AN INTRODUCTION TO
DIFFERENTIAL
EQUATIONS
Stochastic Modeling, Methods and Analysis

Volume
2

AN INTRODUCTION TO
DIFFERENTIAL EQUATIONS
Stochastic Modeling, Methods and Analysis

Volume

2

Anil G Ladde
Chesapeake Capital Corporation, USA

G S Ladde
University of South Florida, USA

World Scientific

NEW JERSEY · LONDON · SINGAPORE · BEIJING · SHANGHAI · HONG KONG · TAIPEI · CHENNAI

Published by

World Scientific Publishing Co. Pte. Ltd.

5 Toh Tuck Link, Singapore 596224

USA office: 27 Warren Street, Suite 401-402, Hackensack, NJ 07601

UK office: 57 Shelton Street, Covent Garden, London WC2H 9HE

British Library Cataloguing-in-Publication Data
A catalogue record for this book is available from the British Library.

AN INTRODUCTION TO DIFFERENTIAL EQUATIONS
Stochastic Modeling, Methods, and Analysis
(Volume 2)

ISBN 978-981-4390-06-4
ISBN 978-981-4390-07-1 (pbk)

Typeset by Stallion Press
Email: enquiries@stallionpress.com

Printed in Singapore by B & Jo Enterprise Pte Ltd

Preface

For more than half a century, stochastic calculus and stochastic differential equations have been playing a central role in analyzing the dynamic phenomena in the biological, engineering, and physical sciences. The advancement of knowledge about stochastic differential equations is rapidly spreading into graduate programs around the world. Unfortunately, the fruits of this advancement have not reached the undergraduate classrooms or the interdisciplinary graduate/undergraduate programs.

This book (Volume 2) resolves this limitation, and is the first available book that can be used in a stochastic modeling or applied mathematics course for interdisciplinary graduate/undergraduate students and for interdisciplinary young researchers with a minimal academic background. It is a stochastic version of *An Introduction to Differential Equations, Volume 1: Deterministic Modeling, Methods, and Analysis*. The two volumes are similar in spirit but differ in content. In fact, the initial concept focused on an introduction to deterministic and stochastic differential equations with applications. Knowing the role and scope (both academic and pedagogical) of the material, many colleagues recommended creating two volumes (as a reminder, the first volume focuses on the deterministic approach). Both volumes offer unique innovations and style to the presentation of topics, methods, and concepts with adequate preparation in deterministic calculus. Below, we highlight the basic features of the Volume 2: stochastic modeling, methods, and analysis.

1. Stochastic modeling. This book offers easy access to the advanced tools and knowledge about developing stochastic models of dynamic processes under both internal and external random environmental perturbations. The stochastic modeling approach is based on classical and statistical theoretical procedures coupled with the basic laws or principles in biology, chemistry, economics, pharmacokinetics, physics, physiology, sociology, etc. Of course, no attempt is made to teach all of these disciplines, but it tries to reach out to readers with interdisciplinary backgrounds and interests. The modeling of stochastic dynamic processes in these various disciplines is also based on both the conceptual and the computational understanding of stochastic differential equations. This combined understanding allows one to study the advanced topics in stochastic differential equations and stochastic dynamic modeling, which are currently not available to undergraduates. Ensuring that our classrooms are using the latest techniques and material is part of our responsibility, and this book offers both of these.

2. Methods. The incorporation of the latest research ideas and techniques into *An Introduction to Differential Equations, Volume 1: Deterministic Modeling, Methods, and Analysis* was motivated by the methods of finding solutions to stochastic differential equations. Hence, there are only three basic methods of solving stochastic differential equations of the Itô–Doob type, namely (i) the eigenvalue-type method, (ii) the method of variation of constants parameters, and (iii) the energy/Lyapunov function method. In short, this book is a reference for developing the methods of solving a larger class of ordinary differential equations. It will also be a reference for the methods of solving other different types of differential equations.

3. Analysis. The Itô–Doob-type stochastic differential equations can be considered to be under the influence of two continuous timescales. The research also includes the theoretical aspect of stochastic differential equations. This justifies our notion of the competitive–cooperative interactions between computational and conceptual ideas.

4. Presentation and pedagogical styles. The presentation of ideas, objectives, style as well as the pedagogical approach parallels the presentation described in the preface to Volume 1.

5. Organizational style. The book is composed of six chapters. Chapter 1 covers the relevant concepts in probability, stochastic processes (Gaussian processes, including Brownian motion and Wiener processes), the stochastic Itô–Doob integral, and differential calculus. In addition, the Itô–Doob laws of differentials as well as the methods of substitution and integration by parts are presented. The topics covered in Sections 1.3–1.5 of Chapter 1 and Chapters 2–5 can be used in a stochastic modeling or applied mathematics course at the senior/graduate level or for the continuation of a first entry-level course in ordinary differential equations, deterministic mathematical modeling, or applied mathematics. Chapter 6 offers a summary of the current trends in the stochastic modeling of dynamic processes under internal/external random perturbations. It can be used in pursuing various undergraduate/graduate research projects.

Chapters 2–5 begin with a stochastic modeling section and contain several examples and illustrations drawn from the biological, chemical, medical, physical, and social sciences.

As with Volume 1, Chapter 2 lays the foundation for the study of stochastic linear Itô–Doob-type differential equations. The problem of solving stochastic linear homogeneous scalar differential equations of the Itô–Doob type is analogous to the problem of solving linear algebraic equations. Thus, the problem of finding a solution to a stochastic linear homogeneous scalar differential equation reduces to a problem of finding a solution to a linear system of algebraic equations. For this purpose, we decompose stochastic differential equations of the Itô–Doob-type into deterministic and stochastic parts. This is analogous to the decomposition of the unperturbed and perturbed parts of ordinary differential equations from Section 4.4 of Volume 1. Then, we use the deterministic procedure described in that section to find a solution process of the original stochastic linear homogeneous scalar differential equation of the Itô–Doob type. The method of finding a solution to stochastic linear nonhomogeneous scalar differential equations of the Itô–Doob type parallels the method described in Volume 1. Again, this introduces the concept of the method of variation of constants parameters to find a particular solution to a stochastic differential equation.

Chapter 3 focuses on the study of the Itô–Doob type of stochastic nonlinear differential equations. The method of solving stochastic nonlinear scalar differential equations includes the method of solving deterministic nonlinear scalar differential equations (Chapter 3, Volume 1) as a special case. Again, the presented method is unique. Separately, this is the first book that provides the method of solving the Itô–Doob-type modeling problems to undergraduate/graduate students, regardless of their major. This unified method of solving nonlinear scalar differential equations is called the energy/Lyapunov function method. The description of this method parallels the deterministic case.

Chapter 4 deals with the first-order stochastic linear system of differential equations. Extensions of the deterministic multivariate problems do generate a few analogous questions and issues in the context of an Itô–Doob type of stochastic calculus setting. These challenges are resolved by presenting generalizations of deterministic results, namely the principle of superposition, the Abel–Jacobi–Liouville theorem and its extensions to their stochastic version, or further by the clarification/proof of the results/observations presented in Chapter 2. Based on our knowledge, its development is original.

Chapter 5 is devoted to higher-order Itô–Doob type stochastic scalar differential equations. The presentation of this material is parallel to that in the deterministic case. Furthermore, by introducing the concept of the Laplace transform in the context of Itô–Doob-type stochastic differential equations, several results are obtained. The problem of second-order Itô–Doob-type stochastic differential equations reduces to a problem of solving algebraic equations. Then, this leads to finding a solution to initial value problems. Finally, this method is illustrated by solving the Langevin and Chandrasekhar stochastic differential equations.

The problem behind stochastic modeling is to just compute, not only a solution process, but also to provide the qualitative information about a dynamic process and an input to design a dynamic process to meet the specified goals of the systems. For this purpose, the specific knowledge of the properties of a solution is not very essential. Historically, by knowing the fundamental solution process of differential equations, the method of variation of parameters and Lyapunov's second method have played a very significant role in the past century. A brief outline of qualitative properties is also presented. Furthermore, dynamic processes operating under more than one timescale as well as in the presence of random perturbations are current research. These topics are briefly highlighted in Chapter 6.

This book can be used at various levels with diverse backgrounds:

(i) In an interdisciplinary undergraduate/graduate-level course on stochastic dynamic modeling or applied mathematics (usually stochastic modeling for biology, chemistry, business, engineering, economics, mathematics, medical sciences, physics, psychology, social sciences, etc.).

(ii) By juniors and seniors majoring in mathematics, engineering, and in physical, or material sciences.

(iii) By interdisciplinary professionals, since this book would be a good source, a do-it-yourself type of manual.

(iv) Several colleagues from India, the UAE, the USA, and other countries have shown interest in using the two volumes for a class in differential equations or in various mathematical modeling courses (deterministic and stochastic settings).

The first author has encouraged the second author to facilitate this work through his assistance for more than 20 years. Knowing that the 21st century problems are highly nonlinear, stochastic, and interdisciplinary, we believe that this book will play a central role in preparing students for a graduate study, not only in engineering and sciences, but also in various multidisciplinary research areas.

Anil G. Ladde
G. S. Ladde

Acknowledgement
and Dedication

We dedicate this volume to the forces behind the cause and development of the topics and ideals contained in this book. This book also is dedicated to memory of the loving and supportive parents of the second author. Although they never had a formal education, they dedicated their lives to the growth and development of their children so that each child could make the most out of any opportunities that might come their way.

I also would like dedicate this work to various people, friends, professors and teachers. These individuals were catalysts for intellectually challenging and preparing me to be an independent problem solver. This intellectual challenge coupled with my academic nurturing was furthered by my professors during my undergraduate, graduate and postgraduate studies. These include Dr. C. G. Capse (People's College, Nanded), Dr A. A. Kayande, Dr. S. G. Deo, Dr. V. Lakshmikantham (Marathwada University, Aurangabad), and Dr. Dragoslav D. Siljak (Santa Clara University, California). Moreover, my classmates and friends, Professors Mou-Hsiung Chang, Jagdish Chandra, J. U. Chamergore, D. Y. Kasture, B. G. Pachpatte and Gopal R. Shendge have supported me unconditionally.

I truly appreciate the endless and unconditional love, patience, and sacrifice by my wife, Sushila, and my children, Anil, Jay and Nathan. They had their needs, but they understood the importance of my academic and research efforts as I spent long hours with my students and my mathematical science research efforts. Ultimately, without my family's encouragement, in particular, Anil's comment about my writing an undergraduate level book, this work would not exist. Furthermore, I chose to act on his suggestion when he agreed to assist me with this effort. Going back to my wife, I want to extend my deepest gratitude to her for constantly asking me on the status of our work. She greeted my request to proof-read my book with a smile. My other sons, Jay and Nathan, also reviewed parts of this work as their comments proved to be valuable. This work is dedicated to my daughter-in-laws, Gina and Nichole, who were very supportive when I worked on this book during my visits to their respective homes. In addition, this work is also dedicated to the arrival of my first grandchild, Lincoln Jay in May 2012.

Early in my teaching career, I was very fortunate to be associated with a wonderful educator and individual, Professor Clarence F. Stephens at the SUNY-Potsdam.

Professor Stephens' insights and leadership at SUNY-Potsdam's math department helped to create a fruitful academic atmosphere as well as a sound research and teaching environment.

I am also thankful for the continuous encouragement and support from my colleagues: Professors: N. U. Ahmed, Edward Allen, Armando Arciniega, Stephen R. Bernfeld, P. –L Chow, Kumer Das, D. Kannan, Andrzej Korzeniowski, Negash Medhin, M. Sambandham, S. Sathananthan, A. S. Vatsala, Drs: Mohmoud J. Anabtawi, Janusz Golec, Byron L. Griffin, Roger Dale Kirby, Bonita A. Lawrence, Srinivasa G. Rajalakshmi, Ongard Sirisaengtaksin, Michael S. Smith and Ling Wu.

A special thanks goes to Dr. Korzeniowski who reviewed most of the original manuscript. His valuable comments and suggestions have significantly strengthened the presentation of the material. Mr. Bong-Jin Choi and Dr. Ling Wu assisted for drawing many of the figures contained in this book. Since 2007, this material has been regularly used for my course on "Stochastic Modeling of Dynamic Process at the University of South Florida at Tampa. The inputs and suggestions of the students, in particular, Ph. D. students: Bong-Jin Choi, Arnut Paothong, Jean-Claude Pedjeu, Olusegun Otunuga, Daniel Siu, Wanduku Divine, and Tadesse Zerihun were very fruitful.

We greatly appreciate the World Scientific Publishers for their kind assistance and confidence in our efforts. In particular, we would like to recognize Ms. He Yue, Editor-In-House, as she is a very kind, efficient, effective and thoughtful editor.

Finally, we greatly acknowledge the support of the Mathematical Sciences Division, the US Army Research Office over the period of time to the second author, in particular, Grant Numbers: W911NF-07-1-0283 and W911NF-12-1-0090. The many aspects of the research work went into the development and preparation of this volume.

Contents

Chapter 1

Elements of Stochastic Processes and Itô–Doob Calculus

Introduction

This chapter serves as a review where the relevant mathematical concepts and statements are presented. Numerical examples are given to illustrate the concepts and conclusions of the mathematical results. Section 1.1 introduces the basic concepts of probability distributions and properties for elementary probability and statistics. It also touches on probability distributions and the central limit theorems. Section 1.2 details the basics of stochastic functions. The conceptual topics in this section revolve around Gaussian and Markov processes, in particular the Brownian motion process, the Wiener process, and white noise. Section 1.3 is devoted to the integral and differential concepts of the Itô–Doob type. The Itô–Doob composite, addition, product, and quotient differential formulas as well as examples are also detailed. A high degree of mathematical rigor is avoided by stating "in the Itô–Doob sense." For the Itô–Doob integral, the methods of substitution and integration by parts are presented in Sections 1.4 and 1.5, respectively.

1.1 Probabilistic Preliminaries

In this section, we provide some basic mathematical terms that will be used to present the probabilistic preliminary material in the text.

Definition 1.1.1. Any experiment/phenomenon whose outcomes are unpredictable is called a random phenomenon or a random experiment.

Example 1.1.1. The number of cars that will cross a given intersection at 8 a.m.

Definition 1.1.2. The collection of all possible outcomes/observations of a random experiment/phenomenon is called the sample space. It is denoted by Ω.

Example 1.1.2. A random experiment consists in tossing a coin twice. All possible outcomes are $\Omega = \{HH, HT, TH, TT\}$ where H stands for "head" and T for "tail." H and T are all possible outcomes of a single toss of the coin.

Definition 1.1.3. An event is a subset of the sample space Ω.

Example 1.1.3. $A = \{HH, HT\}$ is an event with respect to the sample space Ω in Example 1.1.2.

Definition 1.1.4. A single outcome of a random experiment is called an elementary event. It is represented by ω.

Example 1.1.4. In reference to Example 1.1.2, the single outcomes $\{HH\}$, $\{HT\}$, $\{TT\}$ are elementary events of the sample space $\Omega = \{HH, HT, TH, TT\}$.

Example 1.1.5. Let $\mathcal{R}$ be a given bounded region, and $\Delta\mathcal{R}$ be a subregion of $\mathcal{R}$ with a center C. A collection of all objects in $\mathcal{R}$ is considered to be a sample space Ω. A collection of all objects in the subregion $\Delta\mathcal{R}$ with a center C is denoted by $A \equiv A(C, \Delta\mathcal{R})$. This mathematical description of the event $A \equiv A(C, \Delta\mathcal{R})$ enables us to determine whether or not an object is in the subregion with the center C. Moreover, it assists us in determining whether the object is exactly located at C. This can be determined by defining an elementary event ω of the object at a location Y, in the subregion $\Delta\mathcal{R}$, with a center C, and it is represented by $\omega = \omega(Y, \Delta\mathcal{R})$.

Observation 1.1.1. We note that the outcomes of a random experiment/ phenomenon are practical realizations of elementary events. In general, practical outcomes of random phenomena are represented by a finite or an infinite sequence of integers,

$$\omega = (n_1, n_2, n_3, \ldots, n_1, \ldots), \text{ or real numbers } \omega = (x_1, x_1, \ldots, x_1, \ldots). \quad (1.1.1)$$

Definition 1.1.5. The null event or impossible event is an event with no elements of Ω. It is denoted by $\emptyset$.

Example 1.1.6. In Example 1.1.2, an outcome is neither head nor tail. In general, $\emptyset = \{x : x \neq x\}$.

Definition 1.1.6. A probability space is a triplet $(\Omega, \mathfrak{F}, P)$, where Ω is the sample space, $\mathfrak{F}$ is a collection of subsets of events, and P is a set function defined on $\mathfrak{F}$ with its values in $[0, 1]$ such that the two sets of properties are satisfied:

Axioms

S1. The sample space is in $\mathfrak{F}$.
S2. The null event (empty set) $\emptyset$ is in $\mathfrak{F}$.

S3. If a finite or countable number of events $A_1, A_2, \ldots, A_n, \ldots$ belong to $\mathfrak{F}$, then their union belongs to $\mathfrak{F}(A_1 \cup A_2 \cup A_3 \cdots \in \mathfrak{F})$.

S4. If a finite or countable number of events $A_1, A_2, \ldots, A_n, \ldots$ belong to $\mathfrak{F}$, then their intersection belongs to $\mathfrak{F}(A_1 \cap A_2 \cap A_3 \cdots \in \mathfrak{F})$.

Axioms

P1. For every event $A \in \mathfrak{F}$, $0 \leq P(A) \leq 1$ (non-negativity).

P2. $P(\Omega) = 1$ (normed finiteness).

P3. The probability of the union of a finite or countable number of pairwise exclusive events is the sum of the probabilities of the events, that is, $P(A_1 \cup A_2 \cup A_3 \ldots) = P(A_1) + P(A_2) + P(A_3) + \cdots$, whenever $A_i \cap A_j = \emptyset$ for all i and j, $i \neq j$ (additivity).

Observation 1.1.2

(i) We remark that axiom P3 seems to be restrictive, but it is essential to assume that this axiom must also be valid. In Example 1.1.5, $\mathcal{R}$ is a bounded region and hence it can be covered by at-most-countable subregions. The axiomatic probabilistic approach is applicable to this problem. However, if one modifies Example 1.1.5 by considering that $\Delta\mathcal{R} = \{C\}$ instead of a subregion of $\mathcal{R}$ with the location of the object at a point C in $\mathcal{R}$ as an elementary event. $P(\{C\}) = 0$, and $\mathcal{R}$ is a union of all singleton sets $\{C\}$ in $\mathcal{R}$ which is not a countable union. As a result of this, axiom P3 is not applicable.

(ii) We also note that $\mathfrak{F}$ is considered to be the complete information about a random experiment.

Example 1.1.7. In reference to Example 1.1.2, a triplet $(\Omega, \mathfrak{F}, P)$ is a probability space, where $\mathfrak{F} = \{\{HH, HT, TH, TT\}, \emptyset, \{HH\}, \{HT\}, \{TH\}, \{TT\}, \{HH, HT\}, \{HH, TH\}, \{HH, TT\}, \{HT, TH\}, \{HT, TT\}, \{TH, TT\}, \{HH, HT, TH\}, \{HH, HT, TT\}, \{HH, TH, TT\}, \{HT, TH, TT\}\}$, $P(\omega) = \frac{1}{4}$ for $\omega \in \Omega$.

Definition 1.1.7. $\mathfrak{F}$ is called a σ-algebra if $\mathfrak{F}$ satisfies axioms S1–S4.

Example 1.1.8. $\mathfrak{F} = \{\Omega, \emptyset, \{HH\}, \{HT\}, \{TH\}, \{TT\}, \{HH, HT\}, \{HH, TH\}, \{HH, TT\}, \{HT, TH\}, \{HT, TT\}, \{TH, TT\}, \{HH, HT, TH\}, \{HH, TH, TT\}, \{HT, TH, TT\}, \{HH, HT, TT\}\}$ in Example 1.1.2.

Definition 1.1.8. P is called a probability measure or probability if it satisfies axioms P1–P3.

Example 1.1.9. The probability measure is defined in Example 1.1.7.

Definition 1.1.9. Let A and B be two events in $\mathfrak{F}$. We observe that the occurrence of mutually exclusive events $(A \cap B = \emptyset)$ is connected with the concept of the nonintersection of sets. If $A \cap B$ is nonempty, then $P(A \cap B)$ is called the joint probability of events A and B.

Definition 1.1.10 (Cartesian product probability space). Let $\Omega_1 \equiv (\Omega_1, \mathfrak{F}_1, P_1)$ and $\Omega_2 \equiv (\Omega_2, \mathfrak{F}_2, P_2)$ be probability spaces. The problems of joint probability naturally arise due to thee *Cartesian product probability space*, $(\Omega, \mathfrak{F}, P)$, where the sample space Ω is a Cartesian product of two sample spaces Ω_1 and Ω_2, namely $\Omega = \Omega_1 \times \Omega_2$; the sigma algebra $\mathfrak{F}$ is a Cartesian product of two sigma algebras $\mathfrak{F}_1$ and $\mathfrak{F}_2$, namely $\mathfrak{F} = \mathfrak{F}_1 \times \mathfrak{F}_2$; and the probability measure P is a Cartesian product of two probability measures, namely $P = P_1 \times P_2$, and for $A_1 \in \mathfrak{F}_1$, $B_1 \in \mathfrak{F}_2$, $A_1 \times B_1 = C \in \mathfrak{F}$,

$$P(C) = P(A_1 \times B_1) = P_1(A_1)P_2(B_1). \tag{1.1.2}$$

Observation 1.1.3

(i) **Cartesian product probability space.** The joint probability arises in a natural way in the Cartesian product probability space, $(\Omega, \mathfrak{F}, P)$. For example, here an elementary joint event is $\{\omega\}$, where $\omega = (\omega_1, \omega_2) \in \Omega$. This joint elementary event is represented by $\{\omega\} = C_1 \cap C_2$, where $C_1 = \{\omega_1\} \times \Omega_2$ and $C_2 = \Omega_1 \times \{\omega_2\}$. Hence, the joint probability of $\{\omega\} = C_1 \cap C_2$ is given by $P(\{\omega\}) = P(C_1 \cap C_2) = P_1(C_1)P_2(C_2)$.

(ii) **Time-evolving dynamic systems.** Let us denote by A and B the states (for example: 1 — position vectors of a particle in a physical system; 2 — number of species in an ecological system; 3 — number of molecules of a chemical substance in a chemical system; 4 — amounts of electric charge/current; 5 — amount of prices in an economic system; and so on) of a system at times t_1 and t_2, respectively. The goal is to find a joint probability distribution $P(A \cap B)$ of events A and B.

Definition 1.1.11. A triplet $(\Omega, \mathfrak{F}, P)$ is called a complete probability space if every null event is an event in $\mathfrak{F}$.

In order to study the essential structure and its behavior of a random phenomenon, we represent the outcomes of the random phenomenon by assigning real numbers. In general, explicit knowledge about the underlying sample space Ω is not always feasible. Moreover, the underlying elementary events are too complex to describe. The concept of association of outputs of a random phenomenon (elementary events) with the simpler outputs, namely real numbers, is a powerful one. This leads to the concept of a random variable, which provides information about the random phenomenon. In fact, it is a mathematical model of a random phenomenon.

Observation 1.1.4. If ω is defined as a position of a particle in the air, then the descriptions of ω is complicated. The position of the particle depends on linear velocity, angular velocity, momentum, etc. However, it is easier to specify by random variables. Moreover, time varying dynamic events can easily be described by functions of random variables.

Definition 1.1.12. Let $(\Omega, \mathfrak{F}, P)$ be a complete probability space. A random variable X is a function defined on Ω into $\mathbb{R}$ ($X : \Omega \to \mathbb{R}$ or $X : \Omega \to \mathbb{R}^n$). Then, for every Borel set B in $\mathbb{R}$ (or B in $\mathbb{R}^n$), $X^{-1}(B) \in \mathfrak{F}$. This is denoted by $X \in R[\Omega, R]$ (or $X \in R[\Omega, \mathbb{R}^n]$).

The σ-algebra induced by a random variable X is $\mathfrak{F}_X = \{X^{-1}(B) : B \in \mathfrak{B}\}$, where $\mathfrak{B}$ is the Borel σ-algebra of $\mathbb{R}$ (or $\mathbb{R}^n$). Example: The collection of subsets $(-\infty, a]$ for $a \in \mathbb{R}$ (or $\mathbb{R}^n$) is Borel sets in $\mathbb{R}$ (or $\mathbb{R}^n$). The collection $\mathfrak{F}_X$ is the smallest sigma algebra with respect to which X is a random variable.

Example 1.1.10. Every real number is a random variable.

Solution process. To establish the validity of the given mathematical statement, we employ the deductive reasoning (Chapter 1, Volume 1). Let $(\Omega, \mathfrak{F}, P)$ be a complete probability space and let X be a function defined on $(\Omega, \mathfrak{F}, P)$ into $\mathbb{R}/\mathbb{R}^n$ by $X(\omega) = c$ for each $\omega \in \Omega$, and for arbitrary given $c \in \mathbb{R}/\mathbb{R}^n$. From the definition of X, it is clear that it is a function (constant). Now, we need to show that for every Borel set B in $\mathbb{R}/\mathbb{R}^n$, $X^{-1}(B) \in \mathfrak{F}$. Let B be any Borel set in $\mathbb{R}/\mathbb{R}^n$. Then we have

$$X^{-1}(B) = \begin{cases} \emptyset, & \text{if } c \notin B, \\ \Omega, & \text{if } c \in B. \end{cases} \tag{1.1.3}$$

From (1.1.3) and Definition 1.1.6 (Axioms S1 and S2), we conclude that $X^{-1}(B) \in \mathfrak{F}$. Thus, the constant function X is a random variable (Definition 1.1.12).

Definition 1.1.13. Every random variable induces a measure $F_X \equiv P_X$ on $\mathbb{R}$ (or on $\mathbb{R}^n$). It is defined by $F_X(B) = P(X^{-1}(B))$, and it is termed the distribution function (just distribution) of X. In particular, for $B = (-\infty, x]$, $F_X(B) = P(\{\omega : X(\omega) \le x\})$ is denoted by $F_X(x)$ [or simply $F(x)$]. Moreover, $(\mathbb{R}/\mathbb{R}^n, \mathfrak{B}, F_X)$ is a probability space, where X is a random variable defined on $(\Omega, \mathfrak{F}, P)$ into $\mathbb{R}/\mathbb{R}^n$.

Example 1.1.11. Find the distribution of the random variable X defined in Example 1.1.10. Moreover, determine $(\mathbb{R}/\mathbb{R}^n, \mathfrak{B}, F_X)$.

Solution process. For any given $x \in \mathbb{R}/\mathbb{R}^n$, let $B = (-\infty, x]$ be a Borel set. From Definition 1.1.13 and the argument used in the solution process of Example 1.1.10, we have

$$F_X(x) = P(X^{-1}(B)) = \begin{cases} 0, & \text{if } c \notin B \\ 1, & \text{if } c \in B \end{cases}$$

$$= \begin{cases} 0, & \text{if } x < c \\ 1, & \text{if } x \ge c. \end{cases} \tag{1.1.4}$$

This is the probability distribution (cumulative) function of the random variable in Example 1.1.10. Furthermore, $(\mathbb{R}/\mathbb{R}^n, \mathfrak{B}, F_X)$ is determined, where F_X is as defined in (1.1.4).

Definition 1.1.14. Let X and Y be random variables. They induce a measure $F_{X,Y}$ on $\mathbb{R}$ (or on $\mathbb{R}^n$), defined by $F_{X,Y}(B_1, B_2) = P(X^{-1}(B_1) \cap Y^{-1}(B_2))$. This is termed the joint distribution function (just joint distribution) of X and Y. In particular, for $B_1 = (-\infty, x]$ and $B_2 = (-\infty, y]$, $F_{X,Y}(B_1, B_2) = P(\{\omega : X(\omega) \le x$ and $Y(\omega) \le y\})$ is denoted by $F_{X,Y}(x, y)$ [or simply $F(x, y)$].

Observation 1.1.5

(i) Let $(\Omega, \mathfrak{F}, P)$ be a complete probability space. A random variable X is a function defined on Ω into $\mathbb{R}$ ($X : \Omega \to \mathbb{R}$ or $X : \Omega \to \mathbb{R}^n$). Let $\mathfrak{F}_X = \{X^{-1}(B) : B \in \mathfrak{B}\}$ be the sub-σ-algebra induced by a random variable X, where $\mathfrak{B}$ is the Borel σ-algebra of $\mathbb{R}$ (or $\mathbb{R}^n$), and let $\bar{\mathfrak{F}}_X$ be the sub-σ-algebra generated by $\mathfrak{F}_X$. The restriction of P, $P_X \equiv F_X$ to $\mathfrak{F}_X$ is uniquely determined by the distribution of X. Therefore, the probability space $(\Omega, \bar{\mathfrak{F}}_X, P_X)$ is induced by X with respect to which X is a random variable.

(ii) We note that, in practice, one rarely starts with a probability space and a random variable/vector. Instead, one often starts with a probability distribution function $F_X(x)$ of X, x in $\mathbb{R}$ (or $\mathbb{R}^n$). $F_X(x)$ can be considered to be the realization of $F_X \equiv P_X$. Thus, P_X defines a probability measure on $(\mathbb{R}/\mathbb{R}^n, \mathfrak{B}, F_X)$, as follows:

(a) $X : \mathbb{R} \to \mathbb{R}$ (or $X : \mathbb{R}^n \to \mathbb{R}^n$) is the coordinate function defined by $X(x) = x$ (or $X : \mathbb{R}^n \to \mathbb{R}^n$, $X_i(x) = x_i$, for $i = 1, 2, \ldots, n$).

(b) X is a random variable/vector on $(\mathbb{R}/\mathbb{R}^n, \mathfrak{B}, P_X)$, with the probability measure P_X as the joint distribution, in view of the fact that $P_X(\{x : X(x) < a\}) = F_X(a)$ or $P_X(\{x : X_1(x) < a_1, X_2(x) < a_2, \ldots, X_n(x) < a_n\}) = F_X(a_1, a_2, \ldots, a_n)$.

Definition 1.1.15. Let X and Y be random variables. A density of a random variable X is a function defined on $\mathbb{R}$ by the derivative of its distribution function, provided that it is differentiable on $\mathbb{R}$, and is denoted by f_X or f. Similarly, a joint density of random variables X and Y is defined by $\frac{\partial^2}{\partial x \partial y} F(x, y) = f(x, y)$.

Example 1.1.12. Find the probability density function of the distribution function of the random variable in Example 1.1.10. Moreover, find the probability space induced by the random variable.

Solution process. We recall the cumulative probability distribution function (1.1.4) of the random variable in Example 1.1.11. We note that this function F_X in (1.1.4) is constant on both of the intervals $(-\infty, c)$ and $[c, \infty)$. Moreover, it takes the values 0 and 1 on the intervals $(-\infty, c)$ and $[c, \infty)$, respectively. Therefore, it is continuously differentiable on the intervals $(-\infty, c)$ and (c, ∞). At $x = c$, F_X in (1.1.4) has no derivative. In fact, the left-hand derivative of F_X at $x = c$ does not exist, i.e. for $h < 0$, $c + h < c$,

$$\lim_{h \to 0^-} \left[\frac{F_X(c+h) - F_X(c)}{h} = \frac{0-1}{h} = -\frac{1}{h} \right] = \infty. \tag{1.1.5}$$

F_X has the right-hand derivative at $x = c$, and it is 0, i.e. for $h > 0$, $c + h > c$,

$$\lim_{h \to 0^+} \left[\frac{F_X(c+h) - F_X(c)}{h} = \frac{1-1}{h} = \frac{0}{h} \right] = 0. \tag{1.1.6}$$

The probability space induced by X is $(\Omega, \bar{\mathfrak{F}}_X, P_X)$, where $\bar{\mathfrak{F}}_X = \{\Omega, \emptyset\}$ $P_X = F_X$.

Observation 1.1.6. We observe that the density function f_X or f defined in Definition 1.1.15 has the following properties:

(i) $f_X(x) \geq 0$;

(ii) $\int_a^b f_X(x)dx = \int_a^b dF_X(x) = F_X(b) - F_X(a)$;

(iii) $\int_{-\infty}^{\infty} f(x)dx = 1$.

Here, $\int_a^b g(x)dF_X(x)$ is the Riemann–Stieltjes integral of g with respect to F_X [168].

Example 1.1.13. Using the distribution of the random variable in Example 1.1.11, find

$$\int_a^b dF_X(x).$$

Solution process. Let P be $\{x_0, x_1, \ldots, x_k, \ldots, x_n\}$ a partition of the interval $[a, b]$ and let t_k be an element of the kth subinterval $[x_{k-1}, x_k]$ of the given interval $[a, b]$. The Riemann–Stieltjes sum of the function g ($g \equiv 1$, constant) with respect to F_X is:

$$S(P, g, F_X) = \sum_{k=1}^{n} g(t_k)(F_X(x_k) - F_X(x_{k-1}))$$

$$= \sum_{k=1}^{n} g(t_k)\Delta F_X(x_k)$$

$$= \sum_{k=1}^{n} \Delta F_X(x_k) \quad \text{(from assumption } g \equiv 1\text{)},$$

$$S(P, g, F_X) = \begin{cases} \Delta F_X(x_j), & \text{if } c \in [a, b] \text{ and for some } j, c \in [x_{j-1}, x_j], \\ 0, & \text{if } c < a \text{ or } b < c. \end{cases} \quad (1.1.7)$$

The mesh of partition is the length of the largest subinterval of P, and it is denoted by $\mu(P)$. As $\mu(P) \to 0$,

$$S(P, g, F_X) \to \begin{cases} F_X(c^+) - F_X(c^-) = 1, & \text{if } c \in [a, b], \\ 0, & \text{if } c < a \text{ or } b < c \end{cases} \quad (1.1.8)$$

and

$$S(P, g, F_X) \to \int_a^b dF_x(x). \quad (1.1.9)$$

From (1.1.7) and (1.1.8), we have

$$\int_a^b dF_X(x) = \begin{cases} 1, & \text{if } c \in [a, b], \\ 0, & \text{if } c < a \text{ or } b < c. \end{cases} \quad (1.1.10)$$

This completes the solution process of the example.

Definition 1.1.16. A random variable X is said to be P-integrable if the integral $\int_\Omega X dP$ is finite, and it is called an *expectation/mean/first moment of X*. It is denoted by

$$E[X] = \int_\Omega X dP. \tag{1.1.11}$$

From of the definition of the probability space $(\mathbb{R}/\mathbb{R}^n, \mathfrak{B}, P_X)$ in Definition 1.1.13 and Observation 1.1.5, $E[X]$ can be computed by using its distribution or density function (if it exists) as follows:

$$E[X] = \int_{\mathbb{R}/\mathbb{R}^n} x dF(x) = \int_{\mathbb{R}/\mathbb{R}^n} x f(x) dx. \tag{1.1.12}$$

Moreover, higher moments of X are defined by

$$E[|X|^p] = \int_{-\infty}^{\infty} |x|^p dF(x) = \int_{-\infty}^{\infty} |x|^p f(x) dx, \tag{1.1.13}$$

for $p \geq 1$. In particular, for $p = 2$, it is referred to as a mean-square/quadratic-moment.

Observation 1.1.7. From (1.1.13), we note that for $p \geq 1$ the pth moment is the expected value/mean of a function of random variable $g(X) = |X|^p$. This idea can be extended to a random variable Z which is a function of the given random variable $X, Z = h(X)$ as long as the following improper integral exists:

$$E[Z] = E[h(X)] = \int_{-\infty}^{\infty} h(x) dF(x) = \int_{-\infty}^{\infty} h(x) f(x) dx. \tag{1.1.14}$$

This is the expected value of the function $Z = h(X)$. Sufficient conditions for $Z = h(X)$ to be the random variable are that X is a random variable and $h(x)$ is a continuous or piecewise continuous function of $x \in R$.

Example 1.1.14. Using the distribution of the random variable in Example 1.1.11, find:

(a) $\int_{-\infty}^{\infty} x dF_X(x)$;

(b) $\int_{-\infty}^{\infty} |x|^p dF_X(x)$;

(c) $\int_{-\infty}^{\infty} h(x) dF_X(x)$, where h is as defined in Observation 1.1.7.

Solution process. We note that the integrands in Parts (a) and (b) are special cases of the integrand in Part (c). Moreover, they are continuous functions. Therefore, it is enough to sketch the solution process for Part (c). By replacing g with h and imitating the argument used in the development of the solution process of

Example 1.1.13, we arrive at

$$S(P, h, F_X) = \begin{cases} h(t_j)\Delta F_X(x_j), & \text{if } c \in [a, b] \text{ and for some } j, c \in [x_{j-1}, x_j], \\ 0, & \text{if } c < a \text{ or } b < c. \end{cases}$$
(1.1.15)

The mesh of the partition is the length of the largest subinterval of P, and it is denoted by $\mu(P)$. From the continuity of h, as $\mu(P) \to 0$,

$$S(P, h, F_X) = \begin{cases} h(c)[F_X(c^+) - F_X(c^-)], & \text{if } c \in [a, b], \\ 0, & \text{if } c < a \text{ or } b < c, \end{cases}$$
(1.1.16)

and

$$S(P, h, F_X) \to \int_a^b h(x) dF_X(x).$$
(1.1.17)

From knowing $[F_X(c^+) - F_X(c^-)] = 1$, (1.1.16) and (1.1.17), we have

$$\int_a^b h(x) dF_X(x) = \begin{cases} h(c), & \text{if } c \in [a, b], \\ 0, & \text{if } c < a \text{ or } b < c. \end{cases}$$
(1.1.18)

Thus, from the definitions of the improper integral and the expected value of the random functions, we have $E[X] = c$, $E[|X|^p] = |c|^p$ and $E[h(X)] = h(c)$. This completes the solution process of the example.

Definition 1.1.17. Let X be a random variable. The characteristic function of X is defined by

$$Q(\lambda) = Q_X(\lambda) = E[\exp[i\lambda X]] = \int_{-\infty}^{\infty} \exp[i\lambda x] dF(x),$$
(1.1.19)

where λ is a real number. If f is the density function of X, then the formula (1.1.19) reduces to

$$Q(\lambda) = \int_{-\infty}^{\infty} \exp[i\lambda x] f(x) dx.$$
(1.1.20)

Moreover, f can be obtained by the following inversion formula of the Fourier integral [195]:

$$f(x) = \int_{-\infty}^{\infty} \exp[-i\lambda x] Q(\lambda) d\lambda.$$
(1.1.21)

Example 1.1.15. A random variable X is said to be a *Bernoulli random variable* with parameter p, $0 \le p \le 1$, if its probability distribution is given by

$$p(1) = P(\{X = 1\}) = p, \quad p(0) = P(\{X = 0\}) = 1 - p,$$
(1.1.22)

where the random variable X is defined on probability space $\Omega = \{\omega_1, \omega_2\}$ which consists of two elementary events, classified as either "success $\equiv \omega_1$" or "failure $\equiv \omega_2$"; the value of X at ω_1 is 1 $[X(\omega_1) = 1]$, and at ω_2, it is 0 $[X(\omega_1) = 0]$; p, $0 \le p \le 1$, is probability of an elementary event "success."

Example 1.1.16. A random variable X is said to be a *binomial random variable* with parameters p and n if its probability distribution is given by

$$p(i, n, p) = P(\{X = i\}) = \begin{cases} \dfrac{n!}{(n-i)!i!} p^i (1-p)^{n-i}, & \\ 0 & \text{elsewhere,} \end{cases} \qquad (1.1.23)$$

where X is a random variable defined on a probability space Ω generated by repeating n times a random experiment with two outcomes, independently. It is represented by the number of successes i out of n independent trials with a probability of "success" p, $0 \le p \le 1$, and $i = 0, 1, \dots, n$.

Example 1.1.17. A random variable X is said to be a *normal/Gaussian random variable* with parameters μ and σ^2 if the density of X is given by

$$f(x) = \frac{1}{\sigma\sqrt{2\pi}} e^{-(x-\mu)^2/2\sigma^2}, \quad -\infty < x < \infty. \qquad (1.1.24)$$

We note that the graph of this function f in (1.1.24) is a bell-shaped curve that is symmetric around μ.

Example 1.1.18. A random variable X that takes values in $\{0, 1, 2, \dots, n, \dots\}$ is said to be a *Poisson random variable* with parameter λ if, for some $\lambda > 0$,

$$p(i) = P(\{X = i\}) = e^{-\lambda} \frac{\lambda^i}{i!}, \quad i = 0, 1, 2, \dots, n. \dots \qquad (1.1.25)$$

Example 1.1.19. A random variable X is said to be *uniformly distributed* on the interval (a, b) if its probability density function is given by

$$f(x) = \begin{cases} \dfrac{1}{b-a}, & \text{if } a < x < b, \\ 0, & \text{otherwise.} \end{cases} \qquad (1.1.26)$$

Very frequent and sudden random impulses that we encounter in practice may not be able to describe by usually known random functions/processes. For this purpose, in the following, we briefly formulate a concept of generalized functions in the sense of Schwartz and Gel'fand-Shilov [8].

Definition 1.1.18. Let us consider a class of functions $\mathcal{T} = \{f_\epsilon(t) : \epsilon > 0, t \in R\}$, where the function f_ϵ is defined by

$$f_\epsilon(t) = \begin{cases} 0 & \text{for } |t - a| \ge \dfrac{\epsilon}{2}, \\ \dfrac{1}{\epsilon} & \text{for } |t - a| < \dfrac{\epsilon}{2}. \end{cases}$$

This class of functions possesses the following properties:

(a) It has a nature similar to that of an impulse: it acts for a brief instant of time (of duration), but its value is as large as we please.

(b) It is non-negative and

$$\int_{-\infty}^{\infty} f_\epsilon(t)dt = 1.$$

From (a) and (b), this class of functions can be called a class of probability density functions for arbitrary random impulses. The limit of f_ϵ as $\epsilon \to 0$ is called the generalized function δ_a at the impulse instant $t = a[\lim_{\epsilon \to 0} f_\epsilon(t) = \delta_a(t) = \delta(t-a)]$. Moreover,

$$\delta_a(t) = \begin{cases} 0 & \text{for } t \neq a, \\ \text{``}\infty\text{''} & \text{for } t = a, \end{cases}$$

and

$$\int_{a-h}^{a+h} \delta(t-a)dt = 1, \qquad (1.1.27)$$

where $h > 0$. There is no function in the usual sense of the word "function" with these properties. Therefore, it is called the *generalized function*. The function $\delta_a(t) = \delta(t-a)$ is called *Dirac's delta function*. The Dirac function is used as the probability density function of a real-valued random variable X that characterizes the sudden and frequent impulses on the dynamic systems in the biological, chemical, physical, and social sciences.

Example 1.1.20

(a) Let X be a random variable that takes the finite number of values $x_1, x_2, \ldots, x_i, \ldots, x_n$ at the finite sequence of impulses whose strengths are determined by the probabilities $p_i = P(X = x_i)$ for $i = 1, 2, \ldots, n$. A consistent definition of the density function of X is

$$f_X(x) = \sum_{i=1}^{n} p_i \delta(x - x_i). \qquad (1.1.28)$$

$f_X(x)$ in (1.1.28) is consistent in the sense that it satisfies all the properties that are stated in Observation 1.1.6. Moreover, using the integration process and (1.1.27), we can recover the distribution function F_X of X.

(b) The mean of the random variable X is given by

$$E[X] = \int_{-\infty}^{\infty} x f_X(x)dx = \int_{-\infty}^{\infty} x \sum_{i=1}^{n} p_i \delta(x - x_i)dx$$

$$= \sum_{i=1}^{n} p_i \int_{x_i-h_i}^{x_i+h_i} x\delta(x - x_i)dt = \sum_{i=1}^{n} p_i x_i. \qquad (1.1.29)$$

The properties of the expected value/mean are exactly similar to the properties of the definite integral in the deterministic calculus [127]. They are summarized in the following theorem.

Theorem 1.1.1. *Let X and Y be random variables. Then,*

(i) $E[X+Y] = E[X] + E[Y]$ *(additivity property);*

(ii) $E[cX] = cE[X]$ *(homogeneity property) for any given real number;*

(iii) *if $X \geq Y$, then $E[X] \geq E[Y]$.*

Example 1.1.21. Let X be a binomial random variable with parameters p and n. Then,

$$E[X] = \sum_{i=0}^{n} ip(i,n,p) = \sum_{i=0}^{n} i \frac{n!}{(n-i)!i!} p^i (1-p)^{n-i}$$

$$= \sum_{i=1}^{n} \frac{n!}{(n-i)!(i-1)!} p^i (1-p)^{n-i} = np \sum_{i=1}^{n} \frac{(n-1)!}{(n-i)!(i-1)!} p^{i-1} (1-p)^{n-i}$$

$$= np \sum_{i=0}^{n-1} \frac{(n-1)!}{(n-1-i)!i!} p^i (1-p)^{n-1-i} = np. \tag{1.1.30}$$

Definition 1.1.19. Let X be a random variable. The *variance of a random variable* X is defined by

$$\text{Var}(X) = E[(X - E[X])^2]. \tag{1.1.31}$$

$\text{Var}(X)$ is denoted by σ^2. The positive number $\sigma = (\text{Var}(X))^{1/2}$ is called the standard deviation of the random variable X, and $E[(X - E[X])^p]$ is called the pth central moment of X.

It is important to note certain facts about the specific functions of normally distributed random variables. They are presented in the following lemma.

Lemma 1.1.1. *Let X be a normally distributed random variable with parameters μ and σ^2. Then, $L(X) = aX + b$ is also normally distributed with its distribution $(a\mu+b)$ and $a^2\sigma^2$. Moreover, any linear transformation/function of random variables map Gaussian distributions into Gaussian distributions.*

Example 1.1.22. Let X be a normally distributed random variable with parameters μ and σ^2.

(a) For $a = 1$, $b = -\mu$, $L(X)$ in Lemma 1.1.1 becomes $L(X) = X - \mu$. From the application of Lemma 1.1.1, $L(X) = X - \mu$ has a "zero" mean with variance σ^2.

(b) Moreover, if $a = \frac{1}{\sigma}$, $b = \frac{-\mu}{\sigma}$, then $L(X)$ in Lemma 1.1.1 reduces to $L(X) = \frac{X-\mu}{\sigma}$. Again, from the application of Lemma 1.1.1, $L(X) = \frac{X-\mu}{\sigma}$ has a "zero" mean with variance "1." This is referred to as the standard normalized Gaussian random variable.

Problem 1.1.1

(a) Let X be a Gaussian random variable. Show that $E[X] = \mu$ and $\text{Var}(X) = \sigma^2$.

(b) Let X be a binomial random variable. Show that $\text{Var}(X) = npq$, where $q = 1-p$.

Definition 1.1.20. Let X and Y be random variables. The real number

$$\operatorname{cov}(X, Y) = E[(X - E[X])(Y - E[Y])] \tag{1.1.32}$$

is called the covariance of X and Y.

Observation 1.1.8. It is known that the covariance of the Gaussian random variable is positive. Thus, there exists a Gaussian density function defined by the formula (1.1.24).

Definition 1.1.21. Let us assume that A and B belong to $\mathfrak{F}$ and $P(B) > 0$. Further assume that the event B has occurred. The conditional probability of an event A is defined by $\frac{P(A \cap B)}{P(B)}$. It is denoted by $P(A|B) = \frac{P(A \cap B)}{P(B)}$. Moreover, $\mathfrak{G} \subset \mathfrak{F}$ denotes a sub-σ-algebra of $\mathfrak{F}$. Under the given condition $\mathfrak{G} \subset \mathfrak{F}$, the conditional probability of A is defined by $P(A/\mathfrak{G})$. This means that $P(A|B) = \frac{P(A \cap B)}{P(B)}$ for every B belonging to $\mathfrak{G}$ with $P(B) > 0$.

Observation 1.1.9

(i) We assume that A and B belong to $\mathfrak{F}$, and $P(B) > 0$. From the definition of the conditional probability $P(A|B)$, we infer that $P(A|B)$ is defined only on a collection of events in B. We further note that $P(A|B) = \frac{P(A \cap B)}{P(B)}$ implies that

$$P(A \cap B) = P(A|B)P(B) = P(B|A)P(A). \tag{1.1.33}$$

(ii) From the formula (1.1.33), one can easily infer that the joint probability is symmetric with respect to the conditioned/given event. We note that this aspect of the conditional property sheds a different light on the events that are time-dependent. For example, following Observation 1.1.3, let A and B be the states of a system at times t_1 and t_2, respectively. The probability of A given $B(P(A|B))$, i.e. the probability of a state of the system in A at a present time t_1 given that it was in B at the past time t_2 $(t_2 < t_1)$, is a very natural thing to study in time-evolving dynamic systems. The converse of $P(A|B)$ is $P(B|A)$. $P(B|A)$ is the probability of a state of the system in B at a present time t_2, given that it will be in A at a future time t_1 $(t_2 < t_1)$.

Definition 1.1.22. Let $A_1, A_2, \ldots, A_i, \ldots$ and $B_1, B_2, \ldots, B_i, \ldots$ be sequences (finite/infinite) of events in $\mathfrak{F}$. Further assume that $A_j \cap A_i = \emptyset$, $B_j \cap B_i = \emptyset$, $\bigcup_i A_i = \Omega$ and $\bigcup_i B_i = \Omega$. Define $A_{ij} = A_i \cap B_j$. We observe that $A_{i1}, A_{i2}, \ldots, A_{ij}, \ldots$ satisfy the probability axiom P3, $A_{ik} \cap A_{ij} = \emptyset$ for $j \neq k$, $k, j = 1, 2, 3 \ldots, n \ldots$, and $A_i = \bigcup_j A_i \cap B_j$. From axiom P3, we have

$$P(A_i) = \sum_{j=1} P(A_i \cap B_j), \tag{1.1.34}$$

and, similarly, we have

$$P(B_j) = \sum_{i=1} P(B_j \cap A_i). \tag{1.1.35}$$

The probabilities $P(A_i)$ and $P(B_j)$ defined in (1.1.34) and (1.1.35) are called the marginal probabilities of A_i and B_j with respect to $B_1, B_2, \ldots, B_i, \ldots$ and $A_1, A_2, \ldots, A_i, \ldots$, respectively.

Lemma 1.1.2 (Bayes' formula). *Let* $A_1, A_2, \ldots, A_i, \ldots$ *and* $B_1, B_2, \ldots, B_i \ldots$ *be sequences (finite/infinite) of events in* $\mathfrak{F}$. *Further assume that*

(a) *for each* $i \neq j$, $A_j \cap A_i = \emptyset$ *and* $B_j \cap B_i = \emptyset$,
(b) $\bigcup_i A_i = \Omega$, $\bigcup_i B_i = \Omega$;
(c) *for each* i *and* j, $P(A_i) > 0$ *and* $P(B_j) > 0$.

Define $A_{ij} = A_i \cap B_j$. *We observe that* $A_{i1}, A_{i2}, \ldots, A_{ij}, \ldots$ *satisfy the probability axiom P3,* $A_{ik} \cap A_{ij} = \emptyset$ *for* $j \neq k$, $k, j = 1, 2, 3, \ldots, n, \ldots$, *and* $A_i = \bigcup_j A_i \cap B_j$. *Then,*

$$P(A_i) = \sum_{j=1} P(A_i|B_j)P(B_j). \tag{1.1.36}$$

Moreover,

$$P(B_k|A_i) = \frac{P(A_i \cap B_k)}{P(A_i)} = \frac{P(A_i|B_k)P(B_k)}{\sum_{j=1} P(A_i|B_j)P(B_j)}. \tag{1.1.37}$$

Observation 1.1.10. From (1.1.34) and (1.1.36), we conclude that the joint probabilities of A_{ij}, i.e. events A_i and B_j, for every $j = 1, 2, 3, \ldots$, are eliminated from (1.1.34). This observation plays a very important role in the study of stochastic processes/functions.

Definition 1.1.23. Let $(\Omega, \mathfrak{F}, P)$ be a complete probability space. In addition, let A and B belong to $\mathfrak{F}$. The events A and B are said to independent if $P(A \cap B) = P(A)P(B)$. Moreover, the event A and the sub-σ-algebra $\mathfrak{G}$ of $\mathfrak{F}$ are independent if $P(A \cap B) = P(A)P(B)$ for every B belonging to $\mathfrak{G}$.

Definition 1.1.24. Let X and Y be random variables. X and Y are said to independent if for any $B_1, B_2 \in \mathfrak{B}$ the events $(X^{-1}(B_1))$ and $(Y^{-1}(B_2))$ are independent. Moreover, from Definition 1.1.14 and Observation 1.1.5, $\mathfrak{F}_{X,Y}$ is the Borel σ-algebra of $\mathbb{R}$ (or $\mathbb{R}^n$) and

(a) $F_{X,Y}(x, y) = F_X(x)F_X(x)$,
(b) $f_{X,Y}(x, y) = f_X(x)f_Y(y)$.

Observation 1.1.11. Let X and Y be independent random variables. We note that:

(i) $E[XY] = E[X]E[Y]$;
(ii) $\text{cov}(X, Y) = E[(X - E[X])(Y - E[Y])] = 0$.

Using the definition of conditional probability, one can formulate, analogously, the definitions of conditional distribution, conditional density, and conditional expectations of a random variable. We present the properties of the conditional expectation of a random variable.

Theorem 1.1.2. *Let X be a random variable.*

(i) *For $\mathfrak{G} = \{\emptyset, \Omega\}$, $E[X|\mathfrak{G}] = E[X]$.*

(ii) *If $X \geq 0$, then $E[X|\mathfrak{G}] \geq 0$.*

(iii) *If X is a random variable with respect to $\mathfrak{G}$ (in short, it is $\mathfrak{G}$-measurable), then $E[X|\mathfrak{G}] = X$.*

(iv) *If $X = a$, then $E[X|\mathfrak{G}] = a$.*

(v) *$\mathfrak{G} \supset \mathfrak{F}$, if $E[X]$ exists, then $E[E[X/\mathfrak{G}]] = E[X]$.*

(vi) *If $c_1, c_2, \ldots, c_i, \ldots, c_n$ are constants, then $E[\sum_{i=1}^{n} c_i X_i] = \sum_{i=1}^{n} c_i E[X_i]$.*

(vii) *If $X \leq Y$, then $E[X|\mathfrak{G}] \leq E[Y|\mathfrak{G}]$ and $|E[X|\mathfrak{G}]| \leq E[|X||\mathfrak{G}|]$.*

(viii) *If X and $\mathfrak{G}$ are independent, then $E[X|\mathfrak{G}] = E[X]$.*

(ix) *For $\mathfrak{H} \subset \mathfrak{G} \subset \mathfrak{F}$, $E[E[X|\mathfrak{G}]|\mathfrak{H}] = E[X|\mathfrak{H}] = E[E[X|\mathfrak{H}]|\mathfrak{G}]$.*

(x) *If X and Y are independent random variables, then $E[XY|\mathfrak{G}] = E[X|\mathfrak{G}]E[Y|\mathfrak{G}]$.*

Observation 1.1.12. Let Y be a random variable, and the collection $\mathfrak{F}_Y$ be the smallest sigma algebra with respect to which Y is a random variable. If $\mathfrak{G} = \mathfrak{F}_Y$, then $E[X|\mathfrak{G}]$ is denoted by $E[X|Y]$.

In the following, we present many fundamental and well-known theorems that exhibit the fact that the distribution functions of the sum of random variables converge to the distribution of the normal Gaussian random variable.

Theorem 1.1.3 (DeMoivre–Laplace central limit theorem). *Let X be a binomial random variable with with parameters p and n. Then, the distribution of the random variable*

$$Y_n = \frac{x - np}{\sqrt{np(1 - p)}} \tag{1.1.38}$$

approaches the standard normal distribution ($\mu = 0$ and $\sigma^2 = 1$) as $n \to \infty$. Moreover, the random variable X is approximated by $X \simeq np + \sqrt{np(1 - p)}Y_n$ for sufficiently large n, where Y has a standard normal distribution.

Example 1.1.23. For $n = 1, 2, \ldots$, let $X_1, X_2, \ldots, X_i, \ldots, X_n$ be Bernoulli random variables with parameter p, $0 \leq p \leq 1$. Let us define

$$S_n = \sum_{k=1}^{n} X_k.$$

S_n is a binomial random variable with mean np and variance npq, where $q = 1 - p$.

Theorem 1.1.4 (Simeon–Poisson theorem). *Let X be a binomially distributed random variable with parameters p and n. Assume that for $n = 1, 2, \ldots$, the relation*

$p = \frac{\lambda}{n}$ *is valid for some* $\lambda > 0$. *Then,*

$$\lim_{n \to \infty} \left[\lim_{p \to 0} p(i, n, p) \right] = e^{-\lambda} \frac{\lambda^i}{i!}, \quad i = 0, 1, 2, \ldots, n, \ldots.$$

Example 1.1.24. For $n = 1, 2, \ldots$, let $X_1, X_2, \ldots, X_i, \ldots, X_n$ be *Poisson random* variables with parameter λ for some $\lambda > 0$. In addition, let

$$S_n = \sum_{k=1}^{n} X_k$$

be a binomial random variable with mean λ and variance λ.

In the following, we present a general central limit theorem:

Theorem 1.1.5 (central limit theorem). *Let* $X_1, X_2, \ldots, X_k, \ldots, X_n$ *be a sequence of independent, identically distributed random variables, each with mean* μ *and variance* σ^2. *Then, the distribution of*

$$Y_n = \frac{\sum_{k=1}^{n} X_k - n\mu}{\sigma \sqrt{n}}$$

approaches the standard normal distribution ($\mu = 0$ *and* $\sigma^2 = 1$) *as* $n \to \infty$, *i.e.*

$$P\{\omega : Y_n(\omega) \le a\} \to \frac{1}{\sqrt{2\pi}} \int_{-\infty}^{a} e^{-x^2/2} dx \quad \text{as } n \to \infty.$$

Moreover, the random variable $\sum_{k=1}^{n} X_k$ *is approximated by* $\sum_{k=1}^{n} X_k \simeq n\mu + \sigma\sqrt{n} Y_n$ *for sufficiently large* n *where* Y_n *has a standard normal distribution.*

Exercises

1. Let F be a probability distribution of a real-valued random variable X. Show that:

   (a) $F(x)$ is a non-negative function on R;
   (b) $F(x)$ is a nondecreasing function of x;
   (c) $\lim_{x \to \infty} F(x) = 1$;
   (d) $\lim_{x \to -\infty} F(x) = 0$.

2. Let f be a probability density of a real-valued random variable X. Show that:

   (a) $\int_{-\infty}^{\infty} f(x)dx = 1$;
   (b) $P\{\omega : a \le X(\omega) \le b\} = \int_{a}^{b} f(x)dx$.

3. Let X and Y be two random variables. Show that:

   (a) $E[XY|Y = y] = yE[X|Y = y]$;
   (b) $E[g(X, Y)|Y = y] = E[g(X, y)|Y = y]$;
   (c) $E[XY] = EY[E[X|Y]]$.

4. Let X be a Poisson random variable with parameter λ. Show that $E[X] = \lambda$.

5. Let X be a uniformly distributed random variable. Find:

   (a) the distribution function F of X;

   (b) the pth moment of X for any $p \geq 1$.

6. Let X be a random variable with an arbitrary probability distribution function with mean $E[X] = m$. Show that the random variable $Y = X - m$ has a "zero" mean.

7. *Markov inequality.* Let $(\Omega, \mathfrak{G}, P)$ be a complete probability space, and X be a random variable with a finite pth moment, $p \geq 1$. Then, for any $\epsilon > 0$,

$$P(\omega : |X(\omega)| \geq \epsilon) \leq \frac{E[|X|^p]}{\epsilon^p}.$$

(For $p = 2$, the Markov inequality is reduced to the *Chebyshev inequality*.)

1.2 Stochastic/Random Function

In this section, we introduce the basic terms and definitions with regard to the stochastic processes. Further, in-depth details are left to the reader [28, 42, 162, 195].

Definition 1.2.1. Let $J = [t_0, t_0 + a)$ be a subset of R for some $0 < a < \infty$, and let us define $[t_0, t_0 + a) \times \Omega = \{(t, \omega) : t \in J \text{ and } \omega \in \Omega\}$. A stochastic process x is a function of two variables defined on $J \times \Omega$ with its values $x(t, \omega)$ in R (or R^n). For each $(t, \omega) \in J \times \Omega$ and for each fixed $t \in J$, $x(t, \cdot) = X(t)$ is a random variable with respect to probability space $(\Omega, \mathfrak{F}, P)$, and for each $\omega \in \Omega$ (each observation ω), $x(\cdot, \omega)$ is a real-valued function defined on J into R (or R^n). $x(\cdot, \omega)$ is called a sample realization or trajectory/sample path of the stochastic process x. Hereafter, we denote a stochastic process by $x(t)$ for $t \in J$, or by $x \in R[J, R[\Omega, R]]$ (or $x \in R[J, R[\Omega, R^n]]$).

Example 1.2.1. For $t \in R$, $x(t) = \sin(tX)$ is a stochastic process, where X is a normal/Gaussian random variable defined on $(\Omega, \mathfrak{F}, P)$.

Definition 1.2.2. For $t \in J$, the finite-dimensional distribution of a stochastic process $x(t)$ is defined by $P\{\omega : x(t) < a\} = F_t(a)$. Furthermore, for $t_i \in J$, $a_i \in R$ for $i = 1, 2, \ldots, n$, the joint distribution is defined by

$$P(\{\omega : x(t_1) < a_1, x(t_2) < a_2, \ldots, x(t_n) < a_n\}) = F(a_1 t_1, a_2 t_2, \ldots, a_n t_n).$$

We note that this joint distribution function needs to satisfy symmetry and compatibility conditions.[42] It is denoted by $F(a_1 t_1, a_2 t_2, \ldots, a_n t_n) = F_{T_n}(a)$ and $F_{T_n} \equiv P_{T_n}$, where $T_n = \{t_1, t_2, \ldots, t_n\} \subseteq J$. Moreover, if the finite-dimensional distribution F is differentiable with respect to a, then it is called the finite-dimensional density function $f(a_1 t_1, a_2 t_2, \ldots, a_n t_n)$.

Observation 1.2.1

(i) Let $(\Omega, \mathfrak{F}, P)$ be a complete probability space. A stochastic process x is a function of two variables defined on $J \times \Omega$ with its values $x(t, \omega)$ in R (or R^n). By following the argument used in Observation 1.1.5(i), we have the sub-σ-algebra $\mathfrak{F}_x$ induced by a random variable $x(t)$, for every $t \in J$, and the sub-σ-algebra $\bar{\mathfrak{F}}_x$ generated by $\mathfrak{F}_x$. The restriction of P, $P_x \equiv F_x$ to $\mathfrak{F}_x$ is uniquely determined by the finite-dimensional distributions of x. Therefore, the probability space $(\Omega, \bar{\mathfrak{F}}_x, P_x)$ is induced by stochastic process x.

(ii) An observation similar to Observation 1.1.5(ii) can be formulated with respect to the stochastic process. We note that, in practice, one rarely starts with a probability space and a given family of random variables defined on the probability space. Instead, one often starts with a compatible family of finite-dimensional probability distributions $\{P_{T_n} : T_n \subseteq J\}$. From this, one presents a construction of a probability space $(\Omega, \mathfrak{F}, P)$ and a family of random variables $\{x(t) : t \in J\}$ with the above given function as the finite-dimensional probability distributions. We define $\Omega = \mathbb{R}^J = \{\omega : \omega$ is a real-valued function defined on J into $\mathbb{R}\}$. A stochastic process x defined on $J \times \Omega = J \times \mathbb{R}^J$ into $\mathbb{R}$ as $x(t, \omega) = \omega(t)$. $x(t, \omega)$ is called the *coordinate function*, because it is the tth coordinate of ω. For each t and $a \in \mathbb{R}$, $\{\omega : x(t, \omega) = \omega(t) < a\}$ is called a cylindrical set. $\mathfrak{F}$ is the smallest σ-algebra containing all cylindrical sets in $\Omega = \mathbb{R}^J$. The probability measure P is defined by

$$P(\{\omega : (x(t_1, \omega), x(t_2, \omega), \ldots, x(t_n, \omega)) \in B\}) = \int_B dF(a_1 t_1, a_2 t_2, \ldots, a_n t_n),$$

where B is an n-dimensional Borel set.

Several random phenomena in biological, chemical, physical, and social sciences are approximated by Gaussian stochastic processes. One of the reasons is that the Gaussian processes have distribution functions.

Definition 1.2.3. A stochastic process $x(t)$ for $t \in J$ is said to be a *Gaussian process* if for any finite collection $\{t_1, t_2, \ldots, t_i, \ldots, t_n\}$ in J, every finite linear combination of the form $\sum_{k=1}^{n} \alpha_k x(t_k) = X$ is a Gaussian random variable.

Observation 1.2.2

(i) If a process x is a Gaussian, then it is clear that for each $t \in J$, $x(t)$ is a Gaussian random variable. Moreover, a process $x(t)$ for $t \in J$ is Gaussian if and only if:

(a) $E[x^2(t)] < \infty$ for each $t \in J$.

(b) For every finite collection $\{t_1, t_2, \ldots, t_i, \ldots, t_n\}$ in J, the characteristic function of random variables of $x(t_1), x(t_2), \ldots, x(t_n)$ has the form

$$E\left[\exp\left[i\sum_{k=1}^{n}\alpha_k x(t_k)\right]\right]$$

$$= \exp\left[i\sum_{k=1}^{n}\alpha_k m(t_k) - \frac{1}{2}\sum_{k=1}^{n}\sum_{\ell=1}^{n}\alpha_k\alpha_\ell R(t_k, t_\ell)\right], \quad (1.2.1)$$

where $m(t_k) = E[x(t_k)]$ and $R(t_k, t_\ell) = E[(x(t_k) - m(t_k))(x(t_\ell) - m(t_\ell))]$.

(ii) The discussion in (i) shows that all finite-dimensional distributions of Gaussian processes are completely determined by the mean $m(t) = E[x(t)]$ and the covariance function $R(t, s) = E[(x(t) - m(t))(x(s) - m(s))]$. Moreover, the covariance function $R(t, s)$ is more precisely termed an *autocovariance* function of the stochastic process x.

(iii) In addition, the $n \times n$ matrix $R(t_k, t_\ell)$ in (1.2.1) is non-negative definite, i.e. the quadratic form $\sum_{k=1}^{n}\sum_{\ell=1}^{n}\alpha_k\alpha_\ell R(t_k, t_\ell)$ satisfies relation

$$\sum_{k=1}^{n}\sum_{\ell=1}^{n}\alpha_k\alpha_\ell R(t_k, t_\ell) = E\left[\left|\sum_{k=1}^{n}\alpha_k(x(t_k) - m(t_k))\right|^2\right] \geq 0. \quad (1.2.2)$$

Definition 1.2.4. A stochastic process $x(t)$ defined on J is said to be a (*strictly*) *stationary process* if for any finite collection $\{t_1, t_2, \ldots, t_i, \ldots, t_n\}$ in J, and if for any $t_0 \in R$ $\{t_1 + t_0, t_2 + t_0, \ldots, t_i + t_0, \ldots, t_n + t_0\}$ in J, the joint distribution of $x(t_1 + t_0), x(t_2 + t_0), \ldots, x(t_n + t_0)$ does not depend on t_0, i.e.

$$F(a_1(t_1 + t_0), a_2(t_2 + t_0), \ldots, a_n(t_n + t_0)) = F(a_1 t_1, a_2 t_2, \ldots, a_n t_n). \quad (1.2.3)$$

Moreover,

$$F(a_1 t_1, a_2 t_2, \ldots, a_n t_n) = F(a_1 0, a_2(t_2 - t_1), \ldots, a_n(t_n - t_{n-1})). \quad (1.2.4)$$

Definition 1.2.5. A stochastic process $x(t)$ defined on J is said to be a *wide-sense stationary process* if

(a) $E[x^2(t)] < \infty$ for each $t \in J$;
(b) $E[x(t)] = m$ is a constant;
(c) $E[(x(t) - m)(x(s) - m)] = R(t - s)$.

Example 1.2.2

(a) A *binary noise* process $x(t)$ defined on $t \in J$ as:

   (i) It takes values either $+1$ or -1 throughout successive time intervals of fixed length Δ.
   (ii) The values it takes in one interval is independent of the values taken in any other interval.

(iii) All member functions differing by a shift along the t-axis are equally likely.

A possible representation of $x(t)$ is $x(t) = y(t - A)$ with $y(t) = y_n$ on $(n - 1)\Delta < t < n\Delta$, where the random variable A has a uniform distribution in the interval $[0, \Delta]$, and the random variables y_n, $n \in \{\cdots - 3, -2, -1, 0, 1, 2, 3, \ldots\}$ are independent and identically distributed with the density function $f(y) = \frac{1}{2}[\delta(y + 1) + \delta(y - 1)]$. This is an example of the two-valued process $x_1 = -1$, $x_2 = 1$ with $p_1 = p_2 = P(x(t) = x_i)$. From Example 1.1.16, we have $E[x(t)] = \frac{1}{2} - \frac{1}{2} = 0$ for each $t \in R$. Now, we examine the correlation function $\Gamma(t, s)$. For t, s in R, there are two cases:

(i) $|t - s| > \Delta$;
(ii) $|t - s| \le \Delta$.

In the case of (i), from condition (b), $\Gamma(t, s) = E[x(t)x(s)] = E[x(t)]E[x(s)] = 0$ for $t, s \in R$. In the case of (ii), $x(t)$ and $x(s)$ are located in the same interval depending on the value A takes from $[0, \Delta]$. Therefore, $\Gamma(t, s) = E[x(t)x(s)] = 1 \cdot P(A < \Delta - |t - s|) = 1 - \frac{|t-s|}{\Delta}$ whenever $|t - s| \le \Delta$. We combine the conclusions in the above cited two cases. Thus, we have

$$\Gamma(t, s) = \begin{cases} 0, & \text{for } |t - s| > \Delta, \\ 1 - \dfrac{|t - s|}{\Delta}, & \text{for } |t - s| \le \Delta. \end{cases}$$

This shows that the binary noise process is wide-sense stationary, i.e. $E[x(t)] = 0$ and $\Gamma(t, s) = \Gamma(t - s)$.

(b) In Part (a) of Example 1.2.2, we can generalize the sequence of random variables y_n by modifying two-valued independent random variables with two parameters, namely $E[y_n] = 0$ and variance σ^2. From this, it is obvious that $E[x(t)] = 0$. Now, we examine the correlation function $\Gamma(t, s)$. For t, s in R, there are two cases:

(i) $|t - s| > \Delta$;
(ii) $|t - s| \le \Delta$.

In the case of (i), from condition (b), $\Gamma(t, s) = E[x(t)x(s)] = E[x(t)]E[x(s)] = 0$ for $t, s \in R$. In the case of (ii), $x(t)$ and $x(s)$ are located in the same interval depending on the value A takes from $[0, \Delta]$. Therefore, $\Gamma(t, s) = E[x(t)x(s)] = E[x^2(t)] = E[(x(t) - 0)^2] = \sigma^2 P(A < \Delta - |t - s|) = \sigma^2(1 - \frac{|t-s|}{\Delta})$ for $|t - s| \le \Delta$. By combining these conclusions, we obtain

$$\Gamma(t, s) = \begin{cases} 0, & \text{for } |t - s| > \Delta, \\ \sigma^2 \left(1 - \dfrac{|t - s|}{\Delta}\right), & \text{for } |t - s| \le \Delta. \end{cases}$$

This shows that the binary noise process is wide-sense stationary, i.e. $E[x(t)] = 0$ and $\Gamma(t, s) = \Gamma(t - s)$.

Observation 1.2.3

(i) From Definitions 1.2.4 and 1.2.5, we observe that a strictly stationary process with a finite second moment is also a wide-sense stationary process. However, the converse, in general, is not true. An important exception is the Gaussian process. A wide-sense stationary Gaussian process is also stationary.

(ii) We note that for a wide-sense stationary process $x(t)$ for $t \in J$, the variance σ^2 of $x(t)$ is independent of t in J. That is to say

$$\sigma^2 = R(0) = E[(x(t) - m)^2]. \tag{1.2.5}$$

(iii) We also note that if m in Definition 1.2.5 is zero $(E[x(t)] = 0)$, then the autocovariance $R(t - s)$ reduces to

$$E[x(t)x(s)] = R(t - s) = \Gamma(t - s), \tag{1.2.6}$$

and it is referred to as the *autocorrelation* function of the stochastic process. Moreover, $\Gamma(t - s) = \Gamma(s - t)$, and it satisfies the relation (1.2.2).

(iv) The function Γ defined in (1.2.6) is an even function, i.e. $\Gamma(\tau) = \Gamma(-\tau)$, for $\tau \in R$.

(v) $\Gamma(\tau)$ always exists and is finite. Furthermore, $|\Gamma(\tau)| \leq \Gamma(0) = E[x^2(t)]$.

(vi) If $\Gamma(\tau)$ is continuous at the origin, then it is uniformly continuous in τ. Hint: By Schwarz inequality, we have $|\Gamma(\tau + \epsilon) - \Gamma(\tau)| \leq 2\Gamma(0)[\Gamma(0) - \Gamma(\epsilon)]$.

(vii) $\Gamma(\tau)$ is non-negative definite.

(viii) Let $\Gamma(\tau)$ be continuous at the origin. It can be represented in the form

$$\Gamma(\tau) = \frac{1}{2} \int_{-\infty}^{\infty} e^{iw\tau} d\Phi(w). \tag{1.2.7}$$

If $\Phi(w)$ is absolutely continuous, then we have $S(w) = \frac{d}{dw}\Phi(w)$. Hence,

$$\Gamma(\tau) = \frac{1}{2} \int_{-\infty}^{\infty} e^{iw\tau} S(w) dw. \tag{1.2.8}$$

(ix) From (1.2.8), we note that the functions $\Gamma(\tau)$ and $S(w)$ form a Fourier transform pair. The relations (*Wiener–Khintchine formula*)

$$S(w) = \frac{1}{\pi} \int_{-\infty}^{\infty} e^{-iw\tau} \Gamma(\tau) d\tau \quad \text{and} \quad \Gamma(\tau) = \frac{1}{2} \int_{-\infty}^{\infty} e^{iw\tau} S(w) dw. \tag{1.2.9}$$

Definition 1.2.6. The power spectral density function of a wide-sense stationary stochastic process is defined by

$$S(w) = \frac{1}{\pi} \int_{-\infty}^{\infty} e^{-iw\tau} \Gamma(\tau) d\tau. \tag{1.2.10}$$

Observation 1.2.4. From the nature of the power spectral density function $S(w)$ and the autocorrelation function $\Gamma(\tau)$, the *Wiener–Khintchine formula* may be written as

$$S(w) = \frac{2}{\pi} \int_0^\infty \Gamma(\tau) \cos w\tau \, d\tau \quad \text{and} \quad \Gamma(\tau) = \int_0^\infty S(w) \cos w\tau \, dw. \quad (1.2.11)$$

We note that $S(\omega)$ is also an even function. For $\tau = 0$, (1.2.11) reduces to

$$\Gamma(0) = \int_0^\infty S(w) dw = E[x^2(t)]. \quad (1.2.12)$$

Example 1.2.3. In the following, we present autocorrelation functions and the corresponding spectral density functions of a few well-known wide-sense stationary stochastic processes.

(a) *Binary noise* (Example 1.2.2):

$$\Gamma(t) = \begin{cases} 0, & \text{for } |\tau| > \Delta, \\ \sigma^2 \left(1 - \frac{|\tau|}{\Delta}\right), & \text{for } |\tau| \leq \Delta, \end{cases}$$

$$S(w) = \frac{\sigma^2 \Delta}{\pi} \frac{\sin^2\left(\frac{w\Delta}{2}\right)}{\left(\frac{w\Delta}{2}\right)^2}.$$

(b) *Random telegraph signal:*

$$\Gamma(\tau) = e^{-\lambda|\tau|},$$

$$S(w) = \frac{2\lambda}{\pi(w^2 + \lambda^2)}.$$

Definition 1.2.7. A stochastic process $x(t)$ for $t \in R$ is said to be a *white noise*, if it is usually described as a Gaussian stationary process with a zero mean and a constant spectral density function, i.e.

$$S(w) = S_0 \quad \text{on } R. \quad (1.2.13)$$

Observation 1.2.5

(i) In the case of white noise, from (1.2.12), we have

$$\Gamma(0) = \int_0^\infty S(w) dw = E[x^2(t)] = \infty.$$

Therefore, it is not a second-order process. It cannot be a physical process.

(ii) From (1.2.13), we can draw the following conclusion:

$$\Gamma(\tau) = E[x(t + \tau)x(t)] = \pi S_0 \delta(\tau), \quad (1.2.14)$$

where $\delta(\tau)$ is the Dirac delta function. This is due to the fact (1.2.10).

(iii) Many processes that we encounter in practice are well approximated by white noise. The way to explain white noise is to recall the definition of the Dirac delta function (1.1.27). The delta function is never used outside of an integral. The same is true of white noise.

Definition 1.2.8. A stochastic process $x(t)$ for $t \in J$ is said to be a process of independent increments if for any partition $\mathcal{P}$ of $J : t_0 < t_1 < t_2 \cdots < t_i < \cdots < t_n$, the random variables $x(t_0), x(t_1) - x(t_0), x(t_2) - x(t_1), \ldots, x(t_{i+1}) - x(t_i), \ldots, x(t_n) - x(t_{n-1})$ are mutually independent.

Observation 1.2.6. Let $x(t_0), x(t_1) - x(t_0), \ldots, x(t_{i+1}) - x(t_i), \ldots, x(t_{n-1}) - x(t_n)$ be mutually independent random variables. In addition, let $F(x_0 t_0, (x_1 - x_0)t_0 t_1, \ldots, (x_n - x_{n-1})(t_{n-1}t_n))$ and $f(a_0 t_0, a_1 t_1, a_2 t_2, \ldots, a_n t_n)$ be the joint distribution and density (if it exists) functions of these random variables, respectively. From Definition 1.1.24, we have:

(i) $F(x_0 t_0, (x_1 - x_0)t_0 t_1, \ldots, (x_n - x_{n-1})t_{n-1}t_n) = F(x_0 t_0) \prod_{i=1}^{n} F((x_i - x_{i-2}) t_i t_{i-1})$.

(ii) $f(x_0 t_0, (x_1 - x_0)t_0 t_1, \ldots, (x_n - x_{n-1})t_{n-1}t_n) = f(x_0 t_0) \prod_{i=1}^{n} f((x_i - x_{i-2}) t_i t_{i-1})$, where F and f are the marginal probability distribution and density functions of the random variables $x(t_0), x(t_1) - x(t_0), x(t_2) - x(t_1), \ldots, x(t_{i+1}) - x(t_i), \ldots, x(t_{n-1}) - x(t_n)$, respectively, for $i = 0, 1, \ldots, n - 1$.

Definition 1.2.9. A random process $x \in R[J, R[\Omega, R]]$ is said to be a Markov process if for any increasing collection of partition $\mathcal{P} : t_0 < t_1 < t_2 \cdots < t_i < \cdots < t_n$ of J and $B \in \mathfrak{B}$, $\mathfrak{B}$ is the Borel σ-algebra of $\mathbb{R}$,

$$P(\{x(t_n) \in \mathfrak{B} | x(t_1), x(t_2), \ldots, x(t_{n-1})\}) = P(\{x(t_n) \in \mathfrak{B} | x(t_{n-1})\})$$
$$= F(\mathfrak{B}t_n | x_{n-1}t_{n-1}). \qquad (1.2.15)$$

The following result provides various equivalent formulations of the definition of the Markov process.

Observation 1.2.7

(i) For any partition $\mathcal{P} : t_0 < t_1 < \cdots < t_i < \cdots < t_n$ of J, if the finite-dimensional distribution $F(x_0 t_0, x_1 t_1, x_2 t_2, \ldots, x_n t_n)$ of a Markov process x has density function $f(x_0 t_0, x_1, t_1, x_2 t_2, \ldots, x_n t_n)$, then $f(x_0 t_0, x_1 t_1, x_2 t_2, \ldots, x_n t_n)$ can be written as

$$f(x_0 t_0, x_1 t_1, x_2 t_2, \ldots, x_n t_n)$$
$$= f(x_n t_n | x_0 t_0, x_1 t_1, x_2 t_2, \ldots, x_{n-1}t_{n-1})f(x_0 t_0, x_1 t_1, x_2 t_2, \ldots, x_{n-1}t_{n-1})$$
$$= f(x_n t_n | x_{n-1}t_{n-1})f(x_0 t_0, x_1 t_1, x_2 t_2, \ldots, x_{n-1}t_{n-1}). \qquad (1.2.16)$$

Moreover, by using (1.2.16), we find an expression for $f(x_0 t_0, x_1 t_1, x_2 t_2, \ldots, x_n t_n)$:

$$f(x_0 t_0, x_1 t_1, x_2 t_2, \ldots, x_n t_n) = f(x_0 t_0) \prod_{i=1}^{n} f(x_i t_i | x_{i-1} t_{i-1}). \qquad (1.2.17)$$

(ii) From (1.2.17), we conclude that all finite-dimensional distributions are completely determined by two-dimensional distributions. Moreover, this statement is valid for all Markov processes, and it is independent of the existence of density functions.

(iii) Let $E[x(t)] = m(t)$ be the mean of the Markov process $x(t)$. From Definition 1.2.9, we conclude that $x(t)$ is a Markov process if and only if $x(t) - m(t)$ is Markov. As a result of this, we can always assume (without loss of generality) that the Markov process $x \in R[J, R[\Omega, R]]$ has a zero mean.

(iv) Let $x \in R[J, R[\Omega, R]]$ be a Gaussian–Markov process with mean $E[x(t)] = 0$. In addition, let $R(t, s) = E[x(t)x(s)] = \Gamma(t, s)$ be the covariance (autocorrelation) function of x. Then,

$$E[x(t)|x(s) = x] = \frac{\Gamma(t, s)}{\Gamma(s, s)}x. \tag{1.2.18}$$

We note that $x(s)$ and $x(t) - \frac{\Gamma(t,s)}{\Gamma(s,s)}x(s)$ are Gaussian and uncorrelated.

Theorem 1.2.1. *Each of the following condition is equivalent to the relation* (1.2.15):

(i) *For $t_0 \leq s \leq t$, $s, t \in J$, and $B \in \mathfrak{B}$,*

$$P(\{x(t) \in B|\mathfrak{F}_s\}) = P(\{x(t) \in B|x(s)\}),$$

where $\mathfrak{F}_s = \mathfrak{F}[t_0, s]$ is the smallest sub-σ-algebra of $\mathfrak{F}$ generated by all random variables $x(u)$ for $t_0 \leq u \leq s$.

(ii) *For $t_0 \leq s \leq t \leq \alpha$, $s, t, \alpha \in J$, y is $\mathfrak{F}_\alpha^t$-measurable and integrable,*

$$E[y|\mathfrak{F}_s] = E[y|x(s)],$$

where $\mathfrak{F}_\alpha^t = \mathfrak{F}[t, \alpha]$ is the smallest sub-σ-algebra of $\mathfrak{F}$ generated by all random variables $x(u)$ for $t \leq u \leq \alpha$.

(iii) *For $t_0 \leq s \leq t \leq \alpha$ and $A \in \mathfrak{F}_\alpha^t$,*

$$P(A|\mathfrak{F}_s) = P(A|x(s)).$$

(iv) *For $t_0 \leq t_1 \leq t \leq t_2 \leq \alpha$, $t, \alpha \in J$, $A_1 \in \mathfrak{F}_{t_1}$, $A_2 \in \mathfrak{F}_\alpha^{t_2}$,*

$$P(A_1 \cap A_2|\mathfrak{F}_t) = P(A_1|x(t))P(A_2|x(t)).$$

Observation 1.2.8. The concept of the Markov property states that the knowledge of the future state of the system depends only on the knowledge of the most recent state, and it is independent of the past information of the system. The interpretation of Part (iv) of Theorem 1.2.1 is: knowing the present state of a Markov process, the past and future states of the process are statistically independent.

Lemma 1.2.1 (Chapman–Kolmogorov equation). *Let $x \in R[J, R[\Omega, R]]$ be a Markov process, and let $\mathfrak{F}_t = \mathfrak{F}[t_0, t]$ be the smallest sub-σ-algebra of $\mathfrak{F}$ generated by all random variables $x(t)$ for $t \in J$. Then,*

$$F(Bt|xs) = \int_{-\infty}^{\infty} F(Bt|yu)dF(dy\,u|xs), \qquad (1.2.19)$$

where $F(Bt|xs) = P(\{x(t) \in B|x(s) = x\})$.

Proof. From the definition with probability 1, we have

$$F(Bt|xs) = P(\{x(t) \in B|x(s)\}) = P(\{x(t) \in B|\mathfrak{F}_s\}).$$

From this and the fact that for $t_0 \leq s \leq u$, $\mathfrak{F}_s \subseteq \mathfrak{F}_u$ and hence $\mathfrak{F}_s \cap \mathfrak{F}_u = \mathfrak{F}_s$, and the joint probability formula (1.2.17), Theorem 1.1.2(ix), and Theorem 1.2.1(i), we have

$$F(Bt|xs) = P(\{x(t) \in B|\mathfrak{F}_s\}) = P(\{x(t) \in B|\mathfrak{F}_s \cap \mathfrak{F}_u\})$$

$$= E[P(\{x(t) \in B|\mathfrak{F}_u\})|\mathfrak{F}_s] = E[F(Bt|x(u)u)|\mathfrak{F}_s]$$

$$= E[F(Bt|x(u)u)|x(s)] = \int_{-\infty}^{\infty} F(Bt|yu)dF(yu|xs).$$

This establishes the validity of the Chapman–Kolmogorov equation. $\qquad\square$

Observation 1.2.9. We note that $F(Bt|xs)$ has certain properties. This function is called a transition probability of the Markov process. Moreover, Chapman–Kolmogorov equation (1.2.19) in the context of the transition probability density of the Markov process $f(xt|x_0t_0)$ reduces to

$$f(xt|x_0t_0) = \int_{-\infty}^{\infty} f(xt|zu)f(zu|x_0t_0)dz. \qquad (1.2.20)$$

Let $x \in R[J, R[\Omega, R]]$ be a Gaussian–Markov process with mean $E[x(t)] = 0$. In addition, let $R(t, s) = E[x(t)x(s)] = \Gamma(t, s)$ be the covariance (autocorrelation) function of x. Then,

$$E[x(t)|x(s) = y] = \frac{\Gamma(t, s)}{\Gamma(s, s)}y. \qquad (1.2.21)$$

Moreover,

$$E[x(t)|x(t_0) = x_0] = \int_{-\infty}^{\infty} xdF(xt|x_0t_0),$$

$$E[x(t)|x(t_0)] = x_0] = \frac{\Gamma(t, s)}{\Gamma(s, s)} \int_{-\infty}^{\infty} y\,dF(ys|x_0t_0) = \frac{\Gamma(t, s)}{\Gamma(s, s)}\frac{\Gamma(s, t_0)}{\Gamma(t_0, t_0)}x_0, \qquad (1.2.22)$$

and hence

$$\frac{\Gamma(t, t_0)}{\Gamma(t_0, t_0)} = \frac{\Gamma(t, s)}{\Gamma(s, s)}\frac{\Gamma(s, t_0)}{\Gamma(t_0, t_0)}. \qquad (1.2.23)$$

If $\Gamma(s, s)$ is strictly positive in the interior of J, then any solution to (1.2.23) has the form

$$\Gamma(t, s) = g(\max\{t, s\})h(\min\{t, s\}), \quad \text{for } t, s \in J. \tag{1.2.24}$$

Definition 1.2.10. Let $x \in R[J, R[\Omega, R]]$ be a Markov process. The Markov process x is said to homogeneous with respect to time if its transition probability $P(Bt|xs)$ is stationary, i.e. if it satisfies the following condition:

$$P(B(t + u)|x(s + u)) = P(Bt|xs), \tag{1.2.25}$$

for $t_0 \le s \le t \le \alpha$ and $t_0 \le s + u \le t + u \le \alpha$, $s, t, \alpha \in J$. In this case, the Chapman–Kolmogorov equation (1.2.19) reduces to

$$P(B(t + s)|xs) = \int_{-\infty}^{\infty} F(Bt|yu)dF(yu|xs). \tag{1.2.26}$$

Definition 1.2.11. A process $b(t)$ is said to be a *Brownian motion process* if it is defined for $t \ge 0$, and if it is a Gaussian process with independent increments.

Observation 1.2.10

(i) We observe that for $t > s \ge 0$, the structure of the variance function of the increment process $b(t) - b(s)$ associated with a Brownian motion process $b(t)$ has the form

$$D(s, t) = E[((b(t) - b(s) - m(t) + m(s))^2] = D(s, u) + D(u, t), \tag{1.2.27}$$

where $s < u < t$, with $E[b(t)] = m(t)$ and $E(b(s)) = m(s)$.

(ii) For each $t > s \ge 0$, let $b(t) - b(s)$ be an increment associated with a Brownian motion process $b(t)$ with mean $E[b(t)] = m(t)$. We define a random variable $B(s, t) = b(t) - b(s) - m(t) + m(s)$. We observe that

$$E[B(s, t)] = E[b(t) - b(s) - m(t) + m(s)] = 0, \tag{1.2.28}$$

$$\text{var}(B(s, t)) = E[(b(t) - b(s) - m(t) + m(s))^2] = D(s, t), \tag{1.2.29}$$

and hence the process $B(s, t)$ is a Gaussian with a zero mean and variance $D(s, t)$.

(iii) If for any $t > s \ge 0$, an increment process $b(t) - b(s)$ associated with a Brownian motion process $b(t)$ is stationary process, then $b(t)$ is called a *process with stationary increments*. The variance function of the increment process $b(t) - b(s)$ of a Brownian motion process $b(t)$ with stationary increments has the form $D(s, t) = D(t - s)$. We set $t - s = t - u + u - s$. By using this, the expression $D(s,t)$ in (1.2.27) reduces to

$$D(t - s) = D(t - u) + D(u - s), \tag{1.2.30}$$

which implies that [42]

$$D(t) = \sigma^2 t, \quad \text{and also } E[b(t)] = m(t) = mt. \tag{1.2.31}$$

Moreover, $E[b(t) - b(s)] = m(t - s)$, where m is a constant.

(iv) From (ii) and (iii), for a Brownian motion process $b(t)$ with stationary increments, the relations (1.2.28) and (1.2.29) reduce to

$$E[B(s,t)] = 0$$

and

$$\text{var}(B(s,t)) = E[(b(t) - b(s) - m(t) + m(s))^2] = \sigma^2[t - s]. \tag{1.2.32}$$

From this, we conclude that the probability distribution of an increment process $B(s,t)$ defined by $b(t) - b(s)$ coincides with the probability distribution of $B(\tau)$. Moreover, it is normal with a zero mean and variance $\sigma^2 \tau$, i.e.

$$P(\{\omega : a < B(\tau) < b\}) = \frac{1}{\sigma\sqrt{2\pi\tau}} \int_a^b \exp\left[-\frac{u^2}{2\sigma^2\tau}\right] du, \tag{1.2.33}$$

where $\tau = |t - s|$ and $\tau \geq 0$. This exhibits that a Brownian motion process $b(t)$ with stationary increments induces a Brownian motion process $B(\tau)$ with a zero mean and variance $\sigma^2 \tau$ for $\tau \geq 0$.

(v) Let $b(t)$ be a Brownian motion process with stationary increments, and let $B(\tau)$ be a Brownian motion process induced by stationary increments $b(t) - b(s)$ of $b(t)$ with a zero mean and variance $\sigma^2 \tau$ for $\tau \geq 0$. We observe that for any $\tau, \nu, \tau \geq 0$ and $\nu \geq 0$, we compute $E[B(\tau)B(\nu)]$. For this purpose, for any $\tau, \nu \geq 0$ and $\tau \neq \nu$, we observe that

$$uv = \frac{1}{2}[u^2 + v^2 - (u - v)^2], \tag{1.2.34}$$

$$\tau + \nu = 2\min\{\tau, \nu\} + |\tau - \nu|, \tag{1.2.35}$$

and

$$E[(B(\tau) - B(\nu))^2] = \sigma^2|\tau - \nu|. \tag{1.2.36}$$

We consider

$$E[B(\tau)B(\nu)] = \frac{1}{2}[E[(B(\tau))^2] + E[(B(\nu))^2] - E[(B(\tau) - B(\nu))^2]]$$

$$= \frac{1}{2}[\sigma^2\tau + \sigma^2\nu - \sigma^2|\tau - \nu|] \quad \text{[from (1.2.35)]}$$

$$= \frac{1}{2}[\sigma^2(2\min\{\tau, \nu\} + |\tau - \nu|) - \sigma^2|\tau - \nu|]$$

$$= \frac{1}{2}[\sigma^2(2\min\{\tau, \nu\}]$$

and hence

$$E[B(\tau)B(\nu)] = \sigma^2 \min\{\nu, \tau\} = \Gamma(\tau, \nu) = R(\tau, \nu), \qquad (1.2.37)$$

where $\Gamma(\tau, \nu)$ and $R(\tau, \nu)$ are autocorrelation and autocovariance functions of $B(\tau)$ for $\tau \geq 0$. Thus, the product of two Gaussian stationary processes is not a stationary process.

(vi) We note that a Gaussian process with a zero mean and a positive covariance function determines a Brownian motion process.

(vii) A stationary Brownian motion process is a Markov process. For any partition $\mathcal{P}$ of $J : t_0 < t_1 < t_2 \cdots t_i < \cdots < t_n$, and from (1.2.16) and (1.2.17), we have

$$f(x_0 t_0, x_1 t_1, x_2 t_2, \ldots, x_{n-1} t_{n-1}, x_n t_n)$$

$$= f(x_0 t_0) \prod_{i=1}^{n} f(x_i t_i | x_{i-1} t_{i-1})$$

$$= f(x_0 t_0) \prod_{i=1}^{n} \frac{1}{\sigma \sqrt{2\pi(t_i - t_{i-1})}} \exp\left[-\frac{1}{2} \frac{(x_1 - x_{i-1})^2}{\sigma^2(t_i - t_{i-1})}\right],$$

and

$$P(\{x(t_n) \leq x_n | x(t_i) = x_i, i = 0, 1, \ldots, n-1\})$$

$$= F(x_n t_n | x_0 t_0, x_1 t_1, x_2 t_2, \ldots, x_{n-1} t_{n-1})$$

$$= \int_{-\infty}^{\infty} f(z t_n | x_0 t_0, x_1 t_1, x_2 t_2, \ldots, x_{n-1} t_{n-1}) dz$$

$$= \int_{-\infty}^{\infty} \frac{f(x_0 t_0, x_1 t_1, x_2 t_2, \ldots x_{n-1} t_{n-1} z t_n)}{f(x_0 t_0, x_1 t_1, x_2 t_2, \ldots, x_{n-1} t_{n-1})} dz$$

$$= \int_{-\infty}^{\infty} \frac{1}{\sigma \sqrt{2\pi(t_n - t_{n-1})}} \exp\left[-\frac{1}{2} \frac{(z - x_{n-1})^2}{\sigma^2(t_n - t_{n-1})}\right] dz$$

$$= P(\{x(t_n) \leq x_n | x(t_{n-1}) = x_{n-1}\}.$$

This establishes the fact that a stationary Brownian motion process is a Markov process.

Example 1.2.4

(a) Let $b(t)$ be a Brownian motion process with stationary increments, and for $\tau \geq 0$, let $B(\tau)$ be a Brownian motion process induced by the stationary increment process $b(t) - b(s)$ of $b(t)$, with a zero mean and variance $\sigma^2 \tau$. In addition, let us define a stochastic process by

$$W^{\Delta t}(t) = \frac{B(t + \Delta t) - B(t)}{\Delta t}, \quad \text{for } t \geq 0 \quad \text{and} \quad \Delta t > 0. \qquad (1.2.38)$$

Again, by applying Lemma 1.1.1 to $W^{\Delta t}(t)$, we conclude that $W^{\Delta t}(t)$ is a Brownian motion process with a stationary increment with a zero mean and constant variance $\frac{\sigma^2}{\Delta t}$. By imitating the discussion in Observation 1.2.10(v) and using

$$E\left[\left[\frac{B(t+\Delta t)-B(t)}{\Delta t}\right]\left[\frac{B(t)-B(s)}{\Delta t}\right]\right]=0,$$

the independent increment property, we have

$$E[W^{\Delta t}(t)W^{\Delta t}(s)]$$

$$= \frac{1}{2}E[(W^{\Delta t}(t))^2 + (W^{\Delta t}(s))^2 - (W^{\Delta t}(t) - W^{\Delta t}(s))^2] \quad \text{[from (1.2.34)]}$$

$$= \frac{1}{2}\left[E\left[\frac{B(t+\Delta t)-B(t)}{\Delta t}\right]^2 + E\left[\frac{B(s+\Delta t)-B(s)}{\Delta t}\right]^2\right.$$

$$\left. - E[W^{\Delta t}(t) - W^{\Delta t}(s)]^2\right] \quad \text{[from (1.2.38)]}$$

$$= \frac{1}{2}\left[\frac{\sigma^2\Delta t}{(\Delta t)^2} + \frac{\sigma^2\Delta t}{(\Delta t)^2} - E\left[\left(\frac{B(t+\Delta t)-B(t)}{\Delta t} - \frac{B(s+\Delta t)-B(s)}{\Delta t}\right)^2\right]\right]$$

$$= \frac{1}{2}\left[2\frac{\sigma^2\Delta t}{(\Delta t)^2} - E\left[\left(\frac{B(t+\Delta t)-B(s+\Delta t)}{\Delta t} - \frac{B(t)-B(s)}{\Delta t}\right)^2\right]\right]$$

$$= \frac{1}{2}\left[2\frac{\sigma^2\Delta t}{(\Delta t)^2} - \left[E\left[\frac{B(t+\Delta t)-B(s+\Delta t)}{\Delta t}\right]^2 + E\left[\frac{B(t)-B(s)}{\Delta t}\right]^2\right]\right]$$

$$= \frac{1}{2}\left[2\frac{\sigma^2\Delta t}{(\Delta t)^2} - \left[\frac{\sigma^2|t-s|}{(\Delta t)^2} + \frac{\sigma^2|t-s|}{(\Delta t)^2}\right]\right] = \frac{1}{2}\left[2\frac{\sigma^2\Delta t}{(\Delta t)^2} - 2\frac{\sigma^2|t-s|}{(\Delta t)^2}\right]$$

$$= \frac{\sigma^2}{\Delta t}\left(1 - \frac{|t-s|}{\Delta t}\right).$$

Thus, the autocovariance (autocorrelation) function is given by

$$E[W^{\Delta t}(t)W^{\Delta t}(s)] = \frac{\sigma^2}{\Delta t}\left(1 - \frac{|t-s|}{\Delta t}\right) = \Gamma(\tau,\nu) = R(\tau,\nu). \qquad (1.2.39)$$

(b) From Example 1.2.3(a), we further draw the following conclusion. The spectral density function $S(w)$ of a Brownian motion process $W^{\Delta t}(t)$ defined in (1.2.38) with a stationary increment with a zero mean and constant variance $\frac{\sigma^2}{\Delta t}$ is

$$S(w,\Delta t) = \frac{\sigma^2}{\pi}\frac{\sin^2(\frac{w\Delta t}{2})}{(\frac{w\Delta t}{2})^2}. \qquad (1.2.40)$$

(c) From (1.2.40), we observe that

$$\lim_{\Delta t \to 0} S(w, \Delta t) = \lim_{\Delta t \to 0} \left[\frac{\sigma^2}{\pi} \frac{\sin^2(\frac{w\Delta t}{2})}{(\frac{w\Delta t}{2})^2} \right]$$

$$= \frac{\sigma^2}{\pi} \lim_{\Delta t \to 0} \left[\frac{\sin(\frac{w\Delta t}{2})}{(\frac{w\Delta t}{2})} \right]^2 = \frac{\sigma^2}{\pi}. \tag{1.2.41}$$

From (1.2.41) and Definition 1.2.7, we conclude that a *white noise* process can be considered as a limiting process of a Brownian motion process as $\Delta t \to 0$.

Definition 1.2.12. A process $w(t)$ is said to be a *Wiener process* if it is a defined for $t \geq 0$, and if it is a homogeneous Gaussian process with independent increments for which $w(0) = 0$, $E[w(t)] = 0$ and $V[w(t)] = E[(w(t))^2] = t$. It is also referred to as a *normalized Wiener or Brownian motion process*.

Observation 1.2.11

(i) From Definition 1.2.12, it is known [Observation 1.2.10(iv)] that the probability distribution of $w(t + h) - w(t)$ coincides with the probability distribution of $w(h)$, and it is normal with a zero mean and variance h, i.e.

$$P(\{\omega : a < w(h) < b\}) = \frac{1}{\sqrt{2\pi h}} \int_a^b \exp\left[-\frac{u^2}{2h}\right] du, \tag{1.2.42}$$

and its characteristic function is given by the formula

$$E[\exp[izw(h)]] = \exp\left[-\frac{z^2 h}{2}\right], \tag{1.2.43}$$

where $i = \sqrt{-1}$.

(ii) Moreover, it is obvious from the definition of the Wiener process that all observations stated in Observation 1.2.10 remain valid.

(iii) From Example 1.2.4, we conclude that a stochastic process defined by

$$W^{\Delta t}(t) = \frac{w(t + \Delta t) - w(t)}{\Delta t}, \quad \text{for } t \geq 0 \quad \text{and} \quad \Delta t > 0 \tag{1.2.44}$$

is a Brownian motion process with a stationary increment, a zero mean, and constant variance $\frac{1}{\Delta t}$. It is induced by $w(t + \Delta t) - w(t)$. From (1.2.39) and (1.2.40), it clear that the autocovariance (autocorrelation) and spectral density functions are given by

$$E[W^{\Delta t}(t)W^{\Delta t}(s)] = \frac{1}{\Delta t}\left(1 - \frac{|t - s|}{\Delta t}\right) = \Gamma(\tau, \nu) = R(\tau, \nu), \tag{1.2.45}$$

and

$$S(w, \Delta t) = \frac{1}{\pi} \frac{\sin^2(\frac{w\Delta t}{2})}{(\frac{w\Delta t}{2})^2}, \tag{1.2.46}$$

respectively.

(iv) From (1.2.46), we observe that

$$\lim_{\Delta t \to 0} S(w, \Delta t) = \lim_{\Delta t \to 0} \left[\frac{1}{\pi} \frac{\sin^2(\frac{w\Delta t}{2})}{(\frac{w\Delta t}{2})^2} \right] = \frac{1}{\pi}. \tag{1.2.47}$$

From (1.2.47) and Definition 1.2.7, we conclude that the *white noise* process can also be considered as a limiting process of the Wiener process as $\Delta t \to 0$.

(v) We further remark that the white noise process associated with the limiting spectral density function in (1.2.47) can be considered to be a Gaussian process with a zero mean and covariance $\delta(t - s)$. $\delta(t - s)$ is the covariant function of the white noise. Thus, white noise $\xi(t)$ is the derivative of the Wiener process, and its derivative is a generalized function [8]. This is denoted by

$$\xi(t) = \frac{d}{dt} w(t) = w'(t). \tag{1.2.48}$$

Observation 1.2.12. Let w be a scalar Wiener process defined for $t > 0$. In addition, let $[t, t + \Delta t]$ be an interval with finite length Δt. For any natural number n, we denote

$$\tau = \frac{\Delta t}{n}, \quad t_0 = t, \quad t_1 = t + \frac{\Delta t}{n},$$

$$t_2 = t_1 + 2\frac{\Delta t}{n}, \dots, t_k = t_{k-1} + k\frac{\Delta t}{n}, \dots, t_n = t + \Delta t,$$

where $k = 1, 2, 3, \dots, n$. From the definition of the Wiener process w, the n random variables

$$X_1 = w(t_1) - w(t_0), \quad X_2 = w(t_2) - w(t_1), \dots,$$

$$X_k = w(t_k) - w(t_{k-1}), \dots, X_n = w(t_n) - w(t_{n-1})$$

have a zero mean, and are independent and identically distributed. In fact, each $i, k = 1, 2, 3, \dots, n$:

$$E[X_k] = 0, \quad V[X_k] = E[(w(t_k) - w(t_{k-1})^2] = \frac{\Delta t}{n}$$

and

$$E[X_k X_i] = 0, \quad \text{for } k \neq i.$$

Moreover, for $S_n = \sum_{k=1}^{n} X_k$,

$$E[S_n] = 0, \quad V[S_n] = \sum_{k=1}^{n} \frac{\Delta t}{n} = \Delta t.$$

We define

$$W_n = \frac{\sum_{k=1}^{n} X_k}{\sqrt{\Delta t}}.$$

Now, we apply Theorem 1.1.5, and conclude that the distribution of W_n approaches to the standard normal distribution ($\mu = 0$ and $\sigma^2 = 1$) for $n \geq 1$ as $n \to \infty$, i.e.

$$P\{\omega : W_n(\omega) \leq a\} \to \frac{1}{\sqrt{2\pi}} \int_{-\infty}^{a} \exp\left[\frac{-x^2}{2}\right] dx \quad \text{as } n \to \infty.$$

Moreover, the random variable S_n is approximated by $S_n \simeq \sqrt{\Delta t} W_n$ for sufficiently large n, where W_n has a standard normal distribution. On the other hand, the distribution of S_n approaches to the normal distribution with $\mu = 0$ and $\sigma^2 = \Delta t$ as $n \to \infty$, i.e.

$$P\{\omega : S_n(\omega) \leq a\} \to \frac{1}{\sqrt{2\pi\Delta t}} \int_{-\infty}^{a} \exp\left[\frac{-x^2}{2\Delta t}\right] dx \quad \text{as } n \to \infty.$$

The following theorem by J. L. Doob provides an alternative definition of the Wiener process. It is the basis for the *Itô–Doob stochastic calculus*.

Theorem 1.2.2. *Let $(\Omega, \mathfrak{F}, P)$ be a complete probability space, and let the process $w(t)$ be defined and continuous with probability 1 for $t \geq 0$, $w(0) = 0$, and for all $t \geq 0$. In addition, $\mathfrak{F}_t$ be an increasing family of sub-σ-algebras, of $\mathfrak{F}$ ($\mathfrak{F}_{t_1} \subset \mathfrak{F}_{t_2}$ for $t_1 < t$) satisfying the following assumptions:*

(i) *For all $t \geq 0$, the process $w(t)$ is measurable with respect to $\mathfrak{F}_t$;*
(ii) *$E[(w(t+\Delta t) - w(t))|\mathfrak{F}_t] = 0$, w.p.1 (with probability 1) for all $t \geq 0$ and $\Delta t > 0$;*
(iii) *$E[(w(t + \Delta t) - w(t))^2|\mathfrak{F}_t] = \Delta t$, w.p.1 (with probability one) for all $t \geq 0$ and $\Delta t > 0$.*

Then $w(t)$ is a Wiener process.

In the following, an approximation of the *Wiener process* by means of the *random walk process* is illustrated.

Illustration 1.2.1. First, we construct a random walk process. For this purpose, we consider a motion of a particle in a straight line with a starting time at $t = 0$ at the origin $x = 0$. Under the influence of a single independent random impact, in every time interval of length τ, the particle is displaced either to the right (positive direction — "success") or to the left (negative direction — "failure") with a displacement Z of magnitude of Δx. The particle has undergone n independent and identical random impacts with n displacements (either to the right or to the left) at times $\tau, 2\tau, \ldots, k\tau, \ldots, n\tau$. The positions of the particle at times $\tau, 2\tau, \ldots, k\tau, \ldots, n\tau$ are denoted by $x(\tau), x(2\tau), \ldots, x(k\tau), \ldots, x(n\tau)$. The final position of the particle is $x = x(t) = x(n\tau)$ at time t ($t = n\tau$). For p, $0 < p < 1$, and for each impact, it is assumed that

$$P(\{Z = \Delta x > 0\}) = p \quad \text{or} \quad P(\{Z = -\Delta x < 0\}) = 1 - p = q. \tag{1.2.49}$$

Let m be the number of steps taken by the particle to the right. The number of steps taken by the particle to the left will be $n - m$. From the above description, it is clear that $x(k\tau)$ is a discrete-valued stochastic process which is the sum of k independent random variables Z_i, $i = 1, 2, \ldots, k$ and $k = 1, 2, \ldots, n$. We note that for each k, $x(k\tau)$ is a binomial random variable with (k, p) as parameters. From Example 1.1.16, the probability distribution is given by (1.1.23). This stochastic process $x(k\tau)$ is referred to as a two-sided *random walk process*.

In particular, for $k = n$, $x(n\tau)$, let m be the number of steps taken by the particle to the right, and hence the number of steps taken by the particle to the left will be $n - m$. In this case, the distance $D(t) = D(n\tau)$ covered in n steps is described by

$$D(t) = D(n\tau) = [m - (n - m)]\Delta x = [2m - n]\Delta x. \tag{1.2.50}$$

Moreover, from (1.2.49) and by the application of Theorem 1.1.1 and Example 1.1.21, the mean distance is given by

$$E[D(t)] = E[D(n\tau)] = (2np - n)\Delta x = n(p + p - 1)\Delta x$$
$$= n(p - (1 - p))\Delta = n(p - q)\Delta x \tag{1.2.51}$$

and its variance is

$$\text{Var}(D(t)) = \text{Var}(D(n\tau)) = (\Delta x)^2 \, \text{Var}(2x(n\tau) - n) = 4npq(\Delta x)^2. \tag{1.2.52}$$

From the definition of t, (1.2.51) and (1.2.52) can be rewritten as

$$E[D(t)] = t\frac{p - q}{\Delta x}\frac{(\Delta x)^2}{\tau}, \tag{1.2.53}$$

and

$$\text{Var}(D(t)) = 4tpq\,\frac{(\Delta x)^2}{\tau}, \tag{1.2.54}$$

respectively.

We note that the physical nature of the problem imposes certain restrictions on Δx and τ. Similarly, the parameter p cannot be arbitrary. Moreover, the particle cannot go to "infinity" on an interval whose length is small. In view of these considerations, the following conditions seem to be natural for sufficiently large n: for $x = n\Delta x$, $t = n\tau$,

$$\lim_{\tau \to 0} \left[\frac{(\Delta x)^2}{\tau}\right] = \sigma, \quad \lim_{\Delta x \to 0} \left[\frac{(p - q)}{\Delta x}\right] = \frac{\mu}{\sigma} \quad \text{and} \quad \lim_{n \to \infty} pq = \frac{1}{4}, \tag{1.2.55}$$

where μ and σ are certain constants, the former is called the drift, and the latter, the diffusion coefficient. From (1.2.53)–(1.2.55), we obtain

$$\lim_{\Delta x \to 0} \lim_{\tau \to 0} E[D(t)] = t \lim_{\Delta x \to 0} \lim_{\tau \to 0} \left[\frac{p-q}{\Delta x} \frac{(\Delta x)^2}{\tau} \right] \qquad \text{(by the properties of the limit)}$$

$$= \mu t, \qquad\qquad\qquad\qquad (1.2.56)$$

and

$$\lim_{\Delta x \to 0} \lim_{\tau \to 0} \mathrm{Var}(D(t)) = 4t \lim_{\Delta x \to 0} \lim_{\tau \to 0} \left[pq \frac{(\Delta x)^2}{\tau} \right] \qquad \text{(by the properties of the limit)}$$

$$= \sigma t, \qquad\qquad\qquad\qquad (1.2.57)$$

respectively. Now, we define

$$y(t) = \frac{D(t) - n(p-q)\Delta x}{\sqrt{4npq(\Delta x)^2}} = \frac{D(n\tau) - n(p-q)\Delta x}{\sqrt{4npq(\Delta x)^2}}. \qquad (1.2.58)$$

From (1.2.50) and (1.2.58), we have

$$y(n\tau) = y(t) = \frac{D(t) - n(p-q)\Delta x}{\sqrt{4npq(\Delta x)^2}} = \frac{D(n\tau) - n(p-q)\Delta x}{\sqrt{4npq(\Delta x)^2}}$$

$$= \frac{2x(t)\Delta x - n\Delta x - 2np\Delta x + n\Delta x}{2(\Delta x)\sqrt{npq}} = \frac{2x(n\tau)\Delta x - n\Delta x - 2np\Delta x + n\Delta x}{2(\Delta x)\sqrt{npq}}$$

$$= \frac{x(t) - np}{\sqrt{npq}} = \frac{x(n\tau) - np}{\sqrt{npq}}. \qquad\qquad (1.2.59)$$

By the application of Theorem 1.1.3 (DeMoivre–Laplace central limit theorem), we conclude that the process $y(n\tau) = y(t)$ is approximated by a standard normal distribution (a zero mean and variance 1). Moreover, from (1.2.58), $n = \frac{t}{\tau}$ and (1.2.55), we have

$$D(t) \simeq t \frac{p-q}{\Delta x} \frac{(\Delta x)^2}{\tau} + \sqrt{4tpq \frac{(\Delta x)^2}{\tau}} y(n\tau)$$

$$\simeq \mu t + \sqrt{t}\sigma y(n\tau) \simeq \mu t + \sqrt{\sigma}\sqrt{t} y(n\tau)$$

and hence

$$\lim_{\Delta x \to 0} \lim_{\tau \to 0} \sqrt{t} y(t) = w(t), \qquad\qquad (1.2.60)$$

where $w(t)$ is a Wiener process.

Definition 1.2.13. A stochastic process $n(t)$ is said to be a *counting process* if $n(t)$ represents the total number of "events" that have occurred upto the time t.

Example 1.2.5

(a) Let $n(t)$ be for the number of persons who have entered a particular store at or prior to time t. $n(t)$ is a counting process in which an event corresponds to a person entering the store. On the other hand, if $n(t)$ is equal to the number of persons in the store at time t, $n(t)$ is not a counting process.

(b) Let $n(t)$ stand for the number of home runs that an individual player has hit by time t. $n(t)$ is a counting process. An event of this process will occur whenever the player hits a home run.

Observation 1.2.13. From the definition of the counting process $n(t)$, we note that $n(t)$ must satisfy the following conditions:

(i) $n(t) \geq 0$;
(ii) $n(t)$ is an integer-valued process;
(iii) if $s < t$, then $n(s) \leq n(t)$;
(iv) for $s < t$, $n(t) - n(s)$ equals the number of events that have occurred in the interval (s, t).

Definition 1.2.14. The counting process $n(t)$ is said to be a Poisson process having rate λ, $\lambda > 0$, if:

(a) $n(0) = 0$;
(b) the process has independent increments;
(c) the number of events in any interval of length Δt has a Poisson distribution with mean $\lambda \Delta t$. Hence, for all Δt, $t \geq 0$, we have

$$P\{n(t + \Delta t) - n(t) = j\} = \frac{(\lambda \Delta t)^j}{j!} e^{-\lambda \Delta t}, \quad j = 0, 1, 2, \dots. \tag{1.2.61}$$

An alternative definition of a Poisson process with rate λ, $\lambda > 0$, is as follows:

(i) $n(0) = 0$;
(ii) the process has stationary and independent increments;
(iii) $P\{n(\Delta t) = 1\} = \lambda \Delta t + o(\Delta t)$;
(iv) $P\{n(\Delta t) \geq 2\} = o(\Delta t)$.

Exercises

1. A random process $x \in R[J, R[\Omega, R]]$ is said to be a separable process if there exists a countable set $S \subset J = [a, b]$ and a fixed null event Λ so that for any closed set $C \subset (-\infty, \infty)$ and any open interval I, the two sets $\{\omega : x(t, \omega) \in C, t \in I \cap J\}$ and $\{\omega : x(t, \omega) \in C, t \in I \cap S\}$ differ by a subset of Λ. The set S is called a separating set. Show that:

(a) $\sup_{t \in I \cap S} x(t, \omega) = \sup_{t \in I \cap J} x(t, \omega)$;
(b) $\limsup_{s \to t} x(t, \omega) = \lim_{n \to |s-t| < 1/n} \sup x(t, \omega)$.

2. A random process $x \in R[J, R(\Omega, R)]$ is said to be a martingale if for $t \in J$, $x(t)$ is $\mathfrak{F}_t$-measurable, and for $t > s$, $E[x(t)|x(s)] = x(s)$ almost surely (a.s.). Show that the Wiener process is a martingale.

3. Let $x \in R[J, R[\Omega, R]]$ be a separable martingale with a finite second moment on J. Show that $P(\omega : \sup_{t \in J} |x(t, \omega)| \geq \epsilon) \leq \frac{E[(x(b))^2]}{\epsilon^2}$, where b is the least upper bound of J.

4. Let $x \in R[J, R[\Omega, R]]$ be a Brownian motions process. Prove that:
   (a) $E[(x(t) - x(s))^2] = |t - s|$;
   (b) $P(\omega : |x(t + h, \omega) - x(t, \omega)| \geq \epsilon) \leq \frac{|h|}{\epsilon^2}$.

5. Let $x \in R[J, R[\Omega, R]]$ be a separable process, with J being a finite interval. Further assume that $E[(x(t + h) - x(t))^\gamma] \leq L|h|^{1+\alpha}$ (*Kolmogorov condition*), where L, γ and α are positive numbers. Show that $P(\omega : \lim_{h \to 0} |x(t + h, \omega) - x(t, \omega)| = 0) = 1$ uniformly on a closed and bounded subset of J.

1.3 Itô–Doob Stochastic Calculus

In this section, we briefly outline the Itô–Doob-type stochastic integral and differential with respect to the Wiener process. For this purpose, we state a class of random functions (processes) for which the Itô–Doob integral is defined.

Let us denote by $H_2[0, T]$ the space of all random functions f defined on $[0, T]$ into $\mathbb{R}$ that are $\mathfrak{F}_t$-measurable for each $t \in [0, T]$ and satisfy the condition

$$\int_0^T [f(s)]^2 ds < \infty$$

with probability 1 (in short, w.p.1).

Definition 1.3.1. Let $f \in H_2[0, T]$. In particular, if it is a continuous process, the stochastic integral is defined by

$$\int_0^T f(s)dw(s) = \lim_{\mu(P) \to 0} \sum_{k=0}^{n-1} f(t_k)][w(t_{k+1}) - w(t_k)], \tag{1.3.1}$$

where P is the partition of $[0, T]$, i.e. $P : 0 = t_0 < t_1 < \cdots t_k < \cdots t_n = T$. $\mu(P)$ is the mesh/norm of the partition P defined by $\mu(P) = \max_{0 \leq k \leq n}(t_{k+1} - t_k)$. If this limit exists, then it is referred to as the *Itô–Doob stochastic integral*.

Lemma 1.3.1 (properties of the stochastic integral). *The Itô–Doob stochastic integral possesses the following properties:*

(a) *Let $f_1, f_2 \in H_2[0, T]$, and α_1, α_2 be random variables for which $\alpha_1 f_1 + \alpha_2 f_2 \in H_2[0, T]$. Then,*

$$\int_0^T (\alpha_1 f_1 + \alpha_2 f_2)(s)dw(s)$$

$$= \int_0^T \alpha_1 f_1(s)dw(s) + \int_0^T \alpha_2 f_2(s)dw(s). \tag{1.3.2}$$

(b) *If $\chi_{[\alpha,\beta]}$ is the characteristic function $(\chi_{[\alpha,\beta]}(s) = 1$, if $s \in [\alpha, \beta]$, otherwise zero) of the interval $[\alpha, \beta]$ contained in $[0, T]$, then*

$$\int_0^T \chi_{[\alpha,\beta]}dw(s) = w(\beta) = w(\alpha). \tag{1.3.3}$$

(c) *Let $f \in H_2[0,T]$ and $\int_0^T E[f(s)]^2 ds < \infty$. In addition, let $[\alpha, \beta] \subseteq [0,T]$. Then,*

$$E\left[\int_0^T f(s)dw(s)\right] = 0, \quad E\left[\int_0^T f(s)dw(s)\right]^2 = \int_0^T E[f(s)]^2 ds. \quad (1.3.4)$$

Moreover,

$$E\left[\int_0^T f(s)dw(s)|\mathfrak{F}_a\right] = 0,$$

$$E\left[\left[\int_0^T f(s)dw(s)\right]^2 \Big| \mathfrak{F}_a\right] = \int_0^T E[f^2(s)|\mathfrak{F}_a]ds, \quad (1.3.5)$$

whenever f is nonanticipatory with respect to the increasing family of sub-σ-algebra $\mathfrak{F}_t$.

(d) *For all $f \in H_2[0,T]$,*

$$P\left\{\omega : \left|\int_0^T f(s)dw(s)\right| > c\right\} \leq P\left\{\omega : \int_0^T f(s)dw(s) > N\right\} + \frac{N}{c^2}. \quad (1.3.6)$$

Moreover, for each $t \in [0,T]$,

$$I(t) = \int_0^t f(s)dw(s) \quad (1.3.7)$$

is an indefinite integral on $[0,t]$, $I(t)$ is separable and continuous with probability 1 (w.p.1), and

$$P\left\{\omega : \sup_{t_0 \leq t \leq T} |I(t)| > c\right\} \leq P\left\{\omega : \int_0^T f(s)ds > N\right\} + \frac{N}{c^2}. \quad (1.3.8)$$

(e) *For $f \in H_2[0,T]$ and $0 \leq b \leq T$,*

$$\int_0^T f(s)dw(s) = \int_0^b f(s)dw(s) + \int_b^T f(s)dw(s).$$

In the following, we introduce the Itô–Doob-type stochastic differential calculus [28, 60, 61].

Definition 1.3.2. Let x be a stochastic process/function defined on $[0,T]$ whose values are random variables. We say that x has a stochastic differential on $[0,T]$ if the stochastic process/function x satisfies the expression

$$x(t+\Delta t) - x(t) = \int_t^{t+\Delta t} a(s)ds + \int_t^{t+\Delta t} \sigma(s)dw(s), \quad 0 \leq t < t+\Delta t \leq T, \quad (1.3.9)$$

where $\sqrt{|a|}$, $b \in H_2[0, T]$ and $h > 0$ is very small increment to t. This change in x defined in (1.3.9) is denoted by

$$dx(t) = a(t)dt + \sigma(t)dw(t), \qquad (1.3.10)$$

where $\Delta t = dt$ and $\Delta w(t) = w(t + \Delta t) - w(t) = dw(t)$ are the differentials of t and $w(t)$, respectively.

We note that the first integral in (1.3.9) is the usual deterministic integral (Riemann/Cauchy [127, 168]), and the second integral is the Itô–Doob stochastic integral [7, 8, 43, 56, 71, 98, 141]. We further note that the expression (1.3.9) or (1.3.10) is referred to as the Itô–Doob stochastic differential of x, whenever $\Delta t > 0$ and it is very small.

Observation 1.3.1. We note that the operation of a differential is linear. Therefore, the Itô–Doob differential formulas for the product, quotient and composition functions differ from the usual formulas in the deterministic calculus.

In the following, we present a very fundamental formula for the Itô–Doob stochastic differential of a very general composite function. This rule is useful for finding the stochastic Itô–Doob integral of a process. In addition, it will be frequently used to compute closed-form solutions of the Itô–Doob type of stochastic differential equations.

Theorem 1.3.1 (Itô–Doob stochastic differential formula — I). *Let $V \in C[J \times R, R]$, and let $V(t, x)$ be a continuously differentiable function with respect to t and twice continuously differentiable with respect to x. Then,*

$$dV(t, w(t)) = \left[\frac{\partial}{\partial t}V(t, w(t)) + \frac{1}{2}\frac{\partial^2}{\partial x^2}V(t, w(t))\right]dt + \frac{\partial}{\partial x}V(t, w(t))dw(t), \qquad (1.3.11)$$

where w is the Wiener process defined in Section 1.2.

Proof. We apply the deterministic Taylor's formula to V to keep all terms up to the second-degree polynomial in $\Delta w(t)$. For this purpose, we consider

$$V(t + \Delta t, w(t) + \Delta w(t)) = V(t, w(t)) + \frac{\partial}{\partial t}V(t, w(t))\Delta t + \frac{\partial}{\partial x}V(t, w(t))\Delta w(t)$$

$$+ \frac{1}{2}\frac{\partial^2}{\partial x^2}V(t, w(t))(\Delta w(t))^2 + o(\Delta t, \Delta w(t)),$$

which implies that

$$\Delta V = V(t + \Delta t, w(t) + \Delta w(t)) - V(t, w(t))$$

$$= \text{the change in value of } V(t, w(t))$$

$$= \frac{\partial}{\partial t}V(t, w(t))\Delta t + \frac{\partial}{\partial t}V(t, w(t))\Delta w(t)$$

$$+ \frac{1}{2}\frac{\partial^2}{\partial x^2}V(t, w(t))(\Delta w(t))^2 + o(\Delta t, \Delta w(t)). \qquad (1.3.12)$$

By using $o(\Delta t, \Delta w(t)) \equiv o(\Delta t)^{1+\epsilon}$ for $\epsilon > 0$, and the deterministic differential $dt = \Delta t$, and the stochastic differential $dw(t) = \Delta w(t)$ in the sense of Itô–Doob, the stochastic differential of V is defined in terms of dt and $dw(t)$ by the following equation and discarding all higher-order terms in (1.3.12) as

$$dV(t, w(t)) = \frac{\partial}{\partial t} V(t, w(t)) dt + \frac{\partial}{\partial x} V(t, w(t)) dw(t)$$
$$+ \frac{1}{2} \frac{\partial^2}{\partial x^2} V(t, w(t)) (dw(t))^2$$
$$= \left[\frac{\partial}{\partial t} V(t, w(t)) + \frac{1}{2} \frac{\partial^2}{\partial x^2} V(t, w(t)) \right] dt$$
$$+ \frac{\partial}{\partial x} V(t, w(t)) dw(t).$$

This completes the validity of (1.3.11). We note that $E[\Delta w(t)]^2 | \mathfrak{F}_t| = \Delta t$ (Theorem 1.2.2 and Lemma 1.3.1). If Δt is small, then $V(t + \Delta t, w(t) + \Delta w(t)) - V(t, w(t)) \approx dV(t, w(t))$. This approximation is in the sense of Itô–Doob. □

Definition 1.3.3. The formula

$$dV(t, w(t)) = \frac{\partial}{\partial t} V(t, w(t)) dt + \frac{\partial}{\partial x} V(t, w(t)) dw(t) + \frac{1}{2} \frac{\partial^2}{\partial x^2} V(t, w(t)) (dw(t))^2$$

is referred to as a natural Itô–Doob stochastic differential of $V(t, w(t))$.

Example 1.3.1. Given: $V(t, x) = \exp\left[\left(a - \frac{1}{2}\sigma^2\right) t + \sigma x\right]$. Goal: To find an expression for $dV(t, w(t))$.

Solution procedure. We note that the given function is a continuously differentiable function with respect to t and twice continuously differentiable with respect to x. We apply Theorem 1.3.1 [Itô–Doob stochastic differential formula–I] to the given function V.

Step #1. We apply Theorem 1.3.1 [Itô–Doob stochastic differential formula–I]. First, we find $\frac{\partial}{\partial t} V(t, x)$, $\frac{\partial}{\partial x} V(t, x)$, and $\frac{\partial^2}{\partial x^2} V(t, x)$. These partial derivatives are:

$$\frac{\partial}{\partial t} V(t, x) = \frac{\partial}{\partial t} \exp\left[\left(a - \frac{1}{2}\sigma^2\right) t + \sigma x\right]$$
$$= \exp\left[\left(a - \frac{1}{2}\sigma^2\right) t + \sigma x\right]\left(a - \frac{1}{2}\sigma^2\right) \quad \text{(chain rule)},$$

$$\frac{\partial}{\partial x} V(t, x) = \frac{\partial}{\partial x} \exp\left[\left(a - \frac{1}{2}\sigma^2\right) t + \sigma x\right]$$
$$= \exp\left[\left(a - \frac{1}{2}\sigma^2\right) t + \sigma x\right] \sigma,$$

and

$$\frac{\partial^2}{\partial x^2} V(t, x) = \frac{\partial}{\partial x} \exp\left[\left(a - \frac{1}{2}\sigma^2\right) t + \sigma x\right] \sigma = \exp\left[\left(a - \frac{1}{2}\sigma^2\right) t + \sigma x\right] \sigma^2.$$

Step #2.   We use these partial derivatives and evaluate them at $x = w(t)$. In this case

$$\frac{\partial}{\partial t} V(t, w(t)) = \frac{\partial}{\partial t} \exp\left[\left(a - \frac{1}{2}\sigma^2\right) t + \sigma w(t)\right]$$

$$= \exp\left[\left(a - \frac{1}{2}\sigma^2\right) t + \sigma w(t)\right] \left(a - \frac{1}{2}\sigma^2\right),$$

$$\frac{\partial}{\partial x} V(t, w(t)) = \frac{\partial}{\partial x} \exp\left[\left(a - \frac{1}{2}\sigma^2\right) t + \sigma w(t)\right]$$

$$= \exp\left[\left(a - \frac{1}{2}\sigma^2\right) t + \sigma w(t)\right] \sigma,$$

and

$$\frac{\partial^2}{\partial x^2} V(t, w(t)) = \frac{\partial}{\partial x} \exp\left[\left(a - \frac{1}{2}\sigma^2\right) t + \sigma w(t)\right]$$

$$\sigma = \exp\left[\left(a - \frac{1}{2}\sigma^2\right) t + \sigma w(t)\right] \sigma^2.$$

Step #3.   By applying the Itô–Doob differential formula (1.3.11), we have

$$dV(t, w(t)) = \left[\frac{\partial}{\partial t} V(t, w(t)) + \frac{1}{2}\frac{\partial^2}{\partial x^2} V(t, w(t))\right] dt + \frac{\partial}{\partial x} V(t, w(t)) dw(t),$$

and hence

$$d\left(\exp\left[\left(a - \frac{1}{2}\sigma^2\right) t + \sigma w(t)\right]\right)$$

$$= \left[\exp\left[\left(a - \frac{1}{2}\sigma^2\right) t + \sigma w(t)\right] \left(a - \frac{1}{2}\sigma^2\right)\right.$$

$$+ \frac{1}{2}\exp\left[\left(a - \frac{1}{2}\sigma^2\right) t + \sigma w(t)\right] \sigma^2 \Big] dt$$

$$+ \exp\left[\left(a - \frac{1}{2}\sigma^2\right) t + \sigma w(t)\right] \sigma dw(t).$$

By the simplification of the right-hand side expression, we get

$$d\left(\exp\left[\left(a - \frac{1}{2}\sigma^2\right) t + \sigma w(t)\right]\right)$$

$$= a\exp\left[\left(a - \frac{1}{2}\sigma^2\right) t + \sigma w(t)\right] dt + \exp\left[\left(a - \frac{1}{2}\sigma^2\right) t + \sigma w(t)\right] \sigma dw(t)$$

$$= \exp\left[\left(a - \frac{1}{2}\sigma^2\right) t + \sigma w(t)\right] [a dt + \sigma dw(t)].$$

For the given $V(t, x)$, this is the final expression for $dV(t, w(t))$.

We present another basic Itô–Doob stochastic differential formula for the composite function V and process x generated by the Wiener process in (1.3.10).

Theorem 1.3.2 (Itô–Doob stochastic differential formula — II). *Let $V \in C[J \times R, R]$, and let $V(t, x)$ be a continuously differentiable function with respect to t and twice continuously differentiable with respect to x in (1.3.10). Then,*

$$dV(t, x(t)) = LV(t, x(t))dt + \sigma(t)\frac{\partial}{\partial x}V(t, x(t))dw(t), \tag{1.3.13}$$

where w is the Wiener process defined in Section 1.2, in particular Theorem 1.2.2; $x(t)$ is determined by (1.3.10), and L is a linear differential operator associated with (1.3.10) defined by

$$LV(t, x(t)) = \frac{\partial}{\partial t}V(t, x(t)) + a(t)\frac{\partial}{\partial x}V(t, x(t)) + \frac{1}{2}\sigma^2(t)\frac{\partial^2}{\partial x^2}V(t, x(t)). \tag{1.3.14}$$

Proof. By imitating the proof of Theorem 1.3.1, a natural Itô–Doob stochastic differential of $V(t, x(t))$, Definition 1.3.3 is represented by

$$dV(t, x(t)) = \frac{\partial}{\partial t}V(t, x(t))dt + \frac{\partial}{\partial x}V(t, x(t))dx(t) + \frac{1}{2}\frac{\partial^2}{\partial x^2}V(t, x(t))(dx(t))^2. \tag{1.3.15}$$

We note that

$$\begin{aligned}(dx(t))^2 &= (a(t)dt + \sigma(t)dw(t))^2 \\ &= [a^2(t)(dt)^2 + 2a(t)\sigma(t)dt\,dw(t) + \sigma^2(t)dw(t))^2]. \end{aligned} \tag{1.3.16}$$

The terms $a^2(t)(dt)^2$ and $2a(t)\sigma(t)dt\,dw(t)$ are higher-order terms $[o(\Delta t)^{1+\epsilon}$ for $\epsilon > 0]$. The Itô–Doob differential $(dx(t))^2$ is approximately equal to $\sigma^2(t)(dw(t))^2$. We substitute this and $dx(t)$ in (1.3.10) into (1.3.15), and obtain

$$\begin{aligned}dV(t, x(t)) = &\frac{\partial}{\partial t}V(t, x(t))dt + a(t)\frac{\partial}{\partial x}V(t, x(t))dt \\ &+ \sigma(t)\frac{\partial}{\partial x}V(t, x(t))dw(t) + \frac{1}{2}\frac{\partial^2}{\partial x^2}V(t, x(t))\sigma^2(t)(dw(t))^2,\end{aligned}$$

which gives rise to a *standard differential* of $V(t, x(t))$ in the sense of Itô–Doob calculus

$$\begin{aligned}dV(t, x(t)) = &\frac{\partial}{\partial t}V(t, x(t))dt + a(t)\frac{\partial}{\partial x}V(t, x(t))dt \\ &+ \sigma(t)\frac{\partial}{\partial x}V(t, x(t))dw(t) + \frac{1}{2}\sigma^2(t)\frac{\partial^2}{\partial x^2}V(t, x(t))dt \\ = &\left[\frac{\partial}{\partial t}V(t, x(t))dt + a(t)\frac{\partial}{\partial x}V(t, x(t)) + \frac{1}{2}\sigma^2(t)\frac{\partial^2}{\partial x^2}V(t, x(t))\right]dt \\ &+ \sigma(t)\frac{\partial}{\partial x}V(t, x(t))dw(t),\end{aligned} \tag{1.3.17}$$

in view of the Itô–Doob differential of $(dw(t))^2 \simeq dt$ and regrouping of the terms. This completes the verification of the formula (1.3.13) in the context of the definition of the L operator defined in (1.3.14). This completes the proof of the theorem. $\square$

Example 1.3.2. Given: $V(t, x) = x^2$. Goal: To find an expression for $dV(t, x(t))$ with respect to x in (1.3.10).

Solution procedure. We note that the given function $V(t, x) = x^2$ is a continuously differentiable function with respect to t and twice continuously differentiable with respect to x. In fact, it is constant with respect to the independent variable t. We apply Theorem 1.3.2 [Itô–Doob stochastic differential formula — II] to the given function.

Step #1. We apply Theorem 1.3.2 [Itô–Doob stochastic differential formula — II] to $V(t, x) = x^2$. First, we find $\frac{\partial}{\partial t}V(t, x)$, $\frac{\partial}{\partial x}V(t, x)$, and $\frac{\partial^2}{\partial x^2}V(t, x)$. These partial derivatives are

$$\frac{\partial}{\partial t}V(t, x) = 0, \quad \frac{\partial}{\partial x}V(t, x) = 2x \quad \text{and} \quad \frac{\partial^2}{\partial x^2}V(t, x) = 2.$$

Step #2. We use these partial derivatives in Step #1 as well as the Itô–Doob differential $(dw(t))^2 \simeq dt$ to find $dV(t, x(t))$ in (1.3.13). In fact,

$$dV(t, x(t)) = LV(t, x(t))dt + \sigma(t)\frac{\partial}{\partial x}V(t, x(t))dw(t))$$

$$= (2a(t)x(t) + \sigma^2(t))dt + 2\sigma(t)x(t)dw(t),$$

where w is the Wiener process defined in Section 1.2 and $x(t)$ is determined by (1.3.10). This is the final expression for $dV(t, x(t))$ for the given $V(t, x)$.

Let x_1 and x_2 be two stochastic processes satisfying the following Itô–Doob stochastic differential equations:

$$dx_1 = f_1(t, x_1)dt + \sigma_1(t, x_1)dw(t), \tag{1.3.18}$$

and

$$dx_2 = f_2(t, x_2)dt + \sigma_2(t, x_2)dw(t), \tag{1.3.19}$$

where w is the normalized Wiener process, and f_1, σ_1, f_2, and σ_2 are continuous functions.

In the following, we present the fundamental Itô–Doob stochastic differential calculus rules.

Theorem 1.3.3 (Itô–Doob stochastic differential formula — addition rule). *Let $V_1, V_2 \in C[J \times R, R]$. Assume that $V_1(t, z)$ and $V_2(t, z)$ are continuously differentiable functions with respect to t and twice continuously differentiable with respect to z.*

Define $V = V_1 + V_2$. Then,

$$dV(t, x_1, x_2) = dV_1(t, x_1) + dV_2(t, x_2) \tag{1.3.20}$$

$$= [L_1 V_1(t, x_1)dt + L_2 V_2(t, x_2)]dt$$

$$+ \left[\sigma_1(t, x_1)\frac{\partial}{\partial x_1}V_1(t, x_1) + \sigma_2(t, x_2)\frac{\partial}{\partial x_2}V_2(t, x_2) \right] dw(t), \quad (1.3.21)$$

where w is the Wiener process defined in Section 1.2; the Itô–Doob differentials of V_1 and V_2 with respect to the solution processes of (1.3.18) and (1.3.18) are given by

$$dV_1(t, x_1) = L_1 V_1(t, x_1)dt + \sigma_1(t, x_1)\frac{\partial}{\partial x_1}V_1(t, x_1)dw(t), \tag{1.3.22}$$

and

$$dV_2(t, x_2) = L_2 V_2(t, x_2)dt + \sigma_2(t, x_2)\frac{\partial}{\partial x_1}V_2(t, x_2)dw(t), \tag{1.3.23}$$

respectively, and the linear differential operators L_1 and L_2 associated with the Itô–Doob-type stochastic differential equations (1.3.18) and (1.3.18) are

$$L_1 V_1(t, x_1) = \frac{\partial}{\partial t}V_1(t, x_1) + f_1(t, x_1)\frac{\partial}{\partial x_1}V_1(t, x_1) + \frac{1}{2}\sigma_1^2(t, x_1)\frac{\partial^2}{\partial x_1^2}V_1(t, x_1),$$

$$\tag{1.3.24}$$

and

$$L_2 V_2(t, x_2) = \frac{\partial}{\partial t}V_2(t, x_2) + f_2(t, x_2)\frac{\partial}{\partial x_2}V_2(t, x_2) + \frac{1}{2}\sigma_2^2(t, x_2)\frac{\partial^2}{\partial x_2^2}V_2(t, x_2),$$

$$\tag{1.3.25}$$

respectively.

Proof. First, by considering $x = [x_1, x_2]^T$ as a column vector (2×1 matrix) and using Observation 1.5.3 [80], we have

$$\frac{\partial}{\partial t}V(t, x_1, x_2) = \frac{\partial}{\partial t}V_1(t, x_1) + \frac{\partial}{\partial t}V_2(t, x_2),$$

$$\frac{\partial}{\partial x}V(t, x_1, x_2) = \nabla V(t, x_1, x_2) = \left[\frac{\partial}{\partial x_1}V_1(t, x_1), \frac{\partial}{\partial x_2}V_2(t, x_2) \right],$$

and

$$\frac{\partial^2}{\partial x^2}V(t, x_1, x_2) = \begin{bmatrix} \frac{\partial^2}{\partial x_1^2}V_1(t, x_1) & \frac{\partial^2}{\partial x_1 \partial x_2}V_2(t, x_2) \\ \frac{\partial^2}{\partial x_2 \partial x_1}V_1(t, x_1) & \frac{\partial^2}{\partial x_2^2}V_2(t, x_2) \end{bmatrix}^T = \begin{bmatrix} \nabla\frac{\partial}{\partial x_1}V_1(t, x_1) \\ \nabla\frac{\partial}{\partial x_2}V_2(t, x_2) \end{bmatrix}.$$

By using (1.3.18) and (1.3.18) and imitating the proof of Theorem 1.5.10 [80], a natural Itô–Doob stochastic differential of $V(t, x_1, x_2)$ is represented by

$$dV(t, x_1, x_2) = \frac{\partial}{\partial t} V(t, x_1, x_2) dt + \frac{\partial}{\partial x} V(t, x_1, x_2) dx$$

$$+ \frac{1}{2}(dx)^T \left(\frac{\partial^2}{\partial x^2} V(t, x_1, x_2) \right) dx.$$

Now, the rest of the proof can be completed by following the argument used in Theorem 1.3.2. The details are left to the reader. □

Example 1.3.3. Given: $V_1(t, z) = \sin^2 z$ and $V_2(t, z) = e^z$. Goal: To find an expression for $dV(t, x_1(t), x_2(t))$ with respect to the solution processes of (1.3.18) and (1.3.18), where $V(t, x_1, x_2) = V_1(t, x_1) + V_2(t, x_2) = \sin^2 x_1 + e^{x_2}$.

Solution procedure. We note that the given functions $V_1(t, z) = \sin^2 z$ and $V_2(t, z) = e^z$ are continuously differentiable functions with respect to t and twice continuously differentiable with respect to x. In fact, these functions are constant with respect to the independent variable t. We apply Theorem 1.3.3 (Itô–Doob stochastic differential formula — addition rule).

Step #1. We apply the augment and notations used in the proof of Theorem 1.3.3 (Itô–Doob stochastic differential formula — addition rule), and compute

$$\frac{\partial}{\partial t} V(t, x_1, x_2), \quad \frac{\partial}{\partial x} V(t, x_1, x_2) \quad \text{and} \quad \frac{\partial^2}{\partial x^2} V(t, x_1, x_2).$$

With regard to $V(t, x_1, x_2) = \sin^2 x_1 + e^{x_2}$ in this example, $\frac{\partial}{\partial t} V(t, x_1, x_2) = 0$

$$\frac{\partial}{\partial x} V(t, x_1, x_2) = [2 \sin x_1 \cos x_1, e^{x_2}],$$

and

$$\frac{\partial^2}{\partial x^2} V(t, x_1, x_2) = \begin{bmatrix} 2(\cos^2 x_1 - \sin^2 x_1) & 0 \\ 0 & e^{x_2} \end{bmatrix}^T = \begin{bmatrix} \nabla(2 \sin x_1 \cos x_1) \\ \nabla e^{x_2} \end{bmatrix}.$$

Step #2. Using the expressions in Step #1, the formula (1.3.21) reduces to

$$dV(t, x_1, x_2) = \frac{\partial}{\partial t} V(t, x_1, x_2) dt + \frac{\partial}{\partial x} V(t, x_1, x_2) dx$$

$$+ \frac{1}{2}(dx) \left(\frac{\partial^2}{\partial x^2} V(t, x_1, x_2) \right) (dx)^T$$

$$= [2 \sin x_1 \cos x_1, e^{x_2}](dx)^T$$

$$+ \frac{1}{2}(dx) \begin{bmatrix} 2(\cos^2 x_1 - \sin^2 x_1) & 0 \\ 0 & e^{x_2} \end{bmatrix} (dx)^T$$

$$= 2\sin x_1 \cos x_1 dx_1 + e^{x_2} dx_2$$

$$+ \frac{1}{2} 2(\cos^2 x_1 - \sin^2 x_1)(dx_1)^2 + \frac{1}{2} e^{x_2}(dx_2)^2$$

$$= [2\sin x_1 \cos x_1 f_1(t, x_1) + (\cos^2 x_1 - \sin^2 x_1)\sigma_1^2(t, x_1))$$

$$+ (e^{x_2} f_2(t, x_2) + \frac{1}{2} e^{x_2} \sigma_2^2(t, x_2))] dt$$

$$+ [2\sin x_1 \cos x_1 \sigma_1(t, x_1) + e^{x_2} \sigma_2(t, x_2)] dw(t).$$

This establishes the desired expression for $dV(t, x_1, x_2)$ with respect to the solution processes of (1.3.18) and (1.3.18).

Theorem 1.3.4 (Itô–Doob stochastic differential formula — product rule). *Let* $V_1, V_2 \in C[J \times R, R]$. *Assume that* $V_1(t, z)$ *and* $V_2(t, z)$ *are continuously differentiable functions with respect to* t *and twice continuously differentiable with respect to* z. *Define* $V = V_1 V_2$. *Then,*

$$dV(t, x_1, x_2) = V_2(t, x_2)dV_1(t, x_1) + V_1(t, x_1)dV_2(t, x_2)$$

$$+ \sigma_1(t, x_1)\sigma_2(t, x_2)\frac{\partial}{\partial x_1}V_1(t, x_1)\frac{\partial}{\partial x_2}V_2(t, x_2)dt \qquad (1.3.26)$$

$$= \left[V_2(t, x_2)L_1V_1(t, x_1) + \sigma_1(t, x_1)\sigma_2(t, x_2)\frac{\partial}{\partial x_1}V_1(t, x_1)\frac{\partial}{\partial x_2}V_2(t, x_2) \right.$$

$$\left. + V_1(t, x_1)L_2V_2(t, x_2) \right] dt + \left[\sigma_1(t, x_1)V_2(t, x_2)\frac{\partial}{\partial x_1}V_1(t, x_1) \right.$$

$$\left. + \sigma_2(t, x_2)V_1(t, x_1)\frac{\partial}{\partial x_2}V_2(t, x_2) \right] dw(t), \qquad (1.3.27)$$

where w *is the Wiener process defined in Section 1.2, and* $L_1V_1(t, x_1)$ *and* $L_1V_2(t, x_2)$ *are as defined in* (1.3.24) *and* (1.3.25), *respectively.*

Proof. We compute dV by finding $\frac{\partial}{\partial t}V$, $\frac{\partial}{\partial x}V$, and $\frac{\partial^2}{\partial x^2}V$. From deterministic calculus [127], we have

$$\frac{\partial}{\partial t}V(t, x_1, x_2) = V_2(t, x_2)\frac{\partial}{\partial t}V_1(t, x_1)$$

$$+ V_1(t, x_1)\frac{\partial}{\partial t}V_2(t, x_2) \qquad \text{(by the product rule)},$$

$$\frac{\partial}{\partial x}V(t, x_1, x_2) = V_2(t, x_2)\frac{\partial}{\partial x_1}V_1(t, x_1)$$

$$+ V_1(t, x_1)\frac{\partial}{\partial x_2}V_2(t, x_2) \qquad \text{(by the product rule)},$$

$$\frac{\partial^2}{\partial x^2}V_2(t, x_1, x_2) = V_2(t, x_2)\frac{\partial^2}{\partial x_1^2}V_1(t, x_1) + \frac{\partial}{\partial x_1}V_1(t, x_1)\frac{\partial}{\partial x_2}V_2(t, x_2)$$

$$+ V_1(t, x_1)\frac{\partial^2}{\partial x_2^2}V_2(t, x_2) + \frac{\partial}{\partial x_1}V_1(t, x_1)\frac{\partial}{\partial x_2}V_2(t, x_2).$$

By using these derivatives and following the argument used in the proofs of Theorems 1.3.2 and 1.3.3, the natural Itô–Doob stochastic differential of $V(t, x(t))$ in Definition 1.3.3 is represented by

$$dV(t, x_1, x_2) = \left[V_2(t, x_2)\frac{\partial}{\partial t}V_1(t, x_1) + V_1(t, x_1)\frac{\partial}{\partial t}V_2(t, x_2)\right] dt$$

$$+ \left[V_2(t, x_2)\frac{\partial}{\partial x_1}V_1(t, x_1)dx_1 + V_1(t, x_1)\frac{\partial}{\partial x_2}V_2(t, x_2)dx_2\right]$$

$$+ \frac{1}{2}\left[V_2(t, x_2)\frac{\partial^2}{\partial x_1^2}V_1(t, x_1)(dx_1)^2 + V_1(t, x_1)\frac{\partial^2}{\partial x_2^2}V_2(t, x_2))(dx_2)^2\right.$$

$$\left. + 2\frac{\partial}{\partial x_1}V_1(t, x_1)\frac{\partial}{\partial x_2}V_2(t, x_2)dx_1dx_1\right].$$

By substituting for dx_1, dx_2, $(dx_1)^2$, and $(dx_2)^2$ from (1.3.18) and (1.3.18), in the context of the Itô–Doob differentials $(dx_1)^2 \simeq \sigma_1^2(t, x_1)(dw(t))^2$, $(dx_2)^2 \simeq \sigma_2^2(t, x_2)(dw(t))^2$, $(dx_1)(dx_2) \simeq \sigma_1(t, x_1)\sigma_2(t, x_2)(dw(t))^2$, and $(dw(t))^2 \simeq dt$, we obtain

$$dV(t, x_1, x_2) = \left[V_2(t, x_2)\frac{\partial}{\partial t}V_1(t, x_1) + V_1(t, x_1)\frac{\partial}{\partial t}V_2(t, x_2)\right] dt$$

$$+ V_2(t, x_2)\frac{\partial}{\partial x_1}V_1(t, x_1)[f_1(t, x_1)dt + \sigma_1(t, x_1)dw(t)]$$

$$+ V_1(t, x_1)\frac{\partial}{\partial x_2}V_2(t, x_2)[f_2(t, x_2)dt + \sigma_2(t, x_2)dw(t)]$$

$$+ \frac{1}{2}\left[V_2(t, x_2)\frac{\partial^2}{\partial x_1^2}V_1(t, x_1)\sigma_1^2(t, x_1) + V_1(t, x_1)\frac{\partial^2}{\partial x_2^2}V_2(t, x_2))\sigma_2^2(t, x_2)\right.$$

$$\left. + 2\sigma_1(t, x_1)\sigma_2(t, x_2)\frac{\partial}{\partial x_1}V_1(t, x_1)\frac{\partial}{\partial x_2}V_2(t, x_2)dt\right].$$

By regrouping the terms as the coefficients of $V_1(t, x_1)$ and $V_2(t, x_2)$, we have

$dV(t, x_1, x_2)$

$$= V_2(t, x_2)\left(\left[\frac{\partial}{\partial t}V_1(t, x_1) + f_1(t, x_1)\frac{\partial}{\partial x_1}V_1(t, x_1) + \frac{1}{2}\sigma_1^2(t, x_1)\frac{\partial^2}{\partial x_1^2}V_1(t, x_1)\right]dt\right.$$

$$\left.+ \sigma_1(t, x_1)\frac{\partial}{\partial x_1}V_1(t, x_1)dw(t)\right)$$

$$+ V_1(t, x_1)\left(\left[\frac{\partial}{\partial t}V_2(t, x_2) + f_2(t, x_2)\frac{\partial}{\partial x_2}V_2(t, x_2)\right.\right.$$

$$\left.\left.+ \frac{1}{2}\sigma_2^2(t, x_2)\frac{\partial^2}{\partial x_2^2}V_2(t, x_2)\right]dt + \sigma_2(t, x_2)\frac{\partial}{\partial x_2}V_2(t, x_2)dw(t)\right)$$

$$+ \sigma_1(t, x_1)\sigma_2(t, x_2)\frac{\partial}{\partial x_1}V_1(t, x_1)\frac{\partial}{\partial x_2}V_2(t, x_2)dt. \tag{1.3.28}$$

From (1.3.22) and (1.3.22), (1.3.28) reduces to

$$dV_1(t, x_1) = V_2(t, x_2)dV_1(t, x_1) + V_1(t, x_1)dV_2(t, x_2)$$

$$+ \sigma_1(t, x_1)\sigma_2(t, x_2)\frac{\partial}{\partial x_1}V_1(t, x_1)\frac{\partial}{\partial x_2}V_2(t, x_2)dt.$$

This establishes the formula (1.3.26). The validity of (1.3.27) follows from (1.3.24), (1.3.25), and (1.3.28). This completes the proof of the theorem. □

Example 1.3.4. Given: $V_1(t, x_1) = x_1^2$ and $V_2(t, x_2) = x_2^2$. Goal: To find $dV(t, x_1, x_2)$ with respect to the solution processes of (1.3.18) and (1.3.18), where $x_1^2 x_2^2 = V(t, x_1, x_2) = V_1(t, x_1)V_2(t, x_2)$.

Solution procedure. We note that the given functions $V_1(t, x_1) = x_1^2$ and $V_2(t, x_2) = x_2^2$ are continuously differentiable functions with respect to t and twice continuously differentiable with respect to x. In fact, they are independent of the independent variable t. We apply Theorem 1.3.4 (Itô–Doob stochastic differential formula — product rule) to $V(t, x_1, x_2) = x_1^2 x_2^2$.

Step #1. To apply Theorem 1.3.4, we imitate the argument used in Example 1.3.3 to compute

$$\frac{\partial}{\partial t}V(t, x_1, x_2), \quad \frac{\partial}{\partial x}V(t, x_1, x_2) \quad \text{and} \quad \frac{\partial^2}{\partial x^2}V(t, x_1, x_2).$$

With regard to $V(t, x_1, x_2) = x_1^2 x_2^2$, in this example, $\frac{\partial}{\partial t}V(t, x_1, x_2) = 0$,

$$\frac{\partial}{\partial x}V(t, x_1, x_2) = [2x_1 x_2^2, 2x_1^2 x_2],$$

and

$$\frac{\partial^2}{\partial x^2} V(t, x_1, x_2) = \begin{bmatrix} 2x_2^2 & 4x_1x_2 \\ 4x_1x_2 & 2x_1^2 \end{bmatrix}^T = \begin{bmatrix} \nabla(2x_1x_2^2) \\ \nabla(2x_1^2x_2) \end{bmatrix}.$$

Step #2. Using the expressions in Step #1, the formulas (1.3.26) and (1.3.27) reduce to

$$dV(t, x_1, x_2) = \frac{\partial}{\partial t} V(t, x_1, x_2) dt + \frac{\partial}{\partial x} V(t, x_1, x_2)(dx)^T$$

$$+ \frac{1}{2}(dx) \left(\frac{\partial^2}{\partial x^2} V(t, x_1, x_2) \right) (dx)^T$$

$$= [2x_1x_2^2, 2x_1^2x_2](dx)^T + \frac{1}{2}(dx)^T \begin{bmatrix} 2x_2^2 & 4x_1x_2 \\ 4x_1x_2 & 2x_1^2 \end{bmatrix} (dx)$$

$$= 2x_1x_2^2 dx_1 + 2x_1^2x_2 dx_2$$

$$+ \left[\frac{1}{2} 2x_2^2 dx_1 + \frac{1}{2} 4x_1x_2 dx_2 \quad \frac{1}{2} 4x_1x_2 dx_1 + \frac{1}{2} 2x_1^2 x_2 dx_2 \right] (dx)$$

$$= 2x_1x_2^2 dx_1 + 2x_1^2x_2 dx_2$$

$$+ \frac{1}{2} 2x_2^2 (dx_1)^2 + \frac{1}{2} 4x_1x_2 dx_1 dx_2 + \frac{1}{2} 4x_1x_2 dx_2 dx_1 + \frac{1}{2} 2x_1^2 (dx_2)^2$$

$$= [(2x_1x_2^2 f_1(t, x_1) + x_2^2 \sigma_1^2(t, x_1)) + (2x_1^2 x_2 f_2(t, x_2) + x_1^2 \sigma_2^2(t, x_2))$$

$$+ 4x_1x_2 \sigma_1(t, x_1) \sigma_2(t, x_2)] dt$$

$$+ [2x_1x_2^2 \sigma_1(t, x_1) + 2x_1^2 x_2 \sigma_2(t, x_2)] dw(t)$$

$$= x_2^2 [2x_1 f_1(t, x_1) + \sigma_1^2(t, x_1) + 2x_1 \sigma_1(t, x_1) dw(t)]$$

$$+ x_1^2 [(2x_2 f_2(t, x_2) + \sigma_2^2(t, x_2) + 2x_2 \sigma_2(t, x_2) dw(t)]$$

$$+ 4x_1x_2 \sigma_1(t, x_1) \sigma_2(t, x_2)] dt.$$

This establishes the desired expression for $dV(t, x_1, x_2)$ with respect to the solution processes of (1.3.18) and (1.3.18).

Theorem 1.3.5 (Itô–Doob stochastic differential formula — quotient rule). *Let $V_1, V_2 \in C[J \times R, R]$. Assume that $V_1(t, z)$ and $V_2(t, z)$ are continuously differentiable functions with respect to t and twice continuously differentiable with respect to z. Define $V = \frac{V_1}{V_2}$, where $V_2 \neq 0$. Then,*

$$dV(t, x_1, x_2)$$

$$= \frac{V_2(t, x_2) dV_1(t, x_1) - V_1(t, x_1) dV_2(t, x_2)}{V_2^2(t, x_2)}$$

$$- \frac{\sigma_1(t,x_1)\sigma_2(t,x_2)\frac{\partial}{\partial x_1}V_1(t,x_1)\frac{\partial}{\partial x_2}V_2(t,x_2)dt}{V_2^2(t,x_2)}$$

$$+ \frac{\sigma_1(t,x_1)\sigma_2(t,x_2)V_1(t,x_1)(\frac{\partial}{\partial x_2}V_2(t,x_2))^2dt}{V_2^3(t,x_2)} \tag{1.3.29}$$

$$= \left[\frac{V_2(t,x_2)L_1V_1(t,x_1) - V_1(t,x_1)L_2V_2(t,x_2)}{V_2^2(t,x_2)} \right.$$

$$- \frac{\sigma_1(t,x_1)\sigma_2(t,x_2)\frac{\partial}{\partial x_1}V_1(t,x_1)\frac{\partial}{\partial x_2}V_2(t,x_2)}{V_2^2(t,x_2)}$$

$$+ \left. \frac{\sigma_1(t,x_1)\sigma_2(t,x_2)V_1(t,x_1)(\frac{\partial}{\partial x_2}V_2(t,x_2))^2}{V_2^3(t,x_2)} \right] dt$$

$$+ \left[\frac{V_2(t,x_2)\sigma_1(t,x_1)\frac{\partial}{\partial x_1}V_1(t,x_1) - V_1(t,x_1)\sigma_2(t,x_2)\frac{\partial}{\partial x_2}V_2(t,x_2)}{V_2^2(t,x_2)} \right] dw(t), \tag{1.3.30}$$

where w is the Wiener process defined in Section 1.2, and $L_1V_1(t,x_1)$ and $L_1V_2(t,x_2)$ are as defined in (1.3.24) and (1.3.25), respectively.

Proof. The proof of this theorem follows by applying Theorem 1.3.4 to $V = V_1V_2^{-1}$ as the product of two functions, V_1 and V_2^{-1}. The details are left as an exercise to the reader. □

Example 1.3.5. Given: $V_1(t,x_1) = x_1^2$ and $V_2(t,x_2) = x_2$. Goal: To find $dV(t,x_1,x_2)$ with respect to the solution processes of (1.3.18) and (1.3.18), where $\frac{x_1^2}{x_2} = V(t,x_1,x_2) = \frac{V_1(t,x_1)}{V_2(t,x_2)}$.

Solution procedure. We note that the given functions $V_1(t,x_1) = x_1^2$ and $V_2(t,x_2) = x_2$ are continuously differentiable functions with respect to t and twice continuously differentiable with respect to x's. In fact, they are independent of the independent variable t. Moreover, $\frac{x_1^2}{x_2} = V(t,x_1,x_2)$ is a continuously differentiable function with respect to t and twice continuously differentiable with respect to x_1 and for all x_2 for all $x_2 \neq 0$. We apply Theorem 1.3.4 (Itô–Doob stochastic differential formula — quotient rule) to $\frac{V_1(t,x_1)}{V_2(t,x_2)} = \frac{x_1^2}{x_2}$.

Step #1. To apply Theorem 1.3.5, we imitate the argument used in Examples 1.3.3 and 1.3.4 to compute

$$\frac{\partial}{\partial t}V(t,x_1,x_2), \quad \frac{\partial}{\partial x}V(t,x_1,x_2) \quad \text{and} \quad \frac{\partial^2}{\partial x^2}V(t,x_1,x_2).$$

With regard to $V(t, x_1, x_2) = \frac{x_1^2}{x_2}$, in this example, $\frac{\partial}{\partial t} V(t, x_1, x_2) = 0$,

$$\frac{\partial}{\partial x} V(t, x_1, x_2) = \left[\begin{array}{cc} \dfrac{2x_1}{x_2} & -\dfrac{x_1^2}{x_2^2} \end{array}\right],$$

and

$$\frac{\partial^2}{\partial x^2} V(t, x_1, x_2) = \left[\begin{array}{cc} \dfrac{2}{x_2} & -\dfrac{2x_1}{x_2^2} \\[2mm] -\dfrac{2x_1}{x_2^2} & \dfrac{2x_1 x_2}{x_2^3} \end{array}\right]^T = \left[\begin{array}{c} \nabla\left(\dfrac{2x_1}{x_2}\right) \\[2mm] \nabla\left(-\dfrac{x_1^2}{x_2^2}\right) \end{array}\right].$$

Step #2. Using the expressions in Step #1, the formulas (1.3.29) and (1.3.30) reduce to

$$dV(t, x_1, x_2) = \frac{\partial}{\partial t} V(t, x_1, x_2) dt + \frac{\partial}{\partial x} V(t, x_1, x_2)(dt)^T$$

$$+ \frac{1}{2}(dx)\left(\frac{\partial^2}{\partial x^2} V(t, x_1, x_2)\right)(dx)^T$$

$$= \left[\begin{array}{cc} \dfrac{2x_1}{x_2} & -\dfrac{x_1^2}{x_2^2} \end{array}\right](dx) + \frac{1}{2}(dx)^T \left[\begin{array}{cc} \dfrac{2}{x_2} & -\dfrac{2x_1}{x_2^2} \\[2mm] -\dfrac{2x_1}{x_2^2} & \dfrac{2x_1^2 x_2}{x_2^3} \end{array}\right](dx)$$

$$= \frac{2x_1}{x_2} dx_1 - \frac{x_1^2}{x_2^2} dx_2$$

$$+ \left[\begin{array}{cc} \dfrac{1}{2}\dfrac{2}{x_2} dx_1 + \dfrac{1}{2}\left(-\dfrac{2x_1}{x_2^2}\right) dx_2 & \dfrac{1}{2}\left(-\dfrac{2x_1}{x_2^2}\right) dx_1 + \dfrac{1}{2}\dfrac{2x_1^2 x_2}{x_2^3} dx_2 \end{array}\right](dx)$$

$$= \frac{2x_1}{x_2} dx_1 - \frac{x_1^2}{x_2^2} dx_2 + \frac{1}{2}\frac{2}{x_2}(dx_1)^2$$

$$+ \frac{1}{2}\left(-\frac{2x_1}{x_2^2}\right) dx_1 dx_2 + \frac{1}{2}\left(-\frac{2x_1}{x_2^2}\right) dx_1 dx_2 + \frac{1}{2}\frac{2x_1^2 x_2}{x_2^3}(dx_2)^2$$

$$= \left[\left(\frac{2x_1}{x_2} f_1(t, x_1) + \frac{1}{x_2}\sigma_1^2(t, x_1)\right) + \left(-\frac{x_1^2}{x_2^2} f_2(t, x_2) + \frac{x_1^2}{x_2^2}\sigma_2^2(t, x_2)\right)\right.$$

$$\left. - \frac{2x_1}{x_2^2}\sigma(t, x_1)\sigma_2(t, x_2)\right] dt + \left[\frac{2x_1}{x_2}\sigma_1(t, x_1) - \frac{x_1^2}{x_2^2}\sigma_2(t, x_2)\right] dw(t)$$

$$= \left[\frac{x_2(2x_1 f_1(t, x_1) + \sigma_1^2(t, x_1)) - x_1^2 f_2(t, x_2)}{x_2^2}\right.$$

$$-\frac{2x_1}{x_2^2}\sigma_1(t,x_1)\sigma_2(t,x_2) + \frac{x_1^2}{x_2^2}\sigma_2^2(t,x_2)\bigg]dt$$

$$+ \bigg[\frac{x_2 2x_1\sigma_1(t,x_1) - x_1^2\sigma_2(t,x_2)}{x_2^2}\bigg]dw(t)$$

$$= \frac{1}{x_2}\big[(2x_1 f_1(t,x_1) + \sigma_1^2(t,x_1))dt + 2x_1\sigma_1(t,x_1)dw(t)\big]$$

$$- \frac{x_1^2}{x_2^2}[f_2(t,x_2)dt + \sigma_2(t,x_2)dw(t)] + \frac{x_1^2}{x_2^2}\sigma_2^2(t,x_2)dt$$

$$- \frac{2x_1}{x_2^2}\sigma_1(t,x_1)\sigma_2(t,x_2)dt$$

$$= \frac{x_2}{x_2^2}\big[(2x_1 f_1(t,x_1) + \sigma_1^2(t,x_1))dt + 2x_1\sigma_1(t,x_1)dw(t)\big]$$

$$- \frac{x_1^2}{x_2^2}\big[(f_2(t,x_2) - \sigma_2^2(t,x_2))dt + \sigma_2(t,x_2)dw(t)\big]$$

$$- \frac{2x_1}{x_2^2}\sigma_1(t,x_1)\sigma_2(t,x_2)dt.$$

This establishes the desired expression for $dV(t,x_1,x_2)$ with respect to the solution processes of (1.3.18) and (1.3.18).

Definition 1.3.4. A random variable ζ is called a Markov time with respect to the σ-algebra $\mathfrak{F}$ if for arbitrary $t \in [0,T]$, the event $\{\omega : \zeta(\omega) > t\}$ belongs to the σ-algebra $\mathfrak{F}_t$. This means that $\{\omega : \zeta(\omega) > t\}$ is independent of $w(s)$ for $s \geq t$.

Example 1.3.6. Let $w(t)$ be a Wiener process defined on $[0,T]$ into R. The first passage/stopping time is defined by:

(a) $\zeta_1 = \inf\{t : w(t) = a\}$ — the first time the Wiener process $w(t)$ attains the value a;

(b) $\zeta_2 = \inf\{t : \int_0^t \sigma(u)dw(u) = a\}$ — the first time the process $\int_0^t \sigma(u)dw(u)$ attains the value a. ζ_1 and ζ_2 are Markov times. Moreover, depending on a, ζ_1 and ζ_2 can take values between 0 and T (if $T = \infty$, then they can take values ∞).

Theorem 1.3.6. *Let τ be a Markov time with respect to σ-algebra $\mathfrak{F}_t$, and $w(t)$ be a Wiener process defined on $[0,T]$ into R. Then the process defined by*

$$w^*(t) = w(t+\tau) - w(t)$$

is also a Wiener process and it does not depend on $\mathfrak{F}_\tau$; $\mathfrak{F}_\tau$ is the smallest σ-algebra containing the collection of events $\{A_t \cap \{\omega : \tau(\omega) > t\} : A_t \in \mathfrak{F}_t\}$.

Exercises

1. Find the Itô–Doob differentials of the following functions:

   (a) $V(t, w(t)) = \sin(w(t))$;

   (b) $V(t, w(t)) = \sec(w(t))$;

   (c) $V(t, w(t)) = \tan(w(t))$;

   (d) $V(t, w(t)) = \exp[t^2 - 2w^2(t)]$;

   (e) $V(t, w(t)) = 2w^3(t) - 3w^2(t) + w(t) + t$;

   (f) $V(t, w(t)) = \ln(1 + w^2(t))$;

   (g) $V(t, w(t)) = \cosh(w(t))$;

   (h) $V(t, w(t)) = \cos(t + w(t))$;

   (i) $V(t, w(t)) = \sin t \exp[-2w^2(t)]$;

   (j) $V(t, w(t)) = w^2(t) \cos t$;

   (k) $V(t, w(t)) = t + w(t)$;

   (l) $V(t, w(t)) = 2t + w(t)$;

   (m) $V(t, w(t)) = \frac{t + w(t)}{2t + w(t)}$;

   (n) $V(t, w(t)) = \frac{2t + w(t)}{t + w(t)}$.

2. Find: (1) addition, (2) product, and (3) quotient formula with regard to the following Itô–Doob-type stochastic differential equations with specified functions:

   (a) $V_1(t, x) = \sin^2 x$ and $V_2(t, x) = x^2 : dx = 3x\,dt + 2x\,dw(t)$;

   (b) $V_1(t, x) = \exp[x]$ and $V_2(t, x) = \cosh x : dx = ax\,dw(t)$;

   (c) $V_1(t, x) = \sin^2 x$ and $V_2(t, x) = x^2 : dx = -x\,dt$.

3. Find: (1) addition, (2) product, and (3) quotient formula with regard to the Itô–Doob-type stochastic differential equations

   $$dx_1 = \cos x_1 dt + \sin x_1 dw(t),$$

   $$dx_2 = -\sin x_2 dt + \cos x_2 dw(t),$$

   by utilizing the following functions:

   (a) $V_1(t, x) = \sin^2 x$ and $V_2(t, x) = x^2$;

   (b) $V_1(t, x) = \exp[x]$ and $V_2(t, x) = \cosh x$;

   (c) $V_1(t, x) = \cos^2 x$ and $V_2(t, x) = \sin^2 x$;

   (d) $V_1(t, x) = \ln(1 + x^2)$ and $V_2(t, x) = \exp[x]$.

4. Using Observation 1.5.3 [80] and Theorems 1.5.6 [80] and 1.3.1, find $dA(t)$ and $d(\det(A(t)))$, where

   (a) $A(t) = \begin{bmatrix} \cos w(t) & \sin w(t) \\ -\sin w(t) & \cos w(t) \end{bmatrix}$;

(b) $A(t) = \begin{bmatrix} 1 & \exp[w(t)] & t \\ \cos w(t) & t & 1 \\ t & \sin w(t) & \exp[w(t)] \end{bmatrix}.$

5. Let ϕ be a real-valued function with bounded continuous second derivative ϕ'', and $w(t)$ be a Wiener process. $V(t, w(t)) = \phi(w(t)) - \frac{1}{2} \int_a^t \phi''(w(s))ds$. Find the Itô–Doob differential of $V(t, w(t))$.

6. Complete the proof of Theorem 1.3.5 (quotient rule).

1.4 Method of Substitution

In the following, we present a procedure for computing an Itô–Doob integral of the type

$$I(w(t)) = \int f(w(t))dw(t), \qquad (1.4.1)$$

where $f \in C[D(f), R]$, with $D(f) \subseteq R$ standing for the domain of f; f is assumed to be continuously differentiable on $D(f)$.

Solution procedure

Step 1. We denote $u = w(t)$ and rewrite the integral (1.4.1) as

$$I(u) = \int f(u)du,$$

and apply the deterministic methods (if possible) to compute this integral. The computation is given by

$$I(u) = \int f(u)du = F(u) = F(w(t)), \qquad (1.4.2)$$

where F is any antiderivative of f. This fundamental step can be considered to be the *substitution method* for the computation of the stochastic integral (1.4.1).

Step 2. We use the indefinite integral in Step 1, and compute the Itô–Doob differential of $F(w(t))$. For this purpose, we apply the formula (1.3.11) to $F(w(t))$, and obtain

$$dF(w(t)) = \frac{1}{2}\frac{d^2}{dx^2}F(w(t))dt + \frac{d}{dx}F(w(t))dw(t). \qquad (1.4.3)$$

This is possible because of the fact that F is an antiderivative of f, and f is continuously differentiable on its domain $D(f)$. Hence, (1.4.3) can be rewritten as

$$dF(w(t)) = \frac{1}{2}\frac{d}{dx}f(w(t))dt + f(w(t))dw(t). \qquad (1.4.4)$$

Step 3. Now, we apply the Itô–Doob integral, and we have

$$F(w(t)) = \frac{1}{2}\frac{d}{dx}f(w(t))dt + \int f(w(t))dw(t). \tag{1.4.5}$$

By solving for $\int f(w(t))dw(t)$, we obtain the integral (1.4.1) as follows:

$$\int f(w(t))dw(t) = F(w(t)) - \frac{1}{2}\int \frac{d}{dx}f(w(t))dt. \tag{1.4.6}$$

In the following, this procedure is illustrated by presenting several basic indefinite integrals.

Example 1.4.1. Consider $\int w^n(t)dw(t)$, $n \neq -1$. Here $f(x) = x^n$, and $\frac{d}{dx}f = nx^{n-1}$. Using Step 1 in the above described procedure, we arrive at

$$\int u^n du. \tag{1.4.7}$$

This follows by the elementary deterministic integration process [127]

$$\int u^n du = c + \frac{u^{n+1}}{n+1}.$$

Hence

$$F(w(t)) = c + \frac{w^{n+1}(t)}{n+1}, \tag{1.4.8}$$

where F is an antiderivative of f. Applying Step 2 to $F(w(t))$ in (1.4.8), the relation (1.4.4) becomes

$$dF(w(t)) = \frac{1}{2}\frac{d}{dx}f(w(t))dt + f(w(t))dw(t)$$
$$= \frac{n}{2}w^{n-1}(t)dt + w^n(t)dw(t). \tag{1.4.9}$$

From (1.4.9) and (1.4.8) and using Step 3, the relation (1.4.6) reduces to

$$\int w^n(t)dw(t) = F(w(t)) - \frac{n}{2}\int w^{n-1}(t)dt$$
$$= c + \frac{w^{n+1}(t)}{n+1} - \frac{n}{2}\int w^{n-1}(t)dt. \tag{1.4.10}$$

This completes the solution process.

Example 1.4.2. Now, we consider $\int \sin w(t)dw(t)$. Here $f(x) = \sin x$, $\frac{d}{dx}f(x) = \cos x$. Utilizing Step 1, we obtain

$$\int \sin u \, du. \tag{1.4.11}$$

Again, from the elementary deterministic integration process [127], we obtain

$$\int \sin u \, du = c - \cos u.$$

Hence

$$F(w(t)) = c - \cos w(t), \tag{1.4.12}$$

where F is an antiderivative of f. Applying Step 2 to $F(w(t))$ in (1.4.12), the relation (1.4.4) becomes

$$dF(w(t)) = \frac{1}{2}\frac{d}{dx}\sin w(t)dt + \sin w(t)dw(t)$$

$$= \frac{1}{2}\cos w(t)dt + \sin w(t)dw(t). \tag{1.4.13}$$

From (1.4.12) and (1.4.13) and using Step 3, the relation (1.4.6) reduces to

$$\int \sin w(t)dw(t) = F(w(t)) - \frac{1}{2}\int \cos w(t)dt$$

$$= c - \cos w(t) - \frac{1}{2}\int \cos w(t)dt. \tag{1.4.14}$$

This completes the integration process.

Example 1.4.3. Now, we consider $\int \sec^2 w(t)dw(t)$. Here $f(x) = \sec^2 x$, $\frac{d}{dx}f(x) = \tan x \sec^2 x$. Utilizing Step 1, we have

$$\int \sec^2 u \, du. \tag{1.4.15}$$

By the elementary deterministic integration process, we have

$$\int \sec^2 u \, du = c + \tan u.$$

Hence

$$F(w(t)) = c + \tan w(t), \tag{1.4.16}$$

where F is an antiderivative of f. Applying Step 2 to $F(w(t))$ in (1.4.16), the relation (1.4.4) becomes

$$dF(w(t)) = \frac{1}{2}\frac{d}{dx}f(w(t))dt + f(w(t))dw(t)$$

$$= \sec^2 w(t) \tan w(t)dt + \sec^2 w(t)dw(t). \tag{1.4.17}$$

From (1.4.16) and (1.4.17) and using Step 3, the relation (1.4.6) reduces to

$$\int \sec^2 w(t)dw(t) = F(w(t)) - \int \sec^2 w(t)\tan w(t)dt$$

$$= c + \tan w(t) - \int \sec^2 w(t)\tan w(t)dt. \qquad (1.4.18)$$

The example is completed.

Example 1.4.4. Let us consider $\int e^{w(t)}dw(t)$. Here $f(x) = e^x$, $\frac{d}{dx}f(x) = e^x$. Utilizing Step 1, we have

$$\int e^u du. \qquad (1.4.19)$$

Its integral is

$$\int e^u du = c + e^u.$$

Hence

$$F(w(t)) = c + e^{w(t)}, \qquad (1.4.20)$$

where F is an antiderivative of f. Applying Step 2 to $F(w(t))$ in (1.4.20), the relation (1.4.4) becomes

$$dF(w(t)) = \frac{1}{2}\frac{d}{dx}f(w(t))dt + f(w(t))dw(t)$$

$$= \frac{1}{2}e^{w(t)}dt + e^{w(t)}dw(t). \qquad (1.4.21)$$

From (1.4.20) and (1.4.21) and using Step 3, the relation (1.4.6) is reduced to

$$\int e^{w(t)}dw(t) = F(w(t)) - \frac{1}{2}\int e^{w(t)}dt$$

$$= c + e^{w(t)} - \frac{1}{2}\int e^{w(t)}dt. \qquad (1.4.22)$$

This completes our example.

Example 1.4.5. Compute the integral $\int \exp[at + bw(t)]dw(t)$, for $b \neq 0$.

Set $z(t) = at + bw(t)$. Its differential is $dz(t) = d(at + bw(t)) = adt + bdw(t)$. Now, solve for $dw(t)$; its solution is $dw(t) = \frac{dz(t) - adt}{b}$. With this substitution, the given integral is reduced to

$$\int \exp[at + bw(t)]dw(t) = \int \exp[z] \left(\frac{dz(t) - adt}{b} \right)$$

$$= \frac{1}{b} \int \exp[z]dz(t) - \frac{a}{b} \int \exp[z(t)]dt$$

$$= \frac{1}{b} \exp[z] - \frac{a}{b} \int \exp[z(t)]dt. \qquad (1.4.23)$$

Now, by imitating the argument used in Example 1.4.4, in particular from (1.4.20), we have

$$dF(z(t)) = f(z(t))dz(t) + \frac{1}{2}\frac{d}{dx}f(z(t))(dz(t))^2 = e^{z(t)}dz(t) + \frac{1}{2}e^{z(t)}(dz(t))^2$$

$$= e^{z(t)}dz(t) + \frac{1}{2}e^{z(t)}(adt + bdw(t))^2$$

$$= e^{z(t)}dz(t) + \frac{1}{2}e^{z(t)}[a^2(dt)^2 + 2ab\,dt\,dw(t) + b^2(dw(t))^2]$$

$$= e^{z(t)}dz(t) + \frac{1}{2}e^{z(t)}b^2(dw(t))^2 \quad \text{(by the Itô–Doob calculus)}$$

$$= e^{z(t)}dz(t) + \frac{1}{2}e^{z(t)}b^2dt. \qquad (1.4.24)$$

Hence

$$\int \exp[z]dz(t) = c + e^{z(t)} - \frac{b^2}{2} \int e^{z(t)}dt. \qquad (1.4.25)$$

From (1.4.24) and (1.4.25), the integral (1.4.23) reduces to

$$\int \exp[at + bw(t)]dw(t) = c + \frac{1}{b}\exp[at + bw(t)]$$

$$- \frac{b^2}{2b} \int \exp[at + bw(t)]dt - \frac{a}{b} \int \exp[at + bw(t)]dt$$

$$= c + \frac{1}{b}\exp[at + bw(t)] - \left(\frac{b}{2} + \frac{a}{b} \right) \int [\exp[at + bw(t)]]dt. \qquad (1.4.26)$$

This is the desired integral with an arbitrary constant of integration c.

Exercises

By using the procedure described in this section, evaluate the integral:

1. $\int \cos w(t)dw(t)$,
2. $\int \csc^2 w(t)dw(t)$,

3. $\int \sec w(t) \tan w(t) dw(t)$,

4. $\int \csc w(t) \cot w(t) dw(t)$,

5. $\int \sinh w(t) dw(t)$,

6. $\int \cosh w(t) dw(t)$,

7. $\int \tan w(t) dw(t)$,

8. $\int \cot w(t) dw(t)$,

9. $\int \frac{1}{w^2(t)+a^2} dw(t)$,

10. $\int \frac{1}{w(t)} dw(t)$.

1.5 Method of Integration by Parts

In the following, we present a procedure for computing an Itô–Doob integral of the type

$$I(t, w(t)) = \int f(t, w(t)) dw(t), \tag{1.5.1}$$

where $f \in C[J \times R, R]$; it is continuously differentiable in both variables (t, x) as many times as desired.

Solution procedure

Step 1. We apply the Itô–Doob differential formula (1.3.11) (Theorem 1.3.1) to

$$V(t, w(t)) = w(t) f(t, w(t)). \tag{1.5.2}$$

Using product and chain rules and applying the Itô–Doob differential of $V(t, w(t))$ in (1.3.11), we get

$$
\begin{aligned}
dV(t, w(t)) &= d(w(t) f(t, w(t))) \\
&= w(t) f_t(t, w(t)) dt + f(t, w(t)) dw(t) + w(t) f_w(t, w(t)) dw(t) \\
&\quad + \frac{1}{2} [2 f_w(t, w(t)) + w(t) f_{ww}(t, w(t))] (dw(t))^2 \\
&= \left[w(t) f_t(t, w(t)) + f_w(t, w(t)) + \frac{1}{2} w(t) f_{ww}(t, w(t)) \right] dt \\
&\quad + f(t, w(t)) dw(t) + w(t) f_w(t, w(t)) dw(t).
\end{aligned}
\tag{1.5.3}
$$

Step 2. Now, by applying the stochastic integral formula to (1.5.3), we obtain

$$
\begin{aligned}
V(t, w(t)) = c_1 &+ \int f(t, w(t)) dw(t) + \int w(t) f_w(t, w(t)) dw(t) \\
&+ \int \left[w(t) f_t(t, w(t)) + f_w(t, w(t)) + \frac{1}{2} w(t) f_{ww}(t, w(t)) \right] dt.
\end{aligned}
$$

From this, definition of $V(t, w(t))$ in (1.5.2), we have the expression

$$w(t)f(t, w(t)) = c_1 + \int f(t, w(t))dw(t) + \int w(t)f_w(t, w(t))dw(t)$$

$$+ \int \left[w(t)f_t(t, w(t)) + f_w(t, w(t)) + \frac{1}{2}w(t)f_{ww}(t, w(t)) \right] dt.$$

$$(1.5.4)$$

Step 3. By solving the second term on the right-hand side (RHS) of (1.5.4), we obtain an expression for the integral on the RHS of (1.5.1):

$$\int f(t, w(t))dw(t) = w(t)f(t, w(t)) - c_1 - \int w(t)f_w(t, w(t))dw(t)$$

$$- \int \left[w(t)f_t(t, w(t)) + f_w(t, w(t)) + \frac{1}{2}w(t)f_{ww}(t, w(t)) \right] dt.$$

$$(1.5.5)$$

This expression is analogous to that obtained in the usual method of integration by parts in a deterministic calculus course [127].

Step 4. Now, by substituting the RHS expression in (1.5.5) in the RHS of (1.5.1), we have:

$$I(t, w(t)) = c + w(t)f(t, w(t)) - \int w(t)f_w(t, w(t))dw(t)$$

$$- \int \left[w(t)f_t(t, w(t)) + f_w(t, w(t)) + \frac{1}{2}w(t)f_{ww}(t, w(t)) \right] dt.$$

$$(1.5.6)$$

Step 5. We repeat Steps 1–4 with regard to the first integral on the RHS of (1.5.6). Thus,

$$V(t, w(t)) = \frac{1}{2}w^2(t)f_w(t, w(t)).$$

$$(1.5.7)$$

The expressions analogous (1.5.3)–(1.5.5) with respect to $V(t, w(t))$ in (1.5.7) are

$$dV(t, w(t)) = d\left(\frac{1}{2}w^2(t)f_w(t, w(t)) \right)$$

$$= \frac{1}{2}w^2(t)f_{tw}(t, w(t))dt + w(t)f_w(t, w(t))dw(t)$$

$$+ \frac{1}{2}w^2(t)f_{ww}(t, w(t))dw(t)$$

$$+ \frac{1}{2}\left[f_w(t, w(t)) + 2w(t)f_{ww}(t, w(t)) + \frac{1}{2}w^2(t)f_{www}(t, w(t)) \right] (dw(t))^2$$

$$= \frac{1}{2}\left[w^2(t)f_{tw}(t,w(t))dt + f_w(t,w(t)) + 2w(t)f_{ww}(t,w(t))\right.$$

$$\left. + \frac{1}{2}w^2(t)f_{www}(t,w(t))\right]dt$$

$$+ w(t)f_w(t,w(t))dw(t) + \frac{1}{2}w^2(t)f_{ww}(t,w(t))dw(t), \tag{1.5.8}$$

$$\frac{1}{2}w^2(t)f_w(t,w(t)) = c_1 + \int w(t)f_w(t,w(t))dw(t) + \frac{1}{2}\int w^2(t)f_{ww}(t,w(t))dw(t)$$

$$+ \frac{1}{2}\int\left[w^2(t)f_{tw}(t,w(t)) + f_w(t,w(t)) + 2w(t)f_{ww}(t,w(t))\right.$$

$$\left. + \frac{1}{2}w^2(t)f_{www}(t,w(t))\right]dt, \tag{1.5.9}$$

$$\int w(t)f_w(t,w(t))dw(t) = \frac{1}{2}w^2(t)f_w(t,w(t)) - c_1 - \frac{1}{2}\int w^2(t)f_{ww}(t,w(t))dw(t)$$

$$- \frac{1}{2}\int\left[w^2(t)f_{tw}(t,w(t)) + f_w(t,w(t)) + w(t)f_{ww}(t,w(t))\right.$$

$$\left. + \frac{1}{2}w^2(t)f_{www}(t,w(t))\right]dt, \tag{1.5.10}$$

respectively.

Step 6. Again, we repeat the procedure described in Step 4, i.e. now substituting the right-hand side expression in (1.5.10) for the first integral term on the right-hand side of (1.5.6), we have the form

$$I(t,w(t)) = c + f(t,w(t))w(t) - \frac{1}{2}w^2(t)f_w(t,w(t))$$

$$+ \frac{1}{2}\int w^2(t)f_{ww}(t,w(t))dw(t)$$

$$+ \frac{1}{2}\int\left[w^2(t)f_{tw}(t,w(t)) + f_w(t,w(t))\right.$$

$$\left. + 2w(t)f_{ww}(t,w(t)) + \frac{1}{2}w^2(t)f_{www}(t,w(t))\right]dt$$

$$- \int\left[w(t)f_t(t,w(t)) + f_w(t,w(t)) + \frac{1}{2}w(t)f_{ww}(t,w(t))\right]dt. \tag{1.5.11}$$

Step 7. We continue this integration procedure (Steps 1–4) [i.e. the integration with respect to the w term on the right-hand side in (1.5.11)], until the integral term is either repeated or terminated. This completes the procedure for computing the integral (1.5.1).

Example 1.5.1. Compute $I(t) = \int w(t)dw(t)$. Here, $f(t, w(t)) = w(t)$. From (1.5.2) and by imitating Step 1, we have $V(t, w(t)) = w^2(t)$ and

$$dV(t, w(t)) = d(w(t)w(t))$$

$$= w(t)dw(t) + w(t)dw(t) + \frac{1}{2}2(dw(t))^2$$

$$= 2w(t)dw(t) + dt. \tag{1.5.12}$$

By following Steps 2 and 3, we get

$$\int w(t)dw(t) = w^2(t) - c_1 - \int w(t)dw(t) - \int dt. \tag{1.5.13}$$

The first integral term on the RHS is repeated. Therefore, the integration procedure is terminated. Thus, we obtain

$$\int w(t)dw(t) = c + \frac{1}{2}w^2(t) - \frac{1}{2}t. \tag{1.5.14}$$

From (1.5.14), we conclude that

$$I(t) = c + \frac{1}{2}w^2(t) - \frac{1}{2}t. \tag{1.5.15}$$

This completes the computation of $I(t)$.

Example 1.5.2. Compute $I(t) = \int td + \int w(t)dw(t)$.

Here, the first integral is a deterministic integral [127], and the second integral is the integral in Example 1.5.1. Therefore, the computation of the first integral and the second integral is

$$I(t) = I_d(t) + I_s(t),$$

where

$$I_d(t) = \int t\,dt = \frac{1}{2}t^2 + c_1$$

$$I_s(t) = \int w(t)dw(t) = c_2 + \frac{1}{2}w^2(t) - \frac{1}{2}t.$$

Hence,

$$I(t) = c + \frac{1}{2}t^2 + \frac{1}{2}w^2(t) - \frac{1}{2}t.$$

This completes the computation of the given integral.

Example 1.5.3. Find $I(t) = \int t^4 w^5(t) dw(t)$.

Here, $f(t, w(t)) = t^4 w^5(t)$. We set $V(t, w(t)) = t^4 w^6(t)$, and by imitating Step 1 we have

$$dV(t, w(t)) = d(t^4 w^6(t))$$
$$= 4t^3 w^6(t) dt + 6t^4 w^5(t) dw(t) + 15t^4 w^4(t)(dw(t))^2. \qquad (1.5.16)$$

After integration, we obtain

$$c + t^4 w^6(t) = \int 4t^3 w^6(t) dt + \int 6t^4 w^5(t) dw(t) + \int 15t^4 w^4(t) dt$$

and hence $I(t)$ is given by

$$\int t^4 w^5(t) dw(t) = c + \frac{1}{6} t^4 w^6(t) - \frac{2}{3} \int t^3 w^6(t) dt - \frac{5}{2} \int t^4 w^4(t) dt. \qquad (1.5.17)$$

Example 1.5.4. Find $\int_a^b g(t) dw(t)$, where g is a continuously differentiable function on $[a, b]$. Here $f(t, w(t)) = g(t)$. We set $V(t, w(t)) = g(t) w(t)$ and imitate Step #1, and we have

$$dV(t, w(t)) = d(g(t) w(t)) = g'(t) w(t) dt + g(t) dw(t). \qquad (1.5.18)$$

After integration, we get

$$(g(t) w(t) + c)|_a^b = \int_a^b g'(t) w(t) dt + \int_a^b g(t) dw(t),$$

which implies that

$$\int_a^b g(t) dw(t) = g(b) w(b) - g(a) w(a) - \int_a^b g'(t) w(t) dt. \qquad (1.5.19)$$

Exercises

Compute the following indefinite integrals:

1. $\int dw(t)$;
2. $\int t^2 dt + \int tw(t) dw(t)$;
3. $\int t^m w^n(t) dw(t)$;
4. $\int \cos t \, dt + \int \sin w(t) dw(t)$;
5. $\int t \, dw(t)$;
6. $\int t \, dt + \int dw(t)$;
7. $\int \sin^3 w(t) \cos w(t) dw(t)$;
8. $\int (t \cos w(t) + \sin t) dw(t)$;
9. $\int t^2 \cos^3 w(t) dw(t)$;
10. $\int \cos t \sin w(t) dw(t)$.

1.6 Notes and Comments

The content of Section 1.1 is devoted to the basic concepts and vocabulary with illustrations in probability theory. For details see Refs. 44, 82 and 163. Section 1.2 is centered on the fundamental concepts, definitions, and results with examples in stochastic processes. For further reading, refer to Refs. 7, 8, 28, 32, 38, 43, 162, 182 and 195. The basic material on the Itô–Doob stochastic calculus in Section 1.3 is derived from Refs. 7, 8, 32, 43, 52, 56, 61, 71, 98, 141 and 182. The methods of substitution and the integration by parts for the Itô–Doob integral are presented in Sections 1.4 and 1.5, respectively. Several examples are used to illustrate the techniques. This material is from the second author's class notes. Moreover, at the end of each section, several exercises are provided for mastering the techniques.

Chapter 2

First-Order Differential Equations

Introduction

In this chapter, mathematical modeling, the procedures for solving first-order linear stochastic scalar differential equations of the Itô–Doob type, and their fundamental conceptual analysis are developed. Prior to the presentation of the technical procedures and the conceptual analysis, an attempt will be made to answer questions that are frequently asked by students. These include: Why should I learn this material? How will this help me? What are the sources of random fluctuations? Moreover, the understanding of the deterministic versus stochastic modeling problem in the dynamic process is explained.

Section 2.1, focuses on mathematical modeling. Three probabilistic models, namely random walk, Poisson, and Brownian motion, are presented in this section. Several dynamic processes in biological, chemical, engineering, medical, physical, and social sciences are outlined to illustrate the modeling procedures. Section 2.2 deals with first-order stochastic differential equations whose solutions can be directly found by the methods of Itô–Doob-type integration. This class of first-order differential equations is referred to as integrable differential equations. Moreover, the mathematical models of laminar blood flow in an artery and the motion of particles in the air are presented to illustrate the usage of this class of differential equations. Section 2.3 is devoted to first-order scalar homogeneous Itô–Doob-type stochastic differential equations. The eigenvalue-type method utilized for solving linear scalar deterministic differential equations [80] is the most suitable and simple approach to solving Itô–Doob-type linear stochastic differential equations with constant and variable coefficients. It requires a minimal mathematical background. In fact, this approach was originally initiated by the authors to find solutions to this class of stochastic differential equations. The approach reduces the problem of solving linear homogeneous Itô–Doob-type stochastic scalar differential equations to a problem of solving a linear system of algebraic equations. The reduction process decomposes the Itô–Doob type of stochastic differential equations into its deterministic and

stochastic parts. This is analogous to the idea of decomposition of systems of perturbed differential equations in Section 4.4 of Volume 1 [80] into unperturbed and
perturbed parts. We then imitate the deterministic procedure described in that section to find a solution process of the original linear homogeneous scalar stochastic
differential equations of the Itô–Doob type. The general step-by-step procedures for
finding the general solution and the solution to initial value problems are logically
and clearly outlined. Applied and mechanical/numerical examples as well as some
illustrations are presented to better describe the procedures and the usefulness of
the differential equations. Section 2.4 deals with first-order nonhomogeneous scalar
Itô–Doob-type stochastic differential equations. The method of variation of constant
parameters is used to solve the general and the initial value problems. This method
is very powerful as it can be used to solve a very complex differential equation problems in a systematic way. The usefulness of this class of Itô–Doob-type stochastic
differential equations in the mathematical modeling of single-species dynamic processes, namely population dynamics (immigration), Newton's law of cooling, cell
membrane, central nervous system dynamics, and diffusion processes under random
environmental perturbations, is also exhibited. In addition, several numerical illustrations and examples are provided to illustrate the method. Section 2.5 begins with
ten natural questions in dynamic modeling. Due to the prerequisite, three out of ten
questions will be answered. An attempt is made to provide rigorous justifications
for the validity of conceptual algorithms (theorems/lemma/corollary). Moreover, it
provides a foundation for answering the remaining questions posed at the beginning
of the section.

2.1 Mathematical Modeling

This section is devoted to the development of stochastic mathematical models for
dynamic processes in the biological, chemical, engineering, medical, physical, and
social sciences. This development is based on a theoretical experimental setup, the
fundamental laws in sciences, and engineering, and the basic information about
dynamic processes. The presented approach is in the spirit of the deterministic
modeling approach [80]. An attempt is made to incorporate diverse and apparently
different phenomena to relate to a common conceptual framework. Historically, the
well-known probabilistic models, namely the random walk model, Poisson process
model, and Brownian motion model, are used as the basis to develop the stochastic
models for dynamic processes. Using the existing deterministic, predictable, and
anticipated knowledge (information) as well as the mathematical models, several
illustrations/examples/observations are presented in the framework of stochastic,
unpredictable, and random environmental fluctuations.

Random Walk Model 2.1.1 [162, 167]. In the following, we utilize the *random
walk process* to initiate the scope and development of stochastic models of dynamic

processes in the biological, chemical, engineering, medical, physical, and social sciences. We use a conceptually common description of processes in the sciences and engineering known as a "state" of a given system. Examples are: "distance" traveled by an object in the physical process, "concentration" of a chemical substance in a chemical process, "number of species" in a biological process, and "price" of a commodity/service in a sociological process, all measured by scalar/vector quantities.

Let $x(t)$ be a state of a system at a time $t \in [a, b] \subset R$. The state of the system is observed over an interval of $[t, t + \Delta t]$, where Δt is a small increment in t. Without loss of generality, it is assumed that Δt is positive. The process is under the influence of random perturbations. We experimentally observe states $x(t_0) = x(t), x(t_1), x(t_2), \ldots, x(t_k), \ldots, x(t_n) = x(t + \Delta t)$ of a system at $t_0 = t$, $t_1 = t + \tau$, $t_2 = t + 2\tau, \ldots, t_k = t + k\tau, \ldots, t_n = t + \Delta t = t + n\tau$ over the interval $[t, t + \Delta t]$, where n belongs to $\{1, 2, 3, \ldots\}$ and $\tau = \frac{\Delta t}{n}$. These observations are made under the following random walk model (in short, RWM) conditions:

RWM1. The system is under the influence of independent and identical random impulses that takes place at $t_1, t_2, \ldots, t_k, \ldots, t_n$.

RWM2. The influence of a random impact on the state of the system is observed on every time subinterval of length τ.

RWM3. For each $k \in I(1, n) = \{1, 2, \ldots, k, \ldots, n\}$, it is assumed that the state is either increased by $\Delta x(t_k)$ ["success" — the positive increment $(\Delta x(t_k) > 0)$] or decreased by $\Delta x(t_k)$ ["failure" — the negative increment $(\Delta x(t_k) < 0)$]. We refer to $\Delta x(t_k)$ as a *microscopic/local* experimentally or knowledge-based observed increment of the state of the system per impact on the subinterval of length τ.

RWM4. It is assumed that $\Delta x(t_k)$ is constant for $k \in I(1, n)$ and is denoted by $\Delta x(t_k) \equiv Z_k = Z$ with $|Z_k| = \Delta x > 0$. Thus, for each $k \in I(1, n)$, there is a constant random increment Z of magnitude Δx to the state of the system per impact on the subinterval of length τ.

In short, from RWM1, RWM2, and RWM3, under n independent and identical random impacts, the initial state and n experimental, or knowledge-based observed random increments Z_k of constant magnitude Δx to the state,

$$x(t_0) = x(t),$$
$$x(t_1) - x(t_0) = Z_1,$$
$$x(t_2) - x(t_1) = Z_2,$$
$$\cdots\cdots\cdots$$
$$x(t_k) - x(t_{k-1}) = Z_k, \tag{2.1.1}$$
$$\cdots\cdots\cdots$$
$$x(t_n) - x(t_{n-1}) = Z_n,$$

are at $t_1, t_2, \ldots, t_k, \ldots, t_n$ over the given interval $[t, t + \Delta t]$ of length Δt. From RWM4, random variables in (2.1.1) are mutually independent. Moreover, Z_k is defined by

$$Z_k = \begin{cases} \Delta x, & \text{for positive increment (success)}, \\ -\Delta x, & \text{for negative increment (failure)}. \end{cases} \qquad (2.1.2)$$

From (2.1.1), the state $x(t_k)$ at the kth instance and the final state $x(t + \Delta t) = x(t_n)$ of the process are expressed by

$$x(t + k\tau) = x(t) + \sum_{i=1}^{k} Z_i, \quad x(t + \Delta t) = x(t) + \sum_{i=1}^{n} Z_i, \qquad (2.1.3)$$

where $\sum_{i=1}^{n} Z_i$ is referred to as an *aggregate increment* to the given state $x \equiv x(t)$ of the system at the given time t over the interval $[t, t + \Delta t]$ of length Δt. Hence, the *aggregate change* of the state of the system $x(t + \Delta t) - x(t)$ under n observations of the system over the given interval $[t, t + \Delta t]$ of length Δt is described by

$$x(t + \Delta t) - x(t) = n \frac{[\sum_{i=1}^{n} Z_i]}{n} = \frac{\Delta t}{\tau} S_n, \qquad (2.1.4)$$

where $S_n = \frac{1}{n} [\sum_{i=1}^{n} Z_i]$. S_n is the sample average of the state aggregate increment data.

For each random impact and any real number p satisfying $0 < p < 1$, it is assumed that

$$P(\{Z_k = \Delta x > 0\}) = p, \quad P(\{Z_k = -\Delta x < 0\}) = 1 - p = q. \qquad (2.1.5)$$

From (2.1.3), it is clear that $x(t_k) - x(t)$ is a discrete-time real-valued stochastic process which is the sum of k independent Bernoulli random variables Z_i $(Z_i = Z)$, $i = 1, 2, \ldots, k$ and $k = 1, 2, \ldots, n$. We note that for each k, $x(t_k) - x(t_0)$ is a binomial random variable with parameters (k, p). Moreover, the random variable $x(t_k) - x(t_0)$ takes values from the set $\{-k\Delta x, (2-k)\Delta x, \ldots, (2m-k)\Delta x, \ldots, k\Delta x\}$. From Example 1.1.16, its probability distribution is given by (1.1.23). The stochastic process $x(t_k) - x(t)$ is referred to as a *random walk process*. In particular, for $k = n$, let m be a number of positive increments Δx to the state of the system out of a total of n changes. $n - m$ is the number of negative increments $-\Delta x$ to the state of the system out of a total of n changes. Furthermore, $m \in I(0, n)$, and we note that

$$S_n = \frac{1}{n} \left[m \frac{[\sum_{i \in I_+(0,n)} Z_i]}{m} - (n - m) \frac{[\sum_{i \in I_-(0,n)} |Z_i|]}{(n - m)} \right]$$

$$= \frac{1}{n} \left[m \frac{m\Delta x}{m} - (n - m) \frac{(n - m)\Delta x}{(n - m)} \right] \quad \text{(from RWM4)}$$

$$= \frac{1}{n}[m\Delta x - (n-m)\Delta x] \qquad \text{(by algebraic simplification)}$$

$$= \frac{1}{n}\left[(2m-n)\frac{n\Delta x}{n}\right] \qquad \text{(by further algebraic simplification)}$$

$$= \frac{1}{n}\left[(2m-n)\frac{1}{n}\left[\sum_{i=1}^{n}|Z_i|\right]\right] \qquad \text{(by additional simplification)}$$

$$= \frac{1}{n}[(2m-n)S_n^+], \qquad (2.1.6)$$

where $I_+(0,n)$ and $I_-(0,n)$ are defined by $I_+(0,n) = \{i \in I(0,n) : |Z_i| = Z_i\}$ and $I_-(0,n) = \{i \in I(0,n) : |Z_i| = -Z_i\}$, respectively, and $S_n^+ = \frac{1}{n}[\sum_{i=1}^{n}|Z_i|]$.

From (2.1.4) and (2.2.6), the *aggregate change* of the state of the system $x(t + \Delta t) - x(t)$ over the time interval of length Δt under n identical random impacts on the system over the given interval $[t, t + \Delta t]$ of time is described by

$$x(t + \Delta t) - x(t) = \frac{1}{n}(2m-n)S_n^+ \frac{\Delta t}{\tau} \text{ (by algebraic manipulation)}$$

$$= \frac{1}{n}(2m-n)\frac{S_n^+}{\tau}\Delta t \text{ (by regrouping)}. \qquad (2.1.7)$$

Moreover, from (2.1.7) and (1.1.23), and by the application of Lemma 1.1.1, Problem 1.1.1 and Example 1.1.21, the mean of the aggregate change of the state of the system $x(t + \Delta t) - x(t)$ over the interval $[t, t + \Delta t]$ is given by

$$E[x(t + \Delta t) - x(t)] = \sum_{m=1}^{n}\frac{1}{n}(2m-n)\frac{n!}{m!(n-m)!}p^m(1-p)^{n-m}\frac{S_n^+}{\tau}$$

$$\text{(by Example 1.1.21)}$$

$$= \frac{1}{n}(2np-n)\frac{S_n^+}{\tau}\Delta t \quad \text{(by Problem 1.1.1)}$$

$$= (p-(1-p))\frac{S_n^+}{\tau}\Delta t \quad \text{(by simplification)}$$

$$= (p-q)\frac{S_n^+}{\tau}\Delta t \qquad \text{(by simplification)}. \qquad (2.1.8)$$

By Definition 1.1.19, its variance is

$$\mathrm{Var}(x(t + \Delta t) - x(t))$$

$$= E\left[\left[x(t + \Delta t) - x(t) - (p-q)\frac{S_n^+}{\tau}\Delta t\right]^2\right]$$

$$= \left[\sum_{m=1}^{n}\left(\frac{1}{n}(2m-n)\right)^2 p(m,n,p) - (p-q)^2\right]\left(\frac{S_n^+}{\tau}\Delta t\right)^2$$

$$= \left[\frac{1}{n^2} [4np[(n-1)p+1] - 4n^2p + n^2] - (p-q)^2 \right] \left(\frac{S_n^+}{\tau} \Delta t \right)^2$$

$$= \left[\frac{1}{n^2} [4n^2p^2 - 4n^2p + n^2 + 4np - 4np^2] - (p-q)^2 \right] \left(\frac{S_n^+}{\tau} \Delta t \right)^2$$

$$= \left[\frac{1}{n^2} n^2 (2p-1)^2 + \frac{1}{n^2} 4np(1-p) - (p-q)^2 \right] \left(\frac{S_n^+}{\tau} \Delta t \right)^2$$

$$= \left[(p-q)^2 + \frac{1}{n^2} 4np(1-p) - (p-q)^2 \right] \left(\frac{S_n^+}{\tau} \Delta t \right)^2$$

$$= \frac{4}{n} pq \left(\frac{S_n^+}{\tau} \Delta t \right)^2$$

$$= 4pq \frac{(S_n^+)^2}{\tau} \Delta t. \tag{2.1.9}$$

$\frac{S_n^+}{\tau} \left(\frac{\Delta x}{\tau} \right)$ and $\frac{(S_n^+)^2}{\tau} \left(\frac{(\Delta x)^2}{\tau} \right)$ are, respectively *microscopic or local sample average increments* and *microscopic or local sample average square increments* per unit time over the uniform length of sample subintervals $[t_{k-1}, t_k]$, $k = 1, 2, \ldots, n$, of interval $[t, t + \Delta t]$.

We note that the physical nature of the problem imposes certain restrictions on Δx and τ. Similarly, the parameter p cannot be taken arbitrary. Moreover, the state of the system cannot go to "infinity" on an interval whose length is small. Thus, in view of these microscopic or local sample average increments and the microscopic or local sample average square increments per unit time, we have $x(t + \Delta t) - x(t) = n\Delta x$, $\Delta t = n\tau$, $4pq = (p+q)^2 - (p-q)^2 = 1 - (p-q)^2$, and

$$\lim_{\tau \to 0} \left[\frac{(S_n^+)^2}{\tau} \right] = D \quad \lim_{\Delta x \to 0} \lim_{\tau \to 0} \left[(p-q) \frac{S_n^+}{\tau} \right] = C \text{ and } \lim_{\Delta x \to 0} \lim_{\tau \to 0} 4pq = 1,$$

$$\tag{2.1.10}$$

where C and D are certain constants; the former is called a *drift* and the latter, a *diffusion* coefficient. Moreover, C can be interpreted as the *average/expected/mean rate of change of the state* of the system per unit time, and D can be interpreted as the *mean square rate of change of the system* per unit time over an interval of length Δt. From (2.1.8)–(2.1.10), we obtain

$$\lim_{\Delta x \to 0} \lim_{\tau \to 0} E[x(t + \Delta t) - x(t)]$$

$$= \Delta t \lim_{\Delta x \to 0} \lim_{\tau \to 0} \left[(p-q) \frac{S_n^+}{\tau} \right] \text{ (by the properties of the limit)}$$

$$= C\Delta t, \tag{2.1.11}$$

$$\lim_{\Delta x \to 0} \lim_{\tau \to 0} \text{Var}(x(t + \Delta t) - x(t))$$

$$= \Delta t \lim_{\Delta x \to 0} \lim_{\tau \to 0} \left[4pq \frac{(S_n^+)^2}{\tau} \right] \quad \text{(by the properties of the limit)}$$

$$= D\Delta t. \tag{2.1.12}$$

Now, we define

$$y(t, n, \Delta t) = \frac{x(t + \Delta t) - x(t) - n(p - q)S_n^+}{\sqrt{4npq(S_n^+)^2}}. \tag{2.1.13}$$

By the application of Theorem 1.1.3 (DeMoivre–Laplace central limit theorem), we conclude that the process $y(t, n, \Delta t)$ is approximated by a standard normal random variable for each t (a zero mean and variance 1).

In fact, from (2.1.13), $n = \frac{\Delta t}{\tau}$, and (2.1.10), we have

$$y(t, n, \Delta t) = \frac{x(t + \Delta t) - x(t) - (p - q)\frac{S_n^+}{\tau}\Delta t}{\sqrt{4pq\frac{(S_n^+)^2}{\tau}\Delta t}},$$

and hence

$$\lim_{\Delta x \to 0} \lim_{\tau \to 0} y(t, n, \Delta t) = \lim_{\Delta x \to 0} \lim_{\tau \to 0} \left[\frac{x(t + \Delta t) - x(t) - (p - q)\frac{S_n^+}{\tau}\Delta t}{\sqrt{4pq\frac{(S_n^+)^2}{\tau}\Delta t}} \right]$$

$$= \frac{x(t + \Delta t) - x(t) - C\Delta t}{\sqrt{D\Delta t}}. \tag{2.1.14}$$

For fixed Δt, the random variable $\lim_{\Delta x \to 0} \lim_{\tau \to 0} y(t, n, \Delta t)$ has a standard normal distribution (a zero mean and variance 1). Now, by rearranging the expressions in (2.1.14), we get

$$x(t + \Delta t) - x(t) \approx C\Delta t + \sqrt{D}\sqrt{\Delta t} \left[\lim_{\Delta x \to 0} \lim_{\tau \to 0} y(t, n, \Delta t) \right], \tag{2.1.15}$$

and using $\sqrt{\Delta t} \left[\lim_{\Delta x \to 0} \lim_{\tau \to 0} y(t, n, \Delta t) \right] = \Delta w(t) = w(t + \Delta t) - w(t)$, (2.1.15) can be rewritten as

$$x(t + \Delta t) - x(t) \approx C\Delta t + \sqrt{D}\Delta w(t), \tag{2.1.16}$$

where $w(t)$ is a Wiener process. Thus, the aggregate change of the state of the system $x(t + \Delta t) - x(t)$ in (2.1.16) under independent and identical random impacts over the given interval $[t, t + \Delta t]$ is interpreted as the sum of the average/expected/mean change ($C\Delta t$) and the mean square change ($\sqrt{D}\Delta w(t)$) of the state of the system due to the random environmental perturbations.

If Δt is very small, then its differential is $dt = \Delta t$. Hence, from (2.1.16), the Itô–Doob differential dx is defined by

$$dx = C\,dt + \sqrt{D}\,dw(t), \qquad (2.1.17)$$

where C and D are the drift and the diffusion coefficients, respectively. Equation (2.1.17) is called the Itô–Doob-type stochastic differential equation.

Illustration 2.1.1 (economic dynamics). Random Walk Model 2.1.1 can be applied to formulate mathematical models in dynamic processes in social sciences, namely economics, management and information sciences. Following the discussion in Ref. 9, 33, 137, 37, 178 and 186, let $x(t)$ be either a rate of price/value of an asset/information/product/service per unit item/size per unit time (specific rate, i.e. the per capita growth/decay rate) or a price/value of an asset/information/product/service at a time $t \in J = [a, b]$, $a, b \in R$ and $J \subseteq R$. The specific rate of price/value (or the price/value) of the asset/information/product/service is observed over an interval of $[t, t + \Delta t]$, where Δt is a small increment in t. Without loss of generality, we assume that Δt is positive. The process is under the influence of exogenous or endogenous random perturbations of national/international/commerce/trade/monetary/social welfare policies. As a result of this, the $x(t)$ is affected by these random environmental perturbations. Following the development of Random Walk Model 2.1.1, its mathematical description is

$$x(t + \Delta t) - x(t) \approx C\Delta t + \sqrt{D}\Delta w(t),$$

which implies that

$$dx = C\,dt + \sigma\,dw(t).$$

We note that if $x(t)$ is the specific rate of the price/value at a time t, then C is called a measure of the *average specific rate* (per capita growth/decay rate) of the price/value of the asset/information/product/service at the time t, and $\sigma(\sigma^2 = D)$ is called the volatility, which measures the *spread (standard deviation) of the specific rate* (per capita growth/decay rate) of the price/value at a time t over an interval of small length $\Delta t = dt$; if $x(t)$ is the price/value at a time t, C is called a measure of the *index average rate* of growth/decay of the price/value of the asset/information/product/service at a time t; and σ is called the *spread of the average rate* of price/value, which measures the magnitude of the deviation of the rate of the price/value with its mean/expected value at the time t over an interval of small length $\Delta t = dt$.

Poisson Process Model 2.1.2. The modeling of a dynamic process by using the *random walk process* is good as long as the values of p or q are close to $\frac{1}{2}$. In fact, the Demoivre–Laplace theorem (Theorem 1.1.3) shows that the approximation of the binomial distribution by the means of a normal deteriorating due to the lack of consideration of the smaller value of either p or q.

However, the substantial range of phenomena necessitates finding just the smaller probability (either p or q). The smaller values of either p or q are provided by Simeon Poisson (Theorem 1.1.4). Of course, several natural phenomena in the chemical, biological, medical, physical, and social sciences follow the Poisson law, such as spontaneous disintegration of atoms of a radioactive substance, the immigration process in population dynamics, arrival of electrons at an anode in physics, and certain controlled monitory processes in economic systems. In this case, we pick $p_n = \frac{\lambda \Delta t}{n}$, where λ is a parameter in the Poisson process representing a rate of occurrence or the arrival rate per single event, and n is a natural number as defined in Random Walk Model 2.1.1. Of course, $q_n = 1 - \frac{\lambda \Delta t}{n} \approx 1$. In fact, the Poisson process model is a *"one-sided random walk model."* Following the argument used in Random Walk Model 2.1.1, we assume that the Poisson process model (in short, PPM) satisfies the following conditions:

PPM1. The conditions RWM1 and RWM2 of Random Walk Model 2.1.1 remain valid.

PPM2. For $\Delta x > 0$, it is assumed that the state is either increased only (decreased) by an increment Δx [the positive (negative) in the change of the state — "success") or there is no increment (no change in the state — "failure"). As before, we refer to Δx as a microscopic/local experimental increment or knowledge-based observed increment per single impact on the subinterval of length τ to the state of the system.

PPM3. Thus, depending on the nature of the system, there is a constant one-sided random change $\Delta x(t_k) = Z_k$ (either up or down only), with $Z_k = Z$ and the magnitude Δx, or no change in the state on each subinterval of length τ, i.e. Z takes values Δx (or $-\Delta x$) or 0.

PPM4. As stated before, the binomial distribution to the Poisson process Definition 1.2.14 with parameter λ (jump rate per unit time) is $p_n = \lambda \tau = \frac{\lambda \Delta t}{n}$. Hence, $p_n = \frac{\lambda \Delta t}{n}$ and $q_n = 1 - \frac{\lambda \Delta t}{n}$. Furthermore, in this case, the Z_k's in the relations (2.1.1) and (2.1.2) are modified to

$$Z_k = \begin{cases} \Delta x \text{ (or } -\Delta x), & \text{if there is a jump (success)}, \\ 0, & \text{if there is no jump (failure)}. \end{cases} \tag{2.1.18}$$

Now, following the reasoning used in Random Walk Model 2.1.1, we conclude that $x(t_k) - x(t)$ takes values from either the set $\{0, \Delta x, 2\Delta x, \ldots, m\Delta x, \ldots, k\Delta x\}$ or $\{0, -\Delta x, -2\Delta x, \ldots, -m\Delta x, \ldots, -k\Delta x\}$. This depends on the type of process, namely growth of decay. The discrete-time real-valued stochastic process $x(t_k)$ is referred to as a *one-sided random walk process* (right/left-hand side random walk process). Hereafter, without loss of generality and with m defined in (2.1.5), we consider a *right-hand side random walk process* (one-sided random walk process to the right), $\{0, \Delta x, 2\Delta x, \ldots, m\Delta x, \ldots, k\Delta x\}$. In this case, (2.1.4), (2.1.8), and (2.1.9) reduce to

$$x(t + \Delta t) - x(t) = m\Delta x, \tag{2.1.19}$$

$$E[x(t + \Delta t) - x(t)] = np_n\Delta x \quad \text{(Example 1.1.21)}$$

$$= n\frac{\lambda\Delta t}{n}\Delta x$$

$$= \lambda\Delta x\Delta t, \tag{2.1.20}$$

and

$$\mathrm{Var}(x(t + \Delta t) - x(t)) = \mathrm{Var}(x(t + \Delta t) - x(t))$$

$$= np_nq_n(\Delta x)^2 \quad \text{(by Problem 1.1.1)}$$

$$= n\frac{\lambda\Delta t}{n}\left(1 - \frac{\lambda\Delta t}{n}\right)(\pm\Delta x)^2$$

$$\text{(from the definitions of } p_n \text{ and } q_n\text{)}$$

$$= \lambda\Delta t\left(1 - \frac{\lambda\Delta t}{n}\right)(\Delta x)^2 \quad \text{(by simplification)}, \tag{2.1.21}$$

respectively.

We note that the physical nature of the problem imposes certain restrictions only on Δx and n. Hence, the parameter p_n can be arbitrarily small. Moreover, the state of the system cannot go to "infinity" on an interval whose length is small. In view of these considerations, the following conditions seem to be natural for a sufficiently large n: for $x(t + \Delta t) - x(t) = n\Delta x$, $\Delta t = n\tau$,

$$\lim_{\Delta x\to 0}[\lambda(\Delta x)^2] = D, \quad \lim_{\Delta x\to 0}[\lambda\Delta x] = C \quad \text{and} \quad \lim_{n\to\infty}q_n = 1, \tag{2.1.22}$$

where C and D are certain constants; C and D are as interpreted in (2.1.10).

From (2.1.20)–(2.1.22), we obtain

$$\lim_{n\to\infty}E[x(t + \Delta t) - x(t)] = \Delta t\lim_{n\to\infty}[\lambda\Delta x] \quad \text{(by the properties of limits)}$$

$$= C\Delta t, \tag{2.1.23}$$

and

$$\lim_{n\to\infty}\mathrm{Var}(x(t + \Delta t) - x(t)) = \Delta t\lim_{n\to\infty}\left[\lambda\left(1 - \frac{\lambda\Delta t}{n}\right)(\Delta x)^2\right]$$

$$\text{(by the properties of limits)}$$

$$= D\Delta t, \tag{2.1.24}$$

respectively.

Now, we define

$$y(t, n, \Delta t) = \frac{x(t + \Delta t) - x(t) - \lambda \Delta x \Delta t}{\sqrt{\lambda \Delta t \left(1 - \frac{\lambda \Delta t}{n}\right)(\Delta x)^2}}. \tag{2.1.25}$$

By repeating the rest of the argument used in Random Walk Model 2.1.1, from (2.1.22)–(2.1.25), we have

$$x(t + \Delta t) - x(t) \approx C\Delta t + \sqrt{D}\Delta w(t), \tag{2.1.26}$$

and the following Itô–Doob-type stochastic differential equation:

$$dx = Cdt + \sqrt{D}dw(t), \tag{2.1.27}$$

where $\sqrt{\Delta t}\left[\lim_{n \to \infty} y(t, n, \Delta t)\right] = \Delta w(t) = w(t + \Delta t) - w(t)$, with $w(t)$ being the Wiener process, and C and D are the drift and the diffusion coefficients, respectively.

Illustration 2.1.2 (decay/growth dynamic processes). Poisson Process Model 2.1.2 provides a mathematical model for several decay/growth dynamic processes in the biological, chemical, compartmental, pharmacological, physical, and social sciences. Here, a state of the system $x(t)$ stands for the specific rate of an amount/mass/size (or an amount/mass/size) of a substance/population/charge/ionizing beam of particles/etc. at a t in $J = [a, b]$, $a, b \in R$ and $J \subseteq R$. By imitating the argument used in the illustration in the context of Poisson Process Model 2.1.2, we arrive at

$$x(t + \Delta t) - x(t) \approx C\Delta t + \sqrt{D}\Delta w(t),$$

which implies that

$$dx = C\,dt + \sigma\,dw(t).$$

As before (Illustration 2.1.1), we note that if $x(t)$ is the specific rate of a state of the process, then C and σ are as defined in Illustration 2.1.1 and the Poisson Process Model 2.1.2. The details can be reconstructed by employing the previous argument.

Illustration 2.1.3 (birth and death processes). Poisson Process Model 2.1.2 can also be utilized to formulate a mathematical model for dynamic processes that are composed of both birth and death processes in the biological, chemical, compartmental, physical, social sciences. As before, $x(t)$ is described in Illustration 2.1.2. Due to the nature of random environmental perturbations, the increment to the specific rate $x(t)$ (or an amount) of a substance needs to be decomposed into two parts: (i) birth–growth and (ii) death–decay. The increments corresponding to these birth and death processes are modeled by the *one-sided random walk model* (Poisson Process Model 2.1.2) with Poisson parameters β and λ, respectively. In fact, by following the argument used in Poisson Process Model 2.1.2, increments due to the birth and death processes are described by

$$B(t + \Delta t) - B(t) \approx C_b \Delta t + \sqrt{D_b} \Delta w_1(t),$$

$$D(t + \Delta t) - D(t) \approx -C_d \Delta t + \sqrt{D_d} \Delta w_2(t),$$

over the interval $[t, t + \Delta t]$, respectively. Thus, the overall change in $x(t)$ over the interval of length Δt is

$$x(t + \Delta t) - x(t) \approx B(t + \Delta t) - B(t) + D(t + \Delta t) - D(t)$$

$$\approx (C_b - C_d)\Delta t + \sqrt{(D_b + D_b)}\Delta w(t).$$

Moreover,

$$dx = C\, dt + \sigma\, dw(t),$$

where $\Delta w(t) = \sqrt{D_b}\Delta w_1(t) + \sqrt{D_d}\Delta w_2(t)$, the Wiener processes w_1 and w_2 are mutually independent, $C = C_b - C_d$, and $\sigma^2 = D_b + D_b$.

Brownian Motion Model 2.1.3. Let us consider a physical application of the Brownian motion process. Objects are moving in a given medium. The motion is caused by the exceedingly frequent random environmental impacts on objects. The evolution of the objects is chaotic in character. Therefore, its description can only be described in the sense of probabilistic analysis. In physics, this phenomenon is known as the Brownian motion process (in short, BMM). The system is evolving under the following conditions:

BMM1. It is assumed that the environmental agent impact on the particle in the medium is random.

BMM2. The medium is homogeneous. We assume that random impacts of different agents in the environment on the particle are independent of each other.

Under these conditions, changes in the state of each individual object under the influence of random impacts of the environmental agents in the medium are independent of the motion of all other objects. The changes of the state of the same object from an arbitrary state in different time intervals are independent processes as long as the intervals are not too small. Of course, the distribution may depend on the environmental parameters of the medium.

Again, by following the reasoning in Random Walk Model 2.1.1 and the properties of stochastic processes with independent increments (Observation 1.2.10) for each $k = 1, 2, \ldots, n$, the random changes in the state $x(t_k) - x(t_{k-1}) = Z_k$ in (2.1.2) at $t_1, t_2, \ldots, t_k, \ldots, t_n$ are independent, and identically distributed Gaussian random variables with mean and variance

$$m(t_k - t_{k-1}) = m\frac{\Delta t}{n} \quad \text{and} \quad D(t_k, t_{k-1}) = 2D(t_k - t_{k-1}) = \sigma^2\frac{\Delta t}{n}, \quad (2.1.28)$$

respectively. In this case, (2.1.8) and (2.1.9) reduce to

$$E[x(t + \Delta t) - x(t)] = \sum_{i=1}^{n} m\frac{\Delta t}{n}$$

$$= m\Delta t \qquad \text{(by summation simplification)}, \qquad (2.1.29)$$

and

$$\text{Var}(x(t + \Delta t) - x(t)) = \sum_{i=1}^{n} \sigma^2\frac{\Delta t}{n} \quad \text{(by substitution)}$$

$$= \sigma^2\Delta t \qquad \text{(by summation simplification)}, \qquad (2.1.30)$$

respectively.

The changes of the state of the same object from an arbitrary state in different time intervals are independent processes as long as the intervals are not too small. We set

$$m = C \quad \text{and} \quad \sigma^2 = D. \qquad (2.1.31)$$

Here, as before, C and D are certain constants that are as described before. From (2.1.29)–(2.1.31), we have

$$x(t + \Delta t) - x(t) \approx C\Delta t + \sqrt{D}\Delta w(t), \qquad (2.1.32)$$

which reduces to the following Itô–Doob-type stochastic differential equation:

$$dx = C\,dt + \sqrt{D}\,dw(t), \qquad (2.1.33)$$

where C and D are the drift and the diffusion coefficients, respectively.

Illustration   2.1.4 (diffusion processes).   Brownian Motion Model 2.1.3 is frequently utilized to formulate a mathematical model for the dynamic processes in the sciences and engineering that operate under limited resources/capacity/environmental conditions/control mechanism. As before, a state of the system $x(t)$ stands for either a specific rate or an amount of a substance. The evolution of the system is chaotic in character. Therefore, its description can only be analyzed in the sense of probabilistic analysis, particularly by Brownian Motion Model 2.1.3.

Observation 2.1.1

(i) We recall that the experimentally or knowledge-based observed constant random variables $x(t_0) = x(t)$, $Z_1, Z_2, \ldots, Z_k, \ldots, Z_n$ in (2.1.1) or (2.1.18) or

(2.1.28) are mutually independent. Therefore,

$$E[x(t + \Delta t) - x(t)] \text{ and } E[(x(t + \Delta t) - x(t))^2] = \text{Var}(x(t + \Delta t) - x(t))$$

in (2.1.8) and (2.1.9), or (2.1.20) and (2.1.21), or (2.1.29) and (2.1.30) can be replaced by the conditional expectations:

$$E[x(t + \Delta t) - x(t)] = E[(x(t + \Delta t) - x(t)) \mid x(t) = x], \quad (2.1.34)$$

$$\text{Var}(x(t + \Delta t) - x(t)) = E[(x(t + \Delta t) - x(t))^2 \mid x(t) = x]. \quad (2.1.35)$$

(ii) We further note that based on experimental observations, information and the basic scientific laws/principles in the biological, chemical, engineering, medical, physical, and social sciences, that the magnitude of the microscopic or local increment depends on both the initial time t and the initial state $x(t) \equiv x$ of a system. As a result of this, in general, the drift (C) and the diffusion (D) coefficients defined in (2.1.10) (2.1.22), and (2.1.31) don't need not be absolute constants. They may depend on the initial time t and the initial state $x(t) \equiv x$ of the system, as long as their dependence on t and x is very smooth. The modified descriptions in (2.1.34) and (2.1.35) allow us to incorporate both time- and state-dependent random environmental perturbations. Because of this, (2.1.16), (2.1.26), and (2.1.32) can be reduced to

$$x(t + \Delta t) - x(t) \approx C(t, x)\Delta t + \sigma(t, x)\Delta w(t), \quad (2.1.36)$$

where $C(t, x)$ and $\sigma^2(t, x) = D(t, x)$ are also referred to as the average/expected/mean rate and the mean square rate of the state of the system on the interval of length Δt. Moreover, the Itô–Doob-type stochastic differential equations (2.1.17), (2.1.27), and (2.1.33) reduce to

$$dx = C(t, x)dt + \sigma(t, x)dw(t). \quad (2.1.37)$$

The following examples justify the scope and the significance of Observation 2.1.1(ii) in the modeling of dynamic processes in the biological, chemical, engineering, medical, physical, and social sciences.

Example 2.1.1 (economic dynamics). Let us consider Illustration 2.1.1. Utilizing all the notations, definitions, and conditions outlined in Random Walk Model 2.1.1, we modify by specifying the success probability p:

RWM5. We further assume that $p = \frac{1}{2} + (C/D)x(t)\Delta x$ and $\frac{1}{2} - (C/2)x(t)\Delta x = q$. By imitating the argument used in Random Walk Model 2.1.1, we arrive at

$$x(t + \Delta t) - x(t) \approx Cx(t)\Delta t + \sqrt{D}x(t)\Delta w(t),$$

which implies that

$$dx = Cx(t)dt + \sigma x(t)dw(t).$$

Here, we assume that C and $\sigma = \sqrt{D}$ are constants. Under the above conditions, $Cx(t)$ is called a measure of the *average/expected/mean rate of change in the price*. We refer to C as the *specific rate* (per capita growth/decay rate) of change of the price/value of the asset/information/product/service at the time t over an interval of length Δt. $\sigma x(t)$ ($\sigma^2 = D$) is called the *volatility*, which measures the *standard deviation of the rate of change of the price*, and we call σ the unpredictable/stochastic/nonanticipated specific rate of change generated by random perturbations (per capita growth/decay rate under random fluctuations) of the price/value of the asset/information/product/service at an initial time t over an interval of small length $\Delta t = dt$.

Example 2.1.2 (decay dynamic processes). Let us consider Illustration 2.1.2. Using the notations, definitions, and conditions outlined in Poisson Process Model 2.1.2, we modify the conditions PPM2 and PPM4 to incorporate the dependence of the initial state. For simplicity, we present the modified conditions of PPM2 and PPM4 for the decay processes. However, similar modified conditions for the growth processes can analogously be presented.

PPM2*. For $x(t)\Delta x > 0$, it is assumed that the amount/mass/size is either decreased by $x(t)\Delta x$ (the negative increment in the change of the amount/mass/size — "success") or unchanged (no change in the increment — "failure"). We refer to $x(t)\Delta x$ as a microscopic, local experimental or knowledge-based observed increment per impact on an interval of length τ on the amount/mass/size of the substance/population/charge/ionizing beam of particles/etc. at the initial time t.

PPM3*. Thus, there is a one-sided constant random change Z of magnitude $x(t)\Delta x$, or no change in the amount/mass/size of the substance/population/charge/ionizing beam of particles on each subinterval of length τ.

By imitating the argument used in *one-sided random walk model* (Poisson Process Model 2.1.2), we arrive at

$$x(t + \Delta t) - x(t) \approx -Cx(t)\Delta t + \sqrt{D}x(t)\Delta w(t),$$

which implies that

$$dx = -Cx(t)dt + \sigma x(t)dw(t).$$

Here, we assume that C and $\sigma = \sqrt{D}$ are constants. Under the above conditions, $Cx(t)$ is called a measure of the *average/expected/mean rate of decay* of the amount/mass/size of the substance/population/charge/ionizing beam of particles/etc. and we refer to C as the *average/expected/mean specific rate* (per capita decay rate) of the amount/mass/size of the substance/population/charge/

ionizing beam of particles/etc. at the time t over an interval of length Δt. $\sigma x(t)$ ($\sigma^2 = D$) is the measure of the *spread of the change of the decay rate*, and we call σ the measure of the random magnitude of fluctuations (unpredictable/stochastic/nonanticipated) in the specific rate of change (per capita decay rate under random fluctuations) of the amount/mass/size of the substance/population/charge/ionizing beam of particles/etc. at an initial time t over an interval of small length $\Delta t = dt$.

Example 2.1.3 (birth and death processes). Let us consider Illustration 2.1.3. By using the notations, definitions, and conditions outlined in Poisson Process Model 2.1.2, the modifications to incorporate the dependence of the initial state x with regard to both decay and growth processes are made in a similar way to the modifications for decay processes in Example 2.1.2. In fact, we consider:

BDM1. In general, under the environmental random perturbations, the increment to the amount/mass/size of a substance/population/charge/etc. needs to be decomposed into two processes, namely the birth–growth process (BGP) and the death–decay process (DDP).

BDM2. The birth and death processes are separately treated as the *one-sided random walk model* (Poisson Process Model 2.1.2) with Poisson parameters β and λ, respectively.

BDM3. The net change in population over an interval of length Δt is given by

$$x(t + \Delta t) - x(t) = B(t + \Delta t) - B(t) + D(t + \Delta t) - D(t),$$

where part of the amount/mass/size follows the right-hand side random walk process (birth/growth process: $B(t_k)$, $k = 1, 2, \ldots, n$), $B(t_n) = B(t + \Delta t) - B(t)$, and the remaining part of the amount/mass/size follows the left-hand side random walk process (decay process: $D(t_k)$, $k = 1, 2, \ldots, n$), $D(t_n) = D(t + \Delta t) - D(t)$.

By following the argument used in Poisson Process Model 2.1.2 and BDP, have the change in the amount/mass/size due to the birth/growth process is

$$B(t + \Delta t) - B(t) \approx C_b x(t)\Delta t + \sqrt{D_b}x(t)\Delta w_1(t),$$

and the death/decay process is

$$D(t + \Delta t) - D(t) \approx -C_d x(t)\Delta t + \sqrt{D_d}x(t)\Delta w_2(t)$$

over an interval, respectively, where

$$\beta(x(t)\Delta x)^2 = 2D_b x^2(t), \quad \beta x(t)\Delta x = C_b x(t),$$

and

$$\lambda(x(t)\Delta x)^2 = 2D_d x^2(t), \quad \lambda x(t)\Delta x = C_d x(t).$$

From BDM3, the net change the amount/mass/size over an interval of length Δt is given by

$$x(t + \Delta t) - x(t) = B(t + \Delta t) - B(t) + D(t + \Delta t) - D(t)$$

$$\approx C_b x(t) \Delta t + \sqrt{D_b} x(t) \Delta w_1(t)$$

$$- C_d x(t) \Delta t + \sqrt{D_d} x(t) \Delta w_2(t) \qquad \text{(by substitution)}$$

$$\approx (C_b - C_d) x(t) \Delta t + \sqrt{(D_b + D_b)} x(t) \Delta w(t) \qquad \text{(by regrouping)}.$$

Thus,

$$dx = Cx(t)dt + \sigma x(t)dw(t),$$

where $\Delta w(t) = \sqrt{D_b} x(t) \Delta w_1(t) + \sqrt{D_d} x(t) \Delta w_2(t)$, the Wiener processes and w_1 and w_2 are mutually independent, $C = (C_b - C_d)$, and $\sigma^2 = (D_b + D_b)$.

Here, we assume that C and $\sigma = \sqrt{D}$ are constants. Under the above conditions, $Cx(t)$ is called a measure of the *average/expected/mean rate net growth* of the amount/mass/size of the substance/population/charge/ionizing beam of particles/etc. C is called the *average/expected/mean specific rate* (per capita decay rate) of the amount/mass/size of the substance/population/charge/ionizing beam of particles/etc. at time t over an interval of length Δt. $\sigma x(t)$ ($\sigma^2 = 2D$) measures the spread of the *change of the net rate*. σ is the measure of the specific rate of change (per capita decay rate under random fluctuations) (or the amount/mass/size of the substance/population/charge/ionizing beam of particles/etc.) at an initial time t over an interval of small length $\Delta t = dt$.

Example 2.1.4 (diffusion processes). Let us consider Illustration 2.1.4. Employing all the notations, definitions, and conditions outlined in the Brownian Motion Model 2.1.3, we utilize all the notations and definitions outlined in the Brownian Motion Model 2.1.3 in our subsequent discussion. The mathematical model is based on the following assumption:

It is assumed that the mean and the variance of change of the amount/biomass/mass/size/etc. of the substance/population/charge/etc. over an interval $[t, t + \Delta t]$ are given by

$$E[x(t + \Delta t) - x(t)] = m\Delta t = \bar{r} x(t)(1 - bx(t))\Delta t,$$

and

$$\text{Var}(x(t + \Delta t) - x(t)) = \sigma^2 \Delta t = \sigma^2 x^2(t)(1 - bx(t))^2 \Delta t.$$

This assumption implies that the random environmental fluctuations are introduced into the growth rate r, where

$$E[r] = \bar{r} \quad \text{and} \quad \text{Var}(r) = \sigma^2.$$

Hence, by imitating the argument used in Brownian Motion Model 2.1.3, we arrive at

$$x(t + \Delta t) - x(t) = \bar{r}x(t)(1 - bx)(t))\Delta t + \sigma x(t)(1 - bx(t))\Delta w(t),$$

which implies that

$$dx = \bar{r}x(t)(1 - bx(t))dt + \sigma x(t)(1 - bx(t))dw(t),$$

where $C = \bar{r}x(t)(1 - bx(t))$ is called a Verhulst–Pearl logistic mean growth rate of the amount/mass/size of the substance/population/charge/ionizing/etc. Further details of this example are given in Section 3.1 of Chapter 3.

Observation 2.1.2. From (2.1.16), (2.1.26), (2.1.32) and (2.1.36), we have

$$\frac{x(t + \Delta t) - x(t)}{\Delta t} \approx C + \sqrt{D}\frac{\Delta w(t)}{\Delta t}, \tag{2.1.38}$$

and

$$\frac{x(t + \Delta t) - x(t)}{\Delta t} \approx C(t, x) + \sigma(t, x)\frac{\Delta w(t)}{\Delta t}. \tag{2.1.39}$$

In addition to the interpretations about $C/C(t, x)$ and $D/\sigma(t, x)$, we further shed light on their existence in the context of stochastic dynamic processes.

The left-hand side expression in (2.1.38) or (2.1.39) is the average (time average) rate of change of the state of the system under random environmental perturbations over a time interval of length Δt. The first term in the right-hand side expression in (2.1.38) or (2.1.39) is the expected/mean/average (predictable/ deterministic/anticipated) rate of change of the state of the system at a given time t and state x. The second term in the right-hand side expression is the random (unpredictable/stochastic/nonanticipated) rate of change of the state of the system at a given time t and state x due to random environmental perturbations. In short, the evolution of a dynamic system under random environmental perturbations is composed of two components: (a) predictable/deterministic and (b) unpredictable/stochastic. Moreover, they are characterized by $C/C(t, x)$ and $2D/\sigma(t, x)\frac{\Delta w(t)}{\Delta t}$, respectively.

Observation 2.1.3
  (i) From (2.1.11), (2.1.23), (2.1.29), (2.1.34), and (2.1.36), we conclude that

$$E[x(t + \Delta t) - x(t) \mid x(t) = x] = C\Delta t, \tag{2.1.40}$$

and

$$E[x(t + \Delta t) - x(t) \mid x(t) = x] = C(t, x)\Delta t. \tag{2.1.41}$$

The interpretation of (2.1.40)/(2.1.41) is as follows. The conditional expectation/average/mean of the change of the state of the system under random environmental perturbations over a time interval of length Δt for given "$x(t) = x$" is directly proportional to the length of time interval Δt. Furthermore, the constant of proportionality is $C/C(t, x)$.

(ii) From (2.1.40) and (2.1.41), we have

$$\frac{E[x(t + \Delta t) - x(t) \mid x(t) = x]}{\Delta t} = C, \qquad (2.1.42)$$

$$\frac{E[x(t + \Delta t) - x(t) \mid x(t) = x]}{\Delta t} = C(t, x). \qquad (2.1.43)$$

The interpretation of (2.1.42)/(2.1.43) is as follows. The average (time average) of change of the conditional expectation/average/mean of the change of the state of the system under random environmental perturbations over a time interval of length Δt for given "$x(t) = x$" is $C/C(t, x)$. We note that $C/C(t, x)$ is a stochastic process.

(iii) Moreover, from (2.1.42) and (2.1.43), we have

$$\frac{d}{dt} E[x(t) \mid x(t) = x] = \lim_{\Delta t \to 0} E\left[\frac{x(t + \Delta t) - x(t)}{\Delta t} \mid x(t) = x \right] = C,$$

$$\qquad (2.1.44)$$

$$\frac{d}{dt} E[x(t) \mid x(t) = x] = \lim_{\Delta t \to 0} E\left[\frac{x(t + \Delta t) - x(t)}{\Delta t} \mid x(t) = x \right] = C(t, x),$$

$$\qquad (2.1.45)$$

respectively.

The interpretation of (2.1.44)/(2.1.45) is as follows. The instantaneous rate of change of the conditional expectation/average/mean of the state of the system under random environmental perturbations for given "$x(t) = x$" is $C/C(t, x)$. We note that $C/C(t, x)$ is a stochastic process.

(iv) Similarly, from (2.1.12), (2.1.24), (2.1.30), (2.1.35), and (2.1.36), we have

$$E[(x(t + \Delta t) - x(t) - C\Delta t)^2 \mid x(t) = x] = D\Delta t, \qquad (2.1.46)$$

$$E[(x(t + \Delta t) - x(t) - C(t, x)\Delta t)^2 \mid x(t) = x] = \sigma^2(t, x)\Delta t. \qquad (2.1.47)$$

The interpretation of (2.1.46)/(2.1.47) is as follows. The conditional variance of the change of the state of the system under random environmental perturbations over a time interval of length Δt for given "$x(t) = x$" is directly proportional to the length of the time interval Δt. Furthermore, the constant of proportionality is $2D/\sigma^2(t, x)$.

(v) From (2.1.46) and (2.1.47), we have $C/C(t,x)$

$$\frac{E[(x(t+\Delta t)-x(t)-C\Delta t)^2 \mid x(t)=x]}{\Delta t} = D, \qquad (2.1.48)$$

and

$$\frac{E[(x(t+\Delta t)-x(t)-C(t,x)\Delta t)^2 \mid x(t)=x]}{\Delta t} = \sigma^2(t,x). \qquad (2.1.49)$$

The interpretation of (2.1.48)/(2.1.49) is as follows. The average (time average) of the conditional variance of the change of the state of the system under random environmental perturbations over a time interval of length Δt for given "$x(t)=x$" is $D/\sigma^2(t,x)$. We note that $D/\sigma^2(t,x)$ is a stochastic process.

(vi) Moreover, from (2.1.48) and (2.1.49), we have

$$\frac{d}{dt}E[x(t) \mid x(t)=x] = \lim_{\Delta t \to 0} E\left[\frac{(x(t+\Delta t)-x(t))^2}{\Delta t} \mid x(t)=x\right] = D,$$
$$(2.1.50)$$

and

$$\frac{d}{dt}E[x(t) \mid x(t)=x] = \lim_{\Delta t \to 0} E\left[\frac{(x(t+\Delta t)-x(t))^2}{\Delta t} \mid x(t)=x\right] = \sigma^2(t,x),$$
$$(2.1.51)$$

respectively.

The interpretation of (2.1.50)/(2.1.51) is as follows. The instantaneous rate of change of the conditional variance of the state of the system under random environmental perturbations for given "$x(t)=x$" is $D/\sigma^2(t,x)$. We note that $D/\sigma^2(t,x)$ is a stochastic process.

Example 2.1.5. The significance and the scope of Observation 2.1.3 is exhibited by applying the conclusions in the context of (2.1.41), (2.1.43), (2.1.45), (2.1.47), (2.1.49), and (2.1.51). This is due to the fact that (2.1.40), (2.1.42), (2.1.44), (2.1.46), (2.1.48), and (2.1.50) are special cases of (2.1.41), (2.1.43), (2.1.45), (2.1.47), (2.1.49), and (2.1.51), respectively. The example is given in the context of Examples 2.1.1–2.1.4. In order to minimize the repetition, we just present the results corresponding to (2.1.41), (2.1.43), and (2.1.45), with regard to:

(a) Examples 2.1.1 (economic dynamics) and 2.1.3 (birth and death processes):

$$E[x(t+\Delta t)-x(t) \mid x(t)=x] = Cx(t)\Delta t,$$
$$\frac{E[x(t+\Delta)-x(t) \mid x(t)=x]}{\Delta t} = Cx(t),$$

and

$$\frac{d}{dt}E[x(t) \mid x(t)=x] = Cx(t).$$

(b) Example 2.1.2 (decay processes):

$$E[x(t + \Delta t) - x(t) \mid x(t) = x] = -Cx(t)\Delta t,$$

$$\frac{E[x(t + \Delta) - x(t) \mid x(t) = x]}{\Delta t} = -Cx(t),$$

and

$$\frac{d}{dt}E[x(t) \mid x(t) = x] = -Cx(t).$$

(c) Example 2.1.4 (diffusion processes):

$$E[x(t + \Delta t) - x(t) \mid x(t) = x] = \bar{r}x(t)(1 - bx(t))\Delta t,$$

$$\frac{E[x(t + \Delta) - x(t) \mid x(t) = x]}{\Delta t} = \bar{r}x(t)(1 - bx(t)),$$

and

$$\frac{d}{dt}E[x(t) \mid x(t) = x] = \bar{r}x(t)(1 - bx(t)).$$

The other results are left as exercises.

We further observe that the expressions $Cx(t)$, $-Cx(t)$, $C(x(t)$ and $\bar{r}x(t)(1 - bx(t))$ are the average/expected rates that are defined in Examples 2.1.1–2.1.4, respectively. In addition,

$$\frac{\frac{d}{dt}E[x(t) \mid x(t) = x]}{x(t)} = C \quad \text{and} \quad \frac{\frac{d}{dt}E[x(t) \mid x(t) = x]}{x(t)} = \bar{r}(1 - bx(t))$$

are per capita growth rates. Moreover, we conclude that C, $-C$, C and $\bar{r}(1 - bx(t))$ are average/expected specific rates that are defined in Examples 2.1.1–2.1.4, respectively.

Observation 2.1.4. From Example 1.2.4 and Observation 1.2.11, in particular (ii)–(v), (2.1.38) and (2.1.39) reduce to

$$dx = [C + \sqrt{D}\xi(t)]dt, \tag{2.1.52}$$

and

$$dx = [C(t, x) + \sigma(t, x)\xi(t)]dt, \tag{2.1.53}$$

respectively; $\xi(t)$ is a white noise process. This type of stochastic differential equation is referred to as a *Langevin-type* stochastic differential equation.

Observation 2.1.5. Furthermore, in the absence of the random perturbations, this mathematical model of a dynamic process is described by a deterministic differential equation. Moreover, the differential equations corresponding to (2.1.52) and (2.1.52) reduce to

$$dx = C\, dt, \tag{2.1.54}$$

and

$$dx = C(t, x)dt, \tag{2.1.55}$$

respectively. By comparing (2.1.54) with (2.1.44), and (2.1.55) with (2.1.45), we conclude that the deterministic differential equations (2.1.54) and (2.1.55) can be considered to be the mean of the stochastic differential equations (2.1.33) and (2.1.37), respectively.

The presented probabilistic models provide a motivation to study the following type of first-order stochastic differential equations of the Itô–Doob type:

$$dx = f^d(t, x, w(t))dt + f^s(t, x, w(t))dw(t), \tag{2.1.56}$$

where dx is an Itô–Doob differential, f^d and f^s are defined and continuous on $J \times R \times R$ into R, and w is a normalized Wiener process. The differential equation (2.1.56) can be rewritten as

$$dx = \langle f^d(t, x, w(t)), f^s(t, x, w(t)) \rangle \cdot \langle dt, dw(t) \rangle, \tag{2.1.57}$$

where $J = [a, b]$ and "$\cdot$" stands for the dot product of vectors defined in an analytic geometry course. Note that the methods of finding solutions to this type of differential equations include the methods of finding solutions to the following types of deterministic and stochastic differential equations with stochastic process varying rate functions:

$$dx = f^d(t, x, w(t))dt, \tag{2.1.58}$$

$$dx = f^s(t, x, w(t))dw(t), \tag{2.1.59}$$

respectively, as special cases. This chapter deals with the methods of solving the first-order stochastic differential equations of Itô–Doob.

Exercises

1. *Compound interest problem*
   (a) Let $r(t)$ be a continuously adjustable annual interest rate function per dollar per year at a time $t \in J = [a, b]$ (in years).
   (b) The interest rate adjustment process is under the influence of random perturbations.
   (c) It is assumed that the interest rate adjustment process is described by a stochastic process with independent increments.
   (d) Let $x(t)$ be the number of dollars in a savings account at a time t (in years). Show that:
      (i) $\frac{d}{dt}E[r(t + \Delta t) - r(t) \mid r(t) = r] = \bar{r}$, for some real number $\bar{r}$;
      (ii) $\frac{d}{dt}E[(r(t + \Delta t) - r(t) - \bar{r}\Delta t)^2 \mid r(t) = r] = \sigma^2$, for some positive number σ;

(iii) $r'(t) = \bar{r}$;

(iv) $r'(t) = \bar{r} + \sigma\xi(t)$;

(v) $dr(t) = \bar{r}\,dt + \sigma\,dw(t)$;

(vi) $dx(t) = \bar{r}x\,dt$;

(vii) $dx(t) = \bar{r}x\,dt + \sigma x\,dw(t)$;

(viii) give your interpretations for $\bar{r}$, σ^2, $\xi(t)$, and $w(t)$.

2. **Cell growth problem**

   (a) Let $k(t)$ be a continuously adjustable absolute growth rate function per unit mass per unit time at a time $t \in J$.

   (b) The adjustable absolute growth rate process is under the influence of random environmental perturbations. These random perturbations can alter the rate of passage of the chemicals through the cell wall (for example, the influence of ionization on cell membrane) as well as the metabolic rate inside the cell.

   (c) It is assumed that the absolute growth rate adjustment process is described by a stochastic process with independent increments.

   (d) Let $x(t)$ be the amount of the mass at a time t (in an appropriate unit of time). Show that:

   (i) $\frac{d}{dt}E[k(t + \Delta t) - k(t) \mid k(t) = k] = \bar{k}$, for some real number $\bar{k}$;

   (ii) $\frac{d}{dt}E[(k(t + \Delta t) - k(t) - \bar{k}\Delta t)^2 \mid k(t) = k] = \sigma^2$, for some positive number σ;

   (iii) $k' = \bar{k}$;

   (iv) $k'(t) = \bar{k} + \sigma\xi(t)$;

   (v) $dk(t) = \bar{k}\,dt + \sigma\,dw(t)$;

   (vi) $dx(t) = \bar{k}x\,dt$;

   (vii) $dx(t) = \bar{k}x\,dt + \sigma x\,dw(t)$;

   (viii) give your interpretations for $\bar{k}$, σ^2, $\xi(t)$, and $w(t)$.

3. **Population growth problem** [80, 151]

   (a) Let $\beta(t)$ be a continuously adjustable birth rate function per individual per unit time at a time $t \in J$ (in years). $\beta(t)$ be a continuously adjustable specific birth rate.

   (b) The adjustable birth rate process is under the influence of random environmental perturbations. The randomness can be due the size of an organism/emigration/immigration/resources/etc.

   (c) It is natural to assume that the specific birth rate adjustment process is described by a Poisson process.

   (d) Let $x(t)$ be the size of the species at a time t (in an appropriate unit of time). Show that:

   (i) $\frac{d}{dt}E[\beta(t + \Delta t) - \beta(t) \mid \beta(t) = \beta] = \bar{\beta}$, for some real number $\bar{\beta}$;

   (ii) $\frac{d}{dt}E[(\beta(t + \Delta t) - \beta(t) - \bar{\beta}\Delta t)^2 \mid \beta(t) = \beta] = \sigma^2$, for some positive number σ;

   (iii) $\beta' = \bar{\beta}$;

    (iv) $\beta'(t) = \bar{\beta} + \sigma\xi(t)$;

    (v) $d\beta(t) = \bar{\beta}\,dt + \sigma\,dw(t)$;

    (vi) $dx(t) = \bar{\beta}x\,dt$;

    (vii) $dx(t) = \bar{\beta}x\,dt + \sigma x\,dw(t)$;

    (viii) give your interpretations for $\bar{\beta}$, σ^2, $\xi(t)$, and $w(t)$.

4. **Radiation problem** [1, 49, 80]. An ionizing beam of particles consists of protons, neutrons, deuterons, electrons, γ-ray quanta, etc. It is known that high polymers, namely protein/nucleic acids hit by an ionizing beam, may be irreversibly altered. In short, the polymers are damaged. Let N_0 be the number of undamaged molecules of a specific chemical compound. These molecules are in cells, and are assumed to be susceptible to radiation. Let n be the number of ionizing particles which cross the unit area of the target. This $n \in J$ is called a "dose of radiation."

    (a) Let $\lambda(n)$ be a continuously adjustable rate function of decay of undamaged molecules per undamaged molecule per unit dose at a dose level n (number of ionizing particles per unit area of the target). $\lambda(n)$ is considered as a continuously adjustable *specific decay rate* of the undamaged molecular population of a chemical compound.

    (b) The adjustable decay rate process is under the influence of random environmental perturbations. The randomness can be due to the influence of the ionization on a cell membrane or the effectiveness of the ionizing beam treatment.

    (c) It is natural to assume that the specific decay rate of an adjustment process is described by a stochastic process of independent increments.

    (d) Let $N(n)$ be the number of undamaged molecules after exposure to radiation at the level of dose n. Show that:

        (i) $\frac{d}{dt}E[\lambda(n + \Delta n) - \lambda(n) \mid \lambda(n) = \lambda] = \bar{\lambda}$, for some real number $\bar{\lambda}$;

        (ii) $\frac{d}{dt}E[(\lambda(n + \Delta n) - \lambda(t) - \bar{\lambda}\Delta n)^2 \mid \lambda(n) = \lambda] = \sigma^2$, for some positive number σ;

        (iii) $\lambda' = \bar{\lambda}$;

        (iv) $\lambda'(n) = \bar{\lambda} + \sigma\xi(n)$;

        (v) $d\lambda(n) = \bar{\lambda}\,dn + \sigma\,dw(n)$;

        (vi) $dN(n) = \bar{\lambda}N\,dn$;

        (vii) $dN(n) = \bar{\lambda}N\,dn + \sigma N\,dw(n)$;

        (viii) give your interpretations for $\bar{\lambda}$, σ^2, $\xi(n)$, and $w(n)$.

5. **Compartmental problem** [62, 80, 156, 157]. Let S be a solute. It is distributed in a fluid compartment with a fixed volume V_s. The solute S is removed (cleared) from the compartment by removing a constant fraction of the fluid in the compartment per unit time (volume V_r per unit time). But no additional amount of S is added to the fluid compartment. Under these stated conditions, the concentration of S in the fluid compartment will decrease. Other illustrations of this general situation are the "washout" of a solute from a well-mixed compartment

by a steady inflow of the pure solvent through the compartment, and the removal of the solute by a steady constant rate of turnover.

(a) Let $k(t)$ be a continuously adjustable clearance/volume distribution (turnover) rate function per unit amount of S per unit time at a time $t \in J$.

(b) The adjustable turnover rate process is under the influence of random environmental perturbations. These random perturbations can alter the turnover rate of the solute (for example, the influence of good stirring and good mixing with the temperature variation generates a random motion of thermal agitation).

(c) It is assumed that the turnover rate adjustment process is described by a stochastic process with independent increments.

(d) Let $Q(t)$ be the amount of the mass at the time t (in an appropriate unit of time). Show that:

   (i) $\frac{d}{dt}E[k(t+\Delta t) - k(t) \mid k(t) = k] = \bar{k}$, where $\bar{k} = \frac{V_r}{V_s}$;

   (ii) $\frac{d}{dt}E[k(t+\Delta t) - k(t) - \bar{k}\Delta t)^2 \mid k(t) = k] = \sigma^2$, for some positive number σ;

   (iii) $k' = \bar{k}$;

   (iv) $k'(t) = \bar{k} + \sigma\xi(t)$;

   (v) $dk(t) = \bar{k}\,dt + \sigma\,dw(t)$;

   (vi) $dQ(t) = -\bar{k}Q(t)dt$;

   (vii) $dQ(t) = -\bar{k}Q\,dt + \sigma Q(t)dw(t)$;

   (viii) give your interpretations for $\bar{k}$, σ^2, $\xi(t)$, and $w(t)$.

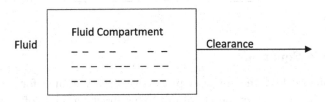

Fig. 2.1.1

6. **Electric circuit problems** [55, 80]. Let us consider a capacitor of fixed capacitance C (farads) carrying an initial charge Q_0 (coulombs). In an RC circuit the voltage difference ΔE [electromotive force/voltage E (volts)] across the capacitor causes the current I (amperes) to flow through a fixed resistance R (ohms). In this circuit, let $Q(t)$ and $\Delta E(t)$ be the charge on the capacitor and the voltage difference, respectively, at any time t (in seconds). It is known that $C\Delta E(t) = Q(t)$. By Ohm's law, we have $\Delta E(t) = I(t)R$. From the definition of current, $I(t) = \frac{d}{dt}Q(t)$. In the RC circuit, we have $I(t) = -\frac{d}{dt}Q(t)$.

(a) Let $k(t) = \frac{1}{CR(t)}$ be a continuously adjustable rate of electric charge (turnover rate) function per unit coulomb per unit time at a time $t \in J$.

(b) The adjustable turnover rate process is under the influence of random electrical fluctuations due to the presence of a register. This can alter the turnover rate of the electric current.

(c) It is observed that the turnover rate adjustment process is described by Poisson Process Model 2.1.2. Show that:

(i) a$\frac{d}{dt}E[k(t+\Delta t)-k(t)\mid k(t)=k]=\bar{k}$, where $\bar{k}=\frac{1}{CR}$;

(ii) $\frac{d}{dt}E[(k(t+\Delta t)-k(t)-\bar{k}\Delta t)^2\mid k(t)=k]=\sigma^2$, for some positive number σ;

(iii) $k'=\bar{k}$;

(iv) $k'(t)=\bar{k}+\sigma\xi(t)$;

(v) $dk(t)=\bar{k}dt+\sigma dw(t)$;

(vi) $dQ(t)=-\bar{k}Q(t)dt$;

(vii) $dQ(t)=-\bar{k}Q(t)dt+\sigma QdQ(t)dw(t)$;

(viii) give your interpretations for $\bar{k}$, σ^2, $\xi(t)$, and $w(t)$.

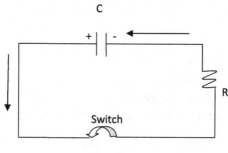

Fig. 2.1.2

2.2 Integrable Equations

In this section, we find the solution process of a very simple form of a class of first-order stochastic differential equations. The direct methods of integration will be used to solve this very simple class of first-order stochastic differential equations, which is called directly *integrable Itô–Doob-type stochastic differential equations*.

2.2.1 *General Problem*

Let us consider the following Itô–Doob type of first-order stochastic differential equation:

$$dx = f^d(t, w(t))dt + f^s(t, w(t))dw(t), \quad t \in J, \tag{2.2.1}$$

where $J = [a, b] \subseteq R$ for $a, b \in R$; f^d and f^s are continuous functions defined on $R \times R$ into R; and the rate functions f^d and f^s are independent of the state variable x (dependent variable) of a dynamic system.

In this subsection, our goal is to discuss a procedure for finding a general solution to (2.2.1). This provides a basis for solving real-world problems.

Definition 2.2.1. Let $J = [a, b]$ for $a, b \in R$, and hence $J \subseteq R$ be an interval. A solution process of a first-order stochastic differential equation of the Itô–Doob-type (2.2.1) is a random process/function x defined on J into R such that it satisfies (2.2.1) on J in the sense of Itô–Doob stochastic calculus. In short, the Itô–Doob-type stochastic differential $dx(t)$ of x is equal to the right-hand side expression in (2.2.1).

Example 2.2.1. Verify that $x(t) = tw + 5$ is a solution to the following differential equation: $dx = w(t)dt + t\,dw(t)$. Moreover, for any arbitrary constant c, show that $x(t) = tw + c$ is also a solution to this differential equation.

Solution procedure. We apply both the deterministic and the Itô–Doob stochastic differential calculus, addition, and product rules (Theorems 1.3.3 and 1.3.4) to $x(t) = tw + 5$, and we have $dx(t) = w(t)dt + t\,dw(t)$. This shows that the given process $x(t) = tw + 5$ satisfies the given differential equations. Furthermore, $x(t) = tw + c$ satisfies the same differential equation. This is due to the fact that the Itô–Doob stochastic differential of a constant is zero. Therefore, by Definition 2.2.1, the given processes are solutions to the given differential equation.

Example 2.2.2. Given that $x(t) = \sin tw^2(t) + c$, where c is an arbitrary constant, prove that $x(t)$ is a solution process of the following differential equation:

$$dx = (\cos tw^2(t) + \sin(t))dt + 2\sin tw(t)dw(t).$$

Solution Procedure. By repeating the argument used in Example 2.2.1, we obtain

$$dx(t) = d(\sin tw^2(t) + c) \qquad \text{(from the given expression)}$$

$$= d(\sin t)w^2(t) + \sin td(w^2(t)) + dc \qquad \text{(by the sum and product rules)}$$

$$= \cos tw^2(t)dt + 2\sin tw(t)dw(t)$$

$$+ \frac{1}{2}2\sin t(dw(t))^2 \qquad \text{(by computations)}$$

$$= \cos tw^2(t)dt + 2\sin tw(t)dw(t) + \frac{1}{2}2\sin t\,dt \quad [\text{by } (dw(t))^2 \simeq dt]$$

$$= (\cos tw^2(t) + \sin(t))dt + 2\sin tw(t)dw(t) \qquad \text{(by regrouping)}.$$

This proves that the given process satisfies the given differential equation.

2.2.2 *Procedure for Finding a General Solution*

The procedure for finding a general solution to the type of stochastic differential equation (2.2.1) is described below. First, we note that Eq. (2.2.1) can be rewritten as

$$dx = \langle f^d(t, w(t)), f^s(t, w(t)) \rangle \cdot \langle dt, dw(t) \rangle. \tag{2.2.2}$$

This representation suggests that the procedure basically depends on the methods of integration in both deterministic calculus [127] and stochastic calculus [8, 43, 98].

Step #1. We decompose (2.2.2) into its deterministic and stochastic parts.

$$dx = f^d(t, w(t))dt, \tag{2.2.3}$$

$$dx = f^s(t, w(t))dw(t), \tag{2.2.4}$$

respectively. The solutions x^d and x^s to (2.2.3) and (2.2.4), respectively, are simply determined by finding the antiderivatives of functions of f^d and f^s with respect to t and w, respectively. In fact, by integrating these functions with respect to t and w, respectively, we have

$$x^d(t) = c^d + \int f^d(t, w(t))dt, \tag{2.2.5}$$

$$x^s(t) = c^s + \int f^s(t, w(t))dw(t), \tag{2.2.6}$$

where c^d and c^s are arbitrary constants (c^d and c^s are either real numbers or real-valued random variables that are independent of the Wiener process w); the integral in (2.2.5) is the usual Cauchy–Riemann integrals [127] and the integral in (2.2.6) is an Itô–Doob-type stochastic integral. These integrals can be computed by employing the methods of deterministic and Itô–Doob stochastic integration (whenever possible).

Step #2. From the application of Theorem 1.3.3, a general solution to (2.2.1) is defined by

$$x(t) = x^d(t) + x^s(t)$$

$$= c + \int f^d(t, w(t))dt + \int f^s(t, w(t))dw(t), \tag{2.2.7}$$

where x^d and x^s are solutions to (2.2.3) and (2.2.4), and are determined in (2.2.5) and (2.2.6), respectively; $c = c^d + c^s$ is an arbitrary constant. The solution (2.2.7) is called a *general solution* to (2.2.1).

In the following, several examples are provided to illustrate the above-stated procedure.

Example 2.2.3. Consider the following differential equation: $dx = w\,dw(t)$. Find its general solution.

Solution procedure. Here, $f^d(t, w(t)) = 0$ and $f^s(t, w(t)) = w(t)$ are the coefficients of dt and $dw(t)$ in (2.2.1) (the presence of the effects of noise only). This is a purely ordinary stochastic differential equation of the Itô–Doob type. In this case, the decompositions (2.2.3) and (2.2.4) reduce to

$$dx^d = 0, \quad dt = 0, \quad dx^s = w(t)dw(t),$$

respectively. Their general solutions are described by general antiderivatives of $f^d(t, w(t)) = 0$ and $f^s(t, w(t)) = w(t)$:

$$x^d(t) = c^d, \quad x^s(t) = c^s + \frac{w^2(t)}{2} - \frac{t}{2},$$

respectively. We note that the antiderivative of $f^s(t, w(t)) = w(t)$ is computed by employing the method of substitution for the Itô–Doob-type stochastic integral described in Section 1.4 of Chapter 1. This completes Step #1. Using Step #2, we obtain

$$x(t) = x^d(t) + x^s(t) = c^d + c^s + \frac{w^2(t)}{2} - \frac{t}{2} = c + \frac{w^2(t)}{2} - \frac{t}{2}.$$

This is the general solution of Example 2.2.3. This completes the above-described procedure.

Example 2.2.4. Consider the following differential equation: $dx = t\,dt + tw\,dw(t)$. Find its general solution.

Solution procedure. Here, $f^d(t, w(t)) = t$ and $f^s(t, w(t)) = tw(t)$ are the coefficients of dt and $dw(t)$ in (2.2.1). This is an ordinary stochastic differential equation of the Itô–Doob-type. In this case, the decompositions (2.2.3) and (2.2.4) are

$$dx^d = t\,dt, \quad dx^s = tw(t)dw(t),$$

respectively. Their general solutions are described by general antiderivatives of $f^d(t, w(t))$ and $f^s t, w(t) = tw(t)$:

$$x^d(t) = c^d + \frac{1}{2}t^2, \quad x^s(t) = c^s + \frac{1}{2}tw^2(t) - \frac{1}{4}t^2 - \frac{1}{2}\int_a^t w^2(s)ds,$$

respectively. We note that the antiderivative of $f^s(t, w(t)) = tw(t)$ is computed by employing the method of integration by parts for the Itô–Doob-type stochastic integral described in Section 1.5 of Chapter 1. This completes Step #1. From

Step #2, we have

$$x(t) = x^d(t) + x^s(t)$$

$$= c^d + \frac{1}{2}t^2 + c^s + \frac{1}{2}tw^2(t) - \frac{1}{4}t^2 - \frac{1}{2}\int_a^t w^2(s)ds$$

$$= c + \frac{1}{4}t^2 + \frac{1}{2}tw^2(t) - \frac{1}{2}\int_a^t w^2(s)ds.$$

This is the general solution of Example 2.2.4. This completes the above-described procedure.

Example 2.2.5. Consider the following differential equation:

$$dx = f^d(t, w(t))dt + t^m w^n dw(t), \quad \text{for } n \neq -1,$$

where f^d is a continuous function defined in (2.2.1). Find a general solution to the given differential equation.

Solution procedure. Here, $f^d(t, w(t))$ and $f^s(t, w(t)) = t^m w^n(t)$ are the coefficients of dt and $dw(t)$ in (2.2.1). This is an ordinary stochastic differential equation of the Itô–Doob-type. The decompositions (2.2.3) and (2.2.4) are

$$dx^d = f^d(t, w(t))dt \quad \text{and} \quad dx^s = t^m w^n(t)dw(t),$$

respectively. We note that the sample paths of the Wiener process w are sample-continuous [8, 43]. Therefore, $f^d(t, w(t))$ is continuous because the composition of two continuous functions is continuous [127]. The function $f^d(t, w(t))$ is integrable with probability 1 in the sense of the Cauchy–Riemann integral [127]. Hence, the general solutions to the above-decomposed differential equations are described by the general antiderivatives of $f^d(t, w(t))$ and $f^s t$, $w(t) = t^m w^n(t)$:

$$x^d(t) = c^d + \int_a^t f^d(s, w(s))ds,$$

$$x^s(t) = c^s + \frac{1}{1+n}t^m w^{n+1}(t) - \frac{m}{n+1}\int_a^t s^{m-1}w^{n+1}(s)ds - \frac{n}{2}\int_a^t s^m w^{n-1}(s)ds,$$

respectively. We note that the antiderivative of $f^s(t, w(t)) = t^m w^n(t)$ is computed by employing the method of integration by parts for the Itô–Doob-type stochastic integral described in Section 1.5 of Chapter 1. This completes Step #1. Using Step #2, we have

$$x(t) = x^d(t) + x^s(t)$$

$$= c^d + \int_a^t f^d(s, w(s))ds$$

$$+ c^s + \frac{1}{1+n}t^m w^{n+1}(t) - \frac{m}{n+1}\int_a^t s^{m-1}w^{n+1}(s)ds$$

$$-\frac{n}{2}\int_a^t s^m w^{n-1}(s)ds$$

$$= c + \frac{1}{1+n}t^m w^{n+1}(t)$$

$$+ \int_a^t \left[f^d(s, w(s)) - \frac{m}{n+1}s^{m-1}w^{n+1}(s) - \frac{n}{2}s^m w^{n-1}(s) \right] ds.$$

This is the general solution of Example 2.2.5. This completes the procedure.

In the following subsection, we discuss a procedure for finding a particular solution to the Itô–Doob-type stochastic integrable differential equation. This idea leads to the formulation of an initial value problem in the study of differential equations and its applications. Moreover, illustrations are provided to exhibit the usefulness of the concept of solving real-world problems.

2.2.3 *Initial Value Problem*

Let us formulate an initial value problem for the Itô–Doob-type stochastic integrable differential equation (2.2.1). We consider

$$dx = f^d(t, w(t))dt + f^s(t, w(t))dw(t), \quad x(t_0) = x_0. \tag{2.2.8}$$

Here x, f^d, f^s, and $w(t)$ are defined in (2.2.1); t_0 is in J; x_0 is a real-valued random variable defined on a complete probability space $(\Omega, \mathfrak{F}, P)$. x_0 is independent of $w(t)$ for all t in J. (t_0, x_0) is called an *initial data* or *initial condition*. In short, x_0 is a random variable taking values in R. It is the value of the solution function at $t = t_0$ (the initial/given time). Moreover, it is an initial/given state of a dynamic process defined on the same probability space $(\Omega, \mathfrak{F}, P)$, and it is independent of $w(t)$ for all t in J. The problem of finding a solution to (2.2.8) is referred to as the *initial value problem* (IVP). Its solution is represented by $x(t) = x(t, t_0, x_0)$ for $t \geq t_0$, and $t, t_0 \in J$.

Definition 2.2.2. A *solution process of the IVP* (2.2.8) is a real-valued random process/function x defined on $[t_0, b] \subseteq J$ such that $x(t)$ and its Itô–Doob differential $dx(t)$ satisfy both (i) the scalar differential equation in (2.2.8) and (ii) a given initial condition (t_0, x_0). In short, (i) $x(t)$ is the solution to (2.2.1) (Definition 2.2.1) and (ii) $x(t_0) = x_0$.

Example 2.2.6. Verify that $x(t) = tw + c$ is a solution to the following differential equation: $dx = w(t)dt + t\,dw(t)$. Then determine a value of the constant c so that $x(t)$ is the solution to $dx = w(t)dt + t\,dw(t)$, $x(0) = 5$.

Solution procedure. By imitating the solution procedure described in Example 2.2.1, we conclude that $x(t) = tw + c$ is the general solutions to the given differential equation. We need to find a constant c so that $x(t) = tw + c$ is the solution process of the given IVP: $dx = w(t)dt + t\,dw(t)$, $x(0) = 5$. For this purpose, we substitute $t = 0$ in $x(t) = tw + c$ and obtain $x(0) = 0w + c$. We know by Definition 2.2.2 that $x(0) = 5$. Hence, $5 = 0w + c$. Now, we solve for c, and obtain $c = 5$. Finally, we substitute for $c = 5$ into the expression $x(t) = tw + c$, and we have $x(t) = tw + 5$. This is the desired solution, $x(t) = tw + 5$, to the given IVP.

Example 2.2.7. Determine the value of a constant c so that $x(t) = \sin tw^2(t) + c$ is the solution to the following IVP:

$$dx = (\cos tw^2(t) + \sin t)dt + 2\sin tw(t)dw(t), \quad x(\pi) = 0,$$

where c is an arbitrary constant.

Solution procedure. Imitating the solution procedure described in Example 2.2.2, we conclude that $x(t) = \sin tw^2(t) + c$ is a solution to

$$dx = (\cos tw^2(t) + \sin t)dt + 2\sin tw(t)dw(t).$$

We follow the procedure outlined in Example 2.2.6 to determine the value of the constant c. Again, for this purpose, we substitute $t = \pi$ into $x(t) = \sin tw^2(t) + c$, and we obtain $x(\pi) = \sin \pi w^2(\pi) + c$. We are given $x(\pi) = 0$. Therefore, we have $0 = \sin \pi w^2(\pi) + c = 0w^2(\pi) + c = c$. Hence, $c = 0$. This completes the goal of the problem.

2.2.4 *Procedure for Solving the IVP*

In the following, we apply the procedure for finding the solution process of the first-order scalar stochastic differential equation (2.2.1). Now, we briefly summarize the procedure for finding a solution to the IVP (2.2.8):

Step #1. We note that the general solution determined in (2.1.7) is represented by:

$$x(t) = c + \int_a^t f^d(s, w(s))ds + \int_a^t f^s(s, w(s))dw(s), \tag{2.2.9}$$

where $a, t \in J = [a, b] \subseteq R$. Here, c is an arbitrary constant. In view of the continuity of rate functions on their respective domains of definitions, the general solution in (2.2.9) of (2.2.1) is defined.

Step #2. In this step, we find an arbitrary constant c in (2.2.9). For this purpose, we utilize the given initial data (t_0, x_0), and solve the following scalar algebraic equation:

$$x(t_0) = c + \int_a^{t_0} f^d(s, w(s))ds + \int_a^{t_0} f^s(s, w(s))dw(s) = x_0, \tag{2.2.10}$$

for any given t_0 in J. This equation has a unique solution, and it is given by

$$c = x(t_0) - \int_a^{t_0} f^d(s, w(s))ds - \int_a^{t_0} f^s(s, w(s))dw(s)$$

$$= x_0 - \int_a^{t_0} f^d(s, w(s))ds - \int_a^{t_0} f^s(s, w(s))dw(s). \qquad (2.2.11)$$

Of course, c depends on f^d, f^s, w, x_0, t_0, and a.

Step #3. The solution to the IVP (2.2.8) is determined by substituting the expression of c in (2.2.11) into (2.2.9). Hence, we have

$$x(t) = c + \int_a^t f^d(s, w(s))ds + \int_a^t f^s(s, w(s))dw(s) \quad \text{[from (2.2.9)]}$$

$$= x_0 - \int_a^{t_0} f^d(s, w(s))ds - \int_a^{t_0} f^s(s, w(s))dw(s)$$

$$+ \int_a^t f^d(s, w(s))ds + \int_a^t f^s(s, w(s))dw(s) \quad \text{[by substitution of } c \text{ in (2.2.11)]}$$

$$= x_0 + \int_a^t f^d(s, w(s))ds - \int_a^{t_0} f^d(s, w(s))dw(s)$$

$$+ \int_a^t f^s(s, w(s))dw(s) - \int_a^{t_0} f^s(s, w(s))dw(s) \quad \text{(by regrouping like terms)}$$

$$= x_0 + \int_{t_0}^t f^d(s, w(s))ds + \int_{t_0}^t f^s(s, w(s))dw(s)$$

[by Lemma 1.3.1, properties (e) of the integral]. $\qquad (2.2.12)$

The solution process of the IVP (2.2.8) is denoted by $x(t) = x(t, t_0, x_0)$. This solution is referred to as a *particular solution* to (2.2.1).

Now, a couple of illustrations are presented to exhibit the usefulness of the concept of the IVP and its applications.

Illustration 2.2.1 (laminar blood flow in an artery [80, 135]). Let us consider a piece of an artery or vein as a cylindrical tube with constant radius R and length ℓ. It is assumed that the laminar blood flow is under the random fluctuations described by the Wiener process. In fact, the assumption characterizes the roughness profile of the inner surface of the artery. This is partially characterized by the state of the vessel which is undergoing an occlusion process. A velocity of the blood flow in the artery/vein is described by

$$dv = -2kr\,dr + \sigma k\,dw(r), \quad v(R) = 0, \quad v(0) = kR^2,$$

where r is the distance of any point of the blood from the axis; $k = P/4\eta\ell$; ℓ, R, and r are measured in centimeters; P stands for the pressure difference between the two ends of the tube dyne/cm^2 = cm^{-1}g s^{-2}) in the CGS system; η is the inner friction of the blood at the artery of the wall and is called the coefficient of viscosity of the blood measured in poise (cm^{-1}g s^{-2}). The speed is highest along the center axis of the tube. It is assumed that the velocity increases from zero at the wall toward the center (the maximum velocity).

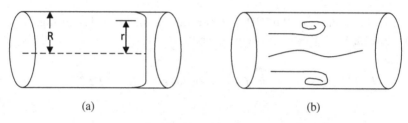

(a)   (b)

Fig. 2.2.1

Solution process. By using the Itô–Doob stochastic integral and the given initial and boundary conditions, we have

$$v(r) = v(0) - 2k \int_0^r s \, ds + \sigma k \int_0^r dw(s)$$

$$= v(0) - kr^2 + \sigma k w(r)$$

$$= kR^2 - kr^2 + \sigma k w(r),$$

and hence

$$v(r) = k(R^2 - r^2) + \sigma k w(r), \quad \text{for } 0 \leq r \leq R,$$

$$= \frac{P(R^2 - r^2) + \sigma P w(r)}{4\eta\ell}, \quad \text{for } 0 \leq r \leq R.$$

This describes the velocity profile of the blood flow through the cylinder-like artery/vein with constant radius R and length ℓ. This is the stochastic version of the Poiseuille laminar flow law. The mean of this velocity process is given by

$$E[v(r)] = k(R^2 - r^2) = \frac{P}{4\eta\ell}(R^2 - r^2), \quad \text{for } 0 \leq r \leq R.$$

The function $E[v(r)]$ was experimentally discovered (the law of laminar flow) by a French physiologist, and physician, Jean Louis Poiseuille. Moreover, the variance of

$v(r)$ is

$$\text{Var}(v(r)) = E[(v(r) - E[v(r)])^2] = E[(\sigma k w(r))^2] = (\sigma k)^2 r$$

$$= \left(\sigma \frac{P}{4\eta\ell}\right)^2 r, \quad \text{for } 0 \le r \le R.$$

Illustration 2.2.2 (vertical motion of particles [55, 80]). A person standing on a cliff throws/sprays a small dust particle upward. It is observed that after $2\,\text{s}$, the particle is at the maximum height in feet. After $5\,\text{s}$, it hits the ground. The motion of the particle is under the gravitational force influenced by random environmental perturbations described by the Wiener process. Find the velocity and the position of the particle at any time t, for the given initial velocity V_0 and position s_0 at the initial time t_0.

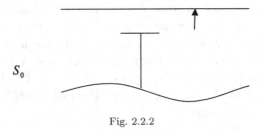

S_0

Fig. 2.2.2

Solution procedure. Let $s(t)$ be a position of the particle at a time t. The motion of the particle is acting under the gravitational force influenced by random environmental perturbations described by the Wiener process. Therefore, its motion is described by the Itô–Doob-type stochastic differential equation

$$dV(t) = -g\,dt + \sigma\,dw(t),$$

where $V(t) = \frac{d}{dt}s(t)$, $g \approx 32\,\text{ft/s}^2$ ($9.81\,\text{m/s}^2$). By using the Itô–Doob stochastic integral, we have

$$V(t) = \frac{d}{dt}s(t) = -gt + \sigma w(t) + c,$$

where c is a constant of integration. We utilize the initial data to determine c, and it is given by $V(t)_{|t=t_0} = V_0 = -gt_0 + \sigma w(t_0) + c$. Hence, $c = V_0 + gt_0 - \sigma w(t_0)$. The velocity $V(t)$ at any time t is given by

$$V(t) = \frac{d}{dt}s(t) = -gt + \sigma w(t) + V_0 + gt_0 - \sigma w(t_0)$$

$$= -g(t - t_0) + \sigma(w(t)) - w(t_0) + V_0.$$

The position of the particle at any time t is given by integrating the above expression over an interval $[t_0, t]$, and we obtain

$$\int_{t_0}^t V(u)du = \int_{t_0}^t \frac{d}{dt}s(u)du = \int_{t_0}^t [-g(u-t_0) + \sigma(w(u)) - w(t_0) + V_0]du$$

$$= s(u)\Big|_{t_0}^t = -\frac{1}{2}g(u-t_0)^2 + (V_0 u - \sigma w(t_0))u\Big|_{t_0}^t + \sigma\int_{t_0}^t w(u)du,$$

$$s(t) - s(t_0) = -\frac{1}{2}g(t-t_0)^2 + (V_0 - \sigma w(t_0))(t-t_0) + \sigma\int_{t_0}^t w(u)du.$$

The position of the particle at any time t is given by

$$s(t) = s_0 - \frac{1}{2}g(t-t_0)^2 + (V_0 - \sigma w(t_0))(t-t_0) + \sigma\int_{t_0}^t w(u)du.$$

The sample paths $w(t, \omega)$ of the Wiener process w are continuous $[t_0, t]$. Therefore, the Riemann–Cauchy integral $\sigma\int_{t_0}^t w(u)du$ of w on the right-hand side exists, and it is a stochastic process with mean zero and variance $\sigma^2\int_{t_0}^t E[(w(u))^2]du = \frac{1}{2}\sigma^2(t-t_0)^2$. Furthermore, the mean velocity and the mean position of the particle at any time t are $E[V(t)] = E[V_0] - g(t-t_0)$ and $E[s(t)] = E[s(t_0)] + E[V_0](t-t_0) - \frac{1}{2}g(t-t_0)^2$. The variance of $V(t)$ and of $s(t)$ are left as an exercise to the reader.

Illustration 2.2.3 (Weber–Fechner law [23, 80]). The Weber–Fechner law says that the increment to the reaction for equal increments in a stimulus decreases as the magnitude of the stimulus increases. It is a fact that the *afferent pathways* transmit the stimuli from the different sensory organs to the brain, and then, the *efferent pathways* transmit the stimuli from the brain to the different end organs. The intensity of the excitation/inhibition in a pathway depends on both the intensity of the stimulus and the number of excited/inhibited nerve fibers. The stimulus to the nerve cell generates an electric current causing the cations to move to the cathode, and the anions to the anode. Moreover, regardless of the background noise, the effects can be very significant. The fluctuations in stimulus intensity are characterized by the Wiener process. Let R be the number of nerve impulses emitted per second by sense organs such as the eyes and nose. Let S be the magnitude of the stimulus, such as the intensity of the light or the odor. Let ΔS be increment to the stimulus, and ΔR be increment to the response induced by the ΔS. From the Weber–Fechner law, we have the mathematical model

$$\Delta R = k\frac{\Delta S}{S} + \sigma\Delta w(S),$$

and hence

$$dR = k\frac{\Delta S}{S} + \sigma dw(S),$$

where k is some positive constant. This is the Itô–Doob-type stochastic integrable differential equation. Its solution is given by using one of the methods of integration. Thus, we have

$$R(S) = k \ln S + \sigma w(S) + c,$$

where c is the constant of integration. This expression is referred to as the Weber–Fechner law. It is the general solution to the derived differential equation. For the IVP, it is assumed that $R(S_0) = 0$. This means that S_0 is the response detection threshold. Thus, the response is described by

$$R(S) = k \ln S - k \ln S_0 + \sigma w(S) = k \ln \left(\frac{S}{S_0} \right) + \sigma w(S).$$

We note that the parameters k and S_0 depend on the type of stimulus and the individual. The curve of $R(S)$ describes the law proposed by Fechner.

Exercises

1. Using substitution, verify that each given random function/process is a solution to the given differential equation:

   (a) $x(t) = \cos 2t + w^3(t)$, $dx = (-\sin 2t + 3w(t))dt + 3w^2(t)dw(t)$;
   (b) $x(t) = t^4 + e^{w(t)}$, $dx = (4t^3 + \frac{1}{2}e^{w(t)})dt + e^{w(t)}dw(t)$.

   Find the general solution process of the following differential equations:

2. $dx = dw(t)$;
3. $dx = w^2(t)dw(t)$;
4. $dx = w^3(t)dw(t)$;
5. $dx = t^2w^2(t)dw(t)$;
6. $dx = t^3w^3(t)dw(t)$;
7. $dx = \frac{1}{1+t^2}dt + \cos w(t)dw(t)$;
8. $dx = \sin t\, dt + \cos t e^{w(t)}dw(t)$;
9. $dx = \ln(t+1)dt + (e^{2t}w^2(t))dw(t)$;
10. $dx = \sec^2 t\, dt + \sin w(t)dw(t)$;
11. $dx = \frac{2t}{1+t^2}dt + (t^3w^3(t) + t^2w^2(t))dw(t)$.
12. Show that the vertical distance traveled by a falling body under environmental fluctuations is given by

$$s(t) = \frac{1}{2}gt^2 + \sigma \int_0^t w(u)du,$$

   where t is time (measured in seconds), $w(t)$ is a Wiener process that characterizes random environmental perturbations, s is the vertical distance traveled by the body, and $g \approx 32\,\text{ft/s}^2$ ($9.81\,\text{m/s}^2$) at the surface of the Earth.

13. In reference to Poiseuille's law (Illustration 2.2.1), let us consider arterial blood with its concentration of O_2-bound hemoglobin. For human blood, its viscosity

is about less than that of venous blood, on the average $\eta = 0.027$ poise (Diem, 1962; p. 548). The blood is flowing through wide an arterial capillary of length $\ell = 2\,\text{cm}$ and $R = 8 \times 10^{-3}\,\text{cm}$. The pressure difference between the two ends of the vessel is $P = 4 \times 10^3\,\text{dyne/cm}^{-2}$ (3 mm mercury). Find: (a) the velocity process $v(r)$; (b) $E[v(r)]$, and (c) $\text{Var}(v(r))$.

14. In a metabolic experiment, it is observed that the rate of change of the speed of glucose metabolism is -0.03 per hour. Assuming that the metabolic activity is under random environmental perturbations described by the Brownian motion process, show that the dynamic of glucose metabolism is described by:

    (a) $dr(t) = -0.06dt + \sigma dw(t)$, $r(0) = 0$, where $r(t) = \frac{d}{dt}M(t)$, with $M(t)$ being the mass of glucose at time, t h; $\sigma > 0$; and $w(t)$ is the Wiener process.
    (b) Find: (i) $r(t)$ and (ii) $E[r(t)]$ and $\text{Var}(r(t))$ at $t = 1, 3$.
    (c) Moreover, as per hemoglobin A 1C test [11], the lower end of the reference interval is given by $M(0) = 4.5$; show that

$$M(t) = 4.5 - 0.03t^2 + \sigma \int_0^t w(s)ds.$$

    (d) Find: (i) $M(t)$ and (ii) $E[M(t)]$ and $\text{Var}(M(t))$ at $t = 13$.

15. In reference to Exercises 2.1, find continuously adjustable specific rates, the mean and variance:

    (a) (iii) and (v) with $r(0) = 0$ (compound interest problem);
    (b) (iii) and (v) with $k(0) = 0$ (cell growth problem);
    (c) (iii) and (v) with $\beta(0) = 0$ (population growth problem);
    (d) (iii) and (v) with $\lambda(0) = 0$ (radiation problem);
    (e) (iii) and (v) with $k(0) = 0$ (compartmental problem);
    (f) (iii) and (v) with $k(0) = 0$ (electric circuit problems).

16. Find the variance of velocity and the position of the particle in Illustration 2.2.2.

2.3 Linear Homogeneous Equations

In this section, we utilize the eigenvalue-type method to compute a solution to a linear first-order scalar homogeneous stochastic differential equation of the Itô–Doob-type. This idea is efficient and useful for solving purely deterministic and stochastic linear first-order scalar homogeneous differential equations. This method can be considered to be an alternative approach to the integrating factor approach to solving deterministic differential equations [80]. Several illustrations are given showing that first-order linear scalar homogeneous stochastic differential equations can describe the mathematical models of several dynamic processes in the biological, chemical, physical, and social sciences.

2.3.1 *General Problem*

Let us consider the following first-order linear scalar homogeneous stochastic differential equation of the Itô–Doob type:

$$dx = f(t)x\,dt + \sigma(t)x\,dw(t), \qquad (2.3.1)$$

where x is a state variable; f and σ are continuous functions defined on an interval $[a,b] = J \subseteq R$ for $a,b \in R$; and w is a scalar normalized Wiener process (Theorem 1.2.2).

In this subsection, we discuss a procedure for finding a general solution to (2.3.1). This provides a basis for analyzing, understanding, and solving various real-world problems in the biological, chemical, physical, and social sciences.

Definition 2.3.1. Let $J = [a,b]$ for $a,b \in R$, and hence $J \subseteq R$ is an interval. A solution process of the first-order linear scalar stochastic differential equation of the Itô–Doob-type (2.3.1) is a random process/function x defined on J into R such that it satisfies (2.3.1) on J in the sense of Itô–Doob stochastic calculus. In short, if we substitute a random function $x(t)$ and its Itô–Doob differential $dx(t)$ into (2.3.1), then the equation remains valid for all t in J.

Example 2.3.1. Show that $x(t) = 5e^{\sigma w(t)}$ is a solution to the differential equation $dx = \frac{1}{2}\sigma^2 x\,dt + \sigma x\,dw(t)$, where σ is any given real number. Moreover, for any arbitrary constant c, show that $x(t) = ce^{\sigma w(t)}$ is also a solution to the given differential equation.

Solution procedure. We apply both the deterministic and the Itô–Doob stochastic differential calculus (Theorems 1.3.1 and 1.3.4) to $x(t) = 5e^{\sigma w(t)}$, and we have

$$dx(t) = 5\sigma e^{\sigma w(t)}dw(t) + \frac{5}{2}\sigma^2 e^{\sigma w(t)}(dw(t))^2 \quad \text{(from Theorems 1.3.2 and 1.3.4)}$$

$$= \frac{5}{2}\sigma^2 e^{\sigma w(t)}dt + 5\sigma e^{\sigma w(t)}dw(t) \qquad \text{[by using } (dw(t))^2 \simeq dt]$$

$$= \frac{1}{2}\sigma^2 x(t)dt + \sigma x(t)dw(t) \qquad \text{(by substitution for } 5e^{\sigma w(t)}).$$

This shows that the given process $x(t) = 5e^{\sigma w(t)}$ satisfies the given differential equation. Furthermore, by imitating the above argument, we can conclude that $x(t) = ce^{\sigma w(t)}$ also satisfies the same differential equation. As noted before, we replace "5" by an arbitrary constant c. From Definition 2.3.1, we conclude that the given processes are solutions to the same given differential equation.

Example 2.3.2. Verify that $x(t) = \exp[\sin t + \sigma w(t)]$ is a solution to the following differential equation: $dx = (\cos t + \frac{1}{2}\sigma^2)x(t)dt + \sigma x(t)dw(t)$. Moreover, for any arbitrary constant c, show that $x(t) = c\exp[\sin t + \sigma w(t)]$ is also a solution to this differential equation.

Solution procedure. We apply the deterministic and the Itô–Doob stochastic differential calculus (Theorems 1.3.1 and 1.3.4) to $\exp[\cos t + \sigma w(t)]$, and obtain

$$dx = d(\exp[\sin t + \sigma w(t)]) \qquad \text{(by the given expression)}$$

$$= d(\exp[\sin t]\exp[\sigma w(t)]) \qquad \text{(by the law of exponents)}$$

$$= d(\exp[\sin t])\exp[\sigma w(t)]$$
$$\quad + \exp[\sin t]d(\exp[\sigma w(t)]) \qquad \text{(by Theorem 1.3.4)}$$

$$= \exp[\sin t]\exp[\sigma w(t)]d(\sin t)$$
$$\quad + \exp[\sin t]\exp[\sigma w(t)]d(\sigma w(t))$$
$$\quad + \frac{1}{2}\exp[\sin t]\exp[\sigma w(t)](\sigma dw(t))^2 \quad \text{(by the chain rule and Theorem 1.3.1)}$$

$$= \exp[\sin t]\exp[\sigma w(t)]\cos t\, dt$$
$$\quad + \exp[\sin t]\exp[\sigma w(t)]\sigma\, dw(t)$$
$$\quad + \frac{1}{2}\sigma^2\exp[\sin t]\exp[\sigma w(t)]dt \qquad [\text{by using } (\sigma dw(t))^2 \simeq \sigma^2 dt]$$

$$= \exp[\sin t + \sigma w(t)]\cos t\, dt$$
$$\quad + \exp[\sin t + \sigma w(t)]\sigma\, dw(t)$$
$$\quad + \frac{1}{2}\sigma^2\exp[\sin t + \sigma w(t)]dt \qquad \text{(by the law of exponents)}$$

$$= \exp[\sin t + \sigma w(t)]\left(\cos t + \frac{1}{2}\sigma^2\right)dt$$
$$\quad + \exp[\sin t + \sigma w(t)]\sigma dw(t) \qquad \text{(by regrouping like terms)}$$

$$= \left(\cos t + \frac{1}{2}\sigma^2\right)x(t)dt + \sigma x(t)dw(t) \quad \text{(by substitution).}$$

Furthermore, by imitating the above argument, we can conclude that $x(t) = c\exp[\sin t + \sigma w(t)]$ also satisfies the same differential equation. As noted before, we replace "1" by an arbitrary constant c. From Definition 2.3.1, we conclude that the given processes are solutions to the given differential equation.

2.3.2 *Procedure for Finding a General Solution*

We introduce an eigenvalue–eigenvector-like approach to the Itô–Doob differential operator. This approach is a natural extension of the approach used for solving the deterministic differential equations of this type [80]. In fact, the problem of finding a solution to a linear homogeneous scalar Itô–Doob-type stochastic differential equation reduces to a problem of its reduction and then solving a corresponding linear system of algebraic equations.

Step #1 (decomposition process). This step is exactly similar to Step #1 described in Section 2.2.2. For the sake of completeness, we repeat it. We decompose (2.3.1) into its deterministic and Itô–Doob-type stochastic parts,

$$dx = f(t)x\,dt, \tag{2.3.2}$$

and

$$dx = \sigma(t)x\,dw(t), \tag{2.3.3}$$

respectively. Our goal of finding the solution to (2.3.1) is decomposed into the following subgoals:

(a) To find the solutions x^d and x^s to (2.3.2) and (2.3.3), respectively.
(b) To create a candidate

$$x(t) \equiv x(t, w(t)) = x^d(t)x^s(t, w(t)), \quad \text{for } t \in J, \tag{2.3.4}$$

for a solution to (2.3.1);
(c) To test the correctness of the candidacy of the solution in (2.3.4).

In the following, a procedure to fulfill the stated subgoals in (a), (b) and (c) are outlined.

Step #2 (reduction process of a system of linear algebraic equations). Let us seek solutions to (2.3.2) and (2.3.3) of the forms

$$x^d(t) = \exp\left[\int_a^t \lambda^d(s)ds\right]c^d, \quad \text{for } t \in J, \tag{2.3.5}$$

$$x^s(t) \equiv x^s(t, w(t)) = \exp\left[\int_a^t \lambda_1^s(s)ds + \int_a^t \lambda_2^s(s)dw(s)\right]c^s, \quad \text{for } t \in J, \tag{2.3.6}$$

respectively, where c^d and c^s are arbitrary unknown constants that are either real numbers or real-valued random variables that are independent of the Wiener process w; and λ^d, λ_1^s, and λ_2^s are unknown continuous functions defined on J into R.

Case 1. We note that if either $c^d = 0$ or $c^s = 0$ (for any λ^d, λ_1^s, and λ_2^s), then either $x^d(t)$ in (2.3.5) or $x^s(t)$ in (2.3.6) is a trivial solution process (zero solution, i.e. zero random process/function on J) of either (2.3.2) or (2.3.3). In this case, the candidate for a solution defined in (2.3.4) is always the trivial solution to (2.3.1).

Case 2. Therefore, our major goal is to seek a nontrivial solution process (nonzero solutions) of (2.3.1). For this purpose, we need to find unknown functions λ^d, λ_1^s, and λ_2^s and unknown nonzero constant numbers c^d and c^s in (2.3.5) and (2.3.6). This is achieved in the following.

Applying the deterministic calculus [127] to (2.3.5) and imitating the procedure described in [80], the problem of finding the solution $x^d(t)$ to (2.3.2) is summarized as follows:

$$dx^d(t) = \lambda^d(t) \exp\left[\int_a^t \lambda^d(s)ds\right] c^d dt, \qquad (2.3.7)$$

$$\exp\left[\int_a^t \lambda^d(s)ds\right] \lambda^d(t)c^d dt = f(t) \exp\left[\int_a^t \lambda^d(s)ds\right] c^d dt, \qquad (2.3.8)$$

$$\exp\left[\int_a^t \lambda^d(s)ds\right] \neq 0, \quad c^d \neq 0, \qquad (2.3.9)$$

implies that

$$\lambda^d(t) = f(t) \qquad (2.3.10)$$

Hence

$$x^d(t) = \exp\left[\int_a^t f(s)ds\right] c^d. \qquad (2.3.11)$$

This completes the determination of the solution process of (2.3.2).

To find a solution to (2.3.3), we imitate the procedure 2.4.2 [80] in the context of the Itô–Doob-type stochastic calculus, particularly the Itô–Doob differential formula (Theorems 1.3.1 and 1.3.4), and we have

$$dx^s(t) = d \exp\left[\int_a^t \lambda_1^s(s)ds + \int_a^t \lambda_2^s(s)dw(s)\right] c^s \qquad \text{[from (2.3.6)]}$$

$$= d\left(\exp\left[\int_a^t \lambda_1^s(s)ds\right] \exp\left[\int_a^t \lambda_2^s(s)dw(s)\right] c^s\right) \qquad \text{(by the law of exponents)}$$

$$= d \exp\left[\int_a^t \lambda_1^s(s)ds\right] \exp\left[\int_a^t \lambda_2^s(s)dw(s)\right] c^s$$

$$+ \exp\left[\int_a^t \lambda_1^s(s)ds\right] d \exp\left[\int_a^t \lambda_2^s(s)dw(s)\right] c^s \qquad \text{(by Theorem 1.3.4)}$$

$$= \lambda_1^s(t) \exp\left[\int_a^t \lambda_1^s(s)ds\right] \exp\left[\int_a^t \lambda_2^s(s)dw(s)\right] c^s dt$$

$$+ \lambda_2^s(t) \exp\left[\int_a^t \lambda_1^s(s)ds\right] \exp\left[\int_a^t \lambda_2^s(s)dw(s)\right] c^s dw(t)$$

$$+ \frac{1}{2}\lambda_2^{s^2}(t) \exp\left[\int_a^t \lambda_1^s(s)ds\right] \exp\left[\int_a^t \lambda_2^s(s)dw(s)\right] c^s dt$$

$$\text{(by Theorems 1.3.1 and 1.3.4)}$$

$$= \left(\lambda_1^s(t) + \frac{1}{2} \lambda_2^{s^2}(t) \right) \exp \left[\int_a^t \lambda_1^s(s) ds \right]$$

$$\times \exp \left[\int_a^t \lambda_2^s(s) dw(s) \right] c^s dt$$

$$+ \lambda_2^s(t) \exp \left[\int_a^t \lambda_1^s(s) ds \right] \exp \left[\int_a^t \lambda_2^s(s) dw(s) \right] c^s dw(t) \qquad \text{(by regrouping)}$$

$$= \left(\lambda_1^s(t) + \frac{1}{2} \lambda_2^{s^2}(t) \right)$$

$$\times \exp \left[\int_a^t \lambda_1^s(s) ds + \int_a^t \lambda_2^s(s) dw(s) \right] c^s dt$$

$$+ \lambda_2^s(t) \exp \left[\int_a^t \lambda_1^s(s) ds + \int_a^t \lambda_2^s(s) dw(s) \right] c^s dw(t)$$

$$\text{(by the law of exponents)}$$

$$= \exp \left[\int_a^t \lambda_1^s(s) ds + \int_a^t \lambda_2^s(s) dw(s) \right]$$

$$\times c^s \left(\lambda_1^s(t) + \frac{1}{2} \lambda_2^{s^2}(t) \right) dt$$

$$+ \exp \left[\int_a^t \lambda_1^s(s) ds + \int_a^t \lambda_2^s(s) dw(s) \right] c^s \lambda_2^s(t) dw(t)$$

$$\text{(by the commutative law).}$$

$$\text{(2.3.12)}$$

From Definition 2.3.1 and the assumption that $x^s(t)$ is a solution to (2.3.3), we conclude that $x^s(t)$ in (2.3.6) and $dx^s(t)$ in (2.3.12) satisfy the stochastic differential equation of the Itô–Doob-type (2.3.3). Thus, we have

$$dx^s = \exp \left[\int_a^t \lambda_1^s(s) ds + \int_a^t \lambda_2^s(s) dw(s) \right] \left(\lambda_1^s(t) + \frac{1}{2} \lambda_2^{s^2}(t) \right) c^s dt$$

$$+ \exp \left[\int_a^t \lambda_1^s(s) ds + \int_a^t \lambda_2^s(s) dw(s) \right] \lambda_2^s(t) c^s dw(t)$$

$$= \sigma(t) \exp \left[\int_a^t \lambda_1^s(s) ds + \int_a^t \lambda_2^s(s) dw(s) \right] c^s dw(t), \qquad \text{(2.3.13)}$$

for any nonzero constant c^s number. Because of the fact that λ_1^s and λ_2^s are arbitrary and unknown functions, by comparing the coefficients of dt and $dw(t)$ on both sides of the equation in (2.3.13) (the method of undetermined coefficients) and $dt \neq dw(t)$ $[\Delta t \neq \Delta w(t) = w(t + \Delta t) - w(t)]$, we obtain

$$\exp \left[\int_a^t \lambda_1^s(s) ds + \int_a^t \lambda_2^s(s) dw(s) \right] \left(\lambda_1^s(t) + \frac{1}{2} \lambda_2^{s^2}(t) \right) = 0, \qquad \text{(2.3.14)}$$

$$\exp\left[\int_a^t \lambda_1^s(s)ds + \int_a^t \lambda_2^s(s)dw(s)\right]\lambda_2^s(t)$$

$$= \sigma(t)\exp\left[\int_a^t \lambda_1^s(s)ds + \int_a^t \lambda_2^s(s)dw(s)\right]. \qquad (2.3.15)$$

For any integrable functions λ_1^s and λ_2^s, we note that

$$\exp\left[\int_a^t \lambda_1^s(s)ds + \int_a^t \lambda_2^s(s)dw(s)\right] \neq 0 \quad \text{on } R.$$

Therefore, from (2.3.14) and (2.3.15), we have

$$\lambda_1^s(t) + \frac{1}{2}\lambda_2^{s^2}(t) = 0, \qquad (2.3.16)$$

and

$$\lambda_2^s(t) = \sigma(t). \qquad (2.3.17)$$

This is the desired system of linear algebraic equations.

Step #3 (method of solving a system of algebraic equations). The goal of finding the solution x^s to (2.3.3) reduces to a problem of solving the algebraic systems (2.3.16) and (2.3.17). The algebraic equation (2.3.17) is completely decoupled. Therefore, the solution $\lambda_2^s(t) = \sigma(t)$ is uniquely determined. By the backward substitution of $\lambda_2^s(t) = \sigma(t)$ into (2.3.16), we solve for $\lambda_1^s(t)$. Thus, $\lambda_1^s(t) = -\frac{1}{2}\lambda_2^{s^2}(t) = -\frac{1}{2}\sigma^2(t)$. In short, the algebraic systems (2.3.16) and (2.3.17) have unique solutions, namely

$$\lambda_2^s(t) = \sigma(t), \quad \lambda_1^s(t) = -\frac{1}{2}\sigma^2(t), \text{ for } t \in J. \qquad (2.3.18)$$

Step #4 (candidate for the solution process). From Step #2 and Step #3, and recalling the fact that the arbitrary constants c^d and c^s are nonzero, the solutions x^d and x^s to (2.3.2) and (2.3.3) are determined. In fact, from (2.3.5), (2.3.6), (2.3.11), and (2.3.18), they are expressed as

$$x^d(t) = \exp\left[\int_a^t f(s)ds\right]c^d, \qquad (2.3.19)$$

and

$$x^s(t) = \exp\left[-\frac{1}{2}\int_a^t \sigma^2(s)ds + \int_a^t \sigma(s)dw(s)\right]c^s, \qquad (2.3.20)$$

for any given arbitrary constants c^d and c^s. This fulfills the stated subgoal (a) in Step #1. From (2.3.19) and (2.3.20), the subgoal (b) immediately follows. In fact, as per (2.3.4), a candidate for the nontrivial solution to (2.3.1) is

$$x(t) = x^d(t)x^s(t) \qquad \text{[from (2.3.4)]}$$

$$= \exp\left[\int_a^t f(s)ds\right]c^d$$

$$\times \exp\left[-\frac{1}{2}\int_a^t \sigma^2(s)ds + \int_a^t \sigma(s)dw(s)\right]c^s \qquad \text{[from (2.3.19) and (2.3.20)]}$$

$$= \exp\left[\int_a^t f(s)ds\right]$$

$$\times \exp\left[-\frac{1}{2}\int_a^t \sigma^2(s)ds + \int_a^t \sigma(s)dw(s)\right]c^d c^s \qquad \text{(by commutativity)}$$

$$= \exp\left[\int_a^t f(s)ds\right]$$

$$\times \exp\left[-\frac{1}{2}\int_a^t \sigma^2(s)ds + \int_a^t \sigma(s)dw(s)\right]c \qquad \text{(by notation)}$$

$$= \exp\left[\int_a^t \left[f(s) - \frac{1}{2}\sigma^2(s)\right]ds + \int_a^t \sigma(s)dw(s)\right]c \quad \text{(by the law of exponents)},$$

$$(2.3.21)$$

where $c = c^d c^s \neq 0$. We note that the last step in (2.3.21) is due to the law of exponents, the commutative properties of multiplication of real numbers/real-valued functions, and the properties of integration.

Step #5 (validity of the solution process). We need to show [the validity of subgoal (c)] that $x(t)$ defined in (2.3.21) is a nontrivial solution to (2.3.1). First, we note that $x(t)$ is a well-defined random process/function defined on J as long as f and σ are defined and continuous or piecewise continuous on J. It is nontrivial solution because $\exp\left[\int_a^t \left[f(s) - \frac{1}{2}\sigma^2(s)\right]ds + \int_a^t \sigma(s)dw(s)\right] \neq 0$ as well as $c \neq 0$. The only thing that remains to be proven is to show that $x(t)$ defined in (2.3.21) and its $dx(t)$ satisfy (2.3.1). For this purpose, we need to find the Itô–Doob differential of $x(t)$ in (2.3.21). By applying Theorem 1.3.4 and the first fundamental theorem of integral calculus [127] to $x(t)$ in (2.3.21), and imitating the derivation procedure of (2.3.12), we conclude that

$$dx(t, w(t)) = d\exp\left[\int_a^t \left[f(s) - \frac{1}{2}\sigma^2(s)\right]ds + \int_a^t \sigma(s)dw(s)\right]c \quad \text{[from (2.3.21)]}$$

$$= \exp\left[\int_a^t \left[f(s) - \frac{1}{2}\sigma^2(s)\right]ds + \int_a^t \sigma(s)dw(s)\right]$$

$$\times \left[\left(f(t) - \frac{1}{2}\sigma^2(t)\right)dt + \sigma(t)dw(t)\right]c$$

$$+ \exp\left[\int_a^t \left[f(s) - \frac{1}{2}\sigma^2(s)\right]ds + \int_a^t \sigma(s)dw(s)\right]\frac{1}{2}\sigma^2(t)dt\,c \quad \text{(by Theorem 1.3.4)}$$

$$= f(t)\exp\left[\int_a^t \left[f(s) - \frac{1}{2}\sigma^2(s)\right]ds + \int_a^t \sigma(s)dw(s)\right]c\,dt$$

$$+ \sigma(t)\exp\left[\int_a^t \left[f(s) - \frac{1}{2}\sigma^2(s)\right]ds + \int_a^t \sigma(s)dw(s)\right]c\,dw(t) \quad \text{(by simplifying)}$$

$$= f(t)x(t, w(t))dt + \sigma(t)x(t, w(t))dw(t) \qquad \text{[from (2.3.21)]}.$$

$$(2.3.22)$$

This proves that the nonzero process defined in (2.3.21) is indeed a solution process of (2.3.1).

Observation 2.3.1. From the deterministic (2.3.2) and stochastic (2.3.3) decomposition of the stochastic differential equation (2.3.1), the corresponding solution processes of (2.3.19) and (2.3.20) can be rewritten as

$$x^d(t) = \exp\left[\int_a^t f(s)ds\right] c^d = \Phi^d(t)c^d, \quad \text{for } t \in J, \qquad (2.3.23)$$

$$x^s(t, w(t)) = \exp\left[-\frac{1}{2}\int_a^t \sigma^2(s)ds + \int_a^t \sigma(s)dw(s)\right] c^s$$

$$= \Phi^s(t, w(t))c^s, \quad \text{for } t \in J, \qquad (2.3.24)$$

respectively, where

$$\Phi^d(t) = \exp\left[\int_a^t f(s)ds\right], \qquad\qquad \text{for } t \in J, \quad (2.3.25)$$

$$\Phi^s(t) \equiv \Phi^s(t, w(t)) = \exp\left[-\frac{1}{2}\int_a^t \sigma^2(s)ds + \int_a^t \sigma(s)dw(s)\right], \quad \text{for } t \in J, \quad (2.3.26)$$

and nonzero arbitrary constants c^d and c^s are as defined in Step #2. Based on this and the deterministic calculus background [127], we are ready to draw the following conclusions:

(i) We observe that both $\Phi^d(t)$ and $\Phi^s(t)$ are nonzero functions for any $a, t \in J$ and $t \geq a$. $\Phi^d(t)$ and $\Phi^s(t)$ have algebraic inverses. In fact,

$$(\Phi^d(t))^{-1} = \exp\left[-\int_a^t f(s)ds\right], \qquad\qquad \text{for } t \in J, \quad (2.3.27)$$

$$(\Phi^s(t))^{-1} \equiv (\Phi^s(t, w(t)))^{-1}$$

$$= \exp\left[\frac{1}{2}\int_a^t \sigma^2(s)ds - \int_a^t \sigma(s)dw(s)\right], \quad \text{for } t \in J. \quad (2.3.28)$$

(ii) The solution $x^d(t)$ defined in (2.3.23) is called the *general solution* of the deterministic component of the differential equation (2.3.2), and $x^s(t)$ defined in (2.3.24) is called the *general solution* of the Itô–Doob-type stochastic component of the differential equation (2.3.3).

(iii) Moreover, $\Phi^d(t)$ defined in (2.3.25) depends on rate function f in the deterministic differential equation (2.3.2) and $a \in J = [a, b]$, and $\Phi^s(t)$ defined in (2.3.26) depends on rate function σ, the Wiener process w in the stochastic differential equation (2.3.3), and $a \in J$. In short, $\Phi^d(t)$ and $\Phi^s(t)$ are uniquely determined by rate functions and the Wiener process w in (2.3.1), and $a \in J = [a, b]$.

(iv) In particular, for $c^d = 1 = c^s$, from (2.3.23) and (2.3.24), we conclude that $\Phi^d(t)$ and $\Phi^s(t)$ are also nontrivial solutions to the differential equations (2.3.2) and (2.3.3), respectively. From (2.3.23) and (2.3.24), we further conclude that any other solutions to (2.3.2) and (2.3.3), respectively, are determined by scalar multiples of $\Phi^d(t)$ and $\Phi^s(t)$, respectively. Therefore, $\Phi^d(t)$ and $\Phi^s(t)$ are called *general fundamental solution processes* of the deterministic differential equation (2.3.2) and the Itô–Doob-type stochastic differential equation (2.3.3), respectively. In fact,

$$d\Phi^d(t) = f(t)\Phi^d(t)dt, \qquad (2.3.29)$$

$$d\Phi^s(t) = \sigma(t)\Phi^s(t)dw(t). \qquad (2.3.30)$$

(v) For $t \in J = [a, b]$, let us define a process

$$\Phi(t) \equiv \Phi(t, w(t)) = \Phi^d(t)\Phi^s(t, w(t)) = \Phi^d(t)\Phi^s(t, w(t)) \equiv \Phi^d(t)\Phi^s(t). \qquad (2.3.31)$$

From (2.3.27) and (2.3.28), we conclude that $\Phi(t)$ is algebraically invertible. Moreover, $\Phi(t)$ is a fundamental solution process of (2.3.1). In fact, $\Phi(t)$ and $d\Phi(t)$ satisfy the stochastic differential equation (2.3.1) on J. This can be justified as follows. We apply Theorem 1.3.4 to $\Phi(t)$, and obtain

$$d\Phi(t)$$
$$= d\Phi^d(t)\Phi^s(t) + \Phi^d(t)d\Phi^s(t) \quad \text{(by Theorem 1.3.4, the product rule)}$$
$$= f(t)\Phi^d(t)dt\Phi^s(t)$$
$$\quad + \Phi^d(t)\sigma(t)\Phi^s(t)dw(t) \qquad \text{[by substitution from (2.3.29) and (2.3.30)]}$$
$$= f(t)\Phi^d(t)\Phi^s(t)dt$$
$$\quad + \sigma(t)\Phi^d(t)\Phi^s(t)dw(t) \qquad \text{[by } \Phi^d(t)\sigma(t) = \sigma(t)\Phi^d(t)]$$
$$= f(t)\Phi(t)dt + \sigma(t)\Phi(t)dw(t) \quad \text{[by substitution (2.3.31)]}. \qquad (2.3.32)$$

This shows that $\Phi(t)$ is a solution process of (2.3.1). Moreover, this random process depends on only rate functions f, $\sigma \in H_2[a, b]$, $a \in J$, and the Wiener process w. In short, Φ is uniquely determined by f, σ and the Wiener process w in (2.3.1) and $a \in J$, and it is referred to as the *general fundamental solution process* of (2.3.1).

(vi) From (v) and (2.3.21), any nontrivial solution to (2.3.1) can be represented by

$$x(t) = \exp\left[\int_a^t \left[f(s) - \frac{1}{2}\sigma^2(s)\right] ds + \int_a^t \sigma(s)dw(s)\right] c = \Phi(t)c, \qquad (2.3.33)$$

where $t \in J = [a, b] \subseteq R$ and $\Phi(t)$ is defined as (2.3.31) on J. Here, c is an arbitrary constant. $x(t) = \Phi(t)c$ is called the *general solution* to (2.3.1) Figure 2.3.1. Process of finding general solution.

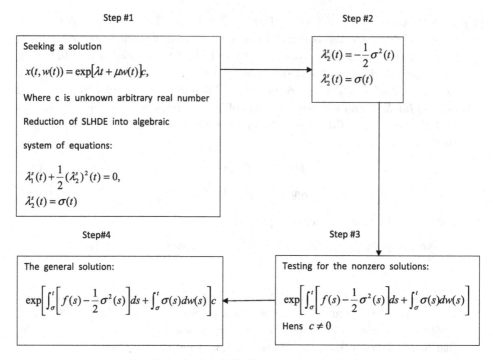

Fig. 2.3.1   Process of finding the general solution.

Remark 2.3.1. Note that the form of the general solution expression (2.3.21) or (2.3.33) for the stochastic differential equation (2.3.1) is an extension of the form of the corresponding deterministic differential equation (2.3.2). In fact, by the choice of $\sigma \equiv 0$, i.e. in the absence of random perturbations/effects, a solution expression in (2.3.33) of (2.3.1) is reduced to a solution expression in (2.3.23).

Theorem 2.3.1. *Let Φ be the fundamental solution process of (2.3.1). Then, Φ is invertible in the algebraic sense, and ϕ^{-1} satisfies the following first-order linear stochastic differential equation of the Itô–Doob type:*

$$d\Phi^{-1}(t) = [-f(t) + \sigma^2(t)]\Phi^{-1}(t)dt - \sigma(t)\Phi^{-1}(t)dw(t). \qquad (2.3.34)$$

Proof. From Observation 2.3.1, the fundamental solution process Φ of (2.3.1) is algebraically invertible. The algebraic inverse is denoted by Φ^{-1}. By using the product rule (Theorem 1.3.4), we compute the following Itô–Doob differential:

$$d(\Phi\Phi^{-1}) = d\Phi\Phi^{-1} + \Phi d\Phi^{-1} + d\Phi d\Phi^{-1} = d1 = 0.$$

From this and (2.3.33), we have

$$d\Phi^{-1} = -\Phi^{-1}[d\Phi\Phi^{-1} + d\Phi d\Phi^{-1}] \qquad \text{(by solving for } d\Phi^{-1})$$

$$= -\Phi^{-1}d\Phi\Phi^{-1} - \Phi^{-1}(t)d\Phi d\Phi^{-1} \qquad \text{(by Theorem 1.2.1M2)}$$

$$= -\Phi^{-1}[f(t)\Phi(t)dt + \sigma(t)\Phi(t)dw(t)]\Phi^{-1}$$

$$-\Phi^{-1}[\sigma(t)\Phi(t)dw(t)]d\Phi^{-1} \qquad \text{[from (2.3.32) \& } dtd\Phi^{-1} \simeq o(dt)]$$

$$= [-f(t)dt - \sigma(t)dw(t)]\Phi^{-1}$$

$$\quad - \sigma(t)dw(t)d\Phi^{-1} \qquad \text{(by the commutative law)}$$

$$= [-f(t)dt - \sigma(t)dw(t)]\Phi^{-1}$$

$$\quad + \sigma^2(t)\Phi^{-1}(dw(t))^2 \qquad \text{(by the Itô–Doob part } d\Phi^{-1})$$

$$= [-f(t)dt - \sigma(t)dw(t)]\Phi^{-1} + \sigma^2(t)\Phi^{-1}dt \quad \text{[by } (dw(t))^2 \simeq dt]$$

$$= [-f(t) + \sigma^2(t)]\Phi^{-1}dt - \sigma(t)\Phi^{-1}dw(t) \quad \text{(by regrouping)}.$$

This completes the proof of the theorem. $\qquad\qquad\qquad\qquad\qquad\qquad\qquad$ □

Observation 2.3.2. From Theorem 2.3.1, we observe that $\Phi^{-1}(t)$ is a fundamental solution process of the following stochastic differential equation:

$$dy = [-f(t) + \sigma^2(t)]y\, dt - \sigma(t)y\, dw(t). \qquad (2.3.35)$$

This differential equation is referred to as the *adjoint* to (2.3.1). This idea is utilized to study both the scalar and systems of stochastic differential equations of the Itô–Doob type. Moreover, $\Phi^{-1}(t)$ is denoted by $\Psi(t)$.

Example 2.3.3. Given: $dx = fx\, dt + \sigma x\, dw(t)$. Find the fundamental and general solutions to the given differential equation.

Solution procedure. Goal of the example: To find the general solution to the given differential equation. Let us compare the coefficients of "dt and $dw(t)$" of the given differential equation with the corresponding coefficients of the differential equation (2.3.1). Here $f(t) \equiv f$ and $\sigma(t) \equiv \sigma$ are constant functions.

Using Step #1, the decomposition of the given differential equation is as follows:

$$dx = fx\, dt, \quad dx = \sigma x\, dw(t).$$

Now, by imitating Step #2 and Step #3, we have

$$\lambda^d(t) = f, \quad \lambda_1^s(t) + \lambda_2^s(t) = 0 \quad \text{and} \quad \lambda_2^s(t) = \sigma,$$

and hence

$$\lambda^d(t) = f, \quad \lambda_1^s(t) = -\frac{1}{2}\sigma^2 \quad \text{and} \quad \lambda_2^s(t) = \sigma.$$

Using these expressions, the representations of solutions in Step #2 are

$$x^d(t) = \exp[f(t-a)]c^d, \quad x^s(t) = \exp\left[-\frac{1}{2}\sigma^2(t-a) + \sigma(w(t) - w(a))\right]c^s.$$

Now, using Step #4 and following the argument used in Step #5, we conclude that

$$x(t) = x^d(t)x^s(t)c \quad \text{(by the law of exponents)}$$

$$= \exp[f(t-a)] \exp\left[-\frac{1}{2}\sigma^2(t-a) + \sigma(w(t) - w(a))\right] c \quad \text{(by regrouping)}$$

$$= \exp\left[\left(f - \frac{1}{2}\sigma^2\right)(t-a) + \sigma(w(t) - w(a))\right] c \qquad \text{(by simplification)},$$

where c is an arbitrary constant. Moreover, from (2.3.31), its fundamental solution is

$$\Phi(t) = \Phi^d(t)\Phi^s(t) = \exp\left[\left(f - \frac{1}{2}\sigma^2\right)(t-a) + \sigma(w(t) - w(a))\right].$$

From Observation 2.3.1(i), (iv), we conclude that

$$x(t) = \exp\left[\left(f - \frac{1}{2}\sigma^2\right)(t-a) + \sigma(w(t) - w(a))\right] c$$

is the desired general solution to the given stochastic differential equation.

Example 2.3.4. Find a general solution to $dx = -fx\,dt + \sigma x\,dw(t)$.

Solution procedure. To avoid the repetition, we just follow the mechanical argument used in the solution process of Example 2.3.3. The Itô–Doob-type stochastic fundamental and general solutions to the given differential equation are

$$\Phi(t) = \Phi^d(t)\Phi^s(t) = \exp\left[\left(-f - \frac{1}{2}\sigma^2\right)(t-a) + \sigma(w(t) - w(a))\right],$$

$$x(t) = \exp\left[\left(-f - \frac{1}{2}\sigma^2\right)(t-a) + \sigma(w(t) - w(a))\right] c,$$

respectively.

Example 2.3.5. Find a general solution to $dx = -fx\,dx - \sigma x\,dw(t)$.

Solution procedure. Again, we just follow the mechanical argument used in the solution process of Example 2.3.3. And again, the Itô–Doob-type stochastic fundamental and general solutions to the given differential equations are

$$\Phi(t) = \Phi^d(t)\Phi^s(t) = \exp\left[\left(-f - \frac{1}{2}\sigma^2\right)(t-a) - \sigma(w(t) - w(a))\right],$$

and

$$x(t) = \exp\left[\left(-f - \frac{1}{2}\sigma^2\right)(t-a) - \sigma(w(t) - w(a))\right] c,$$

respectively.

Example 2.3.6. Given: $\sqrt{1+t^2}\,dx = \sqrt{1+t^2}\sin 2tx\,dt + \sigma x\,dw(t)$. Find the fundamental and general solutions to the given differential equation.

Solution procedure. The goal of the example is to find the general solution to the given Itô–Doob-type stochastic differential equation. Let us compare the coefficients of "dt and $dw(t)$" of the given differential equation with the corresponding coefficients of the differential equation (2.3.1). For this purpose we need to rewrite the differential in a "*standard form* (2.3.1)." It is

$$dx = \sin 2tx \, dt + \sigma \frac{1}{\sqrt{1+t^2}} x \, dw(t).$$

Here $f(t) = \sin 2t$ and $\sigma(t) = \frac{1}{\sqrt{1+t^2}}$.

Using Step #1, the decomposition of the given differential equation is

$$dx = \sin 2tx \, dt, \quad dx = \sigma \frac{1}{\sqrt{1+t^2}} x \, dw(t).$$

Now, by using the logic from Step #2 and Step #3, we have

$$\lambda^d(t) = \sin 2t, \quad \lambda_1^s(t) + \lambda_2^s(t) = 0 \quad \text{and} \quad \lambda_2^s(t) = \sigma \frac{1}{\sqrt{1+t^2}},$$

and hence

$$\lambda^d(t) = \sin 2t, \quad \lambda_1^s(t) = \frac{1}{2}\sigma \frac{1}{1+t^2} = 0, \quad \text{and} \quad \lambda_2^s(t) = \sigma \frac{1}{\sqrt{1+t^2}}.$$

From this information, the representations of solutions in Step #2 are

$$x^d(t) = \exp\left[\int_a^t \sin 2s \, ds\right] c^d,$$

$$x^s(t) = \exp\left[-\frac{1}{2}\sigma^2 \int_a^t \frac{1}{1+t^2} ds + \sigma \int_a^t \frac{1}{\sqrt{1+s^2}} dw(s)\right].$$

We recall that $\int_a^t \sin 2s \, ds = \frac{1}{2}(\cos 2t - \cos 2a)$ and $\int_a^t \frac{1}{1+t^2} ds = (\tan^{-1} t - \tan^{-1} a)$. From this, the above solution processes reduce to

$$x^d(t) = \exp\left[\frac{1}{2}(\cos 2t - \cos 2a)\right] c^d,$$

$$x^s(t) = \exp\left[-\frac{1}{2}\sigma^2(\tan^{-1} t - \tan^{-1} a) + \sigma \int_a^t \frac{1}{\sqrt{1+s^2}} dw(s)\right] c^s.$$

Now, using Step #4 and following the argument used in Step #5, we conclude that

$$x(t) = x^d(t)x^s(t)c \quad \text{(by the law of exponents, regrouping, and simplification)}$$

$$= \exp\left[\frac{1}{2}(\cos 2t - \cos 2a) - \frac{1}{2}\sigma^2(\tan^{-1} t - \tan^{-1} a) + \sigma \int_a^t \frac{1}{\sqrt{1+s^2}} dw(s)\right] c,$$

where c is an arbitrary constant. Moreover, from (2.3.31), its fundamental solution is

$$\Phi(t) = \Phi^d(t)\Phi^s(t)$$

$$= \exp\left[\frac{1}{2}(\cos 2t - \cos 2a) - \frac{1}{2}\sigma^2(\tan^{-1}t - \tan^{-1}a) + \sigma \int_a^t \frac{1}{\sqrt{1+s^2}}dw(s)\right].$$

From Observation 2.3.1(i) and (iv), we conclude that

$$x(t) = \exp\left[\frac{1}{2}(\cos 2t - \cos 2a) - \frac{1}{2}\sigma^2(\tan^{-1}t - \tan^{-1}a) + \sigma \int_a^t \frac{1}{\sqrt{1+s^2}}dw(s)\right]c$$

is the desired general solution to the given stochastic differential equation.

In the following subsection, we discuss a procedure for finding a particular solution to a linear homogeneous Itô–Doob-type stochastic differential equation. This idea leads to the formulation of an IVP in the study of differential equations and its applications. Moreover, illustrations are provided to exhibit the usefulness of the concept of solving real-world problems.

2.3.3 *Initial Value Problem*

Let us formulate an initial value problem for the linear first-order homogeneous stochastic differential equation of the Itô–Doob-type (2.3.1). We consider

$$dx = f(t)x\,dt + \sigma(t)x\,dw(t), \quad x(t_0) = x_0. \tag{2.3.36}$$

Here x, f, σ, and $w(t)$ are as defined in (2.3.1); t_0 is in J; x_0 is a real-valued random variable defined on a complete probability space $(\Omega, \mathfrak{F}, P)$; and x_0 is independent of $w(t)$ for all t in J. (t_0, x_0) is called an *initial data* or *initial condition*. In short, x_0 is a random variable taking values in R. x_0 is the value of the solution function at $t = t_0$ (the initial/given time). Moreover, x_0 is an initial/given state of a dynamic process defined on the same probability space $(\Omega, \mathfrak{F}, P)$, and it is independent of $w(t)$ for all t in J. The problem of finding a solution to (2.3.36) is referred to as the *initial value problem* (IVP). Its solution is represented by $x(t) = x(t, t_0, x_0)$ for $t \geq t_0$ and $t, t_0 \in J$.

Definition 2.3.2. A *solution of the IVP* (2.3.36) is a real-valued random process/function x defined on $[t_0, b) \subseteq J$ such that $x(t)$ and its Itô–Doob differential $dx(t)$ satisfy both the scalar differential equations (2.3.36) and also any given initial condition (t_0, x_0). In short, (i) $x(t)$ is a solution to (2.3.1) (Definition 2.3.1) and (ii) $x(t_0) = x_0$.

Example 2.3.7. Verify that $x(t) = 3\exp[-(t - t_0) + 2(w(t) - w(t_0))]$ is a solution to the following IVP: $dx = x\,dt + 2x\,dw(t)$, $x(t_0) = 3$.

Solution procedure. To show the given process to be a solution process of the given stochastic differential, we find its Itô–Doob differential.

$$dx(t) = d(3\exp[-(t - t_0) + 2(w(t) - w(t_0))]) \qquad \text{(by Theorems 1.3.1 and 1.3.4)}$$

$$= 3\exp[-(t - t_0) + 2(w(t) - w(t_0))]$$

$$\times \, d(-(t - t_0) + 2(w(t) - w(t_0)))$$

$$+ \frac{1}{2}3\exp[-(t - t_0) + 2(w(t) - w(t_0))]$$

$$\times \, d(-(t - t_0) + 2(w(t) - w(t_0)))^2$$

$$= 3\exp[-(t - t_0) + 2(w(t) - w(t_0))](-dt + 2dw(t))$$

$$+ \frac{1}{2}3\exp[-(t - t_0) + 2(w(t) - w(t_0))]$$

$$\times \, (-dt + 2dw(t))^2 \qquad \qquad [\text{by } (dw(t))^2 \simeq dt]$$

$$= 3\exp[-(t - t_0) + 2(w(t) - w(t_0))](-dt + 2dw(t))$$

$$+ \frac{1}{2}3\exp[-(t - t_0) + 2(w(t) - w(t_0))]4dt \quad \text{(by regrouping and simplifying)}$$

$$= 3\exp[-(t - t_0) + 2(w(t) - w(t_0))](dt + 2dw(t)) \qquad \text{(by substitution)}$$

$$= x(t)dt + 2x(t)dw(t)) \qquad \qquad \text{(by substitution)}.$$

This shows that $x(t)$ and $dx(t)$ satisfy the given differential equation. To complete the solution process of the given problem, we need to show that the given solution process satisfies the given initial condition. For this purpose, we substitute $t = t_0$ into $x(t)$ and simplify as

$$x(t_0) = 3\exp[-(t - t_0) + 2(w(t_0) - w(t_0))] \qquad \text{(by evaluation)}$$

$$= 3\exp[0] \qquad \qquad \text{(by simplification)}$$

$$= 3(1) = 3 \qquad \qquad \text{(by definition of an exponential function)}.$$

This shows that the given solution process satisfies the given initial data. Therefore, by Definition 2.3.2, we conclude that the given process $x(t)$ is the solution process of the given IVP.

Example 2.3.8. Verify that $x(t) = 2\exp[\sin w(t) - \sin w(t_0)]$ is a solution to the following IVP:

$$dx = \frac{1}{2}(\cos^2 w(t) - \sin w(t))x \, dt + \cos w(t)x \, dw(t), \quad x(t_0) = 2.$$

Solution procedure.   To show the given process to be a solution process of the given stochastic differential, we find its Itô–Doob differential.

$$dx(t) = d(2\exp[\sin w(t) - \sin w(t_0)]) \qquad \text{(from the given expression)}$$

$$= 2\exp[\sin w(t) - \sin w(t_0)]d(\sin w(t) - \sin w(t_0)) + \exp[\sin w(t) - \sin w(t_0)]$$
$$\times \; d(\sin w(t) - \sin w(t_0))^2 \qquad \text{(by Theorems 1.3.1 and 1.3.4)}$$

$$= x(t)\left(-\frac{1}{2}\sin w(t)dt + \cos w(t)dw(t)\right)$$

$$+ x(t)\frac{1}{2}\left(-\frac{1}{2}\sin w(t)dt + \cos w(t)dw(t)\right)^2 \qquad \text{(by substitution)}$$

$$= x(t)\left(-\frac{1}{2}\sin w(t)dt + \cos w(t)w(t)\right)$$

$$+ \frac{1}{2}x(t)\cos^2 w(t)dt \quad \text{(by } (dw(t))^2 \simeq dt)$$

$$= \frac{1}{2}(\cos^2 w(t) - \sin w(t))x(t)dt$$

$$+ \cos w(t)x(t)dw(t) \qquad \text{(by regrouping and simplifying).}$$

This shows that $x(t)$ and $dx(t)$ satisfy the given differential equation. To complete the solution process of the given problem, we need to show that the given solution process satisfies the given initial condition. For this purpose, we substitute $t = t_0$ into $x(t)$ and simplify as

$$x(t_0) = 2\exp[\sin w(t_0) - \sin w(t_0)] \text{ (by evaluation)}$$

$$= 2\exp[0] \qquad\qquad\qquad \text{(by simplification)}$$

$$= 2(1) = 2 \qquad\qquad\qquad \text{(by definition of an exponential function).}$$

This shows that the given solution process satisfies the given initial data. Therefore, by Definition 2.3.2, we conclude that the given process $x(t)$ is the solution process of the given IVP.

2.3.4 *Procedure for Solving the IVP*

In the following, we develop a procedure for finding a solution process of the linear first-order scalar homogeneous stochastic differential equations of Itô–Doob (2.3.1). The problem of finding a solution to the IVP (2.3.36) is summarized:

Step #1.   From Observation 2.3.1 and (2.3.21) or (2.3.33), the general solution to (2.3.1) is

$$x(t) = \exp\left[\int_a^t \left[f(s) - \frac{1}{2}\sigma^2(s)\right]ds + \int_a^t \sigma(s)dw(s)\right]c, = \Phi(t)c,$$

where $t \in J \subseteq R$, and $\Phi(t)$ is as defined (2.3.31). Here, c is an arbitrary constant.

Step #2. In this step, we need to find an arbitrary constant in (2.3.33). For this purpose, we utilize the given initial given data (t_0, x_0), and solve the scalar algebraic equation

$$x(t_0) = \Phi(t_0)c = x_0, \qquad (2.3.37)$$

for any given t_0 in J. Since $\exp\left[\int_a^{t_0}\left[f(s) - \frac{1}{2}\sigma^2(s)\right]ds + \int_a^{t_0}\sigma(s)dw(s)\right] \neq 0$ for any continuous functions f and σ, the algebraic equation (2.3.37) has a unique solution. This solution is given by

$$c = \Phi^{-1}(t_0)x_0 = \exp\left[-\int_a^{t_0}\left[f(s) - \frac{1}{2}\sigma^2(s)\right]ds + \int_a^{t_0}\sigma(s)dw(s)\right]x_0, \quad (2.3.38)$$

which depends on f, σ, w, x_0, t_0, and a.

Step #3. The solution to the IVP (2.3.36) is determined by substituting the expression (2.3.38) into (2.3.33). Thus, we have

$$\begin{aligned}
x(t) &= \Phi(t)c && \text{[from (2.3.33)]}\\
&= \Phi(t)\Phi^{-1}(t_0)x_0 && \text{[from (2.3.34) and (2.3.38)]}\\
&= \exp\left[\int_{t_0}^t\left[f(s) - \frac{1}{2}\sigma^2(s)\right]ds\right.\\
&\qquad\left. + \int_{t_0}^t\sigma(s)dw(s)\right]x_0 && \text{(from Lemma 1.3.1e)}\\
&= \Phi(t, t_0)x_0 && \text{(by notation)}, && (2.3.39)
\end{aligned}$$

where

$$\Phi(t, w(t), t_0) = \Phi(t, w(t))\Phi^{-1}(t_0, w(t_0)) \equiv \Phi(t)\Phi^{-1}(t_0) \equiv \Phi(t, t_0)$$

$$= \exp\left[\int_{t_0}^t\left[f(s) - \frac{1}{2}\sigma^2(s)\right]ds + \int_{t_0}^t\sigma(s)dw(s)\right]. \qquad (2.3.40)$$

Hence, the solution process $x(t) = x(t, t_0, x_0)$ of the IVP (2.3.36) is represented by

$$x(t) = x(t, t_0, x_0) = \exp\left[\int_{t_0}^t\left[f(s) - \frac{1}{2}\sigma^2(s)\right]ds + \int_{t_0}^t\sigma(s)dw(s)\right]x_0 = \Phi(t, t_0)x_0.$$

This solution is referred to as a *particular solution process* of (2.3.1). $\Phi(t, t_0)$ defined in (2.3.40) is called a *normalized fundamental solution process* of (2.3.1). Because $\Phi(t, t_0)$ has an algebraic inverse $[\Phi(t, t_0) \neq 0$ on $J]$, and $\Phi(t, t_0) = 1$.

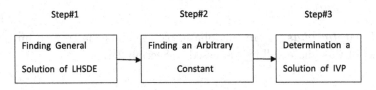

Fig. 2.3.2 Process of finding the solution of the IVP

Now, a few illustrations are presented to exhibit the usefulness of the concept of the IVP, and the procedure for finding the solution to IVPs and its applications. In addition, numerical examples are given.

Illustration 2.3.1 (chemical kinetics under a random environment [80, 183]). We recall the chemical kinetics of nitrogen pentoxide under well-mixed and constant temperature conditions [80]. The decomposition of a gaseous nitrogen pentoxide (N_2O_5) into nitrogen oxide (NO_2) and oxygen (O_2) follows:

$$2N_2O_5 \rightarrow 4NO_2 + O_2.$$

Let $c(t)$ be the concentration of gaseous nitrogen pentoxide measured in moles per liter at a time t.

In general, the controlled, conditions are not completely controllable, due to the external influences. For example, the well-mixed and constant temperature conditions are not perfect. This imperfectness creates the nonhomogeneity and the temperature fluctuations. These slight changes can generate random environmental perturbations. These random perturbations can be described by the Brownian motion process or Poisson process. Following the discussion of Brownian Motion Model 2.1.3, the dynamic of gaseous nitrogen pentoxide under random environmental perturbations is described by the IVP

$$dc(t) = -kc(t)dx + \sigma c(t)dw(x), \quad c_0 = c(0),$$

where $w(t)$ is the normalized Wiener process defined before; $c(t)$, k, and t are as defined in Illustration 2.4.2.[80] By following the solution procedures of Examples 2.3.4 and 2.3.8, we conclude that

$$c(t) = \exp\left[-\left(k + \frac{1}{2}\sigma^2\right)t + \sigma w(t)\right]c_0.$$

This is a stochastic version of the concentration process of gaseous nitrogen pentoxide at a time $t \geq 0$ (Illustration 2.4.2, Volume 1).

Illustration 2.3.2 (stochastic arterial pulse-diastolic process [80, 172]). This illustration is based on Otto Frank (1889)'s pressure–time relationship [172] in the aorta and Illustration 2.2.1 (stochastic version of Poiseuille's law). We recall the basis for the deterministic mathematical model in Illustration 2.4.3 [80].

OFM1 (diastolic phase). During this phase, the rate of inflow from the heart to the aorta is zero.

OFM2 (systolic phase). In this phase, blood enters the aorta from the heart. The model of this phase will be discussed in Section 2.4.

OFM3. In the diastolic phase, it is assumed that: (a) there is a linear relationship between the volume and the pressure; (b) Poiseuille's law is obeyed in this case.

OFM4. Instead of the deterministic Poiseuille's law in OFM3(b), we apply the stochastic version of Poiseuille's law, Illustration 2.2.1.

Let P stand for pressure in the Windkessel, and V be the volume of the blood (fluid). From (a), we have

$$dP = \alpha dV,$$

where α is constant.

If we assume that the pressure at the distal end (venous side) is zero, then Poiseuille's law is applicable, and we have

$$dV = -\frac{P}{\beta}dt + \sigma P dw(t),$$

where $\beta = \frac{8\ell\eta}{\pi r^4}$ is constant and represents the resistance to the blood flow; the minus sign indicates that the greater the pressure, the greater the decrease in the volume per unit time in the Windkessel.

By using the above two expressions, we eliminate V, and we have

$$dP = -\frac{\alpha}{\beta}P\,dt + \sigma P\,dw(t), \quad P(0) = P_0.$$

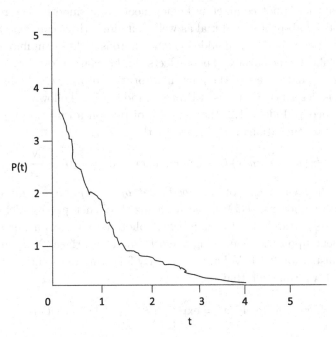

Fig. 2.3.3

Now, we utilize Procedure 2.3.4, and we obtain

$$P(t) = P_0 \exp\left[-\left(\frac{\alpha}{\beta} + \frac{1}{2}\sigma^2\right)t + \sigma w(t)\right].$$

This describes the pressure-time curve during the diastolic phase. Moreover, it signifies the pressure drop in the aorta during the diastolic period. $\frac{2\beta}{2\alpha+\beta\sigma^2} = \tau$ is referred to as the time constant. Finding the mean and variance of pressure is left as an exercise to the reader.

Illustration 2.3.3 (photochemical reaction under a random environment [41, 80]). We recall the Bouguer–Lambert law [41] in photochemistry. This was outlined under the following controlled experimental conditions (Illustration 2.4.1, Volume 1 [80]):

(a) Let A be a surface area cm^2 of a uniform absorbing material.
(b) Let N_0 be the number of uniformly distributed monochromatic photons (the photon is unit of radiant energy moving with the velocity of light and an energy proportional to a frequency called a light quantum) entering in the direction perpendicular to the surface area A of the material in (a).
(c) It is assumed that there is no reflection or scattering of the photons.

The controlled conditions described in Illustration 2.4.1, Volume. 1 [80] are not controllable to the external disturbances (such as water surfaces of lakes and seas). Of course, plant life can exist at the layer that is about a few meters deep. The absorbing material is not completely homogeneous and smooth. The roughness of the surface of the absorbing material as well as its impurities can generate random environmental perturbations. In addition, the photons/light beam may not directly hit perpendicular to the surface. These effects can be characterized by random perturbations. The randomness of the photon absorption process can be like the decay process that is described by Poisson Process Model 2.1.2. Following the discussion of Poisson Process Model 2.1.2, the dynamic of photon absorption under random environmental perturbations is described by the IVP

$$dn(x) = -kn(x)dx + \sigma n(x)dw(x), \quad n_0 = n(0) = \frac{N_0}{A},$$

where k is called the mean of the *coefficient of absorption or the specific rate of absorption* of photons; $w(x)$ is the normalized Wiener process defined before; and $n(x) = \frac{N(x)}{A}$ stands for the number of photons per second per square centimeter incident upon the layer lying x cm below the surface. $N(x)$ and x are as described (Illustration 2.4.1, Volume 1 [80]). By following the solution procedure of Example 2.3.4, we conclude that

$$\Phi(x) = \Phi^d(x)\Phi^s(x) = \exp\left[-\left(k + \frac{1}{2}\sigma^2\right)x + \sigma w(x)\right],$$

$$n(x) = \exp\left[-\left(k + \frac{1}{2}\sigma^2\right)x + \sigma w(x)\right]c,$$

are the fundamental and the general solution to the above-described stochastic differential equation, respectively. Now, by following the solution procedure of Example 2.3.8, we obtain the solution process of the above IVP:

$$n(x) = \exp\left[-\left(k + \frac{1}{2}\sigma^2\right)x + \sigma w(x)\right]n_0.$$

This is a stochastic version of Bouguer's law. Moreover, the stochastic version of Lambert's law is

$$I(x) = I_0 \exp\left[-\left(k + \frac{1}{2}\sigma^2\right)x + \sigma w(x)\right],$$

where $I(x)$ denotes the intensity of the light beam at a depth of x units below the absorbing surface. It is known that the intensity of the light beam is directly proportional to the photon flux $n(x) = \frac{N(x)}{A}$.

Remark 2.3.2 [49, 80]. We remark that the Bouguer–Lambert law is applicable to any homogeneous, transparent substances, such as glass, Plexiglas, liquids, and any thin layers viewed under the microscope. Furthermore, like light waves, other electromagnetic waves, namely X-rays and gamma rays, behave the same way. In fact, the Bouguer–Lambert law is basic in photometry.

Example 2.3.9. Using Illustration 2.3.3, find:

(a) The expected depth of the layer absorbs (i) 50% and (ii) 90% of the photons.
(b) The variance of the depth of the layer absorbs (i) 50% and (ii) 90% of the photons.

Solution procedure. From Illustration 2.3.3, the number of photons $n(x)$ per second per square centimeter incident upon the layer lying x cm below the surface is given by

$$n(x) = \exp\left[-\left(k + \frac{1}{2}\sigma^2\right)x + \sigma w(x)\right]n_0,$$

and hence

$$\frac{n(x)}{n_0} = \exp\left[-\left(k + \frac{1}{2}\sigma^2\right)x + \sigma w(x)\right].$$

Knowing that the function "ln" is an inverse function of "exp," the above mathematical expression can be written as

$$\ln\left[\frac{n(x)}{n_0}\right] = -\left(k + \frac{1}{2}\sigma^2\right)x + \sigma w(x).$$

From this, one can complete the example, by imitiating the deterministic version (Example 2.4.6 [80]).

Exercises

1. Using substitution, verify that each given random process/function is a solution to the given differential equation:

   (a) $(x(t) = \exp[-3t - w(t)]c,\ dx = -\frac{5}{2}x\,dt - x\,dw(t)$;
   (b) $(x(t) = \exp[-3t + w(t)]c,\ dx = -\frac{5}{2}x\,dt + x\,dw(t)$;
   (c) $x(t) = e^{w(t)},\ dx = \frac{1}{2}x\,dt + x\,dw(t)$;
   (d) $x(t) = \frac{1+t^2}{\sqrt{1+\exp[t]}}\exp\left[-\int_0^t \sqrt{\frac{\exp[s]}{1+\exp[s]}}\,dw(s)\right]$;

   $dx = \frac{2t}{1+t^2}x\,dt - \sqrt{\frac{\exp[t]}{1+\exp[t]}}x\,dw(t)$.

2. Find the general and fundamental solutions to the following differential equations:

   (a) $dx = -2x\,dt$;
   (b) $dx = -x\,dw(t)$;
   (c) $dx = -2x\,dt - x\,d\,w(t)$;
   (d) $dx = -2x\,dt$;
   (e) $dx = -x\,dw(t)$;
   (f) $dx = -2x\,dt - x\,dw(t)$;
   (g) $dx = 2x\,dt$;
   (h) $dx = 2x\,dw(t)$;
   (i) $dx = 2x\,dt + 2x\,dw(t)$;
   (j) $dx = bx\,dt$;
   (k) $dx = \sqrt{b}x\,dw(t)$;
   (l) $dx = bx\,dt + \sqrt{b}x\,dw(t)$;
   (m) $dx = -bx\,dt$;
   (n) $dx = -\sqrt{b}x\,dw(t)$;
   (o) $dx = -bx\,dt - \sqrt{b}x\,dw(t)$;
   (p) $dx = 2x\,dt$;
   (q) $dx = \sqrt{6}x\,dw(t)$;
   (r) $dx = 2x\,dt + \sqrt{6}x\,dw(t)$.

3. Find the general solution to the following differential equations:

   (a) $(t + t^2)dx = t^2x\,dt + \sigma x\,dw(t)$;
   (b) $(90 + t)dx = 3x\,dt + \sigma(90 + t)x\,dw(t)$;
   (c) $\sqrt{1 + t^2}dx = (1 + t^2)x\,dt - \sigma x\,dw(t)$;
   (d) $dx = \tan t2x\,dt - \sqrt{b}x\,dw(t)$;
   (e) $(1 + \sin^4 t)dx = \sin 2tx\,dt + \sin t(1 + \sin^4 t)x\,dw(t)$;
   (f) $dx = \ln(1 + t)x\,dt - x\,dw(t)$.

4. Solve the following initial value problems:

   (a) $dx = 4x\,dt,\ x(0) = 2$;
   (b) $dx = 2x\,dw(t),\ x(0) = 2$;
   (c) $dx = 4x\,dt + 2x\,dw(t),\ x(0) = 2$;

(d) $dx = 2x\,dt$, $x(0) = 3$;

(e) $dx = \sqrt{6}x\,dw(t)$, $x(0) = 3$;

(f) $dx = 2x\,dt + \sqrt{6}x\,dw(t)$, $x(0) = 3$;

(g) $(90 + t)dx = 3x\,dt + \sigma(90 + t)x\,dw(t)$, $x(0) = 90$.

5. With reference to Illustrations 2.3.1, 2.3.2, and 2.3.3, find the mean and variance of:

(a) the concentration of nitrogen pentoxide;

(b) the pressure profile of the aorta;

(c) the photon flux.

6. Compare the means in Exercise 5 with the corresponding deterministic expressions in Chapter 2, Volume 1 [80].

7. For $i = 1, 2, \ldots, k$, let f, σ_i be continuous functions defined on J into R, and let $w_i(t)$ be the Wiener process (Theorem 1.2.2) defined on a complete probability space $(\Omega, \mathfrak{F}, P)$. Further, assume that for $i \neq j, i, j-1, 2, \ldots, k$, $w_i(t)$ and $w_j(t)$ are mutually independent on J ($E(dw_i(t)dw_j(t)) = 0$). Show that the process defined by

$$x(t, w_1(t), w_2(t), \ldots, w_k(t))$$

$$= \exp\left[\int_a^t \left[f(s) - \frac{1}{2}\sum_{i=1}^k \sigma_i^2(s)\right]ds + \sum_{k=1}^k \int_a^t \sigma_i(s)dw(s)\right]c$$

is the solution process of the following Itô–Doob stochastic differential equation:

$$dx = f(t)x\,dt + \sum_{i=1}^k \sigma_i(t)x\,dw_i(t).$$

8. Find the general solution to $dx = 2x\,dt + 2x\,dw_1(t) + 2x\,dw_2(t)$.

9. Let F be a continuous cumulative distribution function (cdf) of a random time variable T, and let f, g, and σ be continuous functions. Moreover, $w(t)$ is the Wiener process defined before. Apply the procedure outlined in Section 2.3 to find the general solution process of the following differential equations:

(a) $dx = f(t)x\,dt + g(t)x\,dF(t)$;

(b) $dx = g(t)x\,dF(t) + \sigma(t)x\,dw(t)$.

2.4 Linear Nonhomogeneous Equations

In this section, we extend the *method of variation of constants/parameters* for finding particular solutions to first-order linear nonhomogeneous scalar deterministic differential equations [80] to first-order linear nonhomogeneous scalar stochastic differential equations of the Itô–Doob type. This will further enhance the significance and usefulness of this approach.

2.4.1 *General Problem*

Let us consider the following first-order linear nonhomogeneous scalar stochastic differential equation of the Itô–Doob type:

$$dx = [f(t)x + p(t)]dt + [\sigma(t)x + q(t)]dw(t), \qquad (2.4.1)$$

where x is a state variable; f, p, σ, and q are continuous functions defined on an interval $J = [a, b] \subseteq R$; and $w(t)$ is a normalized scalar Wiener process (Theorem 1.2.2).

In this subsection, we discuss a procedure for finding a general solution to (2.4.1). The concept of the solution process defined in Definition 2.3.1 is directly applicable to (2.4.1). The details are left to the reader. The problem of finding a solution process provides a basis for analyzing, understanding, and solving the real-world problems in the biological, chemical, physical, and social sciences.

2.4.2 *Procedure for Finding a General Solution*

To find a general solution to (2.4.1), we imitate the conceptual ideas of finding a general solution to the linear nonhomogeneous algebraic equations described in Theorem 1.4.7, Volume 1 [80]. The basic ideas are:

(i) To find a general solution process x_c of the first-order linear homogeneous scalar Itô–Doob-type stochastic differential equation corresponding to (2.4.1).

(ii) To find a particular solution process x_p of (2.4.1).

(iii) To set up a candidate, $x = x_0 + x_p$, for testing the general solution to (2.4.1).

Here, we introduce the method of variation of constants/parameters to find a particular solution process x_p of the Itô–Doob stochastic differential equation (2.4.1). This is a very elegant and systematic approach to finding a particular solution to (2.4.1).

Step #1. First, we find a fundamental solution to the first-order linear homogeneous scalar stochastic differential equation corresponding to (2.4.1), i.e.

$$dx = f(t)x\,dt + \sigma(t)x\,dw(t). \qquad (2.4.2)$$

We note that the scalar Itô–Doob-type stochastic differential equation is exactly the same as in (2.3.1). Therefore, we imitate Steps #1 to #5 described in Procedure 2.3.2 for finding a general solution process of (2.3.1), and determine a fundamental solution to (2.4.2) (Observation 2.3.1):

$$\Phi(t) = \Phi^d(t)\Phi^s(t) = \exp\left[\int_a^t \left[f(s) - \frac{1}{2}\sigma^2(s)\right]ds + \int_a^t \sigma(s)dw(s)\right]. \qquad (2.4.3)$$

Step #2 (method of variation of constants parameters). A particular solution process x_p of (2.4.1) is a random function that satisfies the stochastic differential equation (2.4.1). To find a particular solution process of (2.4.1), we define a random function as

$$x_p(t) = \Phi(t)c(t), \quad c(a) = c_a, \tag{2.4.4}$$

where $\Phi(t)$ is as determined in Step #1, and $c(t)$ is an unknown random function with $c(a) = c_a$. Now, our goal of finding a particular solution process $x_p(t)$ of (2.4.1) is equivalent to finding an unknown random process $c(t)$ in (2.4.4) passing through c_a at $t = a$. For this purpose, we assume that $x_p(t)$ is a solution process of (2.4.1). From this assumption, using the Itô–Doob product formula (Theorem 1.3.4), and the fact that $\Phi(t)$ is the fundamental solution to (2.4.2), we have

$$dx_p(t) = d\Phi(t)c(t) + \Phi(t)dc(t) + d\Phi(t)dc(t). \tag{2.4.5}$$

Using the algebraic inverse $\Phi^{-1}(t)$ of $\Phi(t)$, we solve for $dc(t)$, and obtain

$$
\begin{aligned}
dc(t) &= \Phi^{-1}(t)[dx_p(t) - d\Phi(t)c(t) - d\Phi(t)dc(t)] \quad \text{[from (2.4.5)]}\\
&= \Phi^{-1}(t)[[f(t)x_p(t) + p(t)]dt\\
&\quad + [\sigma(t)x_p(t) + q(t)]dw(t)\\
&\quad - d\Phi(t)c(t) - d\Phi(t)dc(t)] \quad \text{(by x_p, the particular solution)}\\
&= \Phi^{-1}(t)[[f(t)x_p(t) + p(t)]dt\\
&\quad + [\sigma(t)x_p(t) + q(t)]dw(t)\\
&\quad - f(t)\Phi(t)c(t)dt - \sigma(t)\Phi(t)c(t)dw(t)\\
&\quad - d\Phi(t)dc(t)] \quad \text{[by (2.3.32)]}\\
&= \Phi^{-1}(t)p(t)dt + \Phi^{-1}(t)q(t)dw(t)\\
&\quad - \Phi^{-1}d\Phi(t)dc(t) \quad \text{(by simplification).} \tag{2.4.6}
\end{aligned}
$$

Now, from (2.3.32) and (2.4.6), we compute $d\Phi(t)dc(t)$ in the last term of (2.4.6):

$$
\begin{aligned}
d\Phi(t)dc(t) &= [f(t)\Phi(t)dt + \sigma(t)\Phi(t)dw(t)]dc(t) \quad \text{[from (2.3.32)]}\\
&= \sigma(t)\Phi(t)dw(t)dc(t) \quad \text{[from $f(t)\Phi(t)dc(t)dt \simeq o(dt)$]}\\
&= (\sigma(t)\Phi(t)dw(t))\Phi^{-1}(t)q(t)dw(t)) \quad \text{[from $dc(t)$ (2.4.6)]}\\
&= (\sigma(t)\Phi(t)\Phi^{-1}(t)q(t)(dw(t))^2 \quad \text{[from $(dw(t))^2 \simeq dt$]}\\
&= \sigma(t)q(t)dt \quad \text{(by simplification).} \tag{2.4.7}
\end{aligned}
$$

Substituting the expression for $d\Phi(t)dc(t)$ in (2.4.7) into (2.4.6), we have

$$
\begin{aligned}
dc(t) &= \Phi^{-1}(t)p(t)dt + \Phi^{-1}(t)q(t)dw(t) - \Phi^{-1}(t)\sigma(t)q(t)dt\\
&= \Phi^{-1}(t)[p(t) - \sigma(t)q(t)]dt + \Phi^{-1}(t)q(t)dw(t). \tag{2.4.8}
\end{aligned}
$$

We recall that the problem of finding a particular solution $x_p(t)$ defined in (2.4.4) reduces to a problem of finding the unknown process $c(t)$ in (2.4.4) passing through c_a at $t = a$. This problem further reduces to a problem of solving the IVP

$$dc(t) = \Phi^{-1}(t)[p(t) - \sigma(t)q(t)]dt + \Phi^{-1}(t)q(t)dw(t), \quad c(a) = c_a. \qquad (2.4.9)$$

By following Procedure 2.2.4 for solving the IVP described in Section 2.2, the solution to the IVP (2.4.9) is given by

$$c(t) = c_a + \int_a^t \Phi^{-1}(s)[p(s) - \sigma(s)q(s)]dt + \int_a^t \Phi^{-1}(s)q(s)dw(s). \qquad (2.4.10)$$

We substitute the expression for $c(t)$ in (2.4.10) into (2.4.4), and we have

$$
\begin{aligned}
x_p(t) &= \Phi(t)\left[c_a + \int_a^t \Phi^{-1}(s)[p(s) - \sigma(s)q(s)]ds + \int_a^t \Phi^{-1}(s)q(s)dw(s)\right] \\
&= \Phi(t)c_a + \int_a^t \Phi(t)\Phi^{-1}(s)[p(s) - \sigma(s)q(s)]ds \\
&\quad + \int_a^t \Phi(t)\Phi^{-1}(s)q(s)dw(s). \qquad (2.4.11)
\end{aligned}
$$

Step #3. From Step #1 and Observation 2.3.1, for any arbitrary constant c, the random function defined by $x_c(t) = \Phi(t)c$ is the general solution to (2.4.2). From the conceptual ideas described in Theorem 1.4.7 [80], we define a random function by $x(t) = x_c(t) + x_p(t)$, where $x_p(t)$ is determined in Step #2. Now, we claim that that the random process $x(t) = x_c(t) + x_p(t)$ is a general solution to (2.4.1). For this purpose, we consider

$$x(t) = x_c(t) + x_p(t)$$

$$= \Phi(t)c + \Phi(t)c_a + \int_a^t \Phi(t)\Phi^{-1}(s)[p(s) - \sigma(s)q(s)]ds$$

$$+ \int_a^t \Phi(t)\Phi^{-1}(s)q(s)dw(s) \qquad \text{[from (2.4.11)]}$$

$$= \Phi(t)(c + c_a) + \int_a^t \Phi(t)\Phi^{-1}(s)[p(s) - \sigma(s)q(s)]ds$$

$$+ \int_a^t \Phi(t)\Phi^{-1}(s)q(s)dw(s) \qquad \text{(by regrouping)}$$

$$= \Phi(t)c + \int_a^t \Phi(t)\Phi^{-1}(s)[p(s) - \sigma(s)q(s)]ds$$

$$+ \int_a^t \Phi(t)\Phi^{-1}(s)q(s)dw(s) \qquad \text{(by notation)}$$

$$= \Phi(t)\Big[c + \int_a^t \Phi^{-1}(s)[p(s) - \sigma(s)q(s)]ds$$

$$+ \int_a^t \Phi^{-1}(s)q(s)dw(s)\Big] \qquad \text{(by Theorem 1.3.1M2 [80])},$$

$$(2.4.12)$$

where $c = c + c_a$ is an arbitrary constant. Because of the fact that c and c_a are arbitrary constants, $\Phi(t)c$ also the general solution to (2.4.2). We show that $x(t)$ is a general solution to (2.4.1).

$$dx(t) = d\Big[\Phi(t)\Big[c + \int_a^t \Phi^{-1}(s)[p(s) - \sigma(s)q(s)]ds + \int_a^t \Phi^{-1}(s)q(s)dw(s)\Big]\Big]$$

$$\text{[from (2.4.12)]}$$

$$= d\Phi(t)\Big[c + \int_a^t \Phi^{-1}(s)[p(s) - \sigma(s)q(s)]ds + \int_a^t \Phi^{-1}(s)q(s)dw(s)\Big]$$

$$+ \Phi(t)d\Big[c + \int_a^t \Phi^{-1}(s)[p(s) - \sigma(s)q(s)]ds + \int_a^t \Phi^{-1}(s)q(s)dw(s)\Big]$$

$$+ d\Phi(t)d\Big[c + \int_a^t \Phi^{-1}(s)[p(s) - \sigma(s)q(s)]ds + \int_a^t \Phi^{-1}(s)q(s)dw(s)\Big]$$

$$\text{(by Theorem 1.3.4)}$$

$$= d\Phi(t)\Big[c + \int_a^t \Phi^{-1}(s)[p(s) - \sigma(s)q(s)]ds + \int_a^t \Phi^{-1}(s)q(s)dw(s)\Big]$$

$$+ \Phi(t)d\Big[\int_a^t \Phi^{-1}(s)[p(s) - \sigma(s)q(s)]ds + \int_a^t \Phi^{-1}(s)q(s)dw(s)\Big]$$

$$+ d\Phi(t)d\Big[\int_a^t \Phi^{-1}(s)[p(s) - \sigma(s)q(s)]ds + \int_a^t \Phi^{-1}(s)q(s)dw(s)\Big]$$

$$\text{(by Theorem 1.3.3)}$$

$$= d\Phi(t)\Big[c + \int_a^t \Phi^{-1}(s)[p(s) - \sigma(s)q(s)]ds + \int_a^t \Phi^{-1}(s)q(s)dw(s)\Big]$$

$$\text{(by Theorem 1.3.3)}$$

$$+ \Phi(t)d\Big[\int_a^t \Phi^{-1}(s)[p(s) - \sigma(s)q(s)]ds\Big] + \Phi(t)d\Big[\int_a^t \Phi^{-1}(s)q(s)dw(s)\Big]$$

$$+ d\Phi(t)d\Big[\int_a^t \Phi^{-1}(s)q(s)dw(s)\Big] + d\Phi(t)d\Big[\int_a^t \Phi^{-1}(s)[p(s) - \sigma(s)q(s)]ds\Big].$$

$$(2.4.13)$$

We recall that

$$d\Phi(t) = f(t)\Phi(t)dt + \sigma(t)\Phi(t)dw(t) \quad \text{[from (2.3.32)]}.$$

From the fundamental theorems of elementary integral calculus [127] and the Itô–Doob calculus, we have

$$d\left[\int_a^t \Phi^{-1}(s)[p(s) - \sigma(s)q(s)ds]\right]$$
$$= \Phi^{-1}(t)[p(t) - \sigma(t)q(t)dt] \qquad \text{(by the fundamental theorem),}$$
$$d\left[\int_a^t \Phi^{-1}(s)q(s)dw(s)\right] = \Phi^{-1}(t)q(t)dw(t) \quad \text{(by the Itô–Doob calculus).}$$

Substituting these expressions into (2.4.13) and using the Itô–Doob calculus, we have

$$dx(t) = [f(t)\Phi(t)dt + \sigma(t)\Phi(t)dw(t)]$$
$$\times \left[c + \int_a^t \Phi^{-1}(s)[p(s) - \sigma(s)q(s)ds] + \int_a^t \Phi^{-1}(s)q(s)dw(s)\right]$$
$$+ \Phi(t)\Phi^{-1}(t)[p(t) - \sigma(t)q(t)]dt + \Phi(t)\Phi^{-1}(t)q(t)dw(t)$$
$$+ [f(t)\Phi(t)dt + \sigma(t)\Phi(t)dw(t)]\Phi^{-1}(t)[p(t) - \sigma(t)q(t)]dt$$
$$+ [f(t)\Phi(t)dt + \sigma(t)\Phi(t)dw(t)]\Phi^{-1}(t)q(t)dw(t) \quad \text{(by multiplying)}$$
$$= f(t)\Phi(t)\left[c + \int_a^t \Phi^{-1}(s)[p(s) - \sigma(s)q(s)ds] + \int_a^t \Phi^{-1}(s)q(s)dw(s)\right]dt$$
$$+ \sigma(t)\Phi(t)\left[c + \int_a^t \Phi^{-1}(s)[p(s) - \sigma(s)q(s)]ds + \int_a^t \Phi^{-1}(s)q(s)dw(s)\right]dw(t)$$
$$+ [p(t) - \sigma(t)q(t)]dt + q(t)dw(t)$$
$$+ [\sigma(t)\Phi(t)dw(t)]\Phi^{-1}(t)q(t)dw(t) + [f(t)\Phi(t)dt]\Phi^{-1}(t)[p(t) - \sigma(t)q(t)]dt$$
$$+ [\sigma(t)\Phi(t)dw(t)]\Phi^{-1}[p(t) - \sigma(t)q(t)]dt + [f(t)\Phi(t)dt]\Phi^{-1}(t)q(t)dw(t)$$
$$= f(t)\Phi(t)\left[c + \int_a^t \Phi^{-1}(s)[p(s) - \sigma(s)q(s)ds] + \int_a^t \Phi^{-1}(s)q(s)dw(s)\right]dt$$
$$+ \sigma(t)\Phi(t)\left[c + \int_a^t \Phi^{-1}(s)[p(s) - \sigma(s)q(s)]ds + \int_a^t \Phi^{-1}(s)q(s)dw(s)\right]dw(t)$$
$$+ [p(t) - \sigma(t)q(t)]dt + q(t)dw(t)$$
$$+ [\sigma(t)\Phi(t)\Phi^{-1}(t)q(t)(dw(t))^2] \qquad \text{(by simplifications)}$$
$$= f(t)\Phi(t)\left[c + \int_a^t \Phi^{-1}(s)[p(s) - \sigma(s)q(s)]ds + \int_a^t \Phi^{-1}(s)q(s)dw(s)\right]dt$$
$$+ \sigma(t)\Phi(t)\left[c + \int_a^t \Phi^{-1}(s)[p(s) - \sigma(s)q(s)]ds + \int_a^t \Phi^{-1}(s)q(s)dw(s)\right]dw(t)$$

$$+ p(t)dt + q(t)dw(t) \quad \text{(from } (dw(t))^2 \simeq dt)$$

$$= f(t)[x_c(t) + x_p(t)]dt + \sigma(t)[x_c(t) + x_p(t)]dw(t)$$

$$+ p(t)dt + q(t)dw(t) \quad \text{[by substitution from } x_c(t) \text{ and } x_p(t)]$$

$$= f(t)x(t)dt + \sigma(t)x(t)dw(t) + p(t)dt + q(t)dw(t) \quad \text{[by } x(t) = x_c(t) + x_p(t)]$$

$$= [f(t)x(t) + p(t)]dt + [\sigma(t)x(t) + q(t)]dw(t) \quad \text{(by regrouping)}. \tag{2.4.14}$$

This shows that $x(t) = x_c(t) + x_p(t)$ is indeed a solution to (2.4.1). This exhibits the claim that $x(t) = x_c(t) + x_p(t)$ is a solution to (2.4.1). Moreover, from (2.3.21) and (2.3.40) the general solution to (2.4.1) is expressed as follows:

$$x(t) = \Phi(t)\left[c + \int_a^t \Phi^{-1}(s)[p(s) - \sigma(s)q(s)]ds\right.$$

$$\left. + \int_a^t \Phi^{-1}(s)q(s)dw(s)\right] \qquad \text{[from (2.4.12)]}$$

$$= \Phi(t)c + \int_a^t \Phi(t,s)[p(s) - \sigma(s)q(s)]ds$$

$$+ \int_a^t \Phi(t,s)q(s)dw(s) \qquad \text{(by Theorem 1.3.1M2 [80])}$$

$$= \exp\left[\int_a^t \left[f(s) - \frac{1}{2}\sigma^2(s)\right]ds + \int_a^t \sigma(s)dw(s)\right]c$$

$$+ \int_a^t \exp\left[\int_s^t \left(f(u) - \frac{1}{2}\sigma^2(u)\right)du + \int_s^t \sigma(u)dw(u)\right][p(s) - \sigma(s)q(s)]ds$$

$$+ \int_a^t \exp\left[\int_s^t \left(f(u) - \frac{1}{2}\sigma^2(u)\right)du\right.$$

$$\left. + \int_s^t \sigma(u)dw(u)\right][q(s)dw(s)]ds \qquad \text{[from (2.3.40)]}. \tag{2.4.15}$$

Observation 2.4.1

(i) From (2.3.33) (Observation 2.3.1), we note that $x_c(t) = \Phi(t)c$ is the general solution to (2.4.2), where $\Phi(t)$ is the fundamental solution to (2.4.2), and c is an arbitrary constant. $x_c(t) = \Phi(t)c$ is also referred to as the *complementary solution* to (2.4.1). However, to find a particular solution to (2.4.1) in Step #2, we assumed that $x_p(t) = \Phi(t)c(t)$ is a particular solution to (2.4.1), where $c(t)$ is an unknown random process. The basic structure of the complementary solution and of the particular solution are the same. In the case of the particular solution, we treated the constant c in the complementary solution as an unknown function of the independent variable t as a parameter. Because of this, the method of finding a particular solution to (2.4.1) is called the *method of variation of constants parameters.*

(ii) Instead of using a general fundamental solution in (2.4.4), one can use the normalized fundamental solution process $\Phi(t, t_0)$ in (2.3.40) corresponding to (2.4.2), and can compute a particular solution to (2.4.1).

(iii) We further observe that the method of variation of constants parameters determines the general solution to (2.4.1). For instance, if $c_a = 0$ in (2.4.9), then the particular solution in (2.4.11) reduces to

$$
x_p(t) = \Phi(t)\left[\int_a^t \Phi^{-1}(s)[p(s) - \sigma(s)q(s)]ds + \int_a^t \Phi^{-1}(s)q(s)dw(s)\right]
$$

$$
= \int_a^t \Phi(t)\Phi^{-1}(s)[p(s) - \sigma(s)q(s)]ds
$$

$$
+ \int_a^t \Phi^1(t)\Phi^{-1}(s)q(s)dw(s) \quad \text{(by Theorem 1.3.1M2)}
$$

$$
= \int_a^t \Phi(t, s)[p(s) - \sigma(s)q(s)]ds
$$

$$
+ \int_a^t \Phi(t, s)(s)q(s)dw(s) \qquad \text{[from (2.3.40)]}. \tag{2.4.16}
$$

If c_a in (2.4.9) is any arbitrary constant c, then the particular solution in (2.4.11) is indeed a general solution to (2.4.1), i.e.

$$
x_p(t) = \Phi(t)\left[c + \int_a^t \Phi^{-1}(s)[p(s) - \sigma(s)q(s)]ds + \int_a^t \Phi^{-1}(s)q(s)dw(s)\right]
$$

$$
= \Phi(t)c + \int_a^t \Phi(t, s)[p(s) - \sigma(s)q(s)]ds + \int_a^t \Phi(t, s)q(s)dw(s). \tag{2.4.17}
$$

This solution process is equivalent to the general solution process described in (2.4.15).

For easy reference, this procedure is summarized in the following flowchart:

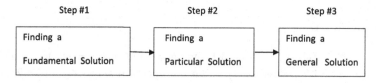

Fig. 2.4.1   Flowchart for finding general solution.

Observation 2.4.2.   In the following, we present an alternative procedure for finding a particular solution to (2.4.1). From the definition of a particular solution in (2.4.4), we have

$$
c(t) = \Phi^{-1}(t)x_p(t). \tag{2.4.18}
$$

Now, by using Theorems 1.3.4 and 2.3.1 and by assuming that $x_p(t)$ is a particular solution to (2.4.1), we have

$$dc(t) = d\Phi^{-1}(t)x_p(t) + \Phi^{-1}(t)dx_p(t) + d\Phi^{-1}(t)dx_p(t) \qquad \text{(by Theorem 1.3.4)}$$

$$= \left[\left[-f(t) + \sigma^2(t)\right]\Phi^{-1}(t)dt - \sigma(t)\Phi^{-1}(t)dw(t)\right]x_p(t)$$

$$+ \Phi^{-1}(t)[[f(t)x_p(t) + p(t)]dt$$

$$+ [\sigma(t)x_p(t) + q(t)]dw(t)] + [-\sigma(t)\Phi^{-1}(t)dw(t)][\sigma(t)x_p(t)$$

$$+ q(t)]dw(t)] \qquad \text{(by simplification of } d\Phi^{-1})$$

$$= \Phi^{-1}(t)[p(t) - \sigma(t)q(t)]dt + \Phi^{-1}(t)q(t)dw(t) \qquad \text{(by simplifications).}$$

$$(2.4.19)$$

This stochastic differential equation is exactly the same as the differential equation (2.4.9). One can imitate the rest of the procedure as described before to determine the particular solution process of (2.4.1).

Example 2.4.1. Given: $dx = (fx + p)dt + (\sigma x + q)dw(t)$. Find: (a) a particular solution and (b) a general solution to the given differential equation.

Solution procedure. First, let us compare the coefficients of "dt and $dw(t)$" of the given differential with the corresponding coefficients of the differential equation (2.4.1). Here, $f(t) \equiv f$, $\sigma(t) \equiv \sigma$, $p(t) \equiv p$, and $q(t) \equiv q$ are constant functions.

From Step #1 of Procedure 2.4.2 and Example 2.3.3, a fundamental solution process of the corresponding first-order linear homogeneous scalar stochastic differential equation

$$dx = fx\,dt + \sigma x\,dw(t)$$

is given by

$$\Phi(t) = \Phi^d(t)\Phi^s(t) = \exp\left[\left(f - \frac{1}{2}\sigma^2\right)(t - a) + \sigma(w(t) - w(a))\right].$$

By applying the method of variation of constants parameters and Observation 2.4.1(iii), the particular solution to the given differential equation is

$$x_p(t) = \Phi(t)\left[\int_a^t \Phi^{-1}(s)(p - \sigma q)ds + \int_a^t \Phi^{-1}(s)q\,dw(s)\right]$$

$$= (p - \sigma q)\int_a^t \exp\left[\left(f - \frac{1}{2}\sigma^2\right)(t - s) + \sigma(w(t) - w(s))\right]ds$$

$$+ q\int_a^t \exp\left[\left(f - \frac{1}{2}\sigma^2\right)(t - s) + \sigma(w(t) - w(s))\right]dw(s) \qquad \text{(by substitution).}$$

Moreover, from (2.4.17), the general solution to the given differential equation is

$$x(t) = x_c(t) + x_p(t)$$

$$x(t) = \exp\left[\left(f - \frac{1}{2}\sigma^2\right)(t-a) + \sigma(w(t) - w(a))\right]c$$

$$+ (p - \sigma q)\int_a^t \exp\left[\left(f - \frac{1}{2}\sigma^2\right)(t-s) + \sigma(w(t) - w(s))\right]ds$$

$$+ q\int_a^t \exp\left[\left(f - \frac{1}{2}\sigma^2\right)(t-s) + \sigma(w(t) - w(s))\right]dw(s) \quad [\text{by } (2.4.15)].$$

This, together with the integral of the third term (Example 1.4.5) in the above equation,

$$q\int_a^t \exp\left[\left(f - \frac{1}{2}\sigma^2\right)(t-s) + \sigma(w(t) - w(s))\right]dw(s)$$

$$= -\frac{q}{\sigma} + \frac{q}{\sigma}\exp\left[\left(f - \frac{1}{2}\sigma^2\right)(t-a) + \sigma(w(t) - w(a))\right]$$

$$+ \frac{q}{\sigma}(\sigma^2 - f)\int_a^t \exp\left[\left(f - \frac{1}{2}\sigma^2\right)(t-s) + \sigma(w(t) - w(s))\right]ds$$

$$(\text{by Example 1.4.5}),$$

yields the solution

$$x(t) = \exp\left[\left(f - \frac{1}{2}\sigma^2\right)(t-a) + \sigma(w(t) - w(a))\right]\left(c + \frac{q}{\sigma}\right) - \frac{q}{\sigma}$$

$$+ \left(p - \sigma q + \frac{q}{\sigma}(\sigma^2 - f)\right)\int_a^t \exp\left[\left(f - \frac{1}{2}\sigma^2\right)(t-s) + \sigma(w(t) - w(s))\right]ds$$

$$= \exp\left[\left(f - \frac{1}{2}\sigma^2\right)(t-a) + \sigma(w(t) - w(a))\right]\left(c + \frac{q}{\sigma}\right) - \frac{q}{\sigma}$$

$$+ \frac{\sigma p - qf}{\sigma}\int_a^t \exp\left[\left(f - \frac{1}{2}\sigma^2\right)(t-s) + \sigma(w(t) - w(s))\right]ds$$

and completes our goal.

Example 2.4.2. Find a general solution to $dx = (-fx + p)dt + (\sigma x + q)dw(t)$.

Solution procedure. To avoid the repetitiveness, we just follow the argument used in the solution process of Example 2.4.1. Here, we have $f(t) \equiv -f$, $\sigma(t) \equiv \sigma$, $p(t) \equiv p$ and $q(t) \equiv q$ as constant functions; moreover, from Example 2.3.4, the fundamental solution corresponding to the homogeneous differential equation

$$dx = -fx\, dt + \sigma x\, dw(t)$$

is

$$\Psi(t) = \Phi^d(t)\Phi^s(t) = \exp\left[\left(-f - \frac{1}{2}\sigma^2\right)(t-a) + \sigma(w(t) - w(a))\right].$$

Imitating the steps outlined in the solution procedure of Example 2.4.1, the particular and the general solution of the given differential are represented by

$$x_p(t) = (p - \sigma q)\int_a^t \exp\left[\left(-f - \frac{1}{2}\sigma^2\right)(t-s) + \sigma(w(t) - w(a))\right]ds$$

$$+\, q\int_a^t \exp\left[\left(-f - \frac{1}{2}\sigma^2\right)(t-s) + \sigma(w(t) - w(a))\right]dw(s)$$

(by substitution),

which gives the solution

$$x(t) = \exp\left[\left(-f - \frac{1}{2}\sigma^2\right)(t-a) + \sigma(w(t) - w(a))\right]\left(c + \frac{q}{\sigma}\right) - \frac{q}{\sigma}$$

$$+\, \frac{\sigma p - qf}{\sigma}\int_a^t \exp\left[\left(-f - \frac{1}{2}\sigma^2\right)(t-s) + \sigma(w(t) - w(s))\right]ds.$$

Example 2.4.3. Let g and h be any given continuous functions defined on J, and let

$$\sqrt{1+t^2}dx = \left[\sqrt{1+t^2}\sin 2tx + g(t)\right]dt + [\sigma x + h(t)]dw(t).$$

Find a general solution to the given differential equation. Moreover, find the general solution to the given differential equation if (a) $g(t)\sqrt{1+t^2} - \sigma h(t) = D(1+t^2)$, where D is a real number, and (b) $D = 0$.

Solution procedure. The goal of the example is to find:

(i) A particular solution.
(ii) A general solution to the given differential equation.
(iii) Solutions for the particular cases (a) and (b).

Following the argument used in Example 2.3.6, we need to rewrite the differential in the "*standard form* (2.4.1)."

$$dx = \left[\sin 2tx + \frac{g(t)}{\sqrt{1+t^2}}\right]dt + \left[\sigma\frac{1}{\sqrt{1+t^2}}x + \frac{h(t)}{\sqrt{1+t^2}}\right]dw(t).$$

Here $f(t) = \sin 2t$, $p(t) = \frac{g(t)}{\sqrt{1+t^2}}$, $\sigma(t) = \frac{\sigma}{\sqrt{1+t^2}}$ and $q(t) = \frac{h(t)}{\sqrt{1+t^2}}$.

From Step #1 of Procedure 2.4.2 and Example 2.3.6, a fundamental solution process of the corresponding first-order linear homogeneous scalar stochastic differential equation

$$dx = \sin 2tx\, dt + \sigma \frac{1}{\sqrt{1+t^2}} x\, dw(t)$$

is

$$\Phi(t) = \exp\left[\frac{1}{2}(\cos 2t - \cos 2a) - \frac{1}{2}\sigma^2(\tan^{-1} t - \tan^{-1} a) + \sigma \int_a^t \frac{1}{\sqrt{1+s^2}} dw(s)\right].$$

By applying the method of variation of constants parameters and Observation 2.4.1(iii), the particular solution to the given differential equation is

$$x_p(t) = \Phi(t)\left[\int_a^t \Phi^{-1}(s)(p(s) - \sigma(s)q(s))ds + \int_a^t \Phi^{-1}(s)q(s)dw(s)\right],$$

where $\Phi(t)$ is as described above, $p(t) = \frac{g(t)}{\sqrt{1+t^2}}$, $\sigma(t) = \frac{\sigma}{\sqrt{1+t^2}}$, and $q(t) = \frac{h(t)}{\sqrt{1+t^2}}$.

Moreover, from (2.4.17), the general solution to the given differential equation is

$$x(t) = x_c(t) + x_p(t)$$

$$= \Phi(t)c + \int_a^t \Phi(t, s)[p(s) - \sigma(s)q(s)]ds + \int_a^t \Phi(t, s)q(s)dw(s),$$

where

$$\Phi(t, s) = \exp\left[\frac{1}{2}(\cos 2t - \cos 2a) - \frac{1}{2}\sigma^2(\tan^{-1} t - \tan^{-1} s) + \sigma \int_a^t \frac{1}{\sqrt{1+s^2}} dw(s)\right].$$

In the case of (a), $g(t)\sqrt{1+t^2} - \sigma h(t) = D(1+t^2)$, where D is a real number, the above general solution reduces to

$$x(t) = \Phi(t)c + D\int_a^t \Phi(t, s)ds + \int_a^t \Phi(t, s)q(s)dw(s),$$

and in the case of (b), $D = 0$, it further reduces to

$$x(t) = \Phi(t)c + \int_a^t \Phi(t, s)q(s)dw(s)$$

Thus, it completes the analysis.

In the following subsection, we discuss a procedure for finding a particular solution to a linear nonhomogeneous stochastic differential equation of the Itô–Doob type. As before, this idea leads to the formulation of an IVP in the study of stochastic differential equations and its applications. Moreover, illustrations are provided to exhibit the usefulness of the concept of solving real-world problems.

2.4.3 *Initial Value Problem*

Let us formulate an IVP for the first-order linear nonhomogeneous stochastic differential equation of the Itô–Doob-type (2.4.1). We consider

$$dx = [f(t)x + p(t)]dt + [\sigma(t)x + q(t)]dw(t), \quad x(t_0) = x_0. \tag{2.4.20}$$

2.4.4 *Procedure for Solving the IVP*

In the following, we apply the procedure for finding a solution process of the first-order linear nonhomogeneous scalar stochastic differential equations of Itô–Doob (2.4.1) to the problem of finding the solution to the stochastic differential equations (2.4.20). Now, we briefly summarize the procedure for finding a solution to the IVP (2.4.20).

Step #1. By following the procedure for finding a general solution to (2.4.1), we have

$$x(t) = \Phi(t)\left[c + \int_a^t \Phi^{-1}(s)[p(s) - \sigma(s)q(s)]ds + \int_a^t \Phi^{-1}(s)q(s)dw(s)\right]$$

$$= \Phi(t)c + \int_a^t \Phi(t, s)[p(s) - \sigma(s)q(s)]ds + \int_a^t \Phi(t, s)q(s)dw(s), \tag{2.4.21}$$

where a, $t \in J \subseteq R$, $\Phi(t)$ is as defined (2.4.3), and $\Phi(t, s) = \Phi(t)\Phi^{-1}(s)$. Here, c is an arbitrary constant.

Step #2. In this step, we need to find an arbitrary constant c in (2.4.21). For this purpose, we utilize the given initial given data $(t_0, x_0) \in J \times R$ and solve the linear scalar algebraic equation

$$x(t_0) = \Phi(t_0)\left[c + \int_a^{t_0} \Phi^{-1}(s)[p(s) - \sigma(s)q(s)]ds + \int_a^{t_0} \Phi^{-1}(s)q(s)dw(s)\right]$$

$$= x_0. \tag{2.4.22}$$

Since $\exp\left[\int_a^{t_0} \left[f(s) - \frac{1}{2}\sigma^2(s)\right]ds + \int_a^{t_0} \sigma(s)dw(s)\right] \neq 0$ for any continuous functions f and σ, the algebraic equation (2.4.22) has a unique solution depending on $t_0, t \in J$, x_0 and rate coefficients in (2.4.1). This solution is given by

$$c = \Phi^{-1}(t_0)x_0 - \left[\int_a^{t_0} \Phi^{-1}(s)[p(s) - \sigma(s)q(s)]ds + \int_a^{t_0} \phi^{-1}(s)q(s)dw(s)\right]. \tag{2.4.23}$$

Step #3. The solution to the IVP (2.4.20) is determined by substituting the expression (2.4.23) into

$$x(t) = \Phi(t)c + \int_a^t \Phi(t,s)[p(s) - \sigma(s)q(s)]ds + \int_a^t \Phi(t,s)q(s)dw(s) \quad \text{[from (2.4.21)]}$$

$$= \Phi(t)\left[\Phi^{-1}(t_0)x_0 - \left[\int_0^{t_0} \Phi^{-1}(s)[p(s) - \sigma(s)q(s)]ds + \int_a^{t_0} \Phi^{-1}(s)q(s)dw(s)\right]\right]$$

$$+ \int_a^t \Phi(t,s)[p(s) - \sigma(s)q(s)]ds + \int_a^t \Phi(t,s)q(s)dw(s) \quad \text{[from (2.4.23)]}$$

$$= \Phi(t,t_0)x_0 - \int_a^{t_0} \Phi(t,s)[p(s) - \sigma(s)q(s)]ds - \int_a^{t_0} \Phi(t,s)q(s)dw(s)$$

$$+ \int_a^t \Phi(t,s)[p(s) - \sigma(s)q(s)]ds$$

$$+ \int_a^t \Phi(t,s)q(s)dw(s) \quad \text{[by Theorem 1.3.1M2 [80] and (2.3.40)]}$$

$$= \Phi(t,t_0)x_0 + \int_{t_0}^t \Phi(t,s)[p(s) - \sigma(s)q(s)]ds$$

$$+ \int_{t_0}^t \Phi(t,s)q(s)dw(s) \quad \text{(by definite integral properties and grouping)},$$

$$(2.4.24)$$

where

$$\Phi(t,w(t),t_0) = \Phi(t,w(t))\Phi^{-1}(t_0,w(t_0)) \equiv \Phi(t)\Phi^{-1}(t_0)$$

$$= \Phi(t,t_0) = \exp\left[\int_{t_0}^t \left[f(s) - \frac{1}{2}\sigma^2\right]ds + \int_{t_0}^t \sigma(s)dw(s)\right]. \quad (2.4.25)$$

Thus, the solution process $x(t) = x(t,t_0,x_0)$ of the IVP (2.4.20) in (2.4.24) is also represented by

$$x(t) = x(t,t_0,x_0) = \exp\left[\int_{t_0}^t \left[f(s) - \frac{1}{2}\sigma^2(s)\right]ds + \int_{t_0}^t \sigma(s)dw(s)\right]x_0$$

$$+ \int_{t_0}^t \exp\left[\int_s^t \left[f(u) - \frac{1}{2}\sigma^2(u)\right]du + \int_s^t \sigma(u)dw(u)\right][p(s) - \sigma(s)q(s)]ds$$

$$+ \int_{t_0}^t \exp\left[\int_s^t \left[f(u) - \frac{1}{2}\sigma^2(u)\right]du + \int_s^t \sigma(u)dw(u)\right]q(s)dw(s). \quad (2.4.26)$$

The solution (2.4.26) is called as a *particular solution* to (2.4.1).

For easy reference, this procedure is summarized in the following flowchart:

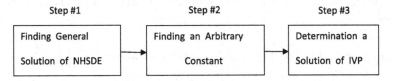

Fig. 2.4.2   Flowchart for finding particular solution

Now, a few examples are given out to demonstrate the procedure. In addition, a few illustrations are presented to exhibit the usefulness of the concept of the IVP and its applications.

Example 2.4.4.   Given: $dx = (fx + p)dt + (\sigma x + q)dw(t)$, $x(t_0) = x_0$. Find the solution to the given IVP: (a) for any f, p, q, and σ such that $\sigma \neq 0$ and (b) for f, p, q, and σ such that $f \neq 0$, $\sigma \neq 0$, and $\frac{p}{f} - \frac{q}{\sigma} = 0$.

Solution procedure.   From Step #1 and Example 2.4.1, the general solution to the given nonhomogeneous differential equation is

$$x(t) = \exp\left[\left(f - \frac{1}{2}\sigma^2\right)(t - a) + \sigma(w(t) - w(a))\right]\left(c + \frac{q}{\sigma}\right) - \frac{q}{\sigma}$$
$$+ \frac{\sigma p - fq}{\sigma}\int_a^t \exp\left[\left(f - \frac{1}{2}\sigma^2\right)(t - s) + \sigma(w(t) - w(s))\right]ds.$$

Following Step #2, the unknown constant is determined by

$$x_0 = \exp\left[\left(f - \frac{1}{2}\sigma^2\right)(t_0 - a) + \sigma(w(t_0) - w(a))\right]\left(c + \frac{q}{\sigma}\right) - \frac{q}{\sigma}$$
$$+ \frac{\sigma p - fq}{\sigma}\int_a^{t_0} \exp\left[\left(f - \frac{1}{2}\sigma^2\right)(t_0 - s) + \sigma(w(t) - w(s))\right]ds,$$

which can be solved for $c + \frac{q}{\sigma}$, and we get

$$c + \frac{q}{\sigma} = \exp\left[\left(f - \frac{1}{2}\sigma^2\right)(a - t_0) + \sigma(w(a) - w(t_0))\right]\left(x_0 + \frac{q}{\sigma}\right)$$
$$- \frac{\sigma p - fq}{\sigma}\int_a^{t_0} \exp\left[\left(f - \frac{1}{2}\sigma^2\right)(a - s) + \sigma(w(t_0) - w(s))\right]ds.$$

Now, we substitute this expression for $c + \frac{q}{\sigma}$ into the general solution. After algebraic simplification, we obtain

$$x(t) = \exp\left[\left(f - \frac{1}{2}\sigma^2\right)(t - t_0) + \sigma(w(t) - w(t_0))\right]\left(x_0 + \frac{q}{\sigma}\right) - \frac{q}{\sigma}$$
$$+ \frac{\sigma p - fq}{\sigma}\int_{t_0}^t \exp\left[\left(f - \frac{1}{2}\sigma^2\right)(t - s) + \sigma(w(t) - w(s))\right]ds.$$

This is the desired solution to the given IVP with respect to (a). The solution to the IVP that satisfies the conditions in (b) is deduced from the above solution expression,

$$x(t) = \exp\left[\left(f - \frac{1}{2}\sigma^2\right)(t - t_0) + \sigma(w(t) - w(t_0))\right]\left(x_0 + \frac{q}{\sigma}\right) - \frac{q}{\sigma}$$
$$+ \frac{\sigma f}{\sigma}\left(\frac{p}{f} - \frac{q}{\sigma}\right)\int_{t_0}^{t}\exp\left[\left(f - \frac{1}{2}\sigma^2\right)(t - s) + \sigma(w(t) - w(s))\right]ds,$$

as follows:

$$x(t) = \exp\left[\left(f - \frac{1}{2}\sigma^2\right)(t - t_0) + \sigma(w(t) - w(t_0))\right]x_0$$
$$+ \frac{q}{\sigma}\left(\exp\left[\left(f - \frac{1}{2}\sigma^2\right)(t - t_0) + \sigma(w(t) - w(t_0))\right] - 1\right).$$

This completes the goal of the problem.

Example 2.4.5. Solve the IVP $dx = (-fx + p)dt + (\sigma x - q)dw(t)$, $x(t_0) = x_0$, (a) for any given f, p, q, and σ such that $\sigma \neq 0$, and (b) for f, p, q, and σ such that $f \neq 0$, $\sigma \neq 0$, and $\frac{p}{f} - \frac{q}{\sigma} = 0$.

Solution procedure. From Step #1 and Example 2.4.2, with a slight modification in sign, the general solution to the given nonhomogeneous differential equation is deduced from

$$x(t) = \exp\left[-\left(f + \frac{1}{2}\sigma^2\right)(t - a) + \sigma(w(t) - w(a))\right]\left(c - \frac{q}{\sigma}\right) + \frac{q}{\sigma}$$
$$+ \left(p + \sigma q - \frac{q}{\sigma}(\sigma^2 + f)\right)\int_{a}^{t}\exp\left[-\left(f + \frac{1}{2}\sigma^2\right)(t - s) + \sigma(w(t) - w(s))\right]ds$$
$$= \exp\left[-\left(f + \frac{1}{2}\sigma^2\right)(t - a) + \sigma(w(t) - w(a))\right]\left(c - \frac{q}{\sigma}\right) + \frac{q}{\sigma}$$
$$+ \frac{\sigma p - fq}{\sigma}\int_{a}^{t}\exp\left[-\left(f + \frac{1}{2}\sigma^2\right)(t - s) + \sigma(w(t) - w(s))\right]ds.$$

Imitating the argument used in Example 2.4.4, we obtain

$$x(t) = \exp\left[\exp\left[-\left(f + \frac{1}{2}\sigma^2\right)(t - t_0) + \sigma(w(t) - w(t_0))\right]\right]\left(x_0 - \frac{q}{\sigma}\right) + \frac{q}{\sigma}$$
$$+ \frac{\sigma p - fq}{\sigma}\int_{t_0}^{t}\exp\left[-\left(f + \frac{1}{2}\sigma^2\right)(t - s) + \sigma(w(t) - w(s))\right]ds.$$

This is the desired solution to the given IVP with respect to (a). The solution to the IVP that satisfies the conditions in (b) is deduced from the solution expression in (a) as:

$$x(t) = \exp\left[\exp\left[-\left(f + \frac{1}{2}\sigma^2\right)(t - t_0) + \sigma(w(t) - w(t_0))\right]\right]\left(x_0 - \frac{q}{\sigma}\right) + \frac{q}{\sigma}$$

$$+ \frac{\sigma f}{\sigma}\left(\frac{q}{f} - \frac{q}{\sigma}\right)\int_{t_0}^{t}\exp\left[-\left(f + \frac{1}{2}\sigma^2\right)(t - s) + \sigma(w(t) - w(s))\right]ds,$$

which implies that

$$x(t) = \exp\left[\exp\left[-\left(f + \frac{1}{2}\sigma^2\right)(t - t_0) + \sigma(w(t) - w(t_0))\right]\right]x_0$$

$$+ \frac{q}{\sigma}\left(1 - \exp\left[-\left(f + \frac{1}{2}\sigma^2\right)(t - t_0) + \sigma(w(t) - w(t_0))\right]\right).$$

This completes the goal of the problem.

Illustration 2.4.1 (Newton's law of cooling under random perturbations [80, 82]). Under the controlled experiment, we recall Newton's law of cooling, which was outlined in Illustration 2.5.2 [80]. In general, the experimental conditions are subject to random fluctuations. These random fluctuations are due to heat transfer near the surface of the body, which creates thermal agitation. As a result of this, Newton's law of the cooling process can be considered to be a decay process that is described by Poisson Process Model 2.1.2. By utilizing this model, the dynamic of a cooling body under random environmental perturbation is described by the following IVP.

$$dT(t) = -k(T(t) - T_s)dt + \sigma(T(t) - T_s)dw(t), \qquad T(t_0) = T_0$$

$$= (-kT(t) + kT_s)dt + (\sigma T(t) - \sigma T_s)dw(t), \quad T(t_0) = T_0,$$

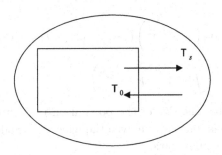

Fig. 2.4.3

where T_s is the constant temperature of the surrounding medium and T_0 is an initial temperature of a body at an initial time t_0 without an internal heating system; k is the coefficient of cooling; and $w(t)$ is the normalized Wiener process defined before.

By following the solution procedure of Example 2.4.5 with $-f = -k$, $p = kT_s$, and $-q = -\sigma T_s$, the temperature process is given by

$$T(t) = \exp\left[-\left(k + \frac{1}{2}\sigma^2\right)(t - t_0) + \sigma(w(t) - w(t_0))\right]\left(T_0 - \frac{\sigma T_s}{\sigma}\right)$$

$$+ \frac{\sigma T_s}{\sigma} + \left(kT_s + \sigma^2 T_s - \frac{\sigma T_s}{\sigma}(\sigma^2 + k)\right)$$

$$\times \int_{t_0}^{t} \exp\left[-\left(k + \frac{1}{2}\sigma^2\right)(t - s) + \sigma(w(t) - w(s))\right]ds$$

$$= \exp\left[-\left(k + \frac{1}{2}\sigma^2\right)(t - t_0) + \sigma(w(t) - w(t_0))\right]T_0$$

$$+ T_s\left(1 - \exp\left[-\left(k + \frac{1}{2}\sigma^2\right)(t - t_0) + \sigma(w(t) - w(t_0))\right]\right).$$

Illustration 2.4.2 (birth–death under random immigration process [80, 82]). From Example 2.1.3, Illustration 2.5.1 [80] and Observation 2.1.5, we have

$$dx = (fx + p)dt + (\sigma x + q)dw(t), \quad x(t_0) = x_0,$$

where x stands for the number of species in a given population. Here, $C(t, x) = fx + p$ and $\sigma(t, x) = \sigma x + q$; f is the expected intrinsic growth rate (birth rate − death rate); p is the expected/deterministic/permitted/legal immigration rate of population; the random fluctuations are characterized by the Wiener process $w(t)$; and σ and q are the magnitudes of the specific growth rate of the population and the constant random/undocumented/illegal immigration rate under random environmental perturbations, respectively. x_0 is the initial size of the population at the initial time t_0. By following the discussion of the solution process of the IVP in Example 2.4.4, we have

$$x(t) = \exp\left[\left(f - \frac{1}{2}\sigma^2\right)(t - t_0) + \sigma(w(t) - w(t_0))\right]x_0$$

$$+ \frac{q}{\sigma}\left(\exp\left[\left(f - \frac{1}{2}\sigma^2\right)(t - t_0) + \sigma(w(t) - w(t_0))\right] - 1\right)$$

$$+ \frac{\sigma p - fq}{\sigma}\int_{t_0}^{t}\exp\left[\left(f - \frac{1}{2}\sigma^2\right)(t - s) + \sigma(w(t) - w(s))\right]ds.$$

This expression describes the size of the population under random environmental perturbations. From this, one can analyze the effects of random and illegal immigration on the size of the population.

Illustration 2.4.3 (stochastic arterial pulse-systolic process [80, 172]). This is a continuation of Illustration 2.3.2, and an extension of Illustration 2.5.6, Volume 1 [80]. We imitate the description of Illustration 2.5.6 and notations in the context of the stochastic version of Poiseuille's law. Let us assume that the volume

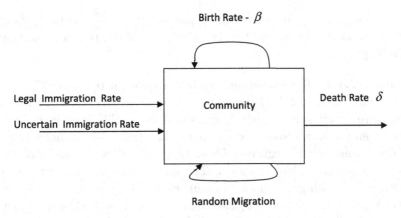

Fig. 2.4.4

of the blood entering the Windkessel during the interval of length Δt is $p(t)\Delta t + q(t)\Delta w(t)$. Here, the functions $p(t)$ and $q(t)$ are continuous on a specified interval. Thus,

$$\text{(Net Change in Volume)} = \text{(Volume Change Due to Inflow)}$$
$$- \text{(Volume Change Due to Outflow)},$$
$$\Delta V = [p(t)\Delta t + q(t)\Delta w(t)] - \frac{P}{\beta}\Delta t + \sigma P\Delta w(t)$$
$$\Delta P = \alpha \Delta V.$$

Using the above two expressions, we eliminate V, and obtain

$$dP = \left(-\frac{\alpha}{\beta}P + \alpha p(t)\right)dt + (\sigma \alpha P + \alpha q(t))dw(t), \quad P(0) = P_0.$$

Now, we utilize Procedure 2.4.4, and we obtain

$$P(t) = P_0 \exp\left[-\left(\frac{\alpha}{\beta} + \frac{1}{2}\sigma^2\alpha^2\right)t + \sigma\alpha w(t)\right]$$
$$+ \int_0^t \exp\left[-\left(\frac{\alpha}{\beta} + \frac{1}{2}\sigma^2\alpha^2\right)(t-s) + \sigma\alpha(w(t)-w(s))\right][\alpha p(s) - \sigma\alpha^2 q(s)]ds$$
$$+ \int_0^t \exp\left[-\left(\frac{\alpha}{\beta} + \frac{1}{2}\sigma^2\alpha^2\right)(t-s) + \sigma\alpha(w(t)-w(s))\right]\alpha q(s)dw(s).$$

This describes the pressure-time curve during the systolic phase. Moreover, it signifies the pressure drop in the aorta during the diastolic period. $\frac{2\beta}{2\alpha+\sigma^2\alpha^2} = \tau$ is referred to as the time constant. The function $e(t) = p(t) + q(t)\xi(t)$ is called the forcing process and $\xi(t)$ the white noise process. $\xi(t) = \frac{dw(t)}{dt}$ is the derivative of the Wiener process. This completes the description of the pressure–time curve in the systolic phase.

Illustration 2.4.4 (cell membrane under random fluctuations [57, 80, 155]). The body fluid is full of both positive- and negative-charged ions. Because of the ionic permeability of the membrane as well as the difference in ionic composition between the intra- and extra-cellular fluids, the potential difference across the resting membrane is negative. These types of differences are controlled by a sodium–potassium exchange system. The ionic permeability of the membrane depends on: (a) opening and closing of ionic channels across the membrane (ionic mobility across the membrane), (b) membrane thickness, and (c) the partition coefficient between the aqueous extra- and intra-cellular fluids and the membrane. It is easier to measure the current carried by ions than the ionic charge. As a result of this, it is natural to consider membrane conductance rather than membrane permeability.

We consider the RC circuit model (Exercise 2.1.6) as a prototype model for the cell membrane. Here, the membrane acts like a resistance (the reciprocal of membrane conductance) and a capacitance in parallel. From Exercise 2.1.6 and Illustration 2.5.4 [80], we have

$$dI_a = -\frac{1}{C_m} I_a dt,$$

where a is the ionic species which can pass through the channel; $I(t) = -\frac{d}{dt} Q(t)$; $C_m \Delta E(t) = Q(t)$; ΔE_m is the voltage difference [electromotive force/voltage E (volts)] across the membrane; C_m is the membrane capacity/per unit area, it is assumed to be constant; $\Delta E(t) = I(t) R_m$ (Ohm's law), with $R_m = \frac{1}{P_a}$, $P_a = \frac{\mu \beta RT}{DF}$, where R is the universal gas constant, T, the absolute temperature, F, Faraday's constant, D, the thickness of the homogeneous membrane, μ, the ionic mobility, β, the partition coefficient, and $\frac{\Delta E}{D}$, the potential constant; and $I_a = P_a \frac{F^2 \Delta E(t)}{RT} \left[\frac{[a]_0 - [a]_i \exp\left[-\frac{F \Delta E}{RT}\right]}{1 - \exp\left[-\frac{F \Delta E}{RT}\right]} \right]$ here, with $[a]_0$ and $[a]_i$ being ionic concentrations outside and inside the cell, respectively. The influence of random fluctuations is due to the changes in ionic conductance. These random perturbations provide insights into the opening and closing of the kinetics of ionic channels. The main ionic species are sodium (Na^+), potassium (K^+), and chloride (Cl^-) ions. The resting membrane potential of the membrane depends on the concentrations of Na^+, K^+, Cl^- and permeability. The types of random fluctuations affecting the membrane potential can be characterized by (a) thermal noise, (b) shot noise, (c) flicker noise, and (d) conductance fluctuations. Most of these random perturbations can be approximated by the Wiener process. With this understanding, the above-described differential equation reduces to the following simplified version of Itô–Doob-type stochastic differential equations:

$$dI_a = (-fI_a + p)dt + (\sigma I_a - q)dw(t),$$

where w is a Wiener process and f, p, σ, and q are rate constants. By imitating the solution procedure of Example 2.4.5, we have

$$I_a(t) = \exp\left[-\left(f + \frac{1}{2}\sigma^2\right)(t - t_0) + \sigma(w(t) - w(t_0))\right]\left(I_{a0} - \frac{q}{\sigma}\right) + \frac{q}{\sigma}$$

$$+ \frac{\sigma f}{\sigma}\left(\frac{p}{f} - \frac{q}{\sigma}\right)\int_{t_0}^{t}\exp\left[-\left(f + \frac{1}{2}\sigma^2\right)(t - s) + \sigma(w(t) - w(s))\right]ds.$$

Under the following conditions on the coefficients, for f, p, q, and σ such that $f \neq 0$, $\sigma \neq 0$, and $\frac{p}{f} - \frac{q}{\sigma} = 0$, we get

$$I_a(t) = \exp\left[-\left(f + \frac{1}{2}\sigma^2\right)(t - t_0) + \sigma(w(t) - w(t_0))\right]I_{a0}$$

$$+ \frac{q}{\sigma}\left(1 - \exp\left[-\left(f + \frac{1}{2}\sigma^2\right)(t - t_0) + \sigma(w(t) - w(t_0))\right]\right).$$

Illustration 2.4.5 (stochastic diffusion process [41, 80]). The mathematical model of the diffusion process described in Illustration 2.5.3, Volume 1 [80] was under a controlled environment. This illustration allows us to incorporate the imperfectness into assumptions. As a result of this, the experimental conditions are subject to random fluctuations. These random fluctuations are due to the imperfect conditions for maintaining the homogeneity of the liquid in the surrounding medium of the cell and the membrane resistance to the solute flow causing generation heat and thermal agitation. This affects the cell membrane permeability (the diffusion coefficient k) and solute flow (Illustration 2.5.3 [80]). As a result, Fick's law of the diffusion process can be considered as a diffusion process that is described by Brownian Motion Model 2.1.3. Utilizing this model, the dynamic of diffusion of the solute in the cell under random environmental perturbation is described by the following IVP:

$$dc(t) = -\frac{kA}{V}(c(t) - C)dt + \sigma(c(t) - C)dw(t), \quad c_0(t_0) = c_0$$

$$= \left(-\frac{kA}{V}c(t) + \frac{kA}{V}C\right)dt + (\sigma c(t) - \sigma C)dw(t), \quad c_0(t_0) = c_0,$$

where $c(t)$ is the concentration of the solute S inside the cell and $w(t)$ is the normalized Wiener process. Following the solution procedure of Example 2.4.5 with

$f = \frac{kA}{V}$, $p = \frac{kA}{V}C$, and $q = \sigma C$, the concentration process is given by

$$c(t) = \exp\left[-\left(\frac{kA}{V} + \frac{1}{2}\sigma^2\right)(t - t_0) + \sigma(w(t) - w(t_0))\right](c_0 - C) + C$$

$$+ \left(\frac{kA}{V}C + \sigma^2 C - \frac{\sigma C}{\sigma}\left(\sigma^2 + \frac{kA}{V}\right)\right)$$

$$\times \int_{t_0}^{t} \exp\left[-\left(\frac{kA}{V} + \frac{1}{2}\sigma^2\right)(t - s) + \sigma(w(t) - w(s))\right] ds$$

$$= \exp\left[-\left(\frac{kA}{V} + \frac{1}{2}\sigma^2\right)(t - t_0) + \sigma(w(t) - w(t_0))\right] x_0$$

$$+ C\left(1 - \exp\left[-\left(\frac{kA}{V} + \frac{1}{2}\sigma^2\right)(t - t_0) + \sigma(w(t) - w(t_0))\right]\right).$$

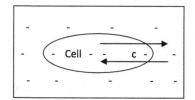

Fig. 2.4.5

Illustration   2.4.6 (central nervous system under random perturbations).   We imitate the description of the modeling of deterministic dynamic of the central nervous system (CNS) in Illustration 2.5.5, Volume 1 [80]. The source of randomness is due to the fluctuations of different physicochemical conditions which affect the transmission at synapses. The protoplasm contains a large number of different ions, cations (positive electric charge) and anions (negative electric charge). The stimulus to the nerve cell generates an electric current, causing the cations to move to the cathode, and the anions to the anode. This leads to the excitation/inhibition of nerve cells at the cathode.

Again, it is assumed that the constant stimulus intensity rate I of excitation/inhibition in a pathway per unit time is measured in the electric current. Let $\epsilon = \epsilon(t)$ be the concentration of the exciting cations and $\iota = \iota(t)$ be the concentration of the inhibiting cations near the cathode at a time t. Based on the theoretical understanding, there is a threshold for the excitation/inhibition process. Let us denote this threshold value by c. Using c, the excitation and the inhibition state of cations are described by

$$\frac{\epsilon}{\iota} \geq c \text{ excitation}, \quad \frac{\epsilon}{\iota} < c \text{ inhibition}.$$

Let ϵ_0 and ι_0 be the concentrations at the rest of the exciting and inhibiting cations, respectively. The dynamic motion of cations and anions can be described by the birth and death processes described in Example 2.1.3,

$$\epsilon(t + \Delta t) - \epsilon(t) = [AI\Delta t - A_1 I \Delta w(t)] + [-a\epsilon(t)\Delta t + a_1\epsilon(t)\Delta w(t)],$$

where A, A_1, a, a_1, and I are positive constants; $w(t)$ is a normalized Wiener process; $[AI\Delta t - A_1 I \Delta w(t)]$ characterizes the increase in the concentration of exciting cations (like the birth process) and $[-a\epsilon(t)\Delta t + a_1\epsilon(t)\Delta w(t)]$ characterizes the increase in the concentration of exciting cations (like the decay process). From this we have

$$d\epsilon(t) = (AI - a\epsilon(t))dt + (a_1\epsilon(t) - A_1 I)dw(t), \quad \epsilon(0) = \epsilon_0.$$

By imitating the above description, the mathematical model for the inhibition process is given by

$$d\iota(t) = (BI - b\iota(t))dt + (b_1\iota(t) - B_1 I)dw(t), \quad \iota(0) = \iota_0,$$

where B, B_1, b, b_1, and I are positive constants; $w(t)$ is a standard Wiener process; $[BI\Delta t - B_1 I \Delta w(t)]$ characterizes the increase in the concentration of inhibiting cations (like the birth process) and $[-b\iota(t)\Delta t + b_1\iota(t)\Delta w(t)]$ characterizes the increase in the concentration of inhibiting cations (like the decay process).

By following the solution procedure of Example 2.4.5, we have

$$\epsilon(t) = \exp\left[-\left(a + \frac{1}{2}a_1^2\right)t + a_1 w(t)\right]\epsilon_0$$
$$- \frac{A_1 I}{a_1}\left(1 - \exp\left[-\left(a + \frac{1}{2}a_1^2\right)t + a_1 w(t)\right]\right)$$
$$+ aI\left(\frac{A}{a} - \frac{A_1}{a_1}\right)\int_{t_0}^t \exp\left[-\left(a + \frac{1}{2}a_1^2\right)(t - s) + a_1(w(t) - w(s))\right]ds,$$

and

$$\iota(t) = \exp\left[-\left(b + \frac{1}{2}b_1^2\right)t + b_1 w(t)\right]\iota_0$$
$$- \frac{B_1 I}{b_1}\left(1 - \exp\left[-\left(b + \frac{1}{2}b_1^2\right)t + b_1 w(t)\right]\right)$$
$$+ bI\left(\frac{B}{b} - \frac{B_1}{b_1}\right)\int_{t_0}^t \exp\left[-\left(b + \frac{1}{2}b_1^2\right)(t - s) + b_1(w(t) - w(s))\right]ds.$$

Under the conditions on the coefficients $\frac{A}{a} - \frac{A_1}{a_1} = \frac{B}{b} - \frac{B_1}{b_1} = 0$, the above expressions for concentration processes of exciting and inhibiting cations reduce to

$$\epsilon(t) = \exp\left[-\left(a + \frac{1}{2}a_1^2\right)t + a_1 w(t)\right]\left[\epsilon_0 + \frac{A_1 I}{a_1}\right] - \frac{A_1 I}{a_1},$$

and

$$\iota(t) = \exp\left[-\left(b + \frac{1}{2}b_1^2\right)t + b_1 w(t)\right]\left[\iota_0 + \frac{B_1 I}{b_1}\right] - \frac{B_1 I}{b_1},$$

respectively.

Exercises

1. Find the general and fundamental solutions to the following differential equations:

   (a) $dx = (-2x + 3)dt$;
   (b) $dx = (-x + 5)dw(t)$;
   (c) $dx = (-2x + 3)dt + (-x + 5)dw(t)$;
   (d) $dx = (2x - 3)dt$;
   (e) $dx = (2x - 3)dt + (x - 5)dw(t)$;
   (f) $dx = (x - 5)dw(t)$;
   (g) $dx = (2x + 3)dt$;
   (h) $dx = (2x + 5)dw(t)$;
   (i) $dx = (2x + 3)dt + (2x - 5)dw(t)$;
   (j) $dx = (bx - q)dt$;
   (k) $dx = (bx - p)dt + \left(\sqrt{b}x - q\right)dw(t)$;
   (l) $dx = \left(\sqrt{b}x - q\right)dw(t)$;
   (m) $dx = (-bx + p)dt$;
   (n) $dx = \left(-\sqrt{b}x - q\right)dw(t)$;
   (o) $dx = (-bx + p)dt - \left(\sqrt{b}x + q\right)dw(t)$;
   (p) $dx = \sqrt{6}xdw(t)$;
   (q) $dx = 2x\,dt + \sqrt{6}x\,dw(t)$.

2. Find the general solution to the following differential equations:

   (a) $(1 + t^2)dx = \left(t^2x + \frac{2\sin t}{1+t^2}\right)dt + \left(2x + \frac{1}{2}\sin t\right)dw(t)$;
   (b) $(90 + t)dx = (3x + p)dt + (90 + t)\left(\sigma x - \frac{q}{90+t}\right)dw(t)$;
   (c) $\sin t\,dx = (\cos tx + \sin 2t)dt$;
   (d) $\sqrt{1 + t^2}dx = (1 + t^2)(x + 2)dt - \left(\sigma x + \sqrt{1 + t^2}\right)dw(t)$;
   (e) $dx = \tan t(2x - b\cos t)dt - \sqrt{b}(x + \sin t)dw(t)$;
   (f) $(1 + \sin^4 t)dx = (\sin 2tx + 1 + \sin^4 t)dt + (1 + \sin^4 t)(\sin tx + \csc t)dw(t)$;
   (g) $dx = (\ln(1 + t)x + p(t))dt - x\,dw(t)$;
   (h) $\frac{1+t}{1+t^2}dx = (x + \sin 2t)dt$.

3. Solve the following initial value problems:

   (a) $dx = (5 - 4x)dt$, $x(0) = 2$;
   (b) $dx = (2x - 1)dw(t)$, $x(0) = 2$;
   (c) $dx = (5 - 4x)dt + (2x - 1)dw(t)$, $x(0) = 2$;
   (d) $dx = \left(\sqrt{6}(x - 1)\right)dw(t)$, $x(0) = 3$;
   (e) $dx = \left(2x + \sqrt{6}p\right)dt + \left(\sqrt{6} + p\right)dw(t)$, $x(0) = 3$;
   (f) $(9 + t)dx = (3x + 9 + t)dt + ((9 + t)^2x + 1)dw(t)$, $x(0) = 1$.

4. Extend the mathematical model described in Problem 5, Section 2.4, Volume 1 under the considerations of random environmental perturbation. Furthermore, find a mathematical expression for:

(a) The amount of wheat y at a level x of the dose of the application of the fertilizer knowing that $y(x_0) = y_0$.

(b) The average/expected amount of wheat y at a level x of the dose of the application of the fertilizer knowing that $y(x_0) = y_0$.

5. Given the RL electric circuit with the externally applied voltage $E_e(t)$, use Exercise 2.1.6 and:

(a) Derive the differential equations with regard to the current.

(b) Write the derived differential equation in the physical standard form.

(c) Find the solution of this example IVP.

(d) Determine solutions of the particular forms: (i) $E_e(t) = E_0$, a constant, (ii) $E_e(t) = E_0 \sin \omega t$.

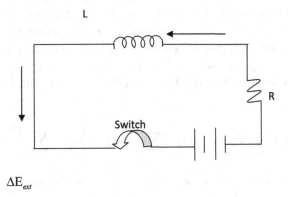

Fig. 2.4.6

6. **Continuous intravenous injection of glucose** [26]. Let e be the rate of infusion (mg/min) of glucose in the blood. Let Q be the quantity (mg) in the blood, and let $c = \frac{Q}{V}$ be the concentration of the glucose in the blood, where V stands for the volume. The injected glucose is eliminated at a rate proportional to the concentration in the blood. Apply Illustration 2.4.5 and find:

(a) The glucose concentration in the blood knowing that $Q(0) = Q_0$.

(b) The glucose concentration in the absence of the infusion.

7. **The heart function test** [26, 80]. The goal of this problem is to find the oxygen debt (x). This is a clinically important problem. During exercise, the body is able to incur an oxygen debt. This means that the muscles are able to work anaerobically, but in the recovery state, oxygen is needed to replenish the energy stores (glucose). It is known that the greater the muscular effort, the greater the amount of oxygen is needed to replenish the energy stores which were used during the exercise. It is assumed that the oxygen debt is proportional to the work done (W). The oxygen debt also depends upon the oxygen supply [amount of oxygen (O) taken up through the lungs]. It is assumed that the oxygen consumption in the muscles is ignored. Show:

(a) $dx = \alpha dW - dO$ particularly.

(b) If the extra oxygen uptake per second (at the lung) is proportional to the oxygen debt at that instant, then $dx = [\alpha P(t) - kx]dt$, where $P(t) = \frac{dW}{dt}$.

(c) Write the differential equation in (b) in the physical standard form.

(d) Formulate the initial value problem for (b) and solve it.

(e) Find the solutions corresponding to the problem (d) for the following cases: (i) $P(t) = 0$; (ii) $P(t) = \beta$ constant.

(f) From the solution expressions in (e) and using the measurable quantity $\frac{dO}{dt}\left(\frac{dO}{dt}\right) = kx$, in particular $\ln\left(\frac{dO}{dt}\right) = \ln\left(\frac{dO}{dt}\right)\Big|_{t=0} - k(t)$, one can find k.

8. With reference to Illustrations 2.4.1–2.4.6, find the mean and variance of the corresponding states of the dynamic processes in the sciences.

9. Compare the means in Exercise 8 with the corresponding deterministic states of the dynamic processes in the sciences (Section 2.5, Volume 1 [80]).

10. For $i = 1, 2, \ldots, k$, let f, σ_i be continuous functions defined on J into R, and let $w_i(t)$ be a Wiener process (Theorem 1.2.2) defined on a complete probability space $(\Omega, \mathfrak{F}, P)$. Further, assume that for $i \neq j, i, j-1, 2, \ldots, k$, $w_i(t)$ and $w_j(t)$ are mutually independent on $J(E(dw_i(t)dw_j(t)) = 0)$. Show that the process defined by

$$x(t) = \Phi(t)c + \int_a^t \Phi(t,s)\left[p(s) - \sum_{i=1}^k \sigma_i(s)q_i(s)\right]ds + \sum_{i=1}^k \int_a^t \Phi(t,s)q_i(s)dw(s)$$

is the general solution process of the following Itô–Doob stochastic differential equation:

$$dx = [f(t)x + p(t)] + \sum_{i=1}^k [\sigma_i(t)x + q_i(t)]dw_i(t),$$

where

$$\Phi(t) \equiv \Phi(t, w_1(t), w_2(t), \ldots, w_k(t))$$

$$= \exp\left[\int_a^t \left[f(s) - \frac{1}{2}\sum_{i=1}^k \sigma_i^2(s)\right]ds + \sum_{i=1}^k \int_a^t \sigma_i(s)dw(s)\right].$$

11. Find the general solution to $dx = [2x + p]dt + [2x + q_1]dw_1(t) + [2x + q_2]dw_2(t)$.

12. Let F be a continuous cumulative distribution function (cdf) of a random time variable T, and let f, g, and σ be continuous functions. Moreover, $w(t)$ is a Wiener process. Apply the procedure outlined in Section 2.4 to find the general solution process of the following differential equations:

(a) $dx = [f(t)x + p(t)]dt + [g(t)x + q(t)]dF(t)$;

(b) $dx = [g(t)x + q_1(t)]dF(t) + [\sigma(t)x + q_2(t)]dw(t)$;

(c) $dx = [f(t)x + p(t)]dt + [g(t)x + q_1(t)]dF(t) + [\sigma(t)x + q_2(t)]dw(t)$.

2.5 Fundamental Conceptual Algorithm and Analysis

In Sections 2.2–2.4, we presented the analytical procedures/methods for finding the solution processes of linear scalar stochastic differential equations of the Itô–Doob type. The presented techniques provide the basis for addressing/raising several important questions in the modeling of dynamic processes in the biological, chemical, engineering, medical, physical, and social sciences. Examples:

1. Are linear differential differential equations adequate mathematical models of dynamic processes?
2. Prior to finding a solution process of a differential equation, how do we know that it has a solution?
3. Are there more than one solutions to a given differential equation?
4. Are the methods of integration sufficient for finding a closed-form solution to every differential equation?
5. Are the procedures for solving differential equations in Sections 2.3 and 2.4 applicable to any other types of differential equations?
6. Do we really need a closed-form solution to a differential equation?
7. In the absence of a closed-form solution to a differential equation, is the modeling of the dynamic process wasteful?
8. Without a closed-form solution, is there any type of qualitative knowledge about a solution process shedding light on the dynamic processes?
9. Is it feasible to study the qualitative properties of solutions to differential equations without the closed-form knowledge about them?
10. Are there any basic universal qualitative properties of solutions to differential equations?

The answers to a few of these questions are outlined in this section. Three of the most basic questions are #2, #3, and #10. These questions generate the three most basic important problems in the theory of differential equations: (i) *existence problem,* (ii) *uniqueness problem,* and (iii) *fundamental properties of solution processes.*

The presented procedures for finding solutions to IVPs in Sections 2.3 and 2.4 justify the existence of solutions to IVPs for linear first-order Itô–Doob-type stochastic differential equations. This can be summarized in the following result. The existence result and the results presented thereafter play a very important role and have a scope for both theoretical and practical points of view, particularly in the study of (a) time-varying and time-invariant (nonstationary and stationary) nonlinear scalar systems, (b) linear time-varying systems, (c) nonhomogeneous systems, and (d) nonlinear systems of stochastic differential equations of the Itô–Doob type.

Theorem 2.5.1 (existence theorem). *Let us consider the IVP*

$$dy = [f(t)y + p(t)]dt + [\sigma(t)y + q(t)]dw(t), \quad y(t_0) = y_0, \qquad (2.5.1)$$

where dx stands for the stochastic differential of the Itô–Doob type; f, p, σ, and q
are any given continuous functions defined on an interval $J = [a, b]$ ($J \subseteq R$) into R;
w is a scalar normalized Wiener process; initial data/conditions $(t_0, y_0) \in J \times R$; and
y_0 is a real-valued random variable defined on a complete probability space $(\Omega, \mathfrak{F}, P)$,
and it is independent of $w(t)$ for all t in J. Then, the IVP (2.5.1) has a solution
$y(t, t_0, y_0) = y(t)$ through the given initial data/conditions (t_0, y_0) for $t \geq t_0$, t_0, and
t in J. Moreover,

$$y(t) = \Phi(t, t_0)y_0 + \int_{t_0}^{t} \Phi(t, s)[p(s) - \sigma(s)q(s)]ds + \int_{t_0}^{t} \Phi(t, s)q(s)dw(s), \quad (2.5.2)$$

where

$$\Phi(t, w(t), t_0) \equiv \Phi(t, t_0) = \exp\left[\int_{t_0}^{t}\left[f(s) - \frac{1}{2}\sigma^2(s)\right]ds + \int_{t_0}^{t}\sigma(s)dw(s)\right] \quad (2.5.3)$$

is the fundamental solution process of (2.3.1), i.e. the first-order linear homogeneous
scalar stochastic differential equation corresponding to (2.5.1):

$$dx = f(t)xdt + \sigma(t)xdw(t), \quad y(t_0) = y_0. \quad (2.5.4)$$

Proof. The justifications for the conclusions of the existence theorem follow from
the computational procedures described in Sections 2.3 and 2.4; Observations 2.3.1,
2.4.1, and 2.4.2; and Theorem 2.3.1. Moreover, the solution expression (2.5.2) is the
byproduct of these procedures.   □

Observation 2.5.1

(i) We assume that the IVP (2.5.1) has a solution. Let us denote it by $y(t) = y(t, t_0, y_0)$. From the definition analogous to Definition 2.3.2 of the IVP
(2.5.4)/(2.3.36), the process $y(t)$ and its Itô–Doob differential must satisfy both
the differential equation (2.5.1) and the given initial conditions/data (t_0, t_0) for
$t \geq t_0$, t_0 and t in J. This means that

$$dy(t) = [f(t)y(t) + p(t)]dt + [\sigma(t)y(t) + q(t)]dw(t), \quad y(t_0) = y_0. \quad (2.5.5)$$

From the concept of the Itô–Doob stochastic integral and differential
(Section 1.3), the solution $y(t)$ of the initial problem (2.5.5) is equivalent to
the following stochastic integral equation:

$$y(t) = y_0 + \int_{t_0}^{t} [f(s)y(s) + p(s)]ds + \int_{t_0}^{t} [\sigma(s)y(s) + q(s)]dw(s). \quad (2.5.6)$$

We recall that the first and second integral terms on the right-hand side
of (2.5.6) are in the sense of Riemann–Cauchy [127] and Itô–Doob calculus,
respectively.

(ii) We remark that the solution to the IVP (2.5.1) has two representations/
expressions: (a) a computationally explicit form (2.5.2); (b) a conceptually
implicit form (2.5.6).

(iii) Under the continuity assumption on the rate coefficients f, p, σ, and q and other standard assumptions, (without the knowledge about the exact expression of the solution) we can, alternatively, establish the existence of the solution process of the IVP (2.5.1). In fact, by using the approximation procedure

$$
y_n(t) = \begin{cases}
y_0 & \text{for } t_0 \leq t \leq t_0 + \dfrac{1}{n}, \\
y_0 + \displaystyle\int_{t_0}^{t-\frac{1}{n}} [f(s)y_n(s) + p(s)]ds + \int_{t_0}^{t-\frac{1}{n}} [\sigma(s)y_n(s) + q(s)]dw(s),
\end{cases}
$$

for $t_0 + \frac{ka}{n} \leq t \leq t_0 + \frac{(k+1)a}{n}$, $k = 1, 2, \ldots, n-1$, one can establish the existence of the solution to the IVP (2.5.1). This procedure utilizes an additional conceptual aspect of the drift $C(t, y) = f(t)y + p(t)$ and the diffusion $F(t, y) = \sigma(t)y + q(t)$ coefficients, namely the following condition:

$$
|C(t, y)| + |F(t, y)| \leq M|y| + N \quad \text{(linear growth condition)}, \qquad (2.5.7)
$$

where M and N are some non-negative real numbers. Of course, it is obvious that $C(t, y) = f(t)y + p(t)$ and $F(t, y) = \sigma(t)y + q(t)$ in (2.5.1) satisfy the growth condition (2.5.7).

(iv) Finally, we observe that (2.5.2) can be rewritten as

$$
y(t) = x(t) + \int_{t_0}^{t} \Phi(t, s)[p(s) - \sigma(s)q(s)]ds + \int_{t_0}^{t} \Phi(t, s)q(s)dw(s), \qquad (2.5.8)
$$

where $x(t)$ is the solution process of the IVP (2.5.4) through (t_0, y_0), i.e.

$$
x(t) = x(t, t_0, y_0).
$$

Our experience of finding a solution to the IVP (2.5.1) is based on the construction of a general fundamental solution (Observation 2.3.1). The construction of a general fundamental solution process depends on finding unknown functions in (2.3.5) and (2.3.6) with arbitrary constants and a choice of "a" in J. As we know, the choice of unknown constants as well as "a" in J depends on the problem-solver and the problem itself. As a result of this, the process of finding a general fundamental solution process of (2.3.1)/(2.5.4) is not unique. Therefore, the presented solution-finding procedures do not address the question of the uniqueness of the solution to the IVP. In short, a general fundamental solution process of (2.3.1)/(2.5.4) is not uniquely determined (see Section 6.1 of Chapter 6). Hence, Theorem 2.5.1 does not guarantee uniqueness of the solution process of (2.5.1). The following result further justifies these statements.

Lemma 2.5.1. *Let Φ_1 be a fundamental solution process of (2.3.1). Φ_2 is any other general fundamental solution process of (2.3.1) if and only if $\Phi_2(t, w(t)) = \Phi_1(t, w(t))C$ on J for some nonzero constant random variable C.*

Proof. First, we assume that $\Phi_1(t, w(t))$ is a given general fundamental solution process of (2.3.1), and $\Phi_2(t, w(t))$ is any general fundamental solution process of (2.3.1). From the definition of the general fundamental solution process of (2.3.1) [Observation 2.3.1(v)], we recall that Φ_1 and Φ_2 will differ only by the choice of "a" in J. First, we prove that $\Phi_2 = \Phi_1 C$ on J for some nonzero constant random variable C. For this purpose, we define $\Psi = \Phi_1^{-1}\Phi_2$ on J. This is defined in view of Observation 2.3.1(v, vi) and the definition of the general fundamental solution process [Observation 2.3.1(v)]. Our goal reduces to showing that Ψ is a nonzero constant random variable. For this purpose, we compute the Itô–Doob stochastic differential of Ψ:

$$d\Psi = d(\Phi_1^{-1}\Phi_2) \quad \text{(by definition of } \Psi\text{)}$$

$$= d(\Phi_1^{-1})\Phi_2 + \Phi_1^{-1}d\Phi_2 + d(\Phi_1^{-1})d\Phi_2 \quad \text{(by Theorem 1.3.4)}$$

$$= [\Phi_1^{-1}[-f(t) + \sigma^2(t)]dt - \Phi_1^{-1}\sigma(t)dw(t)]\Phi_2$$

$$+ \Phi_1^{-1}f(t)\Phi_2dt + \Phi_1^{-1}\sigma(t)\Phi_2dw(t)$$

$$+ [\Phi_1^{-1}[-f(t) + \sigma^2(t)]dt - \Phi_1^{-1}\sigma(t)dw(t)][f(t)\Phi_2dt$$

$$+ \sigma(t)\Phi_2dw(t)] \quad \text{[by Theorem 2.3.1 and from (2.3.32)]}$$

$$= \Phi_1^{-1}[-f(t) + \sigma^2(t)]\Phi_2dt - \Phi_1^{-1}\sigma(t)dw(t)\Phi_2 + \Phi_1^{-1}f(t)\Phi_2dt$$

$$+ \Phi_1^{-1}\sigma(t)\Phi_2dw(t) + \Phi_1^{-1}[-f(t) + \sigma^2(t)]\sigma(t)\Phi_2dw(t)dt$$

$$- \Phi_1^{-1}\sigma^2(t)\Phi_2dt \quad \text{[by simplification and Itô–Doob calculus]}$$

$$= 0 \quad \text{(final simplification).} \tag{2.5.9}$$

From (2.5.9), we conclude that $\Psi(t) = C$ is constant on J. By substitution and simplification, we conclude that $\Phi_2 = \Phi_1 C$ on J. Now, it remains for us to show that Ψ is nonzero on J. From the fact that $\Phi_2 \neq 0 \neq \Phi_1$, $\Phi_2 = \Phi_1 C$ and from elementary algebra, we conclude that $C \neq 0$. Therefore, C is a nonzero constant random variable. This completes the proof of the "if part."

To prove the "only if part," we start with $\Phi_2 = \Phi_1 C$, where C is a nonzero constant random variable, and show that Φ_2 is a general fundamental solution process of (2.3.1). We find the differential (Itô–Doob sense) of both sides of the expression, and obtain

$$d\Phi_2 = d\Phi_1 C \quad \text{(by Theorem 1.5.1P1 [80])}$$

$$= [f(t)\Phi_1dt + \sigma(t)\Phi_1dw(t)]C \quad \text{[by substitution from (2.3.32)]}$$

$$= f(t)\Phi_1Cdt + \sigma(t)\Phi_1Cdw(t) \quad \text{(by rearrangement)}$$

$$= f(t)\Phi_2dt + \sigma(t)\Phi_2dw(t) \quad \text{(by substitution for } \Phi_1C\text{),}$$

and hence, $\Phi_1 C$ is a solution to (2.3.1). Since $\Phi_2 = \Phi_1 C$ for $C \neq 0$. Therefore, we conclude that $\Phi_1 C$ is the general fundamental solution process of (2.3.1). This completes the proof of the theorem. $\qquad\square$

Observation 2.5.2

(i) Obviously, two different linear homogeneous scalar differential equations cannot have the same general fundamental solution process. In fact, from (2.3.32), we note that $f(t)dt + \sigma(t)dw(t) = d\Phi(t)\Phi^{-1}(t)$. This is true in view of Observation 2.3.1(v).

(ii) Let us assume that Φ_1 and Φ_2 are two given general fundamental solution processes of (2.3.1). In addition, let $t_0 \in J$ be given. From Lemma 2.5.1, we have $\Phi_2(t) = \Phi_1(t)C$ on J. For $t_0 \in J$, we have $\Phi_2(t_0) = \Phi_1(t_0)C$. This implies that $C = \Phi_1^{-1}(t_0)\Phi_2(t_0)$. From this, we obtain $\Phi_2(t) = \Phi_1(t)\Phi_1^{-1}(t_0)\Phi_2(t_0)$, which implies that

$$\Phi_2(t, w(t))\Phi_2^{-1}(t_0, w(t_0)) = \Phi_1(t, w(t))\Phi_1^{-1}(t_0, w(t_0)),$$

in view of Observation 2.3.1(i). From this and using the notation of the normalized fundamental solution process of (2.5.4)/(2.3.1), we have

$$\Phi(t, t_0) \equiv \Phi_2(t, w(t), t_0) = \Phi_1(t, w(t), t_0). \qquad (2.5.10)$$

This shows that the normalized fundamental solution process of (2.5.4)/(2.3.1) is uniquely determined by f, σ, w, and t_0. The normalized fundamental solution process $\Phi(t, t_0)$ of (2.5.4)/(2.3.1), defined in (2.5.10), is referred to as the *normalized fundamental solution process of* (2.5.4)/(2.3.1) *at* $t = t_0$.

In the following, we present a few algebraic properties of the normalized fundamental solution process of (2.5.4)/(2.3.1). The proof of the result is based on the observed properties of a general fundamental solution process, Observation 2.3.1, Theorem 2.3.1, and differential/integral calculus.

Lemma 2.5.2. Let $\Phi(t, w(t), t_0, w(t_0)) \equiv \Phi(t, t_0)$ and $\Psi(t, w(t), t_0, w(t_0)) \equiv \Psi(t, t_0)$ be normalized fundamental solutions to (2.5.4)/(2.3.36) and (2.3.35) at $t = t_0$, respectively. Let $\Phi_1(t, t_1)$ be a normalized fundamental solution to (2.5.4)/(2.3.36) at $t = t_1$. Then, for all t_0, t_1, s and t in J,

(a) $\Psi(t, t_0)\Phi(t, t_0) = 1 = \Phi(t_0, t)\Phi(t, t_0),$ *for* $t_0 \leq t$;   (2.5.11)

    where

$\quad \Psi(t, t_0) = \Phi^{-1}(t, t_0) = \Phi(t_0, t),$ *for* $t_0 \leq t$;   (2.5.12)

(b) $\Phi_1(t, t_1) = \Phi(t, t_0)\Phi_1(t_0, t_1),$ *for* $t_0 \leq t$;   (2.5.13)

(c) $\Phi(t, t_0) = \Phi(t, s)\Phi(s, t_0),$ *for* $t_0 \leq t$;   (2.5.14)

(d) $\Phi(t,s) = \Phi(t,t_0)\Psi(s,t_0) = \Phi(t,t_0)\Phi(t_0,s)$,   *for $t_0 \le t$;*   (2.5.15)

(e) $d_s\Phi(t,s) = [-f(s)\Phi(t,s) + \sigma^2(s)\Phi(t,s)]ds - \sigma(s)\Phi(t,s)dw(s))$,   (2.5.16)

where $d_s\Phi(t,s)$ is the Itô–Doob stochastic differential with respect to s.

Proof. To prove (a), we use the Itô–Doob stochastic differential, and compute

$$d(\Psi(t,t_0)\Phi(t,t_0)) = d\Psi(t,t_0)\Phi(t,t_0) + \Psi(t,t_0)d\Phi(t,t_0)$$

$$+ d\Psi(t,t_0)d\Phi(t,t_0) \qquad \text{(by Theorem 1.3.4)}$$

$$= ([-f(t) + \sigma^2(t)]\Psi(t,t_0)dt - \sigma(t)\Psi(t,t_0)dw(t))\Phi(t,t_0)$$

$$+ \Psi(t,t_0)(f(t)\Phi(t,t_0)dt + \sigma(t)\Phi(t,t_0)dw(t))$$

$$+ (-\sigma(t)\Psi(t,t_0)dw(t))\sigma(t)\Phi(t)dw(t) \quad \text{[from (2.3.32) and (2.3.35)]}$$

$$= 0 \qquad \text{(final simplification).}$$

This establishes the fact that $\Psi(t,t_0)\Phi(t,t_0) = c$, where c is an arbitrary constant random variable. This is in view of the fact that if the Itô–Doob differential of a stochastic process is zero on an interval J for $t \ge t_0$, then the process is constant on J. Moreover, $\Psi(t,t_0)$ and $\Phi(t,t_0)$ are normalized solution processes of (2.5.4) and (2.3.35) at $t = t_0$, respectively, and therefore

$$\Psi(t,t_0)\Phi(t,t_0) = c = \Psi(t_0,t_0)\Phi(t_0,t_0) = 1. \qquad (2.5.17)$$

This shows that $\Psi(t,t_0)$ is the algebraic inverse of $\Phi(t,t_0)$, and it is denoted by $\Phi(t_0,t)$. This statement is equivalent to other notations in (2.5.12). Because of these notations and (2.5.17), we have

$$\Psi(t,t_0)\Phi(t,t_0) = 1 = \Phi(t_0,t)\Phi(t,t_0), \quad \text{for } t_0 \le t.$$

This completes the proof of (2.5.11).

To prove (b), from Lemma 2.5.1 and Observation 2.5.2(ii), we have $\Phi_1(t,t_1) = \Phi(t,t_0)C$. By using the assumption of the lemma and $t = t_0$, we get $C = \Phi_1(t_0,t_1)$. After substitution, we obtain

$$\Phi_1(t,t_1) = \Phi(t,t_0)\Phi_1(t_0,t_1) \quad \text{for } t_0 \le t.$$

This establishes the relation (2.5.13).

To prove (c), for $t_0 \le t$ and any s in J, we utilize elementary algebraic properties and the nature of $\Phi(t,t_0)$, and obtain

$$\Phi(t,t_0) = \Phi(t)\Phi^{-1}(t_0) \qquad \text{[by the definition in (2.3.40)]}$$

$$= \Phi(t)\Phi^{-1}(s)\Phi(s)\Phi^{-1}(t_0) \quad \text{[by the existence of } \Phi^{-1}(s)]$$

$$= \Phi(t,s)\Phi(s,t_0) \qquad \text{[by the definitions in (2.3.40) and (2.5.11)].}$$

This completes the proof of (2.5.14).

For the proof of (d), from (2.5.14) and solving for $\Phi(t, s)$, we have

$$\Phi(t, s) = \Phi(t, t_0)\Phi^{-1}(s, t_0) \qquad \text{[by (2.5.14)]}$$

$$= \Phi(t, t_0)\Psi(s, t_0) \qquad \text{[by notation (2.5.12)]}$$

$$= \Phi(t, t_0)\Phi(t_0, s) \quad \text{for } t_0 \le t \quad \text{[by notation (2.5.12)]}.$$

This proves (d).

For the proof of (e), we apply the Itô–Doob differential (Theorem 1.3.4) on both sides with respect to the expression (2.5.15) and obtain

$$d_s\Phi(t, s) = \Phi(t, t_0)[-f(s) + \sigma^2(s)]\Psi(s, t_0)ds - \sigma(s)\Psi(s, t_0)dw(s))$$

$$= ([-\Phi(t, t_0)f(s)\Psi(s, t_0) + \Phi(t, t_0)\sigma^2(s)]\Psi(s, t_0)ds$$

$$- \Phi(t, t_0)\sigma(s)\Psi(t_0, s)dw(s) \quad \text{(by Theorem 1.3.1M2 [80])}$$

$$= ([-f(s) + \sigma^2(s)]\Phi(t, t_0)\Psi(s, t_0)ds$$

$$- \sigma(s)\Phi(t, t_0)\Psi(s, t_0)dw(s) \quad \text{(by the commutative property for scalars)}$$

$$= [-f(s) + \sigma^2(s)]\Phi(t, s)ds$$

$$- \sigma(s)\Phi(t, s)dw(s)) \qquad \text{[by (2.5.15)]}.$$

This establishes the result in (e). This completes the proof of the lemma. $\qquad \square$

Observation 2.5.3

(i) We further emphasize that the proof of Lemma 2.5.2 depends heavily on the algebraic reasoning and the closed-form representation of $\Phi(t, t_0)$ in (2.5.3).

(ii) An alternative proof of these expressions in Lemma 2.5.2 can be given by utilizing the conceptual approach. In fact, the proofs of Parts (a), (c), and (e) will be illustrated at a later stage in the section.

We further remark that (2.3.34) (Theorem 2.3.1) provides another simple calculus approach to finding the inverse of the general fundamental solution to (2.3.1).

For the sake of completeness and prior to the presentation of the uniqueness result, let us define the concept of the uniqueness of the solution process of (2.5.1).

Definition 2.5.1. Let $y_1(t) = y_1(t, t_0, y_0)$ be a solution process of the IVP (2.5.1) for $t \ge t_0$ and t, t_0 in J, and let $y_2(t) = y_2(t, t_0, y_0)$ be any other solution process of (2.5.1). $y_1(t, t_0, y_0)$ is said to be the *unique* solution process of the IVP (2.5.1) if and only if $y_1(t, t_0, y_0) = y_2(t, t_0, y_0)$ for $t \ge t_0$ with probability 1, and t, t_0 in J.

In the following, we are ready to present the uniqueness result for the solution process of (2.5.1).

Theorem 2.5.2 (uniqueness theorem). *Let the hypotheses of Theorem 2.5.1 be satisfied. In addition, let $y(t) = y(t, t_0, y_0)$ be a solution process of the IVP (2.5.1). Then, $y(t, t_0, x_0)$ is unique solution process of (2.5.1) through $(t_0, y_0) \in J \times R$ for $t \geq t_0$ and t, t_0 in J.*

Proof. We assume that the IVP (2.5.1) has two solutions. Let $y_1(t) = y_1(t, t_0, y_0)$ and $y_2(t) = y_2(t, t_0, y_0)$ be the two solutions through the same initial data (t_0, y_0) which correspond to two normalized fundamental solution processes $\Phi_1(t, t_0)$ and $\Phi_2(t, t_0)$ of (2.5.4) at $t = t_0$, respectively. Our goal is to show that $y_1(t) = y_2(t)$, i.e. $y_1(t, t_0, y_0) = y_2(t, t_0, y_0)$ for $t \geq t_0$ and t, t_0 in J. From Observation 2.5.2, we note that $\Phi_2(t, t_0) = \Phi_1(t, t_0)$ for $t \geq t_0$, and hence $\Phi_2(t, t_0)y_0 = \Phi_1(t, t_0)y_0$ is valid for $t \geq t_0$, and t, t_0 in J. From Theorem 2.5.1, we have

$$y_1(t) = \Phi_1(t, t_0)y_0 + \int_{t_0}^{t} \Phi_1(t, s)[p(s) - \sigma(s)q(s)]ds + \int_{t_0}^{t} \Phi_1(t, s)q(s)dw(s)$$

and

$$y_2(t) = \Phi_2(t, t_0)y_0 + \int_{t_0}^{t} \Phi_2(t, s)[p(s) - \sigma(s)q(s)]ds + \int_{t_0}^{t} \Phi_2(t, s)q(s)dw(s),$$

where

$$\Phi_2(t, t_0) = \Phi_1(t, t_0) = \exp\left[\int_{t_0}^{t}\left[f(s) - \frac{1}{2}\sigma^2(s)\right]ds + \int_{t_0}^{t}\sigma(s)dw(s)\right].$$

The expressions on the right-hand side of $y_1(t)$ and $y_2(t)$ are exactly the same. Therefore, $y_1(t, t_0, y_0) = y_2(t, t_0, y_0)$ for $t \geq t_0$ and t, t_0 in J. This establishes the uniqueness of the solution process of the IVP (2.5.1). The proof of the theorem is complete. ☐

Observation 2.5.4

(i) Under the continuity assumption on the rate coefficients f, p, σ, and q and other standard assumptions (without the knowledge about the exact expression of the solution), we can alternatively establish the existence and uniqueness of the solution process of the IVP (2.5.1). In fact, by using the approximation procedure

$$y_n(t) = \begin{cases} y_0 \quad \text{for } n = 0, \text{ on } a \leq t \leq b, \\ y_0 + \int_{a}^{t}[f(s)y_{n-1}(s) + p(s)]ds + \int_{t_0}^{t}[\sigma(s)y_{n-1}(s) + q(s)]dw(s), \end{cases}$$

$$(2.5.18)$$

for $n \geq 1$, we can establish the existence and uniqueness of the solution to the IVP (2.5.1). In addition to the linear growth condition (2.5.7), this procedure utilizes an additional conceptual aspect of the drift $C(t, y) = f(t)y + p(t)$ and the diffusion $F(t, x) = \sigma(t)y + q(t)$ coefficients, namely the following condition:

$$|C(t, x) - C(t, y)| + |F(t, x) - F(t, y)| \leq L|x - y| \quad \text{(Lipschitz condition)},$$
$$(2.5.19)$$

where L is a positive real number. Again, one can easily verify that the functions $C(t, y) = f(t)y + p(t)$ and $F(t, y) = \sigma(t)y + q(t)$ in (2.5.1) satisfy the Lipschitz condition (2.5.19) over the set $J \times R$, $J = [a, b]$ for $a, b \in R$.

(ii) The uniqueness theorem (Theorem 2.5.2) conceptualizes the solution process $y(t) = y(t, t_0, y_0)$ of the IVP (2.5.1) as a function of three variables (t, t_0, y_0). Under conditions of the uniqueness theorem, for given (t, t_0, y_0) and from the definition of function, the output $y(t, t_0, y_0)$ [the value of the solution process at (t, t_0, y_0)] is uniquely determined. This conceptualization plays a very important role in the advanced study of differential equations [8, 32, 43, 71, 98, 100, 114, 141, 182]. Moreover, this notation $y(t, t_0, y_0)$ is just a natural symbol for the solution process of (2.5.1).

(iii) We also observe that the proofs of the relations (2.5.13) and (2.5.14) can be reproduced by applying Theorem 2.5.2. The details are left to the reader.

In the following, we present a few results that shed light on the algebraic properties of solution processes of IVPs corresponding to linear homogeneous scalar differential equations of (2.3.35) and (2.5.4)/(2.3.36). These properties are utilized for studying fundamental properties of the solution process of the IVP (2.5.1). Moreover, these results provide an auxiliary conceptual tool and the insight to undertake the study of both linear and nonlinear systems of differential equations.

Lemma 2.5.3 (principle of superposition). *Let $x_1(t, t_0, x_0) = x_1(t)$ and $x_2(t, t_0, x_0) = x_2(t)$ be solutions to the IVP (2.5.4)/(2.3.36) through (t_0, x_0) and (t_0, y_0) for $t \geq t_0$ and t, t_0 in J. In addition, let α and β be any two scalar real/complex numbers. Then,*

$$\alpha x_1(t, t_0, x_0) + \beta x_2(t, t_0, y_0) \quad (2.5.20)$$

is a solution to (2.5.4) through the initial data $(t, \alpha x_0 + \beta y_0)$ for $t \geq t_0$ and t, t_0 in J.

Proof. From Definition 2.3.2 of the solution to the IVP (2.5.4)/(2.3.36), $x_1(t) = x_1(t, t_0, x_0)$ and $x_2(t) = x_2(t, t_0, y_0)$ being the given solutions to the IVP (2.5.4), we note that for $t = t_0$ the expression $\alpha x_1(t, t_0, x_0) + \beta x_2(t, t_0, y_0)$ reduces to $\alpha x_0 + \beta y_0$. Furthermore, we have the existence of differentials $dx_1(t) = dx_1(t, t_0, x_0)$ and $dx_2(t) = dx_2(t, t_0, y_0)$. Therefore, from Theorems 1.3.3 and 1.3.4 and noting the fact that a differential of a constant is zero, the rest of the proof can be reproduced by imitating the proof of the deterministic version of the lemma [80]. The details are left as an exercise to the reader. $\qquad\square$

Observation 2.5.5

(i) From Illustration 1.4.4 [80], Observation 1.4.5 [80], and the conclusion of Lemma 2.5.3, we conclude that an arbitrary linear combination of solutions to (2.5.4) is also a solution to the IVP (2.5.4).
(ii) In fact, from Observation 1.4.5 [80], Theorem 1.4.5 [80], and Illustration 1.4.5 [80], it implies that the span $S(\{x_1, x_2\})$ of the solution processes x_1 and x_2 of (2.5.4) is a vector subspace of $C[[t_0, b], R]$.
(iii) From Definition 1.4.5 [80], the dimension of the subspace $S(\{x_1, x_2\})$ in (ii) is 1. This is due to the fact that the general solution to (2.5.4)/(2.3.1) depends on only one arbitrary constant. In fact, it is a linear combination of a scalar function.

The following result shows that the normalized solution process $\Phi(t, t_0)$ of (2.5.4)/(2.3.36) possesses certain basic algebraic properties as a function for each fixed (t, t_0) for $t \geq t_0$ and t, t_0 in J.

Lemma 2.5.4. *Let $x(t) = x(t, t_0, x_0)$ be a solution to the IVP (2.5.4) through (t_0, x_0) for $t \geq t_0$ and t, t_0 in J. For fixed (t, t_0), the transformation/mapping $\Phi(t, t_0)$ is defined on R into R by $\Phi(t, t_0)x_0 = x(t, t_0, x_0)$ for x_0 in R. Then, for every x_0, x_1, x_2 and α in R,*

$$\Phi(t, t_0)(x_1 + x_2) = \Phi(t, t_0)x_1 + \Phi(t, t_0)x_2, \tag{2.5.21}$$

$$\Phi(t, t_0)(\alpha x_0) = \alpha \Phi(t, t_0)x_0. \tag{2.5.22}$$

Proof. The proof of the lemma can be reconstructed by imitating the proof of the deterministic version of the lemma [80]. The details are left as an exercise to the reader. □

Observation 2.5.6

(i) We observe that mapping/transformation that satisfies the properties described in (2.5.21) and (2.5.22) is called a *linear transformation/mapping*. For fixed (t, t_0), the normalized fundamental solution $\Phi(t, t_0)$ to (2.5.4)/(2.3.36) at $t = t_0$ is a linear transformation defined on R into itself. In fact, it is polynomial function of degree 1.
(ii) We further remark that the proof of Lemma 2.5.4 is trivial in the light of the algebra of real numbers (Theorem 1.3.1 [80]). However, one is encouraged to use the uniqueness theorem. This line of argument has greater universal appeal for carrying out the study, beyond the study of linear scalar differential equations.

Next, we present a relationship between the solution processes of the linear homogeneous scalar differential equation (2.5.4)/(2.3.36) and its adjoint differential equation (2.3.35).

Lemma 2.5.5. *Let $x(t) = \Phi(t, t_0)x_0$ and $z(t) = x_0\Psi(t, t_0)$ be the solution processes of the IVP (2.5.4)/(2.3.36) and the corresponding IVP associated with its adjoint differential equation (2.3.35) through (t_0, x_0), i.e.*

$$dz = z[-f(t) + \sigma^2(t)]dt - z\sigma(t)dw(t), \quad z(t_0) = x_0, \qquad (2.5.23)$$

respectively, where $\Phi(t, t_0)$ and $\Psi(t, t_0)$ are normalized fundamental solution processes of (2.5.4) and (2.5.23) for $t \geq t_0$ and t, t_0 in J. Then,

$$z(t)x(t) = c, \qquad (2.5.24)$$

where c is a constant number. Moreover, $c = x_0^2$,

$$\Psi(t, t_0)\Phi(t, t_0) = I, \qquad (2.5.25)$$

for all $t \geq t_0$ and t, t_0 in J.

Proof. By imitating the proof of Lemma 2.5.2(a), we compute

$d(z(t)x(t)) = dz(t)x(t) + z(t)dx(t) + dz\,dx$ (by Theorem 1.3.4)

$= ([-f(t) + \sigma^2(t)]z(t)dt - \sigma(t)z(t)dw(t))x(t)$ (by substitution)

$\quad + ([-f(t) + \sigma^2(t)]z(t)dt$

$\quad - \sigma(t)z(t)dw(t))(f(t)x(t)dt + \sigma(t)x(t)dw(t))$

$\quad + z(t)(f(t)x(t)dt + \sigma(t)x(t)dw(t))$ (by linearity, Theorem 1.3.1 [80] and Itô–Doob calculus)

$= [-f(t) + \sigma^2(t)]z(t)x(t)dt - \sigma(t)z(t)x(t)dw(t)$

$\quad + f(t)z(t)x(t)dt + \sigma(t)z(t)x(t)dw(t)$

$\quad - \sigma^2(t)z(t)x(t)dt$ (by simplifications)

$= 0$ (by final simplification).

This establishes the fact that $z(t)x(t) = c$, where c is a constant random variable. This is in view of the fact that if the Itô–Doob differential of a stochastic process is zero on an interval J, then the process is constant on J. This completes the proof of (2.5.24). The proof of $c = x_0^2$ follows from the facts that $z(t_0)x(t_0) = c = x_0^2$, $\Phi(t_0, t_0) = 1 = \Psi(t_0, t_0)$, and $z(t)x(t) = x_0\Psi(t, t_0)\Phi(t, t_0)x_0 = x_0x_0$, for any x_0 in R. The details are left to the reader. Finally, from $x_0\Psi(t, t_0)\Phi(t, t_0)x_0 = x_0x_0$, we have $\Psi(t, t_0)\Phi(t, t_0) = 1$. This completes the proof of (2.5.25). Thus, the proof of the lemma is complete. $\qquad\square$

Observation 2.5.7

(i) The conclusions of Lemma 2.5.5 provide a relationship between the normalized fundamental solution processes of (2.5.4)/(2.3.36) and (2.5.23). Moreover, Lemma 2.5.5 provides an alternative conceptual proof of Lemma 2.5.2(a).

(ii) In addition, Lemma 2.5.5, conceptually, confirms that the solution process provided by Theorem 2.3.1 is indeed the algebraic inverse of a fundamental solution process of (2.5.4)/(2.3.36). Of course, Theorem 2.3.1 provides an analytic method for finding an inverse of any fundamental solution process of (2.5.4)/(2.3.36).

In the following, we give an alternative analytic (conceptual) proof of Lemma 2.5.2(c). This result provides a very general universal argument for establishing the validity of the relation (2.5.14). Moreover, the presented argument provides a conceptual reasoning for undertaking the study of more complex properties of solutions to more general differential equations.

Lemma 2.5.6. *Let $\Phi(t, t_0)$ be a normalized fundamental matrix solution to (2.5.4) at $t = t_0$ and let $x(t, t_0, x_0)$ be the solution process of the IVP (2.5.4) through (t_0, x_0). Then, for all $t \geq t_0$ and t, t_0 in J,*

(a) $$x(t, s, x(s, t_0, x_0)) = x(t, t_0, x_0); \qquad (2.5.26)$$

(b) $$\Phi(t, s)\Phi(s, t_0) = \Phi(t, t_0). \qquad (2.5.27)$$

Proof. The proof of the lemma can be reconstructed by imitating the proof of the deterministic version of the lemma [80]. The details are left as an exercise to the reader. □

A very important byproduct of the uniqueness theorem is that the knowledge of the solution process at a future time depends only on the knowledge of the initial data, and it is independent of its knowledge between the initial time and the future time. This is demonstrated by Lemma 2.5.6 in the context of the solution process of the IVP (2.5.4). The following result extends this idea to the IVP (2.5.1).

Lemma 2.5.7. *Let the hypotheses of Theorem 2.5.2 be satisfied. For $t_0 \leq s \leq t$, let $y(t) = y(t, t_0, y_0)$ and $y(t, s, y(s)) = y(t, s, y(s, t_0, y_0))$ be the solution processes of (2.5.1) through (t_0, y_0) and $(s, y(s))$, respectively. Then,*

$$y(t) = y(t, t_0, y_0) = y(t, s, y(s, t_0, y_0)), \quad \text{for } t_0 \leq s \leq t. \qquad (2.5.28)$$

Proof. The proof of the lemma can be reconstructed by imitating the proof of the deterministic version of the lemma [80]. The details are left as an exercise to the reader. □

Observation 2.5.8

(i) We emphasize that the proof of Lemma 2.5.7 is based on the explicit solution to (2.5.1), algebraic features of the normalized fundamental solution to (2.5.4)/(2.3.1), the computational representation of the solution process of (2.5.1) in (2.5.2), and the properties of Itô–Doob and elementary integrals.

(ii) One can use the integral representation of the solution process in (2.5.6) of (2.5.1) and the properties of integrals to establish the validity of Lemma 2.5.7. We note that we do not require the explicit knowledge of the solution process of (2.5.1).

(iii) The validity of Lemma 2.5.7 also follows from the conceptual reasoning similar to the reasoning presented in the proof of Lemma 2.5.6, and the applications of Theorem 2.5.2. In fact, we observe that $y(t, s, y(s)) = y(t, s, y(s, t_0, y_0))$ (by the notation and definition of the solution process). For $s = t$, we have

$$y(t, s, y(s, t_0, y_0)) = y(t, t, y(t, t_0, y_0)) \quad \text{(by Definition 2.3.2)}$$

$$= y(t, t_0, y_0). \tag{2.5.29}$$

This shows that solution processes $y(t, t_0, y_0)$ and $y(t, s, y(s, t_0, y_0))$ are equal at $s = t$. Hence, by the uniqueness theorem 2.5.2, $y(t, t_0, y_0) = y(t, s, y(s, t_0, y_0))$ is valid for $t_0 \leq s \leq t$.

In the following, we go one step ahead to conclude that the solution process of (2.5.1) has additional basic, natural, and fundamental properties. This statement is illustrated by the following results.

Theorem 2.5.3 (continuous dependence of the solution on initial data). *Let the hypotheses of Theorem 2.5.2 be satisfied. Then, the solution process $y(t) = y(t, t_0, y_0)$ is continuous with respect to (t, t_0, y_0) for $t \geq t_0$ and t, t_0 in J. In particular, it is continuous with respect to the initial conditions/data.*

Proof. First, we note that by virtue of the solution process, $y(t, t_0, y_0)$ is continuous in t for each (t_0, y_0). We need to examine the continuity with respect to (t_0, y_0) for each t. For this purpose, we examine the solution expression (2.5.8). From the expression $y(t) = y(t, t_0, y_0)$ in (2.5.2), it is clear that $y(t)$ is a linear function of y_0, i.e. $y(t)$ is a first-degree polynomial function in y_0 for each fixed (t, t_0) for $t \geq t_0$ and t, t_0 in J. The basic result in elementary calculus states that every polynomial function is continuous. Therefore, we conclude that $x(t)$ is continuous with respect to y_0 for each fixed (t, t_0) for $t \geq t_0$ and t, t_0 in J. To prove the continuity of the solution process $y(t, t_0, y_0)$ of (2.5.1) with respect to t_0 for fixed (t, y_0), we need to examine the continuity of functions in three terms on the right-hand side of (2.5.8). We note that the continuity of the function $\Phi(t, t_0)y_0$ in the first term with respect to t_0 follows from Lemma 2.5.2(e). This is because of the fact that $d_{t_0}\Phi(t, t_0)$ satisfies the adjoint differential equation (2.5.23). This implies that $\Phi(t, t_0)$ is continuous in t_0, and hence the function $\Phi(t, t_0)y_0$ in the first term is continuous with respect to t_0. The continuity of functions in the second and third terms on the right-hand side with respect to t_0 follows from the continuity of the rate coefficient functions of the differential equations and their properties of indefinite integrals as a function of t_0. From this elementary argument, we conclude that the solution process $y(t, t_0, y_0)$

is continuous with respect to the initial data (t_0, y_0). This completes the proof of the theorem. □

In fact, in the following we present a result that exhibits the differentiability of the solution process of (2.5.1) in the sense of Itô–Doob calculus with respect to all three variables (t, t_0, y_0). Prior to this result, we first prove a result concerning the differentiability of the solution process of (2.5.4)/(2.3.36) in the sense of Itô–Doob with respect to all three variables (t, t_0, x_0).

Lemma 2.5.8. *Let $x(t, t_0, x_0)$ be the solution process of the IVP (2.5.4) through (t_0, x_0). Then, for all $t \geq t_0$ and t, t_0 in J,*

(a) *$\frac{\partial}{\partial x_0} x(t) = \frac{\partial}{\partial x_0} x(t, t_0, x_0)$ exists and it satisfies the differential equation*

$$dx = f(t)x \, dt + \sigma(t)x \, dw(t), \qquad \frac{\partial}{\partial x_0} x(t_0) = 1, \qquad (2.5.30)$$

where $\frac{\partial}{\partial x_0} x(t, t_0, x_0)$ stands for a deterministic partial derivative of the solution process $x(t, t_0, x_0)$ of the IVP (2.5.4) with respect to x_0 at (t, t_0, x_0) for fixed (t, t_0);

(b) *$\frac{\partial}{\partial t_0} x(t) = \frac{\partial}{\partial t_0} x(t, t_0, x_0)$ exists (in the generalized sense, this is beyond the level of the course) and it satisfies the differential equation*

$$dx = f(t)x \, dt + \sigma(t)x \, dw(t), \qquad \frac{d}{dt_0} x(t_0) = z_0, \qquad (2.5.31)$$

where $\partial_{t_0} x(t, t_0, x_0) = \frac{\partial}{\partial t_0} x(t, t_0, x_0) dt_0$ stands for the Itô–Doob-type stochastic partial differential of the solution process $x(t, t_0, x_0)$ of (2.5.4) with respect to t_0 at (t, t_0, x_0) for fixed (t, x_0); its initial differential condition $\partial x(t_0)$ $\left(\frac{dx_0(t_0)}{dt_0} = \frac{\partial}{\partial t_0} x(t_0) = z_0 \right)$ satisfies the differential equation

$$\frac{\partial}{\partial t_0} x(t_0, t_0, x_0) dt_0 \equiv \partial x(t_0) = [-f(t_0) + \sigma^2(t_0)]x_0 dt_0 - \sigma(t_0)x_0 dw(t_0),$$

$$(2.5.32)$$

or

$$\frac{d}{dt_0} x(t_0) \equiv \frac{d}{dt_0} x(t_0, t_0, x_0) = z_0 = [-f(t_0) + \sigma^2(t_0)]x_0 - \sigma(t_0)x_0 \frac{d}{dt_0} w(t_0),$$

where $\frac{d}{dt_0} w(t_0)$ is the white noise (Observation 1.2.11) process of the Wiener process.

Proof. From Lemma 2.5.4, the solution $x(t, t_0, x_0) = \Phi(t, t_0)x_0$ to the IVP (2.5.4) through (t_0, x_0) is a linear function of x_0, i.e. $x(t, t_0, x_0)$ is a first-degree polynomial in x_0 for each fixed (t, t_0) for $t \geq t_0$ and t, t_0 in J. The basic result in elementary calculus states that every polynomial function is continuously differentiable. Therefore, the solution process $x(t) = x(t, t_0, x_0)$ is differential with respect to x_0 for each fixed (t, t_0). In fact, we have

$$\frac{\partial}{\partial x_0} x(t) = \frac{\partial}{\partial x_0} x(t, t_0, x_0) = \Phi(t, t_0). \tag{2.5.33}$$

Moreover, $\frac{\partial}{\partial x_0} x(t, t_0, x_0)$ is a continuous function in x_0. This is because of the fact that $\frac{\partial}{\partial x_0} x(t, t_0, x_0) = \Phi(t, t_0)$ is independent of x_0 (i.e. constant) for each fixed (t, t_0) for $t \geq t_0$ and t, t_0 in J. From (2.5.33), it is clear that $\frac{\partial}{\partial x_0} x(t) = \frac{\partial}{\partial x_0} x(t, t_0, x_0)$ is the normalized fundamental solution process of (2.5.4)/(2.3.36). Therefore, the validity of the statement (a) follows, immediately.

To prove the statement (b), again from the definition of the solution $x(t, t_0, x_0) = \Phi(t, t_0) x_0 = \Phi(t, w(t)) \Phi^{-1}(t_0, w(t_0)) x_0$ to the IVP (2.5.4), Theorem 2.3.1, and applying the Itô–Doob formula (Theorem 1.3.4), we have

$$\partial_{t_0} x(t, t_0, x_0) = \Phi(t, w(t)) d_{t_0} \Phi^{-1}(t_0, w(t_0)) x_0 \quad \text{[from (2.3.40)]}$$

$$= \Phi(t, w(t))[\Phi^{-1}(t_0, w(t_0))[-f(t_0) + \sigma^2(t_0)] dt_0$$

$$- \Phi^{-1}(t_0, w(t_0)) \sigma(t_0) dw(t_0)] x_0 \quad \text{[from (2.3.35)]}$$

$$= \Phi(t, t_0)[-f(t_0) + \sigma^2(t_0)] x_0 dt_0$$

$$- \Phi(t, t_0) \sigma(t_0) x_0 dw(t_0)] \quad \text{(by simplification)}$$

$$= \Phi(t, t_0) \partial x(t_0) \quad \text{[from (2.5.32)], \tag{2.5.34}}$$

where $\partial_{t_0} x(t, t_0, x_0)$ denotes the Itô–Doob partial differential of the solution process $x(t, t_0, x_0)$ of (2.5.4) with respect to t_0 at (t, t_0, x_0); $\partial x(t_0)$ is as defined in the lemma, and it satisfies the differential equation (2.5.32).

From (2.5.32), it is clear that $\partial_{t_0} x(t, t_0, x_0)$ satisfies the IVP (2.5.31) with initial data $\partial x(t_0) = \partial_{t_0} x(t, t_0, x_0)$, which satisfies the differential equation (2.5.32). $\square$

Observation 2.5.9

(i) From the proof of Lemma 2.5.8, we observe that $\frac{\partial}{\partial x_0} x(t, t_0, x_0)$ is independent of x_0, i.e. $\frac{\partial}{\partial x_0} x(t, t_0, x_0) = \Phi(t, t_0)$. It is obvious that $\frac{\partial}{\partial x_0} x(t, t_0, x_0)$ is continuous in (t, t_0, x_0). Moreover, its second partial derivative of $x(t, t_0, x_0)$ is $\frac{\partial^2}{\partial x_0^2} x(t, t_0, x_0)$ exists for fixed (t, t_0) and $\frac{\partial^2}{\partial x_0^2} x(t, t_0, x_0) = 0$.

(ii) By using (i) and Lemma 2.5.8, we introduce the Itô–Doob partial differentials of the solution process $x(t, t_0, x_0)$ of (2.5.4) with respect to x_0 and t_0 at (t, t_0, x_0), respectively, as follows:

$$\partial_{x_0} x(t, t_0, x_0) = \frac{\partial}{\partial x_0} x(t, t_0, x_0) dx_0 = \Phi(t, t_0) dx_0, \tag{2.5.35}$$

$$\partial_{t_0} x(t, t_0, x_0) = \frac{\partial}{\partial x_0} x(t, t_0, x_0) \partial x_0(t_0) = \Phi(t, t_0) \partial x_0(t_0), \tag{2.5.36}$$

where dx_0 is the Itô–Doob partial differential of $x(t, t_0, x_0)$ at t_0 for fixed (t, t_0), i.e.

$$dx(t_0) = dx_0 = f(t_0) x_0 dt_0 + \sigma(t_0) x_0 dw(t_0), \tag{2.5.37}$$

and $\frac{\partial}{\partial t_0} x(t_0, t_0, x_0) dt_0 = \partial x(t_0)$ is the partial differential of $x(t, t_0, x_0)$ in (2.5.32) with respect to t_0 at t_0 for fixed x_0. Note the difference between dx_0 and $\partial x(t_0)$.

(iii) From notations (2.5.35) and (2.5.36), we compute the Itô–Doob-type second mixed partial differentials of the solution process $x(t, t_0, x_0)$ of (3.5.4) as

$$\partial^2_{t_0 x_0} x(t, t_0, x_0) = \partial_{t_0}(\partial_{x_0} x(t, t_0, x_0)) \qquad \text{(by the definition of the partial differential)}$$

$$= \partial_{t_0} \left(\frac{\partial}{\partial x_0} x(t, t_0, x_0) dx_0 \right) \qquad \text{[from (2.5.35)]}$$

$$= \Phi(t, t_0)[-f(t_0) + \sigma^2(t_0)] dt_0 dx_0$$

$$\quad - \Phi(t, t_0)\sigma(t_0) dw(t_0) dx_0 \qquad \text{[from (2.5.16)]}$$

$$= -\Phi(t, t_0)\sigma^2(t_0) x(t_0) dt_0 \qquad \text{[from (2.5.37)].} \tag{2.5.38}$$

On the other hand,

$$\partial^2_{x_0 t_0} x(t, t_0, x_0) = \partial_{x_0}(\partial_{t_0} x(t, t_0, x_0)) \qquad \text{(by the definition of the partial differential)}$$

$$= \partial_{x_0}(\Phi(t, t_0) \partial x(t_0)) \qquad \text{[from (2.5.36)]}$$

$$= \Phi(t, t_0) \partial_{x_0}(\partial x(t_0)) \qquad \text{(from Theorem 1.5.1P2 [80])}$$

$$= \Phi(t, t_0)[-f(t_0) + \sigma^2(t_0)] dt_0$$

$$\quad - \Phi(t, t_0)\sigma(t_0) dw(t_0) \partial_{x_0} x_0 \qquad \text{[from (2.5.32)]}$$

$$= \Phi(t, t_0)[-f(t_0) + \sigma^2(t_0)] dt_0 dx_0$$

$$\quad - \Phi(t, t_0)\sigma(t_0) dw(t_0) dx_0 \qquad \text{(from the notation)}$$

$$= -\Phi(t, t_0)\sigma^2(t_0) x(t_0) dt_0 \qquad \text{[from (2.5.37)].} \tag{2.5.39}$$

Hence, we conclude that its mixed second differential exist and $\partial^2_{x_0 t_0} x(t, t_0, x_0) = \partial^2_{t_0 x_0} x(t, t_0, x_0)$, and $\left(\frac{\partial^2}{\partial t_0 \partial x_0} x(t, t_0, x_0)\right)$ is the normalized fundamental solution process of the adjoint differential equation (2.5.23).

(iv) We further observe that Lemma 2.5.8(b) provides a conceptual proof for Lemma 2.5.2(e). In fact, from (2.5.32) and (2.5.34), we have

$$\frac{\partial}{\partial x_0} \left(\partial_{t_0} x(t, t_0, x_0) \right)$$

$$= \frac{\partial}{\partial x_0} (\Phi(t, t_0) \partial x(t_0)) \quad \text{[from (2.5.34)]}$$

$$= \frac{\partial}{\partial x_0} (\Phi(t, t_0)[-f(t_0) + \sigma^2(t_0)]dt_0 - \Phi(t, t_0)\sigma(t_0)dw(t_0)]x_0)$$

$$\text{[from (2.5.32)]}$$

$$= \Phi(t, t_0) \left([-f(t_0) + \sigma^2(t_0)]dt_0 - \Phi(t, t_0)\sigma(t_0)dw(t_0)]\frac{\partial}{\partial x_0} x_0 \right)$$

$$\text{(from Theorem 1.5.1P2)}$$

$$= \Phi(t, t_0) \left([-f(t_0) + \sigma^2(t_0)]dt_0 - \Phi(t, t_0)\sigma(t_0)dw(t_0)] \right)$$

$$\left(\text{from } \frac{\partial}{\partial x_0} x_0 = I_{n \times n} \right).$$

On the other hand,

$$\partial_{t_0} \left(\frac{\partial}{\partial x_0} x(t, t_0, x_0) \right) = \partial_{t_0} (\Phi(t, t_0)) \quad \text{[from (2.5.33)]}$$

$$= \partial_{t_0} \Phi(t, t_0) \quad \text{(by the notation)}.$$

From the above discussion and (iii), the validity of (2.5.16) immediately follows.

Theorem 2.5.4 (differentiability of solutions with respect to initial conditions). *Let the hypotheses of Theorem 2.5.2 be satisfied. In addition, let $y(t, t_0, y_0)$ be the solution process of the IVP (2.5.1). Then, for all $t \geq t_0$ and t, t_0 in J.*

(i) $\frac{\partial}{\partial y_0} y(t) = \frac{\partial}{\partial y_0} y(t, t_0, y_0)$ *exists and it satisfies the IVP (2.5.30)*;

(ii) $\frac{\partial}{\partial t_0} y(t) = \frac{\partial}{\partial t_0} y(t, t_0, y_0)$ *exists (in the generalize sense) and it satisfies the differential equation*

$$dx = f(t)x \, dt + \sigma(t)x \, dw(t), \quad \partial_{t_0} y(t, t_0, y_0) = \partial y(t_0), \quad (2.5.40)$$

where $\partial y(t_0)$ satisfies the differential equation

$$\partial y(t_0) = [(-f(t_0) + \sigma^2(t_0))y_0 - p(t_0) + \sigma(t_0)q(t_0)]dt_0$$

$$- [\sigma(t_0)y_0 + q(t_0)]dw(t_0) \quad (2.5.41)$$

or

$$\frac{d}{dt_0}y(t_0) = z_0 = [(-f(t_0) + \sigma^2(t_0))y_0 - p(t_0) + \sigma(t_0)q(t_0)]$$

$$- [\sigma(t_0)y_0 + q(t_0)]\frac{d}{dt_0}w(t_0),$$

where $\frac{d}{dt_0}w(t_0)$ is the white noise (Observation 1.2.11) process of the Wiener process.

Proof. From the direct observation of the solution expression of the IVP (2.5.1) in (2.5.2), we conclude that $y(t, t_0, y_0)$ is a first-degree polynomial in y_0. Because of Observation 2.5.1(iv), this first term is indeed $x(t, t_0, y_0) = \Phi(t, t_0)y_0$. By imitating the argument used in the proof of Lemma 2.5.8, we conclude that $\frac{\partial}{\partial y_0}y(t, t_0, y_0)$ exists and it is equal to $\Phi(t, t_0)$, with $\Phi(t, t_0)$ being the normalized fundamental solution process of (2.5.4)/(2.3.1). As a result of this, $\frac{\partial}{\partial y_0}y(t, t_0, y_0) = \Phi(t, t_0) = \frac{\partial}{\partial y_0}x(t, t_0, y_0)$, and it satisfies the corresponding (*variational differential equation*) first-order linear homogeneous scalar stochastic differential equation (2.5.4) with the initial data $(t_0, 1)$. Hence, $\frac{\partial}{\partial y_0}y(t, t_0, y_0)$ is the solution of the IVP (2.5.30). This completes the proof of (i).

To prove (ii), again from the solution expression (2.5.8) of the IVP (2.5.1),

$$y(t, t_0, y_0) = x(t, t_0, y_0) + \int_{t_0}^{t} \Phi(t, s)[p(s) - \sigma(s)q(s)]ds + \int_{t_0}^{t} \Phi(t, s)q(s)dw(s).$$

$$(2.5.42)$$

The first term $x(t, t_0, y_0) = \Phi(t, t_0)y_0$ on the right-hand side has the Itô–Doob differential with respect to t_0. Moreover, from the application of Lemma 2.5.8, in particular (2.5.33), we can conclude that

$$\partial_{t_0}x(t, t_0, y_0) = \Phi(t, t_0)\partial x(t_0),$$

where $\partial x(t_0) = [-f(t_0) + \sigma^2(t_0)]y_0dt_0 - \sigma(t_0)y_0dw(t_0)$.

By using Lemma 2.5.2(d) and applying the fundamental theorem of integral calculus to the second and third terms on the right-hand side of (2.5.42), we get

$$\partial_{t_0}\left(\int_{t_0}^{t} \Phi(t, s)[p(s) - \sigma(s)q(s)]ds\right)$$

$$= \Phi(t, a)\partial_{t_0}\left(\int_{t_0}^{t} \Psi(s, a)[p(s) - \sigma(s)q(s)]ds\right)$$

$$= \Phi(t, a)\left[\partial_{t_0}\left(\int_{a}^{t} \Psi(s, a)[p(s) - \sigma(s)q(s)]ds\right)\right. \quad \text{[by Lemma 2.5.2(d)]}$$

$$- \partial_{t_0} \left(\int_a^{t_0} \Psi(s,a)[p(s) - \sigma(s)q(s)]ds \right) \Bigg]$$

$$= \Phi(t,a)\Psi(t_0,a)[-p(t_0) + \sigma(t_0)q(t_0)]dt_0 \qquad \text{[by Lemma 2.5.2(d)]}$$

$$= \Phi(t,t_0)[-p(t_0) + \sigma(t_0)q(t_0)]dt_0, \qquad (2.5.43)$$

and

$$\partial_{t_0} \left(\int_{t_0}^t \Phi(t,s)q(s)dw(s) \right)$$

$$= \Phi(t,a)\partial_{t_0} \left(\int_{t_0}^t \Psi(s,a)q(s)dw(s) \right) \qquad \text{[by Lemma 2.5.2(d)]}$$

$$= \Phi(t,a) \left[\partial_{t_0} \left(\int_a^t \Psi(s,a)q(s)dw(s) \right) - \partial_{t_0} \left(\int_a^{t_0} \Psi(s,a)q(s)dw(s) \right) \right]$$

$$= -\Phi(t,a)\Psi(s,a)q(t_0)dw(t_0)) \qquad \text{[by Lemma 2.5.2(d)]}$$

$$= -\Phi(t,t_0)q(t_0)dw(t_0). \qquad (2.5.44)$$

From (2.5.42)–(2.5.44), we have

$$\partial_{t_0} y(t,t_0,y_0)$$

$$= \partial_{t_0} x(t,t_0,y_0) + \partial_{t_0} \left(\int_{t_0}^t \Phi(t,s)[p(s) - \sigma(s)q(s)]ds \right)$$

$$+ \partial_{t_0} \left(\int_{t_0}^t \Phi(t,s)q(s)dw(s) \right) \qquad \text{[by Theorem 1.3.3 (sum formula)]}$$

$$= \Phi(t,t_0)[[-f(t_0) + \sigma^2(t_0)]y_0 dt_0 - \sigma(t_0)y_0 dw(t_0)] - \Phi(t,t_0)q(t_0)dw(t_0)$$

$$+ \Phi(t,t_0)[-p(t_0) - \sigma(t_0)q(t_0)]dt_0 \qquad \text{[from (2.5.43) and (2.5.44)]}$$

$$= \Phi(t,t_0)[[-f(t_0) + \sigma^2(t_0)]y_0 - p(t_0) + \sigma(t_0)q(t_0)]dt_0$$

$$- [\sigma(t_0)y_0 + q(t_0)]dw(t_0)] \qquad \text{(by regrouping)}$$

$$= \Phi(t,t_0)\partial_{t_0} y(t_0), \qquad (2.5.45)$$

where $\partial_{t_0} y(t_0)$ is as defined (2.5.41), i.e.

$$\partial_{t_0} y(t_0) = [(-f(t_0) + \sigma^2(t_0))y_0 - p(t_0) + \sigma(t_0)q(t_0)]dt_0$$

$$- [\sigma(t_0)y_0 + q(t_0)]dw(t_0). \qquad (2.5.46)$$

From the representation of the solution to the IVP (2.3.36)/(2.5.4) and (2.5.45), it is clear that $\partial_{t_0} y(t,t_0,y_0)$ is the solution to the IVP (2.5.40). This completes the proof of the theorem. □

Observation 2.5.10. We note that the proof of Theorem 2.5.4 can be reconstructed by utilizing the conceptual aspect of Lemma 2.5.8. For the sake of simplicity, we chose to prove it by utilizing algebraic properties coupled with the solution representation of (2.5.2).

In the following, we present a very important result that connects the solution process of (2.5.1) with the corresponding homogeneous differential equation (2.5.4). This result is the Alekseev-type variation of constants/parameters formula [80, 98]. It provides a conceptual tool and insight for undertaking the study and for developing the results concerning linear/nonlinear and time-varying/time-invariant scalar/systems of differential equations. In addition, it provides an alternative method for investigating the qualitative properties of more complex differential equations [98, 112, 178].

Theorem 2.5.5 (variation of constants formula). *Let the hypotheses of Theorem 2.5.2 be satisfied. In addition, let $y(t) = y(t, t_0, y_0)$ and $x(t) = x(t, t_0, y_0)$ be the solution processes of the IVPs (2.5.1) and (2.5.4) through the initial data (t_0, y_0), respectively. Then,*

$$y(t, t_0, y_0) = x(t, t_0, y_0) + \int_{t_0}^{t} \Phi(t, s)[p(s) - \sigma(s)q(s)]ds + \int_{t_0}^{t} \Phi(t, s)q(s)dw(s).$$

$$(2.5.47)$$

Proof. From the assumption of the theorem, for $t_0 \leq s \leq t$, we define a solution to the IVP (2.5.4) through s, $y(s)$, as follows:

$$x(t, s, y(s, t_0, y_0)) = x(t, s, y(s)), \qquad (2.5.48)$$

where $y(s) = y(s, t_0, y_0)$ a solution to the IVP (2.5.1) through (t_0, y_0). From (2.5.48) and the definition of the IVP, we observe that for $s = t_0$

$$x(t, t_0, y(t_0, t_0, y_0)) = x(t, t_0, y_0). \qquad (2.5.49)$$

From the application of Lemma 2.5.8, the Itô–Doob differential formula (Theorem 1.3.4), the chain rule, and Observation 2.5.9 with respect to s for $t_0 \leq s \leq t$, we have

$$d_s x(t, s, y(s)) = \partial_{t_0} x(t, s, y(s)) + \partial_{x_0} x(t, s, y(s)) + \partial_{t_0 x_0}^2 x(t, s, y(s))$$

[by Theorem 1.3.4 (sum rule)]

$$= \frac{\partial}{\partial x_0} x(t, s, y(s))\partial y(s) + \frac{\partial}{\partial x_0} x(t, s, y(s))dy(s)$$

$$+ \partial_{t_0} \frac{\partial}{\partial x_0} x(t, s, y(s))dy(s) \qquad \text{[from (2.5.35)–(2.5.37)]}$$

$$= \Phi(t, s)\partial y(s) + \Phi(t, s)dy(s)$$

$$+ \Phi(t, s)\partial y(s)dy(s) \qquad \text{[from (2.5.35)–(2.5.37)]}$$

$$= \Phi(t,s)([-f(s) + \sigma^2(s)]y(s)ds - \sigma(s)y(s)dw(s))$$
$$+ \Phi(t,s)([f(s)y(s) + p(s)]ds + [\sigma(s)y(s) + q(s)]dw(s))$$
$$+ \Phi(t,s)([-f(s) + \sigma^2(s)]y(s)ds$$
$$- \sigma(s)y(s)dw(s))([f(s)y(s) + p(s)]ds$$
$$+ [\sigma(s)y(s) + q(s)]dw(s))]$$

[by Lemma 2.5.8, (2.5.31), and (2.5.32)]

$$= \Phi(t,s)[-f(s) + \sigma^2(s)]y(s)ds - \Phi(t,s)\sigma(s)y(s)dw(s)$$
$$+ \Phi(t,s)[f(s)y(s) + p(s)]ds + \Phi(t,s)[\sigma(s)y(s) + q(s)]dw(s)$$
$$- \Phi(t,s)[\sigma^2(s)y(s) + \sigma(s)q(s)]ds \qquad \text{(by Itô–Doob calculus)}$$

$$= \Phi(t,s)\Big[[(p(s) - \sigma(s)q(s))]ds + q(s)dw(s)\Big] \qquad \text{(by simplifying)}.$$

$$(2.5.50)$$

By integrating (2.5.50) on both sides with respect to s from t_0 to t, using Lemma 2.5.6, and the notations and the uniqueness (Theorem 2.5.2) of the solution process of (2.5.4), we get

$$x(t,t,y(t,t_0,y_0)) - x(t,t,y(t,t_0)) = y(t,t_0,y_0) - x(t,t_0,y_0)$$
$$= \int_{t_0}^t \Phi(t,s)[p(s) - \sigma(s)q(s)]ds + \int_{t_0}^t \Phi(t,s)q(s)dw(s),$$

which can be written as

$$y(t,t_0,y_0) = x(t,t_0,y_0) + \int_{t_0}^t \Phi(t,s)[p(s) - \sigma(s)q(s)]ds + \int_{t_0}^t \Phi(t,s)q(s)dw(s).$$

This completes the proof of the theorem. Moreover, from (2.5.2) and (2.5.8), we have

$$y(t,t_0,y_0) = \Phi(t,t_0)y_0 + \int_{t_0}^t \Phi(t,s)[p(s) - \sigma(s)q(s)]ds$$

$$+ \int_{t_0}^t \Phi(t,s)q(s)dw(s). \qquad (2.5.51)$$

$$\square$$

Observation 2.5.11

(i) From the conclusion of Theorem 2.5.5, it is obvious that this theorem provides an alternative conceptual approach to the closed-form representation of the solution to (2.5.1). In fact, one can compare the solution expressions on the right-hand sides of (2.5.8)/(2.5.2) and (2.5.47)/(2.5.51). These expressions are exactly the same.

(ii) We further observe that the solution representation of (2.5.1) in either (2.5.2) or (2.5.8) was derived based on the conceptual knowledge of the corresponding homogeneous scalar differential equation (2.3.1) or (2.5.4). In particular, the smoothness (conceptual) properties of the solution process with respect to its initial data/conditions outlined in Observation 2.5.9 were utilized.

(iii) For the validity of Theorem 2.5.5, we merely needed the existence of the solution process of (2.5.1).

Exercises

1. Using Uniqueness Theorem 2.5.2 and Observation 2.5.4(iii), prove the relations (2.5.13) and (2.5.14).
2. Complete the proof of Lemma 2.5.3.
3. Construct the proof of Lemma 2.5.4.
4. Given: $x\Psi(t, t_0)\Phi(t, t_0)x = xx$, for any $x \neq 0$. Show that $\Psi(t, t_0)\Phi(t, t_0) = 1$.
5. Prove Lemmas 2.5.6 and 2.5.7.
6. Using the integral representation of the solution process of (2.5.1) in (2.5.6), show that the conclusion of Lemma 2.5.7 remains true.

2.6 Notes and Comments

The initiation and development of the material in this chapter is original. The development of the stochastic modeling procedure in Section 2.1 is a modified form of the classical modeling processes in sciences and engineering. It is based on the elementary descriptive statistical approach which includes the classical approach in unified way. The examples, exercises, and illustrations are based on the second author's classroom experience over a period of more than 40 years. The method of solving integrable differential equations is the topic of Section 2.2. The applied Illustrations 2.2.1–2.2.3 are derived from [11, 80], [55, 80] and [23, 80], respectively. The initiation of an eigenvalue and eigenvector–type approach is the most suitable approach for the introduction of stochastic modeling, methods, and the motivation for an advanced study at the undergraduate and early graduate levels with a minimal mathematical background. The concepts of fundamental, general, and particular solutions are introduced and determined, systematically. Several stochastic models for dynamic processes in the biological, chemical, physiological, and physical sciences are developed to illustrate the role of stochastic modeling and the methods of finding closed-form solutions of the processes. Numerous examples with analytic closed-form solutions are presented to illustrate these concepts. In addition, the applied Illustrations 2.3.1 and 2.3.2 from chemistry [11, 41, 80, 125, 143, 183] and Illustration 2.3.3 [1, 6, 26, 54, 80, 154, 155, 160, 166, 172, 174, 188, 189] from physiology are outlined. Section 2.4 deals with the method of variation of parameters for solving first-order linear nonhomogeneous stochastic scalar differential equations of the Itô–Doob type. Several numerical and applied examples are presented to illustrate the method and concepts. In addition to the classical applied Illustrations 2.4.1 and 2.4.2, Illustrations 2.4.3–2.4.6 in the biological, physical, and physiological sciences [80] are outlined. Several numerical and applied exercises are formulated and presented at the end of the sections. The material in Section 2.5 outlines the basis and scope of conceptual algorithms. It is based on the second author's class experience and notes over a period of more than 40 years. The objective is to generate curiosity, to motivate, and to challenge the undergraduate to expand the knowledge beyond the material presented here.

Chapter 3

First-Order Nonlinear Differential Equations

Introduction

The mathematical modeling and procedures for solving first-order nonlinear Itô–Doob-type stochastic scalar differential equations are the objectives of this chapter. The general step-by-step procedure for finding the general solution and the solution to IVPs are outlined. We begin with the mathematical modeling of dynamic processes in sciences and engineering (in Section 3.1). Two probabilistic models, namely the modified Brownian motion and sequential colored noise models are outlined. As in Chapter 2, several dynamic processes in the biological, chemical, engineering, medical, physical, and social sciences are presented to illustrate the modeling procedures. Section 3.2 details a five-step general algorithm for finding either an implicit or an explicit form of the solution process for a class of first-order nonlinear Itô–Doob-type stochastic scalar differential equations. This problem-solving algorithm is called the energy/Lyapunov function method. The five-step general algorithm is applied logically to a particular subclasses of first-order Itô–Doob-type stochastic scalar nonlinear differential equations in Sections 3.3 and 3.4. Section 3.3 focuses on a subclass of first-order nonlinear stochastic scalar differential equations that are reducible to Itô–Doob-type stochastic integrable differential equations. On the other hand, Section 3.4 deals with a subclass of first-order nonlinear Itô–Doob-type stochastic scalar differential equations that are reducible to Itô–Doob-type stochastic scalar linear nonhomogeneous differential equations. The remainder of the chapter is devoted to showing the usefulness and scope of the general theoretical algorithms developed in Sections 3.3 and 3.4. The concepts of the Itô–Doob-type stochastic variable separable, homogeneous, Bernoulli, and essentially time-invariant differential equations are introduced in Sections 3.5–3.8, respectively. The class of separable, homogeneous, Bernoulli-type, and essentially time-invariant Itô–Doob-type stochastic differential equations belong to the subclass determined in Sections 3.3 and 3.4. The solution procedure for Itô–Doob-type stochastic variable separable differential equations is presented in Section 3.5. Moreover, the mathematical models for enzymatic reactions,

175

acid-catalyzed hydrolysis, concurrent chemical reactions, single-species ecosystems, liquid leakage problem, etc. are also presented. The methods for finding solutions to homogeneous differential equations, Bernoulli-type differential equations, and essentially time-invariant differential equations are outlined in Sections 3.6, 3.7, and 3.8, respectively. Mathematical models of relative growth-allometry and the dynamic of swarms are described in Section 3.6, and mathematical models of economic growth–Solow's model and population extinction–Richard Levin's model are presented in Section 3.7.

3.1 Mathematical Modeling

The material in this section parallels Section 2.1 of Chapter 2. We briefly outline the development of stochastic mathematical models for dynamic processes in the biological, chemical, engineering, medical, physical and social sciences. In Chapter 2, the well-known probabilistic models, namely random walk, Poisson process, and Brownian motion models are used to derive the Itô–Doob-type mathematical description of dynamic processes. In this section, we focus on a generalized version of the Brownian motion model to derive the Itô–Doob-type mathematical description of dynamic processes. This brings to light another existing model-building approach for the classrooms of the 21st century. Several illustrations, examples, and observations are made to establish the relationship between the existing deterministic, predictable, and anticipated knowledge and information based in the framework of stochastic, unpredictable, and random environmental fluctuations.

The Gaussian white noise modeling processes. We recall Observations 2.1.1, 2.1.3–2.1.5, which were based on the dynamic and modeling processes generated by the random walk, Poisson process, and Brownian motion models. In particular, the mathematical models described in (2.1.37), (2.1.43), (2.1.53), and (2.1.55) are as follows:

$$dx = C(t, x)dt + \sigma(t, x)dw(t), \tag{3.1.1}$$

$$\frac{d}{dt}E[x(t) \mid x(t) = x] = \lim_{\Delta t \to 0} E\left[\frac{x(t + \Delta t) - x(t)}{\Delta t} \mid x(t) = x\right] = C(t, x), \tag{3.1.2}$$

$$dx = C(t, x)dt + \sigma(t, x)\xi(t)dt, \tag{3.1.3}$$

$$dx = C(t, x)dt, \tag{3.1.4}$$

where $w(t)$ is a Wiener process (Definition 1.2.12) and $\xi(t)$ is a Gaussian white noise process (Definition 1.2.7). The observations provide insights for a model-building process of dynamic processes in the biological, chemical, engineering, medical, physical, and social sciences. Further details are given in the following observations.

Observation 3.1.1. (i) The differential equation (3.1.4) is the deterministic component of either (3.1.1) or (3.1.3). This is obtained by ignoring environmental or internal random fluctuations. This implies that all system and environmental parameters in $C(t, x)$ are assumed to be under ideally fixed and controlled conditions. Moreover, the system in (3.1.4) is called the deterministic unperturbed system of (3.1.1) or (3.1.3).

(ii) The differential equation (3.1.1) or (3.1.3) is considered to be a stochastic perturbation of the deterministic differential equation (3.1.4). The random terms $\sigma(t, x)dw(t)$ and $\sigma(t, x)\xi(t)$ on the right-hand sides of (3.1.1) and (3.1.3), respectively, are usually interpreted as random perturbations caused by the presence of microscopic and/or the imperfections of the controlled conditions, either known or unknown and/or either environmental or internal fluctuations in the parameters of $C(t, x)$. Moreover, they exhibit the magnitude of the influence of random fluctuations on the state of the dynamic process.

(iii) Based on our model-building processes in Section 2.1 of Chapter 2, we recall that the Gaussian white noise driven differential equation (3.1.3) was derived from the stochastic differential equation of the Itô–Doob-type (3.1.1) (Example 1.2.4 and Observation 1.2.11).

(iv) The observations (i)–(iii) provide an idea about building a more general and feasible stochastic mathematical model for dynamic processes in the biological, chemical, engineering, medical, physical, and social sciences. The idea is to start with the deterministic mathematical model (3.1.4). From (i) and (ii), one can identify the parameter(s) and the source of random internal or environmental perturbations of the parameter(s) of the mathematical model (3.1.4), and formulate a stochastic mathematical model as

$$dx = F(t, x, \xi(t))dt, \quad x(t_0) = x_0, \tag{3.1.5}$$

and, in particular,

$$dx = C(t, x)dt + \sigma(t, x)\xi(t)dt, \quad x(t_0) = x_0, \tag{3.1.6}$$

where ξ is a stochastic process that belongs to $R[[a, b], R[\Omega, R]]$; rate functions F, $C(t, x)$, and $\sigma(t, x)$ are sufficiently smooth, and are defined on $[a, b] \times R$ into R, $x_0 \in R$ and $t_0 \in [a, b]$. Whenever the sample paths $\xi(t, \omega)$ of $\xi(t)$ are smooth functions (sample continuous), one utilizes the deterministic calculus [168], and can look for the solution process determined by (3.1.5)/(3.1.6). We note that such a solution process is a random function with all sample paths starting at x_0.

(v) Of course, there is an exception to the situation described in (iv). For example, if $\xi(t)$ in (3.1.5) or (3.1.6) is a Gaussian process, then the discussion (iv) may not be applicable. For instance, the smoothness of the sample paths of the

Gaussian process may cause a contrast with certain statistical properties of the solution process of (3.1.1) (namely the Markovian property). The lack of correlation may create serious mathematical difficulties regarding the interpretation of (3.1.6). This is because the sample paths of $x(t)$ may not be differentiable in the usual sense. In addition, there is no one-to-one correspondence between the solution process $x(t)$ of (3.1.5)/(3.1.6) and a stationary Gaussian process $\xi(t)$ with $E[\xi(t)] = 0$. But there is a rather narrow high-peaked correlation function, $E[\xi(t)\xi(s)] = \Gamma(t - s)$. Thus, $\Gamma(\tau)$ is nonzero in a neighborhood of $\tau = 0$ with a very sharp maximum at $\tau = 0$. In fact, it creates ambiguity/well-definedness in the modeling process.

In the following, we present a Gaussian white noise modeling process that will overcome the above (v) and other difficulties. We summarize the Gaussian white noise modeling process to arrive at the special form of the mathematical model (3.1.6). The basic ideas are outlined below.

Modified Brownian Motion Model 3.1.1 ([158]). From Brown Motion Model 2.1.3 in Section 2.1 of Chapter 2, we note that certain dynamic models can be directly described by a stationary Gaussian process with independent increments. For example, in the biological or chemical dynamic processes: (r_1) the sizes of species are integral values, (r_2) the successive generations may not overlap in time, and (r_3) breeding as well as environmental parameters may vary in discrete time. In order to model these types of processes, we first must modify the development of Brownian Motion Model 2.1.3, and then provide the derivation of the Gaussian white noise driven differential equation in Observation 2.1.4, (2.1.53)/(3.1.3).

Let $x(t)$ be a state of a dynamic system at a time t. The state of the system is observed over an interval $[t, t + \Delta t]$, where Δt is a small increment in t. Without loss of generality, we assume that Δt is positive. Let $C(t, x)$ and $\sigma(t, x)$ be smooth deterministic rate functions over an interval $[t, t + \Delta t]$. Here, $x(t_0) = x(t) \equiv x$. The dynamic process is under the influence of random perturbations. We observe states $x(t_0) = x(t)$, $x(t_1), x(t_2), \ldots, x(t_k), \ldots, x(t_n) = x(t + \Delta t)$ of a system at $t_0 = t$, $t_1 = t + \tau, t_2 = t + 2\tau + \cdots t_k = t + k\tau, \ldots, t_n = t + \Delta t = t + n\tau$ over the interval $[t, t + \Delta t]$, where n belongs to $\{1, 2, 3, \ldots\}$ and $\tau = \frac{\Delta t}{n}$.

From the above-experimental setup, the microscopic changes in the state of the system over n subintervals are:

$$x(t_0) = x(t),$$

$$x(t_1) - x(t_0) = \frac{\Delta t}{n}C(t, x) + \sigma(t, x)Z_1,$$

$$x(t_2) - x(t_1) = \frac{\Delta t}{n}C(t, x) + \sigma(t, x)Z_2,$$

$$\cdots\cdots\cdots\cdots$$

$$x(t_k) - x(t_{k-1}) = \frac{\Delta t}{n} C(t, x) + \sigma(t, x) Z_k,$$

$$\cdots\cdots\cdots\cdots$$

$$x(t_n) - x(t_{n-1}) = \frac{\Delta t}{n} C(t, x) + \sigma(t, x) Z_n, \tag{3.1.7}$$

where $k = 1, 2, \ldots, n$, Z_k are independent, identically distributed Gaussian random variables with mean $E[Z_k] = 0$, and without loss of generality, its variance is $\mathrm{Var}(Z_k) = \frac{\Delta t}{n}$ (Example 1.1.22). Hence, the changes in the state of the system at $t_1, t_2, \ldots t_k, \ldots, t_n$ are independent. From (3.1.7), the change in the state over the interval $[t, t + \Delta t]$ is given by

$$x(t + \Delta t) - x(t)$$

$$= [x(t_1) - x(t_0)] + [x(t_2) - x(t_1)] + \cdots + [x(t_k) - x(t_{k-1})] + [x(t_n) - x(t_{n-1})]$$

$$= \frac{\Delta t}{n} C(t, x) + \sigma(t, x) Z_1 + \frac{\Delta t}{n} C(t, x) + \sigma(t, x) Z_2 + \cdots + \frac{\Delta t}{n} C(t, x) + \sigma(t, x) Z_n$$

$$= \left[\frac{\Delta t}{n} C(t, x) + \cdots + \frac{\Delta t}{n} C(t, x) \right] + [\sigma(t, x) Z_1 + \sigma(t, x) Z_2 + \cdots + \sigma(t, x) Z_n]$$

$$= C(t, x) \Delta t + \sigma(t, x) \sum_{i=1}^{n} Z_i \quad \text{(by properties of summation)}, \tag{3.1.8}$$

where $C(t, x) \Delta t + \sigma(t, x) \sum_{i=1}^{n} Z_i$ is termed an *aggregate increment* to the given state $x \equiv x(t)$ of the system at the given time t over the interval $[t, t + \Delta t]$ of length Δt.

We recall that

$$W_n = \frac{1}{\sqrt{\Delta t}} \sum_{i=1}^{n} Z_i \to Z \quad \text{as } n \to \infty \quad \text{(by Theorem 1.1.5)},$$

where Z is a standard Gaussian random variable with zero mean and variance 1, and hence

$$\sum_{i=1}^{n} Z_i = \sqrt{\Delta t} W_n \to \sqrt{\Delta t} Z \quad \text{as } n \to \infty. \tag{3.1.9}$$

From Observation 1.2.11, we conclude that the distribution of $\sqrt{\Delta t} Z$ is equal to the distribution of the increment process $\Delta w(t) = w(t + \Delta t) - w(t)$, where w is the Wiener process defined in Definition 1.2.12. From this discussion and (3.1.9), (3.1.8) can be approximated by

$$x(t + \Delta t) - x(t) \simeq C(t, x) \Delta t + \sigma(t, x) \Delta w(t).$$

This yields the Itô–Doob-type stochastic differential equation (Itô–Doob SDE or SDE)

$$dx = C(t, x) dt + \sigma(t, x) dw(t) \quad \text{(Definition 1.3.2)}. \tag{3.1.10}$$

This type of stochastic differential is considered to be equivalent to the Gaussian white noise driven differential equation

$$dx = C(t, x)dt + \sigma(t, x)\xi(t)dt, \qquad (3.1.11)$$

where $C(t, x)$ and $\sigma(t, x)$ are the drift and the diffusion coefficients, respectively; $\xi(t) = \frac{d}{dt}w(t)$, i.e.

$$w(t) = \int_0^t \xi(s)ds \qquad (3.1.12)$$

in a certain sense. We further emphasize that the development of the mathematical model (3.1.11) is based on the Itô–Doob-type integral calculus.

Sequential Colored Noise Model 3.1.2. The practical importance of modeling by the smooth correlated processes [Observation 3.1.1(iv)] is a significant step in the modeling of dynamic processes. In order to utilize the more feasible stochastic modeling process (the usage of phenomenological/biological/chemical/medical/physical social laws) and the knowledge of the system/environmental parameter(s) [Observation 3.1.1(iv)], one needs to modify the one-shot modeling approach into a sequential modeling approach. The colored noise modeling (in short, CNM) approach alleviates the limitations raised in Observation 3.1.1(v). The basic ideas are outlined below:

CNM1. Let us start with a sequence $\{\xi_n(t)\}_{n=1}^\infty$ of sufficiently smooth (sample-pathwise continuous in the usual calculus sense [98, 168]) Gaussian processes which converges in some sense to a Gaussian white noise process $\xi(t)$ in (3.1.6). For each n, we associate a sequence of stochastic differential equations with the smooth random process, as follows:

$$dx_n = C(t, x_n)dt + \sigma(t, x_n)\xi_n(t)dt, \quad x_n(t_0) = x_0, \qquad (3.1.13)$$

where $C(t, x)$ and $\sigma(t, x)$ are as described in (3.1.6).

CNM2. We assume that the IVP (3.1.13) has a unique solution process. In fact, sufficient conditions are given in Section 3.2 to insure the existence and the uniqueness of the solution process of (3.1.13). The IVP (3.1.13) generates a sequence $\{x_n(t)\}_{n=1}^\infty$ of solution processes corresponding to the Gaussian sequence $\{\xi_n(t)\}_{n=1}^\infty$ in CNM1.

CNM3. Under reasonable conditions on the rate functions $C(t, x)$ and $\sigma(t, x)$ in (3.1.6) and a suitable convergent sequence of Gaussian processes $\{\xi_n(t)\}_{n=1}^\infty$ in CNM1, it is shown that the sequence of solution processes $\{x_n(t)\}_{n=1}^\infty$ determined by (3.1.13) converges almost surely or in a quadratic mean or even in probability to a process $x(t)$. Moreover, $x(t)$ is the solution process of (3.1.6).

CNM4. The above-described ideas in CNM1–CNM3 make up a precise mathematical interpretation of (3.1.6). However, we still need to show that (3.1.6) can be modeled by an Itô–Doob form of the stochastic differential equation (3.1.10). Moreover, we need to shed light on the concept of convergence of $\{\xi_n(t)\}_{n=1}^{\infty}$ to the white noise process in (3.1.6). For this purpose, we define

$$w_n(t) - w_n(t_0) = \int_{t_0}^{t} \xi_n(s)ds, \tag{3.1.14}$$

where the integral on the right-hand side of (3.1.14) is an integral in the sense of sample-pathwise [98], since for $\omega \in (\Omega, \mathcal{F}, P)$ and each sample path $\xi_n(s, \omega)$ of $\xi_n(s)$ is continuous in the elementary calculus sense. From (3.1.14), we have

$$\frac{d}{dt}w_n(t) = \frac{d}{dt}\left[\int_{t_0}^{t} \xi_n(s)ds\right] \quad \text{(by differentiating both sides)}$$

$$= \xi_n(t) \quad \text{(by fundamental theorem of integral calculus [127]).} \tag{3.1.15}$$

Moreover, the differential of $w_n(t)$ in (3.1.14) is given by

$$dw_n(t) = \xi_n(t)dt. \tag{3.1.16}$$

The integral representation of the solution process of (3.1.13) can be rewritten as

$$x_n(t) = x_n(t_0) + \int_{t_0}^{t} C(s, x_n(s))ds + \int_{t_0}^{t} \sigma(s, x_n(s))\xi_n(s)ds$$

$$= x_n(t_0) + \int_{t_0}^{t} C(s, x_n(s))ds + \int_{t_0}^{t} \sigma(s, x_n(s))dw_n(s) \quad \text{[from (3.1.16)].}$$

$$\tag{3.1.17}$$

The integrals on the right-hand side of (3.1.17) are the Cauchy–Riemann integrals [127]. From (3.1.14), we make a precise notion of convergence of $\{\xi_n(t)\}_{n=1}^{\infty}$ to the Gaussian white noise by using the result of advanced calculus [168] as

$$\lim_{n\to\infty}[w_n(t) - w_n(t_0)]$$

$$= \lim_{n\to\infty}\left[\int_{t_0}^{t} \xi_n(s)ds\right] \quad \text{(by the limit in the "almost sure" sense)}$$

$$= \int_{t_0}^{t} \lim_{n\to\infty}[\xi_n(s)]ds$$

(by the interchange of the limit with the integral theorem [168])

$$= \nu\left[\int_{t_0}^{t} [\xi(s)]ds\right] = \nu[w(t) - w(t_0)], \tag{3.1.18}$$

for some constant ν. $w(t)$ is a normalized Wiener process. In this discussion, without loss of generality, we assume that $\nu = 1$. Otherwise, ν can be absorbed into σ in (3.1.17).

CNM5.   To conclude the convergence of $\{x_n(t)\}_{n=1}^{\infty}$, we need to show the convergence of the two terms on the right-hand side of (3.1.17). The procedure for showing this convergence amounts to the steps below:

Step #1.   This step is concerned with establishing the following:

$$\lim_{n\to\infty} [y_n(t)] = \lim_{n\to\infty} \left[\int_{t_0}^t \phi(s, w_n(s))dw_n(s) \right]$$

$$= \int_{t_0}^t \phi(s, w(s))dw(s) + \frac{1}{2}\int_{t_0}^t \frac{\partial}{\partial z}\phi(s, w(s))ds, \qquad (3.1.19)$$

where ϕ is a known smooth function of two variables. This is achieved by considering a deterministic partial indefinite integral of a given smooth deterministic function ϕ:

$$\psi(t, x) = \int_0^x \phi(t, z)dz. \qquad (3.1.20)$$

Now, we find the total deterministic differential of the deterministic function ψ in (3.1.20). We use the fundamental theorem of integral calculus and the chain rule [127], and obtain

$$d\psi(t, x) = \frac{\partial}{\partial x}\psi(t, x)dx + \frac{\partial}{\partial t}\psi(t, x)dt \qquad \text{(from Calculus III: differential)}$$

$$= \phi(t, x)dx + \left[\int_0^x \frac{\partial}{\partial t}\phi(t, z)dz \right]dt \quad \text{[from (3.1.20)]}. \qquad (3.1.21)$$

Assuming suitable domains and the ranges of functions, we evaluate both sides of the mathematical expressions along $w_n(t)$, and we have

$$d\psi(t, w_n(t)) = \phi(t, w_n(t))dw_n(t) + \frac{\partial}{\partial t}\psi(t, w_n(t))dt. \qquad (3.1.22)$$

We note that the left-hand side of (3.1.22) is the total derivative of the composite function $g_n(t) = \psi(t, w_n(t))$. Thus, we have

$$\int_{t_0}^t dg_n(s)ds = g_n(t) - g_n(t_0) = \psi(t, w_n(t)) - \psi(t_0, w_n(t_0)) = \int_{t_0}^t d\psi(s, w_n(s))ds.$$

From this, the definition of $y_n(t)$ in (3.1.19), (3.1.20), and (3.1.22), we get

$$y_n(t) = \int_{t_0}^t \phi(s, w_n(s))dw_n(s)$$

$$= \psi(t, w_n(t)) - \psi(t_0, w_n(t_0)) - \int_{t_0}^t \frac{\partial}{\partial t}\psi(s, w_n(s))ds. \qquad (3.1.23)$$

Again, under the smoothness assumptions on ϕ, we apply the argument used in the proof of (3.1.18) to (3.1.23), and we arrive at

$$\lim_{n\to\infty}[y_n(t)] = \lim_{n\to\infty}\left[\psi(t,w_n(t)) - \psi(t_0,w_n(t_0)) - \int_{t_0}^t \frac{\partial}{\partial t}\psi(s,w_n(s))ds\right]$$

$$= \lim_{n\to\infty}\psi(t,w_n(t)) - \lim_{n\to\infty}\psi(t_0,w_n(t_0))$$

$$- \lim_{n\to\infty}\left[\int_{t_0}^t \frac{\partial}{\partial t}\psi(s,w_n(s))ds\right] \quad \text{(by sum property of the limit)}$$

$$= \psi(t,w(t)) - \psi(t_0,w(t_0)) - \int_{t_0}^t \frac{\partial}{\partial t}\psi(s,w(s))ds. \tag{3.1.24}$$

Now, applying the Itô–Doob differential formula to $\psi(t,w(t))$ (Theorem 1.3.1), we obtain

$$d\psi(t,w(t)) = \frac{\partial}{\partial t}\psi(t,w(t))dt + \frac{\partial}{\partial x}\psi(t,w(t))dw(t)$$

$$+ \frac{1}{2}\frac{\partial^2}{\partial x^2}\psi(t,w(t))dt \quad \text{(by Theorem 1.3.1)}$$

$$= \frac{\partial}{\partial t}\psi(t,w(t))dt + \phi(t,w(t))dw(t)$$

$$+ \frac{1}{2}\frac{\partial}{\partial x}\phi(t,w(t))dt \quad \text{[by (3.1.20) and (3.1.21)].} \tag{3.1.25}$$

This implies that

$$\phi(t,w(t))dw(t) + \frac{1}{2}\frac{\partial}{\partial z}\phi(t,w(t))dt = d\psi(t,w(t)) - \frac{\partial}{\partial t}\psi(t,w(t))dt.$$

Integrating both sides, we arrive at

$$\int_{t_0}^t \phi(s,w(s))dw(s) + \frac{1}{2}\int_{t_0}^t \frac{\partial}{\partial x}\phi(s,w(s))ds$$

$$= \psi(t,w(t)) - \psi(t_0,w(t_0)) - \int_{t_0}^t \frac{\partial}{\partial t}\psi(s,w(s))ds. \tag{3.1.26}$$

We note that the right-hand side expressions in (3.1.24) and (3.1.26) are the same. Therefore, the left-hand side expressions in (3.1.24) and (3.1.26) must be equal. This establishes the validity of (3.1.19).

Step #2. This step deals with the procedure for finding a limit of the sequence of the solution process $\{x_n(t)\}_{n=1}^\infty$ determined by (3.1.13) or its equivalent stochastic differential equation [using (3.1.16) and (3.1.17)] as

$$dx_n = C(t, x_n)dt + \sigma(t, x_n)dw_n(t), \quad x_n(t_0) = x_0, \tag{3.1.27}$$

where $w_n(t)$ is defined in (3.1.14). For this purpose, we assume that $\sigma(t, z)$ in (3.1.6) satisfies the conditions $\sigma(t, z) \neq 0$, and it is continuously differentiable. We set $\phi(t, z) = \frac{1}{\sigma(t,z)}$ in (3.1.20). Under the smoothness conditions on the rate functions C and σ, and imitating the procedure outlined in Step #1, one can conclude that $\{x_n(t)\}_{n=1}^{\infty}$ converges to a process $x(t)$ on $[t_0, b]$. The final conclusion is to show that $x(t)$ satisfies the following Itô–Doob-type stochastic differential equation:

$$dx = \left[C(t, x) + \frac{1}{2}\sigma(t, x)\frac{\partial}{\partial x}\sigma(t, x) \right] dt + \sigma(t, x)dw(t), \quad x(t_0) = x_0. \tag{3.1.28}$$

This is achieved by the procedure for solving the Itô–Doob-type stochastic differential equation in the form of (3.1.27). The procedure is to reduce the differential equation (3.1.27) into the following reduced stochastic integrable differential equation (reduced SDE):

$$dm = f(t)dt + g(t)dw(t), \tag{3.1.29}$$

where $f(t)$ and $g(t)$ are suitable stochastic processes determined by the rate functions C and σ in (3.1.6). The detailed procedure is discussed in Section 3.3. The extra term $\frac{1}{2}\sigma(t, x)\frac{\partial}{\partial x}\sigma(t, x)$ in (3.1.28) is referred to as the *correction term* generated by the sequential colored noise modeling process.

In summary, we have shown that if we interpret the Gaussian white noise driven differential equation (3.1.6) by the limit of a sequence of stochastic differential equations (3.1.13) with colored noise processes, then the Gaussian white noise driven differential equation (3.1.6) is equivalent to the Itô–Doob-type stochastic differential equation (3.1.28).

Observation 3.1.2

(i) We present a few approximations $\{w_n(t)\}_{n=1}^{\infty}$ to the Brownian motion $b(t)$ process (Definition 1.2.11 and Observation 1.2.10) [195]:

   (a_1) For each t, $w_n(t) \rightarrow b(t)$ as $n \rightarrow \infty$ almost surely. For each n, and almost all ω belonging to the given complete probability space $(\Omega, \mathcal{F}, P)$, $\{w_n(t)\}_{n=1}^{\infty}$ is sample continuous and of bounded variation on $[a, b]$.

   (a_2) Let us assume that a_1 is valid and further assume that, for almost ω, the sequence $\{w_n(t)\}_{n=1}^{\infty}$ is uniformly bounded, i.e. $\sup_n[\sup_{t \in [a,b]} w_n(t)] < \infty$.

   (a_3) In addition to a_2, it is assumed that, for almost ω, $w_n(t)$ has a continuous derivative $\frac{d}{dt}w_n(t)$ for each n.

(a$_4$) For each n, $w_n(t)$ is a polynomial approximation of the Wiener process $w(t)$ defined by

$$w_n(t) = w_n(t_j^n) + \left[w_n(t_{j+1}^n) - w_n(t_j^n)\right] \frac{(t - t_j^n)}{(t_{j+1}^n - t_j^n)}, \quad t_j^n \le t \le t_{j+1}^n,$$

(3.1.30)

where a partition of $[a, b]$ $P_n : a = t_0^n < t_1^n < \cdots < t_j^n < \cdots < t_n^n = b$ and $\mu(P_n)$ is a measure of a partition P_n defined by

$$\mu(P_n) = \max_j \{(t_1^n - t_0^n), \ldots, (t_{j+1}^n - t_j^n), \ldots, (t_n^n - t_{n-1}^n)\} \to 0 \text{ as } n \to \infty.$$

(3.1.31)

(ii) We present a few convergence results concerning (3.1.19). Let $\phi(t, z)$ be a real-valued function having continuous partial derivatives $\frac{\partial}{\partial t}\phi(t, z)$ and $\frac{\partial}{\partial z}\phi(t, z)$ on its domain $[a, b] \times R$. Let $\{w_n(t)\}_{n=1}^\infty$ be as defined in a$_2$. Then,

$$\lim_{n\to\infty} \left[\int_{t_0}^t \phi(s, w_n(s))dw_n(s)\right] = \int_{t_0}^t \phi(s, w(s))dw(s) + \frac{1}{2}\int_{t_0}^t \frac{\partial}{\partial z}\phi(s, w(s))ds,$$

(3.1.32)

most of the time. Furthermore, if $\phi(t, z) \equiv \phi(z)$ and a$_2$ is replaced by a$_1$, then the relation (3.1.32) remains true.

(iii) We present a few convergence results concerning the solution process of (3.1.27). For this purpose, we assume that the rate functions $C(t, x)$ and $\sigma(t, x)$ are defined on $[a, b] \times R$ into R, and satisfy the following conditions:

(1) $C(t, x)$, $\sigma(t, x)$, $\frac{\partial}{\partial t}\sigma(t, x)$, and $\frac{\partial}{\partial x}\sigma(t, x)$ are continuous on $[a, b] \times R$.

(2) $C(t, x)$, $\sigma(t, x)$, $\frac{\partial}{\partial t}\sigma(t, x)$, and $\frac{\partial}{\partial x}\sigma(t, x)$ satisfy a uniform Lipschitz condition in x.

(3) $x(t)$ is a solution process of (3.1.28).

(4) $x_n(t_0) = x(t_0) = x_0$, and it is independent of $w(t) - w(t_0)$ for all $t \ge t_0$.

(5) Under these hypotheses and a$_4$ in (i) with $E[x_0^4] < \infty$, this implies that

$$\lim_{n\to\infty} x_n(t) \to x(t) \text{ as } n \to \infty \text{ in the mean square sense.}$$

(6) Under these hypotheses and a$_3$ in (i) with $|\sigma(t, x)| > c > 0$ and $|\sigma(t, x)| < L\sigma^2(t, x)$, this implies that

$$\lim_{n\to\infty} x_n(t) \to x(t) \text{ as } n \to \infty \text{ in the "almost sure" sense.}$$

Example 3.1.1. Let us consider a linear differential equation with the Gaussian noise perturbations:

$$dx = [f(t)x + \sigma x\xi(t)]dt, \quad x(t_0) = x_0,$$

where $\xi(t)$ is Gaussian white noise; it is approximated by a sequence of colored noise processes $\{\xi_n(t)\}_{n=1}^\infty$, $\xi_n(t) = \frac{d}{dt}w_n(t)$, and $\{w_n(t)\}_{n=1}^\infty$ (is as described in Observation 3.1.2(i)(a$_4$)).

Solution procedure. We associate the given differential equation and the sequence of smooth random functions with a sequence of differential equations as in (3.1.13), and we write

$$dx_n = [f(t) + \sigma\xi_n(t)]x\,dt, \quad x_n(t_0) = x_0.$$

For each n, this is like a deterministic differential equation with smooth random rate coefficients. We apply Procedure 2.3.4 to solve this IVP, and we have

$$x_n(t) = \exp\left[\int_{t_0}^t (f(s) + \sigma\xi_n(s))ds\right] x_0$$

$$= \exp\left[\int_{t_0}^t \left(f(s) + \sigma\frac{d}{dt}w_n(s)\right)ds\right] x_0$$

$$\text{[from (3.1.30) and (3.1.15)]}$$

$$= \exp\left[\int_{t_0}^t f(s)ds + \sigma\int_{t_0}^t \frac{d}{dt}w_n(s)ds\right] x_0$$

$$\text{(by deterministic integral properties)}$$

$$= \exp\left[\int_{t_0}^t f(s)ds + \sigma(w_n(t) - w_n(t_0))\right] x_0$$

$$\text{(from antiderivative concept)}.$$

From the knowledge of the exponential function and the assumption [Observation 3.1.2(i)(a₄)], we conclude that

$$\lim_{n\to\infty} x_n(t) \to x(t)$$

$$= \exp\left[\int_{t_0}^t f(s)ds + \sigma(w(t) - w(t_0))\right] x_0 \quad \text{(in the ``almost sure'' sense)}.$$

Moreover, by the Itô–Doob formula (Theorem 1.3.1), we infer that

$$x(t) = \exp\left[\int_{t_0}^t f(s)ds + \sigma(w(t) - w(t_0))\right] x_0,$$

which satisfies the following Itô–Doob-type stochastic differential equation:

$$dx = \left[f(t) + \frac{1}{2}\sigma^2\right] x\,dt + \sigma x\,dw(t).$$

This completes the solution procedure of the given example.

In the following, we present a few illustrations and examples of mathematical modeling of the dynamic process by Modified Brownian Motion Model 3.1.1 and Sequential Colored Noise Model 3.1.2.

Illustration 3.1.1 (chain reactions: Rice–Herzfeld mechanism). Imitating the deterministic version of the Rice–Herzfed mechanism [80, 143], we have

$$R_1 + R_1 \rightarrow P_1(k_{4a}) \text{ (termination)},$$

$$R_2 + R_2 \rightarrow P_2(k_{4b}) \text{ (termination)},$$

$$R_1 + R_2 \rightarrow P_3(k_{4c}) \text{ (termination)}.$$

We note that the radical R_1 propagates the chain by the bimolecular process, and R_2 is assumed to be the first-order.

Utilizing the law of mass action and repeating the detailed mathematical description of Rice–Herzfeld decomposition of the single chemical species (Steps #1 to #3 [80]), the mathematical description of the above-described single species chemical reaction system is:

$$\frac{d}{dt}[M] = -[M](k_1 + k_2[R_1]),$$

$$\frac{d}{dt}[R_1] = k_1[M] + k_3[R_2] - k_2[R_1][M] - k_{4a}[R_1]^2,$$

$$\frac{d}{dt}[R_2] = k_2[R_1][M] - k_3[R_2], \qquad (3.1.33)$$

where $[S]$ (the number of moles of S per unit volume) is the concentration of a substance S, and $\frac{d}{dt}[S]$ is the instantaneous rate of change of the concentration (velocity). By following the standard stationary state (a concept introduced by Bodenstein, 1913) approximation scheme (referred to as the *stationary state hypothesis* [80, 175]) with algebraic simplifications, we arrive at

$$d[M] = -\left[k_1[M] + k_2\sqrt{\frac{k_1}{k_{4a}}}[M]^{3/2}\right] dt \quad \text{(by algebraic simplification)}. \quad (3.1.34)$$

The source of random perturbations can be due to the formation of free radicals. Therefore, the rate parameters k_1, k_2, k_{4a}, and the parameter induced by the steady-state hypothesis about $[R_1]$ and $[R_2]$ can be subject to random perturbations. Moreover, $[R_1]$ and $[R_2]$ appear in (3.1.33) as full-fledged state variables. Here, it is assumed that the parameters k_1 and k_{4a} are subject to the random fluctuations, with $\frac{k_1}{k_{4a}}$ being almost constant. Moreover, we assume that these random perturbations are described by the stationary Gaussian process with independent increments. As a result of this consideration, Modified Brownian Motion Model 3.1.1 is applicable to these types of chain reaction systems. In this case, the deterministic model (3.1.34) reduces to the Itô–Doob-type stochastic differential equation

$$d[M] = -[M]\left[\bar{k}_1 + k_2\sqrt{\frac{k_1}{k_{4a}}}[M]^{1/2}\right] dt + \sigma[M]dw(t),$$

$$[M(t_0)] = [M_0], \qquad (3.1.35)$$

where the smooth deterministic rate functions in (3.1.10) are $C(t, [M]) = [M]$ $[\bar{k}_1 + \sqrt{\frac{k_2}{k_1 k_{4a}}}[M]^{1/2}]$, $k_1 = \bar{k}_1 + \xi(t)$, and $\sigma(t, [M]) = \sigma[M]$ for some $\sigma \neq 0$; $\sigma \, dw(t) = \xi(t)dt$ characterizes the effects of the environmental random fluctuations (thermal) with $E[\xi(t)] = 0$, and $\xi(t)$ is a stationary Gaussian process with independent increments as defined in (3.1.12).

Example 3.1.2. The Rice–Herzfeld mechanism of a chain reaction is exhibited by the thermal decomposition of acetaldehyde (CH_3CHO) (Example 3.2.1 [80]). The thermal decomposition of acetaldehyde produces two free radicals, namely CH_3 and CHO (chain carriers). In this case, the three-state process is as follows [126]:

Initiation: $CH_3CHO \rightarrow CH_3 + CHO$ (k_1) (initiation)

Propagation: $CH_3 + CH_3CHO \rightarrow CH_3 + CH_4 + CO$ (k_2) (propagation)

Termination: $CH_3 + CH_3 \rightarrow C_2H_6$ (k_{4a}) (termination)

$CHO + CHO \rightarrow CH_4 + CO$ (k_{4b}) (termination)

and the overall reaction is

$$CH_3CHO \rightarrow CH_4 + CO$$

with the traces of C_2H_6 and H_2.

We set $[M] = [CH_3CHO]$, $[R_1] = [CH_3]$, $[R'_1] = [CHO]$, $[R_2] = [CH_4] + [CO]$, $[P_1] = [C_2H_6]$, and $[P_2] = [CH_4] + [CO]$. Employing the argument used in Example 3.2.1 [80] and Illustration 3.1.1, the system of differential equations (3.1.33)–(3.1.35) reduces to

$$\frac{d}{dt}[M] = -[M](k_1 + k_2[R_1]),$$

$$\frac{d}{dt}[R_1] = k_1[M] - k_2[R_1][M] + k_2[R_1][M] - k_{4a}[R_1]^2,$$

$$\frac{d}{dt}[R'_1] = k_1[M] - k_{4b}[R'_1]^2,$$

$$\frac{d}{dt}[R_2] = k_1[M][R_1]m,$$

$$d[M] = -k_1[M] + k_2\sqrt{\frac{k_1}{k_{4a}}}[M]^{3/2}dt,$$

$$d[M] = -[M]\left[\bar{k} + k_2\sqrt{\frac{k_1}{k_{4a}}}[M]^{1/2}\right]dt + \sigma[M]dw(t), \quad [M(t_0)] = [M_0],$$

respectively.

Illustration 3.1.2 (chemical process: enzymatic reaction). Recalling the detailed derivation of the deterministic dynamic model of the enzymatic reaction process (ERP1 to ERP7 and Steps #1 to #4 in Illustration 3.2.2 [80]), under the

Briggs–Haldane approach [175], the *Michaelis–Menten rate equation* and the *Henry–Michaelis–Menten rate equation* are:

$$d[S] = -\frac{k_p[E^*][S]}{K_m + [S]}dt, \quad [S(t_0)] = [S_0], \quad (3.1.36)$$

$$d[S] = -\frac{V_{\max}[S]}{K_m + [S]}dt, \quad [S(t_0)] = [S_0], \quad (3.1.37)$$

respectively, where $K_m = \frac{([E^*]-[ES])[S]}{[ES]} = \frac{(k_{-1}+k_p)}{k_1}$, $[E^*] = [E] + [ES]$; $V_{\max} = k_p[E^*]$; $[S]$, $[E]$, and $[ES]$ are concentrations of the substrate (S), enzyme (E), enzyme-substrate complex (ES), respectively; and k_p stands for the rate constant for the breakdown of ES to $E + P$. For more detailed description, see Refs. 80 and 175.

ERP8. The source of the random perturbations can be centered upon the enzyme-substrate complex formation process. In particular, the parameter k_p (or $[E^*]$ or both k_p and $[E^*]$) can be considered under the influence of the random perturbations. The natural mathematical model of this type of enzymatic reaction system can be described by the stationary Gaussian process with independent increments.

Under the considerations of ERP8, Modified Brown Motion Model 3.1.1 is directly applicable to this enzyme and substrate reaction system. Therefore, from (3.1.10), the deterministic models (Illustration 3.2.2 in Ref. 80) (3.1.36) and (3.1.37) reduce to the following respective Itô–Doob-type stochastic differential equations:

$$d[S] = -\frac{\overline{k_p[E^*]}[S]}{K_m + [S]}dt + \frac{\sigma[S]}{K_m + [S]}dw(t), \quad [S(t_0)] = [S_0], \quad (3.1.38)$$

$$d[S] = -\frac{V_{\max}[S]}{K_m + [S]}dt + \frac{\sigma[S]}{K_m + [S]}dw(t) \quad [S(t_0)] = [S_0], \quad (3.1.39)$$

where the smooth deterministic rate functions in (3.1.10) are $C(t, [S]) = \frac{-\overline{k_p[E^*]}[S]}{K_m + [S]}$, $k_p[E^*] = \overline{k_p[E^*]} + \xi(t)$, and $\sigma(t, [S]) = \frac{\sigma[S]}{K_m + [S]}$ for some constant $\sigma \neq 0$; and $\sigma dw(t) = \xi(t)dt$ characterizes the effects of the environmental random fluctuations with $E[\xi(t)] = 0$, and $\xi(t)$ is a stationary Gaussian process with independent increments as defined (3.1.12).

Illustration 3.1.3 (ecological process). The biological processes have a natural tendency to maintain steady/equilibrium states, or the gradual adjustment to the environmental changes. The birth and death processes in the biological systems play a significant role. For the development of a mathematical model of single-species ecological processes, we augment the following assumption to the assumptions SEP1–SEP5 in Illustration 3.2.3 [80]:

SEP6. In open communities, the sizes of species are integral values. The successive generations may not overlap in time. The breeding as well as environmental parameters may vary in discrete time. Because of this, it is difficult to control the population growth by only the birth and death processes. For example, migration is a natural process to be taken into account for the growth rate. The environmental random fluctuations can also modify the resources and cause the migration process. The natural mathematical model of these types of ecosystems can be described by the stationary Gaussian process with independent increments. Because of these considerations, the Modified Brownian Motion Model 3.1.1 is directly applicable to these types of species in the communities. Therefore, from (3.1.10), the deterministic model described in SEP5 in Illustration 3.2.3 [80] reduces to the Itô–Doob-type stochastic differential equation

$$dN = \alpha N(\bar{\kappa} - N)dt + \sigma N dw(t), \quad N(t_0) = N_0, \qquad (3.1.40)$$

where the smooth deterministic rate functions in (3.1.10) are $C(t, N) = \alpha N(\kappa - N)$, $\kappa = \bar{\kappa} + \xi(t)$, and $\sigma(t, N) = \sigma N$ for some $\sigma \neq 0$; $\sigma\, dw(t) = \xi(t)dt$ characterizes the effects of random environmental fluctuations with $E[\xi(t)] = 0$. It incorporates the random growth rate due to the migration of the population, and $\xi(t)$ is a stationary Gaussian process with independent increments as defined in (3.1.12). The solution process of (3.1.40) is $N_0(\alpha N_0 + (\kappa - N_0)\exp[-(\alpha\kappa - \frac{1}{2}\sigma^2)(t - t_0) + \sigma(w(t) - w(t_0))])^{-1}$. For details, see Example 3.7.4.

SEP7. We note that the stochastic model described in SEP6 is limited to a class of single-species processes in the biological sciences. Moreover, the more natural way of building a stochastic model is to start from the deterministic model SEP3, Illustration 3.2.3 [80]. From SEP4 and SEP5 [80], one can conclude that α is a natural parameter that is sensitive to the internal or environmental random perturbations. Therefore, one can modify this parameter to: $\alpha \equiv \lambda p(t, N)$, where $p(t, N)$ regulates the growth process under the influence of a Gaussian process and λ is a positive constant. In particular, $p(t, N) \equiv \bar{p}(N) + \Lambda(N)\xi(t)$, where $\xi(t)$ is a Gaussian white noise process. With this consideration, the deterministic mathematical model in SEP3 reduces to

$$dN = \lambda N[\bar{p}(N) + \Lambda(N)\xi(t)]dt, \quad N(t_0) = N_0, \qquad (3.1.41)$$

where $\bar{p}(N)$ and $\Lambda(N)$ are smooth functions to assure the existence and uniqueness of the sequence of IVPs:

$$dN_n = \lambda N_n[\bar{p}(N_n) + \Lambda(N_n)\xi_n(t)]dt, \quad N_n(t_0) = N_0,$$

where $\{\xi_n(t)\}_{n=1}^{\infty}$ is a sequence of sufficiently smooth (sample-pathwise continuous in the usual calculus sense) Gaussian processes which converges in some sense to the above Gaussian white noise process $\xi(t)$.

Now, one can imitate the argument used in the development of Sequential Colored Noise Model 3.1.2 and conclude that the solution process of (3.1.41) is equivalent to the solution process of the following Itô–Doob-type stochastic differential equation:

$$dN = \lambda N \left[\bar{p}(N) + \frac{\nu^2}{2} \Lambda(N) \left[N \frac{d}{dN} \Lambda(N) + \Lambda(N) \right] \right] dt,$$

$$+ \lambda \nu N \Lambda(N) dw(t), \quad N(t_0) = N_0 \tag{3.1.42}$$

where ν is as defined in (3.1.18).

Example 3.1.3. From SEP7, Illustration 3.1.3, we can present several well-known mathematical models as special cases:

(a) For the choice of functions $\bar{p}(N) = \kappa - N$ and $\Lambda(N) = \lambda^{-1}$, (3.1.41) and (3.1.42) become

$$dN = \lambda N[(\kappa - N) + \lambda^{-1} \xi(t)] dt$$

$$= [\lambda N(\kappa - N) + N \xi(t)] dt, \quad N(t_0) = N_0,$$

and

$$dN = \lambda N \left(\kappa + \frac{\nu^2}{2\lambda} - N \right) dt + \nu N dw(t), \quad N(t_0) = N_0, \tag{3.1.43}$$

respectively.

(b) If we substitute the choice of functions $\bar{p}(N) = \kappa - N$ and $\Lambda(N) = \lambda^{-1} N^{-1/2}$ into (3.1.41) and (3.1.42), then we have

$$dN = \lambda N[(\kappa - N) + \lambda^{-1} N^{-1/2} \xi(t)] dt$$

$$= [\lambda N(\kappa - N) + N^{1/2} \xi(t)] dt, \quad N(t_0) = N_0,$$

and

$$dN = \left[\lambda N(\kappa - N) + \frac{\nu^2}{4} \right] dt + \nu N^{1/2} dw(t), \quad N(t_0) = N_0, \tag{3.1.44}$$

respectively.

(c) For the choice of functions $\bar{p}(N) = \kappa - N$ and $\Lambda(N) = \lambda^{-1}(\kappa - N)$ from (3.1.41) and (3.1.42), we have

$$dN = \lambda N[(\kappa - N) + \lambda^{-1}(\kappa - N) \xi(t)] dt$$

$$= [\lambda N(\kappa - N) + N(\kappa - N) \xi(t)] dt, \quad N(t_0) = N_0,$$

and

$$dN = N(\kappa - N) \left[\lambda + \frac{\nu^2}{2} (\kappa - 2N) \right] dt + \nu N(\kappa - N) dw(t), \quad N(t_0) = N_0. \tag{3.1.45}$$

(d) For the choice of functions $\bar{p}(N) = \kappa - \ln N$ and $\Lambda(N) = \lambda^{-1}$, (3.1.41) and (3.1.42) reduce to

$$dN = \lambda N[(\kappa - \ln N) + \lambda^{-1}\xi(t)]dt$$
$$= [\lambda N(\kappa - \ln N) + N\xi(t)]dt, \quad N(t_0) = N_0,$$

and

$$dN = \lambda N\left(\kappa + \frac{\nu^2}{2\lambda} - \ln N\right)dt + \nu N dw(t), \quad N(t_0) = N_0, \qquad (3.1.46)$$

respectively.

(e) For functions $\bar{p}(N) = (\kappa - \ln N)$ and $\Lambda(N) = \lambda^{-1}(\kappa - \ln N)$, (3.1.41) and (3.1.42) become

$$dN = \lambda N[(\kappa - \ln N) + \lambda^{-1}(\kappa - \ln N)\xi(t)]dt$$
$$= [\lambda N(\kappa - \ln N) + N(\kappa - \ln N)\xi(t)]dt, \quad N(t_0) = N_0,$$

and

$$dN = N(\kappa - \ln N)\left[\lambda - \frac{\nu^2}{2} + \frac{\nu^2}{2}(\kappa - \ln N)\right]dt$$
$$+ \nu N(\kappa - \ln N)dw(t), \quad N(t_0) = N_0, \qquad (3.1.47)$$

respectively.

Example 3.1.4. Let us continue with the genetic population dynamic (GPD) mathematical model described in Example 3.2.2 [80] under the assumptions GPD1–GDP3. Here, we augment the following assumption:

GPD4. The randomness is one of the sources causing the frequency of alleles in gene population changes. The random fluctuations are due to evolutionary pressures — in particular, selection intensities due to environmental fluctuations and also random sampling of gametes ("gametes" means the gem cell carrying one set of chromosomes from the parent to the offspring) in a finite population.

In the following, we consider a particular case of random fluctuations in the selection intensities. In order to observe the effects of this type of random perturbations, it is assumed that the population size is very large. This implies that the effects of random sampling can be neglected. The following deterministic gene frequency logistic growth model of the A_1 allele (Example 3.2.2 [80]) is recast as

$$dx = sx(1 - x)dt, \quad x(t_0) = x_0,$$

where x is the frequency of the A_1 allele; s is the selective advantage of A_1 over A_2, and it is under random perturbations, and $s = \bar{s} + \xi(t)$ with $\xi(t)$ being the Gaussian white noise process in (3.1.6) that satisfies (3.1.18). With this consideration and following the argument used in SEP7, Illustration 3.1.3, the stochastic version of the above deterministic differential is

$$dx = (\bar{s} + \xi(t))x(1 - x)dt$$

$$= \bar{s}x(1 - x)dt + x(1 - x)\xi(t)dt, \quad x(t_0) = x_0.$$

From (3.1.45), we conclude that the above stochastic differential is equivalent to the following Itô–Doob-type stochastic differential equation:

$$dx = x(1 - x)\left[\bar{s} + \frac{\nu^2}{2}(1 - 2x)\right]dt + \nu x(1 - x)dw(t), \quad x(t_0) = x_0.$$

Illustration 3.1.4 (epidemiological process). Most epidemiological phenomena are very complex. Mathematical models provide information about communicable epidemic diseases to the international, national, state, and local health departments for their planning and decision-making processes. The deterministic mathematical model

$$dI = \alpha I(N - I)dt, \quad I(0) = 1 \tag{3.1.48}$$

under conditions EDP1–EDP6 in Illustration 3.2.4 [80] is overly simplified. In this illustration, we modify EDP2 and EDP4 by introducing the randomness in the parameters as follows:

EDP7. In the open community/system, it is difficult to control the "well-mixed" condition (EDP2). In addition, the migration process is possible by holding the total size fixed N. Because of this, the parameter N in (3.1.48) is under random environmental fluctuations. N can be treated as a random parameter, $N = \bar{N} + \xi(t)$ with $E[\xi(t)] = 0$, and it is a stationary Gaussian process with independent increments. Moreover, in view of the modified Brownian motion model, $\alpha^{-1}\sigma\,dw(t) = \xi(t)dt$ for some $\sigma \neq 0$, where $w(t)$ is a Wiener process. Under EDP1–EDP7, the deterministic model reduces to the Itô–Doob-type stochastic differential equation

$$dI = \alpha I(\bar{N} - I)dt + \sigma I\,dw(t), \quad I(0) = 1. \tag{3.1.49}$$

Illustration 3.1.5 (economic process). We continue to study the economic growth models due to Harrod–Domar and Solow [185]. The development of the deterministic single composite commodity model described under conditions EGM1–EGM4 in Illustration 3.2.5 [80] is extended to the stochastic model. This is achieved by incorporating the randomness recognized by Solow in the dynamics.

The deterministic single composite commodity model described under conditions EGM1–EGM4 (Illustration 3.2.5) is

$$dr = sf(r, 1)dt - nr\,dt, \tag{3.1.50}$$

where f, $r = \frac{K}{L}$, n, and s are as defined in Illustration 3.2.5 [80].

It is known and recognized that random fluctuations in economic dynamics do arise in a natural way (see Illustration 2.1.1). In addition, according to N. Kadlor and Solow [184], the source of the random fluctuations is described below:

EGM5. The rate of growth of labor n is not constant, but it is under the random perturbations due to the influence of demographic factors and resource limitations. These factors have already been emphasized in Illustration 3.1.3 and Example 3.1.3 in the context of the models related to population dynamic processes. Moreover, the constant saving rate s is subject to random fluctuations. Thus, the ratio of capital to output is very volatile in any fluctuating economy. In short, all of the constant parameters n, s, and k in the Harrod–Domar model are subject to random perturbations.

In the light of EGM5 (Illustration 3.2.5, Volume 1), we utilize and apply the type of parametric random perturbations described in SEP6, Illustration 3.1.3 to the parameters n and s. Hence, from (3.1.10), the deterministic model (3.1.50) described in EGM4 takes the following different forms of the Itô–Doob-type stochastic differential equations:

$$dr = [sf(r,1) - \bar{n}r]dt + \sigma r\, dw(t), \quad r(t_0) = r_0, \tag{3.1.51}$$

$$dr = [\bar{s}f(r,1) - nr]dt + \sigma f(r,1)dw(t), \quad r(t_0) = r_0, \tag{3.1.52}$$

$$dr = [\bar{s}f(r,1) - \bar{n}r]dt + \sigma[sf(r,1) - nr]dw(t), \quad r(t_0) = r_0, \tag{3.1.53}$$

where $n = \bar{n} + \sigma\xi(t)$ and $s = \bar{s} + \sigma\xi(t)$, with $\xi(t)$ being defined as in (3.1.12).

In addition, we can employ the type of parametric random perturbations described in SEP7, Illustration 3.1.3 to the parameters n and s, and obtain

$$dr = [sf(r,1) - \bar{n}r]dt + r\xi(t)dt, \quad r(t_0) = r_0, \tag{3.1.54}$$

$$dr = [\bar{s}f(r,1) - nr]dt + f(r,1)\xi(t)dt, \quad r(t_0) = r_0, \tag{3.1.55}$$

$$dr = [\bar{s}f(r,1) - \bar{n}r]dt + [f(r,1) - r]\xi(t)dt, \quad r(t_0) = r_0, \tag{3.1.56}$$

where $\xi(t)$ is a Gaussian white noise process as described in (3.1.6). Hence, by applying Sequential Colored Noise Model 3.1.2 and using the argument of Illustration 3.1.3, the solution processes of the stochastic differential equations (3.1.54)–(3.1.56) are equivalent to the solution processes of the following Itô–Doob-type stochastic differential equations:

$$dr = \left[sf(r,1) - \bar{n}r + \frac{\nu^2}{2}\right]dt + \nu r\, dw(t), \quad r(t_0) = r_0, \tag{3.1.57}$$

$$dr = \left[\bar{s}f(r,1)\left(1 + \frac{\nu^2}{2}\frac{d}{dr}f(r,1)\right) - nr\right] \tag{3.1.58}$$

$$\times\, dt + \nu f(r,1)dw(t), \quad r(t_0) = r_0,$$

$$dr = \left[\bar{s} f(r,1)\left(1 + \frac{\nu^2 \bar{s}}{2}\frac{d}{dr}f(r,1)\right) - \bar{n}r + \frac{\nu^2}{2} \right]$$

$$\times\, dt + \nu[f(r,1) - r]dw(t), \quad r(t_0) = r_0. \tag{3.1.59}$$

Example 3.1.5 (Cobb–Douglas function [185]). The production function is defined by

$$Y = K^a L^{1-a} = f(K,L),$$

where a is a parameter and $a < 1$, $a \neq 0$. From Example 3.2.3 [80], we have:

$$dr = (sr^a - nr)dt.$$

We imitate the procedure outlined in Illustration 3.1.5, and finally conclude that the stochastic mathematical models with respect to (3.1.51)–(3.1.59) in the context of the deterministic Cobb–Douglas production function reduce to

$$dr = [sr^a - nr]dt + \sigma r\, dw(t), \quad r(t_0) = r_0,$$

$$dr = [\bar{s}r^a - nr]dt + \sigma r^a dw(t), \quad r(t_0) = r_0,$$

$$dr = [\bar{s}r^a - \bar{n}r]dt + \sigma[sr^a - nr]dw(t), \quad r(t_0) = r_0,$$

where $n = \bar{n} + \sigma\xi(t)$ and $s = \bar{s} + \sigma\xi(t)$, with $\xi(t)$ being defined as in (3.1.12);

$$dr = [sr^a - \bar{n}r]dt + r\xi(t)dt, \quad r(t_0) = r_0,$$

$$dr = [\bar{s}r^a - nr]dt + r^a\xi(t)dt, \quad r(t_0) = r_0,$$

$$dr = [\bar{s}r^a - \bar{n}r]dt + [r^a - r]\xi(t)dt, \quad r(t_0) = r_0, \tag{3.1.60}$$

where $\xi(t)$ is a Gaussian white noise process as described in (3.1.6), and their corresponding equivalent differential equations in the sense of Itô–Doob-type stochastic differential equations are

$$dr = \left[sr^a - \bar{n}r + \frac{\nu^2}{2} \right]dt + \nu r\, dw(t), \quad r(t_0) = r_0,$$

$$dr = \left[\bar{s}r^a\left(1 + \frac{a\nu^2\bar{s}}{2}r^{a-1}\right) - nr \right]dt + \nu r^a dw(t), \quad r(t_0) = r_0,$$

$$dr = \left[\bar{s}r^a\left(1 + \frac{a\nu^2\bar{s}}{2}r^{a-1}\right) - \bar{n}r + \frac{\nu^2}{2} \right]dt + \nu[r^a - \nu r]dw(t), \quad r(t_0) = r_0.$$

$$\tag{3.1.61}$$

Exercises

1. Under the conditions of Illustration 3.1.1, formulate the stochastic mathematical model corresponding to (3.1.34) by treating the specified parameters under the influence of:

   (a) The Gaussian white noise process with independent increments, with the parameter k_2 only.
   (b) The Gaussian white noise process, with the parameter k_2 only.
   (c) The Gaussian white noise process, with the parameter k_1 only.

2. Using Illustration 3.1.1, find:

   (a) The mathematical description of Rice–Herzfeld decomposition of the single chemical species.
   (b) The steady states of decomposition of the single chemical species.
   (c) The steady states of intermediate substrates.
   (d) The instantaneous rate of change of the concentration (velocity) $\frac{d}{dt}[M]$.
   (e) The stochastic mathematical model analogous to (3.1.35) in the case of:

       (i) The terminal reaction (k_{4b}).
       (ii) The terminal reaction (k_{4c}).

3. Use the mathematical models developed in Exercise 2 to formulate appropriate mathematical models analogous to the models of Exercise 1.

4. Let $[A_0] = a$ and $[B_0] = b$ be the given concentration of n-amyl florid $(n - C_5H_{11}F)$ and sodium ethoxide $(NaOC_2H_5)$, and $a \neq b$. It is known that one mole of n-amyl florid reacts with exactly one mole of sodium ethoxide to form one mole of sodium florid and one mole of n-amyl ethoxide. The reaction mechanism is described by

$$n - C_5H_{11}F + NaOC_2H_5 \to NaF + n - C_5H_{11}OC_2H_5.$$

   Let $x \equiv x(t)$ be the number of moles of n-amyl florid that has reacted with the moles of sodium ethoxide. We note that a and b decrease by the same number of moles. This reaction requires collision of molecules of n-amyl florid with molecules of sodium ethoxide. Find an expression for $x(t)$ at any time t.

5. **Unimolecular reaction** [183]. Let M be a normal molecule, and let M^* be its activated molecule. The problem of the origin of the activation energy in a unimolecular gas reaction is well known. A reaction in the gas phase is assumed to be unimolecular, provided that it:

   (i) follows the first-order rate law;
   (ii) is homogeneous;
   (iii) is not a chain reaction;
   (iv) changes order from 1 to 2 at a pressure of a few millimeters.

   In 1923, F. A. Lindemann pointed out that it is possible for normal molecules to receive their energy of activation by collision. The possible time delay between the activation and reaction processes is due to the deactivation of activated molecules by the collision with the normal molecules. The Lindemann mechanism may be described as follows:

$$M + M \rightarrow M^* + M \quad \text{(activation: reaction rate } k_1\text{)},$$

$$M^* + M \rightarrow M + M \quad \text{(deactivation: reaction rate } k_2\text{)},$$

$$M^* \rightarrow P \qquad \text{(product formation: reaction rate } k_3\text{)},$$

where M and M^* stand for an unimolecular reactant and unimolecular activated molecule, respectively.

(a) Derive a system of differential equations that represents the given Lindemann mechanism.

(b) Determine the reaction rate of M at any pressure.

(c) Using the reaction rate in (b), justify the fact that the order of reaction changes from 1 to 2.

6. Under the conditions of Illustration 3.1.1, formulate the stochastic mathematical model of the unimolecular reaction system in Exercise 5 by treating the specified parameters under the influence of:

(a) The Gaussian white noise process with independent increments, with the parameter k_1 only.

(b) The Gaussian white noise process, with the parameter k_1 only.

(c) The Gaussian white noise process, with the parameter $[E^*]$ only.

(d) The Gaussian white noise process with independent increments, with the parameter $[E^*]$ only.

7. Under the conditions of Illustration 3.1.2 and the representation K_m, formulate the stochastic mathematical model corresponding to (3.1.36) and (3.1.37) by treating the specified parameters under the influence of:

(a) The Gaussian white noise process with independent increments, with the parameter k_1 only.

(b) The Gaussian white noise process, with the parameter k_1 only.

(c) The Gaussian white noise process, with the parameter $[E^*]$ only.

(d) The Gaussian white noise process with independent increments, with the parameter $[E^*]$ only.

8. In Example 3.1.4, we assumed that $s = \xi(t)$ (i.e. A_1 is selectively neutral on the average); $\xi(t)$ is the Gaussian white noise process defined in (3.1.6) which satisfies (3.1.18). Under this modification, derive a stochastic model.

9. In Example 3.1.4, we neglected the random environmental perturbations in both the selection intensities and the sampling of gametes, and allowed migration under random perturbations. Under this modification, derive the stochastic mathematical models described by both the modified Brownian motion process and the sequential colored noise modeling process.

10. Formulate the stochastic mathematical model corresponding to the deterministic model of EDP7, Illustration 3.2.4 [80] by treating the parameter α under the influence of:

   (a) The Gaussian white noise process with independent increments.
   (b) The Gaussian white noise process.

11. ***Fixed proportion (Harrod–Domar model)*** [80, 185]. Let a stand for the number of units of capital for producing a unit of output, and let b signify the number of workers for producing a unit of output. Of course, a unit of output can be produced with more capital and/or labor than this. Let us define a production function f as

$$Y = f(K, L) = \min\left(\frac{K}{a}, \frac{L}{b}\right).$$

   (a) Show that the production function f satisfies the neoclassical conditions, i.e. it is homogeneous of degree 1, and it admits an unlimited substitutability property between capital and labor.
   (b) Find (i) the domain and range of f, and (ii) make a sketch of f.
   (c) Show that $dr = [s\min(\frac{r}{a}, \frac{1}{b}) - nr]$, where $r = \frac{K}{L}$, and s and n are as defined in Illustration 3.1.5.
   (d) Find differential equations corresponding to (3.1.57)–(3.159).

12. A family of functions is defined by $Y = f(K, L) = (aK^p + L^p)^{1/p}$, for $p > 0$.

   (a) Show that f is a homogeneous function of degree 1.
   (b) For what values of p does f satisfy the "unlimited" substitutability property?
   (c) For $p = \frac{1}{2}$, show that $Y = f(K, L) = a^2 K + 2a\sqrt{KL} + L$.
   (d) Show that $dr = s(A\sqrt{r}+1)(B\sqrt{r}+1)$, where $A = a - \sqrt{n/s}$, $B = a + \sqrt{n/s}$, s and n are as defined in Illustration 3.1.5.
   (e) Find differential equations corresponding to (3.1.57)–(3.1.59).

3.2 Method of Energy Functions

In this section, we present a very general conceptual algorithm for finding a solution process of a first-order scalar nonlinear Itô–Doob-type stochastic differential equation. The method seeks an energy function associated with a given dynamic process. Knowing the existence of the solution process, we assume that there is an energy function associated with a given dynamic system. The basic ideas are: (1) seeking an unknown energy function, (2) associating a conceptual and simpler Itô–Doob-type stochastic differential equation with an unknown energy function and the original nonlinear Itô–Doob-type stochastic differential equation, (3) determining the energy function and the rates coefficients of the simpler differential equation in the context of its conceptual form and the original nonlinear Itô–Doob-type stochastic differential equation, and (4) finding a representation of the solution to the original differential equation in the context of the energy function and the solution to Itô–Doob-type reduced stochastic differential equation. Knowing the original rate functions, the reduction process determines the rate functions of the simpler Itô–Doob-type stochastic differential equation and the energy function. A solution to an original nonlinear Itô–Doob-type stochastic differential equation is recasted in the context of the energy function and the solution to solvable reduced differential equations, such as: (a) the directly integrable Itô–Doob-type stochastic differential equation, and (b) a first-order linear stochastic differential equation of the Itô–Doob-type.

3.2.1 General Problem

Let us consider the following first-order nonlinear Itô–Doob-type of scalar stochastic differential equation (Itô–Doob-SDE or SDE):

$$dx = f(t,x)dt + \sigma(t,x)dw(t), \qquad (3.2.1)$$

where f and σ are continuous functions defined on $J \times R$ into R, $J = [a,b]$.

In the following, let us present a definition of the solution process of (3.2.1).

Definition 3.2.1. Let $J = [a,b]$ for $a, b \in R$, and hence $J \subseteq R$ be an interval. In addition, $\mathfrak{F}_t$, $a \leq t \leq b$, be the minimal σ-algebra generated by the Wiener process (Definition 1.2.12) $w(s)$ for $a \leq s \leq t \leq b$, and let $H_2[a,b]$ be the class of processes defined in Definition 1.3.1. A random process/function x defined on J is said to be a solution process of a first-order nonlinear stochastic differential equation of the Itô–Doob-type (3.2.1) if it satisfies the following conditions: (a) the process $x(s)$ for $a \leq s \leq t \leq b$ is a random variable, (b) the processes $|f(t,x(t))|^{1/2}$ and $\sigma(t,x(t))$ belong to $H_2[a,b]$, and (c) the random process/function $x(t)$ and its Itô–Doob-type stochastic differential $dx(t)$ satisfy (3.2.1) on J in the sense of the Itô–Doob stochastic calculus. In short, if we substitute a random function x and its Itô–Doob-type stochastic differential $dx(t)$ into (3.2.1), then the equation remains valid on J.

Now, we present a result that provides sufficient conditions for the existence and the uniqueness of the solution process corresponding to (3.2.1).

Theorem 3.2.1. *Assume that f and σ are defined and continuous functions on $J \times R$ into R, $J = [a,b]$. Further assume that:*

(H$_{3.2}$). *The rate functions f and σ in (3.2.1) satisfy the following conditions:*

$$|f(t,x)|^2 + |\sigma(t,x)|^2 \leq K^2(1 + |x|^2) \qquad (growth\ condition),$$

and

$$|f(t,x) - f(t,y)| + |\sigma(t,x) - \sigma(t,y)| \leq L|x - y| \quad (Lipschitz\ condition),$$

for $(t,x) \in J \times R$, where K and L are some positive numbers. Then, the IVP

$$dx = f(t,x)dt + \sigma(t,x)dw(t), \quad x(t_0) = x_0, \qquad (IVP)$$

has a unique solution $x(t) = x(t, t_0, x_0)$ through (t_0, x_0) for $t \geq t_0$, $t, t_0 \in J$ for any random variable x_0 that is independent of the Wiener process $w(t)$ for all $t \geq t_0 \geq a$, and the second moment of x_0 is finite.

In this subsection, we discuss a general procedure for finding a representation of a general solution to (3.2.1) (explicit/implicit). This provides a basis for solving the real-world problems.

3.2.2 *Procedure for Finding a General Solution Representation*

The procedure for finding a representation of a general solution to the stochastic differential equation (3.2.1) is described below. A method of the energy/Lyapunov function is used to determine a closed-form representation of a solution to (3.2.1).

Step #1 (seeking an energy function). Let us assume that $x(t)$ is a solution process of (3.2.1). To find an explicit/implicit representation of $x(t)$ as a function of t and $w(t)$, we seek an unknown function $V(t, x)$ (energy/Lyapunov function) defined on $J \times R$ into R which possesses the following properties:

(a) $V(t, x)$ is continuous on $J \times R$ ($V \in C[J \times R, R]$).
(b) For $(t, x) \in J \times R$, $V(t, x)$ is monotonic in x for each t.
(c) V is continuously differentiable with respect to t for each fixed x and twice continuously differentiable with respect to x for each fixed t.
(d) For each $t \in J$, $V(t, x)$ has an inverse function $E(t, x)$ defined on $J \times R$ into R, i.e. $V(t, E(t, x)) = x = E(t, V(t, x))$.

Step #2 [differential of the energy function along the differential equation field (3.2.1)]. We find a differential of $V(t, x)$ along the Itô–Doob-type stochastic differential equation field (3.2.1) in R^2. We simply assume that (3.2.1) has a solution process $x(t)$ defined on J with an unknown closed-form representation. We recall that our main goal is to find an explicit/implicit representation of $x(t)$ on J (if possible). For this purpose, we apply the Itô–Doob stochastic differential formula (Theorem 1.3.2) to $V(t, x(t))$ with respect to the solution process of (3.2.1), and we have

$$dV(t, x(t)) = LV(t, x(t))dt + \sigma(t, x)\frac{\partial}{\partial x}V(t, x(t))dw(t), \tag{3.2.2}$$

where w is the Wiener process defined in Definition 1.2.12 (Theorem 1.2.2), $x(t)$ is the solution process of (3.2.1), and L is a linear differential operator associated with (3.2.1), and it is defined by

$$LV(t, x(t)) = \frac{\partial}{\partial t}V(t, x(t)) + f(t, x(t))\frac{\partial}{\partial x}V(t, x(t)) + \frac{1}{2}\sigma^2(t, x)\frac{\partial^2}{\partial x^2}V(t, x(t)). \tag{3.2.3}$$

Step #3 (idea of a simpler differential equation). We denote $m(t) = V(t, x(t))$ as a composite function of $V(t, x)$ and $x(t)$. With this notation, from (3.2.2) and (3.2.3), we make a conceptual choice of convenient functions F and Λ so that the process $m(t)$ satisfies the following differential equation:

$$dm = F(t, m)dt + \Lambda(t, m)dw(t), \tag{3.2.4}$$

where

$$F(t, V(t, x)) = \frac{\partial}{\partial t} V(t, x) + f(t, x) \frac{\partial}{\partial x} V(t, x) + \frac{1}{2}\sigma^2(t, x) \frac{\partial^2}{\partial x^2} V(t, x), \tag{3.2.5}$$

and

$$\Lambda(t, V(t, x)) = \sigma(t, x) \frac{\partial}{\partial x} V(t, x). \tag{3.2.6}$$

Conditions (3.2.5) and (3.2.5) allow us to choose an unknown energy function in Step #1(a) depending on the conceptual choice of the convenient functions F and Λ in (3.2.4). "Convenient" means allowing one to find an explicit/implicit (if possible) form of the solution process $x(t)$ of (3.2.1) for the class of rate functions f and σ (as large as possible). The choice of the functions F and Λ in (3.2.4) satisfying the conditions (3.2.5) and (3.2.5) depends on the class of rate functions f and σ and the ability to find a closed-form solution to (3.2.4). The differential equation (3.2.4) is referred to as the *Itô–Doob-type stochastic reduced differential equation.*

Step #4 (determination of energy and convenient choice of functions). Depending on the class of functions determined by the conditions (3.2.5) and (3.2.5), Step #3 allows us to determine: (i) an unknown energy function described in Step #1, and (ii) the conceptual convenient choice rate functions F and Λ in (3.2.4). In short, an energy function and the reduced differential equation (3.2.4) are determined with respect to the class of nonlinear Itô–Doob-type stochastic differential equations determined in Step #3.

Step #5 [determination of the solution representation of (3.2.1)]. The energy function and reduced differential equation determined in Step #4 with respect to the class of nonlinear differential equations in Step #3 are used for finding a representation of the solution to (3.2.1). This is achieved by using the closed-form solution $m(t)$ to (3.2.4) and the property (d) of the energy function $V(t, x)$ in Step #1 (if it exists). The representation of the original solution process is given by

$$x(t) = E(t, m(t)) = E(t, V(t, x(t))). \tag{3.2.7}$$

This is the desired closed-form conceptual solution representation of the given solution $x(t)$ to (3.2.1). This completes the most general and a broad procedure for finding a representation of a general solution to a given problem.

Brief Summary. The following flowchart summarizes the basic steps described in the energy function method.

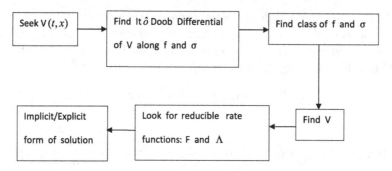

Fig. 3.2.1 Energy function method.

Observation 3.2.1. In the next two sections, we determine the classes of nonlinear stochastic differential equations (3.2.1) which are reducible to (a) integrable and (b) first-order linear stochastic differential equations. We further note that this technique is not limited to these classes. It is an open-ended technique, and it can be extended to reducible nonlinear stochastic differential equations that can be solved in closed forms. In short, the energy function method is a source of research topics for undergraduate/graduate students and interested interdisciplinary researchers.

3.3 Integrable Reduced Equations

In this section, we make a particular choice of convenient functions F and Λ in (3.2.4) as follows: $F(t, m) = p(t)$ and $\Lambda(t, m) = q(t)$, where p and q are unknown continuous functions. We assume that p and q satisfy the conditions (3.2.5) and (3.2.5) in Section 3.2. Using this "conceptual and convenient choice of functions," we determine the class of differential equations (3.2.1) as per Step #3 of Procedure 3.2.2. From the knowledge of this class, we then determine an energy/Lyapunov function and the expressions for the conceptual choice of functions p and q. In short, the energy function and rate functions F and Λ depend on (3.2.1). The determined class of differential equations is reducible to Itô–Doob-type stochastic integrable differential equations. The method of solving the differential equations of Section 2.2 of Chapter 2 is used to find a solution representation of (3.2.1).

3.3.1 *General Problem*

We consider a reduction problem for a broad class of nonlinear stochastic differential equations (3.2.1). This class of differential equations is reducible to the following Itô–Doob-type of first-order integrable reduced stochastic differential equation (reduced SDE) [43, 71]:

$$dm = p(t)dt + q(t)dw(t), \qquad (3.3.1)$$

where p and q are continuous functions defined on R into R; the rate functions p and q are independent of the state variable m (dependent variable) of a transformed dynamic system.

(H$_{3.3}$). In addition to hypothesis (H$_{3.2}$), we assume that f and σ are continuously differentiable with respect to both t and x variables. Furthermore, $\sigma(t, x) \neq 0$ and $G(t, x) = \int_c^x \frac{dy}{\sigma(t,y)}$ is invertible in x for each t.

3.3.2 *Procedure for Finding a General Solution Representation*

To find an expression for a solution process of (3.2.1), we describe a brief procedure: (ISDE$_1$) to determine a class of rate functions f and σ in (3.2.1), (ISDE$_2$) to find the energy/Lyapunov function, and (ISDE$_3$) to look for the convenient choice of functions p and q in the context of the class and the given rate functions f and σ in (3.2.1).

Step #1. We imitate Steps #1 and #2 outlined in of Procedure 3.2.2, and arrive at (3.2.2) and (3.2.3).

Step #2 (Step #3 of General Procedure 3.2.2). Now, by using the conceptual choice of functions $F(t, m) = p(t)$ and $\Lambda(t, m) = q(t)$ in (3.3.1), we repeat Step #3 of General Procedure 3.2.2. In this case, (3.2.5) and (3.2.5) become

$$p(t) = \frac{\partial}{\partial t}V(t, x) + f(t, x)\frac{\partial}{\partial x}V(t, x) + \frac{1}{2}\sigma^2(t, x)\frac{\partial^2}{\partial x^2}V(t, x), \qquad (3.3.2)$$

and

$$q(t) = \sigma(t, x)\frac{\partial}{\partial x}V(t, x), \qquad (3.3.3)$$

respectively. Under the conditions (3.3.2) and (3.3.2), we find first, a sufficient condition for the class of rate functions f and σ of (3.2.1). For this purpose, we partially differentiate both sides of (3.3.2) with respect to x and t, and obtain

$$0 = \frac{\partial}{\partial x}\left(\sigma(t, x)\frac{\partial}{\partial x}V(t, x)\right) \text{ [from (3.3.2)]}$$

$$= \frac{\partial}{\partial x}\sigma(t, x)\frac{\partial}{\partial x}V(t, x) + \sigma(t, x)\frac{\partial^2}{\partial x^2}V(t, x) \text{ (by the product formula)},$$

$$(3.3.4)$$

and

$$\frac{d}{dt}q(t) = \frac{\partial}{\partial t}\left(\sigma(t, x)\frac{\partial}{\partial x}V(t, x)\right) \text{ [from (3.3.2)]}$$

$$= \frac{\partial}{\partial t}\sigma(t, x)\frac{\partial}{\partial x}V(t, x) + \sigma(t, x)\frac{\partial^2}{\partial t\partial x}V(t, x) \text{ (by the product formula)}.$$

$$(3.3.5)$$

From (3.3.2)–(3.3.4), we solve for $\frac{\partial^2}{\partial x^2}V(t,x)$ and $\frac{\partial^2}{\partial t\partial x}V(t,x)$, and we have

$$\frac{\partial^2}{\partial x^2}V(t,x) = -\frac{\frac{\partial}{\partial x}\sigma(t,x)}{\sigma(t,x)}\frac{\partial}{\partial x}V(t,x) \quad \text{[from (3.3.4) solving for } \frac{\partial^2}{\partial x^2}V(t,x)\text{]}$$

$$= -\frac{q(t)\frac{\partial}{\partial x}\sigma(t,x)}{\sigma^2(t,x)} \quad \text{[by using (3.3.2)],} \qquad (3.3.6)$$

and

$$\frac{\partial^2}{\partial t\partial x}V(t,x) = \frac{\frac{d}{dt}q(t)}{\sigma(t,x)} - \frac{\frac{\partial}{\partial t}\sigma(t,x)}{\sigma(t,x)}\frac{\partial}{\partial x}V(t,x) \quad \text{[from (3.3.4)]}$$

$$= \frac{\sigma(t,x)\frac{d}{dt}q(t) - q(t)\frac{\partial}{\partial t}\sigma(t,x)}{\sigma^2(t,x)} \quad \text{[by using (3.3.2)].} \quad (3.3.7)$$

We partially differentiate both sides of (3.3.2) with respect to x, and obtain

$$0 = \frac{\partial}{\partial x}\left[\frac{\partial}{\partial t}V(t,x) + f(t,x)\frac{\partial}{\partial x}V(t,x) + \frac{1}{2}\sigma^2(t,x)\frac{\partial^2}{\partial x^2}V(t,x)\right] \quad \text{[from (3.3.2)]}$$

$$= \frac{\partial^2}{\partial t\partial x}V(t,x) + \frac{\partial}{\partial x}\left[f(t,x)\frac{\partial}{\partial x}V(t,x) + \frac{1}{2}\sigma^2(t,x)\frac{\partial^2}{\partial x^2}V(t,x)\right] \quad \text{(by the sum rule).}$$

$$(3.3.8)$$

We substitute the expressions in (3.3.2), (3.3.6), and (3.3.6) into Equation (3.3.8), and obtain

$$0 = \frac{\sigma(t,x)\frac{d}{dt}q(t) - q(t)\frac{\partial}{\partial t}\sigma(t,x)}{\sigma^2(t,x)} + \frac{\partial}{\partial x}\left[f(t,x)\frac{q(t)}{\sigma(t,x)} - \frac{1}{2}\sigma^2(t,x)\frac{q(t)\frac{\partial}{\partial x}\sigma(t,x)}{\sigma^2(t,x)}\right]$$

$$= \frac{\sigma(t,x)\frac{d}{dt}q(t) - q(t)\frac{\partial}{\partial t}\sigma(t,x)}{\sigma^2(t,x)}$$

$$+ \frac{\partial}{\partial x}\left[f(t,x)\frac{q(t)}{\sigma(t,x)} - \frac{1}{2}q(t)\frac{\partial}{\partial x}\sigma(t,x)\right] \quad \text{(by simplifying).}$$

Hence,

$$\frac{\frac{d}{dt}q(t)}{\sigma(t,x)} = q(t)\frac{\frac{\partial}{\partial t}\sigma(t,x)}{\sigma^2(t,x)} - q(t)\frac{\partial}{\partial x}\left[\frac{f(t,x)}{\sigma(t,x)} - \frac{1}{2}\frac{\partial}{\partial x}\sigma(t,x)\right]$$

$$= q(t)\left[\frac{\frac{\partial}{\partial t}\sigma(t,x)}{\sigma^2(t,x)} - \frac{\partial}{\partial x}\left(\frac{f(t,x)}{\sigma(t,x)}\right) + \frac{1}{2}\frac{\partial^2}{\partial x^2}\sigma(t,x)\right] \quad \text{(by regrouping).}$$

This implies that

$$\frac{\frac{d}{dt}q(t)}{q(t)} = \sigma(t,x)\left[\frac{\frac{\partial}{\partial t}\sigma(t,x)}{\sigma^2(t,x)} - \frac{\partial}{\partial x}\left(\frac{f(t,x)}{\sigma(t,x)}\right) + \frac{1}{2}\frac{\partial^2}{\partial x^2}\sigma(t,x)\right]. \tag{3.3.9}$$

Now, we know that the left-hand side of (3.3.9) is independent of x. We again implicitly differentiate partially [127] with respect to x, and we have

$$0 = \frac{\partial}{\partial x}\left(\sigma(t,x)\left[\frac{\frac{\partial}{\partial t}\sigma(t,x)}{\sigma^2(t,x)} - \frac{\partial}{\partial x}\left(\frac{f(t,x)}{\sigma(t,x)}\right) + \frac{1}{2}\frac{\partial^2}{\partial x^2}\sigma(t,x)\right]\right). \tag{3.3.10}$$

This is a sufficient condition for the reducibility of (3.2.1) to the integrable type of stochastic differential equation (3.3.1) [43, 71]. The relation (3.3.10) determines the class of differential equations (3.2.1) which is reducible to the (3.3.1) type.

Step #3 (Step #4 of General Procedure 3.2.2). Now, we are ready to enter into the task of Step #4 of Procedure 3.2.2. First, we note that a "convenient choice of the function" $q(t)$ in (3.3.1) can be determined by the right-hand side expression in (3.3.9) and is in terms of rate functions $f(t,x)$ and $\sigma(t,x)$ of (3.2.1). Therefore, we denote this known function in (3.3.9) by

$$\alpha(t) = \sigma(t,x)\left[\frac{\frac{\partial}{\partial t}\sigma(t,x)}{\sigma^2(t,x)} - \frac{\partial}{\partial x}\left(\frac{f(t,x)}{\sigma(t,x)}\right) + \frac{1}{2}\frac{\partial^2}{\partial x^2}\sigma(t,x)\right]. \tag{3.3.11}$$

Using (3.3.11) and (3.3.9), we have

$$\frac{\frac{d}{dt}q(t)}{q(t)} = \alpha(t).$$

This is a first-order linear homogeneous deterministic differential equation [80] with unknown function $q(t)$:

$$\frac{d}{dt}q(t) = \alpha(t)q(t). \tag{3.3.12}$$

Applying the method of Section 2.3 of Chapter 2 [80], the general solution to (3.3.12) is

$$q(t) = \exp\left[\int_a^t \alpha(s)ds\right]C, \tag{3.3.13}$$

where C is an arbitrary constant of integration. For the nontrivial function $q(t)$, we choose $C \neq 0$. This determines a choice function q in (3.3.1).

To find a second choice function $p(t)$ and an energy function $V(t, x)$, from (3.3.13), we rewrite (3.3.2) as follows:

$$\sigma(t, x)\frac{\partial}{\partial x}V(t, x) = q(t) = \exp\left[\int_a^t \alpha(s)ds\right]C. \tag{3.3.14}$$

Now, by using assumption (H$_{3.3}$), we solve for $\frac{\partial}{\partial x}V(t, x)$, and obtain

$$\frac{\partial}{\partial x}V(t, x) = \frac{1}{\sigma(t, x)}\exp\left[\int_a^t \alpha(s)ds\right]C.$$

We partially integrate the above relation on both sides with respect to x for fixed t, and obtain

$$V(t, x) = \int_c^x \left(\frac{\partial}{\partial x}V(t, y)\right)dy = C\exp\left[\int_a^t \alpha(s)ds\right]\int_c^x \frac{dy}{\sigma(t, y)} + B(t), \tag{3.3.15}$$

where $B(t)$ is an arbitrary constant of integration which depends on t. The dependence of $B(t)$ on t is due to the partial integration in (3.3.15) with respect to x for fixed t. We need to find B subject to the other conditions on $V(t, x)$. For this purpose, we find $\frac{\partial}{\partial t}V(t, x)$, $\frac{\partial}{\partial x}V(t, x)$ and $\frac{\partial^2}{\partial x^2}V(t, x)$ as

$$\frac{\partial}{\partial t}V(t, x) = \frac{\partial}{\partial t}\left[\exp\left[\int_a^t \alpha(s)ds\right]C\int_c^x \frac{dy}{\sigma(t, y)} + B(t)\right]$$

$$= \frac{\partial}{\partial t}\left[\exp\left[\int_a^t \alpha(s)ds\right]C\int_c^x \frac{dy}{\sigma(t, y)}\right] + \frac{\partial}{\partial t}B(t) \quad \text{(by the sum rule)}$$

$$= \exp\left[\int_a^t \alpha(s)ds\right]C\left[\alpha(t)\int_c^x \frac{dy}{\sigma(t, y)} - \int_c^x \frac{\frac{\partial}{\partial t}\sigma(t, y)}{\sigma^2(t, y)}dy\right]$$

$$+ \frac{d}{dt}B(t) \quad \text{(by the product and chain rules)} \tag{3.3.16}$$

$$\frac{\partial}{\partial x}V(t, x) = \frac{1}{\sigma(t, x)}\exp\left[\int_a^t \alpha(s)ds\right]C, \tag{3.3.17}$$

and

$$\frac{\partial^2}{\partial x^2}V(t, x) = -\frac{\frac{\partial}{\partial x}\sigma(t, x)}{\sigma^2(t, x)}\exp\left[\int_a^t \alpha(s)ds\right]C. \tag{3.3.18}$$

To find $B(t)$ and $p(t)$, we substitute the expressions (3.3.16)–(3.3.17) into (3.3.2), and we have

$$p(t) = \frac{\partial}{\partial t}V(t, x) + f(t, x)\frac{\partial}{\partial x}V(t, x) + \frac{1}{2}\sigma^2(t, x)\frac{\partial^2}{\partial x^2}V(t, x)$$

$$= \exp\left[\int_a^t \alpha(s)ds\right]C\left[\alpha(t)\int_c^x \frac{dy}{\sigma(t, y)} - \int_c^x \frac{\frac{\partial}{\partial t}\sigma(t, y)}{\sigma^2(t, y)}dy\right]$$

$$+ \frac{d}{dt}B(t) + \frac{f(t,x)}{\sigma(t,x)} \exp\left[\int_a^t \alpha(s)ds\right] C$$

$$- \frac{1}{2}\sigma^2(t,x)\frac{\frac{\partial}{\partial x}\sigma(t,x)}{\sigma^2(t,x)} \exp\left[\int_a^t \alpha(s)ds\right] C \qquad \text{(by substitution)}$$

$$= \exp\left[\int_a^t \alpha(s)ds\right] C \left[\alpha(t)\int_c^x \frac{dy}{\sigma(t,y)} + \frac{f(t,x)}{\sigma(t,x)} - \frac{1}{2}\frac{\partial}{\partial x}\sigma(t,x)\right.$$

$$\left. - \int_c^x \frac{\frac{\partial}{\partial t}\sigma(t,y)}{\sigma^2(t,y)}dy\right] + \frac{d}{dt}B(t) \qquad \text{(by regrouping)}.$$

We solve for $\frac{d}{dt}B(t)$ and obtain

$$\frac{d}{dt}B(t) = p(t) - \left[\alpha(t)\int_c^x \frac{dy}{\sigma(t,y)} - \int_c^x \frac{\frac{\partial}{\partial t}\sigma(t,y)}{\sigma^2(t,y)}dy\right.$$

$$\left. + \frac{f(t,x)}{\sigma(t,x)} - \frac{1}{2}\frac{\partial}{\partial x}\sigma(t,x)\right]\exp\left[\int_a^t \alpha(s)ds\right]C. \qquad (3.3.19)$$

Hence,

$$\frac{\frac{d}{dt}B(t) - p(t)}{\exp\left[\int_a^t \alpha(s)ds\right]C} = -\left[\alpha(t)\int_c^x \frac{dy}{\sigma(t,y)} + \frac{f(t,x)}{\sigma(t,x)}\right.$$

$$\left. - \frac{1}{2}\frac{\partial}{\partial x}\sigma(t,x) - \int_c^x \frac{\frac{\partial}{\partial t}\sigma(t,y)}{\sigma^2(t,y)}dy\right], \qquad (3.3.20)$$

provided that $C \neq 0$.

For $C \neq 0$, from (3.3.11), we observe that the left-hand side expression in (3.3.20) is independent of x, and the right-hand expression in (3.3.20) depends only on the rate functions in (3.2.1). Therefore, we set by a known function $\beta(t)$ as follows:

$$\beta(t) = -\left[\alpha(t)\int_c^x \frac{dy}{\sigma(t,y)} + \frac{f(t,x)}{\sigma(t,x)} - \frac{1}{2}\frac{\partial}{\partial x}\sigma(t,x) - \int_c^x \frac{\frac{\partial}{\partial t}\sigma(t,y)}{\sigma^2(t,y)}dy\right]. \qquad (3.3.21)$$

From (3.3.21), (3.3.19) reduces to

$$\frac{d}{dt}B(t) = \beta(t)\exp\left[\int_a^t \alpha(s)ds\right]C + p(t). \qquad (3.3.22)$$

From (3.3.22), we integrate both sides, and obtain

$$B(t) = \int_a^t \left[\beta(s) \exp\left[\int_a^s \alpha(u)du \right] C + p(s) \right] ds + C_1, \qquad (3.3.23)$$

where C_1 is another arbitrary constant of integration. We note that $B(t)$ is an arbitrary known constant of an integration function that depends on the known functions $\alpha(t)$ in (3.3.11), $\beta(t)$ in (3.3.21), and an arbitrary chosen function $p(s)$ in (3.3.1). Thus, from (3.3.15) and (3.3.23), the unknown function $V(t,x)$ in Step #1 is almost completely determined by

$$V(t,x) = \exp\left[\int_a^t \alpha(s)ds \right] C \int_c^x \frac{dy}{\sigma(t,y)} + B(t)$$

$$= \exp\left[\int_a^t \alpha(s)ds \right] C \int_c^x \frac{dy}{\sigma(t,y)} \qquad \text{(by substitution)}$$

$$+ \int_a^t \left[\beta(s) \exp\left[\int_a^s \alpha(u)du \right] C + p(s) \right] ds + C_1, \qquad (3.3.24)$$

where $C \neq 0$, C and C_1 are arbitrary constants of integration, the functions $\beta(t)$ and $\alpha(t)$ depend on the rate functions f and σ in (3.2.1), and $p(t)$ is an arbitrary choice of function in (3.3.1). This completes the step of determining an energy function as well as the rate functions of the reduced differential equation (3.3.1).

Moreover, from (3.3.11), (3.3.13), (3.3.16)–(3.3.17), (3.3.21), and (3.3.22), we have

$$dV(t,x(t)) = \left[\frac{\partial}{\partial t}V(t,x(t)) + f(t,x(t))\frac{\partial}{\partial x}V(t,x(t)) + \frac{1}{2}\sigma^2(t,x)\frac{\partial^2}{\partial x^2}V(t,x(t)) \right] dt$$

$$+ \sigma(t,x)\frac{\partial}{\partial x}V(t,x(t))dw(t) \quad \text{(by Theorem 1.3.2)}$$

$$= \exp\left[\int_a^t \alpha(s)dy \right] C \left[\alpha(t) \int_c^{x(t)} \frac{dy}{\sigma(t,y)} - \int_c^{x(t)} \frac{\frac{\partial}{\partial t}\sigma(t,y)}{\sigma^2(t,y)}dy \right]$$

$$+ \frac{d}{dt}B(t) + \frac{f(t,x(t))}{\sigma(t,x(t))} \exp\left[\int_a^t \alpha(s)ds \right] C$$

$$- \frac{1}{2}\sigma^2(t,x(t))\frac{\frac{\partial}{\partial x}\sigma(t,x(t))}{\sigma^2(t,x(t))} \exp\left[\int_a^t \alpha(s)ds \right] C\,dt$$

$$+\frac{\sigma(t,x(t))}{\sigma(t,x(t))}\exp\left[\int_a^t \alpha(s)ds\right]Cdw(t) \quad \text{(by substitution from above)}$$

$$= \exp\left[\int_a^t \alpha(s)ds\right]C\left[\alpha(t)\int_c^x \frac{dy}{\sigma(t,y)} - \int_c^x \frac{\frac{\partial}{\partial t}\sigma(t,y)}{\sigma^2(t,y)}dy + \beta(t)\right.$$

$$\left.+\frac{f(t,x(t))}{\sigma(t,x(t))} - \frac{1}{2}\frac{\partial}{\partial x}\sigma(t,x(t))\right]dt + p(t)dt$$

$$+ \exp\left[\int_a^t \alpha(s)ds\right]Cdw(t) \quad \text{(by regrouping and simplifying)}$$

$$= p(t)dt + q(t)dw(t) \quad \text{[by substitution for } \beta(t) \text{ and } q(t)\text{].} \quad (3.3.25)$$

This shows that $m(t) = V(t, x(t))$ satisfies (3.3.1), where $q(t)$ is as defined in (3.3.13); $p(t)$ and $C \neq 0$ are an arbitrary function and a constant, respectively. Thus, $p(t)$ can be chosen by a problem-solver. Hence, a general solution process of (3.3.1) is given by

$$m(t) = V(t, x(t)) = \int_a^t p(s)ds + C\int_a^t \exp\left[\int_a^s \alpha(u)du\right]dw(s). \quad (3.3.26)$$

Step #4 (Step #5 of General Procedure 3.2.2). The class of differential equations (3.2.1) is determined by the condition (3.3.10). The energy function in (3.3.24) is utilized for finding a representation of the solution process of (3.2.1). By comparing the right-hand side expressions in (3.3.24) and (3.3.26), we have

$$\int_a^t p(s)ds + C\int_a^t \exp\left[\int_a^s \alpha(u)du\right]dw(s)$$

$$= \exp\left[\int_a^t \alpha(s)ds\right]C\int_c^{x(t)} \frac{dy}{\sigma(t,y)}$$

$$+ \int_a^t \left[\beta(s)\exp\left[\int_a^s \alpha(u)du\right]C + p(s)\right]ds + C_1. \quad (3.3.27)$$

From (3.3.27), we solve for $x(t)$. First, we solve for $\exp[\int_a^t \alpha(s)ds]C\int_c^{x(t)}\frac{dy}{\sigma(t,y)}$ as follows:

$$\exp\left[\int_a^t \alpha(s)ds\right]C\int_c^{x(t)} \frac{dy}{\sigma(t,y)}$$

$$= -C\int_a^t \exp\left[\int_a^s \alpha(u)du\right]\beta(s)ds - \int_a^t p(s)ds + C_1$$

$$+ \int_a^t p(s)ds + C \int_a^t \exp\left[\int_a^s \alpha(u)du\right] dw(s) \quad \text{(by simplifying)}$$

$$= -C \int_a^t \exp\left[\int_a^s \alpha(u)du\right] \beta(s)ds + C \int_a^t \exp\left[\int_a^s \alpha(u)du\right] dw(s) + C_1.$$

This implies that

$$G(t, x(t)) - G(t, c) = \int_c^{x(t)} \frac{dy}{\sigma(t, y)}$$

$$= \frac{\exp\left[-\int_a^t \alpha(s)ds\right]}{C}\left[-C \int_a^t \exp\left[\int_a^s \alpha(u)du\right] \beta(s)ds + C_1\right.$$

$$\left. + C \int_a^t \exp\left[\int_a^s \alpha(u)du\right] dw(s)\right] \quad [\text{by solving for } \int_c^{x(t)} \tfrac{dy}{\sigma(t,y)}]$$

$$= \exp\left[-\int_a^t \alpha(s)ds\right]\left[-\int_a^t \exp\left[\int_a^s \alpha(u)du\right] \beta(s)ds + \frac{C_1}{C}\right.$$

$$\left. + \int_a^t \exp\left[\int_a^s \alpha(u)du\right] dw(s)\right] \quad \text{(by algebra)}$$

$$= \exp\left[-\int_a^t \alpha(s)ds\right] C_1 - \int_a^t \exp\left[-\int_s^t \alpha(u)du\right] \beta(s)ds$$

$$+ \int_a^t \exp\left[-\int_s^t \alpha(u)du\right] dw(s) \quad \text{(by integration property).}$$

$$(3.3.28)$$

From the hypothesis (H$_{3.3}$), $G(t, x(t))$ is invertible for each t, and therefore we have

$$x(t) = G^{-1}\left(t, G(t,c) + \frac{C_1}{C} \exp\left[-\int_a^t \alpha(s)ds\right] - \int_a^t \exp\left[-\int_s^t \alpha(u)du\right] \beta(s)ds\right.$$

$$\left. + \int_a^t \exp\left[-\int_s^t \alpha(u)du\right] dw(s)\right) = E(t, m(t)), \quad (3.3.29)$$

where $m(t)$ is a general solution process of (3.3.1) as determined in Section 2.2 of Chapter 2, and $x(t)$ is a general solution process of the Itô–Doob-type stochastic differential equation (3.2.1) which belongs to the class determined in Step #3 (General Procedure 3.2.2). We note that the representation of the solution process is independent of an arbitrary choice function $p(t)$. This completes the solution representation process of the class of differential equations (3.2.1) determined by the condition (3.3.10).

Observation 3.3.1

(i) From (3.3.15), (3.3.23), (3.3.26), and (3.3.29), we recall that $V(t, x)$ depends on an arbitrary constant of integration function $B(t)$ in (3.3.15) and $B(t)$ depends on an arbitrary choice function $p(t)$ in (3.3.23). Moreover, the functions $\alpha(t)$ in (3.3.11), $\beta(t)$ in (3.3.22), $B(t)$, and $\int_a^t p(s)ds$ are related by (3.3.23). In fact, $p(t)$ is determined by (3.3.22)

(ii) We also note that by setting

$$z(t, x) = \left[\frac{f(t, x)}{\sigma(t, x)} - \frac{1}{2} \frac{\partial}{\partial x} \sigma(t, x) - \int_c^x \frac{\frac{\partial}{\partial t} \sigma(t, y)}{\sigma^2(t, y)} dy \right], \tag{3.3.30}$$

we have

$$\frac{\partial}{\partial x} z(t, x) = \frac{\partial}{\partial x} \left[\frac{f(t, x)}{\sigma(t, x)} - \frac{1}{2} \frac{\partial}{\partial x} \sigma(t, x) \right] - \frac{\frac{\partial}{\partial t} \sigma(t, x)}{\sigma^2(t, x)}. \tag{3.3.31}$$

Thus, from (3.3.31), the condition (3.3.10) which determines the class can be rewritten as

$$0 = \frac{\partial}{\partial x} \left(\sigma(t, x) \left[\frac{\partial}{\partial x} \left(\frac{f(t, x)}{\sigma(t, x)} \right) - \frac{1}{2} \frac{\partial^2}{\partial x^2} \sigma(t, x) - \frac{\frac{\partial}{\partial t} \sigma(t, x)}{\sigma^2(t, x)} \right] \right)$$

$$= \frac{\partial}{\partial x} \left(\sigma(t, x) \frac{\partial}{\partial x} z(t, x) \right). \tag{3.3.32}$$

We further note that from (3.3.30) and (3.3.31), $\alpha(t)$ in (3.3.11) and $\beta(t)$ in (3.3.21) can be rewritten as

$$\alpha(t) = -\sigma(t, x) \frac{\partial}{\partial x} z(t, x), \tag{3.3.33}$$

and

$$\beta(t) = - \left[\alpha(t) \int_c^x \frac{dy}{\sigma(t, y)} + z(t, x) \right], \tag{3.3.34}$$

respectively.

(iii) From (3.3.32), we remark that the reducible class of Itô–Doob SDEs (3.2.1) to a scalar integrable stochastic differential equation (reduced ISDE or ISDE) is defined and denoted by

$$\text{ISDE-}(f, \sigma) = \left\{ (f, \sigma) \text{ in } (3.2.1) : \frac{\partial}{\partial x} \left(\sigma(t, x) \frac{\partial}{\partial x} z(t, x) \right) = 0, \text{ with } \sigma \neq 0 \right\}. \tag{3.3.35}$$

Moreover, the condition defining the class provides a conceptual structure for the drift rate in (3.2.1). This structure can be employed for finding solution by using the knowledge of f and an algebraic method of undetermined coefficients.

(iv) From the definition of the class ISDE-(f, σ), we can construct a conceptual representation of this class with arbitrary functions/constants. In fact, from the defining property of this class and using the method of Section 2.2 of Chapter 2 [80], we solve the deterministic differential equation

$$\frac{\partial}{\partial x}\left(\sigma(t, x)\frac{\partial}{\partial x}z(t, x)\right) = 0. \tag{3.3.36}$$

We integrate both sides repeatedly with respect to x for fixed t, and obtain

$$\sigma(t, x)\frac{\partial}{\partial x}z(t, x) = f_1(t). \tag{3.3.37}$$

From (3.3.37), by solving for $\frac{\partial}{\partial x}z(t, x)$ and integrating partially with respect to x for fixed t, we obtain

$$z(t, x) = f_1(t)\int_c^x \frac{dy}{\sigma(t, y)} + f_2(t), \tag{3.3.38}$$

where $f_1(t)$ and $f_2(t)$ are an arbitrary continuous functions which are constants of integration. From (3.3.30) and (3.3.38), we can find an expression for $f(t, x)$ as follows:

$$z(t, x) = \frac{f(t, x)}{\sigma(t, x)} - \frac{1}{2}\frac{\partial}{\partial x}\sigma(t, x) - \int_c^x \frac{\frac{\partial}{\partial t}\sigma(t, y)}{\sigma^2(t, y)}dy = f_1(t)\int_c^x \frac{dy}{\sigma(t, y)} + f_2(t),$$

which implies that

$$\frac{f(t, x)}{\sigma(t, x)} = \frac{1}{2}\frac{\partial}{\partial x}\sigma(t, x) + \int_c^x \frac{\frac{\partial}{\partial t}\sigma(t, y)}{\sigma^2(t, y)}dy + f_1(t)\int_c^x \frac{dy}{\sigma(t, y)} + f_2(t). \tag{3.3.39}$$

Hence,

$$f(t, x) = \sigma(t, x)\left[\frac{1}{2}\frac{\partial}{\partial x}\sigma(t, x) + \int_c^x \frac{\frac{\partial}{\partial t}\sigma(t, y)}{\sigma^2(t, y)}dy + f_1(t)\int_c^x \frac{dy}{\sigma(t, y)} + f_2(t)\right]. \tag{3.3.40}$$

This is the conceptual structural representation of the member (f, σ) of the class ISDE-(f, σ). Moreover, from (3.3.33), (3.3.33), (3.3.37), and (3.3.38), the arbitrary continuous functions $f_1(t)$ and $f_2(t)$ are

$$f_1(t) = -\alpha(t) \text{ and } f_2(t) = -\beta(t), \tag{3.3.41}$$

respectively.

(v) In particular, if $\frac{\partial}{\partial x}(z(t,x)) = 0$ implies that $z(t,x)$ in (3.3.30) must be a function of t only. Moreover, in this case, from (3.3.37), (3.3.39) reduces to

$$f(t,x) = \sigma(t,x)\left[\frac{1}{2}\frac{\partial}{\partial x}\sigma(t,x) + \int_c^x \frac{\frac{\partial}{\partial t}\sigma(t,y)}{\sigma^2(t,y)}ds + f_2(t)\right]. \tag{3.3.42}$$

f in (3.3.42) together with $\sigma(t,x)$ satisfying the condition (3.3.36) represents the membership of the class ISDE-(f,σ).

(vi) If the rate functions f and σ are stationary (autonomous), i.e. $f(t,x) \equiv f(x)$ and $\sigma(t,x) \equiv \sigma(x)$, then the class of SDEs in (3.2.1) is reducible to scalar integrable stochastic differential equations with constant coefficients (reduced ISDECC). This class is defined and denoted by [43]

$$\text{ISDECC-}(f,\sigma) = \left\{(f,\sigma) \text{ in } (3.2.1) : \frac{d}{dx}\left(\sigma(x)\frac{d}{dx}z(x)\right) = 0, \text{ with } \sigma \neq 0\right\}, \tag{3.3.43}$$

where

$$z(x) = \left[\frac{f(x)}{\sigma(x)} - \frac{1}{2}\frac{\partial}{\partial x}\sigma(x)\right]. \tag{3.3.44}$$

Moreover, the corresponding energy-like function in (3.3.24) and the rate function f in (3.3.40) are described by

$$V(t,x) = \exp[\alpha(t-a)]C\int_c^x \frac{dy}{\sigma(y)} + \left[\frac{\beta}{\alpha} - \frac{\beta}{\alpha}\exp[-\alpha(t-a)]\right]C + \int_a^t p(s)ds + C_1, \tag{3.3.45}$$

$$f(x) = \sigma(x)\left[\frac{1}{2}\frac{\partial}{\partial x}\sigma(x) + f_1\int_c^x \frac{dy}{\sigma(y)} + f_2\right]. \tag{3.3.46}$$

f in (3.3.46) together with $\sigma(x)$ satisfying the condition (3.3.43) represents the membership of class ISDECC-(f,σ).

Observation 3.3.2. In the following, we provide a few tips that can be useful for solving the SDE (3.2.1):

T_1: **Identification.** Identify the rate coefficients in the given Itô–Doob SDE.

T_2: **Rewriting.** Using algebra and deterministic calculus [127], rewrite the given SDE in standard form as in (3.2.1).

T_3: **Comparison.** Using the tip T_2, rewrite the SDE in a comparable form.

T_4: **Construction of the class condition.** Again, using algebraic and calculus tricks, create a class-defining condition from the rate coefficients (f,σ).

T_5: *Class identification.*   Identify the class membership of the given SDE.

T_6: *Reduced SDE.*   From T_4 and T_5, determine the rate coefficients of reduced SDE.

T_7: *Energy function.*   Knowing T_4–T_6, recognize the energy function.

T_8: *Solving the reduced SDE.*   Solve the reduced SDE and equate with the energy function evaluated along the solution to the given SDE.

T_9: *Step #5.*   Step #5 of General Procedure 3.2.2.

In the following, we present a few illustrations to exhibit the procedure described in this section.

Illustration 3.3.1. Given:

(a) $dx = \sigma(t,x)\left[\left[\frac{1}{2}\frac{\partial}{\partial x}\sigma(t,x) + \int_c^x \frac{\frac{\partial}{\partial t}\sigma(t,y)}{\sigma^2(t,y)}dy + F(t)\int_c^x \frac{dy}{\sigma(t,y)} + f(t)\right]dt + dw(t)\right],$

(b) $dx = \sigma(t,x)\left[\left[\frac{1}{2}\frac{\partial}{\partial x}\sigma(t,x) + \int_c^x \frac{\frac{\partial}{\partial t}\sigma(t,y)}{\sigma^2(t,y)}dy + F(t)\int_c^x \frac{dy}{\sigma(t,y)}\right]dt + dw(t)\right],$

(c) $dx = \sigma(t,x)\left[\left[\frac{1}{2}\frac{\partial}{\partial x}\sigma(t,x) + \int_c^x \frac{\frac{\partial}{\partial t}\sigma(t,y)}{\sigma^2(t,y)}dy + f(t)\right]dt + dw(t)\right],$

(d) $dx = \sigma(t,x)\left[\left[\frac{1}{2}\frac{\partial}{\partial x}\sigma(t,x) + \int_c^x \frac{\frac{\partial}{\partial t}\sigma(t,y)}{\sigma^2(t,y)}dy\right]dt + dw(t)\right],$

(e) $dx = \sigma(x)\left[\left[\frac{1}{2}\frac{\partial}{\partial x}\sigma(x) + A\int_c^x \frac{dy}{\sigma(y)} + B\right]dt + dw(t)\right],$

(f) $dx = \sigma(x)\left[\left[\frac{1}{2}\frac{\partial}{\partial x}\sigma(x)\right]dt + dw(t)\right],$

where σ satisfies the hypothesis $(H_{3.3})$, F and f are given continuous functions, and A and B are arbitrary given constants. Find the general solution to the given differential equation.

Solution procedure.   Let us solve Illustration 3.3.1(a). For this purpose, we follow the tips provided in Observation 3.3.2. From T_1, we have

$$f(t,x) = \sigma(t,x)\left[\frac{1}{2}\frac{\partial}{\partial x}\sigma(t,x) + \int_c^x \frac{\frac{\partial}{\partial t}\sigma(t,y)}{\sigma^2(t,y)}dy + F(t)\int_c^x \frac{dy}{\sigma(t,y)} + f(t)\right].$$

$$(3.3.47)$$

The standard representation of (a) is $dx = f(t,x)dt + \sigma(t,x)dw(t)$, where the drift and diffusion coefficients are as identified. This utilizes the tip T_2. The illustration is very general; we cannot do any further algebraic or calculus simplifications. We skip

T_3. From T_4, we play with the drift coefficient (here algebraically only), and obtain

$$\frac{f(t,x)}{\sigma(t,x)} = \frac{1}{2}\frac{\partial}{\partial x}\sigma(t,x) + \int_c^x \frac{\frac{\partial}{\partial t}\sigma(t,y)}{\sigma^2(t,y)}dy + F(t)\int_c^x \frac{dy}{\sigma(t,y)} + f(t). \qquad (3.3.48)$$

Using this, knowing the rate functions and the class structure (3.3.32) along with the objective of T_5, we have

$$\frac{f(t,x)}{\sigma(t,x)} - \frac{1}{2}\frac{\partial}{\partial x}\sigma(t,x) - \int_c^x \frac{\frac{\partial}{\partial t}\sigma(t,y)}{\sigma^2(t,y)}dy = F(t)\int_c^x \frac{dy}{\sigma(t,y)} + f(t) \qquad (3.3.49)$$

From (3.3.30) and (3.3.31), we have

$$\frac{\partial}{\partial x}z(t,x) = \frac{\partial}{\partial x}\left[\frac{f(t,x)}{\sigma(t,x)} - \frac{1}{2}\frac{\partial}{\partial x}\sigma(t,x)\right] - \frac{\frac{\partial}{\partial t}\sigma(t,x)}{\sigma^2(t,x)} = \frac{F(t)}{\sigma(t,x)}. \qquad (3.3.50)$$

This together with simplifications gives us

$$\sigma(t,x)\frac{\partial}{\partial x}z(t,x) = \sigma(t,x)\frac{\partial}{\partial x}\left[\frac{f(t,x)}{\sigma(t,x)} - \frac{1}{2}\frac{\partial}{\partial x}\sigma(t,x)\right] - \frac{\frac{\partial}{\partial t}\sigma(t,x)}{\sigma^2(t,x)} = F(t).$$
$$(3.3.51)$$

Again, we partially differentiate implicitly with respect to x for fixed t, and conclude that

$$\frac{\partial}{\partial x}\left(\sigma(t,x)\frac{\partial}{\partial x}z(t,x)\right) = 0$$

$$= \frac{\partial}{\partial x}\left(\sigma(t,x)\left[\frac{\partial}{\partial x}\left(\frac{f(t,x)}{\sigma(t,x)}\right) - \frac{1}{2} - \frac{\partial^2}{\partial x^2}\left(\sigma(t,x) - \frac{\frac{\partial}{\partial t}\sigma(t,x)}{\sigma^2(t,x)}\right)\right]\right).$$

This is the condition that determines the class ISDE-(f, σ). This completes the task of T_5. Performing partial differentiation in (3.3.51) and making a comparison with (3.3.11), we have $\alpha = -F(t)$. This determines $q(t)$ in (3.3.13). From (3.3.13), (3.3.15), (3.3.19), and (3.3.49), $p(t) = f(t)\exp\left[\int_a^t \alpha(s)ds\right] + \frac{d}{dt}B(t)$ and $B(t)$ is determined by the knowledge of rate functions in these problems. This information determines the energy function and reducible SDE. The tips T_6 and T_7 are fulfilled. Following tips T_8 and T_9, we have

$$x(t) = G^{-1}\left(t, G(t,c) + \frac{C_1}{C}\exp\left[-\int_a^t \alpha(s)ds\right] - \int_a^t \exp\left[-\int_s^t \alpha(u)du\right]\beta(s)ds\right.$$

$$\left. + \int_a^t \exp\left[-\int_s^t \alpha(u)du\right]dw(s)\right).$$

This completes the solution process of (a).

Following the argument similar to (a), we conclude that (f, σ) belongs to the class ISDE-(f, σ). In the case of (b), from (3.3.41), $\alpha(t) = -F(t)$ and $\beta(t) = 0$. Hence,

$$x(t) = G^{-1}\left(t, G(t, C) + \frac{C_1}{C} \exp\left[-\int_a^t \alpha(s)ds\right] + \int_a^t \exp\left[-\int_s^t \alpha(u)du\right] dw(s)\right).$$

(c) In this case, $\alpha(t) = 0$ and $\beta(t) = -f(t) = p(t)$. Hence, from (3.3.14), we have

$$V(t, x) = C \int_c^x \frac{dy}{\sigma(t, y)}, \quad \text{for } C \neq 0.$$

$$x(t) = G^{-1}\left(t, G(t, C) + \frac{C_1}{C} - \int_a^t \beta(s)ds + w(t) - w(a)\right).$$

(d) In this case, $\alpha(t) = 0$ and $\beta(t) = 0 = p(t)$. Therefore,

$$x(t) = G^{-1}\left(t, G(t, c) + \frac{C_1}{C} + \int_a^t dw(s)\right) = G^{-1}\left(t, G(t, c) + \frac{C_1}{C} + w(t) - w(a)\right).$$

(e) This differential equation belongs to ISDECC-(f, σ). Moreover, this case is very similar to case (a). Here, $\alpha(t) = -a$ and $\beta(t) = b$.

(f) This case is similar to cases (d) and (c).

Illustration 3.3.2. Given:

$$dx = k(t)\left[-g(t)k(t)\sin\left(g(t)x + \theta(t)\right)\cos^3\left(g(t)x + \theta(t)\right)\right.$$

$$+ F(t)\sin\left(g(t)x + \theta(t)\right)\cos\left(g(t)x + \theta(t)\right) - \frac{x\frac{d}{dt}g(t) + \frac{d}{dt}\theta(t)}{k(t)g(t)}$$

$$\left. + f(t)\cos^2\left(g(t)x + \theta(t)\right)\right]dt + k(t)\cos^2\left(g(t)x + \theta(t)\right)dw(t),$$

where $f(t)$, $F(t)$, $g(t)$, $k(t)$, and $\theta(t)$ are any given continuous functions with $k(t)g(t) \neq 0$. Find the general solution to the given differential equation for: (a) any $F(t)$, (b) $F(t) = \frac{k(t)\frac{d}{dt}g(t) + g(t)\frac{d}{dt}k(t)}{k^2(t)g^2(t)}$, and (c) $F(t) = 0$.

Solution procedure. The goal is to find a general solution to the given differential equation under conditions (a)–(c).

Prior to utilizing the tips in Observation 3.3.2, first, we recall that $\sin 2\phi = 2\sin\phi\cos\phi$, $\cos\phi\sec\phi = 1$ and $\sin\phi = \cos\phi\tan\phi$. Using the tips T_1–T_3, we have

$$f(t, x) = k(t)\cos^2\left(g(t)x + \theta(t)\right)\left[-\frac{1}{2}g(t)k(t)\sin 2\left(g(t)x + \theta(t)\right) + f(t)\right.$$

$$\left. - \frac{x\frac{d}{dt}g(t) + \frac{d}{dt}\theta(t)}{k(t)g(t)}\sec^2\left(g(t)x + \theta(t)\right) + F(t)\tan\left(g(t)x + \theta(t)\right)\right],$$

$$\sigma(t, x) = k(t)\cos^2\left(g(t)x + \theta(t)\right).$$

Hence,

$$\frac{f(t,x)}{\sigma(t,x)} = -\frac{1}{2}k(t)g(t)\sin 2(g(t)x + \theta(t)) + F(t)\tan(g(t)x + \theta(t))$$

$$- \frac{x\frac{d}{dt}g(t) + \frac{d}{dt}\theta(t)}{k(t)g(t)}\sec^2(g(t)x + \theta(t)) + f(t). \qquad (3.3.52)$$

Here, using algebra and calculus (tip T_4), we find a comparable membership structure for the given drift rate function in the context of the conceptual structure:

$$\frac{\partial}{\partial x}\sigma(t,x) = -2k(t)g(t)\cos(g(t)x + \theta(t))\sin(g(t)x + \theta(t)) \quad \text{(by the chain rule)}$$

$$= -k(t)g(t)\sin 2\,(g(t)x + \theta(t)) \qquad \text{(from trig identity)},$$

$$\frac{\partial}{\partial t}\sigma(t,s) = \frac{d}{dt}k(t)\cos^2(g(t)x + \theta(t))$$

$$- 2k(t)\cos(g(t)x + \theta(t))\sin(g(t)x + \theta(t))\left[x\frac{d}{dt}g(t) + \frac{d}{dt}\theta(t)\right],$$

and

$$\int_c^x \frac{\frac{\partial}{\partial t}\sigma(t,y)}{\sigma^2(t,y)}dy = \int_c^x \frac{\frac{d}{dt}k(t)\cos^2(g(t)y + \theta(t))}{k^2(t)\cos^4(g(t)y + \theta(t))}dy$$

$$- \int_c^x \frac{2k(t)\left[y\frac{d}{dt}g(t) + \frac{d}{dt}\theta(t)\right]\cos(g(t)y + \theta(t))\sin(g(t)y + \theta(t))}{k^2(t)\cos^4(g(t)y + \theta(t))}dy$$

$$= \int_c^x \frac{\frac{d}{dt}k(t)\cos^2(g(t)y + \theta(t))}{k^2(t)\cos^4(g(t)y + \theta(t))}dy$$

$$+ \int_c^x \frac{-2k(t)\left[y\frac{d}{dt}g(t) + \frac{d}{dt}\theta(t)\right]\cos(g(t)y + \theta(t))\sin(g(t)y + \theta(t))}{k^2(t)\cos^4(g(t)y + \theta(t))}dy$$

$$= \frac{\frac{d}{dt}k(t)}{k^2(t)}\int_c^x \sec^2(g(t)y + \theta(t))dy$$

$$- \frac{2k(t)\frac{d}{dt}\theta(t)}{k^2(t)}\int_c^x \sec^2(g(t)y + \theta(t))\tan(g(t)y + \theta(t))dy$$

$$- 2\frac{k(t)\frac{d}{dt}g(t)}{k^2(t)}\int_c^x \sec^2(g(t)y + \theta(t))\tan(g(t)y + \theta(t))y\,dy$$

$$\text{(by substitution)}$$

$$= \frac{\frac{d}{dt}k(t)}{k^2(t)g(t)}\tan(g(t)x + \theta(t)) - \frac{k(t)\frac{d}{dt}\theta(t)}{k^2(t)g(t)}\sec^2(g(t)x + \theta(t)) + b(t)$$

$$- 2\frac{k(t)\frac{d}{dt}g(t)}{k^2(t)}\int_c^x \sec^2(g(t)y + \theta(t))\tan(g(t)y + \theta(t))y\,dy$$

(by integration)

$$= \frac{\frac{d}{dt}k(t)}{k^2(t)g(t)}\tan(g(t)x + \theta(t)) - \frac{k(t)\frac{d}{dt}\theta(t)}{k^2(t)g(t)}\sec^2(g(t)x + \theta(t)) + b(t)$$

$$- \frac{k(t)\frac{d}{dt}g(t)}{k^2(t)g(t)}x\sec^2(g(t)x + \theta(t))$$

$$+ \frac{k(t)\frac{d}{dt}g(t)}{k^2(t)g(t)}\int_c^x \sec^2(g(t)y + \theta(t))dy \quad \text{(by integration)}$$

$$= \frac{\frac{d}{dt}k(t)}{k^2(t)g(t)}\tan(g(t)x + \theta(t)) - \frac{k(t)\frac{d}{dt}\theta(t)}{k^2(t)g(t)}\sec^2(g(t)x + \theta(t)) + b(t)$$

$$- \frac{k(t)\frac{d}{dt}g(t)}{k^2(t)g(t)}x\sec^2(g(t)x + \theta(t)) + \frac{k(t)\frac{d}{dt}g(t)}{k^2(t)g^2(t)}\tan(g(t)x + \theta(t))$$

(by regrouping)

$$= \frac{g(t)\frac{d}{dt}k(t) + k(t)\frac{d}{dt}g(t)}{k^2(t)g^2(t)}\tan(g(t)x + \theta(t))$$

$$- \frac{x\frac{d}{dt}g(t) + \frac{d}{dt}\theta(t)}{k(t)g(t)}\sec^2(g(t)x + \theta(t)) + b(t),$$

where $b(t) \equiv b(t,c)$ is a constant of integration for some given c.

Now, we construct the class-defining condition for the given rate functions (f, σ) as follows:

$$z(t,x) = \frac{f(t,x)}{\sigma(t,x)} - \frac{1}{2}\frac{\partial}{\partial x}\sigma(t,x) - \int_c^x \frac{\frac{\partial}{\partial t}(\sigma, y)}{\sigma^2(t,y)}dy$$

(by given f, σ and substitution)

$$= -\frac{1}{2}k(t)g(t)\sin 2(g(t)x + \theta(t)) + F(t)\tan(g(t)x + \theta(t))$$

$$- \frac{x\frac{d}{dt}g(t) + \frac{d}{dt}\theta(t)}{k(t)g(t)}\sec^2(g(t)x + \theta(t)) + f(t)$$

$$+ \frac{1}{2}k(t)g(t)\sin 2(g(t)x + \theta(t)) + \frac{x\frac{d}{dt}g(t) + \frac{d}{dt}\theta(t)}{k(t)g(t)}\sec^2(g(t)x + \theta(t))$$

$$- \frac{g(t)\frac{d}{dt}k(t) + k(t)\frac{d}{dt}g(t)}{k^2(t)g^2(t)}\tan(g(t)x + \theta(t)) + b(t)$$

(by regrouping and simplifying)

$$= f(t) + b(t) + \left[F(t) - \frac{g(t)\frac{d}{dt}k(t) + k(t)\frac{d}{dt}g(t)}{k^2(t)g^2(t)}\right]\tan(g(t)x + \theta(t)).$$

Using this representation and making a comparison with Illustration 3.3.1, we conclude that the condition of the pair (f, σ) belongs to the class ISDE-(f, σ) with $\sigma \neq 0$. We follow tips T_5–T_7 to find the energy function and rate functions of reduced SDE. We start with

$$\frac{\partial}{\partial x}z(t, x) = \frac{\partial}{\partial x}\left[f(t) + b(t) + \left[F(t) - \frac{g(t)\frac{d}{dt}k(t) + k(t)\frac{d}{dt}g(t)}{k^2(t)g^2(t)}\right]\right.$$

$$\left. \times \ \tan(g(t)x + \theta(t))\right]$$

$$= \left[F(t) - \frac{g(t)\frac{d}{dt}k(t) + k(t)\frac{d}{dt}g(t)}{k^2(t)g^2(t)}\right]\frac{\partial}{\partial x}\tan(g(t)x + \theta(t))$$

$$= g(t)\left[F(t) - \frac{g(t)\frac{d}{dt}k(t) + k(t)\frac{d}{dt}g(t)}{k^2(t)g^2(t)}\right]\sec^2(g(t)x + \theta(t)).$$

From (3.3.33), (3.3.33), and the above expressions of $z(t, x)$ and $\frac{\partial}{\partial x}z(t, x)$, we have

$$\alpha(t) = -\sigma(t, x)\frac{\partial}{\partial x}z(t, x)$$

$$= -k(t)\cos^2(g(t)x + \theta(t))g(t)$$

$$\times \left[F(t) - \frac{g(t)\frac{d}{dt}k(t) + k(t)\frac{d}{dt}g(t)}{k^2(t)g^2(t)}\right]\sec^2(g(t)x + \theta(t))$$

$$= -k(t)g(t)\left[F(t) - \frac{k(t)\frac{d}{dt}g(t) + g(t)\frac{d}{dt}k(t)}{k^2(t)g^2(t)}\right]$$

$$= \frac{g(t)\frac{d}{dt}k(t) + k(t)\frac{d}{dt}g(t)}{k(t)g(t)} - k(t)g(t)F(t).$$

$$= \frac{g(t)\frac{d}{dt}k(t)}{k(t)g(t)} + \frac{k(t)\frac{d}{dt}g(t)}{k(t)g(t)} - k(t)g(t)F(t)$$

$$= \frac{\frac{d}{dt}k(t)}{k(t)} + \frac{\frac{d}{dt}g(t)}{g(t)} - k(t)g(t)F(t),$$

$$\beta(t) = -\left[\alpha(t)\int_c^x \frac{dy}{\sigma(t,y)} + z(t,x)\right]$$

$$= -\left[-\frac{k(t)g(t)}{g(t)k(t)}\left[F(t) - \frac{g(t)\frac{d}{dt}k(t) + k(t)\frac{d}{dt}g(t)}{k^2(t)g^2(t)}\right]\tan(g(t)x + \theta(t))\right.$$

$$\left. + f(t) + b(t) + \left[F(t) - \frac{g(t)\frac{d}{dt}k(t) + k(t)\frac{d}{dt}g(t)}{k^2(t)g^2(t)}\right]\tan(g(t)x + \theta(t))\right]$$

$$= -(f(t) + b(t)).$$

Thus from (3.3.22), $p(t) = (f(t) + b(t))\exp[\int_a^t \alpha(s)ds] + \frac{d}{dt}B(t)$. Now, by imitating Step #4 of General Procedure 3.3.2, particularly from (3.3.27) and (3.3.28), we have

$$G(t, x(t)) = \frac{1}{k(t)g(t)}\tan(g(t)x + \theta(t)) \equiv \int_c^{x(t)} \frac{dy}{\sigma(t,y)} + G(t,c)$$

$$= -\int_a^t \exp\left[-\int_s^t \alpha(u)du\right]\beta(s)ds + \int_a^t \exp\left[-\int_s^t \alpha(u)du\right]dw(s)$$

$$+ C_1\exp\left[-\int_a^t \alpha(s)ds\right] + G(t,c),$$

for C_1 and c arbitrary constants. From the invertibility property of $G(t, x(t))$ $(H_{3.3})$, we have

$$x(t) = \frac{1}{g(t)}\left[\tan^{-1}\left(k(t)g(t)\left[-\int_a^t \exp\left[-\int_s^t \alpha(u)du\right]\beta(s)ds\right.\right.\right.$$

$$+ \int_a^t \exp\left[-\int_s^t \alpha(u)du\right]dw(s)$$

$$\left.\left.\left. + C_1\exp\left[-\int_a^t \alpha(s)ds\right] + G(t,c)\right]\right) - \theta(t)\right].$$

This is the expression for the general solution process of problem (a).

With respect to the general solution process representation of the problem (b), we observe that

$$\alpha(t) = -k(t)g(t)\left[F(t) - \frac{k(t)\frac{d}{dt}g(t) + g(t)\frac{d}{dt}k(t)}{k^2(t)g^2(t)}\right] \quad \text{(by substitution)}$$

$$= k(t)g(t)\left[\frac{k(t)\frac{d}{dt}g(t) + g(t)\frac{d}{dt}k(t)}{k^2(t)g^2(t)} - \frac{k(t)\frac{d}{dt}g(t) + g(t)\frac{d}{dt}k(t)}{k^2(t)g^2(t)}\right] = 0.$$

From this and the application of Illustration 3.3.1(c) with $p(t) = f(t) + b(t) + \frac{d}{dt}B(t)$ together with Part (a), the general solution representation of the problem (b) is

$$x(t) = \frac{1}{g(t)}\left[\tan^{-1}\left(k(t)g(t)\left[\int_a^t f(s)ds\right] + (w(t) - w(a)) + C_1 + G(t,c)\right) - \theta(t)\right],$$

for C_1 and c arbitrary constants. This is the expression for the general solution process of problem (b).

For the case of the problem (c), we observe that

$$\alpha(t) = -k(t)g(t)\left[-\frac{k(t)\frac{d}{dt}g(t) + g(t)\frac{d}{dt}k(t)}{k^2(t)g^2(t)}\right] \quad \text{[by substitution for } F(t) = 0\text{]}$$

$$= k(t)g(t)\frac{k(t)\frac{d}{dt}g(t) + g(t)\frac{d}{dt}k(t)}{k^2(t)g^2(t)} \quad \text{(by simplifying)}$$

$$= \frac{k(t)\frac{d}{dt}g(t) + g(t)\frac{d}{dt}k(t)}{k(t)g(t)}$$

$$= \left(\frac{\frac{d}{dt}g(t)}{g(t)} + \frac{\frac{d}{dt}k(t)}{k(t)}\right) = \frac{d}{dt}\ln(|g(t)|) + \frac{d}{dt}\ln(|k(t)|).$$

Hence,

$$\int_a^t \alpha(s)ds = \int_a^t \left[\frac{\frac{d}{ds}g(s)}{g(s)} + \frac{\frac{d}{ds}k(s)}{g(s)}\right]ds$$

$$= \ln|g(t)| - \ln|g(a)| + \ln|k(t)| - \ln(|k(a)|).$$

We substitute this expression into the general solution representation problem (a), and we have

$$x(t) = \frac{1}{g(t)}\left[\tan^{-1}\left(k(t)g(t)\left[\int_a^t \left[(f(s) + b(s))\frac{|g(s)k(s)|}{|g(t)k(t)|}ds\right.\right.\right.\right.$$

$$\left.\left.\left.+ \int_a^t \frac{|g(s)k(s)|}{|g(t)k(t)|}dw(s) + C_1\frac{|g(a)k(a)|}{|k(t)g(t)|}\right) - \theta(t)\right]\right.$$

$$= \frac{1}{g(t)}\left[\tan^{-1}\left(\left[\int_a^t (f(s) + b(s))g(s)k(s)ds\right.\right.\right.$$

$$\left.\left.\left.+ \int_a^t g(s)k(s)dw(s) + C_1\right]\right) - \theta(t)\right],$$

for C_1 arbitrary generic constant and $|k(t)g(t)| = k(t)g(t)$. This is the expression for the general solution to problem (c).

In the following, we present a few examples to illustrate the procedure described in this section.

Example 3.3.1. Given: $dx = -\frac{b^2}{2}\csc^2 x \cot x \, dt + b \csc x \, dw(t)$. Find the general solution to the given differential equation.

Solution procedure. From Illustration 3.3.1(f), we conclude that this differential equation belongs to class ISDECC-(f, σ). From (3.3.28), its general solution is given by

$$G(x(t)) = \int_c^{x(t)} \frac{\sin s}{b} ds = \frac{-\cos x}{b} = (w(t) - w(a)) + \frac{G(c)}{b}.$$

This implies that $\cos x(t) = (\cos c - b(w(t) - w(a)))$. Thus,

$$x(t) = \cos^{-1}(\cos c - b(w(t) - w(a))).$$

This is the general solution to the given differential equation.

Example 3.3.2. Solve the IVP

$$dx = [\sec^2 2x \tan 2x + f(t)\sec 2x + F(t)\tan 2x]dt + \sec 2x \, dw(t), \quad x(t_0) = x_0.$$

Solution procedure. The goal is to find the solution to the given IVP. For this purpose, by imitating the argument of Illustration 3.3.2, we need to find a general solution to it. First, we observe that $k(t) = 1$, $g(t) = 2$, $k(t)g(t) = 2$, $k^2(t)g(t) = 2$, $k^2(t)g^2(t) = 4$, $\frac{d}{dt}k(t) = 0$,

$$\frac{d}{dt}g(t) = 0, \quad \frac{\frac{d}{dt}g(t)}{g(t)} = 0, \quad \frac{\frac{d}{dt}g(t)}{k(t)g(t)} = 0, \quad \frac{k(t)\frac{d}{dt}g(t) + g(t)\frac{d}{dt}k(t)}{k^2(t)g^2(t)} = 0,$$

$$f(t,x) = \sec^2 2x \tan 2x + f(t) \sec 2x + F(t) \tan 2x$$

$$= \sec 2x[\sec 2x \tan 2x + f(t) + F(t) \sin 2x],$$

$$\sigma(t,x) = \sec 2x.$$

Now, we show that the pair (f,σ) belongs to the class ISDE-(f,σ). For this purpose, we further note that

$$\frac{f(t,x)}{\sigma(t,x)} = \sec 2x \tan 2x + f(t) + F(t) \sin 2x,$$

$$\frac{\partial}{\partial x}\sigma(t,x) = 2 \sec 2x \tan 2x, \qquad \frac{\partial}{\partial t}\sigma(t,x) = 0,$$

$$z(t,x) = \frac{f(t,x)}{\sigma(t,x)} - \frac{1}{2}\frac{\partial}{\partial x}\sigma(t,x) - \int_c^x \frac{\frac{\partial}{\partial t}\sigma(t,y)}{\sigma^2(t,y)} dy$$

$$= \sec 2x \tan 2x + f(t) + F(t)\sin 2x - \sec 2x \tan 2x \quad \text{(by substitution)}$$

$$= f(t) + F(t)\sin 2x,$$

$$\frac{\partial}{\partial x}z(t,x) = 2F(t)\cos 2x, \quad \alpha(t) = -\sigma(t,x)\frac{\partial}{\partial x}z(t,x) = -2F(t),$$

and $\beta(t) = -f(t)$. From (3.3.28), the general solution is given by

$$G(t,x(t)) - G(c) \equiv \int_c^{x(t)} \frac{ds}{\sigma(t,s)} = \frac{1}{2}\sin 2x(t) - \frac{1}{2}\sin 2c$$

$$= \int_a^t \exp\left[-\int_s^t \alpha(u)du\right] f(s)ds + \int_a^t \exp\left[-\int_s^t \alpha(u)du\right] dw(s)$$

$$+ C_1 \exp\left[-\int_a^t \alpha(u)du\right].$$

Hence,

$$\exp\left[\int_a^t \alpha(u)du\right]\sin 2x(t) = 2\left[\int_a^t \exp\left[\int_a^s \alpha(u)du\right] f(s)ds\right.$$

$$\left. + \int_a^t \exp\left[\int_a^s \alpha(u)du\right] dw(s) + C_1 + \frac{1}{2}\sin 2c\right].$$

By substituting $t = t_0$, $x(t_0) = x_0$ into the above expression, we have

$$\exp\left[\int_a^{t_0} \alpha(u)du\right]\sin 2x_0 = 2\left[\int_a^{t_0} \exp\left[\int_a^s \alpha(u)du\right] f(s)ds\right.$$

$$\left. + \int_a^{t_0} \exp\left[\int_a^s \alpha(u)du\right] dw(s) + C_1 + \frac{1}{2}\sin 2c\right].$$

Solving for C_1, we get

$$C_1 + \frac{1}{2}\sin 2c = \exp\left[\int_a^{t_0} \alpha(u)du\right]\sin 2x_0 - 2\left[\int_a^{t_0}\exp\left[\int_a^s \alpha(u)du\right]f(s)ds\right.$$

$$\left. + \int_a^{t_0}\exp\left[\int_a^s \alpha(u)du\right]dw(s)\right].$$

Now, we substitute this value of C_1 into the above expression, and obtain

$$\exp\left[\int_a^t \alpha(u)du\right]\sin 2x(t)$$

$$= 2\left[\int_a^t \exp\left[\int_a^s \alpha(u)du\right]f(s)ds + \int_a^t \exp\left[\int_a^s \alpha(u)du\right]dw(s)\right]$$

$$- 2\left[\int_a^{t_0}\exp\left[\int_a^s \alpha(u)du\right]f(s)ds + \int_a^{t_0}\exp\left[\int_a^s \alpha(u)du\right]dw(s)\right]$$

$$+ \exp\left[\int_a^{t_0}\alpha(u)du\right]\sin 2x_0$$

$$= 2\left[\int_a^t \exp\left[\int_a^s \alpha(u)du\right]f(s)ds - \int_a^{t_0}\exp\left[\int_a^s \alpha(u)du\right]f(s)ds\right]$$

$$+ 2\left[\int_a^t \exp\left[\int_a^s \alpha(u)du\right]dw(s) - \int_a^{t_0}\exp\left[\int_a^s \alpha(u)du\right]dw(s)\right]$$

$$+ \exp\left[\int_a^{t_0}\alpha(u)du\right]\sin 2x_0$$

$$= 2\int_{t_0}^t \exp\left[\int_a^s \alpha(u)du\right]f(s)ds + 2\int_{t_0}^t \exp\left[\int_a^s \alpha(u)du\right]dw(s)$$

$$+ \exp\left[\int_a^{t_0}\alpha(u)du\right]\sin 2x_0.$$

Thus,

$$x(t) = \frac{1}{2}\sin^{-1}\left(2\int_{t_0}^t \exp\left[-\int_s^t \alpha(u)du\right]f(s)ds\right.$$

$$\left. + 2\int_{t_0}^t \exp\left[-\int_s^t \alpha(u)du\right]dw(s) + \exp\left[-\int_{t_0}^t \alpha(u)du\right]\sin 2x_0\right).$$

This is the solution to the given IVP.

Observation 3.3.3. General Procedure 3.3.2 reduces the problem of finding the solution to (3.2.1) to a problem of solving the directly integrable stochastic differential equation (3.3.1), where $m(t) = V(t, x(t))$ (the energy function to be found). Here, $m(t)$ is as in (3.3.15) with known $\alpha(t) = \alpha(t, f, \sigma)$ in (3.3.11) and $B(t) = B(t, \alpha(t), f, \sigma)$ in (3.3.23). Then, $x(t)$ is extracted from $V(t, x(t))$ as

$$G(x(t)) = \int^{x(t)} \frac{dy}{\sigma(t, y)} = \frac{m(t) - B(t)}{C} \exp\left[-\int^t \alpha(s)ds\right]$$

in Step #4. Since $m(t) = \int_0^t p(s)ds + \int_0^t q(s)dw(s)$ and $z(t) = \int_0^t q(s)dw(s)$ is uniquely determined by the Gaussian process $E[z(s)] = 0$, $\mathrm{cov}(z(s), z(t)) = \int_0^{\min(t,s)} q^2(u)du)$, it follows that $x(t) = \Psi((z(t)) + \nu(t))$ for some function Ψ, where $z(t)$ is Gaussian and $\nu(t)$ is a deterministic process.

Brief Summary. A summary of the basic mechanical steps needed to solve nonlinear Itô–Doob-type stochastic scalar differential equations is presented in the following flowchart:

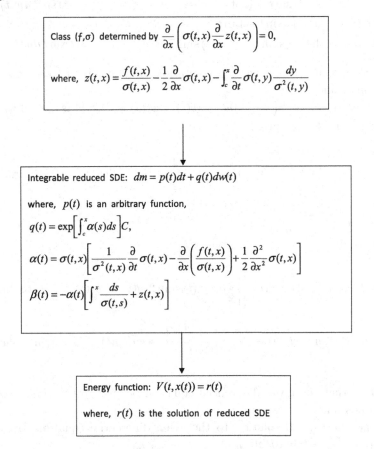

Fig. 3.3.1 Summary rediciable ISDE and the solution process.

Exercises

1. Show that the following differential equations belong to the class of ISDECC-(f, σ):

   (a) $dx = (1 + \exp[-2x] - \exp[-4x])dt + \exp[-2x]dw(t)$;

   (b) $dx = -\exp[-4x]dt + \exp[-2x]dw(t)$;

   (c) $dx = \left[-\frac{1}{2}x + \sqrt{1-x^2}(1 + \sin^{-1} x)\right] dt + \sqrt{1-x^2}dw(t)$;

   (d) $dx = [\sec^2 2x \tan 2x + \sec 2x + 3\tan 2x]dt + \sec 2x\, dw(t)$;

   (e) $dx = [-\csc^2 2x \cot 2x - \csc 2x + 2\cot 2x]dt + \csc 2x\, dw(t)$.

2. Show that the following differential equations belong to the class of ISDE-(f, σ):

   (a) $dx = \left(\frac{x - (1+t)^2}{1+t} - \frac{x}{2(1+t)^2}\exp\left[-\frac{x^2}{1+t^2}\right]\right) dt + \exp\left[-\frac{x^2}{2(1+t)^2}\right] dw(t)$;

   (b) $dx = [-\csc^2 2x \cot 2x - \sin t \csc 2x + \cot t \cot 2x]dt + \csc 2x\, dw(t)$;

   (c) $dx = [x(1+x^2)\sin t + (\cot t - \cos t)\tan^{-1} x]dt + (1+x^2)\sin t\, dw(t)$;

   (d) $dx = [-x\cos t + (\sin t - \tan t)\sinh^{-1} x]dt + \sqrt{1+x^2}\cos t\, dw(t)$;

   (e) $dx = [x\cosh t + (\sinh t - \tanh t)\sinh^{-1} x]dt - (1+x^2)\cosh t\, dw(t)$.

3. Find the general solution representation of the differential equations in Problems 1 and 2.

4. Show that $x(t) = G^{-1}\left(\int_{t_0}^{t} f(s)ds + w(t) - w(t_0) + G(x_0)\right)$ is the solution to the following initial value problem:

$$dx = \left[f(t)h(x) + \frac{1}{2}h(x)\frac{d}{dx}h(x)\right] dt + h(x)dw(t), \quad x(t_0) = x_0,$$

   where $G(x) = \int_c^x \frac{dy}{h(y)} + G(c)$.

5. Given:

$$dx = k(t)\cosh^2(g(t)x + \theta(t))\left[\left[\frac{1}{2}g(t)k(t)\sinh 2(g(t)x + \theta(t)) + F(t)\tanh(g(t)x\right.\right.$$

$$\left.\left. + \theta(t)\tanh + f(t)) - \frac{x\frac{d}{dt}g(t) + \frac{d}{dt}\theta(t)}{k(t)g(t)}\operatorname{sech}^2(g(t)x + \theta(t))\right] dt + dw(t)\right].$$

   (a) Show that the given differential equation belongs to the class ISDE-(f, σ) (if possible).

   (b) Find the general solution to the given differential equation for: (i) $F(t)$, (ii) $F(t) = \frac{k(t)\frac{d}{dt}g(t) + g(t)\frac{d}{dt}k(t)}{k^2(t)g^2(t)}$, and (iii) $F(t) = 0$.

6. Given:

$$dx = k(t)\sin^2(g(t)x + \theta(t))\left[\left[\frac{1}{2}g(t)k(t)\sin^2(g(t)x + \theta(t)) + F(t)\cot(g(t)x + \theta(t))\right.\right.$$

$$\left.\left. + f(t) - \frac{x\frac{d}{dt}g(t) + \frac{d}{dt}\theta(t)}{k(t)g(t)}\csc^2(g(t)x + \theta(t))\right]dt + dw(t)\right].$$

(a) Show that the given differential equation belongs to the class ISDE-(f,σ) (if possible).
(b) Find the general solution to the given differential equation for:
(i) $F(t)$, (ii) $F(t) = \frac{k(t)\frac{d}{dt}g(t) + g(t)\frac{d}{dt}k(t)}{k^2(t)g^2(t)}$, and (iii) $F(t) = 0$.

7. Solve the following initial value problems:

(a) $dx = (1+t)\cos^2\left(\frac{x}{2(1+t)^2}\right)\left[\left[\left(\frac{1}{4(1+t)} + 1\right)\sin\left(\frac{x}{(1+t)^2}\right) - \frac{2}{x(1+t)^2}\right]dt + dw(t)\right]$,
$x(0) = 0$;
(b) $dx = \sin^2(x + \sin t)\left[\left(\frac{1}{2}\sin 2(x + \sin t) - \frac{1}{2}\cos t\right)dt + dw(t)\right]$, $x(0) = 1$;
(c) $dx = \frac{1}{2}\tan^2 t \sec x\, dt + \sec x\, dw(t)$, $x(t_0) = x_0$;
(d) $dx = -\frac{a^2}{2}\sin x \cos^3 xx\, dt + a\cos^2 x\, dw(t)$, $x(t_0) = x_0$;
(e) $dx = \frac{a^2}{2}x(2x^2 - 1)dt + x\sqrt{(x^2 - 1)}dw(t)$;
(f) $dx = a^2\sinh x \cosh^3 x\, dt + a\cosh^2 x\, dw(t)$, $x(t_0) = x_0$.

3.4 Linear Nonhomogeneous Reduced Equations

In this section, we find another very broad class of Itô–Doob-type stochastic differential equations (3.2.1). For this purpose, we make another special conceptual choice of the convenient functions F and Λ in (3.2.4) of the types $F(t, m) = \alpha(t)m + p(t)$ and $\Lambda(t, m) = \beta(t)m + q(t)$. The class of differential equations (3.2.1), reducible to first-order nonhomogeneous scalar stochastic differential equations of the Itô–Doob-type with variable coefficients, is presented.

3.4.1 *General Problem*

We consider a reduction problem for another broad class of nonlinear Itô–Doob-type stochastic differential equations (3.2.1). This class of differential equations (3.2.1) is reduced to the following Itô–Doob-type of first-order nonhomogeneous scalar stochastic differential equation with variable coefficients [43, 71]:

$$dm = [\mu(t)m + p(t)]dt + [\nu(t)m + q(t)]dw(t), \qquad (3.4.1)$$

where μ, ν, p, and q are continuous functions defined on R into R and w is a scalar normalized Wiener process.

(H$_{3.4}$). In addition to hypothesis $(H_{3.2})$, we assume that $\sigma(t,x) = g(t)h(x)$, f and h are continuously differentiable with respect to both variables (t,x). Furthermore, $h(x) \neq 0$, $G(x) = \int_c^x \frac{dy}{h(y)}$ is invertible, and for any arbitrary real numbers γ and δ the functions ν and q satisfy conditions $q(t) = \gamma g(t)$ and $\nu(t) = \delta g(t)$.

3.4.2 *Procedure for Finding a General Solution Representation*

A procedure for finding a solution representation for a broader class of differential equations (3.2.1) is presented. We outline the method in the same spirit as in Section 3.3. The procedure consists of the following: (LSDE$_1$) to determine a class of rate functions f and σ in (3.2.1), (LSDE$_2$) to find the energy/Lyapunov function, and (LSDE$_3$) to look for the convenient choice of functions μ, ν, p, and q, in particular μ, p, γ, and δ and [in view of the hypothesis (H$_{3.4}$)]. From this reduction procedure and using a general solution to (3.4.1), a general solution expression of the Itô–Doob-type stochastic differential equation (3.2.1) is determined in either explicit or implicit form.

To fulfill the goal of finding an expression for the solution process of (3.2.1), we describe the following brief procedure for determining the solution representation of the nonlinear Itô–Doob-type stochastic differential equation (3.2.1).

Step #1. We imitate Steps #1 and #2 of General Procedure 3.2.2 outlined in Section 3.2, and arrive at (3.2.2) and (3.2.3).

Step #2 (Step #3 of General Procedure 3.2.2). By using the conceptual choice of functions $F(t,V(t,x)) = \mu(t)V(t,x) + p(t)$ and $\Lambda(t,V(t,x)) = \nu(t)V(t,x) + q(t)$, we repeat Step #3 of General Procedure 3.2.2. Again, our goal is to find a broader class of differential equations (3.2.1) that is reducible to (3.4.1). For this purpose, we use this choice of functions to rewrite (3.2.5) and (3.2.5) as

$$\mu(t)V(t,x) + p(t) = \frac{\partial}{\partial t}V(t,x) + f(t,x)\frac{\partial}{\partial x}V(t,x) + \frac{1}{2}\sigma^2(t,x)\frac{\partial^2}{\partial x^2}V(t,x), \quad (3.4.2)$$

and

$$\nu(t)V(t,x) + q(t) = \sigma(t,x)\frac{\partial}{\partial x}V(t,x), \tag{3.4.3}$$

respectively. Under the conditions (3.4.2) and (3.4.3), we find a sufficient condition for the class of rate functions f and σ of (3.2.1). First, using (3.4.3), we find a candidate for $V(t,x)$. For this purpose, from (3.4.3) and in view of the hypothesis (H$_{3.4}$), $\sigma(t,x) = g(t)h(x)$, $q(t) = \gamma g(t)$, $\nu(t) = \delta g(t)$ and $\sigma(t,x) \neq 0$ on $J \times R$, we have

$$\frac{\partial}{\partial x}V(t,x) = \frac{\nu(t)V(t,x) + q(t)}{\sigma(t,x)} = \frac{\delta g(t)V(t,x)}{g(t)h(x)} + \frac{\gamma g(t)}{g(t)h(x)} = \frac{\delta V(t,x)}{h(x)} + \frac{\gamma}{h(x)}$$

$$= \frac{\delta V(t,x)}{h(x)} + \frac{\gamma}{h(x)}. \tag{3.4.4}$$

From this and applying the method of variation of constants parameters in Section 2.4 of Chapter 2 [80] for each t in J, we have

$$V(t,x) = \exp\left[\int_c^x \frac{\delta}{h(y)} dy\right] C + \int_c^x \exp\left[\int_y^x \frac{\delta}{h(u)} du\right] \frac{\gamma}{h(y)} dy$$

$$= \exp\left[\int_c^x \frac{\delta}{h(y)} dy\right] C - \frac{\gamma}{\delta}$$

$$= \exp[\delta G(x)]C - \frac{\gamma}{\delta} \quad \text{(from H}_{3.4}\text{)}, \tag{3.4.5}$$

where C is an arbitrary nonzero positive function (constant of integration) on J. We note that $V(t,x)$ depends on the constant parameters δ and γ. Our goal is to find them (if possible).

For $V(t,x)$ in (3.4.5), we compute $\frac{\partial}{\partial t}V(t,x)$, $\frac{\partial}{\partial x}V(t,x)$, and $\frac{\partial^2}{\partial x^2}V(t,x)$

$$\frac{\partial}{\partial t}V(t,x) = \exp\left[\int_c^x \frac{\delta}{h(y)} dy\right] \frac{d}{dt}C, \tag{3.4.6}$$

$$\frac{\partial}{\partial x}V(t,x) = \frac{\delta}{h(x)} \exp\left[\int_c^x \frac{\delta}{h(y)} dy\right] C, \tag{3.4.7}$$

and

$$\frac{\partial^2}{\partial x^2}V(t,x) = \frac{\delta^2}{h^2(x)} \exp\left[\int_c^x \frac{\delta}{h(y)} dy\right] C - \frac{\delta \frac{d}{dx}h(x)}{h^2(x)} \exp\left[\int_c^x \frac{\delta}{h(y)} dy\right] C$$

$$= \left[\frac{\delta^2}{h^2(x)} - \frac{\delta \frac{d}{dx}h(x)}{h^2(x)}\right] \exp\left[\int_c^x \frac{\delta}{h(y)} dy\right] C. \tag{3.4.8}$$

By using (3.4.6)–(3.4.7), we compute the right-hand side expression in (3.4.2):

$$\frac{\partial}{\partial t}V(t,x) + f(t,x)\frac{\partial}{\partial x}V(t,x) + \frac{1}{2}\sigma^2(t,x)\frac{\partial^2}{\partial x^2}V(t,x)$$

$$= \exp\left[\int_c^x \frac{\delta}{h(y)} dy\right] \frac{d}{dt}C + f(t,x)\frac{\delta}{h(s)} \exp\left[\int_c^x \frac{\delta}{h(y)} dy\right] C$$

$$+ \frac{1}{2}\sigma^2(t,x)\left[\frac{\delta^2}{h^2(x)} - \frac{\delta \frac{d}{dx}h(x)}{h^2(x)}\right] \exp\left[\int_c^x \frac{\delta}{h(y)} dy\right] C$$

(by substituton and rearrangement)

$$= \left[\frac{d}{dt}C + \left[f(t,x)\frac{\delta}{h(x)} + \frac{1}{2}g^2(t)h^2(x)\left[\frac{\delta^2}{h^2(x)} - \frac{\delta\frac{d}{dx}h(x)}{h^2(x)} \right] \right] C \right]$$

$$\times \exp\left[\int_c^x \frac{\delta}{h(y)}dy \right] \quad \text{(by substitution)}$$

$$= \left[\frac{d}{dt}C + \left[\frac{1}{2}g^2(t)\delta^2 + f(t,x)\frac{\delta}{h(x)} - \frac{1}{2}\delta g^2(t)\frac{d}{dx}h(x) \right] C \right] \exp\left[\int_c^x \frac{\delta}{h(y)}dy \right]$$

$$= \left[\frac{d}{dt}C + \left[\frac{1}{2}g^2(t)\delta^2 + \delta\left[\frac{f(t,x)}{h(x)} - \frac{1}{2}g^2(t)\frac{d}{dx}h(x) \right] \right] C \right] \exp\left[\int_c^x \frac{\delta}{h(y)}dy \right].$$

$$(3.4.9)$$

From (3.4.5) and (3.4.9), (3.4.2) can be rewritten as

$$\frac{\partial}{\partial t}V(t,x) + f(t,x)\frac{\partial}{\partial x}V(t,x) + \frac{1}{2}\sigma^2(t,x)\frac{\partial^2}{\partial x^2}V(t,x) = \mu(t)V(t,x) + p(t),$$

$$\left[\frac{d}{dt}C + \left[\frac{1}{2}g^2(t)\delta^2 + \delta\left[\frac{f(t,x)}{h(x)} - \frac{1}{2}g^2(t)\frac{d}{dx}h(x) \right] \right] C \right] \exp\left[\int_c^x \frac{\delta}{h(y)}dy \right]$$

$$= \mu(t)\left[\exp\left[\int_c^x \frac{\delta}{h(y)}dy \right] C - \frac{\gamma}{\delta} \right] + p(t).$$

This relation can be further rewritten as

$$\left[\frac{d}{dt}C + \left[\delta\left[\frac{f(t,x)}{h(x)} - \frac{1}{2}g^2(t)\frac{d}{dx}h(x) \right] + \frac{1}{2}g^2(t)\delta^2 - \mu(t) \right] C \right]$$

$$\times \exp\left[\int_c^x \frac{\delta}{h(y)}dy \right] = p(t) - \frac{\mu(t)\gamma}{\delta}. \quad (3.4.10)$$

Now, we set

$$z(t,x) = \frac{f(t,x)}{h(x)} - \frac{1}{2}g^2(t)\frac{d}{dx}h(x). \quad (3.4.11)$$

By substituting this $z(t,x)$ into (3.4.10), we obtain

$$\left[p(t) - \frac{\mu(t)\gamma}{\delta} \right] = \left[C\left[\delta z(t,x) + \frac{1}{2}g^2(t)\delta^2 - \mu(t) \right] + \frac{d}{dt}C \right] \exp\left[\int_c^x \frac{\delta}{h(y)}dy \right].$$

$$(3.4.12)$$

The left-hand side of (3.4.12) is a function of t only. Therefore, by differentiating both sides with respect to x, we obtain

$$0 = \frac{\partial}{\partial x}\left(\left[C\left[\delta z(t,x) + \frac{1}{2}g^2(t)\delta^2 - \mu(t)\right] + \frac{d}{dt}C\right]\exp\left[\int_c^x \frac{\delta}{h(y)}dy\right]\right)$$

$$= \left[\frac{\partial}{\partial x}(C\delta z(t,x)) + \frac{\partial}{\partial x}\left(C\left[\frac{1}{2}g^2(t)\delta^2 - \mu(t)\right] + \frac{d}{dt}C\right)\right]\exp\left[\int_c^x \frac{\delta}{h(y)}dy\right]$$

$$+ \left[C\left[\delta z(t,x) + \frac{1}{2}g^2(t)\delta^2 - \mu(t)\right] + \frac{d}{dt}C\right]\frac{\partial}{\partial x}\left(\exp\left[\int_c^x \frac{\delta}{h(y)}dy\right]\right)$$

(by laws of derivatives)

$$= \left[C\left[\delta z(t,x) + \frac{1}{2}g^2(t)\delta^2 - \mu(t)\right] + \frac{d}{dt}C\right]\frac{\partial}{\partial x}\left(\exp\left[\int_c^x \frac{\delta}{h(y)}dy\right]\right)$$

$$+ C\frac{\partial}{\partial x}(\delta z(t,x))\exp\left[\int_c^x \frac{\delta}{h(s)}ds\right]\quad\left[\text{from } \frac{\partial}{\partial x}\left(C\left[\frac{1}{2}g^2(t)\delta^2 - \mu(t)\right] + \frac{d}{dt}C\right) = 0\right]$$

$$= \left[C\left[\delta z(t,x) + \frac{1}{2}g^2(t)\delta^2 - \mu(t)\right] + \frac{d}{dt}C\right]\frac{\delta}{h(x)}\exp\left[\int_c^x \frac{\delta}{h(y)}dy\right]$$

$$+ C\frac{\partial}{\partial x}([\delta z(t,x)])\exp\left[\int_c^x \frac{\delta}{h(y)}dy\right]\left[\text{by } \frac{\partial}{\partial x}\left(\int_c^x \frac{\delta}{h(y)}dy\right) = \frac{\delta}{h(x)}\right.$$

and regrouping$\Big]$

$$= \left\{C\left[\delta z(t,x) + \frac{1}{2}g^2(t)\delta^2 - \mu(t)\right] + \frac{d}{dt}C\right.$$

$$\left. + Ch(x)\frac{\partial}{\partial x}z(t,x)\right\}\frac{\delta}{h(x)}\exp\left[\int_c^x \frac{\delta}{h(y)}dy\right]\quad\text{(by simplifying)}$$

$$= Ch(x)\frac{\partial}{\partial x}z(t,x) + C\left[\delta z(t,x) + \frac{1}{2}g^2(t)\delta^2 - \mu(t)\right] + \frac{d}{dt}C. \tag{3.4.13}$$

Now, using (3.4.13), we obtain

$$C\left[h(x)\frac{\partial}{\partial x}z(t,x) + \delta z(t,x) + \frac{1}{2}g^2(t)\delta^2\right] = \mu(t)C - \frac{d}{dt}C$$

$$C\beta(t) = \mu(t)C - \frac{d}{dt}C, \tag{3.4.14}$$

where

$$\beta(t) = h(x)\frac{\partial}{\partial x}(z(t,x)) + \delta z(t,x) + \frac{1}{2}g^2(t)\delta^2.$$

We choose $\mu(t) = \beta(t)$, and hence $C(t)$ is a constant function. (Otherwise, $C(t) = \exp[\int_a^t(\mu(s)-\beta(s))ds]c$). Again, by differentiating both sides of (3.3.14) with respect to x, we have

$$0 = \frac{\partial}{\partial x}\left(h(x)\frac{\partial}{\partial x}z(t,x) + \delta z(t,x)\right) \quad \text{(by the sum formula of derivatives)}$$

$$= \frac{\partial}{\partial x}\left(h(x)\frac{\partial}{\partial x}z(t,x)\right) + \delta\frac{\partial}{\partial x}z(t,x). \tag{3.4.15}$$

For $\frac{\partial}{\partial x}z(t,x) \neq 0$ and solving for δ, we have

$$\delta = -\frac{\frac{\partial}{\partial x}\left(h(x)\frac{\partial}{\partial x}z(t,x)\right)}{\frac{\partial}{\partial x}z(t,x)}. \tag{3.4.16}$$

This determines $\nu(t)$ [hypothesis (H$_{3.4}$)]; $\nu(t) = \delta g(t)$.

    Again, we note that the left-hand side of (3.4.16) is independent of both t and x. Therefore, by differentiating both sides of (3.4.16) with respect to both t and x, we have

$$\frac{\partial}{\partial t}\left[\frac{\frac{\partial}{\partial x}\left(h(x)\frac{\partial}{\partial x}z(t,x)\right)}{\frac{\partial}{\partial x}z(t,x)}\right] = \frac{\partial}{\partial x}\left[\frac{\frac{\partial}{\partial x}\left(h(x)\frac{\partial}{\partial x}z(t,x)\right)}{\frac{\partial}{\partial x}z(t,x)}\right] = 0. \tag{3.4.17}$$

This is a sufficient condition for the reducibility of (3.2.1) to the first-order non-homogeneous scalar Itô–Doob-type stochastic differential equation with variable coefficients (3.4.1).

    In summary, (3.4.17) determines the class of differential equations (3.2.1) reducible to (3.4.1) by the energy function defined in (3.4.5).

Step #3 (Step #4 of General Procedure 3.2.2).   Now, we are ready to enter into the task of Step #4 of Procedure 3.2.2 to find $V(t,x)$ as well as the convenient choice of functions μ, ν, p, and q in (3.4.1). For this purpose, we utilize the sufficient condition (3.4.17). This condition gives rise to (3.4.16). From (3.4.11), we conclude that δ in (2.4.16) is determined by the rate functions $f(t,x)$ and $\sigma(t,x)$ of (3.2.1). We also note that the convenient choice of functions μ, ν, p, and q in (3.4.1) are also determined in terms of the rate functions $f(t,x)$ and $\sigma(t,x)$ of (3.2.1). In fact, from the hypothesis (H$_{3.4}$), $\sigma(t,x) = g(t)h(x)$, $\nu(t) = \delta g(t)$, and $q(t) = \gamma g(t)$. Thus, $\nu(t)$ and $q(t)$ are determined in terms of the rate function $\sigma(t,x)$ of (3.2.1) and (3.4.16). From (3.4.14) C is positive constant if and only if $\beta(t) = \mu(t)$ is determined in (3.4.14). Now, using the expression $\mu(t)$ and (3.4.16), $p(t)$ is determined from (3.4.12). In fact, from (3.4.14) and (3.4.16), (3.4.12) reduces to

$$p(t) = \frac{\mu(t)\gamma}{\delta} - C\left[h(x)\frac{\partial}{\partial x}(z(t,x))\right]\exp\left[\int_c^x \frac{\delta}{h(y)}dy\right]$$

$$= \frac{\mu(t)\gamma}{\delta} - Cf_1(t), \quad \text{for } \delta \neq 0.$$

Hence, $\delta p(t) = \mu(t)\gamma - C\delta f_1(t)$. Of course, we note that $p(t)$ depend on γ and δ, C, $q(t)$ depend on γ. These parameters are determined from the rate functions $f(t,x)$ and $\sigma(t,x)$ of (3.2.1) and the method of finding undetermined coefficients.

Step #4 (Step #5 of General Procedure 3.2.2). Under the condition (3.4.17), the nonlinear Itô–Doob-type stochastic differential equation (3.2.1) reduces to (3.4.1). This condition determines the coefficients of the first-order linear nonhomogeneous Itô–Doob-type stochastic differential equation (3.4.1). The general solution representation of (3.2.1) is given by the inverse of the energy function $V(t,x)$ in (3.2.7):

$$x(t) = E(t, m(t)), \tag{3.4.18}$$

where $m(t)$ is a general solution process of (3.4.1) determined in Chapter 2 of Section 2.4; and $x(t)$ is a general solution process of the stochastic differential equation (3.2.1) that belongs to the class determined by (3.4.17). In fact from (3.4.5) and (2.4.15), we have

$$m(t) = V(t, x(t)) = C\exp\left[\int_c^{x(t)} \frac{\delta}{h(s)}ds\right] - \frac{\gamma}{\delta}C\exp[\delta G(x(t))] - \frac{Y}{\delta} \quad \text{(from H}_{3.4}\text{)}$$
$$\tag{3.4.19}$$

$$m(t) = \Phi(t)c + \int_a^t \Phi(t,s)[p(s) - \nu(s)q(s)]ds + \int_a^t \Phi(t,s)q(s)dw(s),$$

where

$$\Phi(t, w(t), a) \equiv \Phi(t) = \exp\left[\int_a^t \left[\mu(s) - \frac{1}{2}\nu^2(s)\right]ds + \int_{t_0}^t \nu(s)dw(s)\right].$$

Thus,

$$C\exp[\delta G(x(t))] - \frac{\gamma}{\delta}$$

$$= \Phi(t)c + \int_a^t \Phi(t,s)[p(s) - \nu(s)q(s)]ds + \int_a^t \Phi(t,s)q(s)dw(s). \tag{3.4.20}$$

This together with (3.4.5), the hypothesis (H$_{3.4}$), $q(t) = \gamma g(t)$ and $\nu(t) = \delta g(t)$ implies that

$$
x(t) = G^{-1}\left[\ln\left[\Phi(t)c + \int_a^t \Phi(t,s)[p(s) - \nu(s)q(s)]ds\right.\right.
$$

$$
\left.\left. + \int_a^t \Phi(t,s)q(s)dw(s) + \frac{\gamma}{\delta}\right] - \ln C\right]
$$

$$
= G^{-1}\left[\ln\left[\Phi(t)c + \int_a^t \Phi(t,s)[p(s) - \delta\gamma g^2(s)]ds\right.\right.
$$

$$
\left.\left. + \int_a^t \Phi(t,s)\gamma g(s)dw(s) + \frac{\gamma}{\delta}\right] - \ln C\right], \tag{3.4.21}
$$

where δ and γ are arbitrary constant parameters, and

$$
G(x(t)) = \int_c^{x(t)} \frac{\delta}{h(y)}dy + G(c). \tag{3.4.22}
$$

This completes the order reduction and the solution representation process of (3.2.1) under the specified conditions on f and σ in (3.4.17).

Observation 3.4.1

(i) We remark that the class of nonlinear Itô–Doob-type stochastic differential equations reducible to linear nonhomogeneous Itô–Doob-type stochastic scalar differential equations with time-varying coefficients is defined and denoted by

$$
\text{LSDEVC-}(f,\sigma) = \left\{(f,\sigma) \text{ in } (3.2.1): \delta\frac{\partial}{\partial x}z(t,x)\right.
$$

$$
\left. + \frac{\partial}{\partial x}\left(h(x)\frac{\partial}{\partial x}z(t,x)\right) = 0, \quad \text{with } \sigma = gh, \sigma \neq 0\right\}.
$$

Moreover, the condition describing the class provides a conceptual structure of the drift rate f in (3.2.1).

(ii) From the definition of the class LSDEVC-(f,σ), for arbitrary γ, δ with $\delta \neq 0$, we can construct a representative of this class. The property defining this class,

$$
\delta\frac{\partial}{\partial x}z(t,x) + \frac{\partial}{\partial x}\left(h(x)\frac{\partial}{\partial x}z(t,x)\right) = 0, \tag{3.4.23}
$$

is rewritten as

$$
\frac{\partial}{\partial x}\left(h(x)\frac{\partial}{\partial x}z(t,x)\right) = -\frac{\delta}{h(x)}\left(h(x)\frac{\partial}{\partial x}z(t,x)\right). \tag{3.4.24}
$$

This is a linear homogeneous deterministic differential equation with variable coefficients. We apply the method of Section 2.4 of Chapter 2 [80]. Its general solution is

$$h(x)\frac{\partial}{\partial x}z(t,x) = \exp\left[-\delta\int_c^x \frac{dy}{h(y)}\right]f_1(t), \qquad (3.4.25)$$

where $f_1(t)$ is an arbitrary continuous function which is a constant of integration. We note that Equation (3.4.25) is a deterministic integrable differential equation; see Section 2.3 of Chapter 2 [80]. Its general solutions is

$$z(t,x) = \int_c^x \frac{1}{h(u)}\exp\left[-\delta\int_c^u \frac{dy}{h(y)}\right]f_1(t)du + f_2(t)$$

$$= -\frac{f_1(t)}{\delta}\exp\left[-\delta\int_c^x \frac{dy}{h(y)}\right] + f_2(t) \quad \text{(by the substitution method),}$$

$$(3.4.26)$$

where $f_2(t)$ is another arbitrary continuous function which is a constant of integration. From (3.4.11) and (3.4.26), we can find a conceptual expression for $f(t,x)$:

$$z(t,x) = \frac{f(t,x)}{h(x)} - \frac{1}{2}g^2(t)\frac{d}{dx}h(x) = -\frac{f_1(t)}{\delta}\exp\left[-\delta\int_c^x \frac{dy}{h(y)}\right] + f_2(t),$$

which implies that

$$\frac{f(t,x)}{h(x)} = \frac{1}{2}g^2(t)\frac{d}{dx}h(x) - \frac{f_1(t)}{\delta}\exp\left[-\delta\int_c^x \frac{dy}{h(y)}\right] + f_2(t). \qquad (3.4.27)$$

Thus,

$$f(t,x) = h(x)\left[\frac{1}{2}g^2(t)\frac{d}{dx}h(x) - \frac{f_1(t)}{\delta}\exp\left[-\delta\int_c^x \frac{dy}{h(y)}\right] + f_2(t)\right]. \qquad (3.4.28)$$

This is the structural representation of a drift component of the member pair (f,σ) of the class LSDEVC-(f,σ) of nonlinear Itô–Doob-type stochastic differential equations in (3.2.1). It will be used to classify and/or solve a given nonlinear Itô–Doob-type stochastic differential equation.

Now, from (3.4.12) and (3.4.28), $\mu(t)$ and $p(t)$ in (3.4.1) are as follows:

$$\mu(t) = \frac{1}{2}g^2(t)\delta^2 + \delta f_2(t) \quad \text{and} \quad p(t) = \frac{1}{2}g^2(t)\delta\gamma + \gamma f_2(t) - Cf_1(t). \quad (3.4.29)$$

The f in (3.4.28), together, with σ satisfying the condition (3.4.23), represents the membership of the class LSDEVC-(f,σ).

(iii) If the rate functions f and σ are stationary (autonomous), i.e. $f(t,x) \equiv f(x)$ and $\sigma(t,x) \equiv \sigma(x)$, then the nonlinear Itô–Doob-type stochastic differential equation (3.2.1) is

$$dx = f(x)dt + \sigma(x)dw(t), \qquad (3.4.30)$$

which is reducible to the stochastic differential equation (3.4.1) of the type

$$dm = [\mu m + p]dt + [\nu m + q]dw(t), \qquad (3.4.31)$$

where μ, ν, p, and q are arbitrary constants.

In this case, the respective expressions (3.4.5), (3.4.15), and (3.4.17) reduce to

$$V(x) = C \exp\left[\int_c^x \frac{\delta}{\sigma(s)} ds\right] - \frac{\gamma}{\delta} C \exp[\delta G(x(t))] - \frac{\gamma}{\delta} \quad \text{(from H}_{3.4}) \tag{3.4.32}$$

$$\frac{d}{dx}\left(\sigma(x)\frac{d}{dx}z(x)\right) + \delta \frac{d}{dx} z(x) = 0, \qquad (3.4.33)$$

$$\frac{d}{dx}\left[\frac{\frac{d}{dx}\left(\sigma(x)\frac{d}{dx}z(x)\right)}{\frac{d}{dx}z(x)}\right] = 0 \quad \text{for} \quad \frac{d}{dx}z(x) \neq 0, \qquad (3.4.34)$$

where $V(t,x) \equiv V(x)$, $z(t,x) \equiv z(x)$, and with these notational changes δ is determined as in (3.4.16). Moreover, from Observation 3.4.1(i), the class of differential equations (3.4.30) is reducible to linear nonhomogeneous Itô–Doob-type scalar stochastic differential equations with constant coefficients. It is denoted by

$$\text{LSDECC-}(f,\sigma) = \left\{ (f,\sigma) \text{ in (3.2.1): } \delta\frac{\partial}{\partial x}z(x) \right.$$

$$\left. + \frac{\partial}{\partial x}\left(\sigma(x)\frac{\partial}{\partial x}z(x)\right) = 0, \quad \text{with } \sigma = h, \sigma \neq 0 \right\},$$

where $z(x) = \frac{f(x)}{\sigma(x)} - \frac{1}{2}\frac{d}{dx}\sigma(x)$. Furthermore, (3.4.26)–(3.4.28) reduce to

$$z(x) = -\frac{f_1}{\delta} \exp\left[-\delta \int_c^x \frac{dy}{\sigma(y)}\right] + f_2, \qquad (3.4.35)$$

which implies that

$$\frac{f(x)}{\sigma(x)} = \frac{1}{2}\frac{d}{dx}\sigma(x) - \frac{f_1}{\delta} \exp\left[-\delta \int_c^x \frac{dy}{\sigma(y)}\right] + f_2, \qquad (3.4.36)$$

and hence

$$f(x) = \sigma(x)\left[\frac{1}{2}\frac{d}{dx}\sigma(x) - \frac{f_1}{\delta} \exp\left[-\delta \int_c^x \frac{dy}{\sigma(y)}\right] + f_2\right]. \qquad (3.4.37)$$

From (3.4.12), (3.4.37), $g(t) = 1$, and nonzero constant C, we have

$$p(t) - \frac{\mu(t)\gamma}{\delta} = C\left[\delta z(x) + \frac{1}{2}\delta^2 - \mu(t)\right]\exp\left[\int_c^x \frac{\delta}{h(y)}dy\right]$$

$$= C\left[\delta\left[-\frac{f_1}{\delta}\exp\left[-\delta\int_c^x \frac{dy}{\sigma(y)}\right] + f_2\right]\right.$$

$$\left. + \frac{1}{2}\delta^2 - \mu(t)\right]\exp\left[\int_c^x \frac{\delta}{h(y)}dy\right]$$

$$= C\left[-f_1\exp\left[-\delta\int_c^x \frac{dy}{\sigma(y)}\right] + \delta f_2\right.$$

$$\left. + \frac{1}{2}\delta^2 - \mu(t)\right]\exp\left[\int_c^x \frac{\delta}{h(y)}dy\right],$$

for all t and x. Now, we use the method of undetermined coefficients to find $\mu(t)$ and $p(t)$ as

$$\mu(t) = \frac{1}{2}\delta^2 + \delta f_2 \quad \text{and} \quad p(t) = \frac{1}{2}\delta^2\gamma + \gamma f_2 - Cf_1, \tag{3.4.38}$$

respectively, where f_1, f_2, γ, and C arbitrary parameters/constants determined by the rate functions f and σ, otherwise γ and C are chosen as $\gamma = 0$ and $C = 1$. In this case, f in (3.4.37) together with $\sigma(x)$ satisfies the condition (3.4.33). Hence, it represents the membership of the class LSDECC-(f, σ).

(iv) The structures of the drift coefficients in (3.4.28) and (3.4.37) play a mechanical tool for solving this type of Itô–Doob SDE (3.2.1). These structural representations of the drift component of the member pair (f, σ) of either the class LSDEVC-(f, σ) or LSDECC-(f, σ) of the Itô–Doob SDE (3.2.1) are used to classify and/or to solve a given SDE.

Observation 3.4.2

(i) We further note that in Steps #2–#4 of Procedure 3.4.2, the conclusions and discussions in Observation 3.4.1(ii)–(iv) are valid for arbitrary γ, δ with $\delta \neq 0$. In the case of $\delta = 0$, $V(t) = 0$, and $q(t) = \gamma g(t)$, (3.4.4), (3.4.5), and (3.4.9) become

$$\frac{\partial}{\partial x}V(t, x) = \frac{\gamma}{h(x)}, \tag{3.4.39}$$

$$V(t, x) = \gamma\int_c^x \frac{dy}{h(y)} + C, \tag{3.4.40}$$

$$\frac{\partial}{\partial t}V(t,x) + f(t,x)\frac{\partial}{\partial x}V(t,x) + \frac{1}{2}\sigma^2(t,x)\frac{\partial^2}{\partial x^2}V(t,x)$$

$$= f(t,x)\frac{\gamma}{h(x)} - \frac{1}{2}\sigma^2(t,x)\frac{\gamma}{h^2(x)}\frac{d}{dx}h(x)$$

(by substitution and rearrangment)

$$= f(t,x)\frac{\gamma}{h(x)} - \frac{1}{2}g^2(t)h^2(x)\frac{\gamma}{h^2(x)}\frac{d}{dx}h(x) \quad \text{(by simplification)}$$

$$= \left[\gamma\frac{f(t,x)}{h(x)} - \frac{1}{2}\gamma g^2(t)\frac{d}{dx}h(x)\right], \tag{3.4.41}$$

respectively.

From (3.4.40) and (3.4.41), (3.4.2) can be rewritten as

$$\frac{\partial}{\partial t}V(t,x) + f(t,x)\frac{\partial}{\partial x}V(t,x) + \frac{1}{2}\sigma^2(t,x)\frac{\partial^2}{\partial x^2}V(t,x) = \mu(t)V(t,x) + p(t),$$

which implies that

$$\gamma\left[\frac{f(t,x)}{h(x)} - \frac{1}{2}g^2(t)\frac{d}{dx}h(x)\right] = \mu(t)\left[\int_c^x \frac{\gamma}{h(y)}dy + C\right] + p(t).$$

This relation can be further rewritten as

$$\gamma\left[\frac{f(t,x)}{h(x)} - \frac{1}{2}g^2(t)\frac{d}{dx}h(x) - \mu(t)\int_c^x \frac{dy}{h(y)}\right] = p(t) + \mu(t)C. \tag{3.4.42}$$

Now, we set

$$z(t,x) = \frac{f(t,x)}{h(x)} - \frac{1}{2}g^2(t)\frac{d}{dx}h(x). \tag{3.4.43}$$

By substituting this $z(t,x)$ into (3.4.42), we obtain

$$\frac{p(t) + \mu(t)C}{\gamma} = z(t,x) - \mu(t)\int_c^x \frac{dy}{h(y)} \tag{3.4.44}$$

for C is an arbitrary constant of integration (3.4.40).

Now, imitating the argument used in Step #2 of Procedure 3.4.2, we arrive at an expression similar to (3.4.14) and (3.4.15), i.e.

$$h(x)\frac{\partial}{\partial x}z(t,x) = \mu(t), \tag{3.4.45}$$

and

$$\frac{\partial}{\partial x}\left(h(x)\frac{\partial}{\partial x}z(t,x)\right) = 0. \tag{3.4.46}$$

This is the sufficient condition for the reducibility of (3.4.1) to the first-order nonhomogeneous scalar stochastic differential equation with variable coefficients (3.4.1). For $\delta = 0$, the relation (3.4.45) determines the class of SDEs (3.2.1) that is reducible to (3.4.1).

(ii) In this case of $\delta = 0$, (3.4.28) and (3.4.37) reduce to

$$f(t,x) = h(x)\left[\frac{1}{2}g^2(t)\frac{d}{dx}h(x) + f_1(t)\int_c^x \frac{dy}{h(y)} + f_2(t)\right], \qquad (3.4.47)$$

and

$$f(x) = \sigma(x)\left[\frac{1}{2}\frac{d}{dx}\sigma(x) + f_1\int_c^x \frac{dy}{h(y)} + f_2\right], \qquad (3.4.48)$$

respectively. Again, from (3.4.45), using these structures of drifts in (3.4.47) and (3.4.47) and the method of undetermined coefficients, $\mu(t)$ and $p(t)$ in (3.4.1) are given by

$$\mu(t) = f_1(t) \text{ and } p(t) = f_2(t), \qquad (3.4.49)$$

and

$$\mu(t) = f_1 \text{ and } p(t) = f_2, \qquad (3.4.50)$$

respectively.

(iii) For $\delta = 0$, the structures (3.4.47) and (3.4.47) of drift coefficients constitute a mechanical tool for solving the nonlinear Itô–Doob-type stochastic differential equation (3.2.1). These structural representations of the drift component of the member pair (f,σ) of either the class LSDEVC-(f,σ) or LSDECC-(f,σ) of the nonlinear Itô–Doob SDE (3.2.1) are used to classify and/or to solve a given nonlinear Itô–Doob-type stochastic differential equation.

In the following, we present a few illustrations to exhibit the procedure described in this section.

Illustration 3.4.1. Given:

(a) $dx = h(x)\left[\frac{1}{2}g^2(t)\frac{d}{dx}h(x) - \frac{f(t)}{\delta}\exp\left[-\delta\int_c^x \frac{ds}{h(s)}\right] + F(t)\right]dt + g(t)h(x)dw(t)$,

(b) $dx = h(x)\left[\frac{1}{2}g^2(t)\frac{d}{dx}h(x) + F(t)\right]dt + g(t)h(x)dw(t)$,

(c) $dx = h(x)\left[\frac{1}{2}g^2(t)\frac{d}{dx}h(x) + f(t)\int_c^x \frac{ds}{h(s)} + F(t)\right]dt + g(t)h(x)dw(t)$,

(d) $dx = \sigma(x)\left[\frac{1}{2}\frac{d}{dx}\sigma(x) - \frac{a}{\delta}\exp\left[-\delta\int_c^x \frac{ds}{h(s)}\right] + b\right]dt + \sigma(x)dw(t)$,

(e) $dx = \sigma(x)\left[\frac{1}{2}\frac{d}{dx}\sigma(x) + a\int_c^x \frac{ds}{h(s)} + b\right] + \sigma(x)dw(t)$,

(f) $dx = \sigma(x)\left[\frac{1}{2}\frac{d}{dx}\sigma(x) + b\right] + \sigma(x)dw(t)$,

where σ satisfies the hypothesis (H3.4), f and F are given continuous functions, and a and b are arbitrary given constants. Find the general solution to the given differential equation.

Solution procedure. Following the tips T_1–T_9, outlined in Observation 3.3.2, and imitating the argument used Illustration 3.3.1, one establishes the membership of the class of Illustration 3.4.1(a)–(f). Moreover, the rate coefficients of the corresponding reduced SDEs, are as follows:

(a) $\nu(t) = \delta g(t), \gamma = 0, q(t) = \gamma g(t) = 0, \mu(t) = \frac{1}{2}g^2(t)\delta^2 + \delta F(t)$ and $p(t) = \frac{1}{2}g^2(t)\delta\gamma + \gamma F(t) - Cf(t)$ [by (3.4.29)];

(b) $\delta = 0 = \gamma, q(t) = g(t), \mu(t) = 0$ and $p(t) = F(t)$ [by (3.4.49)];

(c) $\delta = 0, q(t) = \gamma g(t), \gamma = 1, \mu(t) = f(t)$ and $p(t) = F(t)$ [by (3.4.49)];

(d) $\nu(t) = \delta g(t), \gamma = 0, q(t) = \gamma g(t) = 0, \mu(t) = \frac{1}{2}\delta^2 + \delta b$ and $p(t) = \frac{1}{2}\delta\gamma + \gamma b - Ca$ [by (3.4.38)];

(e) $\delta = 0, \gamma = 1, q(t) = 1 = g(t), \mu(t) = a$ and $p(t) = b$ [by (3.4.49)];

(f) $\delta = 0, \gamma = 0, q(t) = 1, \mu(t) = 0$ and $p(t) = b$ [by (3.4.49)].

From this, we conclude the pair of rate functions (f, σ) of problems: (i) (a)–(c) belong to the class LSDEVC-(f, σ) and (ii) (d)–(f) belong to LSDECC-(f, σ). Thus, by following Steps #1–#4 of the preceding procedure, the representation of the general solution to the given differential equation can be determined. The details are left to the reader. This completes the discussion of the illustration. Further detailed discussion can be reformulated in conjunction with Illustration 3.3.1.

Illustration 3.4.2. Given:

$$dx = \left[\frac{\nu}{2}g^2(t)\exp[2\nu x] - \frac{f(t)}{\delta}\exp\left[\nu x + \frac{\delta}{\nu}\exp[-\nu x]\right]\right.$$

$$\left. + F(t)\exp[\nu x]\right]dt + g(t)\exp[\nu x]dw(t),$$

where $f(t)$, $F(t)$, and $g(t)$ are any given continuous functions, and $\delta > 0$ and $\nu < 0$. Find the general solution to the given differential equation for: (a) any functions f and F, (b) $F(t) = 0$, (c) $f = 0$, and (d) $F(t) = 0 = f$.

Solution procedure. The goal is to find a general solution to the given Itô–Doob-type stochastic differential equation. First, depending on the nature of the functions $f(t)$ and $F(t)$, we need to show that the pair of rate functions (f, σ) defined by

$$f(t, x) = \frac{\nu}{2}g^2(t)\exp[2\nu x] - \frac{f(t)}{\delta}\exp\left[\nu x + \frac{\delta}{\nu}\exp[-\nu x]\right] + F(t)\exp[\nu x]$$

$$= \exp[\nu x]\left[\frac{\nu}{2}g^2(t)\exp[\nu x] - \frac{f(t)}{\delta}\exp\left[\frac{\delta}{\nu}\exp[-\nu x]\right] + F(t)\right],$$

$$\sigma(t, x) = g(t)\exp[\nu x], \text{ and } h(x) = \exp[\nu x]$$

belongs to either the class LSDEVC-(f, σ) or LSDECC-(f, σ). To fulfill the objective of T_5, we first observe that

$$\frac{d}{dx} h(x) = \nu \exp[\nu x], \quad \text{and} \quad \exp\left[-\delta \int_c^x \frac{dy}{\exp[\nu y]}\right] = \exp\left[\frac{\delta}{\nu} \exp[-\nu x]\right],$$

and hence

$$\frac{f(t, x)}{h(x)} = \left[\frac{\nu}{2} g^2(t) \exp[\nu x] - \frac{f(t)}{\delta} \exp\left[\frac{\delta}{\nu} \exp[-\nu x]\right] + F(t)\right]$$

$$= \left[\frac{1}{2} g^2(t) \frac{d}{dx} h(x) - \frac{f(t)}{\delta} \exp\left[-\delta \int_c^x \frac{dy}{\exp[\nu y]}\right] + F(t)\right].$$

Now, by comparing the rate functions with Illustration 3.4.1(a), we conclude that it belongs to the class LSDEVC-(f, σ). Moreover, we have

$$\mu(t) = \frac{1}{2} g^2(t) \delta^2 + \delta F(t) \quad \text{and} \quad p(t) = \frac{1}{2} g^2(t) \delta \gamma + \gamma F(t) - C f(t),$$

where $0 \neq C$ and γ is an arbitrary constant. We choose $\gamma = 0$ and $C = 1$ because the SDE is identical with the conceptual representation of the given drift coefficient. Now, by imitating Step #4 of Procedure 3.4.2, we have

$$x(t) = G^{-1}\left(\ln\left[\Phi(t)c - \int_a^t \Phi(t, s) p(s) ds\right] + G(c)\right),$$

$$G(x(t)) - G(c) = \int_c^{x(t)} \frac{\delta}{h(s)} ds = \delta \int_c^{x(t)} \exp[-\nu y] dy = -\frac{\delta}{\nu} \exp[-\nu x(t)].$$

Thus,

$$x(t) = -\frac{1}{\nu} \ln\left[-\frac{\nu}{\delta}\left(\ln\left[\Phi(t)c - \int_a^t \Phi(t, s) f(s) ds\right]\right) + G(c)\right].$$

This completes the solution representation of Part (a). The solution representations of (b), (c), and (d) are

$$x(t) = -\frac{1}{\nu} \ln\left(-\frac{\nu}{\delta}\left(\ln\left[\Phi(t)c - \int_a^t \Phi(t, s) f(s) ds\right] + G(c)\right)\right)$$

with $\mu(t) = \frac{1}{2} g^2(t) \delta^2$,

$$x(t) = -\frac{1}{\nu} \ln\left(\int_a^t F(s) ds + \int_a^t g(s) dw(s) + c\right)$$

with $\gamma = 1, \mu(t) = 0$ and $p(t) = F(t)$ and

$$x(t) = -\frac{1}{\nu} \ln \left(c + \int_a^t g(s)dw(s) \right)$$

with $\gamma = 1, \mu(t) = 0$ and $p(t) = 0$, respectively. This completes the illustration.

Illustration 3.4.3. Given:

$$dx = \left[-\nu g^2(t) \sin(\nu x) \cos^3(\nu x) - \frac{f(t)}{\delta} \cos^2(\nu x) \exp\left[-\frac{\delta}{\nu} \tan(\nu x) \right] \right.$$

$$\left. + F(t) \cos^2(\nu x) \right] dt + g(t) \cos^2(\nu x) dw(t),$$

where $f(t)$, $F(t)$, and $g(t)$ are any given continuous functions, $\delta > 0$, and ν is an arbitrary given real number. Find the general solution to the given differential equation for: (a) any functions f and F, (b) $F(t) = 0$, (c) $f = 0$, and (d) $F(t) = 0 = f$.

Solution procedure. Following the argument used in Illustration 3.4.2, we have

$$f(t,x) = -\nu g^2(t) \sin(\nu x) \cos^3(\nu x) - \frac{f(t)}{\delta} \cos^2(\nu x)$$

$$\times \exp\left[-\frac{\delta}{\nu} \tan(\nu x) \right] + F(t) \cos^2(\nu x)$$

$$= \cos^2(\nu x) \left[-\nu g^2(t) \sin(\nu x) \cos(\nu x) - \frac{f(t)}{\delta} \exp\left[-\frac{\delta}{\nu} \tan(\nu x) \right] + F(t) \right],$$

$$\sigma(t,x) = g(t) \cos^2(\nu x), \quad h(x) = \cos^2(\nu x), \quad \frac{d}{dx} h(x) = -2\nu \sin(\nu x) \cos(\nu x),$$

and $\exp[-\delta \int_c^x \frac{dy}{\cos^2(\nu y)}] = \exp[-\frac{\delta}{\nu} \tan(\nu x)]$. Thus,

$$\frac{f(t,x)}{h(x)} = \left[-\nu g^2(t) \sin(\nu x) \cos(\nu x) - \frac{f(t)}{\delta} \exp\left[-\frac{\delta}{\nu} \tan(\nu x) \right] + F(t) \right]$$

$$= \left[\frac{\nu}{2} g^2(t) \frac{d}{dx} h(x) - \frac{f(t)}{\delta} \exp\left[-\delta \int_c^x \frac{dy}{h(y)} \right] + F(t) \right].$$

Now, by comparing the rate functions with Illustration 3.4.1(a), we conclude that the given SDE belongs to the class LSDEVC-(f, σ). Moreover,

$$\mu(t) = \frac{1}{2} g^2(t) \delta^2 + \delta F(t) \text{ and } p(t) = \frac{1}{2} g^2(t) \delta\gamma + \gamma F(t) - Cf(t),$$

where $0 \neq C$ and γ is an arbitrary constant. We choose $\gamma = 0$ and $C = 1$ because the SDE is identical with the conceptual representation of the given drift coefficient. Furthermore, we have

$$x(t) = G^{-1}\left[\ln\left[\Phi(t)c - \int_a^t \Phi(t,s)p(s)ds\right] + G(c)\right].$$

Hence,

$$G(x(t)) - G(c) = \int^{x(t)} \frac{\delta}{h(y)}dy = \delta \int_c^{x(t)} \sec^2(\nu y)dy = \frac{\delta}{\nu}[\tan(\nu x) - \tan(\nu c)].$$

Thus,

$$x(t) = \frac{1}{\nu}\left[\tan^{-1}\left(\frac{\nu}{\delta}\left[\ln\left[\Phi(t)c - \int_a^t \Phi(t,s)f(s)ds\right] + G(c)\right]\right)\right].$$

This completes the solution representation of Part (a). The solution representations of (b), (c), and (d) are

$$x(t) = \frac{1}{\nu}\left[\tan^{-1}\left(\frac{\nu}{\delta}\left[\ln\left[\Phi(t)c - \int_a^t \Phi(t,s)f(s)ds\right] + G(c)\right]\right)\right]$$

with $\mu(t) = \frac{1}{2}g^2(t)\delta^2$,

$$x(t) = \frac{1}{\nu}\tan^{-1}\left(\int_a^t F(s)ds + \int_a^t \nu g(s)dw(s) + c\right)$$

with $\delta = 0, \gamma = 1, \mu(t) = 0$ and $p(t) = F(t)$ and

$$x(t) = \frac{1}{\nu}\tan^{-1}\left(\nu \int_a^t g(s)dw(s) + c\right)$$

with $\delta = 0, \gamma = 0, \mu(t) = 0, q(t) = g(t)$, and $p(t) = 0$, respectively. This completes the illustration.

Illustration 3.4.4. Given:

$$dx = e(x)E(x)\left[\frac{1}{2}g^2(t)\left(E(x)\frac{d}{dx}e(x) + 1\right)\right.$$

$$\left. - \frac{f(t)}{\partial[E(x)]^\delta} + F(t)\right]dt + g(t)e(x)E(x)dw(t),$$

where $e(x)$, $E(x)$, $g(t)$, and $\delta > 0$ satisfy the hypothesis (H$_{3.4}$) with $E(x) = \int^x \frac{ds}{e(s)} > 0$, and $f(t)$ and $F(t)$ are as defined in Illustration 3.4.1. Find the general solution to the given differential equation for: (a) any functions f and F, (b) $F(t) = 0$, (c) $f = 0$, and (d) $F(t) = 0 = f$.

Solution procedure. First, again by imitating the argument used in the preceding illustrations, we have

$$f(t,x) = e(x)E(x)\left[\frac{1}{2}g^2(t)\left(E(x)\frac{d}{dx}e(x)+1\right) - \frac{f(t)}{\delta[E(x)]^\delta} + F(t)\right],$$

$$\sigma(t,x) = g(t)e(x)E(x), \quad h(x) = e(x)E(x),$$

$$\frac{d}{dx}h(x) = \frac{d}{dx}[e(x)E(x)] = e(x)\frac{d}{dx}E(x) + E(x)\frac{d}{dx}e(x) \quad \text{(by the product rule)}$$

$$= e(x)\frac{1}{e(x)} + E(x)\frac{d}{dx}e(x)$$

(by the mean value theorem for integral calculus)

$$= 1 + E(x)\frac{d}{dx}e(x),$$

$$\exp\left[-\delta\int_c^x \frac{dy}{e(y)E(y)}\right]$$

$$\left[\text{by the substitution method: } u = E(y), du = \frac{1}{e(y)}dy\right]$$

$$= \exp\left[-\delta\int_c^x \frac{du}{u}\right] = \exp[-\delta\ln|u|].$$

Thus,

$$\exp\left[-\delta\int_c^x \frac{dy}{e(y)E(y)}\right] = \frac{1}{[E(x)]^\delta} \quad \text{if } E(x) > 0,$$

Hence,

$$\frac{f(t,x)}{h(x)} = \frac{1}{2}g^2(t)\left(E(x)\frac{d}{dx}e(x)+1\right) - \frac{f(t)}{\delta[E(x)]^\delta} + F(t) \quad \text{(by substitution)}$$

$$= \left[\frac{1}{2}g(t)\frac{d}{dx}h(x) - \frac{f(t)}{\delta}\exp\left[-\delta\int_c^x \frac{ds}{h(s)}\right] + F(t)\right].$$

Now, by comparing the rate functions with Illustration 3.4.1(a), we conclude that

$$\mu(t) = \frac{1}{2}g^2(t)\delta^2 + \delta F(t) \text{ and } p(t) = \frac{1}{2}\delta\gamma g^2(t) + \gamma F(t) - Cf(t),$$

for any γ, and the given differential equation belongs to the class LSDEVC-(f, σ). Furthermore, we have $\gamma = 0$, and its solution process is

$$x(t) = G^{-1}\left[\ln\left[\Phi(t)c + \int_a^t \Phi(t, s)p(s)ds\right] + G(c)\right],$$

where

$$G(x(t)) = \int_c^{x(t)} \frac{\delta}{h(y)}dy = \delta \int_c^{x(t)} \frac{dy}{e(y)E(y)}dy = \delta \ln\left[\frac{E(x)}{C}\right].$$

Thus, we have

$$E(x(t)) = \exp\left(\frac{1}{\delta}\left[\ln\Phi(t)c - \int_a^t \Phi(t, s)f(s)ds\right]\right) C$$

$$= \exp\left(\ln\Phi(t)c - \int_a^t \Phi(t, s)f(s)ds\right)^{1/\delta} C$$

$$= \left[\Phi(t)c - \int_a^t \Phi(t, s)f(s)ds\right]^{1/\delta} C.$$

Hence,

$$x(t) = E^{-1}\left[\left(\Phi(t)c - \int_a^t \Phi(t, s)f(s)ds\right)^{1/\delta} C\right].$$

This completes the solution representation of Part (a). The solution representations of (b), (c), and (d) are given by

$$x(t) = E^{-1}\left[\left(\Phi(t)c - \int_a^t \Phi(t, s)f(s)ds\right)^{1/\delta} C\right]$$

with $\mu(t) = \frac{1}{2}g^2(t)\delta^2$,

$$x(t) = E^{-1}\left(\int_a^t F(s)ds + \int_a^t g(s)dw(s) + C\right)$$

with $\delta = 0, \gamma = 1, \mu(t) = 0, q(t) = g(t)$ and $p(t) = F(t)$, and

$$x(t) = E^{-1}\left(\int_a^t g(s)dw(s) + C\right)$$

with $\delta = 0, \gamma = 1, \mu(t) = 0, q(t) = g(t)$ and $p(t) = 0$, respectively. This completes the illustration.

In the following, we present a few examples to illustrate the procedure described in this section.

Example 3.4.1. Given: $dx = \left(-\frac{\kappa^2}{2}\csc^3 x \cos x + b\cot x\right)dt + \kappa\csc x\, dw(t)$. Find the general solution to the given differential equation.

Solution procedure. From Illustration 3.4.1(e) and (3.4.49), we conclude that it belongs to the class LSDECC-(f,σ). Thus, from (3.4.49), we conclude that $g(t) = \kappa$, $\gamma = 1$, $\delta = 0$, $\gamma = 1$, $q(t) = \kappa$, $\nu(t) = 0$, $\mu(t) = b$, and $p(t) = 0$. The Itô–Doob-type reduced stochastic differential equation is

$$dm = -bm\,dt + \kappa\,dw(t),$$

and its general solution is

$$m(t) = \exp[-b(t-a)]c + \kappa\int_a^t \exp[-b(t-s)]dw(s),$$

where c is an arbitrary nonzero constant. The general solution to the given example is given by

$$G(x(t)) - G(c) = \int_c^{x(t)} \sin s\,ds = \exp[-b(t-a)]c + \kappa\int_a^t \exp[-b(t-s)]dw(s).$$

This implies that

$$-\cos x(t) = \exp[-b(t-a)]c + \kappa\int_a^t \exp[-b(t-s)]dw(s) - \cos c.$$

Hence,

$$x(t) = \cos^{-1}\left(\cos c - \exp[-b(t-a)]c - \kappa\int_a^t \exp[-b(t-s)]dw(s)\right).$$

This is the general solution to the given differential equation.

Example 3.4.2. Given:

$$dx = \frac{\lambda x + \mu}{\eta x + \kappa}\left[-\frac{1}{2}g^2(t)\frac{\eta\mu - \kappa\lambda}{(\eta x + \kappa)^2} - \frac{f(t)}{\delta}[|\lambda x + \mu|]^{\frac{(\eta\mu - \kappa\lambda)\delta}{\lambda^2}}\right.$$

$$\left.\times \exp\left[-\frac{\eta}{\lambda}\delta x\right] + F(t)\right]dt + g(t)\frac{\lambda x + \mu}{\eta x + \kappa}dw(t),$$

where η, κ, λ, and μ are given real numbers, and either η or κ is different from zero; similarly, either λ or μ is different from zero; g, f, and F are continuous functions as defined in Illustration 3.4.1. Find the general solution to the given differential equation for: (a) any functions f and F, (b) $F(t) = 0$, (c) $f = 0$, and (d) $F(t) = 0 = f$.

Solution procedure. The goal is to find a general solution to the given differential equation. We repeat the solution tips outlined in Observation 3.3.2. The rate functions in the example are identified below. Employing the argument used in Illustrations 3.4.1 and 3.4.2, we set

$$f(t,x) = \frac{\lambda x + \mu}{\eta x + \kappa}\left[-\frac{1}{2}g^2(t)\frac{\eta\mu - \kappa\lambda}{(\eta x + \kappa)^2} - \frac{f(t)}{\delta}[|\lambda x + \mu|]^{(\eta\mu - \kappa\lambda)\frac{\delta}{\lambda^2}}\right.$$

$$\left. \times \exp\left[-\frac{\eta}{\lambda}\delta x\right] + F(t)\right]$$

$$= \frac{\lambda x + \mu}{\eta x + \kappa}\left[-\frac{1}{2}g^2(t)\frac{\eta\mu - \kappa\lambda}{(\eta x + \kappa)^2} - \frac{f(t)}{\delta}\right.$$

$$\left. \times \exp\left[-\frac{\delta}{\lambda}\left(\eta x + \frac{\kappa\lambda - \eta\mu}{\lambda}\ln(|\lambda x + \mu|)\right)\right] + F(t)\right],$$

$$\sigma(t,x) = g(t)\frac{\lambda x + \mu}{\eta x + \kappa}, \quad h(x) = \frac{\lambda x + \mu}{\eta x + \kappa}.$$

Thus,

$$\frac{f(t,x)}{\frac{\lambda x + \mu}{\eta x + \kappa}} = \left[-\frac{1}{2}g^2(t)\frac{\eta\mu - \kappa\lambda}{(\eta x + \kappa)^2} - \frac{f(t)}{\delta}\right.$$

$$\left. \times \exp\left[-\frac{\delta}{\lambda}\left(\eta x + \frac{\kappa\lambda - \eta\mu}{\lambda}\ln(|\lambda x + \mu|)\right)\right] + F(t)\right].$$

Using algebra and calculus, we compute

$$\frac{d}{dx}h(x) = \frac{d}{dx}\left(\frac{\lambda x + \mu}{\eta x + \kappa}\right) = -\frac{(\lambda x + \mu)\eta - (\eta x + \kappa)\lambda}{(\eta x + \kappa)^2} = -\frac{\eta\mu - \kappa\lambda}{(\eta x + \kappa)^2},$$

$$G(x) - G(c) = \int_c^x \frac{dy}{h(y)} = \int_c^x \left(\frac{1}{\frac{\lambda y + \mu}{\eta y + \kappa}}\right)dy = \int_c^x \left(\frac{\eta y + \kappa}{\lambda y + \mu}\right)dy$$

$$= \int_c^x \left(\frac{\eta}{\lambda} + \frac{\kappa\lambda - \eta\mu}{\lambda(\lambda y + \mu)}\right)dy$$

$$= \left[\frac{\eta}{\lambda}x + \frac{\kappa\lambda - \eta\mu}{\lambda^2}\ln(|\lambda x + \mu|)\right] - \left[\frac{\eta}{\lambda}c + \frac{\kappa\lambda - \eta\mu}{\lambda^2}\ln(|\lambda c + \mu|)\right],$$

and

$$\exp\left[-\delta\int_c^x \frac{\delta}{d(y)}dy\right] = \exp\left[-\delta\left[\frac{\eta}{\lambda}x + \frac{\kappa\lambda - \eta\mu}{\lambda^2}\ln(|\lambda x + \mu|)\right]\right]C$$

$$= [|\lambda s + \mu|]^{(\eta\mu - \kappa\lambda)\frac{\delta}{\lambda^2}}\exp\left[-\delta\frac{\eta}{\lambda}x\right]C.$$

Hence,

$$\frac{f(t,x)}{\sigma(t,x)} = \left[-\frac{1}{2}g^2(t)\frac{\eta\mu - \kappa\lambda}{(\eta x + \kappa)^2} - \frac{f(t)}{\delta} \right.$$

$$\times \exp\left[-\frac{\delta}{\lambda}\left(\eta x + \frac{\kappa\lambda - \eta\mu}{\lambda}\ln(|\lambda x + \mu|) \right) \right] + F(t) \Bigg]$$

$$= \left[-\frac{1}{2}g^2(t)\frac{d}{dx}h(x) - \frac{f(t)}{\delta}\exp\left[-\delta\int_c^x \frac{dy}{h(y)} \right] + F(t) \right].$$

This expression is identical with Illustration 3.4.1(a). Therefore, we conclude that (f, σ) belongs to the class LSDEVC-(f, σ). We imitate the rest of the argument presented in the preceding illustrations, in particularly Illustration 3.4.1. The solution representation is described by

$$x(t) = G^{-1}\left[\ln\left[\Phi(t)c - \int_a^t \Phi(t,s)p(s)ds \right] + C \right]$$

where the nonzero constant C. The details are left to the reader.

Example 3.4.3. Solve the IVP

$$dx = (1 + x^2)\tan^{-1}x[2(1 + 2x\tan^{-1}x) - f(t)\tan^{-1}x + F(t)]dt$$

$$+ 2(1 + x^2)\tan^{-1}x\, dw(t), \quad x(t_0) = x_0.$$

Solution procedure. Goal: To find the solution to the given IVP. For this purpose, we imitate the arguments of Example 3.4.2, and Illustrations 3.4.1 and 3.4.4. First, we need to find a general solution. We observe that

$$f(t,x) = (1 + x^2)\tan^{-1}x[2(1 + 2x\tan^{-1}x) - f(t)\tan^{-1}x + F(t)],$$

$$\sigma(t,x) = 2(1 + x^2)\tan^{-1}x, \quad h(x) = (1 + x^2)\tan^{-1}x = e(x)E(x),$$

$$e(x) = 1 + x^2, \quad E(x) - E(c) = \int_c^x \frac{dy}{1 + y^2} = \tan^{-1}x - \tan^{-1}c,$$

$$\frac{d}{dx}h(x) = \frac{d}{dx}(e(x)E(x)) = e(x)\frac{d}{dx}E(x)$$

$$+ E(x)\frac{d}{dx}e(x) = 1 + 2x\tan^{-1}x,$$

$$G(x) - G(c) = \int_c^x \frac{ds}{h(s)} = \int_c^x \frac{dy}{(1 + y^2)\tan^{-1}y} = \ln\left(\frac{|\tan^{-1}x|}{|\tan^{-1}c|} \right),$$

$$\exp\left[-\delta\int_c^x \frac{dy}{h(y)} \right] = \frac{1}{(\tan^{-1}x)^\delta}$$

Now, we compare the given and the computational representations as

$$\frac{f(t,x)}{\sigma(t,x)} = \left[\frac{1}{2}g^2(t)\frac{d}{dx}h(x) - \frac{f(t)}{\delta}\right]$$

$$\times \exp\left[-\delta\int^x \frac{dy}{h(y)}\right] + F(t)\right] \quad \text{(conceptual representation)}$$

$$= \left[2(1 + 2x\tan^{-1}x) - f(t)\tan^{-1}x + F(t)\right] \quad \text{(given representation)}$$

$$= \left[2(1 + 2x\tan^{-1}x) - \frac{f(t)}{\delta}[\tan^{-1}x]^{-\delta} + F(t)\right] \quad \text{(computational form)}.$$

These representations are conceptually identical. Therefore, applying Illustration 3.4.1, we conclude that (f,σ) belongs to the class LSDEVC-(f,σ). Again, we compare and determine the parameters and rate functions in the reduced SDE. They are $\delta = -1$, $g(t) = 2$, $C = 1$, $\gamma = 0$, $q(t) = 0$, $-2 = \nu(t)$, $\mu(t) = 2 - F(t)$, and $p(t) = f(t)$.

Following the argument used in Illustration 3.4.3, we have the general solution representation of the problem (a):

$$x(t) = \tan\left(\Phi(t)G(c) + \int_a^t \Phi(t,s)f(s)ds\right).$$

The solution to the given IVP is given by

$$\tan^{-1}x = \Phi(t)G(c) + \int_a^t \Phi(t,s)f(s)ds,$$

$$\tan^{-1}x_0 = \Phi(t_0)G(c) + \int_a^{t_0} \Phi(t_0,s)f(s)ds.$$

This implies that

$$G(c) = \Phi^{-1}(t_0)\tan^{-1}x_0 - \int_a^{t_0} \Phi(s)f(s)ds.$$

We substitute this value into the above solution expression, and get

$$\tan^{-1}x = \Phi(t)\Phi^{-1}(t_0)G(c)\tan^{-1}x_0 - \Phi(t)\int_a^{t_0}\Phi^{-1}(s)f(s)ds + \int_a^t\Phi(t,s)f(s)ds$$

$$= \Phi(t,t_0)\tan^{-1}x_0 - \int_a^{t_0}\Phi(t,s)f(s)ds + \int_a^t\Phi(t,s)f(s)ds$$

$$= \Phi(t,t_0)\tan^{-1}x_0 + \int_{t_0}^t\Phi(t,s)f(s)ds,$$

and hence

$$x(t, x_0, t_0) = \tan\left(\Phi(t, t_0)\tan^{-1} x_0 + \int_{t_0}^{t}\Phi(t, s)f(s)ds\right).$$

This is the solution process of the given IVP.

Observation 3.4.3. An observation similar to Observation 3.3.3 can be made. The details are left to the reader.

Brief Summary. A summary of the basic mechanical steps needed to solve non-linear stochastic scalar differential equations is presented in the following flowchart:

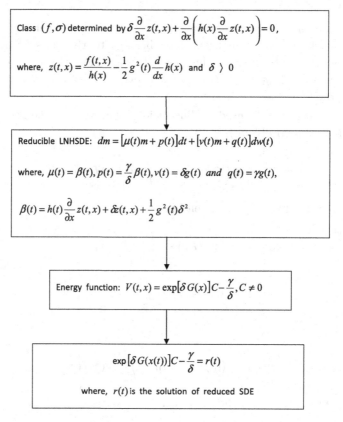

Fig. 3.4.1 Summary of reducible LSDE and the solution processes.

Exercises

1. Show that the following differential equations belong to the class LSDECC-(f, σ):

   (a) $dx = (-50\exp[-8x] - \exp[-(4x + \exp[4x])])dt + 5\exp[-4x]dw(t)$;

(b) $dx = -50 \exp[-4x]dt + 5 \exp[-4x]dw(t);$

(c) $dx = [-\sinh 2x - \cosh x \exp[-\tan^{-1}(\sinh x)] + 4 \cosh x]dt + 2 \cosh x \, dw(t);$

(d) $dx = x\sqrt{1-x^2}\left(\left[\frac{2x^2-1}{2\sqrt{1-x^2}} - \frac{x}{1+\sqrt{1-x^2}}\right] dt - dw(t)\right);$

(e) $dx = \left[\frac{x}{2} + \sqrt{1+x^2}\right] dt + \sqrt{1+x^2}\,dw(t).$

Hint: $\int \frac{dx}{\sqrt{1+x^2}} = \ln\left(x + \sqrt{1+x^2}\right)$ and $\int \frac{dx}{x\sqrt{1-x^2}} = -\ln\left(\frac{1+\sqrt{1-x^2}}{x}\right).$

2. Show that the following differential equations belong to the class LSDEVC-(f,σ):

(a) $dx = [-\sin^2 t \exp[-4x] - \exp[-2x(x + \exp[2x]] + \cos^2 t \exp[-2x])dt + \sin t \exp[-2x]dw(t);$

(b) $dx = \left[-\frac{1}{2}\cos^2 t(2x^2+1)x + \frac{\cos^2 tx^2}{1+\sqrt{1+x^2}}\right] dt - \cos t \, dw(t);$

(c) $dx = \left[-\frac{1}{2}\cosh^2 \exp[x](1+\exp[x]) + \frac{(1+\exp[x])^2}{\exp[x]}\right] dt + \cosh t(1+\exp[x])dw(t);$

(d) $dx = \left[\frac{1}{2}\exp[2t]\,\mathrm{csch}^2\, x \coth x + \ln(t+1)\,\mathrm{csch}\, x\right] dt - \exp[t]\coth x \, dw(t);$

(e) $dx = \left[-\frac{1}{4}\sinh^2 t \sinh 2x + \frac{1}{2}\cosh^2 t \sinh x \coth\left(\frac{x}{2}\right)\right] dt - \sinh t \sinh x \, dw(t).$

Hint: $\int \mathrm{csch}\, x \, dx = \ln\left(\tan\left(\frac{x}{2}\right)\right)$, $\int \tanh x \, dx = \ln(\cosh x)$, and $\int \frac{dx}{1+\exp[x]} = \ln\left(\frac{\exp[x]}{1+\exp[x]}\right).$

3. Find the general solution representation of the differential equations in Problems #1 and #2.

4. Given:

$$dx = \frac{H(x)}{\frac{d}{dx}H(x)}\left[\left[\frac{1}{2}g^2(t)\left(1 - \frac{H(x)\frac{d^2}{dx^2}H(x)}{\left(\frac{d}{dx}H(x)\right)^2}\right)\right.\right.$$
$$\left.\left. - \frac{f}{\delta}[H(x)]^{-\delta} + F\right] dt + g(t)dw(t)\right],$$

where H is any given twice continuously differentiable positive function; g, f, and F are any given constants; δ is a positive number; and $\frac{H(x)}{\frac{d}{dx}H(x)}$ is not constant. Further assume that $\gamma = 0$ in the hypothesis (H3.4), and the number C (3.4.5) is equal to 1.

(a) Show that the given differential equation belongs to the class LSDECC-(f,σ) (if possible).

(b) In this case, the energy function is $V(t,x) = [H(x)]^\delta.$

(c) Find the general solution to the given differential equation for:

    (i) any $F(t)$ and $f(t);$

    (ii) $f(t) = 0;$

    (iii) $F = 0$ and $\delta = 1;$

    (iv) $H(x) = 1 + \exp[x]$ with $\delta = 1$, $f(t) = 4$ and $g(t) = 1;$

    (v) $H(x) = \frac{x}{\sqrt{1+x^2}}$ with $\delta = 1$, $f(t) = 4$ and $g(t) = 1.$

5. Solve the following initial value problems:

(a) $dx = \left[\frac{ax}{2(1+t^2)} + \sqrt{a}\left[x\sqrt{a^2x^2 + ac} + ax^2 + c\right] - \left(1 + \frac{1}{2(t+1)^2}\right)\sqrt{ax^2 + c}\right] dt$
$+ \frac{1}{1+t}\sqrt{ax^2 + c}\, dw(t)$, $x(0) = 1$, $a > 0$;

(b) $dx = \left[-\frac{1}{4}\sin 4ax + \cos ax + \sin 2ax\right] dt + \frac{1}{2}\sin 2ax\, dw(t)$, $x(t_0) = x_0$, $a \neq 0$;

(c) $dx = \left[\frac{1}{2}ax - \frac{1}{2}\sqrt{ax^2 + c}\right] dt + \sqrt{ax^2 + c}\, dw(t)$, $x(0) = 1$, $a > 0$;

(d) $dx = \left[-\sin 4x + \sin 2x\right] dt + \sin 2x\, dw(t)$, $x(t_0) = x_0$;

(e) $dx = \left[\frac{1}{2}x \ln x(\ln x + 1) + x \tan t - \frac{3}{2}x \ln x\right] dt + x \ln x\, dw(t)$, $x(0) = 1$.

Hint: $\int \frac{dx}{\cos ax \sin ax} = \frac{1}{a}\ln\tan ax$, $\int \frac{dx}{\sqrt{ax^2+c}} = \frac{1}{\sqrt{a}}\ln(x\sqrt{a} + \sqrt{ax^2 + c})$.

3.5 Variable Separable Equations

In this section, we apply the energy function method developed in Section 3.3 to a particular class of differential equations (3.2.1). This class of differential equations is easily reducible to (3.3.1). Its structure is characterized by the rate functions $f(t,x)$ and $\sigma(t,x)$ in (3.2.1). In fact, the nonlinear rate functions are decomposed into a product of two functions in which one of them is a function of an independent variable, and the other is a function of a dependent variable.

Definition 3.5.1. A function F defined on $J \times R$ into R is said to be a *separable function* in t and x variables if $F(t,x) = H(t)G(x)$.

Definition 3.5.2. A differential equation (3.2.1) is said to be a *separable differential equation of the Itô–Doob-type* if its rate functions $f(t,x)$ and $\sigma(t,x)$ are separable functions in t and x variables.

(H$_{3.5}$). In addition to the hypothesis (H$_{3.2}$), we assume that $f(t,x)$ and $\sigma(t,x)$ are separable functions in t and x variables: $f(t,x) = a(t)b(x)$, $\sigma(t,x) = g(t)h(x)$ and b and h are continuously differentiable with respect to both (t,x). Furthermore, $h(x) \neq 0$ and $G(x) = \int_c^x \frac{dy}{h(y)}$ is invertible.

3.5.1 *General Problem*

In the light of the hypothesis (H$_{3.5}$), (3.2.1) reduces to

$$dx = a(t)b(x)dt + g(t)h(x)dw(t), \tag{3.5.1}$$

where h and b satisfy the conditions in (H$_{3.5}$).

3.5.2 *Procedure for Finding a General Solution Representation*

The procedure for representing the solution process of (3.5.1) is similar to the deterministic case with a few modifications [80].

Step #1. In this case, from (3.3.2) and by imitating the procedures in the deterministic case (Section 3.6 [80]) and Section 3.3, we conclude that $q(t)$ in (3.3.1) and $V(t,x)$ are determined. Moreover, their expressions are given by

$$q(t) = g(t), \quad V(t,x) \equiv V(x) = \int_c^x \frac{dy}{h(y)}. \tag{3.5.2}$$

Step #2. We need to determine $p(t)$. For this purpose, from (3.5.1) and (3.5.2), (3.3.2) can be reduced to

$$p(t) = a(t)b(x)\frac{\partial}{\partial x}V(t,x) + \frac{1}{2}g^2(t)h^2(x)\frac{\partial^2}{\partial x^2}V(t,x)$$

$$= \frac{a(t)b(x)}{h(x)} - \frac{1}{2}g^2(t)h^2(x)\frac{\frac{d}{dx}h(x)}{h^2(x)}$$

$$= \frac{a(t)b(x)}{h(x)} - \frac{1}{2}g^2(t)\frac{d}{dx}h(x). \tag{3.5.3}$$

We differentiate both sides of (3.5.3) with respect to x for fixed t, and obtain

$$0 = \frac{\partial}{\partial x}\left[\frac{a(t)b(x)}{h(x)} - \frac{1}{2}g^2(t)\frac{d}{dx}h(x)\right] = a(t)\frac{\partial}{\partial x}\left[\frac{b(x)}{h(x)}\right] - \frac{1}{2}g^2(t)\frac{\partial}{\partial x}\left[\frac{d}{dx}h(x)\right]. \tag{3.5.4}$$

By algebraic simplification, assuming that $\frac{d}{dx}[\frac{d}{dx}h(x)] \neq 0$, and from (3.5.4), we get

$$\frac{g^2(t)}{a(t)} = 2\frac{\frac{d}{dx}\left[\frac{b(x)}{h(x)}\right]}{\frac{d}{dx}\left[\frac{d}{dx}h(x)\right]} = C, \tag{3.5.5}$$

for some constant C which is determined by the given rate functions b and h. From (3.5.5), we have

$$\frac{1}{2}C\frac{d}{dx}\left[\frac{d}{dx}h(x)\right] = \frac{d}{dx}\left[\frac{b(x)}{h(x)}\right], \quad Ca(t) = g^2(t), \tag{3.5.6}$$

which can be integrated with regard to x, and we arrive at

$$\frac{1}{2}C\frac{d}{dx}h(x) + C_1 = \frac{b(x)}{h(x)}, \tag{3.5.7}$$

where C_1 is another constant of integration and is determined by (3.5.7).

From (3.5.6) and (3.5.7) and multiplying both sides by $h(x)$, we obtain the following expression for $b(x)$:

$$b(x) = \frac{1}{2}Ch(x)\frac{d}{dx}h(x) + C_1h(x). \tag{3.5.8}$$

From (3.5.6), we have the following representation for the drift rate function in (3.5.1):

$$f(t,x) = a(t)b(x) = \frac{1}{2}g^2(t)h(x)\frac{d}{dx}h(x) + \frac{C_1}{C}g^2(t)h(x). \tag{3.5.9}$$

We substitute the expression (3.5.9) into (3.5.3), and solve for $p(t)$:

$$p(t) = \frac{a(t)b(x)}{h(x)} - \frac{1}{2}g^2(t)\frac{d}{dx}h(x)$$

$$= \frac{a(t)b(x)}{h(x)} - \frac{a(t)b(x)}{h(x)} + \frac{C_1}{C}g^2(t) \quad \text{[by using (3.5.8)]}$$

$$= \frac{C_1}{C}g^2(t), \tag{3.5.10}$$

where arbitrary constants C and C_1 depend on h and b in (3.5.1) [(3.5.7)]. Thus, the function $p(t)$ is determined. We note that (3.5.5) determines the class of separable differential equations (3.5.1) that is reducible to the Itô–Doob-type stochastic integrable differential equation (3.3.1) with $p(t) = \frac{C_1}{C}g^2(t)$ and $q(t) = g(t)$.

Step #3. By applying Step #3 of General Procedure 3.3.2 and using the hypothesis $(H_{3.5})$, we get

$$V(t,x(t)) \equiv V(x(t)) = G(x(t)) = \int_c^x \frac{dy}{h(y)} = \int_a^t p(s)ds + \int_a^t g(s)dw(s) + G(c),$$
$$\tag{3.5.11}$$

where c is an arbitrary constant.

Step #4. From the hypothesis $(H_{3.5})$ and Step #4 of General Procedure 3.3.2, we have

$$x(t) = G^{-1}\left(\frac{C_1}{C}\int_a^t g^2(s)ds + \int_a^t g(s)dw(s) + G(c)\right). \tag{3.5.12}$$

This is the general solution representation of (3.5.1). This completes the procedure.

Observation 3.5.1. Like Observation 3.3.1, we have already found the representation in (3.5.9) for a class of separable stochastic differential equations (3.5.1). Furthermore, from (3.5.1), (3.5.9), along with the definition of $z(t,x)$ in (3.3.30) (Observation 3.3.1), we have

$$z(t,x) = \left[\frac{f(t,x)}{\sigma(t,x)} - \frac{1}{2}\frac{\partial}{\partial x}\sigma(t,x) - \int_c^x \frac{\frac{\partial}{\partial t}\sigma(t,y)}{\sigma^2(t,y)}dy\right]$$

$$= \left[\frac{a(t)b(x)}{g(t)h(x)} - \frac{1}{2}g(t)\frac{d}{dx}h(x) - \frac{\frac{dg(t)}{dt}}{g^2(t)}\int_c^x \frac{dy}{h(y)}\right] \quad \text{[from (3.5.1) and (3.5.9)]}$$

$$= \left[\frac{1}{2} g(t) \frac{d}{dx} h(x) + \frac{C_1}{C} g(t) - \frac{1}{2} g(t) \frac{d}{dx} h(x) - \frac{\frac{dg(t)}{dt}}{g^2(t)} \int_c^x \frac{dy}{h(y)} \right]$$

(by substitution)

$$= \left[\frac{C_1}{C} g(t) - \frac{\frac{dg(t)}{dt}}{g^2(t)} \int_c^x \frac{dy}{h(y)} \right] \quad \text{(by simplification)}.$$

$$(3.5.13)$$

Thus, from (3.5.13), we have $0 = \frac{\partial}{\partial x} [\sigma(t, x) \frac{\partial}{\partial x} z(t, x)]$. From Observation 3.3.1(ii), we conclude the separable differential equations belong to the class ISDE-(f, σ). Furthermore, we note that from (3.5.1) and (3.5.13), $\alpha(t)$ in (3.3.11) and $\beta(t)$ in (3.3.21) are:

$$\alpha(t) = -\sigma(t, x) \frac{\partial}{\partial x} z(t, x) = \frac{\frac{dg(t)}{dt}}{g(t)}, \quad \beta(t) = \frac{C_1}{C} g(t), \qquad (3.5.14)$$

and we note that $q(t) = \exp \left[\int_a^t \alpha(s) ds \right] = g(t)$.

Brief Summary. The following flowchart, Fig. 3.5.1, summarizes the steps described in finding the general solution of a class of variable separable stochastic nonlinear scalar differential equations.

Example 3.5.1. Given: $dx = k^2(t) \cos^2 ax \left[-\frac{a}{2} \sin 2ax + f \right] dt + k(t) \cos^2 ax \, dw(t)$, where f is a constant, $k(t)$ any given continuous function and a is a nonzero constant. Find the general solution to the given differential equation for: (a) any $k(t)$ and f; (b) $f = 0$.

Solution procedure. First, we need to show that the given differential equation is separable. For this purpose, we note that

$$f(t, x) = k(t) \cos^2 ax \left(-\frac{a}{2} k(t) \sin 2ax + f k(t) \right) \quad \text{and} \quad \sigma(t, x) = k(t) \cos^2 ax$$

satisfy the hypothesis (H3.5) with $g(t) = k(t)$, $h(x) = \cos^2 ax$, and $G(x) = \frac{1}{a} \tan ax$. To apply Procedure 3.5.2, we need to check the condition (3.5.6). We compute

$$\frac{1}{2} \frac{d}{dx} \left[\frac{d}{dx} h(x) \right] = \frac{1}{2} \frac{d}{dx} \left[\frac{d}{dx} \cos^2 ax \right] = \frac{a}{2} \frac{d}{dx} [-2 \cos ax \sin ax] = -a^2 \cos 2ax,$$

and

$$\frac{d}{dx} \left[\frac{b(x)}{h(x)} \right] = \frac{d}{dx} \left[\frac{\cos^2 ax \left(-\frac{a}{2} \sin 2ax + f \right)}{\cos^2 ax} \right]$$

$$= \frac{d}{dx} \left[\left(-\frac{a}{2} \sin 2ax + f \right) \right] = -a^2 \cos 2ax.$$

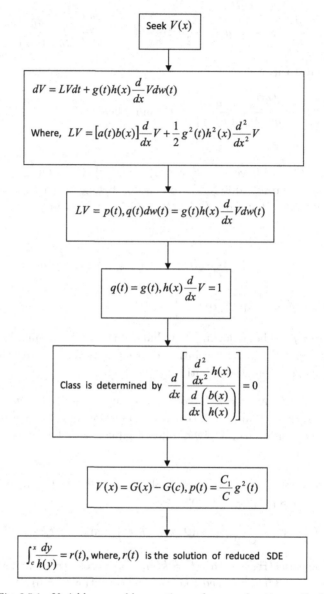

Fig. 3.5.1 Variable separable equation and energy function method.

From this we conclude that

$$\frac{1}{2}\frac{d}{dx}\left[\frac{d}{dx}h(x)\right] = \frac{d}{dx}\left[\frac{b(x)}{h(x)}\right],$$

which shows that the conditions (3.5.6) and (3.5.7) are satisfied with $C = 1$ and $C_1 = f$. Therefore, the given SDE belongs to the class of separable differential equations. Moreover, from Observation 3.5.1, the given differential equation belongs to the class ISDE-(f, σ).

Now, we follow Steps #3 and #4 of Procedure 3.5.2, and we have

$$V(x) = G(x(t)) \equiv \int_c^{x(t)} \frac{dy}{h(y)} = \frac{1}{a} \tan ax$$

$$= \left[f \int_a^t k^2(s)ds + \int_a^t k(s)dw(s) + G(c) \right],$$

and hence

$$x(t) = \frac{1}{a} \tan^{-1} \left(\left[f \int_a^t k^2(s)ds + \int_a^t k(s)dw(s) + G(c) \right] \right).$$

This is the expression for the general solution process of problem (a). In the case of (b), we have

$$x(t) = \frac{1}{a} \tan^{-1} \left(\int_a^t k(s)dw(s) + G(c) \right).$$

By following the argument used in Illustrations 3.3.2 and 3.4.2 together with Part (a), the general solution representation of problem (b) can be reconstructed. Details are left to the reader.

Illustration 3.5.1 (hydrolysis of ester). Let us consider the reaction mechanism of acid-catalyzed hydrolysis, Illustration 3.6.1 [80]:

$S + HA \rightleftarrows SH^+ + A^-$ (reaction rate: k_1 "$\rightarrow$" and k_{-1} "$\leftarrow$" — fast reaction),

$SH^+ + H_2O \rightarrow P_1 + P_2$ (reaction rate: k_2 "$\rightarrow$" — slow reaction),

where S, HA, SH^+, A^-, P_1 and P_2 stand for ester ($R'CO_2R$), acid, base ($R'C^+OOHR$), acid ($R'COOH$), and products (ROH and H^+), respectively. Let $x = [RCO_2R']$ be the concentration reduction of the ester. The ionic mobility in the reaction system can characterized by the Gaussian white noise process. Thus, the rate coefficient $k(k = k_1 k_2 k_{-1}^{-1})$ is under random perturbations.

(a) Show that the dynamic of the hydrolysis of ester is described by

$$dx = \left[k(a-x)(b+x) + \frac{1}{2}(a-b-2x) \right] dt$$

$$+ \sigma(a-x)(b+x)dw(t), \quad x(0) = 0,$$

    where $x = [H^+]$.

(b) Find the solution process of x.

(c) Make a sketch of the solution process.

Solution procedure. Imitating the solution process of Illustration 3.6.1 [80], we get:

$$dx = k(a - x)(b + x)dt, \quad x(0) = 0.$$

Due to the ionically charged chemical reactants, the parameter k is under random perturbations. In particular, $k \equiv \bar{k} + \sigma\xi(t)$, where $\xi(t)$ is the Gaussian white noise process. Therefore, by the application of Sequential Colored Noise Model 3.1.2, we have

$$dx = \left[\bar{k}(a - x)(b + x) + \frac{1}{2}(a - b - 2x)\right]dt + \sigma(a - x)(b + x)dw(t), \quad x(0) = 0.$$

This completes the procedure for the derivation of the differential equation.

To find the solution process of the IVP in (a), we follow Procedure 3.5.2 and the argument used in Example 3.5.1, and get:

$$1 = k(a - x)(b + x)\frac{\partial}{\partial x}V(x) \quad \text{and} \quad dV(x) = \frac{1}{(a - x)(b + x)}dx.$$

This is an integrable form of the deterministic differential equation, and its general solution is given by

$$V(x) = G(x) = \int_c^x \frac{1}{(a - y)(b + y)}dy + G(c)$$

$$= \frac{1}{a + b}\int_c^x \left[\frac{1}{a - y} + \frac{1}{b + y}\right]dy + G(c)$$

$$= \frac{1}{a + b}[\ln(b + x) - \ln(a - x)] + G(c) \quad \text{(by integration by partial fractions)}$$

$$= \frac{1}{a + b}\ln\left(\frac{b + x}{a - x}\right) + G(c) \quad \text{(by algebra and } a - x > 0\text{)}.$$

Hence,

$$V(x) - V(x_0) = \frac{1}{a + b}\ln\left(\frac{a(b + x)}{b(a - x)}\right) = \ln\left[\left(\frac{a(b + x)}{b(a - x)}\right)\right]^{1/a + b}.$$

On the other hand, from (3.5.10) and (3.5.11), we have

$$V(x) - V(x_0) = \int_{t_0}^t \bar{k}\,ds + \int_{t_0}^t \sigma\,dw(s) = \bar{k}(t - t_0) + \sigma(dw(t) - w(t_0)).$$

Noting that $V(x) = G(x)$ and using the above two expressions for $V(x) - V(x_0)$, we obtain

$$\ln\left[\left(\frac{a(b + x)}{b(a - x)}\right)\right]^{1/a + b} = \bar{k}(t - t_0) + \sigma(dw(t) - w(t_0)).$$

From this and the fact that exponential function has an inverse (ln), we obtain

$$\left[\left(\frac{a(b+x)}{b(a-x)}\right)\right]^{1/a+b} = \exp[\bar{k}(t-t_0) + \sigma(dw(t) - w(t_0))].$$

By algebraic simplification, we get

$$\frac{a(b+x)}{b(a-x)} = \exp\left[\frac{\bar{k}(t-t_0) + \sigma(dw(t) - w(t_0))}{a+b}\right].$$

Hence,

$$a(b+x) = b(a-x)\exp\left[\frac{\bar{k}(t-t_0) + \sigma(dw(t) - w(t_0))}{a+b}\right].$$

Thus,

$$x(t)\left(a + b\exp\left[\frac{\bar{k}(t-t_0) + \sigma(dw(t) - w(t_0))}{a+b}\right]\right)$$

$$= ab\left(\exp\left[\frac{\bar{k}(t-t_0) + \sigma(dw(t) - w(t_0))}{a+b}\right] - 1\right).$$

Therefore,

$$x(t) = \frac{ab\left(\exp\left[\frac{\phi(t,t_0,w(t))}{a+b}\right] - 1\right)}{a + b\exp\left[\frac{\phi(t,t_0,w(t))}{a+b}\right]} = \frac{ab\left(1 - \exp\left[-\frac{\phi(t,t_0,w(t))}{a+b}\right]\right)\exp\left[\frac{\phi(t,t_0,w(t))}{a+b}\right]}{\left(\exp\left[-\frac{\phi(t,t_0,w(t))}{a+b}\right]a + b\right)\exp\left[\frac{\phi(t,t_0,w(t))}{a+b}\right]}$$

$$= \frac{ab\left(1 - \exp\left[-\frac{\phi(t,t_0,w(t))}{a+b}\right]\right)}{\exp\left[-\frac{\phi(t,t_0,w(t))}{a+b}\right]a + b},$$

where $\phi((t, t_0, w(t))) = \bar{k}(t-t_0) + \sigma(w(t) - w(t_0))$.

This is the solution to the above given IVP. This completes the solution process of (b).

To make the sketch (Figure 3.5.2), we note that:

(i) $x(0) = 0$, and its horizontal asymptote is a, i.e. $x(t) \to a$ and $\frac{\sigma(w(t)-w(t_0))}{(t-t_0)} \to 0$ (by the strong law of large numbers) as $(t - t_0) \to \infty$ provided that $t \to \infty$ (with probability 1).

(ii) Moreover, its domain and range are $[0, \infty)$ and $[0, a)$, respectively.

(iii) Its sample path sketch is a sigmoid curve. This completes the solution process of the given IVP.

Example 3.5.2 (Pearl–Verhulst logistic stochastic model). We consider the stochastic mathematical model for a class of single-species processes in biological

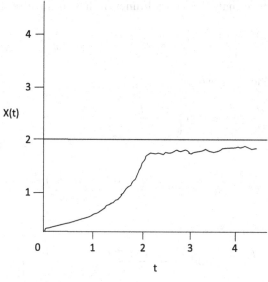

Fig. 3.5.2

sciences described in SEP7, Illustration 3.1.3. In particular, we consider the stochastic model described in (3.1.45) [Example 3.1.3(c)], i.e.

$$dN = N(\kappa - N)\left[\lambda + \frac{\nu^2}{2}(\kappa - 2N)\right]dt + \nu N(\kappa - N)dw(t), \quad N(t_0) = N_0,$$

where N, λ, ν, and κ are as defined in (3.1.43). Find the expression for the size of the species at any time t.

Solution procedure. We imitate Procedure 3.5.2 and the arguments used in Illustration 3.5.1. This gives us:

$$1 = N(\kappa - N)\frac{\partial}{\partial N}V(N), \text{ and hence } dV(N) = \frac{1}{N(\kappa - N)}dN.$$

This is an integrable form of the differential equation, and its general solution is given by

$$V(N) = G(N) = \int^N \frac{1}{y(\kappa - y)}dy + C = \frac{1}{k}\int_c^N \left[\frac{1}{y} + \frac{1}{\kappa - y}\right]ds + C$$

$$= \frac{1}{\kappa}\left[\ln(N) - \ln(|\kappa - N|)\right] + C \quad \text{(by integration by partial fractions)}$$

$$= \frac{1}{\kappa}\ln\left(\left|\frac{N}{\kappa - N}\right|\right) + C \quad \text{(by algebra and } |\kappa - N| > 0\text{)}.$$

Hence,

$$V(N) - V(N_0) = \frac{1}{\kappa}\ln\left(\left|\frac{N(\kappa - N_0)}{N_0(\kappa - N)}\right|\right).$$

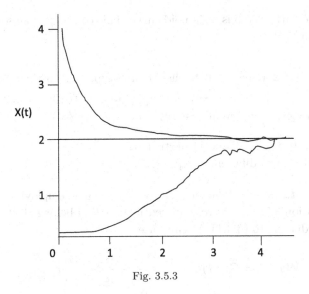

Fig. 3.5.3

From the above V, (3.5.10), and (3.5.11), we have

$$V(N(t)) - V(N_0) = \int_{t_0}^t \lambda \, ds + \int_a^t \nu \, dw(s) = \lambda(t - t_0) + \nu(w(t) - w(t_0)).$$

Again, noting that $V(x) = G(x)$ and using the above two expressions for $V(N(t)) - V(N_0)$, we obtain

$$\ln\left[\left|\frac{N(\kappa - N_0)}{N_0(\kappa - N)}\right|\right] = \kappa\lambda(t - t_0) + \kappa\nu(w(t) - w(t_0)).$$

Hence,

$$\left|\frac{N(\kappa - N_0)}{N_0(\kappa - N)}\right| = \exp[\kappa\lambda(t - t_0) + \kappa\nu(w(t) - w(t_0))].$$

We solve this equation for $N \equiv N(t)$ and obtain

$$N(t) = \frac{\kappa N_0 \exp[\kappa\lambda(t - t_0) + \kappa\nu(w(t) - w(t_0))]}{(\kappa - N_0) + N_0 \exp[\kappa\lambda(t - t_0) + \kappa\nu(w(t) - w(t_0))]}$$

$$= \frac{\kappa N_0}{N_0 + (\kappa - N_0) \exp[-(\kappa\lambda(t - t_0) + \kappa\nu(w(t) - w(t_0)))]}.$$

This is solution to the above given IVP.

To make the sketch, we note that:

(a) $N(t_0) = N_0$, and its horizontal asymptote is κ, i.e. $N(t) \to \kappa$, as $\frac{\sigma(w(t) - w(t_0))}{(t - t_0)}$
   $\to 0$ (by the strong law of large numbers) as $(t - t_0) \to \infty$ provided that $t \to \infty$
   (with probability 1).

(b) For $N_0 > 0$, its domain and range are $[0, \infty)$ and $[N_0, \kappa)$ or $(\kappa, N_0]$, respectively.

(c) The sample path sketch is a sigmoid curve. This completes the solution process of the given IVP.

Illustration   **3.5.2** (Henry–Michaelis–Menten-type stochastic model).   Let us consider $dx = \frac{\lambda x + \mu}{\nu x + \psi}\left[\epsilon + \frac{e\eta^2}{2(\nu x + \psi)^2}\right]dt + \eta \frac{\lambda x + \mu}{\nu x + \psi}dw(t)$, $x(t_0) = x_0$, where ϵ, λ, μ, ν, and ψ are any given real numbers with $e = \lambda\psi - \nu\mu \neq 0 \neq \lambda$. Find:

(a) the solution to the IVP, and in particular;
(b) $x_0 = 0$ of the given differential equation.

Solution procedure.   By following Steps #1, #2 and employing the argument used in Illustration 3.5.1, the given differential equation belongs to the class ISDE-(f, σ) and is reducible to (3.3.1). In this case,

$$V(x(t)) = G(x(t)) = \int_c^{x(t)} \frac{\nu y + \psi}{\lambda y + \mu}dy = \frac{\nu}{\lambda}x$$

$$+ \frac{e}{\lambda^2}\ln(\lambda x(t) + \mu) = \epsilon t + \eta w(t) + G(c).$$

To solve the IVP, we need to find $V(x(t)) - V(x_0)$. Thus,

$$V(x(t)) - V(x_0) = \frac{e}{\lambda^2}\ln\left[\frac{(\lambda x(t) + \mu)}{\lambda x_0 + \mu}\right] + \frac{\nu}{\lambda}(x - x_0).$$

Moreover,

$$V(x(t)) - V(x_0) = \epsilon(t - t_0) + \eta(w(t) - w(t_0)).$$

Thus,

$$\frac{e}{\lambda^2}\ln\left[\frac{(\lambda x(t) + \mu)}{\lambda x_0 + \mu}\right] + \frac{\nu}{\lambda}(x - x_0) = \epsilon(t - t_0) + \eta(w(t) - w(t_0)).$$

Hence,

$$\frac{e}{\lambda^2}\ln\left[\frac{(\lambda x(t) + \mu)}{\lambda x_0 + \mu}\right] = \epsilon(t - t_0) + \eta(w(t) - w(t_0)) - \frac{\nu}{\lambda}(x - x_0).$$

This implies (by properties of ln and algebraic simplification) that

$$\lambda x(t) + \mu = (\lambda x_0 + \mu)\exp\left[\frac{\epsilon\lambda^2}{e}(t - t_0) + \frac{\eta\lambda^2}{e}(w(t) - w(t_0)) - \frac{\lambda\nu}{e}(x - x_0)\right].$$

Thus,

$$x(t) = \frac{1}{\lambda}\left[(\lambda x_0 + \mu)\exp\left[\frac{\epsilon\lambda^2}{e}(t - t_0) + \frac{\eta\lambda^2}{e}(w(t) - w(t_0)) - \frac{\lambda\nu}{e}(x - x_0)\right] - \mu\right].$$

This is the solution to the IVP in the implicit form. In particular, for $x_0 = 0$, we have

$$x(t) = \frac{\mu}{\lambda} \exp\left[\frac{\epsilon\lambda^2}{e}(t - t_0) + \frac{\eta\lambda^2}{e}(w(t) - w(t_0)) - \frac{\lambda\nu}{e}x\right] - \frac{\mu}{\lambda}.$$

This completes the solution process.

Example 3.5.3 (Michaelis–Menten-type stochastic model). From Illustration 3.1.2, let $x = [S]$ be a concentration of a reactant (substrate). It decomposes under the presence of an enzyme E. Let x_0 be the initial concentration of the substrate at the initial time $t = t_0$. In addition, let the $[E^*]$ be the total enzyme concentration in the steady state. From Illustration 3.5.1 and Exercise 7(c) (Section 3.1), we have

$$dx = \frac{\lambda x}{\psi + x}\left[-1 + \frac{e\eta^2}{(\psi + x)^2}\right]dt + \eta\frac{\lambda x}{\psi + x}dw(t), \quad x(t_0) = x_0,$$

where λ, η, and ψ are as defined in Illustration 3.5.2; and $e = \lambda\psi$. Find the solution to the IVP and in particular, for $x_0 = 0$.

Solution procedure. From direct usage of the solution process and the solution representation of Illustration 3.5.2, we have

$$x(t) = x_0 \exp\left[-\frac{\lambda}{\psi}(t - t_0) + \frac{\eta\lambda}{\psi}(w(t) - w(t_0)) - \frac{1}{\psi}(x - x_0)\right].$$

This is the solution to the IVP.

Example 3.5.4. Given: $dx = 18x\,dt + 3\sqrt{4x^2 + 25}\,dw(t)$. Find the general solution to the given differential equation.

Solution procedure. From Observation 3.5.1, we conclude that this differential equation belongs to the class ISDECC-(f, σ). Its general solution is given by

$$G(x(t)) = \int_c^{x(t)} \frac{dy}{\sqrt{4y^2 + 25}} = \frac{1}{2}\ln\left(2x + \sqrt{4x^2 + 25}\right) + G(c) = 3(w(t) - w(a)).$$

This implies that

$$\ln\left(2x + \sqrt{4x^2 + 25}\right) = 6(w(t) - w(a)) + C,$$

$$2x(t) + \sqrt{4x^2 + 25} = \exp[6(w(t) - w(a)) + C]$$

$$= \exp[6(w(t) - w(a))]C.$$

This is the general solution to the given SDE.

Illustration 3.5.3 (stochastic version of Torricelli's law). Let us examine Illustration 3.6.2 [80]. We assume that the area around the bottom hole of the tank is subject to thermal noise random perturbations. Because of this, the liquid flow rate is under the influence of the colored noise perturbations. We recall the deterministic mathematical model of the leakage of the liquid [80]:

$$\rho \frac{d}{dt} V(t) = -a\sqrt{2\rho gh}, \quad A(h)\frac{dh}{dt} = -k\sqrt{h}, \quad h(0) = h_0,$$

where $k = a\sqrt{\frac{2g}{\rho}}$ is replaced by

$$dh = -k\frac{h^{1/2}}{A(h)} dt$$

This is a stochastic version of Torricelli's law and

$$\frac{dh}{dt} = \left[-k\frac{h^{1/2}}{A(h)} + \frac{\sigma^2(A(h) - 2h\frac{dA}{dt}(h))}{4A^3(h)} \right] dt + \sigma \frac{h^{1/2}}{A(h)} dw(t), \quad h(0) = h_0,$$

respectively. By following Procedure 3.6.2 [80], we can conclude that the general solution to the SDE is given by

$$V(h(t)) - V(h_0) = -kt + \sigma w(t),$$

where

$$V(h(t)) = G(h(t)) = \int_{h_0}^{h(t)} \frac{A(s)}{\sqrt{s}} ds + G(h_0).$$

If G satisfies the hypothesis (H$_{3.5}$), the solution process is given by $h(t) = G^{-1}(-kt + \sigma w(t) + V(h_0))$. This completes the solution process of the illustration.

Example 3.5.5. Suppose that a spherical shape tank with diameter $4\,\text{m}$ has a circular hole with radius $1\,\text{cm}$. Let $g = 10\,\text{m/s}^2$. The tank is half-full of water, and it is under colored noise perturbations.

(a) Show that the IVP associated with the example is

$$dh = \left[-0.0001\frac{\sqrt{20h}}{A(h)} + \frac{\sigma^2(A(h) - 2h\frac{dA}{dt}(h))}{4A^3(h)\sqrt{h}} \right] \pi dt$$

$$+ 0.000\sigma\pi\frac{\sqrt{20h}}{A(h)} dw(t), \quad h(0) = 2.$$

(b) Find the solution representation of the IVP in (a).
(c) How long will it take to drain the water completely?
(d) Find the mean and the variance of the time in (c).

Solution procedure. Let $h(t)$ be the height of the water in the tank. The area of the hole is 0.0001 m^2. From Illustration 3.6.2 [80] with $A(h) = \pi r^2 = \pi[4-(2-h)^2] = \pi(4h - h^2)$, the solution to the IVP is given by

$$G(h(t)) = \int_2^h \left[\frac{(4s - s^2)}{\sqrt{s}}\right] ds = \int_2^h (4s^{1/2} - s^{3/2})ds = \left(\frac{8}{3}s^{3/2} - \frac{2}{5}s^{5/2}\right)\Big|_2^h$$

$$= \left(\frac{8}{3}h(t)^{3/2} - \frac{2}{5}h(t)^{5/2}\right) - \left(\frac{16}{3}\sqrt{2} - \frac{8}{5}\sqrt{2}\right)$$

$$= \left(\frac{8}{3}h(t)^{3/2} - \frac{2}{5}h(t)^{5/2}\right) - \frac{56}{15}\sqrt{2}$$

$$= -0.0004\sqrt{5}t + 0.000\sigma w(t).$$

Hence, the solution representation is given in the following implicit form:

$$\left(\frac{8}{3}h(t)^{3/2} - \frac{2}{5}h(t)^{5/2}\right) - \frac{56}{15}\sqrt{2} = -0.0004\sqrt{5}t + 0.000\sigma w(t).$$

The height of the water tank at any time is given by the above implicit expression of h. The tank will be empty when $h(t) = 0$. From this, the solution expression reduces to

$$0.0004\sqrt{5}\,t = \frac{56}{15}\sqrt{2} + 0.000\sigma w(t).$$

Now, we solve for t, and get

$$t = \frac{\frac{56}{15}\sqrt{2} + 0.000\sigma w(t)}{0.0004\sqrt{5}} = 10000\frac{14}{75}\sqrt{10} + \frac{\sigma\sqrt{5}}{20}w(t).$$

This is the desired time to empty the tank. This completes the solution process of the problem. The mean and the variance of the time to empty the tank are $10000\frac{14}{75}\sqrt{10}$ and $\frac{\sigma^2}{80}$.

Brief Summary. A summary of the basic mechanical steps needed to solve nonlinear stochastic scalar differential equations is presented in Figure 3.5.4 flowchart:

Exercises

Determine if the given differential equation is separable. If so, then show that it belongs to the appropriate class LSDECC-(f, σ), and find the general solution to the given differential equation:

1. (a) $dx = (1+t)^2 a^{-x}\left[-\frac{\ln a}{2}a^{-x} + 2\right] dt + (1+t)a^{-x}dw(t)$, $a > 0$;

   (b) $dx = (1+t)^2 a^{-x}\left[-\frac{(\ln a)}{2}a^{-x} + 1\right] dt + (1+t)a^{-x}dw(t)$, $x(t_0) = x_0$, $a > 0$.

2. (a) $dx = \sec t\left[\frac{b}{4}\sec t + 2(a + bx) + \sqrt{a + bx}\right] dt + \sec t\sqrt{a + bx}\, dw(t)$, $b \neq 0$;

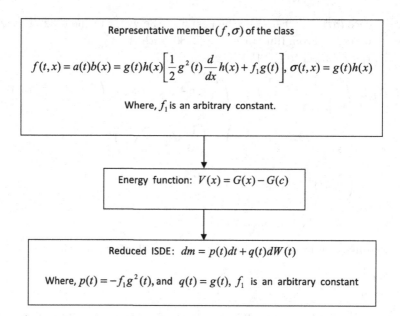

Fig. 3.5.4   Summary of the solution process.

(b) $dx = \sec t \left[\frac{b}{4}\sec t + 2(a + bx)\right] dt + \sec t\sqrt{a + bx}\, dw(t),\ b \neq 0;$

(c) $dx = \sec t \left[\frac{b}{4}\sec t + \sqrt{a + bx}\right] dt + \sec t\sqrt{a + bx}\, dw(t),\ x(t_0) = x_0,\ b \neq 0;$

3. (a) $dx = e^t \left[-\frac{(2a+bx)e^t}{4x^3} + \frac{\sqrt{a+bx}}{x}\right] dt + e^t \frac{\sqrt{a+bx}}{x}\, dw(t),\ x(t_0) = x_0,\ b \neq 0;$

   (b) $dx = -\frac{(2a+bx)e^{2t}}{4x^3} dt + e^t \frac{\sqrt{a+bx}}{x}\, dw(t),\ x(t_0) = x_0,\ b \neq 0.$

4. (a) $dx = \frac{\sqrt{(a^2-x^2)^3}}{x} \left[-\frac{(a^2+2x^2)(a^2-x^2)}{2x^2\sqrt{a^2-x^2}} + 1\right] dt + \frac{\sqrt{(a^2-x^2)^3}}{x}\, dw(t);$

   (b) $dx = \frac{\sqrt{(a^2-x^2)^3}}{x} \left[-\frac{(a^2+2x^2)(a^2-x^2)}{2x^2\sqrt{a^2-x^2}}\right] dt + \frac{\sqrt{(a^2-x^2)^3}}{x}\, dw(t),\ x(t_0) = x_0.$

5. (a) $dx = (1+t)^2 \left[-\frac{\ln 3}{2}3^{-2x} + 23^{-x}\right] dt + (1+t)3^{-x}\, dw(t);$

   (b) $dx = [1 + 2(1 + 2x)]dt + \sqrt{1 + 2x}\, dw(t);$

   (c) $dx = \left[-\frac{8+3x}{4x^3} + 2(4 - 3x)(4 + 3x) + \frac{\sqrt{4+3x}}{x}\right] dt + \frac{\sqrt{4+3x}}{x}\, dw(t),\ b \neq 0;$

   (d) $dx = \frac{\sqrt{(1-x^2)^3}}{x} \left[-\frac{(1+3x^2)(1-x^2)}{2x^2\sqrt{1-x^2}} + 1\right] dt + \frac{\sqrt{(1-x^2)^3}}{x}\, dw(t).$

6. (a) $dx = (1+t)^2 \left[-\frac{\ln 3}{2}3^{-2x} + \ln 3\right] dt + (1+t)3^{-x}\, dw(t),\ x(0) = 1;$

   (b) $dx = \left[\frac{1}{2} + \sqrt{1 + 2x}\right] dt + \sqrt{1 + 2x}\, dw(t),\ x(0) = 0;$

   (c) $dx = \left[-\frac{8+3x}{x^3} + \frac{\sqrt{4+3x}}{x}\right] dt + \frac{\sqrt{4+3x}}{x}\, dw(t),\ x(0) = \frac{5}{3};$

   (d) $dx = \frac{\sqrt{(1-x^2)^3}}{x} \left[-\frac{(1+2x^2)(1-x^2)}{2x^2\sqrt{1-x^2}} + 1\right] dt + \frac{\sqrt{(1-x^2)^3}}{x}\, dw(t),\ x(0) = 0.$

7. (a) $dx = -\frac{\ln 3}{2}(1+t)^2 3^{-2x} dt + (1+t)3^{-x} dw(t)$, $x(0) = 1$;

   (b) $dx = \frac{1}{2}dt + \sqrt{1+2x}\, dw(t)$, $x(0) = 0$;

   (c) $dx = -\frac{(8+3x)}{x^3} dt + \frac{\sqrt{4+3x}}{x}\, dw(t)$, $x(0) = \frac{5}{3}$;

   (d) $dx = -\frac{(1+2x^2)(1-x^2)}{2x^3} dt + \frac{\sqrt{(1-x^2)^3}}{x}\, dw(t)$, $x(0) = 0$.

8. Find the solution process of the following IVPs:

   (a) (3.1.47);
   (b) Example 3.1.4;
   (c) (3.1.57);
   (d) (3.1.58);
   (e) (3.1.59);
   (f) (3.1.61).

3.6 Homogeneous Equations

In this section, we present a subclass of differential equations (3.2.1) that is reducible to (3.4.1)/(3.3.1). This class of differential equations is referred to as Itô–Doob-type stochastic homogeneous differential equations. Prior to introducing the definition of a homogeneous differential equation, we need a mathematical concept of a homogeneous function of degree n, where n is a real number.

Definition 3.6.1. Let $F(t, x)$ be defined on $J \times R$ into R. A function F is said to be *homogeneous in (t, x) with degree n* if for $t \neq 0$ and $x = tv$, $F(t, tv) = t^n F(1, v)$, for some real number n. In particular, for $n = 0$, F is simply called a *homogeneous function*.

Definition 3.6.2. A differential equation (3.2.1) is said to be an Itô–Doob-type stochastic *homogeneous differential equation* if the rate functions $f(t, x)$ and $\sigma(t, x)$ in (3.2.1) are homogeneous functions of degree 0, i.e. $f(t, x)$ and $\sigma(t, x)$ are homogeneous functions in (t, x).

3.6.1 *General Problem*

We imitate the Procedure 3.4.2 with regard to the homogeneous stochastic differential equation of the Itô–Doob-type (3.2.1), i.e.

$$dx = f(t, x)dt + \sqrt{t}\sigma(t, x)dw(t), \tag{3.6.1}$$

where f and σ are defined in (3.2.1), and $t \neq 0$. This procedure leads to the reduction of the differential equation (3.6.1) to (3.4.1).

(H$_{3.6}$). In addition to hypothesis (H$_{3.2}$), we assume that the rate functions $f(t, x)$ and $\sigma(t, x)$ in (3.2.1) are homogeneous functions of degree 0; $\sigma(1, v) \neq 0$. Moreover, $H(x) = \int^x \frac{dy}{\sigma(1,y)}$ is defined and invertible.

3.6.2 *Procedure for Finding a General Solution Representation*

The procedure described in Section 3.4.2 for reducing a subclass of a class of stochastic differential equations (3.2.1) to (3.4.1) is repeated [Observation 3.4.1(i)]. With this subclass of Itô–Doob-type stochastic homogeneous differential equations, we associate a suitable natural energy/Lyapunov function in a unified and coherent way.

Step #1. Using the nature of the subclass of differential equations described by (3.6.1), we associate an unknown energy/Lyapunov function in Step #1 of Procedure 3.2.2 as

$$V(t, x) = G\left(\frac{x}{t}\right), \quad \text{for } t \neq 0 \tag{3.6.2}$$

where G is an unknown energy/Lyapunov function. The problem of seeking energy function V is equivalent to the problem of seeking unknown function G.

Step #2. The computation of the differential $dV(t, x)$ of V in (3.2.2) is achieved by using the following substitution:

$$v = \frac{x}{t}. \tag{3.6.3}$$

We obtain

$$\frac{\partial}{\partial t} V(t, x) = \frac{\partial}{\partial t} G\left(\frac{x}{t}\right) \qquad \text{(by definition)}$$

$$= \frac{d}{dv} G(v) \frac{\partial}{\partial t} v \qquad \text{(by the chain rule)}$$

$$= \frac{d}{dv} G(v) \left(-\frac{x}{t^2}\right) \quad \left[\text{from (3.6.3)}, \frac{\partial}{\partial t} v = \frac{\partial}{\partial t}\left(\frac{x}{t}\right) = -\frac{x}{t^2}\right], \tag{3.6.4}$$

$$\frac{\partial}{\partial x} V(t, x) = \frac{\partial}{\partial x} G\left(\frac{x}{t}\right) \qquad \text{(by definition)}$$

$$= \frac{d}{dv} G(v) \frac{\partial}{\partial x} v \qquad \text{(by the chain rule)}$$

$$= \frac{d}{dv} G(v) \left(\frac{1}{t}\right) \quad \left[\text{from (3.6.3)}, \frac{\partial}{\partial x} v = \frac{\partial}{\partial x}\left(\frac{x}{t}\right) = \frac{1}{t}\right], \tag{3.6.5}$$

$$\frac{\partial^2}{\partial x^2} V(t, x) = \frac{\partial^2}{\partial x^2} G\left(\frac{x}{t}\right) = \frac{\partial}{\partial x}\left(\frac{\partial}{\partial x} G\left(\frac{x}{t}\right)\right)$$

$$\text{(by definition)}$$

$$= \frac{d^2}{dv^2} G(v) \frac{1}{t} \frac{\partial}{\partial x} v \qquad \text{(by the chain rule)}$$

$$= \frac{d^2}{dv^2} G(v) \left(\frac{1}{t^2}\right) \quad \left[\text{from (3.6.5)}, \frac{\partial}{\partial x} v = \frac{\partial}{\partial x}\left(\frac{x}{t}\right) = \frac{1}{t}\right]. \tag{3.6.6}$$

Hence, from (3.2.2), (3.2.3), (3.6.1)–(3.6.6), we have

$$dV(t,x) = dG\left(\frac{x}{t}\right) \quad \text{(by definition)}$$

$$= LG(v)dt + \sqrt{t}\sigma(t,x)\frac{d}{dv}G(v)\left(\frac{1}{t}\right)dw(t)$$
[by (3.6.1), (3.2.2), and (3.6.5)]

$$= LG(v)dt + \sqrt{t}\sigma(t,tv)\frac{d}{dv}G(v)\left(\frac{1}{t}\right)dw(t) \quad \text{(by substitution)}$$

$$= LG(v)dt + \left[\frac{\sigma(1,v)}{\sqrt{t}}\right]\frac{d}{dv}G(v)dw(t) \quad \text{(by homogeneity of } f \text{ and } \sigma),$$

$$(3.6.7)$$

where w is the Wiener process defined in Definition 1.2.12, $x(t)$ is determined by (3.6.1), and L in (3.2.3) is a linear operator associated with (3.6.1)–(3.6.4), (3.6.5), and (3.6.6). That is,

$$LV(t,x) = \frac{d}{dv}G(v)\left(-\frac{x}{t^2}\right) + f(t,x)\frac{d}{dv}G(v)\left(\frac{1}{t}\right)$$

$$+ \frac{t}{2}\sigma^2(t,x)\frac{d^2}{dv^2}G(v)\left(\frac{1}{t^2}\right) \quad \text{[from (3.6.3)]}$$

$$= \frac{d}{dv}G(v)\left(-\frac{tv}{t^2}\right) + f(t,tv)\frac{d}{dv}G(v)\left(\frac{1}{t}\right)$$

$$+ \frac{t}{2}\sigma^2(t,tv)\frac{d^2}{dv^2}G(v)\left(\frac{1}{t^2}\right) \quad \text{[from (3.6.3)]}$$

$$= \frac{d}{dv}G(v)\left(-\frac{tv}{t^2}\right) + f(t,tv)\frac{d}{dv}G(v)\left(\frac{1}{t}\right)$$

$$+ \frac{t}{2}\sigma^2(t,tv)\frac{d^2}{dv^2}G(v)\left(\frac{1}{t^2}\right) \quad \text{(by homogeneity of } f \text{ and } \sigma)$$

$$= \frac{d}{dv}G(v)\left(-\frac{v}{t}\right) + f(1,v)\frac{d}{dv}G(v)\left(\frac{1}{t}\right)$$

$$+ \frac{1}{2}\sigma^2(1,v)\frac{d^2}{dv^2}G(v)\left(\frac{1}{t}\right) \quad \text{(by regrouping).}$$

Thus, from this and (3.6.2),

$$LG(v) = \left[\frac{f(1,v)-v}{t}\right]\frac{d}{dv}G(v) + \frac{1}{2t}\sigma^2(1,v)\frac{d^2}{dv^2}G(v). \quad (3.6.8)$$

In this case, from (3.6.2), (3.6.3), (3.6.7), and (3.6.8), the expressions (3.4.2) and (3.4.3) reduce to

$$\mu(t)G(v) + p(t) = \left[\frac{f(1,v) - v}{t} \right] \frac{d}{dv}G(v) + \frac{1}{2t}\sigma^2(1,v)\frac{d^2}{dv^2}G(v), \quad (3.6.9)$$

and

$$\nu(t)G(v) + q(t) = \left[\frac{\sigma(1,v)}{\sqrt{t}} \right] \frac{d}{dv}G(v), \quad (3.6.10)$$

respectively.

Step #3. Now, by using (3.6.9), we find a candidate for function $G(v)$. For this purpose, from (3.6.9), it is clear that $\frac{\sigma(1,v)}{\sqrt{t}} = g(t)h(v)$, where $g(t) = \frac{1}{\sqrt{t}}$ and $h(v) = \sigma(1,v)$; by knowing this structure, we choose q, ν, μ, and p in (3.4.1) as follows: $q(t) = \gamma g(t)$, $\nu(t) = \delta g(t)$, $t\mu(t) = \epsilon$, and $tp(t) = \eta$, where δ, γ, ϵ, and η are arbitrary constants to be determined; from the hypothesis (H$_{3.6}$), we have $\sigma(1,v) \neq 0$ on $J \times R$. Therefore, Step #2 of Procedure 3.4.2 is satisfied. Thus, from (3.6.10), we have

$$\frac{d}{dv}G(v) = \frac{\sqrt{t}\,[\nu(t)G(v) + q(t)]}{\sigma(1,v)} = \frac{\delta\sqrt{t}G(v)}{\sqrt{t}\sigma(1,v)} + \frac{\sqrt{t}\gamma}{\sqrt{t}\sigma(1,v)} = \frac{\delta G(v)}{\sigma(1,v)} + \frac{\gamma}{\sigma(1,v)}.$$
$$(3.6.11)$$

Hence, (3.4.5), (3.4.10), (3.4.11), and (3.4.17) reduce to

$$G(v) = \exp[\delta H(\nu)]\,C - \frac{\gamma}{\delta}, \quad \text{(from H3.6)} \quad (3.6.12)$$

$$\left[\delta \left[\frac{f(1,v) - v}{\sigma(1,v)} - \frac{1}{2}\frac{d}{dv}\sigma(1,v) \right] + \frac{1}{2}\delta^2 - \varepsilon \right] \exp\left[\int_c^v \frac{\delta}{\sigma(1,y)}dy \right] = \frac{\delta\eta - \epsilon\gamma}{\delta C},$$
$$(3.6.13)$$

$$z(v) = \frac{f(1,v) - v}{\sigma(1,v)} - \frac{1}{2}\frac{d}{dv}\sigma(1,v) \quad (3.6.14)$$

and, for $\frac{d}{dv}z(v) \neq 0$,

$$\frac{d}{dv}\left[\frac{\frac{d}{dv}\left(\sigma(1,v)\frac{d}{dv}z(v) \right)}{\frac{d}{dv}z(v)} \right] = 0, \quad (3.6.15)$$

respectively, where $C \neq 0$.

Step #4. Finally, we repeat Steps #3 and #4 of Procedure 3.4.2. This completes the procedure for solving the homogeneous stochastic differential equation of the Itô–Doob-type (3.6.1).

Observation 3.6.1

(i) In (3.6.11), if $\delta = 0$ (Observation 3.4.2(i)), then (3.6.11), (3.6.12), (3.6.13), and (3.6.15) reduce to

$$\frac{d}{dv}G(v) = \frac{\gamma}{\sigma(1,v)}, \tag{3.6.16}$$

$$G(v) = \gamma \int_c^x \frac{dy}{\sigma(1,y)} + G(c), \tag{3.6.17}$$

$$\gamma \left[\frac{f(1,v) - v}{\sigma(1,v)} - \frac{1}{2}\frac{d}{dv}\sigma(1,v) \right] = t\mu(t)G(v) + tp(t) = \gamma \epsilon G(v) + \eta + \epsilon G(c), \tag{3.6.18}$$

and

$$\frac{d}{dv}\left(\sigma(1,v)\frac{d}{dv}(z(v)) \right) = 0, \tag{3.6.19}$$

respectively, where γ, ϵ, η, and $G(c)$ are arbitrary constants.

(ii) We further observe that under the transformation $x = tv$, the differential equation (3.6.1) reduces to its equivalent form as follows:

$$dx = f(t,x)dt + \sqrt{t}\sigma(t,x)dw(t) = v\,dt + t\,dv,$$

$$dx = f(t,tv)dt + \sqrt{t}\sigma(t,tv)dw(t) = v\,dt + t\,dv,$$

$$dx = f(1,v)dt + \sqrt{t}\sigma(1,v)dw(t) = v\,dt + t\,dv,$$

and hence

$$dv = \frac{f(1,v) - v}{t}dt + \frac{\sigma(1,v)}{\sqrt{t}}dw(t). \tag{3.6.20}$$

Thus, the homogeneous SDE (3.6.1) is reduced to the separable SDE (3.5.1).

(iii) From Observation 3.4.1(ii), for $\delta \neq 0$, a representative of this subclass is determined by

$$\frac{f(1,v) - v}{t} = \frac{1}{t}\sigma(1,v)\left[\frac{1}{2}\frac{d}{dv}\sigma(1,v) - \frac{f_1}{\delta}\exp\left[-\delta\int_c^v \frac{dy}{\sigma(1,y)} \right] + f_2 \right]. \tag{3.6.21}$$

Now, from (3.4.12) and (3.4.37), $\mu(t)$ and $p(t)$ in (3.4.1) are as follows:

$$\mu(t) = \frac{1}{t}\left[\frac{1}{2}\delta^2 + \delta f_2 \right] \quad \text{and} \quad p(t) = \frac{1}{t}\left[\frac{1}{2}\delta\gamma + \gamma f_2 - Cf_1 \right], \tag{3.6.22}$$

where δ, f_1, and f_2 are arbitrary constants of integration.

In the case of $\delta = 0$ and from Observation 3.4.2(ii), we have

$$\frac{f(1,v) - v}{t} = \frac{1}{t}\sigma(1,v)\left[\frac{1}{2}\frac{d}{dv}\sigma(1,v) + f_1 \int_c^v \frac{dy}{\sigma(1,y)} + f_2\right], \qquad (3.6.23)$$

and from (3.4.49), $\mu(t)$ and $p(t)$ in (3.4.42) reduce to

$$\mu(t) = \frac{1}{t}f_1, \quad p(t) = \frac{1}{t}f_2. \qquad (3.6.24)$$

(iv) The relationship expressed in (i) leads to the enlargement of the separable SDE determined in Observation 3.5.1. We recall that $q(t) = g(t)$ and $G(t) = \int^x \frac{dy}{h(y)}$. We start with the unknown energy function $V(x)$, apply the argument used in Steps #1–#3 of Procedure 3.2.2, and arrive at

$$LV(x) = a(t)b(x)\frac{d}{dx}V(x) + \frac{1}{2}g^2(t)h^2(x)\frac{d^2}{dx^2}V(x), \qquad (3.6.25)$$

$$\mu(t)V(x) + p(t) = a(t)b(x)\frac{d}{dx}V(x) + \frac{1}{2}g^2(t)h^2(x)\frac{d^2}{dx^2}V(x), \qquad (3.6.26)$$

and

$$\nu(t)V(x) + q(t) = g(t)h(x)\frac{d}{dx}V(x). \qquad (3.6.27)$$

For $q(t) = g(t) = \nu(t)$, V is determined as in (3.4.5):

$$V(x) = \exp\left[\int_c^x \frac{\delta}{h(y)}dy\right]C - \frac{\gamma}{\delta},$$
$$= \exp[\delta H(\nu)]C - \frac{\gamma}{\delta}, \quad \text{(from H3.6)} \qquad (3.6.28)$$

where δ, γ, and $C \neq 0$ are arbitrary constants and are determined by diffusion rate function σ. Using (3.6.26) and (3.6.28) and repeating the argument in Procedure 3.4.2, we arrive at (3.4.12):

$$z(t,x) = \frac{a(t)b(x)}{h(x)} - \frac{1}{2}g^2(t)\frac{d}{dx}h(x), \qquad (3.6.29)$$

$$\left[p(t) = \frac{\mu(t)\gamma}{\delta}\right] = \left[C\left[\delta z(t,x) + \frac{1}{2}g^2(t)\delta^2 - \mu(t)\right] + \frac{d}{dx}C\right]\exp\left[\int_c^x \frac{\delta}{h(s)}ds\right]. \qquad (3.6.30)$$

Brief Summary. The following flowchart, Fig. 3.6.1, summarizes the steps of finding the general class of Itô–Doob stochastic homogeneous stochastic nonlinear scalar differential equations via the energy function method.

In the following, we present a few illustrations to exhibit the procedure described in this section.

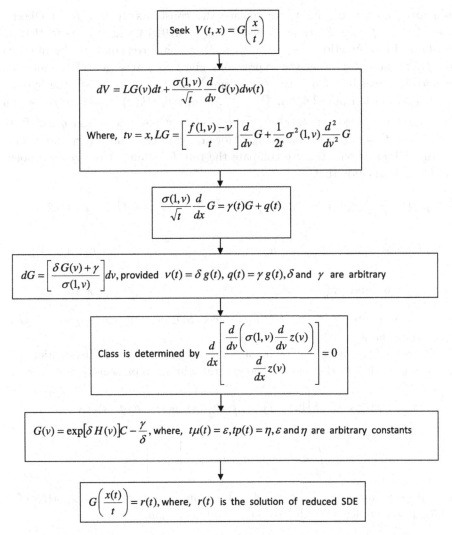

Fig. 3.6.1   Homogeneous equations and energy function method.

Illustration 3.6.1. Given:

(a) $dv = \frac{1}{t}\sigma(1,v)\left[\frac{1}{2}\frac{d}{dv}\sigma(1,v) - \frac{f}{\delta}\exp\left[-\delta\int_c^v\frac{ds}{\sigma(1,s)}\right] + F\right]dt + \frac{\sigma(1,v)}{\sqrt{t}}dw(t),\ \delta \neq 0;$

(b) $dx = \frac{1}{t}\sigma(1,v)\left[\frac{1}{2}\frac{d}{dv}\sigma(1,v) + f\int_c^v\frac{ds}{\sigma(1,s)} + F\right]dt + \frac{\sigma(1,v)}{\sqrt{t}}dw(t),\ \delta = 0;$

(c) $dv = \frac{1}{t}\sigma(1,v)\left[\frac{1}{2}\frac{d}{dx}\sigma(1,v) + F\right] + \frac{\sigma(1,v)}{\sqrt{t}}dw(t),\ \delta = 0;$

where $v = tx$, σ satisfies the hypothesis (H$_{3.6}$), and f and F are given constants.
Find the general solution to the given differential equations.

Solution procedure. First, we replace the constants f_1 and f_2 in Observation 3.6.1 by f and F, respectively. From (3.6.20)–(3.6.22), we note that the transformed rate function $\frac{f(1,v)-v}{t}$ in (3.6.20), which corresponds to the rate function $f(t,x)$ in (3.6.1), has the expressions given in Parts (a)–(c). As noted in Step #3 of Procedure 3.6.2, the transformed rate functions satisfy the hypothesis $(H_{3.4})$. In fact, from (3.6.20), $\frac{\sigma(1,v)}{\sqrt{t}} \equiv \Lambda(t,v) = g(t)h(v)$, where $g(t) = \frac{1}{\sqrt{t}}$ and $h(v) = \sigma(1,v)$. Moreover, $G(x) = \int_c^x \frac{dy}{h(y)}$ is invertible, and ν and q satisfy the conditions $q(t) = \gamma g(t)$ and $\nu(t) = \delta g(t)$ for arbitrary real numbers γ and δ. Then, by using Observation 3.6.1, we compare the rate functions of the given problems (a)–(c), and conclude that

$$\mu(t) = \frac{1}{t}\left[\frac{1}{2}\delta^2 + \delta F\right] \quad \text{and} \quad p(t) = \frac{1}{t}\left[\frac{1}{2}\delta + \gamma F - Cf\right] \quad \text{[by (3.6.22)]},$$

$$\mu(t) = \frac{1}{t}f \quad \text{and} \quad p(t) = \frac{1}{t}F \quad\quad\quad \text{[by (3.6.24)]},$$

$$\mu(t) = 0 \quad \text{and} \quad p(t) = \frac{1}{t}F \quad\quad\quad \text{[by (3.6.24)]}$$

are the choice rate functions of the reduced SDE (3.4.1) corresponding to the differential equations (a)–(c), respectively.

The rest of the argument is similar to the argument used in Illustration 3.4.1. The details are left to the reader. The general solution representation is given by

$$x(t) = tG^{-1}\left[\ln\left[\Phi(t)c + \int_a^t \Phi(t,s)\left[p(s) - \delta\gamma g^2(s)\right]ds\right.\right.$$

$$\left.\left. + \int_a^t \Phi(t,s)\gamma g(s)dw(s) + \frac{\gamma}{\delta}\right] + C\right],$$

where μ, p, ν, and q are as specified in Procedure 3.6.2 and $C = \exp[\delta H(c)]C$ in (3.4.5). This completes the discussion of the illustration.

Illustration 3.6.2 (stochastic allometric law). In the study of relative growth-allometry [59, 187], it has been well recognized that the relationship between the pair of objects is stochastic in nature. In fact, by using regression analysis, an attempt has been made to incorporate the randomness character in a biometric point of view.

We extend the deterministic allometric law described in Illustration 3.7.1 [80] to the stochastic framework. We assume that all of the assumptions CGM1–CGM4 of Illustration 3.7.1 remain valid, except that we decompose the effects that are characterized by an environmental common factor E. These effects are incorporated through the specific growth rates. We assume that the specific rate parameters are subject to random environmental perturbations and the density-dependent effects due to the organ–body interactions on the dynamic (Illustration 3.2.3 [80]).

Under these considerations, the allometric dynamic model (Illustration 3.7.1 [80]) reduces to

$$dy = \bar{\alpha}\frac{y}{x}dx + \sigma(x,y)dw(x).$$

Here, $C(x,y) = \kappa\frac{y}{x}$ and the parameter κ is considered to be under random perturbations; $\kappa = \bar{\alpha} + \xi(x)$ and $\sigma(x,y) = \nu\frac{y}{\sqrt{x}}$; x and y are sizes/weights of the body and an organ; and $\nu\sqrt{x}\,dw(x) = \xi(x)dx$ characterizes the effects of random fluctuations with $E[\xi(x) = 0]$. $\xi(x)$ is the stationary Gaussian white noise process with independent increments as defined in (3.1.12).

We note that this is an Itô–Doob-type stochastic homogeneous differential equation. By following the solution procedure and using the transformation $y = xv$, we have

$$dv = \frac{v(\bar{\alpha}-1)}{x}dx + \frac{\nu v}{\sqrt{x}}dw(x).$$

By using the method for finding the solution procedure in Section 2.3 of Chapter 2, we obtain

$$v = \exp\left[\left(\bar{\alpha}-1-\frac{1}{2}\nu2\right)(\ln x - \ln c) + \nu\int_c^x \frac{1}{\sqrt{s}}dw(s)\right]C.$$

It is written as

$$y(x) = xv = x\exp\left[\left(\bar{\alpha}-1-\frac{1}{2}\nu2\right)\ln x + \nu\int_c^x \frac{1}{\sqrt{s}}dw(s)\right]C$$

$$= x^{(\bar{\alpha}-\frac{1}{2}\nu^2)}\exp\left[\nu\int_c^x \frac{1}{\sqrt{s}}dw(s)\right]C.$$

This represents a pattern of growth of an organ relative to the entire body, whose specific rates are under random environmental perturbations. It is the stochastic analog of deterministic *allometry*.

Example 3.6.1. Given:

$$dx = \frac{11x^4 - 8x^2t^2 - 16t^4}{8x^3t}dt + \frac{4t^2 + x^2}{2x\sqrt{t}}dw(t).$$

(a) Verify that the given differential equation is homogeneous.
(b) Find the general solution to it.

Solution procedure. Goal: To verify that the given differential equation is homogeneous, and then find a general solution to it. We utilize the tips provided in Observation 3.3.2 to solve the problem. Without further details, we identify and write the SDE in a standard form. Here the pair of rate functions (f, σ) are defined by

$$dx = \frac{11x^4 - 8x^2t^2 - 16t^4}{8x^3t} dt + \sqrt{t}\frac{4t^2 + x^2}{2xt} dw(t) \quad \text{(standard form)},$$

where

$$f(t, x) = \frac{11x^4 - 8x^2t^2 - 16t^4}{8x^3t} \quad \text{and} \quad \sqrt{t}\sigma(t, x) = \frac{4t^2 + 3x^2}{2x\sqrt{t}} = \sqrt{t}\frac{4t^2 + 3x^2}{2xt}.$$

They are homogeneous functions of degree 0. In fact, using the transformation $x = vt$, we obtain

$$f(t, tv) = \frac{11(vt)^4 - 8(vt)^2t^2 - 16t^4}{8(vt)^3t} = \frac{11v^4t^4 - 8v^2t^4 - 16t^4}{8v^3t^4}$$

$$= \frac{11v^4 - 8v^2 - 16}{8v^3} = f(1, v),$$

and

$$\sigma(t, tv) = \frac{4t^2 + (tv)^2}{2(tv)t} = \frac{4t^2 + t^2v^2}{2vt^2} = \frac{4 + v^2}{2v} = \sigma(1, v).$$

This shows that the rate functions are homogeneous functions of degree 0. Therefore, the given differential equation is homogeneous. Now, we need to verify that the pair (f, σ) also belongs to the class LSDECC-(f, σ). For this purpose, we compute the terms $\frac{d}{dv}\sigma(1, v)$, $\int^v \frac{ds}{\sigma(1.s)}$, $\exp\left[-\delta\int_c^v \frac{ds}{\sigma(1,s)}\right]$ and the constants δ, f, F, and γ in the context of Illustrations 3.3.1 and 3.6.1 (if possible).

$$\frac{f(1, v) - v}{t} = \frac{\frac{11v^4 - 8v^2 - 16}{8v^3} - v}{t} = \frac{11v^4 - 8v^2 - 16 - 8v^4}{8tv^3}$$

$$= \frac{3v^4 - 8v^2 - 16}{8tv^3},$$

$$\frac{d}{dv}\sigma(1, v) = \frac{d}{dv}\left[\frac{4 + v^2}{2v}\right] = \frac{2v\frac{d}{dv}(4 + v^2) - (4 + v^2)\frac{d}{dv}(2v)}{(2v)^2}$$

$$= \frac{4v^2 - 8 - 2v^2}{4v^2} = \frac{2v^2 - 8}{4v^2} = \frac{v^2 - 4}{2v^2},$$

$$\int^v \frac{ds}{\sigma(1, s)} = \int^v \frac{2s\,ds}{4 + s^2} = \ln(4 + v^2)$$

$$\text{(by the choice of } C = 1\text{),}$$

$$\exp\left[-\delta\int^v \frac{ds}{\sigma(1, s)}\right] = \exp[-\delta\ln(4 + v^2)] = \exp\left[\ln(4 + v^2)^{-\delta}\right] = \frac{1}{(4 + v^2)^\delta}.$$

By using Illustration 3.6.1, in particular (3.6.21) and above computed components, we write the given drift function in its conceptual structural representation form

$$\frac{3v^4 - 8v^2 - 16}{8v^3} = f(1, v) - v \quad \text{(from the above computation)}$$

$$= \sigma(1, v)\left[\frac{1}{2}\frac{d}{dv}\sigma(1, v) - \frac{f(t)}{\delta}\exp\left[-\delta\int_c^v \frac{ds}{\sigma(1, v)}\right] + F(t)\right]$$

$$\text{[from (3.6.21)]}$$

$$= \frac{4 + v^2}{2v}\frac{1}{2}\frac{v^2 - 4}{2v^2} - \frac{f(t)}{\delta}\frac{4 + v^2}{2v}\frac{1}{(4 + v^2)^\delta} + F(t)\frac{4 + v^2}{2v}$$

$$\text{(by substitution)}$$

$$= \frac{v^4 - 16}{8v^3} - f(t)\frac{4 + v^2}{2\delta v}\frac{1}{(4 + v^2)^\delta} + F(t)\frac{4 + v^2}{2v}$$

$$\text{(by simplification)}$$

$$= \frac{v^4 - 16}{8v^3} + \left[F(t) - \frac{f(t)}{\delta}\frac{1}{(4 + v^2)^\delta}\right]\frac{4 + v^2}{2v}$$

$$\text{(by regrouping)}.$$

Since the expression $f(1, v) - v$ (left-hand side) is independent of t, the right-hand side expression must be independent of t for the arbitrary independent functions $f(t)$ and $F(t)$. This implies that $f(t)$ and $F(t)$ must be arbitrary constants; they are denoted by A and B, respectively. With this argument, the last expression in the above can be written as

$$\frac{4 + v^2}{2v}\left[B - \frac{A}{\delta}\frac{1}{(4 + v^2)^\delta}\right] = \frac{3v^4 - 8v^2 - 16}{8v^3} - \frac{v^4 - 16}{8v^3} \quad \text{(by simplifying)}$$

$$= \frac{3v^4 - 8v^2 - 16 - v^4 + 16}{8v^3} = \frac{2v^4 - 8v^2}{8v^3} = \frac{v^2 - 4}{4v} \quad \text{(by simplifying)}.$$

This implies that

$$B - \frac{A}{\delta}\frac{1}{(4 + v^2)^\delta} = \frac{v^2 - 4}{2(4 + v^2)} = \frac{v^2 + 4 - 8}{2(4 + v^2)} = \frac{1}{2} + \frac{-4}{4 + v^2}.$$

From the above expression, by comparing the coefficients of like terms (the method of undetermined coefficients), we conclude that

$$F(t) = B = \frac{1}{2}, \quad \frac{f(t)}{\delta} = \frac{A}{\delta} = 4, \quad \frac{1}{(4 + v^2)^\delta} = \frac{1}{4 + v^2}.$$

Thus, we finally have

$$\delta = 1, \quad A = 4 = f(t), \quad F(t) = B = \frac{1}{2}.$$

In summary, we have computed the unknown parameters and established the structural representations in the conceptual setting as follows:

$$f(1,v) - v = \sigma(1,v)\left[\frac{1}{2}\frac{d}{dv}\sigma(1,v) - \frac{f(t)}{\delta}\exp\left[-\delta\int_c^v\frac{ds}{\sigma(1,v)}\right] + F(t)\right]$$

$$= \frac{4+v^2}{2v}\left[\frac{v^2-4}{4v^2} - \frac{4}{4+v^2} + \frac{1}{2}\right] = \frac{3v^4 - 8v^2 - 16}{8v^3}.$$

This exhibits the validity of the representation of $f(1,v) - v$ in (3.6.21). Therefore, Illustration 3.6.1(a) is applicable. Hence, the given SDE belongs to the class LSDECC(f,σ). Moreover, the choice of rate functions of the reduced SDE (3.4.1) corresponding to this given SDE is $\delta = 1$, $\nu(t) = \frac{1}{\sqrt{t}} = g(t)$, $h(v) = \sigma(1,v)$, $C = 1$, $\gamma = 0$, $q(t) = 0$, $\epsilon = 1$, $\mu(t) = \frac{1}{t}$, $\eta = -4$, and $p(t) = \frac{-4}{t}$. In this case, the reduced SDE is

$$dm = \frac{1}{t}(m-4)dt + \frac{1}{\sqrt{t}}m\,dw(t),$$

and its general solution (Section 2.4) is

$$m(t) = \exp\left[\frac{1}{2}\ln t + \int_a^t\frac{1}{\sqrt{s}}dw(s)\right]c$$

$$- 4\exp\left[\frac{1}{2}\ln t + \int_a^t\frac{1}{\sqrt{s}}dw(s)\right]\int_a^t\Phi^{-1}(s)ds$$

$$= t^{1/2}\exp\left[\int_a^t\frac{1}{\sqrt{s}}dw(s)\right]c$$

$$- 4t^{1/2}\exp\left[\int_a^t\frac{1}{\sqrt{s}}dw(s)\right]\int_a^t s^{-1/2}\exp\left[-\int_a^s\frac{1}{\sqrt{u}}dw(u)\right]ds$$

$$= t^{1/2}\exp\left[\int_a^t\frac{1}{\sqrt{s}}dw(s)\right]\left[c - 4\int_a^t s^{-1/2}\exp\left[-\int_a^s\frac{1}{\sqrt{u}}dw(u)\right]ds\right].$$

From (3.4.5) and Step #4 of Procedure 3.4.2, the general solution to the given differential is

$$m(t) = V(t,v(t)) = \frac{4+c^2}{4+v^2(t)}$$

$$= t^{1/2}\exp\left[\int_a^t\frac{1}{\sqrt{s}}dw(s)\right]m(a) - 4\int_a^t t^{1/2}s^{-1/2}\exp\left[\int_s^t\frac{1}{\sqrt{u}}dw(u)\right]ds,$$

where $V(t, v(t)) = \exp\left[-\int_c^v \frac{\delta}{h(y)}dy\right] = \exp\left[-\int_c^v \frac{2ydy}{4+y^2}\right] = \frac{4+c^2}{4+v^2}$ and $v(t) = tx(t)$, and hence, one can solve for $x(t)$. The details are left to the reader as an algebraic problem. This completes the solution representation process of the given differential equation.

Example 3.6.2. Solve the IVP

$$dx = \frac{1}{2t}\left(x + \sqrt{t^2 - x^2}\right) dt + \sqrt{\frac{t^2 - x^2}{t}}\,dw(t), \quad x(t_0) = x_0, \text{ for } t \geq t_0 > 0.$$

Solution procedure. Goal: To find the solution to the given IVP. For this purpose, we imitate the argument of Example 3.6.1. First, we need to find the general solution. We observe that

$$dx = \frac{1}{2t}\left(x + \sqrt{t^2 - x^2}\right) dt + \sqrt{t}\sqrt{\frac{t^2 - x^2}{t^2}}\,dw(t),$$

and

$$f(t, x) = \frac{1}{2t}\left(x + \sqrt{t^2 - x^2}\right) = f(t, tv) = \frac{1}{2t}\left(tv + \sqrt{t^2 - t^2v^2}\right)$$

$$= \frac{1}{2}\left(v + \sqrt{1 - v^2}\right) = f(1, v),$$

$$\sqrt{t}\sigma(t, x) = \sqrt{\frac{t^2 - x^2}{t}} = \sqrt{\frac{t(t^2 - x^2)}{tt}} = \sqrt{t}\sqrt{\frac{t^2 - x^2}{t^2}}$$

[by comparison with (3.6.1)],

$$\sigma(t, x) = \sqrt{\frac{t^2 - x^2}{t^2}} = \sqrt{\frac{t^2 - t^2v^2}{t^2}} = \sqrt{1 - v^2} = \sigma(1, v).$$

This shows that f and σ are homogeneous functions of degree 0. Therefore, by Definition 3.6.2, the given differential equation is homogeneous. In addition, we compute

$$\frac{d}{dv}\sigma(1, x) = \frac{d}{dv}\sqrt{1 - v^2} = \frac{-2v}{2\sqrt{1 - v^2}} = \frac{-v}{\sqrt{1 - v^2}}, \quad \int^v \frac{ds}{\sqrt{1 - s^2}} = \sin^{-1} v,$$

$$\frac{f(1, v) - v}{t} = \frac{\frac{1}{2}\left(v + \sqrt{1 - v^2}\right) - v}{t} = \frac{\left(v + \sqrt{1 - v^2}\right) - 2v}{2t}$$

$$= \frac{-v + \sqrt{1 - v^2}}{2t} = \frac{\sqrt{1 - v^2}}{t}\left[-\frac{v}{2\sqrt{1 - v^2}} + \frac{1}{2}\right],$$

$$\frac{f(1, v) - v}{t} = \frac{1}{t}\sigma(1, v)\left[\frac{1}{2}\frac{d}{dv}\sigma(1, v) + F\right] = \frac{\sqrt{1 - v^2}}{t}\left[\frac{-v}{\sqrt{1 - v^2}} + F\right]$$

$$= \frac{\sqrt{1 - v^2}}{t}\left[-\frac{v}{\sqrt{1 - v^2}} + \frac{1}{2}\right].$$

From the last mathematical expression, we conclude that $F = \frac{1}{2}$. This shows that this example is in the form of Illustration 3.6.1(c). Hence, the differential equation belongs to the class LSDEVC-(f, σ). Moreover, the choice of rate functions of the reduced SDE (3.4.1) corresponding to the given SDE and Observation 3.6.1(iii) is $\delta = 0$, $\nu(t) = 0$, $g(t) = \frac{1}{\sqrt{t}}$, $h(v) = \sigma(1, v)$, $C = 0$, $\gamma = 1$, $q(t) = \frac{1}{\sqrt{t}}$, $p(t) = g^2(t)F = \frac{1}{2t}$, and $\mu(t) = 0$. Thus, the reduced SDE is

$$dm = \frac{1}{2t}dt + \frac{1}{\sqrt{t}}dw(t), \quad m(t_0) = m_0.$$

The general solution to this reduced differential equation (Section 2.2) is

$$m(t) = \frac{1}{2}\ln t + \int_a^t \frac{1}{\sqrt{s}}dw(s) + C.$$

The solution to the IVP associated with the reduced SDE is

$$m(t) = m_0 + \frac{1}{2}\ln t - \frac{1}{2}\ln t_0 + \int_{t_0}^t \frac{1}{\sqrt{2}}dw(s) = m_0 + \frac{1}{2}\ln\left[\frac{t}{t_0}\right] + \int_{t_0}^t \frac{1}{\sqrt{s}}dw(s).$$

From (3.4.5) and Step #4 of Procedure 3.4.2, the solution to the given IVP is:

$$G(v(t)) = G(v(t_0)) + \frac{1}{2}\ln\left[\frac{t}{t_0}\right] + \int_{t_0}^t \frac{1}{\sqrt{s}}dw(s),$$

and

$$G(v(t)) - G(v(t_0)) = \int_{v(t_0)}^{v(t)} \frac{\gamma}{\sigma(1, y)}dy = \int_{v(t_0)}^{v(t)} \frac{ds}{\sqrt{1 - s^2}} = \sin^{-1}v(t) - \sin^{-1}v(t_0).$$

By equating the left-hand side expressions, we have

$$\sin^{-1}v(t) = \sin^{-1}v(t_0) + \frac{1}{2}\ln\left[\frac{t}{t_0}\right] + \int_{t_0}^t \frac{1}{\sqrt{s}}dw(s).$$

The solution to the given IVP is

$$x(t) = t\sin\left(\sin^{-1}\left(\frac{x_0}{t_0}\right) + \frac{1}{2}\ln\left[\frac{t}{t_0}\right] + \int_{t_0}^t \frac{1}{\sqrt{s}}dw(s)\right).$$

This completes the solution representation process of the given IVP.

Example 3.6.3. Given:

$$dx = \left[\frac{x}{t} - \frac{1}{2}\exp\left[-\frac{2x}{t}\right]\right]dt + \sqrt{t}\exp\left[-\frac{x}{t}\right]dw(t), \quad x(t_0) = x_0, \text{ for } t \geq t_0 > 0.$$

(a) Verify that the given differential equation is homogeneous.
(b) Find: (i) the general solution; (ii) the IVP.

Solution procedure. Goal: To find the solution to the given IVP. For this purpose, we imitate the arguments used in Examples 3.6.1 and 3.6.2. Again, we observe that

$$f(t, x) = \frac{x}{t} - \frac{1}{2} \exp\left[-\frac{2x}{t}\right] = f(t, tv) = \frac{vt}{t} - \frac{1}{2} \exp\left[-\frac{2vt}{t}\right]$$

$$= v - \frac{1}{2} \exp[-2v] = f(1, v),$$

$$\sqrt{t}\sigma(t, x) = \sqrt{t} \exp\left[-\frac{x}{t}\right],$$

$$\sigma(t, x) = \exp\left[-\frac{x}{t}\right] = \exp\left[-\frac{x}{t}\right] = \exp\left[-\frac{vt}{t}\right] = \exp[-v] = \sigma(1, v).$$

This shows that f and σ are homogeneous functions of degree 0. Therefore, from $x = tv$, $t \neq 0$, we conclude that the given SDE is homogeneous. Moreover, we have

$$\frac{d}{dv}\sigma(1, v) = \frac{d}{dv} \exp[-v] = -\exp[-v],$$

$$\int_c^v \frac{dy}{\exp[-y]} = \int_c^v \exp[y]dy = \exp[v] - \exp[c],$$

$$\frac{f(1, v) - v}{t} = \frac{2v - \exp[-2v] - 2v}{2t} = \frac{-\exp[-2v]}{2t} = \frac{\exp[-v]}{t}\left[-\frac{1}{2}\exp[-v]\right],$$

$$\frac{f(1, v) - v}{t} = \frac{1}{t}\sigma(1, v)\left[\frac{1}{2}\frac{d}{dx}\sigma(1, v) + F\right] = \frac{\exp[-v]}{t}\left[-\frac{1}{2}\exp[-v] + F\right].$$

From the last mathematical expression, we conclude that $F = 0$. This shows that this example is a particular case of Illustration 3.6.1(c). Hence, the SDE belongs to the class LSDEVC-(f, σ). Moreover, the choice of rate functions of the reduced SDE (3.4.1) corresponding to this given SDE is $\delta = 0$, $\nu(t) = 0$, $q(t) = g(t) = \frac{1}{\sqrt{t}}$, $C = 0$, $\gamma = 1$, $h(v) = \sigma(1, v) = \exp[-v]$, $p(t) = 0$, $\mu(t) = 0$. Therefore, the reduced SDE is

$$dm = \frac{1}{\sqrt{t}}dw(t), \quad m(t_0) = m_0$$

and its general solution (Section 2.2) is $m(t) = \int_a^t \frac{1}{\sqrt{s}}dw(s) + c$. Hence, the solution to the IVP is $m(t) = m_0 + \int_{t_0}^t \frac{1}{\sqrt{s}}dw(s)$. From (3.4.5) and Step #4 of Procedure 3.4.2, the general solution to the given differential is

$$G(v(t)) = G(v(t_0)) + \int_{t_0}^t \frac{1}{\sqrt{s}}dw(s),$$

and

$$G(v(t)) - G(v(t_0)) = \int_{v(t_0)}^{v(t)} \frac{\gamma}{\sigma(1, y)}dy = \int_{v(t_0)}^{v(t)} \exp[y]dy$$

$$= \exp[v(t)] - \exp[v(t_0)].$$

Hence,

$$\exp[v(t)] = \exp[v(t_0)] + \int_{t_0}^{t} \frac{1}{\sqrt{s}} dw(s).$$

The solution to the given IVP is

$$x(t) = t \ln\left(\exp[v(t_0)] + \int_{t_0}^{t} \frac{1}{\sqrt{s}} dw(s) \right).$$

This is the solution process of the given IVP.

Example 3.6.4 (dynamics of swarms under random perturbations). We assume that all the assumptions regarding swarm dynamics in Illustration 3.7.2 [80] are satisfied. We extend this model under random environmental perturbations.

In the formulation of the deterministic dynamic model of swarms of birds (Illustration 3.7.2 [80]), it is assumed that the speed of swarms is constant relative to the constant speed of the wind. In fact, this assumption is subject to random perturbations due to the unpredictable atmospheric changes. As a result of this, the parameter α in Illustration 3.7.2 is subject to random perturbations. Therefore, by following the development of stochastic models of Section 3.1, we conclude that the swarm dynamic model in Illustrations 3.7.2 [80] can be recasted as

$$dy = \frac{y}{x} + (\bar{\alpha} + \xi(x))\frac{\sqrt{x^2 + y^2}}{x} = \left(\frac{y}{x} + \bar{\alpha}\frac{\sqrt{x^2 + y^2}}{x} \right) dx + \frac{\sqrt{x^2 + y^2}}{x}\xi(x)dx,$$

where $\xi(x)$ is a Gaussian white noise process as described in (3.1.6). Hence, by applying of Sequential Colored Noise Model 3.1.2 and using the argument of Illustration 3.1.3, the solution process of the above-described stochastic differential equation is described by the following Itô–Doob-type stochastic differential equation is:

$$dy = \left(\frac{\left(1 + \frac{\eta^2}{2}\right)y}{x} + \bar{\alpha}\frac{\sqrt{x^2 + y^2}}{x} \right) dx + \eta\frac{\sqrt{x^2 + y^2}}{x}\sqrt{x}\, dw(x), \quad y(x_0) = 0,$$

where $w(x) = \eta \int_0^x [\frac{\xi(s)}{\sqrt{s}}]ds$ is a Wiener process. We note that this is a stochastic homogeneous differential equation (Definition 3.6.2 and Illustration 3.6.1).

To find the solution to stochastic homogeneous differential equation, first, we need to show that it belongs to the class LSDECC-(f,σ). We observe that, for $y = xv, x \neq 0$,

$$\frac{d}{dv}\sigma(1,v) = \frac{d}{dv}\left(\eta\sqrt{1 + v^2} \right) = \frac{1}{2\sqrt{1 + v^2}}2\eta v = \frac{\eta v}{\sqrt{1 + v^2}},$$

$$\int_c^v \frac{dy}{\sqrt{1 + y^2}} = \ln\left(v + \sqrt{1 + v^2} \right) - \ln\left(c + \sqrt{1 + c^2} \right),$$

$$\frac{f(1,v)-v}{x} = \frac{\left(1+\frac{\eta^2}{2}\right)v+\bar{\alpha}\sqrt{1+v^2}-v}{x} = \frac{\frac{\eta^2}{2}v+\bar{\alpha}\sqrt{1+v^2}}{x},$$

$$\frac{f(1,v)-v}{x} = \frac{1}{x}\sigma(1,v)\left[\frac{1}{2}\frac{d}{dv}\sigma(1,v)+F\right]$$

$$= \frac{\eta\sqrt{1+v^2}}{x}\left[\frac{\eta v}{2\sqrt{1+v^2}}+F\right] = \frac{\eta\sqrt{1+v^2}F+\frac{\eta^2}{2}v}{x}.$$

From the last mathematical expression, we conclude that $\eta F = \bar{\alpha}$. This shows that it is an example of Illustration 3.6.1(c). Hence, the differential equation belongs to the class LSDECC-(f,σ). Moreover, the choice of rate functions of the reduced differential equation (3.4.1) corresponding to this given differential equation is

$$\delta = 0, \quad \nu(x) = 0, \quad g(x) = \frac{1}{\sqrt{x}},$$

$$\eta h(v) = \sigma(1,v) = \eta\sqrt{1+v^2}, \quad C \neq 0,$$

$$\gamma = \eta, \quad q(x) = \frac{\eta}{\sqrt{x}}, \quad p(x) = \frac{\bar{\alpha}}{x}, \quad \mu(x) = 0.$$

In this case, the reduced differential equation is

$$dm = \frac{\bar{\alpha}}{x}dx + \frac{\eta}{\sqrt{x}}dw(x), \quad m(x_0) = 0,$$

its general solution (Section 2.2) is

$$m(x) = \bar{\alpha}\ln(x) + \int_a^x \frac{1}{\sqrt{s}}dw(s) + C,$$

and the solution to the IVP is

$$0 = m(x_0) = \bar{\alpha}\ln(x_0) + \int_{t_0}^{t_0} \frac{1}{\sqrt{s}}dw(s) + C.$$

Hence, $C = -\bar{\alpha}\ln(x_0)$. From the above discussion, we get

$$\ln\left(v+\sqrt{1+v^2}\right) = \bar{\alpha}\left[\ln(x)-\ln(x_0)\right] + \int_0^x \frac{1}{\sqrt{s}}dw(s)$$

$$= \ln\left(\frac{x}{x_0}\right)^{\bar{\alpha}} + \int_0^x \frac{1}{\sqrt{s}}dw(s)$$

and thus

$$\left(v+\sqrt{1+v^2}\right) = \left(\frac{x}{x_0}\right)^{\bar{\alpha}}\exp\left[\int_0^x \frac{1}{\sqrt{s}}dw(s)\right].$$

By using the algebraic logic utilized in Illustration 3.7.2 [80], we obtain

$$y(x) = \frac{1}{2}x\left[\left(\frac{x}{x_0}\right)^{\bar{\alpha}}\exp\left[\int_0^x \frac{1}{\sqrt{s}}dw(s)\right] - \left(\frac{x}{x_0}\right)^{-\bar{\alpha}}\exp\left[-\int_0^x \frac{1}{\sqrt{s}}dw(s)\right]\right]$$

$$= \frac{x_0}{2}\left[\left(\frac{x}{x_0}\right)^{\bar{\alpha}+1}\exp\left[\int_0^x \frac{1}{\sqrt{s}}dw(s)\right] - \left(\frac{x}{x_0}\right)^{1-\bar{\alpha}}\exp\left[-\int_0^x \frac{1}{\sqrt{s}}dw(s)\right]\right].$$

This is the phase plane trajectory of the motion of the swarm under the influence of random perturbations. Again, the interpretation of this mathematical description is left as an exercise.

Brief Summary. A summary of the basic mechanical steps needed to solve non-linear stochastic scalar differential equations is presented in Figure 3.6.2 flowchart.

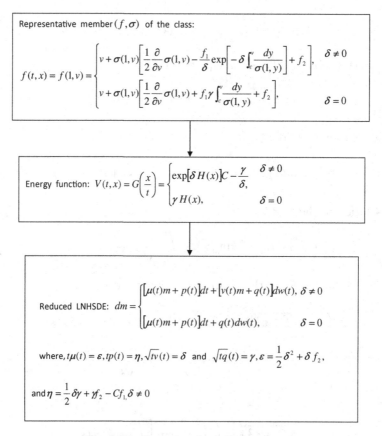

Fig. 3.6.2   Summary of solution processes.

Exercises

1. Let A and B be any given real numbers. Decide which SDEs are homogeneous and which are not:

   (a) $dx = \frac{8(1+B)x^4+(4B-2A)x^2t^2-t^4}{8x^3t}\,dt + \frac{t^2+2x^2}{2x\sqrt{t}}\,dw(t)$;

   (b) $dx = \left[\frac{x}{t} + \frac{t^3+xt^2}{(t-x)^3} + \frac{At}{(t-x)}\right]\exp\left[\frac{x}{2t}\right] + B\frac{t+x}{t-x}\,dt + \sqrt{t}\frac{t+x}{t-x}\,dw(t)$;

   (c) $dx = \frac{(3+2B)x^4+2(B+Ad^2)x^2t^2-d^4t^4}{3x^3t}\,dt + \frac{x^2+d^2t^2}{x\sqrt{t}}\,dw(t)$, for any real number d;

   (d) $dx = \left[\frac{3x+dt+B\sqrt{2dxt+x^2}}{2t}\right]\,dt + \frac{\sqrt{2dxt+x^2}}{t}\,dw(t)$.

   Hint: $\int \frac{ds}{\sqrt{2ds+s^2}} = \ln\left(s+d+\sqrt{2ds+s^2}\right)$ and $\int \frac{(1-s)}{(1+s)}\,ds = 2\ln(1+s) - s$.

2. By using $C = 1$ in (3.4.5), for differential equations in Exercise 1, find:

   (a) The feasible values of δ in the differential equations (a), (b), (c), and (d), respectively;

   (b) The general solution representation of the differential equations (b)–(d):

      (i) For any arbitrary constants A and B.

      (ii) For any arbitrary constants A with $B = 0$.

      (iii) For any arbitrary constants B with $A = 0$.

      (iv) For constants $B = 0$ and $A = 0$.

3. By using $C = 1$ in (3.4.5) and the value of δ's in Exericse 2, find the general solution representation of the differential equations in Exercise 1(a):

   (a) For constants A and B that satisfy the condition: $A = B = 1$.

   (b) For constants A and B that satisfy the condition: $A = 0$ and $B = -1$.

   (c) For constants A and B that satisfy the condition: $A = -1$ and $B = -1$.

4. By using $C = 1$ in (3.4.5) and the value of δ's in Exercise 2, find the general solution representation of the differential equations in Exercise 1(b):

   (a) For constants A and B that satisfy the condition: $A = B$ and $B = 1$.

   (b) For constants A and B that satisfy the condition: $A = 0$ and $B = 1$.

   (c) For constants A and B that satisfy the condition: $A = 1$ and $B = 0$.

5. By using $C = 1$ in (3.4.5) and the value of δ's in Exercise 2, find the general solution representation of the differential equations in Exercise 1(c):

   (a) For constants A and B that satisfy the condition: $A = 0$ and $B = -\frac{3}{2}$.

   (b) For constants A and B that satisfy the condition: $B = -Ad^2$ and $d = 2$.

   (c) For constants A and B that satisfy the condition: $B = -Ad^2$ and $d = 2$.

6. By using $C = 1$ in (3.4.5) and the value of δ's in Exercise 2, find the general solution representation of the differential equations in Exercise 1(d):

   (a) For constants A and B that satisfy the condition: $A = -\frac{1}{2}d$ and $B = 1$.

   (b) For constants A and B that satisfy the condition: $B = A = 0$ and $d = 2$.

   (c) For constants A and B that satisfy the condition: $B = 0$, $A = -1$, and $d = 2$.

7. Let P and Σ be continuous functions. Further assume that λ, μ, ν, ψ, ρ, and ϕ are any given real numbers with $e = \lambda\psi - \nu\mu \neq 0$ and constants ν and λ are both not zero.

(a) $dx = P\left(\frac{\lambda x + \mu t + \rho}{\nu x + \psi t + \phi}\right) dt;$

(b) $dx = P\left(\frac{\lambda x + \mu t + \rho}{\nu x + \psi t + \phi}\right) dt + \sqrt{t - \frac{\nu\rho - \lambda\phi}{e}}\, \Sigma\left(\frac{\lambda x + \mu t + \rho}{\nu x + \psi t + \phi}\right) dw(t).$

   (i) By using a transformation,

$$x = y + k, \quad dx = dy,$$
$$t = z + \ell, \quad dt = dz,$$

show that the given differential equations in (a) and (b) are reducible to:

(a) $dy = P\left(\frac{\lambda y + \mu z}{\nu y + \psi z}\right) dz,$

(b) $dy = P\left(\frac{\lambda y + \mu z}{\nu y + \psi z}\right) dz + \sqrt{z}\, \Sigma\left(\frac{\lambda y + \mu z}{\nu y + \psi z}\right) dw(z),$

where k and ℓ are unknown real numbers that are determined by the following linear nonhomogeneous system of algebraic equations:

$$\lambda k + \mu\ell + \rho = 0,$$
$$\nu k + \psi k + \phi = 0.$$

   (ii) Show that the SDE is reducible to a homogeneous SDE.

8. In addition to the assumptions on the rate function of SDEs in Exercise 7, we further assume that the rate functions P and Σ have the following forms:

$$P(r) = \frac{H(r)}{\frac{d}{dr}H(r)\frac{d}{dv}r}\left(-\frac{H(r)}{2}\left[\frac{\frac{d^2}{dr^2}H(x)}{\left(\frac{d}{dr}H(r)\right)^2} + \frac{\frac{d^2}{dr^2}r}{\frac{d}{dr}H(r)\left(\frac{d}{dv}r\right)^2}\right]\right.$$
$$\left. + \frac{A}{\delta}[H(r)]^{-\delta} + B\right),$$

and

$$\Sigma(r) = \frac{H(r)}{\frac{d}{dr}H(r)\frac{d}{dv}r}, \quad r(x,t) = \frac{\lambda x + \mu t}{\nu x + \psi t},$$

where $r(x,t)$ is a homogeneous function of degree 0. Under this description, find a general solution expression of the differential equation

$$dv = \frac{H(r(v))}{\frac{d}{dr}H(r(v))\frac{d}{dv}r(v)}\left(-\frac{H(r(v))}{2}\left[\frac{\frac{d^2}{dr^2}H(r(v))}{\left(\frac{d}{dr}H(r(v))\right)^2} + \frac{\frac{d^2}{dr^2}r(v)}{\frac{d}{dr}H(r(v))\left(\frac{d}{dv}r(v)\right)^2}\right]\right.$$
$$\left. + \frac{A}{\delta}[H(v)]^{-\delta} + B\right) dt + \frac{H(r(v))}{\frac{d}{dr}H(r(v))\frac{d}{dv}r(v)}\, dw(t).$$

(a) For constants A and B that satisfy the condition: $r(v) = \frac{\lambda v + \mu}{\nu v + \psi}$, $A = 0$, and $B = 0$.

(b) For constants A and B that satisfy the condition: $r(v) = \frac{v + \mu}{-v + \psi}$, $A = 0$, and $B = -5$.

(c) For constants A and B that satisfy the condition: $r(v) = \frac{v + \mu}{-v + \psi}$, $A = 0$, and $B = 0$.

3.7 Bernoulli Equations

In this section, we present another subclass of the differential equations (3.2.1) that is reducible to (3.4.1). This class of differentials is referred as Itô–Doob-type stochastic Bernoulli differential equations. First, we introduce a definition of the Bernoulli differential equation.

Definition 3.7.1. A differential equation (3.2.1) is said to be an Itô–Doob-type stochastic Bernoulli differential equation if the rate functions $f(t, x)$ and $\sigma(t, x)$ in (3.2.1) are of the following forms: $f(t, x) = P(t)x + Q(t)x^n + \frac{n}{2}\Upsilon(t)x^{2n-1}$ and $\sigma(t, x) = \Sigma(t)x + \Upsilon(t)x^n$ for any real number $n \neq 1$.

3.7.1 *General Problem*

We imitate Procedure 3.4.1 with regard to the Itô–Doob-type stochastic Bernoulli differential equation (3.2.1), i.e. for any $n \neq 1$,

$$dx = [P(t)x + Q(t)x^n + \frac{n}{2}\Upsilon^2(t)x^{2n-1}]dt + [\Sigma(t)x + \Upsilon(t)x^n]dw(t), \qquad (3.7.1)$$

where P, Q, Σ, and Υ are continuous functions; moreover, (3.7.1) has a solution process. The following procedure leads to reduction of the differential equation (3.7.1) to (3.4.1).

(H₃.₇). It is assumed that P, Q, Σ, and Υ are continuous functions; moreover, (3.7.1) has a solution process. In addition, we assume that either $\Sigma(t)$ or $\Upsilon(t)$ is different from zero.

3.7.2 *Procedure for Finding a General Solution Representation*

By Procedure 3.4.2 and Observation 3.4.1(i), the process for the reduction of a subclass of the stochastic differential equations (3.2.1) to (3.4.1) is repeated. Again, for this subclass of Bernoulli-type SDEs, we associate a suitable energy/Lyapunov function in a unified way.

Step #1. We repeat Steps #1 and #2 of Procedure 3.4.2. In the context of (3.7.1), (3.4.2) and (3.4.3) reduce to

$$\mu(t)V(t, x) + p(t) = \frac{\partial}{\partial t}V(t, x) + \left[P(t)x + Q(t)x^n + \frac{n}{2}\Upsilon(t)x^{2n-1}\right]\frac{\partial}{\partial x}V(t, x)$$

$$+ \frac{1}{2}[\Sigma(t)x + \Upsilon(t)x^n]^2 \frac{\partial^2}{\partial x^2}V(t, x), \qquad (3.7.2)$$

and

$$\nu(t)V(t,x) + q(t) = [\Sigma(t)x + \Upsilon(t)x^n]\frac{\partial}{\partial x}V(t,x). \tag{3.7.3}$$

Step #2. Again, by using (3.7.3), we find a candidate for $V(t,x)$. For this purpose, from (3.7.3) and in view of the structure of rate functions in the Bernoulli differential equation (3.7.1), $f(t,x) = P(t)x + Q(t)x^n + \frac{n}{2}\Upsilon(t)x^{2n-1}$ and $\sigma(t,x) = \Sigma(t)x + \Upsilon(t)x^n$ for $n \neq 1$, we have two different ways of choosing $V(t,x)$:

$$\text{(i)} \quad \Sigma(t)x\frac{\partial}{\partial x}V(t,x) = \nu(t)V(t,x), \quad \text{(ii)} \quad \Upsilon(t)x^n\frac{\partial}{\partial x}V(t,x) = q(t). \tag{3.7.4}$$

Choice (i). From the hypothesis (H$_{3.7}$), $\Sigma(t)x \neq 0$, we choose $V(t,x)$ which satisfies (i) in (3.7.4) with a choice of $\nu(t) = \delta\Sigma(t)$ for an unknown nonzero constant δ ($\delta \neq 0$). The goal is to find $V(t,x)$, rate functions $\nu(t)$, $q(t)$, $\mu(t)$, and $p(t)$ of the reduced SDE (3.4.1) [(3.7.2) and (3.7.3)]. We note that to find $\nu(t)$, it is enough to find δ. For this purpose, we rewrite (i):

$$\frac{\partial}{\partial x}V(t,x) = \frac{\nu(t)}{\Sigma(t)x}V(t,x) \quad [\text{by the assumption } \Sigma(t)x \neq 0]$$

$$= \frac{\delta}{x}V(t,x) \qquad [\text{from the hypothesis } [(\text{H}_{3.7})]. \tag{3.7.5}$$

From this, it is clear that the quotient of $\frac{\partial}{\partial x}V(t,x)$ with $V(t,x)$ is independent of t. Therefore, it follows that $V(t,x) \equiv V(x)$, i.e. $V(t,x)$ is independent of t.

By applying the method of Section 2.3 of Chapter 2 [80], for each t in J, we have

$$V(t,x) \equiv V(x) = \exp\left[\delta\int_c^x \frac{ds}{s}\right]C$$

$$= \exp\left[\delta\ln\left(\frac{|x|}{|c|}\right)\right]C \qquad \left[\text{from } \int_c^x \frac{ds}{s} = \ln(|x|) - \ln(|c|) = \left(\frac{|x|}{|c|}\right)\right]$$

$$= \exp\left[\ln\left(\left(\frac{|x|}{|c|}\right)^\delta\right)\right]C \quad [\text{from } \delta\ln M = \ln(M)^\delta]$$

$$= x^\delta C \quad (\text{from: exp is the inverse of ln, and a generic constant}), \tag{3.7.6}$$

where C is a nonzero arbitrary constant of integration. Otherwise, if $C = 0$, $V(x) \equiv 0$. This is not useful for finding a nontrivial solution representation of (3.7.1). Now, we compute $\frac{\partial}{\partial x}V(t,x)$ and $\frac{\partial^2}{\partial x^2}V(t,x)$.

$$\frac{d}{dx}V(x) = \delta x^{\delta-1}C, \tag{3.7.7}$$

and

$$\frac{d^2}{dx^2}V(x) = \delta(\delta-1)x^{\delta-2}C. \tag{3.7.8}$$

From (3.7.4) and (3.7.7), we have

$$q(t) = \Upsilon(t)x^n \frac{\partial}{\partial x} V(t, x)$$

$$= \Upsilon(t)x^n \delta x^{\delta-1} C \quad \text{[by substitution from (3.7.7)]}$$

$$= \delta\Upsilon(t)x^{n+\delta-1} C \quad \text{(by simplification).} \tag{3.7.9}$$

By using the assumption (H$_{3.7}$), the method of undetermined coefficients, and (3.7.9), we have

$$q(t) - \delta\Upsilon(t)Cx^{n+\delta-1} = 0. \tag{3.7.10}$$

Therefore, we choose $q(t)$ and δ so that

$$x^{n+\delta-1} = x^0 = 1, \tag{3.7.11}$$

and

$$q(t) = \delta\Upsilon(t)C. \tag{3.7.12}$$

From (3.7.11), we have

$$n + \delta - 1 = 0,$$

and hence

$$\delta = 1 - n. \tag{3.7.13}$$

Moreover,

$$q(t) = (1 - n)\Upsilon(t)C. \tag{3.7.14}$$

By using (3.7.6)–(3.7.8) and (3.7.13), we compute the right-hand expression of (3.7.2):

$$\frac{d}{dt}V(x) + \left[P(t)x + Q(t)x^n + \frac{n}{2}\Upsilon^2(t)x^{2n-1} \right] \frac{d}{dx}V(x)$$

$$+ \frac{1}{2}[\Sigma(t)x + \Upsilon(t)x^n]^2 \frac{d^2}{dx^2}V(x)$$

$$= \delta \left[P(t)x + Q(t)x^n + \frac{n}{2}\Upsilon^2(t)x^{2n-1} \right] x^{\delta-1}C$$

$$+ \frac{1}{2}\delta(\delta - 1)[\Sigma(t)x + \Upsilon(t)x^n]^2 x^{\delta-2}C \quad \text{(by substitution)}$$

$$= \frac{1}{2}\delta(\delta - 1)\left[\Sigma^2(t)x^2 + 2\Sigma(t)\Upsilon(t)x^{n+1} + \Upsilon^2(t)x^{2n} \right] x^{\delta-2}C$$

$$+ \delta \left[P(t)x + Q(t)x^n + \frac{n}{2}\Upsilon^2(t)x^{2n-1} \right] x^{\delta-1}C \quad \text{(by binomial expansion)}$$

$$= \frac{1}{2}\delta(\delta-1)\left[\Sigma^2(t)x^\delta + 2\Sigma(t)\Upsilon(t)x^{n+\delta-1} + \Upsilon^2(t)x^{2n+\delta-2}\right]C$$

$$+\delta\left[P(t)x^\delta + Q(t)x^{n+\delta-1} + \frac{n}{2}\Upsilon^2(t)x^{2n+\delta-2}\right]C \quad \text{(by simplifying)}$$

$$= \delta\Bigg(\left[P(t) + \frac{1}{2}(\delta-1)\Sigma^2(t)\right]x^\delta + [Q(t) + (\delta-1)\Sigma(t)\Upsilon(t)]x^{n+\delta-1}$$

$$+ \left[\frac{n}{2}\Upsilon^2(t) + \frac{1}{2}(\delta-1)\Upsilon^2(t)\right]x^{2n+\delta-2}\Bigg)C \quad \text{(by regrouping)}$$

$$= (1-n)\Bigg(\left[P(t) - \frac{n}{2}[\Sigma^2(t)]\right]x^\delta + [Q(t) - n\Sigma(t)\Upsilon(t)]$$

$$+ \left[\frac{n}{2}\Upsilon^2(t) - \frac{n}{2}\Upsilon^2(t)\right]x^{2n+\delta-2}\Bigg)C \quad \text{[from (3.7.13)]}$$

$$= (1-n)\Bigg(\left[P(t) - \frac{n}{2}\Sigma^2(t)\right]x^\delta + [Q(t) - n\Sigma(t)\Upsilon(t)]\Bigg)C \quad \text{(by simplifying)}$$

$$= (1-n)\left[P(t) - \frac{n}{2}\Sigma^2(t)\right]x^\delta C + (1-n)[Q(t) - n\Sigma(t)\Upsilon(t)]C \quad \text{(reorganizing)}.$$

$$(3.7.15)$$

From (3.7.6) and (3.7.15), (3.7.2) can be rewritten as

$$\mu(t)V(t,x) + p(t) = \frac{\partial}{\partial t}V(t,x) + \left[P(t)x + Q(t)x^n + \frac{n}{2}\Upsilon^2(t)x^{2n-1}\right]\frac{\partial}{\partial x}V(t,x)$$

$$+ \frac{1}{2}[\Sigma(t)x + \Upsilon x^n]^2 \frac{\partial^2}{\partial x^2}V(t,x) \quad \text{(by substitution)},$$

$$\mu(t)Cx^\delta + p(t) = (1-n)\left[P(t) - \frac{n}{2}\Sigma^2(t)\right]Cx^\delta + (1-n)[Q(t) - n\Sigma(t)\Upsilon(t)]C.$$

$$(3.7.16)$$

Now, from (3.7.16) and using the method of undetermined coefficients, we compare the coefficients of x, and conclude that

$$\mu(t)C = (1-n)\left[P(t) - \frac{n}{2}\Sigma^2(t)\right]C,$$

$$p(t) = (1-n)[Q(t) - n\Sigma(t)\Upsilon(t)]C.$$

Hence,

$$\mu(t) = (1-n)\left[P(t) - \frac{n}{2}\Sigma^2(t)\right] \tag{3.7.17}$$

$$p(t) = (1-n)[Q(t) - n\Sigma(t)\Upsilon(t)]C, \tag{3.7.18}$$

where C is a nonzero arbitrary constant of integration in (3.7.6). In particular, one can pick $C = 1$.

From (3.7.4), (3.7.17), and (3.7.18) with $C = 1$, the reduced SDE (3.4.1) is:

$$dm = (1 - n)\left[\left[P(t) - \frac{n}{2}\Sigma^2(t)\right]m + [Q(t) - n\Sigma(t)\Upsilon(t)]\right]dt$$

$$+ (1 - n)[\Sigma(t)m + \Upsilon(t)]dw(t). \tag{3.7.19}$$

From the above discussion, we note that the Case of $\delta = 0$ is not feasible. In fact, the presented procedure does not provide the existence of a desired energy function.

Choice (ii).   Under the hypothesis (H3.7), $\Upsilon(t)x \neq 0$, and we choose $V(t, x)$ so that it satisfies (ii) in (3.7.4) with $q(t) = \gamma\Upsilon(t)$ for some unknown nonzero constant γ ($\gamma \neq 0$). The goal is to find $V(t, x)$, rate functions $q(t)$, $\nu(t)$, $\mu(t)$, and $p(t)$, in the reduced differential equation (3.4.1) [(3.7.2) and (3.7.3)]. We note that to find $q(t)$, it is enough to find γ. For this purpose, we consider (ii):

$$\Upsilon(t)x^n \frac{\partial}{\partial x}V(t, x) = q(t). \tag{3.7.20}$$

From (3.7.20) and the choice of $q(t)$, we solve for $\frac{\partial}{\partial x}V(t, x)$, and obtain

$$\frac{\partial}{\partial x}V(t, x) = \frac{q(t)}{\Upsilon(t)x^n}$$

$$= \frac{\gamma\Upsilon(t)}{\Upsilon(t)x^n} \quad \text{(by substitution)}$$

$$= \frac{\gamma}{x^n} \quad \text{(by simplification).} \tag{3.7.21}$$

Again, the right-hand side is independent of t. Therefore, the left-hand side must also be independent of t [because $V(t, x)$ is an arbitrary function]. Thus, by integrating both sides of this expression with respect to x, we get

$$V(t, x) \equiv V(x) = \frac{\gamma}{(1 - n)x^{n-1}} + C$$

$$= x^{1-n} + C \quad \text{(by the choice of } \gamma \text{ and the law of exponents),} \tag{3.7.22}$$

where $\gamma = 1 - n$, and C is an arbitrary constant of integration. Again, we compute $\frac{\partial}{\partial x}V(t, x)$ and $\frac{\partial^2}{\partial x^2}V(t, x)$.

$$\frac{d}{dx}V(x) = \frac{d}{dx}(x^{1-n} + C) = (1 - n)x^{-n}, \tag{3.7.23}$$

$$\frac{d^2}{dx^2}V(x) = -(1 - n)nx^{-n-1}. \tag{3.7.24}$$

Now, by using (3.7.21), (3.7.23), and (3.7.24) and repeating the above argument, we compute the right-hand expression in (3.7.2). Here, we get the same expression as in (3.7.15) except for the appearance of C, an additive arbitrary constant, instead of a multiplicative constant:

$$\frac{d}{dt}V(x) + \left[P(t)x + Q(t)x^n + \frac{n}{2}\Upsilon^2(t)x^{2n-1}\right]\frac{d}{dx}V(x)$$

$$+ \frac{1}{2}[\Sigma(t)x + \Upsilon(t)x^n]^2\frac{d^2}{dx^2}V(x)$$

$$= (1-n)\left(\left[P(t) - \frac{n}{2}\Sigma^2(t)\right]x^{1-n} + [Q(t) - n\Sigma(t)\Upsilon(t)]\right). \qquad (3.7.25)$$

From (3.7.22) and (3.7.25), (3.7.2) can be rewritten as

$$\mu(t)V(t,x) + p(t) = \frac{\partial}{\partial t}V(t,x) + [P(t)x + Q(t)x^n$$

$$+ \frac{n}{2}\Upsilon^2(t)x^{2n-1}\right]\frac{\partial}{\partial x}V(t,x)$$

$$+ \frac{1}{2}[\Sigma(t)x + \Upsilon x^n]^2\frac{\partial^2}{\partial x^2}V(t,x) \quad \text{(by substitution)},$$

$$\mu(t)x^{1-n} + p(t) + \mu(t)C = (1-n)\left[P(t) - \frac{n}{2}\Sigma^2(t)\right]x^{1-n}$$

$$+ (1-n)[Q(t) - n\Sigma(t)\Upsilon(t)]. \qquad (3.7.26)$$

Now, from (3.7.4), (3.7.26), (3.7.23) and by using the method of undetermined coefficients, we compare the coefficients of x, and conclude that

$$\nu(t)V(t,x) = \Sigma(t)x\frac{d}{dx}V(x)$$

$$\nu(t)(x^{1-n} + C) = \Sigma(t)x(1-n)x^{-n} \quad \text{[from (3.7.22) and (3.7.23)]}$$

$$\nu(t) = (1-n)\Sigma(t), \quad C = 0 \text{ [C: the constant of integration in (3.7.22)]}$$

$$\mu(t) = (1-n)\left[P(t) - \frac{n}{2}\Sigma^2(t)\right], \quad \nu(t) = \gamma\Sigma(t), \quad q(t) = \gamma\Upsilon(t),$$

$$\qquad (3.7.27)$$

$$p(t) = (1-n)[Q(t) - n\Sigma(t)\Upsilon(t)], \qquad (3.7.28)$$

where C is an arbitrary constant of integration, and one can pick $C = 0$.

In this case, from (3.7.4), (3.7.24), (3.7.27), and (3.7.28) with $C = 0$, we have (3.7.19).

Step #4: Under the condition (3.7.4), the nonlinear stochastic Bernoulli differential equation (3.7.1) reduces to (3.7.19). Thus, the condition (3.7.4) determines the subclass of functions f and σ in (3.2.1) which are described in (3.7.1). Moreover, for this subclass of differential equations (3.2.1), its general solution representation is given by the inverse of function $V(x) = x^{1-n}$ either in (3.7.6) with $C = 1$ or in (3.7.22) with $C = 0$:

$$x(t) = E(t, m(t)), \tag{3.7.29}$$

where $m(t)$ is the general solution process of (3.7.19), which can be determined by the method of Section 2.4 of Chapter 2, and $x(t)$ is the general solution process of the stochastic differential equation (3.7.1), which belongs to the above-determined subclass. This completes the reduction and the solution representation process (3.2.1) under the specified conditions of f and σ in the Bernoulli-type SDE (3.7.1).

Brief Summary. The following flowchart, 3.7.1, summarizes the steps of finding the general solution of a class of Itô–Doob-type stochastic Bernoulli-type nonlinear scalar differential equations via the energy function method.

Observation 3.7.1. We recall that any one of the two relations in (3.7.4) plays a role in determining the candidate for the energy function. Moreover, from Procedure 3.7.2, it is clear that the arguments used in finding a candidate in the procedure provide different energy functions, namely (i) $V_\delta(x) = x^{1-n}C$ and (ii) $V_\gamma(x) = x^{1-n} + C$, depending on the choices (i) and (ii) in (3.7.4), where C's are arbitrary constants of integration, respectively. We note that the arbitrary constant C in (i) is nonzero, and that in (ii) it is any arbitrary constant C. It can be proved that any other combination of the relations in (3.7.4) does not provide the feasible energy function.

Next, we present a handful of examples to illustrate the procedure described in this section.

Illustration 3.7.1. Given: $dx = \left[Px + Qx^n + \frac{n}{2}\Upsilon^2 x^{2n-1}\right]dt + [\Sigma x + \Upsilon x^n]dw(t)$, where P, Q, Σ, and Υ are arbitrary given real numbers, and n is any real number that satisfies $n \neq 1$. Find a general solution representation of the solution to this differential equation.

Solution procedure. From Steps #1 and #2 of the procedure for solving the Bernoulli-type SDE and Choice (i), for any real number $n \neq 1$ and $\delta \neq 0$, the candidate for the energy function in (3.7.6) with $C = 1$ reduces to $V(x) = x^\delta$. From (3.7.13), (3.7.14), (3.7.17), and (3.7.18), $V(t, x)$ in (3.7.6) and the rate functions in (3.4.1) [(3.7.2) and (3.7.3)] are $V(t, x) = x^{1-n}$ ($\delta = 1 - n$), $\nu(t) = (1 - n)\Sigma$, $q(t) = (1 - n)\Upsilon$, $\mu(t) = (1 - n)(P - \frac{n}{2}\Sigma^2)$, and $(1 - n)(Q - n\Sigma\Upsilon) = p(t)$.

Using the method of Section 2.4 of Chapter 2, the general solution to the reduced SDE (3.7.19) in the context of the given problem is

$$V(x(t)) = m(t)$$

$$= \Phi(t)m(a) + [(1 - n)[Q - \Sigma\Upsilon] \int_a^t \Phi(t, s)ds + (1 - n)\Upsilon \int_a^t \Phi(t, s))dw(s)],$$

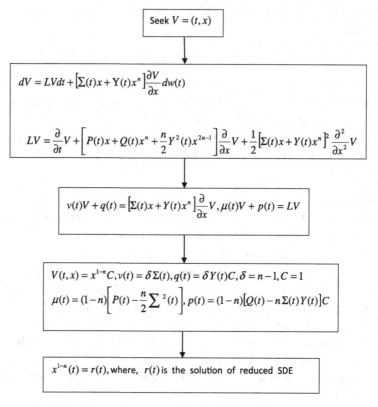

Fig. 3.7.1 Energy function method and Bernoulli-type equations.

where

$$\Phi(t) = \exp\left[(1-n)\left(P - \frac{1}{2}\Sigma^2\right)t + (1-n)\Upsilon w(t)\right] \text{ and } V(x(t)) = (x(t))^{1-n}.$$

The general solution to the original given differential equation is

$$x(t) = \left(\Phi(t)c + [(1-n)[Q - \Sigma\Upsilon]\int_a^t \Phi(t,s)ds\right.$$

$$\left. + (1-n)\Upsilon\int_a^t \Phi(t,s))dw(s)]\right)^{-1/(n-1)}.$$

Illustration 3.7.2 (economic growth–Solow model [80, 185]). Let us consider the growth model of Solow under the Cobb–Douglas production function described in Example 3.1.5 under random environmental perturbations:

$$dr = \left[r^a\left(\bar{s} + \frac{a\nu^2}{2}r^{a-1}\right) - nr\right]dt + \nu r^a dw(t), \quad r(t_0) = r_0.$$

Find the expression for the capital per worker in the above model.

Solution procedure. We note that this is the Bernoulli type differential equation. To find the expression for the capital per worker at any time t, we compare this equation with the one in Illustration 3.7.1. Thus, we have $V(t, r) = r^{1-a}$ $(\delta = 1-a)$, $P = -n$, $Q = \bar{s}$, $\Sigma = 0$, and $\Upsilon = \nu$. In this case, $\nu(t) = (1-a)\Sigma = 0$, $q(t) = (1-a)\nu$, $\mu(t) = (1-a)(-n-0) = -(1-a)n$, and $p(t) = (1-a)(\bar{s}-0) = (1-a)\bar{s}$. The reduced differential equation (3.7.19) is

$$dm = [-1(1-a)nm + (1-a)\bar{s}]dt + (1-a)\nu \, dw(t).$$

By using the method in Section 2.4 of Chapter 2, the general solution to this differential equation is

$$V(r(t)) = m(t)$$

$$= \Phi(t)m(a) + (1-a)\bar{s} \int_a^t \Phi(t, s)ds + (1-a)\nu \int_a^t \Phi(t, s))dw(s)$$

$$= \Phi(t)[m(a) + (1-a)\bar{s} \int_a^t \Phi^{-1}(s)ds + (1-a)\nu \int_a^t \Phi^{-1}(s)dw(s)],$$

where

$$\Phi(t) = \exp[-(1-a)nt] \text{ and } V(r(t)) = [r(t)]^{1-a}.$$

To find the solution to the given IVP, we need to compute the constant of integration c. For this purpose, we use $r(t_0) = r_0$, and we have

$$V(r_0) = [r_0]^{1-a} = \Phi(t_0)m(a) + (1-a)\bar{s} \int_a^{t_0} \Phi(t_0, s)ds$$

$$+ (1-a)\nu \int_a^{t_0} \Phi(t, s))dw(s).$$

We solve for $m(a)$, and obtain

$$m(a) = \Phi^{-1}(t_0) \left[[r_0]^{1-a} - (1-a)\bar{s} \int_a^{t_0} \Phi(t_0, s)ds - (1-a)\nu \int_a^{t_0} \Phi(t_0, s))dw(s) \right]$$

$$= \Phi^{-1}(t_0)[r_0]^{1-a}$$

$$+ \Phi^{-1}(t_0) \left[-(1-a)\bar{s} \int_a^{t_0} \Phi(t_0, s)ds - (1-a)\nu \int_a^{t_0} \Phi(t_0, s))dw(s) \right]$$

$$= \Phi^{-1}(t_0)[r_0]^{1-a}$$

$$+ \Phi^{-1}(t_0)\Phi(t_0) \left[-(1-a)\bar{s} \int_a^{t_0} \Phi^{-1}(s)ds - (1-a)\nu \int_a^{t_0} \Phi^{-1}(s)dw(s) \right]$$

$$= \Phi^{-1}(t_0)[r_0]^{1-a} - (1-a)\bar{s} \int_a^{t_0} \Phi^{-1}(s)ds - (1-a)\nu \int_a^{t_0} \Phi^{-1}(s)dw(s),$$

We substitute this in the above equation,

$$V(r(t)) = [r(t)]^{1-a} = \Phi(t)\left[m(a) + (1-a)\bar{s}\int_a^t \Phi^{-1}(s)ds\right.$$

$$\left. + (1-n)\nu\int_a^t \Phi^{-1}(s)dw(s)\right]$$

$$= \Phi(t)\left[\Phi^{-1}(t_0)[r_0]^{1-a} - (1-a)\bar{s}\int_a^{t_0}\Phi^{-1}(s)ds\right.$$

$$- (1-n)\nu\int_a^{t_0}\Phi^{-1}(s)dw(s)$$

$$\left. + (1-a)\bar{s}\int_a^t \Phi^{-1}(s)ds + (1-n)\nu\int_a^t \Phi^{-1}(s)dw(s)\right]$$

(by substitution)

$$= \Phi(t)\left[\Phi^{-1}[r_0]^{1-a} + (1-a)\bar{s}\int_a^t \Phi^{-1}(s)ds - (1-a)\bar{s}\int_a^{t_0}\Phi(t_0,s)ds\right.$$

$$\left. + (1-n)\nu\int_a^t \Phi^{-1}(s)dw(s) - (1-n)\nu\int_a^{t_0}\Phi^{-1}(s)dw(s)\right]$$

(by regrouping)

$$= \Phi(t,t_0)[r_0]^{1-a} + (1-a)\bar{s}\int_{t_0}^t \Phi(t,s)ds$$

$$+ (1-a)\nu\int_a^{t_0}\Phi(t,s)dw(s) \quad \text{(by simplification and properties of } \Phi\text{)}.$$

The capital per worker $r(t)$ at any time t is given by

$$r(t) = \left(\Phi(t,t_0)[r_0]^{1-a} + (1-a)\bar{s}\int_{t_0}^t \Phi(t,s)ds\right.$$

$$\left. + (1-a)\nu\int_a^{t_0}\Phi(t,s)dw(s)\right)^{1/(1-a)}.$$

This completes the solution process of the given problem.

Example 3.7.1. Given:

$$dx = [x\sin^2 t + x^2\sin 2t + x^5\cos 2t]dt + [x\sin t + x^2\cos t]dw(t), \quad x(t_0) = x_0.$$

Find: (a) the general solution and (b) the particular solution to the IVP.

Solution procedure. We observe that this is the Bernoulli-type SDE with $n = 2$, $P(t) = \sin^2 t$, $Q(t) = \sin 2t$, $\Sigma(t) = \sin t$, and $\Upsilon(t) = \cos t$. The candidate for the energy function in (3.7.6) with $C = 1$ is $V(x) = x^\delta$. From (3.7.13), (3.7.14), (3.7.17), and (3.7.18), $V(t, x)$ in (3.7.6) and the rate functions in (3.4.1) [(3.7.2) and (3.7.3)] are $V(t, x) = x^{-1}$ ($\delta = -1$), $\nu(t) = -\sin t$, $q(t) = -\cos t$, $\mu(t) = -(\sin^2 t - \sin^2 t) = 0$, and $p(t) = -(\sin 2t - 2\sin t \cos t) = 0$. The reduced SDE (3.7.19) is

$$dm = [-\sin tm - \cos t]dw(t),$$

and its solution is

$$x^{-1}(t) = V(x(t)) = m(t)$$

$$= \Phi(t, a)m(a) + \int_a^t \Phi(t, r)[p(r) - \nu(r)q(r)]ds + \int_a^t q(r)dw(r)$$

$$= \Phi(t, a)m(a) - \int_a^t \Phi(t, r)[\sin r \cos r]dr - \int_a^t \Phi(t, r)\cos r \, dw(r),$$

where

$$\Phi(t, a) = \exp\left[-\frac{1}{2}\int_a^t \sin^2 u \, du - \int_a^t \sin u \, dw(u)\right].$$

Thus by setting $m(a) = c$, the general solution to the given SDE is

$$x(t) = \left(\Phi(t, a)c - \frac{1}{2}\int_a^t \Phi(t, r)\sin 2r \, dr - \int_a^t \Phi(t, r)\cos r \, dw(r)\right)^{-1}.$$

For given $t = t_0$ and $x(t_0) = x_0 \neq 0$, the solution to the IVP is

$$\Phi^{-1}(t_0, a)x_0^{-1} = c - \frac{1}{2}\int_a^{t_0} \Phi^{-1}(r, a)\sin 2r \, dr - \int_a^{t_0} \Phi^{-1}(r, a)\cos r \, dw(r)$$

and hence

$$c = \Phi^{-1}(t_0, a)x_0^{-1} + \frac{1}{2}\int_a^{t_0} \Phi^{-1}(r, a)\sin 2r \, dr + \int_a^{t_0} \Phi^{-1}(r, a)\cos r \, dw(r).$$

By substituting this into the above expression of $x(t)$ and using the properties of the integral, we get

$$x(t) = \left(\Phi(t, t_0)x_0^{-1} - \frac{1}{2}\int_{t_0}^t \Phi(t, r)\sin 2r \, dr - \int_{t_0}^t \Phi(t, r)\cos r \, dw(r)\right)^{-1}.$$

This is the desired solution to the IVP.

Example 3.7.2. Given:

$$dx = [4x + 5x^{2/3} + x^{1/3}]dt + \sqrt{3}[x + x^{2/3}]dw(t).$$

Find the general solution to the given differential equation.

Solution procedure. We observe that this is the Bernoulli-type SDE with $n = \frac{2}{3}$, $P(t) = 4$, $Q(t) = 5$, and $\Sigma(t) = \sqrt{3} = \Upsilon(t)$. The candidate for the energy function in (3.7.6) with $C = 1$ is $V(x) = x^\delta$. In this case, from (3.7.13), (3.7.14), (3.7.17), and (3.7.18), $V(t,x)$ in (3.7.6) and the rate functions in (3.4.1) [(3.7.2) and (3.7.3)] are $V(t,x) = x^{1/3}$ $(\delta = \frac{1}{3})$, $\nu(t) = \frac{1}{\sqrt{3}}$, $q(t) = \frac{1}{\sqrt{3}}$, $\mu(t) = \frac{1}{3}(4-1) = 1$, and $p(t) = \frac{1}{3}(5-2) = 1$. The reduced differential equation (3.7.19) is

$$dm = (m+1)dt + \frac{1}{\sqrt{3}}(m+1)dw(t),$$

and its solution is given by

$$x^{1/3} = V(x(t)) = m(t)$$

$$= \Phi(t,a)m(a) + \int_a^t \Phi(t,s)\left[1 - \frac{1}{3}\right]ds + \int_a^t \frac{1}{\sqrt{3}}dw(s)$$

$$= \Phi(t,a)m(a) + \frac{2}{3}\int_a^t \Phi(t,r)dr + \frac{1}{\sqrt{3}}\int_a^t \Phi(t,r)dw(r),$$

where

$$\Phi(t,a) = \exp\left[\frac{5}{6}(t-a) + \frac{1}{\sqrt{3}}(w(t) - w(a))\right].$$

The general solution to the given differential equation is

$$x(t) = \left[\Phi(t,a)m(a) + \frac{2}{3}\int_a^t \Phi(t,r)dr + \frac{1}{\sqrt{3}}\int_a^t \Phi(t,r)dw(r)\right]^3.$$

Example 3.7.3. Given: $dx = [1x + 4x^{4/5} + 2x^{3/5}]dt + \sqrt{5}[x + x^{4/5}]dw(t)$. Find the general solution to the given differential equation.

Solution procedure. We observe that this is the Bernoulli-type differential equation with $n = \frac{4}{5}$, $P(t) = 1$, $Q(t) = 4$, and $\Sigma(t) = \sqrt{5} = \Upsilon(t)$. The candidate for the energy function in (3.7.6) with $C = 1$ is $V(x) = x^\delta$. In this case, from (3.7.13), (3.7.14), (3.7.17), and (3.7.18), $V(t,x)$ in (3.7.6) and the rate functions in (3.4.1) [(3.7.2) and (3.7.3)] are $V(t,x) = x^{1/5}$ $(\delta = \frac{1}{5})$, $\nu(t) = \frac{1}{\sqrt{5}}$, $q(t) = \frac{1}{\sqrt{5}}$, $\mu(t) = \frac{1}{5}(1-2) = -\frac{1}{5}$, and $p(t) = \frac{1}{5}(4-4) = 0$. The reduced differential equation (3.7.19) is

$$dm = -\frac{1}{5}m\,dt + \frac{1}{\sqrt{5}}(m+1)dw(t),$$

and its solution is

$$x^{1/5} = V(x(t)) = m(t)$$

$$= \Phi(t,a)m(a) + \int_a^t \Phi(t,s)\left[-\frac{1}{5}\right]ds + \int_a^t \frac{1}{\sqrt{5}}dw(s)$$

$$= \Phi(t,a)m(a) - \frac{1}{5}\int_a^t \Phi(t,r)dr + \frac{1}{\sqrt{5}}\int_a^t \Phi(t,r)dw(r),$$

where

$$\Phi(t, a) = \exp\left[-\frac{3}{10}(t - a) + \frac{1}{\sqrt{5}}(w(t) - w(a))\right].$$

This is the general solution to the given differential equation.

$$x(t) = \left[\Phi(t, a)m(a) - \frac{1}{5}\int_a^t \Phi(t, r)dr + \frac{1}{\sqrt{5}}\int_a^t \Phi(t, r)dw(r)\right]^5.$$

This is the desired general solution.

Example 3.7.4 (epidemic model (Illustration 3.1.4)). Given: $dI = \alpha I(\bar{N} - I)dt + \sigma I dw(t)$, $I(0) = 1$. Find the general solution, and the particular solution to the IVP.

Solution procedure. We observe that this is the Bernoulli-type differential equation with $n = 2$, $P(t) = \alpha\bar{N}$, $Q(t) = -\alpha$, $\Sigma(t) = \sigma$, and $\Upsilon(t) = 0$. The candidate for the energy function in (3.7.6) with $C = 1$ is $V(x) = x^\delta$. From (3.7.13), (3.7.14), (3.7.17), and (3.7.18), $V(t, x)$ in (3.7.6) and the rate functions in (3.4.1) [(3.7.2) and (3.7.3)] are $V(t, x) = x^{-1}$ ($\delta = -1$), $\nu(t) = -\sigma$, $q(t) = 0$, $\mu(t) = -(\alpha\bar{N} - \sigma^2)$, and $p(t) = -(-\alpha) = \alpha$. The reduced differential equation (3.7.19) is

$$dm = [-(\alpha\bar{N} - \sigma^2)m + \alpha]dt - \sigma m\, dw(t),$$

and its solution is

$$I^{-1}(t) = V(I(t)) = m(t)$$

$$= \Phi(t, a)c + \int_a^t \Phi(t, s)\alpha\, ds$$

$$= \Phi(t, a)c + \alpha\int_a^t \Phi(t, s)ds,$$

where $c = m(a)$ and

$$\Phi(t, a) = \exp\left[-\left(\alpha\bar{N} - \frac{1}{2}\sigma^2\right)(t - a) - \sigma(w(t) - w(a))\right].$$

Thus, the general solution to the given differential equation is

$$I(t) = \left(\Phi(t, a)c + \alpha\int_a^t \Phi(t, s)ds\right)^{-1}.$$

For $t = t_0$ and $I(0) = I_0 = 1$, the solution to the given IVP

$$\Phi^{-1}(t_0, a)I_0^{-1} = c + \alpha\int_a^{t_0} \Phi^{-1}(r, a)dr$$

and hence

$$c = \Phi^{-1}(t_0, a)I_0^{-1} - \alpha\int_a^{t_0} \Phi^{-1}(r, a)dr.$$

By substituting this into the above expression of $I(t)$ and using the properties of the integral, we get

$$I(t) = \left(\Phi(t, t_0) I_0^{-1} + \alpha \int_{t_0}^t \Phi(t, r) dr \right)^{-1} = \left(\frac{\Phi(t, t_0) + \alpha I_0 \int_{t_0}^t \Phi(t, r) dr}{I_0} \right)^{-1}$$

$$= \frac{I_0 \exp\left[\left(\alpha \bar{N} - \frac{1}{2}\sigma^2 \right)(t - t_0) + \sigma(w(t) - w(t_0)) \right]}{1 + \alpha I_0 \int_{t_0}^t \exp\left[\left(\alpha \bar{N} - \frac{1}{2}\sigma^2 \right)(s - t_0) + \sigma(w(s) - w(t_0)) \right] ds}.$$

This is the desired solution to the IVP.

Illustration 3.7.3. (population extinction–Richard Levin's model [128]). In evolutionary biology, the dynamic approach has attracted some interested in the study of the extinction of species. The study of extinction plays an important role in the area of economic entomology and the theory of pest control. It has been suggested that extinction is affected by the physical environment. The species adaptability to the environmental changes has also some limit. Richard Levin's demographic equilibrium of population extinction model (PEM) [128] is based on the following considerations:

PEM1. A species is growing in a large number of subcommunities. Each subcommunity is well-established.

PEM2. The members of a subcommunity are subject to migration, and eventually go extinct.

PEM3. The species in each subcommunity satisfy the Pearl–Verhulst logistic equation (Illustration 3.2.3 [80]).

Under these considerations, the mathematical model for each subcommunity of a given species is described by

$$dN = [mN(1 - N) - eN]dt, \quad N(0) = N_0,$$

where N is the size of the species in a given subcommunity of the community, and e stands for the average extinction rate per species.

We observe that $mN(1 - N) - eN = mN[(1 - \frac{e}{m}) - N]$. By setting $g(t) = m$, $k = 1 - \frac{e}{m}$ and imitating the solution procedure of Example 3.6.2 [80], the solution to the above IVP is given by

$$N(t) = \frac{k N_0 \exp[mkt]}{(k - N_0) + N_0 \exp[mkt]} = \frac{k N_0}{N_0 + (k - N_0) \exp[-mkt]}$$

$$= \frac{N_0}{\frac{N_0}{k} + \left(1 - \frac{N_0}{k}\right) \exp[-(m - e)t]} = \frac{N_0}{\frac{mN_0}{m-e} + \left(1 - \frac{mN_0}{m-e}\right) \exp[-(m - e)t]}.$$

From the above expression for the size of species, we remark that for $e < m$, the species will persist, and for $e > m$, the species goes toward extinction at the rate of $e - m$.

PEM4. Here, the extinction rate parameter can be treated as a source of random environmental perturbations. It is characterized by the $\xi(x)$ stationary Gaussian white noise process in (3.1.6). The presented deterministic model can be reduced to

$$dN = mN \left[\left(1 - \frac{e}{m} \right) - N \right] dt + \sigma N \, dw(t), \quad N(0) = N_0.$$

By repeating the argument used in Example 3.7.4, the solution process of the above stochastic differential equation is

$$N(t) = \frac{N_0 \exp \left[\left(m - e - \frac{1}{2}\sigma^2 \right) (t - t_0) + \sigma(w(t) - w(t_0)) \right]}{1 + mN_0 \int_{t_0}^t \exp \left[\left(m - e - \frac{1}{2}\sigma^2 \right) (s - t_0) + \sigma(w(s) - w(t_0)) \right] ds}.$$

Brief Summary. A summary of the basic mechanical steps needed to solve nonlinear stochastic scalar differential equations of the Bernoulli-type is given in the following flowchart:

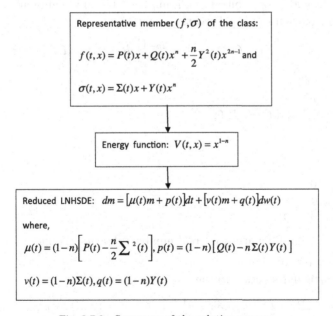

Fig. 3.7.2   Summary of the solution process.

Exercises

1. Find the general solution to the following SDE:

(a) $dx = (4x + 5x^{2/3})dt + \sqrt{3}x \, dw(t)$;

(b) $dx = 5x^{2/3}dt + \sqrt{3}x\,dw(t)$;

(c) $dx = x^{1/3}dt + \sqrt{3}(x + x^{2/3}dw(t)$;

(d) $dx = 2x^{4/5}dt + \sqrt{5}x\,dw(t)$;

(e) $dx = (1x + 2x^{4/5})dt + \sqrt{5}x\,dw(t)$;

(f) $dx = 2x^{3/5}dt + \sqrt{5}[x + x^{4/5}]dw(t)$;

(g) $dx = [\cos^2 tx + \sin 2xtx^3]dt + \cos tx\,dw(t)$;

(h) $dx = [\sin 2tx^2 + \sin^2 tx^5]dt + [\cos tx + \sin tx^2]dw(t)$;

(i) $dx = [\cos^2 tx + \sin 2tx^2 + \sin^2 tx^3]dt + [\cos tx + \sin tx^2]dw(t)$;

(j) $dx = [(\sec^2 t + \tan^2 t)x + \frac{1}{2}x^5 + \frac{5}{2}\cot^2 tx^9]dt + \tan t(x + \cot tx^5)dw(t)$;

(k) $dx = [(\sec^2 t + \tan^2 t)x + \frac{1}{2}x^5]dt + \tan tx\,dw(t)$;

(l) $dx = \left[\frac{3}{2(1+t)}x + \frac{1}{2}x^{1/2} + \frac{1+t}{4}\right]dt + \left[\frac{1}{\sqrt{1+t}}x + \sqrt{(1+t)}x^{1/2}\right]dw(t)$.

2. Find the solution to the following IVPs:

(a) $dx = (1x + 2x^{4/5})dt + \sqrt{5}x\,dw(t)$, $x(0) = 1$;

(b) $dx = [\cos^2 tx + \sin 2tx^{-2}]dt + \cos tx\,dw(t)$, $x(0) = 1$;

(c) $dx = [\sin^2 tx + \sin 2tx^3]dt + \cos tx\,dw(t)$.

3. Let P, Q, Σ, and Υ be arbitrary real numbers. For any real number $n \neq 1$, find the general solution to the differential equation

$$dx = \left[Px + Qx^n + \frac{n}{2}\Upsilon^2 x^{2n-1}\right]dt + [\Sigma x + \Upsilon x^n]dw(t)$$

(a) for $\Sigma^2 = P$;

(b) for $Q = n\Sigma\Upsilon$;

(c) for $\Sigma^2 = P$ and $Q = n\Sigma\Upsilon$;

(d) for $\Upsilon = 0$;

(e) for $\Sigma = 0$;

(f) for $P = 0$;

(g) for $Q = 0$;

(h) for $\Sigma = 0 = \Upsilon$;

(i) for $P = 0 = Q$.

4. Let P, Q, Σ, and Υ be arbitrary real numbers. For $n = 2$, find:

(a) The general solution;

(b) The particular solution to the initial value problems

$$dx = [Px + Qx^{1.1} + .55\Upsilon^2 x^{1.2}]dt + [\Sigma x + \Upsilon x^{1.1}]dw(t), \quad x(t_0) = x_0.$$

   (i) for $\Sigma^2 = P$;

   (ii) for $Q = 2\Sigma\Upsilon$;

   (iii) for $\Sigma^2 = P$ and $Q = 2\Sigma\Upsilon$;

   (iv) for $\Upsilon = 0$;

   (v) for $\Sigma = 0$;

   (vi) for $P = 0$;

(vii) for $Q = 0$;

(viii) for $\Sigma = 0 = \Upsilon$;

(ix) for $P = 0 = Q$.

5. Let P, Q, Σ, and Υ be arbitrary real numbers. For $n = 3$, find:

(a) The general solution;

(b) The particular solution to the initial value problems

$$dx = \left[Px + Qx^{1.4} + .7\Upsilon^2 x^{1.8}\right] dt + [\Sigma x + \Upsilon x^{1.4}]dw(t), \quad x(t_0) = x_0.$$

(i) for $\Sigma^2 = P$;

(ii) for $Q = 3\Sigma\Upsilon$;

(iii) for $\Sigma^2 = P$ and $Q = 3\Sigma\Upsilon$;

(iv) for $\Upsilon = 0$;

(v) for $\Sigma = 0$;

(vi) for $P = 0$;

(vii) for $Q = 0$;

(viii) for $\Sigma = 0 = \Upsilon$;

(ix) for $P = 0 = Q$.

6. Find the solution process of the IVPs associated with differential equations in Exercises 3–5 with the initial conditions at $t = t_0$ and $x(t_0) = x_0 \neq 0$.

3.8 Essentially Time-Invariant Equations

In this section, we present another subclass of differential equations (3.2.1) that is reducible to (3.4.1). This class of differential equations is equivalent to time-invariant differential equations.

Definition 3.8.1. A differential equation (3.2.1) is said to be an essentially time-invariant differential equation if the rate functions $f(t, x)$ and $\sigma(t, x)$ in (3.2.1) are of the following forms: $f(t, x) = F(ax + bt + c)$ and $\sigma(t, x) = \Sigma(ax + bt + c)$.

3.8.1 *General Problem*

Procedure 3.5.2 is applied to find a solution representation of the following essentially time-invariant stochastic differential equation of the Itô–Doob-type (3.2.1):

$$dx = F(ax + bt + c)dt + \Sigma(ax + bt + c)dw(t), \tag{3.8.1}$$

where F and Σ are smooth enough to assure the existence of a solution process of (3.8.1); and a, b, and c are arbitrary given real numbers. The described procedure leads to the reduction of the SDE (3.4.1).

(H$_{3.8}$). In addition to hypothesis (H$_{3.2}$), we assume that $F(v)$ and $\Sigma(v)$ are continuously differentiable with respect to v. In addition, $a\Sigma(v) \neq 0$, and $H(v) = \int_c^v \frac{dy}{a\bar{z}(y)}$ exists and invertible.

3.8.2 *Procedure for Finding a General Solution Representation*

Using Procedure 3.4.2 and Observation 3.4.1(i), the process for reduction of a subclass of stochastic differential equations (3.2.1) in (3.4.1) is repeated. For this subclass, we find a suitable energy/Lyapunov function in a unified and systematic way.

Step #1. Utilizing the nature of the subclass of SDEs described by (3.8.1), we associate an unknown energy/Lyapunov function in Step #1 of General Procedure 3.2.2:

$$V(t,x) = G(ax + bt + c), \tag{3.8.2}$$

where G is an unknown energy/Lyapunov function. The problem of seeking energy function V is equivalent to the problem of seeking unknown function G, and a, b, and c are arbitrary given real numbers in (3.8.1).

Step #2. The computation of the differential $dV(t,x)$ in (3.2.2) is achieved by using substitution

$$v = ax + bt + c. \tag{3.8.3}$$

From (3.8.2) and (3.8.3), we have

$$\frac{\partial}{\partial t} V(t, x(t)) = \frac{\partial}{\partial t} G(ax + bt + c) \quad \text{(by the chain rule)}$$

$$= \frac{d}{dv} G(v) \frac{\partial}{\partial t} v \quad \left[\text{from (3.8.3)}, \frac{\partial}{\partial t} v = \frac{\partial}{\partial t}(ax + bt + c) = b\right]$$

$$= \frac{d}{dv} G(v) b, \tag{3.8.4}$$

$$\frac{\partial}{\partial x} V(t, x(t)) = \frac{\partial}{\partial x} G(ax + bt + c) \quad \text{(by the chain rule)}$$

$$= \frac{d}{dv} G(v) \frac{\partial}{\partial x} v \quad \left[\text{from (3.8.3)}, \frac{\partial}{\partial x} v = \frac{\partial}{\partial x}(ax + bt + c) = a\right]$$

$$= \frac{d}{dv} G(v) a, \tag{3.8.5}$$

$$\frac{\partial^2}{\partial x^2} V(t, x(t)) = \frac{\partial^2}{\partial x^2} G(ax + bt + c) \quad \text{(by the chain rule)}$$

$$= \frac{d}{dv}a\frac{d}{dv}G(v)\frac{\partial}{\partial x}v \quad \left[\text{from (3.8.5)}, \frac{\partial}{\partial x}v = \frac{\partial}{\partial x}(ax+bt+c) = a\right]$$

$$= \frac{d^2}{dv^2}G(v)a^2. \tag{3.8.6}$$

Hence, from (3.2.2), (3.2.3), (3.8.1)–(3.8.6), we have

$$dV(t, x(t)) = dG(ax(t) + bt + c) \quad \text{[from (3.8.1), (3.2.2), and (3.8.5)]}$$

$$= LG(ax + bt + c)dt$$

$$+ \Sigma(ax(t) + bt + c)\frac{d}{dv}G(v)dw(t)$$

[from (3.8.1), (3.2.2), and (3.8.5)]

$$= LG(v)dt + a\Sigma(v)\frac{d}{dv}G(v)dw(t) \quad \text{(by substitution)}$$

$$= LG(v)dt + a\Sigma(v)\frac{d}{dv}G(v)dw(t) \quad \text{[by (3.8.3) and (3.8.5)]}, \tag{3.8.7}$$

where w is the Wiener process defined in Definition 1.2.12, $x(t)$ is determined by (3.8.1), and L is a linear operator associated with (3.8.1)–(3.8.7):

$$LV(t, x(t)) = \frac{d}{dv}G(v)b + F(ax(t) + bt + c)\frac{d}{dv}G(v)a$$

$$+ \frac{a^2}{2}\Sigma^2(ax(t) + bt + c)\frac{d^2}{dv^2}G(v)$$

$$= b\frac{d}{dv}G(v) + aF(v)\frac{d}{dv}G(v) + \frac{a^2}{2}\Sigma^2(v)\frac{d^2}{dv^2}G(v) \quad \text{[from (3.8.3)]}.$$

Thus, from this and (3.8.2),

$$LG(v) = [aF(v) + b]\frac{d}{dv}G(v) + \frac{a^2}{2}\Sigma^2(v)\frac{d^2}{dv^2}G(v). \tag{3.8.8}$$

From (3.8.2), (3.8.3), (3.8.7), and (3.8.8), the expressions (3.4.2) and (3.4.3) reduce to

$$\mu(t)G(v) + p(t) = [aF(v) + b]\frac{d}{dv}G(v) + \frac{a^2}{2}\Sigma^2(v)\frac{d^2}{dv^2}G(v), \tag{3.8.9}$$

$$\nu(t)G(v) + q(t) = a\Sigma(v)\frac{d}{dv}G(v), \tag{3.8.10}$$

respectively.

Step #3. Using (3.8.10), we find a candidate for $G(v)$. For this purpose, from (3.8.10), and by knowing this structure, we choose q, ν, μ, and p in (3.4.1) as follows: $q(t) = \gamma$, $\nu(t) = \delta \neq 0$, $\mu(t) = \epsilon$, and $p(t) = \eta$, where δ, γ, η, and ϵ are arbitrary constants; from the hypothesis (H$_{3.8}$), we have $a\Sigma(v) \neq 0$ on R. Therefore, Step #3 of Procedure 3.4.2 is satisfied. Thus, we have

$$\frac{d}{dv}G(v) = \frac{\nu(t)G(v) + q(t)}{a\Sigma(v)} = \frac{\delta G(v)}{a\Sigma(v)} + \frac{\gamma}{a\Sigma(v)} = \frac{\delta G(v)}{a\Sigma(v)} + \frac{\gamma}{a\Sigma(v)}. \tag{3.8.11}$$

In this case, (3.4.5), (3.4.10), (3.4.11), and (3.4.17) reduce to

$$G(v) = \exp\left[\delta H(v)\right]C - \frac{\gamma}{\delta}, \quad \text{(from (H}_{3.8}\text{))} \tag{3.8.12}$$

$$\left[\delta\left[\frac{aF(v) + b}{a\Sigma(v)} - \frac{a^2}{2}\frac{d}{dx}\Sigma(v)\right] + \frac{1}{2}\delta^2 a^2 - \epsilon\right]\exp\left[\int_c^x \frac{\delta}{a\Sigma(y)}dy\right] = \frac{\delta\epsilon - \eta\gamma}{\delta C},$$
$$\tag{3.8.13}$$

where $C \neq 0$ is an arbitrary constant.

$$z(v) = \frac{aF(v) + b}{a\Sigma(v)} - \frac{a^2}{2}\frac{d}{dx}\Sigma(v), \tag{3.8.14}$$

and for $\frac{d}{dx}(z(v)) \neq 0$

$$\frac{d}{dv}\left[\frac{\frac{d}{dv}(a\Sigma(v)\frac{d}{dv}z(v))}{\frac{d}{dv}z(v)}\right] = 0, \tag{3.8.15}$$

respectively.

Step #4. Finally, we repeat Steps #3 and #4 of Procedure 3.4.2. This completes the procedure for solving the essentially time-invariant stochastic differential equation of the Itô–Doob-type (3.8.1).

Observation 3.8.1

(i) In (3.8.11), if $\delta = 0$, then (3.8.11), (3.8.12), (3.8.13), and (3.8.15) reduce to

$$\frac{d}{dx}H(v) = \frac{\gamma}{a\Sigma(v)}, \tag{3.8.16}$$

$$H(v) = \gamma\int_c^x \frac{dy}{a\Sigma(y)} + H(c), \tag{3.8.17}$$

$$\gamma\left[\frac{aF(v) + b}{a\Sigma(v)} - \frac{a^2}{2}\frac{d}{dx}\Sigma(v)\gamma\right] = \mu H(v) + p = \gamma\mu\frac{1}{a\Sigma(v)} + p + \mu H(c),$$
$$\tag{3.8.18}$$

$$\frac{d}{dv}\left(a\Sigma(v)\frac{d}{dv}z(v)\right) = 0, \tag{3.8.19}$$

respectively.

(ii) From Observation 3.4.1(iv) and for $\delta \neq 0$, the representative of this subclass is determined by

$$aF(v) + b = a\Sigma(v)\left[\frac{1}{2}a\frac{d}{dv}\Sigma(v) - \frac{f_1}{b}\exp\left[-\delta\int_c^v \frac{dy}{a\Sigma(y)}\right] + f_2\right]. \qquad (3.8.20)$$

From (3.8.20) and knowing the rate functions along with using the method of undetermined coefficients, δ, f_1, and f_2 need to be determined. Now, from (3.4.12) and (3.4.41), $\mu(t)$ and $p(t)$ in (3.4.1) are as follows:

$$\mu(t) = \frac{1}{2}\delta^2 + \delta f_2, \quad p(t) = \frac{1}{2}\delta\gamma + \gamma f_2 - G(c)f_1, \qquad (3.8.21)$$

where f_1 and f_2 are arbitrary constants of integration. Again, we use the method of undetermined coefficients to find f_1 and f_2. In the case of $\delta = 0$,

$$aF(v) + b = a\Sigma(v)\left[\frac{1}{2}\frac{d}{dv}a\Sigma(v) + f_1\int_c^x \frac{dy}{a\Sigma(y)} + f_2\right], \qquad (3.8.22)$$

and from (3.4.49), $\mu(t)$ and $p(t)$ in (3.4.42) reduce to

$$\mu(t) = f_1 \text{ and } p(t) = f_2. \qquad (3.8.23)$$

(iii) We further observe that under the transformation, $v = ax + bt + c$ in (3.8.3) the differential equation (3.8.1) reduces to its equivalent form:

$dv = a\,dx + b\,dt \quad$ [from (3.8.1)]

$\quad = aF(ax + bt + c)dt + a\Sigma(ax + bt + c)dw(t) + b\,dt \quad$ (by substitution)

$\quad = aF(v)dt + a\Sigma(v)dw(t) + b\,dt \quad\quad\quad\quad$ (by regrouping)

$\quad = [aF(v) + b]dt + a\Sigma(v)dw(t). \qquad\qquad\qquad (3.8.24)$

(iv) All outlined statements of Observations 3.4.1(vi) are applicable to (3.8.24). In the light of this observation, illustrations and examples can be constructed accordingly.

(v) By following the argument used in (iii), one can extend this approach for more general differential equations than (3.8.1). For instance, by using the transformation in (iii), the class of SDEs can reduced to variable separable and Bernoulli-type SDEs. Let us consider the following SDEs:

$$dx = \left[\kappa(t)F(ax + bt + c) - \frac{b}{a}\right]dt + g(t)\Sigma(ax + bt + c)dw(t), \qquad (3.8.25)$$

$$dx = \left[P(t)(ax + bt + c) + Q(t)(ax + bt + c)^n\right.$$

$$+ \frac{n}{2} \Upsilon^2(t)(ax + bt + c)^{2n-1} - \frac{b}{a} \bigg] dt$$

$$+ [\Sigma(t)(ax + bt + c) + \Upsilon(t)(ax + bt + c)^n] dw(t) \qquad (3.8.26)$$

are reduced to

$$dv = a\kappa(t)F(v)dt + ag(t)\Sigma(v)dw(t), \qquad (3.8.27)$$

$$dv = \left[aP(t)v + aQ(t)v^n + \frac{n}{2} a\Upsilon^2(t)v^{2n-1} \right] dt + [a\Sigma(t)v + a\Upsilon(t)v^n] dw(t),$$

$$(3.8.28)$$

respectively. With this discussion, one can apply the procedures described in Sections 3.5 and 3.7 to find the solution representations of the class of differential equations defined by (3.8.25) and (3.8.26).

Brief Summary. The flowchart (Figure 3.8.1) summarizes the steps of finding the general of a class of essentially time-invariant stochastic nonlinear scalar differential equations via the energy.

In the following, we present a few examples to demonstrate the procedure described in this section.

Example 3.8.1. Given:

$$dx = \frac{1}{a} \left[\frac{1}{2} \sin 4(ax + bt + c) + \cos^2(ax + bt + c) - b + \sin 2(ax + bt + c) \right] dt$$

$$+ \frac{1}{a} \sin 2(ax + bt + c)dw(t), \quad x(t_0) = x_0,$$

where a, b, and c are any given real numbers, and $a \neq 0$. Find: (a) the general solution, and (b) the solution process of the given initial value problem.

Solution procedure. We observe that this SDE is of type (3.8.1) with

$$\Sigma(ax + bt + c) = \frac{1}{a} \sin 2(ax + bt + c),$$

and

$$F(ax + bt + c) = \frac{1}{a} \left[\frac{1}{2} \sin 4(ax + bt + c) + \cos^2(ax + bt + c) \right.$$

$$\left. - b + \sin 2(ax + bt + c) \right].$$

By following the steps in Procedure 3.8.2 and Observation 3.8.1, we compute the expressions (3.8.20) and the differential equation (3.8.24) in the context of the given

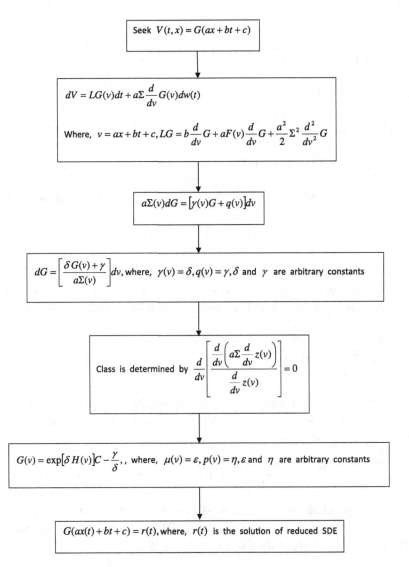

Fig. 3.8.1 Energy function method and essentially time-invariant equations.

example as:

$$aF(v) + b = \frac{1}{2}\sin 4(ax + bt + c) + \cos^2(ax + bt + c) + \sin 2(ax + bt + c)$$

$$= \frac{1}{2}\sin 4v + \cos^2 v + \sin 2v,$$

$$a\Sigma(v) = \sin 2v, \quad a\frac{d}{dv}\Sigma(v) = 2\cos 2v, \quad \int_c^v \frac{dy}{a\Sigma(y)}$$

$$= \int \frac{dy}{\sin 2y} = \frac{1}{2}[\ln\tan v - \ln\tan c],$$

and

$$\exp\left[-\delta\int_c^v \frac{dy}{a\Sigma(y)}\right] = \exp\left[-\delta\left[\frac{1}{2}\ln(\tan v)\right]\right] = [\tan v]^{-\delta/2}.$$

Thus,

$$aF(v) + b = a\Sigma(v)\left[\frac{1}{2}a\frac{d}{dv}\Sigma(v) - \frac{f_1}{\delta}\exp\left[-\delta\int_c^v \frac{dy}{a\Sigma(y)}\right] + f_2\right]$$

(by substitution)

$$= \sin 2v\left[\frac{1}{2}2\cos 2v - \frac{f_1}{\delta[\tan v]^{\delta/2}} + f_2\right] \quad \text{(by simplification)}$$

$$= \sin 2v\cos 2v - \frac{2f_1\sin v\cos v}{\delta[\tan v]^{\delta/2}} + f_2\sin 2v,$$

for unknown constants δ, f_1, and f_2. The subgoal is to determine these constants for the given expression $\frac{1}{2}\sin 4v + \cos^2 v + \sin 2v$. For this purpose, we use the above two expressions of $aF(v) + b$ and the method of undetermined coefficients as

(a) $\sin 2v\cos 2v = \frac{1}{2}\sin 4v$;
(b) $-\frac{2f_1\sin v\cos v}{\delta[\tan v]^{\delta/2}} = \cos^2 v$;
(c) $f_2\sin 2v = \sin 2v$.

We note that (a) is valid from the trigonometric function "double-angle formula." Equation (c) is valid for $f_2 = 1$. Equation (b) is valid if $\delta = 2$, $\gamma = 0$, and $f_1 = -1$. Let us verify this statement:

$$-\frac{2f_1\sin v\cos v}{2[\tan v]} = \frac{\sin v\cos v}{\tan v} = \sin v\cos v\cot v = \sin v\frac{\cos^2 v}{\sin v} = \cos^2 v.$$

From this, we conclude that the given SDE satisfies the condition (3.4.37). Therefore, the given SDE belongs to the class LSDECC-(f,σ) defined in Observation 3.4.1(iv). In this case, the reducible SDE (3.4.1) is

$$dm = (4m + 1)dt + 2m\,dw(t).$$

This is in view of Step #3 of Procedure 3.8.2, (3.8.21), $q(t) = 0$, $v(t) = 2$, $C = 1$, $\mu(t) = 2$, and $p(t) = 1$. The solution representation of the given differential equation is

$$V(t, v(t)) = \exp\left[\delta\left[\frac{1}{2}\ln(\tan v)\right]\right] = [\tan v]^{2/2} = \tan v = m(t)$$

$$= \Phi(t)c + \int_a^t \Phi(t, r)dr,$$

where c is an arbitrary constant of integration and $\Phi(t) = \exp[2t + 2w(t)]$. The general solution to the given SDE is provided by

$$\tan(ax(t) + bt + c) = \Phi(t)c + \int_a^t \Phi(t, r)dr.$$

The solution to the IVP is determined by

$$\tan(ax(t_0) + bt_0 + c) = \Phi(t_0)c + \int_a^{t_0} \Phi(t_0, r)dr,$$

and hence

$$\tan(ax_0 + bt_0 + c) = \Phi(t_0)\left[c + \int_a^{t_0} \Phi^{-1}(t_0, r)dr\right],$$

which implies that

$$c = \Phi^{-1}(t_0)\tan(ax_0 + bt_0 + c) - \int_a^{t_0} \Phi^{-1}(r)dr.$$

Substituting this value of an arbitrary constant of integration into the above expression, we get

$$\tan(ax(t) + bt + c) = \Phi(t)c + \int_a^t \Phi(t, r)dr \quad \text{(by substitution)}$$

$$= \Phi(t)\left[\Phi^{-1}(t_0)\tan(ax_0 + bt_0 + c) - \int_a^{t_0} \Phi^{-1}(r)dr \right.$$

$$\left. + \int_a^t \Phi(t, r)dr\right] \quad \text{(by properties of } \Phi)$$

$$= \Phi(t, t_0)\tan(ax_0 + bt_0 + c)$$

$$- \int_a^{t_0} \Phi(t, r)dr + \int_a^t \Phi(t, r)dr$$

$$\text{(by properties of the integral)}$$

$$= \Phi(t, t_0)\tan(ax_0 + bt_0 + c) + \int_{t_0}^t \Phi(t, r)dr.$$

Hence,

$$ax(t) + bt + c = \tan^{-1}\left(\Phi(t, t_0)\tan(ax_0 + bt_0 + c) + \int_{t_0}^t \Phi(t, r)dr\right).$$

Thus,

$$x(t) = \frac{1}{a}\left[\tan^{-1}\left(\Phi(t, t_0)\tan(ax_0 + bt_0 + c) + \int_{t_0}^t \Phi(t, r)dr\right) - bt - c\right].$$

This completes the process of solving the given IVP.

Example 3.8.2. Given:

$$dx = \left[\frac{ax + bt + c}{1 + (ax + bt + c)^2}\left[1 - \frac{a}{2}\frac{(ax + bt + c)^2 - 1}{[1 + (ax + bt + c)^2]^2}\right] - \frac{b}{a}\right]dt$$

$$+ \frac{ax + bt + c}{1 + (ax + bt + c)^2}dw(t), \quad x(t_0) = x_0.$$

For $a \neq 0 \neq b$ and $1 > -\frac{1}{b}$, solve the given initial value differential equations.

Solution procedure. By applying Observation 3.8.1(iii), we have

$$dv = \frac{v}{1 + v^2}\left[a - \frac{a^2}{2}\frac{(v^2 - 1)}{(1 + v^2)^2}\right]dt + a\frac{v}{1 + v^2}dw(t).$$

This is a variable separable differential equation. Now, we use Procedure 3.5.2 or Procedure 3.8.2. Here,

$$V(t, v) - V(t_0, v_0) \equiv G(v) - G(v_0) = \int_{v_0}^{v}\frac{(1 + y^2)dy}{y} = \frac{1}{2}(v^2 - v_0^2) + \ln\left[\frac{v}{v_0}\right].$$

We note that

$$dV(t, v(t)) = \frac{1 + v^2}{v}dv + \frac{1}{2}\frac{v^2 - 1}{v^2}(dv)^2$$

$$= \frac{1 + v^2}{v}\left[\frac{v}{1 + v^2}\left[a - \frac{a^2}{2}\frac{(v^2 - 1)}{(1 + v^2)^2}\right]dt + a\frac{v}{1 + v^2}dw(t)\right]$$

$$+ \frac{1}{2}\frac{(v^2 - 1)}{v^2}\frac{v^2 a^2}{[1 + v^2]^2}dt$$

$$= a\,dt + a\,dw(t).$$

In this case, the reduced SDE (3.3.1) is $dm = a\,dt + a\,dw(t)$, and hence

$$G(v(t)) - G(v_0) = a(t - t_0) + a(w(t) - w(t_0)),$$

$$v^2(t) + 2\ln v = v_0^2 + 2\ln v_0 + a(t - t_0) + a(w(t) - w(t)),$$

where $v = ax + bt + c$ is the implicit solution representation of the given differential equation. This completes the solution process.

Brief Summary. A summary of the basic steps needed to solve nonlinear stochastic scalar differential equations is presented in the flowchart:

Exercises

1. Find the general solution to the following differential equations:

   (a) $dx = [4(x + 2t + 1) + 5(x + 2t + 1)^{2/3} - 2]dt + \sqrt{3}(x + 2t + 1)dw(t);$
   (b) $dx = [(x + 2t + 1)^{1/3} - 2]dt + \sqrt{3}[x + 2t + 1 + (x + 2t + 1)^{2/3}]dw(t).$

Representative member (f,x) of the class:†

$$f(t,x) = F = \begin{cases} \dfrac{1}{a}\left[-b+a\Sigma\left[\dfrac{a}{2}\dfrac{d}{dv}\Sigma(v)-\dfrac{f_1}{\delta}\exp\left[-\delta\int_c^v\dfrac{dy}{a\Sigma(y)}\right]+f_2\right]\right] & \delta \neq 0 \\[4mm] \dfrac{1}{a}\left[-b+a\Sigma\left[\dfrac{a}{2}\dfrac{d}{dv}\Sigma(v)+f_1\left[\delta\int_c^v\dfrac{dy}{a\Sigma(y)}\right]+f_2\right]\right], & \delta = 0 \end{cases}$$

where, $v = ax + bt + c, a \neq 0, f_1$ and f_2 are arbitrary constants

---

Energy function: $V(t,x) = G(ax+bt+c) = \begin{cases} \exp[\delta H(x)] - \dfrac{\gamma}{\delta}, & \delta \neq 0 \\[3mm] \gamma H(x), & \delta = 0 \end{cases}$

Reduced LNHSDE: $dm = \begin{cases} [\mu(t)m + p(t)]dt + [v(t)m + q(t)]dw(t), & \delta \neq 0 \\[2mm] [\mu(t)m + p(t)]dt + q(t)dw(t), & \delta = 0 \end{cases}$

where, $\mu(t) = \dfrac{1}{2}\delta^2 + \delta f_2, p(t) = \dfrac{1}{2}\delta\gamma + \gamma f_2 - Cf_1, v(t) = \delta$ and

$q(t) = \gamma$

Fig. 3.8.2   Summary of the solution process.

2. Find the solution to the following initial value problems:
   (a) $dx = [5(2x + t - 1)^{1/3} - 1]dt + \sqrt{3}(2x + t - 1)dw(t), x(0) = 1$;
   (b) $dx = [\cos^2 t(x + t + 1) + \sin 2t(x + t + 1)^{-2} - 1]dt + \cos t(x + t + 1)dw(t),$
   $x(0) = 1$.

3.9 Notes and Comments

Historically, little attention has been given to the development of methods for solving nonlinear stochastic differential equations of the Itô–Doob-type. The idea of closed-form solutions and sufficient conditions for finding the closed-form solutions was very briefly introduced by Gikhman and Skorokhod [pp. 34–39, 42]. A few techniques for solving linear and certain classes of nonlinear stochastic differential equations were presented by Kloeden and Platen [Chapter 4, 71]. This would be the first book available for the use of stochastic modeling and method research topics for interdisciplinary graduate/undergraduate students and interdisciplinary researchers with a minimal academic background. The two approaches to stochastic modeling of dynamic processes are outlined in Section 3.1 [78, 158, 195]. Illustration 3.1.1 and Example 3.1.2 are based on the fundamental ideas in chemical kinetics [80, 99]. The randomness is due to the formation of free radicals. The discussion in Illustration 3.1.2 was extracted from Refs. 80 and 175. The outlines of Illustration 3.1.3 [131, 158] and Example 3.1.3 [80, 132, 148] exhibit ecological models. The random perturbations are due to environmental perturbations. Furthermore, the population genetics model is presented in Example 3.1.4 [45, 67, 68, 69, 80, 132], and the random fluctuations are due to evolutionary processes. Illustration 3.1.4 is summarized on the basis of the epidemiological models in Refs. 10, 12, and 80. A single composite commodity model of R. M. Solow [80, 185] and its special cases [185] are discussed in Illustration 3.1.5 and Example 3.1.5. The constant saving rate, demographic factors, and resource limitations are subject to random fluctuations. A very general outline of the 21st century approach, namely "the energy or Lyapunov-like function method," is given in Section 3.2. By employing the knowledge gained in Chapter 2, the feasibility of the general procedure described in Section 3.2 is further justified in Sections 3.3 and 3.4. This is achieved by classifying nonlinear scalar Itô–Doob-type stochastic differential equations that are reducible to integrable and linear scalar Itô–Doob-type stochastic differential equations. Finally, it is shown that variable separable differential equations (Section 3.5), homogeneous differential equations (Section 3.6), and essentially time-invariant differential equations (Section 3.8) belong to the class of integrable reduced stochastic differential equations, while the Bernoulli-type differential equations (Section 3.7) belong to the class of linear reduced differential equations. The energy/Lyapunov-like function method in Section 3.2 and its special cases in Sections 3.3–3.8 are new and extensions of deterministic work [80]. Illustrations 3.5.1 and 3.5.2 are based on Ref. 80. Illustration 3.6.2 is a stochastic version of the deterministic allometric law [59, 80, 159]. Illustration 3.7.2 is a stochastic version of the economic growth–Solow model [80, 185], Example 3.7.4 is a stochastic version of the epidemic model [80, 194], and Illustration 3.7.3 is a stochastic extinction model of the Levin type [128].

Chapter 4

First-Order Systems of Linear Differential Equations

Introduction

The mathematical modeling procedures for solving first-order linear Itô–Doob-type stochastic systems differential equations and theoretical analysis are the premise of this chapter. It is also an extension of deterministic multivariate problems. In addition, efforts are made to contrast the scalar-versus-system and deterministic-versus-stochastic differential equations. The format and the presentation of this chapter are similar to Chapters 2 and 4 of Volume 1. In Section 4.1, several dynamic processes in the biological, chemical, engineering, medical, physical, and social sciences are discussed to illustrate the multivariate stochastic modeling procedures. Section 4.2 begins with first-order linear systems of stochastic differential equations with constant coefficients. The eigenvalue type of method for solving stochastic scalar and deterministic systems of homogeneous differential equations is extended to solve stochastic systems of homogeneous differential equations with constant coefficients. By considering the idea of two different timescales $[t, w(t)]$, systems of algebraic equations are generated. Again, the method for solving stochastic systems of linear homogeneous differential equations reduces to the problem of solving linear systems of algebraic equations with constant coefficients. By introduction a stochastic version of the principle of superposition, the fundamental matrix, and the Abel–Jacobi–Liouville theorem, their role and scope in finding the IVPs are completely addressed. In fact, the general step-by-step procedures for finding the general solution and the solution to the IVPs are systematically extended to linear Itô–Doob-type stochastic systems of differential equations with constant coefficient matrices. The deterministic procedure for finding the fundamental matrix solution is extended to linear Itô–Doob-type stochastic systems of differential equations in Section 4.3. The byproduct is utilized to solve the IVPs. In addition, several applied and mechanical examples seeking illustrations are given to demonstrate the procedures and the related mathematics. In Section 4.4, the procedures for seeking solutions to linear systems of homogeneous differential equations are extended to a general linear stochastic system of homogeneous differential equations with constant coefficients. The general step-by-step procedures for finding the general solution and the solution to the IVPs for linear Itô–Doob-type stochastic systems of differential equations are logically outlined. The limitation on the technique for finding a closed-form solution, even with constant coefficients, is emphasized. Numerous applied and mechanical examples with illustrations are given to explain the procedures. Section 4.5 deals with systems of first-order nonhomogeneous Itô–Doob-type stochastic differential equations. The method of variation of constants parameters is used to find the general solution and the solution to the IVPs. Section 4.6 offers further insights. The presented material is needed for studying systems of linear stochastic differential equations with time-varying coefficient matrix functions. Again, this section's material provides a foundation for answering the main questions. Moreover, this can be used for studying the higher-order differential equations that are discussed in Chapter 5.

4.1 Mathematical Modeling

The material in this section is a natural and straightforward extension of Sections 2.1 and 3.1 of Chapters 2 and 3, respectively. In general, dynamic processes in the biological, chemical, engineering, medical, physical, and social sciences are highly complex, with several interacting and interconnected subcomponents. In fact, for a better understanding of the multispecies dynamic process, a single-state-variable description of the multispecies dynamic system is inadequate. It is natural to expect the development in a several-state-variable description. Again, the development is based on a theoretical experimental setup, the fundamental laws in science and engineering, and the knowledge-based information about dynamic processes. An attempt is made to incorporate diverse and apparently different phenomena to relate them to a common conceptual/computational framework and ideas. As a result of this, the well-known probabilistic models described in Sections 2.1 and 3.1 are utilized to generate the Itô–Doob-type mathematical description framework for dynamic processes.

Multistate Probabilistic Modeling Procedure 4.1.1.   For the development of the mathematical models of dynamic processes, we employ the random walk, Poisson process, and Brownian motion models of Chapter 2. The simultaneous usage of these probabilistic models incorporates the different types of interactions between the components of multistate dynamic processes. We use a conceptually common description of multispecies processes in sciences and engineering consisting of d components $[x_1, x_2, \ldots, x_\ell, \ldots, x_d]^T \in R^d$, as a "$d$-dimensional state vector" of the system, such as the "concentration vector" of chemical reactants in a chemical reaction process, the "size/biomass vector of d-species" in a biological process, or the "price vector" of d commodities/services in business and sociological processes. It is measured as a vector quantity.

Let $x(t)$ be a d-dimensional state vector of a system at a time t. The state of the system is observed over an interval $[t, t + \Delta t]$, where Δt is a small increment in t. Without loss of generality, we assume that Δt is positive. For the sake of simplicity, we assume that the process is under the influence of random perturbations. We experimentally observe the states of a system: $x(t_0) = x(t)$, $x(t_1)$, $x(t_2), \ldots, x(t_k), \ldots, x(t_n) = x(t + \Delta t)$ at $t_0 = t$, $t_1 = t + \tau$, $t_2 = t + 2\tau, \ldots, t_k = t + k\tau, \ldots, t_n = t + \Delta t = t + n\tau$ over the interval $[t, t + \Delta t]$, where n belongs to $\{1, 2, 3 \ldots, \}$ and $\tau = \frac{\Delta t}{n}$. With these observations, for the sake of simplicity, we consider a modeling procedure of a state of the dynamic process under the influence of a single random perturbation (SRP):

SRP1. The system is under the influence of independent and identical random impulses that are occurring at $t_1, t_2, \ldots, t_k, \ldots, t_n$.

SRP2. The influence of a single independent random impact on the state of the system is observed on every time subinterval of length τ.

SRP3. For each $\ell \in \{1, 2, \ldots, d\}$ and for each $k = 1, 2, \ldots, n$, the ℓth component $x_\ell(t_k)$ of the state vector $x(t_k)$ at the time t_k is either increased or decreased, or there is no change by the presence of the gth and rth components of the dynamic process for any $g, r \in \{1, 2, \ldots, d\}$. From the description in Section 2.1, $G_{\ell\ell}$, $G_{g\ell}$, and $R_{\ell r}$ ($R_{\ell\ell} = 0$) stand for experimentally or knowledge-based observed *microscopic or local* increments per impact at t_k to the ℓth component influenced by itself, by the ℓth component to the gth component, and by the rth component to the ℓth component of the system, respectively, over the subinterval of length τ. Moreover, we further note that for a pair of pairs (g, ℓ) and (ℓ, r), the nature of increments depends on the peculiar qualities of the ℓth, gth, and rth components of the multicomponent process, and the influence of the cross-interactions between components. We define $d \times d$ matrices $G = (G_{g\ell})$ and $R = (R_{\ell r})_{d \times d}$, whose entries represent increments to the state of a d-dimensional multicomponent dynamic process.

SRP4. Following the modeling description in Section 2.1 of Chapter 2, we assume that all elements $G_{g\ell}$ and $R_{\ell r}$ of the matrices G and R defined in SRP3 are Bernoulli, Poisson or Gaussian normal random variables. For all pairs of ordered pairs (g, ℓ) and (ℓ, r), depending on the nature of the ℓth, gth, and rth components of the multicomponent process and the influence of the cross-interactions between components of the process, we further assume that $G_{g\ell} = \Delta\alpha_{g\ell} z$ and $R_{\ell r} = \Delta x_{\ell r} z$, and satisfy any one of the following conditions:

(i) $\Delta\alpha_{g\ell}$ and $\Delta x_{\ell r}$ are the magnitudes of the random variables $G_{g\ell}$ and $R_{\ell r}$, and z is a Bernoulli random variable that takes values ± 1 with the same parameter, p.

(ii) $\Delta\alpha_{g\ell}$ and $\Delta x_{\ell r}$ are the jumps of the random variables $G_{g\ell}$ and $R_{\ell r}$, and z is a Poisson random variable that takes values either $z = 0$ or 1, or $z = -1$ or 0 with a parameter λ;

(iii) $\Delta\alpha_{g\ell}$ and $\Delta x_{\ell r}$ are coefficients of the random variables $G_{g\ell}$ and $R_{\ell r}$, and z is a Gaussian normal random variable with parameters $(\mu\tau, \sigma^2\tau)$. In short, the multicomponent dynamic process is under the influence of a single random perturbation.

In summary, for each $\ell \in \{1, 2, \ldots, d\}$, under n independent and identical impacts of a single random perturbation that took place at $t_1, t_2, \ldots, t_k, \ldots, t_n$ over the given interval $[t, t + \Delta t]$ of length Δt with an initial state and $2n$ experimentally or knowledge-based observed random increments $G_{g\ell}(t_i) = G_{g\ell}^i = \Delta\alpha_{g\ell}u^i$ and $R_{\ell r}(t_i) = R_{\ell r}^i = \Delta x_{\ell r}v^i$ for any $g, r \in \{1, 2, \ldots, d\}$ and $i \in \{1, 2, \ldots, n\}$, we have

$$x_\ell(t_0) = x_\ell(t),$$

$$x_\ell(t_1) - x_\ell(t_0) = G_{1\ell}^1 + \cdots + G_{g\ell}^1 + \cdots + G_{d\ell}^1 + R_{\ell 1}^1 + \cdots + R_{\ell\ell-1}^1$$

$$+ \cdots + R_{\ell\ell+1}^1 + R_{\ell d}^1 = \sum_{j=1}^{d} G_{j\ell}^1 + \sum_{j \neq \ell}^{d} R_{\ell j}^1,$$

$$x_\ell(t_2) - x_\ell(t_1) = G_{1\ell}^2 + \cdots + G_{g\ell}^2 + \cdots + G_{d\ell}^2 + R_{\ell 1}^2 + \cdots + R_{\ell\ell-1}^2$$

$$+ \cdots + R_{\ell\ell+1}^2 + R_{\ell d}^2 = \sum_{j=1}^{d} G_{j\ell}^2 + \sum_{j \neq \ell}^{d} R_{\ell j}^2,$$

$$\ldots \quad \ldots \quad \ldots \quad \ldots$$

$$x_\ell(t_k) - x_\ell(t_{k-1}) = G_{1\ell}^k + \cdots + G_{g\ell}^k + \cdots + G_{d\ell}^k + R_{\ell 1}^k + \cdots + R_{\ell\ell-1}^k$$

$$+ \cdots + R_{\ell\ell+1}^k + R_{\ell d}^k = \sum_{j=1}^{d} G_{j\ell}^k + \sum_{j \neq \ell}^{d} R_{\ell j}^k,$$

$$\ldots \quad \ldots \quad \ldots \quad \ldots$$

$$x_\ell(t_n) - x_\ell(t_{n-1}) = G_{1\ell}^n + \cdots + G_{g\ell}^n + \cdots + G_{d\ell}^n + R_{\ell 1}^n + \cdots + R_{\ell\ell-1}^n$$

$$+ \cdots + R_{\ell\ell+1}^n + R_{\ell d}^n = \sum_{j=1}^{d} G_{j\ell}^n + \sum_{j \neq \ell}^{d} R_{\ell j}^n. \tag{4.1.1}$$

From (4.1.1), for each $\ell \in \{1, 2, \ldots, d\}$, the ℓth component of the state at the final impulse is expressed by

$$x_\ell(t + \Delta t) = x_\ell(t) + \sum_{i=1}^{n}\sum_{s=1}^{d} G_{s\ell}^i + \sum_{i=1}^{n}\sum_{j \neq \ell}^{d} R_{\ell j}^i$$

$$= x_\ell(t) + \sum_{j=1}^{d}\sum_{i=1}^{n} G_{j\ell}^i + \sum_{j \neq \ell}^{d}\sum_{i=1}^{n} R_{\ell j}^i \quad \text{(double sum property and index change)}$$

$$= x_\ell(t) + \sum_{i=1}^{n} G_{\ell\ell}^i + \sum_{j\neq\ell}^{d}\sum_{i=1}^{n}(G_{j\ell}^i + R_{\ell j}^i) \qquad \text{(by regrouping)}$$

$$= x_\ell(t) + \sum_{i=1}^{n} \Delta a_{\ell\ell}u^i + \sum_{j\neq\ell}^{d}\sum_{i=1}^{n}(\Delta\alpha_{j\ell}u^i + \Delta x_{\ell j}v^i) \qquad \text{(by substitution)},$$

$$(4.1.2)$$

where $\sum_{j=1}^{d}\sum_{i=1}^{n} G_{j\ell}^i$ and $\sum_{j\neq\ell}^{d}\sum_{i=1}^{n} R_{\ell j}^i$ are referred to as *global aggregate increments* to the ℓth component influence by the ℓth component itself and by all other components of the dynamic process, respectively, over the interval $[t, t+\Delta t]$ of length Δt. The increments $\sum_{i=1}^{n} G_{g\ell}^i$ and $\sum_{i=1}^{n} R_{\ell r}^i$ are called *local aggregate increments* to the ℓth component influeced by the ℓth component to the gth component, and the rth component to the ℓth component of the process, respectively, over the interval $[t, t + \Delta t]$ of length Δt. These local increments are denoted by $\Delta_{g\ell}^n(\ell) = \sum_{i=1}^{n} G_{g\ell}^i$ and $\Delta_{\ell r}^n(r) = \sum_{i=1}^{n} R_{\ell r}^i$.

From the conditions SRP1–SRP4 and (4.1.2), it is clear that for each $\ell \in \{1, 2, \ldots, d\}$, $x_\ell(t_n) - x_\ell(t)$ is a discrete-time-real-valued stochastic process consisting of $2d - 1$ terms in which each term is the sum of n independent and identical random variables $G_{g\ell}^i$, $R_{\ell r}^i$ with identical parameters ($G_{g\ell}^i = G_{g\ell} = \Delta\alpha_{g\ell}z$ and $R_{\ell r}^i = R_{\ell r} = \Delta x_{\ell r}z$), $i = 1, 2, \ldots, n$. Moreover, for each n and the pairs (g, ℓ) and (ℓ, r), depending on the nature of the ℓth, gth, and rth components, the influence of the ℓth component on the gth component and the influence of the rth component on the ℓth component of the system, $\sum_{i=1}^{n} G_{g\ell}^i = \sum_{i=1}^{n} \Delta\alpha_{g\ell}z = \Delta\alpha_{g\ell}\sum_{i=1}^{n} z$ and $\sum_{i=1}^{n} R_{\ell r}^i = \sum_{i=1}^{n}\Delta x_{\ell r}z = \Delta x_{\ell r}\sum_{i=1}^{n} z$, are either binomial random variables with the same parameters (n, p) and (n, p) or $(n, \frac{\lambda\Delta t}{n})$ and $(n, \frac{\lambda\Delta t}{n})$, or Gaussian normal random variables with parameters $(\mu\frac{\Delta t}{n}, \sigma^2\frac{\Delta t}{n})$. From the expression (4.1.2), the *aggregate change* of the ℓth component $x_\ell(t + \Delta t) - x_\ell(t)$ of the state of the system under n identical random impacts of single random perturbations over the given time interval $[t, t+\Delta t]$ of length Δt is given by

$$x_\ell(t + \Delta t) - x_\ell(t) = \sum_{i=1}^{n} \Delta\alpha_{\ell\ell}u^i + \sum_{j\neq\ell}^{d}\sum_{i=1}^{n}(\Delta\alpha_{j\ell}u^i + \Delta x_{\ell j}v^i) \qquad \text{(by substitution)}$$

$$= \left[\Delta\alpha_{\ell\ell}\sum_{i=1}^{n} Z^i + \sum_{j\neq\ell}^{d}\left(\Delta\alpha_{j\ell}\sum_{i=1}^{n} Z^i + \Delta x_{\ell j}\sum_{i=1}^{n} Z^i\right)\right], \qquad (4.1.3)$$

where $Z^i = z$, for $i = 1, 2, \ldots, n$.

From the above discussion and by following the arguments used in the random walk, Poisson, and Brownian motion modeling in Section 2.1 of Chapter 2, depending on the nature of random variables $\Delta_{g\ell}^n(\ell)$ and $\Delta_{\ell r}^n(r)$, and using (2.1.8), (2.1.20), and (2.1.29), we have the corresponding one of the following relations:

$$E[\Delta_{g\ell}^n(\ell)] = \Delta t(p-q)\frac{\Delta\alpha_{g\ell}}{\tau}, \quad E[\Delta_{\ell r}^n(r)] = \Delta t(p-q)\frac{\Delta x_{\ell r}}{\tau}, \tag{4.1.4}$$

$$E[\Delta_{g\ell}^n(\ell)] = \pm\lambda\Delta\alpha_{g\ell}\Delta t, \quad E[\Delta_{\ell r}^n(r)] = \pm\lambda\Delta x_{\ell r}\Delta t, \tag{4.1.5}$$

$$E[\Delta_{g\ell}^n(\ell)] = \mu\Delta\alpha_{g\ell}\Delta t, \quad E[\Delta_{\ell r}^n(r)] = \mu x_{\ell r}\Delta t \tag{4.1.6}$$

respectively, where $\frac{\Delta\alpha_{g\ell}}{\tau}$ and $\frac{\Delta x_{\ell r}}{\tau}$ are called the *average microscopic or local increments* per impact to the ℓth component influenced by the ℓth component to the gth component and the rth component to the ℓth component of the process, respectively, over a microscopic interval $[t_{k-1}, t_k]$, $k = 1, 2, \ldots, n$ and for any $g, r \in \{1, 2, \ldots, d\}$.

From (4.1.3) and Theorem 1.1.1, we obtain

$$E[x_\ell(t+\Delta t) - x_\ell(t)]$$

$$= E\left[\sum_{i=1}^n G_{\ell\ell}^i + \sum_{j\neq\ell}^d\sum_{i=1}^n (G_{j\ell}^i + R_{\ell j}^i)\right] \quad \text{[from (4.1.3) and Definition 1.1.16]}$$

$$= E\left[\sum_{i=1}^n \Delta_{\ell\ell}^n(\ell) + \sum_{j\neq\ell}^d\sum_{i=1}^n (\Delta_{j\ell}^n(\ell) + \Delta_{\ell j}^n(j))\right] \quad \text{(by notation)}$$

$$= \sum_{i=1}^n E[\Delta_{\ell\ell}^n(\ell)] + \sum_{j\neq\ell}^d\sum_{i=1}^n (E[\Delta_{j\ell}^n(\ell)] + E[\Delta_{\ell j}^n(j)]) \quad \text{(by Theorem 1.1.1).}$$

$$\tag{4.1.7}$$

In the case of (4.1.4), (4.1.7) reduces to

$$E[x_\ell(t+\Delta t) - x_\ell(t)]$$

$$= \Delta t(p-q)\frac{\Delta\alpha_{\ell\ell}}{\tau} + \sum_{j\neq\ell}^d\left[\Delta t(p-q)\frac{\Delta\alpha_{j\ell}}{\tau} + \Delta t(p-q)\frac{\Delta\alpha_{\ell j}}{\tau}\right]$$

$$= (p-q)\frac{1}{\tau}\left[\Delta\alpha_{\ell\ell} + \sum_{j\neq\ell}^d(\Delta\alpha_{j\ell} + \Delta x_{\ell j})\right]\Delta t \quad \text{(algebraic simplifications).}$$

$$\tag{4.1.8}$$

The derivations similar to (4.1.8) with respect to the cases (4.1.5) and (4.1.6) are

$$E[x_\ell(t + \Delta t) - x_\ell(t)] = \pm\lambda \left[\Delta\alpha_{\ell\ell} + \sum_{j\neq\ell}^{d}(\Delta\alpha_{j\ell} + \Delta x_{\ell j}) \right] \Delta t, \qquad (4.1.9)$$

$$E[x_\ell(t + \Delta t) - x_\ell(t)] = \mu \left[\Delta\alpha_{\ell\ell} + \sum_{j\neq\ell}^{d}(\Delta\alpha_{j\ell} + \Delta x_{\ell j}) \right] \Delta t, \qquad (4.1.10)$$

respectively.

Thus, the mean of the *overall aggregate change* $x(t + \Delta t) - x(t)$ of the state of the system under n identical random impacts of single random perturbations, over the given time interval $[t, t + \Delta t]$ of length Δt, is given by

$$E[x(t + \Delta t) - x(t)] = \begin{cases} \dfrac{1}{\tau}(p - q)\Delta X e \Delta t & \text{if random walk model,} \\[2mm] \pm\lambda\Delta X e \Delta t & \text{if Poisson process model,} \\[2mm] \mu\Delta X e \Delta t & \text{if Brownian motion model,} \end{cases} \qquad (4.1.11)$$

where e, $x(t)$ and ΔX are $d\times 1$, $d\times 1$ and $d\times d$ matrices defined by $e = [1, 1, \ldots, 1]^T$, $x(t) = [x_1(t), \ldots, x_\ell(t), \ldots, x_d(t)]^T$ and $\Delta X = (\Delta X_{\ell s})_{d\times d}$, where

$$\Delta X_{\ell s} = \begin{cases} \Delta x_{\ell s} & \text{for } s \neq \ell, \\[2mm] \displaystyle\sum_{j=1}^{d}\Delta\alpha_{j\ell} & \text{for } s = \ell. \end{cases} \qquad (4.1.12)$$

The matrix ΔX is referred to as the increment matrix associated with the theoretical experimental setup (4.1.1). Furthermore, we note that we have employed Definition 1.3.1 and the argument used in (1.1.8)–(1.1.10) [80]. Moreover, by using the above notation, (4.1.3) can be rewritten as

$$\Delta x \equiv x(t + \Delta t) - x(t) = \Delta X \xi = \Delta X e z, \qquad (4.1.13)$$

where ξ is a $d \times 1$ random matrix (vector) defined by $\xi = [z, z, \ldots, z]^T$.

Now, we need to find the covariance of $\Delta x \equiv x(t + \Delta t) - x(t)$. The extension of covariance of two random variables (Definition 1.1.20) to two random vectors is

$$E[(\Delta x - E[\Delta x])(\Delta x - E[\Delta x])^T]$$

$$= E[(\Delta x - E[\Delta x])(\Delta x^T - E[\Delta x^T])] \quad \text{(from Theorem 1.3.2(1) [80])}$$

$$= E[\Delta x \Delta x^T - E[\Delta x]\Delta x^T - \Delta x E[\Delta x^T]$$

$$+ E[\Delta x]E[\Delta x^T]] \quad \text{(Theorem 1.3.1M2 [80])}$$

$$= E[\Delta x \Delta x^T] - E[E[\Delta x]\Delta x^T] - E[\Delta x E[\Delta x^T]]$$

$$+ E[E[\Delta x]E[\Delta x^T]] \quad \text{(by Theorem 1.1.1)}$$

$$= E[\Delta x \Delta x^T] - E[\Delta x]E[\Delta x^T] - E[\Delta x]E[\Delta x^T]$$

$$+ E[\Delta x]E[\Delta x^T]] \quad \text{(by Theorem 1.1.1)}$$

$$= E[\Delta x \Delta x^T] - E[\Delta x]E[\Delta x^T] \quad \text{(by simplification)}$$

$$= E[(x(t + \Delta t) - x(t))(x(t + \Delta t) - x(t))^T]$$

$$- E[(x(t + \Delta t) - x(t))]E[(x(t + \Delta t) - x(t))^T] \quad \text{(by substitution)}$$

$$= E[\Delta X e z (\Delta X e z)^T]$$

$$- E[(x(t + \Delta t) - x(t))]E[(x(t + \Delta t) - x(t))^T] \quad \text{[from (4.1.13)]}$$

$$= \Delta X e (\Delta X e)^T E[z^2] - E[(x(t + \Delta t) - x(t))]E[(x(t + \Delta t) - x(t))^T].$$

$$\text{(4.1.14)}$$

From (4.1.11), (4.1.14), (2.1.9), (2.1.24), and (2.1.30), we have

$$E[(\Delta x - E[\Delta x])(\Delta x - E[\Delta x])^T] = (E[(\Delta x_\ell - E[\Delta x_\ell])(\Delta x_r - E[\Delta x_r])^T])_{d \times d}$$

$$= \begin{cases} \left(4pq\Delta t \left(\sum_{j=1}^{d} \frac{\Delta X_{\ell j}}{\tau} \right) \left(\sum_{j=1}^{d} \Delta X_{rj} \right) \right)_{d \times d} \\[2em] \left(\lambda \Delta t \left(1 - \frac{\lambda \Delta t}{n} \right) \left(\sum_{j=1}^{d} \Delta X_{\ell j} \right) \left(\sum_{j=1}^{d} \Delta X_{rj} \right) \right)_{d \times d} \\[2em] \left(\sigma^2 \Delta t \left(\sum_{j=1}^{d} \Delta X_{\ell j} \right) \left(\sum_{j=1}^{d} \Delta X_{rj} \right) \right)_{d \times d} , \end{cases}$$

$$\text{(4.1.15)}$$

where $(\sum_{j=1}^{d} \Delta X_{\ell j})(\sum_{j=1}^{d} \Delta X_{rj}) = (\sum_{j=1}^{d} \Delta X_{rj})(\sum_{j=1}^{d} \Delta X_{\ell j})$, i.e. the matrix is symmetric. This covariance matrix describes the measure of the single random perturbation effects on the d interacting components of the dynamic process. We recall that the variance of the random variable Y measures (spread or dispersion) the expected square of the deviation of Y from its expected value. The concept of the covariance of two random variables is a measure of dependence between them. Of course, if the random variables are independent, the covariance is identically zero. However, if the covariance is zero, the random variables are called uncorrelated.

From (4.1.8)–(4.1.10) and following the arguments used in modeling Section 2.1 of Chapter 2, we have

$$
\lim_{\Delta x \to 0} \lim_{\tau \to 0} E[x_\ell(t + \Delta t) - x_\ell(t)] = \left[\alpha_{\ell\ell} + \sum_{j \neq \ell}^{d} (\alpha_{j\ell} + x_{\ell j}) \right] \nu \Delta t = \sum_{j=1}^{d} c_{\ell j} \nu \Delta t,
$$

(4.1.16)

with

$$
c_{\ell r} = \begin{cases} \Delta x_{\ell r} & \text{for } r \neq \ell, \\ \displaystyle\sum_{j=1}^{d} \alpha_{j\ell} & \text{for } r = \ell, \end{cases}
$$

(4.1.17)

where $c_{\ell\ell}\nu$, $c_{g\ell}\nu$, and $c_{\ell r}\nu$ are *microscopic or local drift coefficients* of the ℓth component state influenced by the ℓth component to it-self, to the gth component, and the rth component to the ℓth component of the process, respectively, over an interval of length Δt; ν is either 1 or $\pm\lambda$ or μ, depending on the nature of z in SRP4 (Bernoulli or Poisson or Gaussian normal random variable). Moreover, $c_{\ell\ell}\nu$, $c_{g\ell}\nu$, and $c_{\ell r}\nu$ are interpreted as the *microscopic or local average/mean/expected rates of change of state* of the ℓth component of the process per unit time due to the influence of itself, the influence of the ℓth component on the gth component and the influence of the rth component of the process, respectively.

In addition, from (4.1.15) and the fact that

$$
\lim_{\Delta x \to 0} \lim_{\tau \to 0} \left[4pq \left(\sum_{j=1}^{d} \frac{\Delta X_{\ell j}}{\tau} \right) \left(\sum_{s=1}^{d} \Delta X_{rs} \right) \right] = D_{\ell r} \quad \text{(by the limit property)}
$$

(4.1.18)

we have

$$\lim_{\Delta x \to 0} \lim_{\tau \to 0} E[(\Delta x - E[\Delta x])(\Delta x - E[\Delta x])^T]$$

$$= \begin{cases} \Delta t \left(\left(\sum_{j=1}^{d} c_{\ell j} \right) \left(\sum_{j=1}^{d} c_{rj} \right) \right)_{d \times d} \\[2em] \lambda \Delta t \left(\left(\sum_{j=1}^{d} c_{\ell j} \right) \left(\sum_{j=1}^{d} c_{rj} \right) \right)_{d \times d} \\[2em] \Delta t \left(\sigma^2 \left(\sum_{j=1}^{d} c_{\ell j} \right) \left(\sum_{j=1}^{d} c_{rj} \right) \right)_{d \times d} \end{cases}$$

$$= \begin{cases} \Delta t (D_{\ell r})_{d \times d} \\[0.5em] \pm \lambda \Delta t (D_{\ell r})_{d \times d} \\[0.5em] \Delta t (\sigma^2 D_{\ell r})_{d \times d} \end{cases}$$

$$= \gamma \Delta t D, \tag{4.1.19}$$

where $\ell, r \in \{1, 2, \ldots, d\}$, $D_{r\ell} = D_{\ell r} = (\sum_{j=1}^{d} c_{\ell j})(\sum_{j=1}^{d} c_{rj})$ are called *microscopic or local covariance coefficients*, where γ is either 1 or λ or σ^2, depending on z being a Bernoulli random variable or a Poisson random or Gaussian normal random variable. It measures the single random perturbation effects on the pairs of ordered pairs (ℓ, r) and (r, ℓ) components of the process over an interval of length Δt. For $\ell = r$, $D_{\ell\ell} = (\sum_{j=1}^{d} c_{\ell j})(\sum_{j=1}^{d} c_{\ell j})$ is called the *microscopic or local variance coefficient*. It measures the single random perturbation effects of the ℓth component over an interval of length Δt. Moreover, under the single random perturbation, $D_{\ell\ell}$ and $D_{r\ell}$ are interpreted as the *microscopic or local mean square rates of change of state* of the ℓth component of the process per unit time due to the influence of itself and the influence of the rth component of the process, respectively.

Furthermore, from (4.1.19), $\sum$ is defined by:

$$\sum = \gamma (D_{\ell r})_{d \times d}, \tag{4.1.20}$$

where $D_{\ell r}$'s are defined in (4.1.19), and the matrix Σ is symmetric and nonsingular.

From (4.1.16), (4.1.17), (4.1.19), (4.1.20) and extending the argument used in Section 2.1 of Chapter 2, we have

$$x(t + \Delta t) - x(t) \approx \nu C e \Delta t + \sqrt{\Sigma} e \Delta w(t), \tag{4.1.21}$$

where matrices C, e, Σ, and $\Delta w(t)$ are defined by $C = (c_{\ell r})_{d \times d}$, $e = [1, 1, \ldots, 1, \ldots, 1]^T$, $\Sigma = \gamma(D_{r\ell})_{d \times d}$; ν, $c_{\ell r}$, and $D_{r\ell}$ are as defined in (4.1.16), (4.1.17), and (4.1.19), respectively; and $\Delta w(t)$ is defined by

$$\lim_{\Delta t \to 0} \left[(\Sigma)^{-1/2} \left[x(t + \Delta t) - x(t) - E[x(t + \Delta t) - x(t)] \right] \right]$$

$$= ez\sqrt{\Delta t} = e\Delta w(t), \tag{4.1.22}$$

where z is a standard Gaussian normal random variable, and $w(t)$ is a Wiener process. For small Δt, from (4.1.22), (4.1.21) reduces to an Itô–Doob-type stochastic system of differential equations:

$$dx(t) = \nu C e \, dt + \sqrt{\Sigma} e \, dw(t), \tag{4.1.23}$$

where νC and $\sqrt{\Sigma}$ are the drift and the diffusion coefficient matrices, respectively.

Observation 4.1.1. We note that the above-described modeling procedure of the multispecies dynamic process under a single random perturbation can easily be extended to the modeling procedure under simultaneous multiple random disturbances. In this case, the conditions SRP1–SRP4 ("S" stands for "single") are replaced by MRP1–MRP4 ("M" stands for "multiple"). In addition, the experimentally or knowledge-based observed *microscopic or local* increments $G_{g\ell}$ and $R_{\ell r}$ defined in SRP3 and SRP4 belong to BPW$[(\Omega, \mathcal{F}, P), R]$, which is a union of arbitrary subcollections of Bernoulli random variables with a parameter $0 < p < 1$, Poisson random variables with a parameter λ, and Gaussian normal random variables with parameters (μ, σ^2). Depending on the nature of the ℓth and rth components of the multicomponent process and the influence of the cross-interactions between components of the process, we further assume that $G_{g\ell} = \Delta\alpha_{g\ell}z_\ell$ and $R_{\ell r} = \Delta x_{\ell r}z_r$, where z_ℓ, $z_r \in$ BPW$[(\Omega, \mathcal{F}, P), R]$ are mutually independent for $\ell \neq r$, and have characteristics similar to z in SRP4. Again, following the above description, we have (4.1.2), where n independent and identical random variables $G_{g\ell}^i(t_i) = G_{g\ell}^i$ and $R_{\ell r}^i(t_i) = R_{\ell r}^i$ with an identical parameter ($G_{g\ell}^i = G_{g\ell} = \Delta\alpha_{g\ell}z_\ell$ and $R_{\ell r}^i = R_{\ell r} = \Delta x_{\ell r}z_r$), $i = 1, 2, \ldots, n$. In this case,

$$x_\ell(t + \Delta t) - x_\ell(t) = \left(\Delta\alpha_{\ell\ell} + \sum_{j \neq \ell}^{d} \Delta\alpha_{j\ell} \right) G_\ell^i + \sum_{j \neq \ell}^{d} \sum_{i=1}^{n} \Delta x_{\ell j} R_j^i, \tag{4.1.24}$$

where $G_\ell^i = z_\ell$ and $Z_j^i = z_j$ for $i = 1, 2, \ldots, n$; moreover, (4.1.13) and (4.1.14) reduce to

$$\Delta x \equiv x(t + \Delta t) - x(t) = \Delta X z, \tag{4.1.25}$$

$$E[(\Delta x - E[\Delta x])(\Delta x - E[\Delta x])^T]$$

$$= \Delta X E[zz^T]\Delta X^T - E[(x(t + \Delta t) - x(t))]E[(x(t + \Delta t) - x(t))^T], \tag{4.1.26}$$

where z is a $d \times 1$ random matrix (vector) defined by $z = [z_1, \ldots, z_r, \ldots z_d]^T$; $z_r \in$ BPW$[(\Omega, \mathcal{F}, P), R]$ and the components of z are mutually independent. Depending on the nature of the components of z, repeating the remainder of the argument used in a single random perturbation in a multicomponent process and using the same notations, (4.1.16) and (4.1.19) remain valid. Thus, (4.1.21) and (4.1.22) reduce to

$$x(t + \Delta t) - x(t) \approx Cm\Delta t + \sqrt{\Sigma}\Delta w(t), \qquad (4.1.27)$$

where matrices C and Σ are defined by $C = (c_{\ell r})_{d \times d}$ and $\Sigma = (D_{r\ell})_{d \times d}$, respectively; and $\Delta w(t)$ is a $d \times 1$ vector defined by

$$\lim_{\Delta t \to 0} [(\Sigma)^{-1/2}[x(t + \Delta t) - x(t) - E[x(t + \Delta t) - x(t)]] = \xi\sqrt{\Delta t} = \Delta w(t), \quad (4.1.28)$$

where $w(t) = [w_1(t), \ldots, w_r(t), \ldots, w_d(t)]_{d \times 1}^T$ is a d-dimensional normalized Wiener process with zero mean and variance $I_{d \times d}$; $\Delta w(t) = \xi\sqrt{\Delta t}$, with $\xi = [\xi_1, \ldots, \xi_r, \ldots, \xi_d]^T$; the rth component of ξ is a standard Gaussian normal random variable, and for $r \neq j$, ξ_r and ξ_j are mutually independent ($E[\xi_r\xi_j] = 0$); $\Lambda = \text{diag}(\nu_1, \ldots, \nu_r, \ldots, \nu_d)$, whose components ν_r can take on any value, 1 or $\pm\lambda$ or μ, depending on z_r being a Bernoulli random variable or Poisson random or Gaussian normal random variable for $r \in \{1, 2, \ldots, r, \ldots d\}$; $m = \Lambda[1, 1, \ldots, 1, \ldots, 1]^T$. Thus, the change $[x(t+\Delta t) - x(t)]$ of the state of the system in (4.1.27) is interpreted as the sum of the average/expected/mean rate of change of the system $Cm\Delta t$ and the mean square rate of change of state of the system $\sqrt{\Sigma}\Delta w(t)$ over the interval of length Δt, due to the multiple random environmental perturbations on the system with its d interacting components.

Finally, (4.1.27) reduces to Itô–Doob-type stochastic system differential equations:

$$dx(t) = C\Lambda e \, dt + \sqrt{\Sigma} \, dw(t), \qquad (4.1.29)$$

where C and $\sqrt{\Sigma}$ are as defined in (4.1.23), and $w(t)$ is a d-dimensional normalized Wiener process with zero mean and variance $I_{d \times d}$.

Illustration 4.1.1 (multieconomy processes [9, 80, 178]). Single state probabilistic (Illustration 2.1.1) and multicomponent deterministic (Illustration 4.2.1 [80]) models are extended to multistate probabilistic dynamic processes in economics, management and information sciences. Here, state vector $x(t)$ of the system stands for either a rate or a specific rate of the price or value vector of a multivariate entity belonging to E, which is a collection of multivariate assets, information, markets, products, services per unit item or size vector $\bar{1} = [1, \ldots, 1, \ldots, 1]_{1 \times d}$ per unit time at time t in $J = [a, b]$, $a, b \in R$ and $J \subseteq R$. The rate or specific rate of price or value of the entity is observed over an interval of $[t, t + \Delta t]$, where Δt is a small increment in t. The process is under the influence of exogenous or endogenous (or both exogenous and endogenous) random perturbations generated by uncertainties such as national and international commerce, trade, monetary, social welfare policies and goals. As a result of this, $x(t)$ is affected by these random environmental perturbations. Its mathematical description can be recasted by following the development of Multistate Probabilistic Model 4.1.1 as

$$x(t + \Delta t) - x(t) \approx \nu Ce\Delta t + \sqrt{\Sigma}e\Delta w(t),$$

which implies that

$$dx(t) = \nu Ce \, dt + \sqrt{\Sigma}e \, dw(t).$$

We note that if $x(t)$ is the specific rate of the price or value vector at time t, then the coefficient matrix νC is called a measure of the instantaneous *average specific rate* of the multiple component price or value of the entity at time t over an interval of small length Δt; and $\sqrt{\Sigma}$ is called the volatility matrix of the multiple economy, which measures the *magnitude of interactions* of the price or value vector at time t over an interval of small length $\Delta t = dt$; if $x(t)$ is the price or value vector at time t, the matrix νC is called a measure of the *average rate* of growth/decay of the price or value vector of the entity at time t over an interval of small length $\Delta t = dt$; and $\sqrt{\Sigma}$ measures the magnitude of the deviation of the rate of the price or value vector with its mean/expected value at time t over an interval of small length $\Delta t = dt$.

Illustration 4.1.2 (multispecies decay/growth processes [62, 96, 160, 178]). Illustration 2.1.2 and multicomponent deterministic (Illustration 4.2.2 [80]) models are extended to the probabilistic models of several multicomponent decay/growth dynamic processes in the biological, chemical, compartment, pharmacological, physical, and social sciences. Again, here, a state vector of the system $x(t)$ stands for either a specific rate of an amount (mass/size) or an amount (mass/size) of an entity that belongs to E; E is a collection of multicomponent states (substance, population, charge, ionizing beam of particles, etc.) of the system at t in $J = [a, b]$, $a, b \in R$ and $J \subseteq R$. The specific rate of an amount (size) or an amount (size) of the entity is observed over an interval of $[t, t + \Delta t]$, where Δt is a small increment in t. Volume, weight, diameter, number, etc. are examples of size. Without loss of generality, we assume that Δt is positive. The process is under the influence of internal structural or external environmental random perturbations, such as temperature, electric, voltage, shot effects, thermal noise, or a large class of impulse noises. By following the above-described Poisson process generated Model 4.1.1, we have

$$x(t + \Delta t) - x(t) \approx \nu Ce\Delta t + \sqrt{\Sigma}e\Delta w(t),$$

and hence

$$dx(t) = \nu Ce \, dt + \sqrt{\Sigma}e \, dw(t).$$

As before (Illustration 4.1.1), we note that if $x(t)$ is the specific rate of the multistate entity vector at time t, then the coefficient matrix νC is called a measure of the *average/expected/mean specific rate/specific drift rate* of the multiple component amount (size) vector of the multistate entity at time t over an interval of small length $\Delta t = dt$; $\sqrt{\Sigma}$ is called the *diffusion coefficient* matrix, which measures the magnitude of the fluctuation of $x(t + \Delta t) - x(t)$ about its mean value of the specific rate of the multiamount (size) of the multistate entity at time t over an interval of small length $\Delta t = dt$; if $x(t)$ is the amount or size of the multivariate entity at time t over an interval of small length Δt, then the matrix νC is called a measure of the *average/expected/mean rate* of growth/decay of the entity vector at time t, and $\sqrt{\Sigma}$ measures the deviation from the mean of the amount or size at time t over an interval of small length $\Delta t = dt$.

Illustration 4.1.3 (multicomponent diffusion processes [80, 158, 166]). This illustration generalizes Illustration 2.1.4 and multivariate deterministic Illustration 4.2.3 [80] to the multivariate diffusion processes in the biological, biochemical, biophysical, chemical, compartmental, epidemiological, pharmacological, physical, physiological, and social sciences, and systems under internal structural or external environmental random perturbations, such as limited resources, capacity, environmental conditions, uncontrolled forces or mechanisms. As before, a state vector $x(t)$ of the system can be defined as in Illustration 4.1.2. The evolution of the system is chaotic in character, and it is described by (4.1.23)/(4.1.29).

Observation 4.1.2

(i) We recall that the experimentally or knowledge-based observed constant random vectors associated with matrices in (4.1.1), $x(t_0) = x(t)$, $[(G_1^i)^T, \ldots, (G_j^i)^T, \ldots, (G_d^i)^T]$, $[R^{i1}, R^{i2}, \ldots, R^{ij}, \ldots, R^{id}]$, are mutually independent of $x(t) = [x_1(t), \ldots, x_\ell(t), \ldots, x_d(t)]^T$, $G_{g\ell}^i = \Delta \alpha_{g\ell} u^i$ and $R_{\ell r}^i = \Delta x_{\ell r} v^i$ for each $g, r \in \{1, 2, \ldots, d\}$ and $i \in \{1, 2, \ldots, n\}$. Therefore, expectations in (4.1.11) and (4.1.15) can be replaced by the conditional expectations

$$E[x(t + \Delta t) - x(t)] = E[x(t) + \Delta t) - x(t) \mid x(t) = x], \qquad (4.1.30)$$

$$\mathrm{Var}(x(t + \Delta t) - x(t)) = E[[(\Delta x - E[\Delta x])(\Delta x - E[\Delta x])^T \mid x(t) = x], \qquad (4.1.31)$$

respectively, where $\Delta x = x(t + \Delta t) - x(t)$.

(ii) We further note that, in general, the magnitude of the microscopic or local increment depends on both the initial time t and the initial state $x(t) \equiv x$ of a system. In the light of this, the drift (νC) and the diffusion ($\sqrt{\Sigma}$) coefficient matrices defined in (4.1.16), (4.1.17) and (4.1.20) may depend on both the initial time t and the initial state $x(t) \equiv x$ of the system, as long as it dependence on t and x is very smooth. Under this consideration, (4.1.30) and (4.1.31) allow us to incorporate both time- and state-dependent random environmental perturbations. Thus, (4.1.21) reduces to

$$x(t + \Delta t) - x(t) \approx C(t, x)x\Delta t + \sigma(t, x)\Delta w(t), \qquad (4.1.32)$$

where $C(t, x)x$ and $\sigma(t, x) = \sqrt{\Sigma(t, x)}x$ are also referred to as the average/expected/mean rate and the mean square rate of the state of the system on the interval of length Δt. Moreover, the Itô–Doob-type system of stochastic differential equations (4.1.23) and (4.1.29) reduce to

$$dx(t) = C(t, x)x\,dt + \sigma(t, x)dw(t). \qquad (4.1.33)$$

(iii) Furthermore, in the absence of the random perturbations, the mathematical model of the multivariate dynamic process is described by the corresponding deterministic system of differential equations. In fact, the system of differential equations corresponding to (4.1.33) is

$$dx = C(t, x)x\,dt, \qquad (4.1.34)$$

particularly

$$dx = [C(t)x + c(t)]dt, \qquad (4.1.35)$$

where C is a $d \times d$ multicomponent state dynamic rate matrix function, and $c(t)$ is deterministic d-dimensional input rate function.

In the following, we utilize the Gaussian white noise modeling process of Section 3.1. We recall Observation 3.3.1. We extend the Gaussian white noise modeling process to the multivariate processes. The ideas follow.

Modified Brownian Motion Model 4.1.2. The basic idea is to start with the deterministic mathematical model (4.1.35)/(4.1.34). From Observation 3.3.1(i), (ii), one can identify the system parameter(s) and the source of random internal or environmental perturbations of the parameter(s) of the mathematical model (4.1.35)/(4.1.34). Using this, we reformulate a stochastic mathematical model by imitating Modified Brownian Motion Model 3.1.1 of Chapter 3 and the recently developed work [2]. We recall that this dynamic model was directly described by the stationary Gaussian process with independent increments (GPI). For example, in the biological or chemical dynamic processes: (GPI1) the size of species are integral values, (GPI2) the successive generations may not overlap in time, and (GPI3) breeding as well as environmental parameters may vary in a discrete time. For the sake of completeness, we briefly recast the description as follows.

Let $x(t)$ be a d-dimensional state vector of a dynamic system at time t. The state of the system is observed over an interval $[t, t + \Delta t]$, where Δt is a small increment in t. Without loss of generality, we assume that Δt is positive. We assume that the deterministic multistate $d \times d$ matrix rate function C and the d-dimensional rate vector function $c(t)$ contain system parameters that are subject to random environmental fluctuations. In fact, we assume that $C(t)$ and $c(t)$ are perturbed by independent increment random processes described by a $d \times d$ matrix $\Delta(t)$ and a d-dimensional vector $\kappa(t)$, respectively, as

$$C(t) = A(t) + \Delta(t) \quad \text{and} \quad c(t) = p(t) + \kappa(t). \tag{4.1.36}$$

Here, the elements $\Delta_{\ell j}(t)$ and $\kappa_\ell(t)$ of $\Delta(t) = (\Delta_{\ell j}(t))_{d \times d}$ and $\kappa(t) = (\kappa_\ell(t))_{d \times 1}$, respectively, are correlated Gaussian white noise processes with time-varying intensity parameters as $E[\Delta_{ir}(t)\Delta_{\ell j}(s)] = m_{ir,\ell j}(t)\delta(t - s)$,

$$E[\Delta_{ir}(t)\kappa_\ell(s)(s)] = d_{ir,\ell}(t)\delta(t - s) \quad \text{and} \quad E[\kappa_i(t)\kappa_\ell(s)] = e_{ik}(t)\delta(t - s).$$

From Observation 1.2.11(v), (4.1.35) can be rewritten as

$$dx = [A(t) + \Delta(t)]x + [p(t) + \kappa(t)]dt.$$

Hence,

$$dx = [A(t) + \Delta(t))x]dt + [p(t) + \kappa(t)]dt. \tag{4.1.37}$$

We define

$$dZ(t) = [(\Delta^1(t))^T \cdots (\Delta^r(t))^T \cdots (\Delta^d(t))^T \cdots (\kappa(t))^T]^T dt, \tag{4.1.38}$$

and hence

$$Z(t) = \int_{t_0}^t [(\Delta^1(s))^T \cdots (\Delta^r(s))^T \cdots (\Delta^d(s))^T \cdots (\kappa(s))^T]ds,$$

where $(\Delta^1(t))^T, \ldots, (\Delta^r(t))^T, \ldots, (\Delta^d(t))^T$ are the transpose of the column vectors of the $d \times d$ matrix $\Delta(t)$. $Z(t)$ is $m = (d^2 + d)$-dimensional Gaussian column vector with independent increments and $E[Z(t)] = 0$. An $m \times m$ covariance matrix,

$$E[Z(t)Z^T(s)] = E\left[\int_{t_0}^{\min(t,s)} \Sigma(u)du\right],$$

where $\Sigma(t)$ is a block matrix defined by

$$\Sigma(t) = \begin{bmatrix} M(t) & D(t) \\ D^T(t) & E(t) \end{bmatrix},$$

with $M(t) = (m_{ir,\ell j}(t))_{d^2 \times d^2}$, $D(t) = (d_{ir,\ell}(t))_{d^2 \times d}$ and $E(t) = (e_{ik}(t))_{d \times d}$. From the property (1.3.5) of the Itô–Doob integral, we have

$$Z(t) = \int_{t_0}^t \sqrt{\Sigma(t)}dw(s) \text{ implies } dZ(t) = \sqrt{\Sigma(t)}dw(s), \tag{4.1.39}$$

where $w(t)$ is an m-dimensional Wiener process. We note that if $\Sigma(t)$ is a singular matrix, then any $m \times n$ matrix Γ $(n < m)$ that satisfies the equation $\Gamma\Gamma^T = \Sigma(t)$

can be used. Now, we decompose $\sqrt{\Sigma(t)}$ as

$$\sqrt{\Sigma(t)} = \begin{bmatrix} B_1^*(t) & \cdots & B_m^*(t) \\ q_1(t) & \cdots & q_m(t) \end{bmatrix} = \begin{bmatrix} B_1(t) & \cdots & B_m(t) \\ q_1(t) & \cdots & q_m(t) \end{bmatrix}, \tag{4.1.40}$$

where $q_\ell \in R^d$, $B_\ell^*(t) \in R^{d^2}$ and $B_\ell(t) \in R^{d \times d}$, for $\ell \in \{1, 2, \ldots, m\}$; moreover, the $B_\ell(t)$ matrix is formed from $B_\ell^*(t)$ in a natural way. From (4.1.39) and (4.1.40), we have

$$dZ(t) = \begin{bmatrix} \sum_{i=1}^{m} B_\ell(t) dw_\ell(t) \\ \sum_{i=1}^{m} q_\ell(t) dw_\ell(t) \end{bmatrix}. \tag{4.1.41}$$

From (4.1.38) and (4.1.41), (4.1.37) reduces to

$$dx = [A(t)dt + \Delta(t)dt]x + [p(t)dt + \kappa(t)dt]$$

$$= \left[A(t)dt + \sum_{i=1}^{m} B_\ell(t) dw_\ell(t) \right] x + \left[p(t)dt + \sum_{i=1}^{m} q_\ell(t) dw_\ell(t) \right] \quad \text{(by substitution)}$$

$$= [A(t)x + p(t)]dt + \left[\sum_{i=1}^{m} B_\ell(t)x \, dw_\ell(t) + \sum_{i=1}^{m} q_\ell(t) dw_\ell(t) \right] \quad \text{(by regrouping)}$$

$$= [A(t)x + p(t)]dt + \sum_{i=1}^{m} [B_\ell(t)x + q_\ell(t)]dw_\ell(t) \quad \text{(by simplification)}. \tag{4.1.42}$$

The following specific illustrations and examples justify the scope and the significance of Observation 4.1.2 in the modeling of the dynamic processes in engineering and sciences. They are in the framework of the above-presented mathematical model building procedures of in this section.

Illustration 4.1.4 (electric circuit network [55, 80]). We assume that all of the conditions ECN1–ECN5 in Illustration 4.2.4 [80] remain unchanged. Repeating the argument, and utilizing the notations and definitions, we obtain the following deterministic mathematical model [80]:

$$dI = CI \, dt + P \, dt, \tag{4.1.43}$$

where

$$I = [I_2 \ I_3]^T, \quad C = \begin{bmatrix} -\dfrac{R_1}{L_2} & -\dfrac{R_1}{L_2} \\ -\dfrac{R_1}{L_3} & -\dfrac{R_1 + R_3}{L_3} \end{bmatrix} \quad \text{and} \quad P = \begin{bmatrix} \dfrac{E}{L_2} & \dfrac{E}{L_3} \end{bmatrix}^T. \tag{4.1.44}$$

Now, we introduce a source of random perturbations.

ECN6. Source of random fluctuations. In the electric circuit, the randomness can be due to the presence of registers as well as external electromotive force E. The random fluctuations in the network can be characterized by:

(i) thermal noise at the junctions;
(ii) shot noise;
(iii) flicker noise;
(iv) Conductance fluctuations.

Most of these random perturbations can be described by the stationary Gaussian process with independent increments $\kappa(t)$.

Step #4 (derivation of a stochastic model). From ECN6, Modified Brownian Motion Model 4.1.2 is directly applicable. By following (4.1.36) and (4.1.37) and using (4.1.42), we have

$$dI = [A + \Delta(t)]I\, dt + [p + \kappa(t)]dt = [A\, dt + \Delta(t)dt]I + [p\, dt + \kappa(t)dt]$$
$$= [AI + p]dt + [\Delta(t)I\, dt + \kappa(t)dt], \qquad (4.1.45)$$

where $C = A + \Delta(t)$ and $P = [p + \kappa(t)]$, with

$$A = \begin{bmatrix} -\dfrac{\bar{R}_1}{L_2} & -\dfrac{\bar{R}_1}{L_2} \\[2mm] -\dfrac{\bar{R}_1}{L_3} & -\dfrac{\bar{R}_1 + \bar{R}_3}{L_3} \end{bmatrix} \quad \text{and} \quad P = \left[\dfrac{\bar{E}}{L_2} \; \dfrac{\bar{E}}{L_3} \right]^T.$$

Moreover, (4.1.45) reduces to

$$dI = [AI + p]dt + \sum_{\ell=1}^{2}[B_\ell x + q_\ell]dw_\ell(t). \qquad (4.1.46)$$

This completes the stochastic modeling process corresponding to the basic electric network.

Illustration 4.1.5 (cellular dynamics [40, 175]). This illustration is a continuation of a deterministic version of the cellular dynamic model (Illustration 4.2.5 [80]). We assume that all assumptions, definitions, and notations remain true. The development of a prototype deterministic model of a cellular dynamics governed by a two-species monomolecular chemical reaction in the cell is summarized by

$$\frac{d}{dt}[A_1] = -k_{21}[A_1] + k_{12}[A_2] + (k_{1e} - k_{10}),$$
$$\frac{d}{dt}[A_2] = k_{21}[A_1] - k_{12}[A_2] + (k_{2e} - k_{20}). \qquad (4.1.47)$$

Source of random fluctuations. For the source of random perturbations, we recall Exercise 2 (Section 2.1) and Illustrations 2.4.4 and 2.4.5; in particular, the influence of ionization of cell membrane as well as the Brownian motion of metabolites inside and outside the cell. Thus, the rate parameters k_{21}, k_{12}, k_{1e}, k_{2e}, k_{10}, and k_{20} are subject to the random fluctuations. Moreover, we assume that these random perturbations are described by the stationary Gaussian process with independent increments. Under this consideration, Modified Brownian Motion Model 4.1.2 is directly applicable to these types of chain reaction systems. Therefore, the deterministic model described in (4.1.47) reduces to the Itô–Doob-type stochastic differential equation

$$dx = [Ax + p]dt + [Bx + q]dw(t), \qquad (4.1.48)$$

$$A = \begin{bmatrix} -\bar{k}_{21} & \bar{k}_{12} \\ \bar{k}_{21} & -\bar{k}_{12} \end{bmatrix}, \quad B = \begin{bmatrix} -\sigma\bar{k}_{21} & \sigma\bar{k}_{12} \\ \sigma\bar{k}_{21} & -\sigma\bar{k}_{12} \end{bmatrix}, \quad x = [[A_1] \quad [A_2]]^T,$$

$$p = [\bar{k}_{1e} - \bar{k}_{10} \quad \bar{k}_{2e} - \bar{k}_{20})]^T, \quad q = [\sigma(\bar{k}_{1e} - \bar{k}_{10}) \quad \sigma(\bar{k}_{2e} - \bar{k}_{20})]^T,$$

and $w(t)$ is the Wiener process.

Illustration 4.1.6 (multispecies ecosystems [130, 134]). In the following, a stochastic version of a deterministic mathematical model (Illustration 4.2.6 [80]) is presented. Employing the definitions, notations, and argument used in the deterministic illustration, we arrive at

$$dN_\ell = \left(e_\ell + \sum_{j=1}^{n} \alpha_{\ell j} N_j \right) dt.$$

Hence,

$$dN = (CN + e)dt, \qquad (4.1.49)$$

where $C = (\alpha_{\ell j})_{n \times n}$ community matrix. $\alpha_{\ell j}$ characterizes the influence of the jth-species on the ℓth-species in the community. The sign of $\alpha_{\ell j}$ determines the nature of the interactions between the (ℓ, j)th pair of species. For details, see Ref. 80.

Source of random fluctuations. In the open communities/systems, the sizes of species are integral values, the successive generations may not overlap in time, and breeding as well as environmental parameters may vary in discrete time. It is difficult to control the population growth by only the birth and death processes. For example, migration is a natural process to be taken into account for the growth rate. The environmental random fluctuations can also modify the resources as well as the cause of the migration process. The natural mathematical model of these types of ecosystems can be described by the stationary Gaussian process with independent increments. As a result of these considerations, Modified Brownian Motion

Model 4.1.2 is directly applicable to these types of species in the communities. For further details, see Illustration 3.1.3. Therefore, the deterministic model described in (4.1.49) reduces to the Itô–Doob-type stochastic differential equation

$$dN = (CN + a)dt + (BN + b)dw(t), \quad N(t_0) = N_0. \tag{4.1.50}$$

This completes the modeling of the multispecies process.

Observation 4.1.3. Imitating the steps, an observation parallel to Observation 4.2.2 [80] can be reconstructed. The further details are left to the reader.

Example 4.1.1. A stochastic version of a classical Lotka–Volterra model with the one-predator–one-prey system (Example 4.2.1 [80]) is presented. Imitating the development of the deterministic model under the same assumptions, notations, and definitions [80], we have

$$dN_1 = (\alpha_0 N_1 - \alpha_1 N_1 N_2)dt,$$

$$dN_2 = (-\beta_0 N_2 + \beta_1 N_1 N_2)dt.$$

For this system, applying Observation 4.1.3, we obtain the algebraic system of equations

$$\begin{cases} \alpha_0 N_1 - \alpha_1 N_1 N_2 = 0, \\ -\beta_0 N_2 + \beta_1 N_1 N_2 = 0, \end{cases} \quad \text{which implies} \quad \begin{cases} \alpha_0 - \alpha_1 N_2 = 0, \\ -\beta_0 + \beta_1 N_1 = 0, \end{cases}$$

and solve it. Its nonzero solution is $N^* = [N_1^* N_1^*]^T = \left[\frac{\beta_0}{\beta_1} \frac{\alpha_0}{\alpha_1} \right]^T$.

$$dN = CN \, dt,$$

where $N = [N_1 \ N_2]^T$ and

$$C = \begin{bmatrix} \alpha_0 - \alpha_1 N_2^* & -\alpha_1 N_1^* \\ \beta_1 N_2^* & -\beta_0 + \beta_1 N_2^* \end{bmatrix} = \begin{bmatrix} 0 & -\frac{\alpha_1 \beta_0}{\beta_1} \\ \frac{\beta_1 \alpha_0}{\alpha_1} & 0 \end{bmatrix}.$$

Moreover, the stochastic version of this linearized model is

$$dN = CN \, dt + BN \, dw(t).$$

The solution processes of the stochastic model will be discussed in Sections 4.2 and 4.3.

Illustration 4.1.7 (multicompartment system). We assume that all of the conditions MCS1–MCS6, and the notations and definitions of Illustration 4.2.7 [80], remain intact. Utilizing the development of the deterministic model of the multi-compartmental system (Illustration 4.2.7 [80]), we have

$$dq_\ell = \left(e_\ell - \left(\alpha_{0\ell} + \sum_{\substack{j \neq \ell}}^{n} \alpha_{j\ell} \right) q_\ell + \sum_{\substack{j \neq \ell}}^{n} \alpha_{\ell j} q_j \right) dt,$$

and hence

$$dq = (Qq + e)dt, \qquad (4.1.51)$$

where $a = [e_1, e_2, \ldots, e_\ell, \ldots, e_n]^T$, $A = (a_{\ell j})_{n \times n}$ and

$$q_{\ell j} = \begin{cases} -\left(\alpha_{0\ell} + \sum_{\substack{j \neq \ell}}^{n} \alpha_{j\ell} \right) & \text{for } j = \ell, \\ \alpha_{\ell j} & \text{for } j \neq \ell. \end{cases}$$

Following the reasoning used in Ref. 80, we obtain

$$dc = (Ac + a)dt, \qquad (4.1.52)$$

where $a = \left[\frac{e_1}{V_1}, \frac{e_2}{V_2}, \ldots, \frac{e_\ell}{V_\ell} \ldots, \frac{e_n}{V_n} \right]^T$, $A = (a_{\ell j})_{n \times n}$, and

$$a_{\ell j} = \begin{cases} -\dfrac{1}{V_\ell} \left(\alpha_{0\ell} + \sum_{\substack{j \neq \ell}}^{n} \alpha_{j\ell} \right) & \text{for } j = \ell, \\ \dfrac{\alpha_{\ell j} V_j}{V_\ell} & \text{for } j \neq \ell. \end{cases}$$

We note that the above transformation leads to a coefficient matrix in (4.1.52) without biophysical significance (unless it is regarded as a volume), such as the fractional clearance rates in (4.1.51). Hereafter, we will be using (4.1.52) in our discussion.

Source of random fluctuations. From MCS6 [80], we note that the transfer of the material from one region to another is under random perturbations. The permeability of the cell membrane is affected by the environmental changes surrounding the membrane. Moreover, the biological cells are identical in nature. As a result of this, the rate parameters A are subject to the random parameters described by the Gaussian process with independent increments. By following the argument used in Section 4.1.2 and employing (4.1.52), we have

$$dc = (Ac + a)dt + (Bc + b)dw(t). \qquad (4.1.53)$$

This completes the stochastic version of the compartmental modeling process.

Example 4.1.2 (pharmacokinetics [80, 157, 188, 189]). We assume that all of the conditions TPK1–TPK3 as well as the definitions and notations are true. Here, a stochastic version of the deterministic pharmacokinetics model (Example 4.2.2 [80]) is formulated. Following the development of the deterministic model building approach, we obtain a mathematical model with the substance kinetics being a two-compartmental system consisting of blood circulation and tissues

$$da_1 = [-(k_{21} + k_{31})a_1 + k_{12}a_2 + k_{10}M_0[-k_{10}t])]dt,$$

$$da_2 = [(k_{21}a_1 - (k_{12} + k_{20})a_2])dt,$$

which can be rewritten as

$$da = [Ka + k(t)]dt, \quad a_0 = [0, 0],$$

where

$$K = \begin{bmatrix} -k_{21} + k_{31} & k_{12} \\ k_{21} & -(k_{12} + k_{20}) \end{bmatrix} \quad \text{and} \quad k(t) = k_{10}M_0[-k_{10}t \; 0].$$

Source of random fluctuations. For the source of random perturbations, we recall Exercises 5 and 6 (Section 2.1) and Illustrations 2.4.5 and 2.4.6; in particular, the influence of ionization of cell membrane on the removal of the hypothesis PKT3, causes the Brownian-motion-type activities of the metabolites inside and outside the cell. As a result of this, the rate parameters k_{21}, k_{12}, k_{1e}, k_{2e}, k_{10} and k_{20} are subject to the random fluctuations. Moreover, we assume that these random perturbations are described by the stationary Gaussian process with independent increments. Because of this consideration, Modified Brownian Motion Model 4.1.2 is directly applicable to these types of chain reaction systems. Therefore, the above-described Teorell deterministic model reduces to the Itô–Doob-type system of stochastic differential equations

$$da = [Aa + p(t)]dt + [Ba + q(t)]dw(t),$$

$$A = \begin{bmatrix} -(\bar{k}_{21} + \bar{k}_{31}) & \bar{k}_{12} \\ \bar{k}_{21} & -(\bar{k}_{12} - \bar{k}_{20}) \end{bmatrix}, \quad B = \begin{bmatrix} -\sigma\bar{k}_{21} & \sigma\bar{k}_{12} \\ \sigma\bar{k}_{21} & -\sigma\bar{k}_{12} \end{bmatrix},$$

$$p(t) = [\bar{k}_{21}M_0[-k_{10}t] \; 0]^T, \quad q(t) = [\sigma M_0[-k_{10}] \; 0]^T, \quad x = [a_1 \; a]^T,$$

and $w(t)$ is the Wiener process.

Illustration 4.1.8 (group dynamics [58, 180]). It is assumed that all of the conditions HGD1–HGD3, the notations, and definitions are valid. The mathematical model of the group dynamic system (Illustration 4.2.8 [80]) is extended to the stochastic settings. The derived deterministic model is

$$da = [-\alpha_{11}a + \alpha_{12}f + \gamma_2 e]dt,$$

$$df = [\alpha_{21}a - \alpha_{22}f]dt,$$

which can be rewritten as

$$dx = (Ax + a)dt, \tag{4.1.54}$$

where $\alpha_{11} = \gamma_1\beta_1 + \gamma_2$, $\alpha_{12} = \gamma_1$, and $\alpha_{21} = \beta_0\alpha_1$ are positive numbers, and $\alpha_{22} = \alpha_0\alpha_1 - \beta_1$; $x = [a, f]^T$; A is a 2×2 constant matrix; and $a = [\gamma_2 e, 0]^T$ is an input to the group dynamic.

Source of random fluctuations. The group dynamic system is an open system subject to internal as well as external random perturbations. All the above-specified parameters are subject to random perturbations. The sizes of species are integral values, the successive activities do not overlap in time, and the environmental parameters vary discretely in time. In the light of this, the parameters can be considered under the stationary Gaussian process with independent increments. From these considerations, Modified Brownian Motion Model 4.1.2 is directly applicable to the group dynamic system. Therefore, the deterministic model described in (4.1.54) reduces to the Itô–Doob-type system of stochastic differential equations

$$dx = (Ax + a)dt + (Bx + b)dw(t), \quad x(t_0) = x_0. \tag{4.1.55}$$

This completes the modeling process of group/individual social interaction dynamics.

Example 4.1.3 (Kendall's mathematical marriage model). All of the conditions KMM1–KMM6, the notations and definitions in Example 4.2.3 [80] are assumed to be true. We utilize the Kendall's simple deterministic mathematical model described by

$$dF = [-(\delta + \rho)F + (\beta + \delta)C]dt,$$

$$dC = [-2\delta C + \rho F]dt,$$

$$dM = [-\delta M + (\beta + \delta)C - \rho F]dt.$$

We note that the first two equations are independent of M (decoupled from M). Therefore, one can solve this system by solving first the system of two differential coupled equations

$$dF = [-(\delta + \rho)F + (\beta + \delta)C]dt,$$

$$dC = [-2\delta C + \rho F]dt,$$

i.e

$$dx = A\,x\,dt, \quad x(0) = x_0,$$

where $x = [F, C]^T$, $A = \begin{bmatrix} -(\delta + \rho) & (\beta + \delta) \\ \rho & -2\delta \end{bmatrix}$, and $x_0 = [F(0), C(0)]^T$. Then, using this solution and applying Procedure 2.4.3 of Chapter 2, one can solve the equation

$$dM = [-\delta M + (\beta + \delta)C - \rho F]dt.$$

From Illustrations 4.1.4, 4.1.7, and 4.1.8, one can present a stochastic mathematical model for the marriage system, and it is represented by the Itô–Doob-type stochastic differential equation

$$dx = (Ax + a)dt + (Bx + b)dw(t), \quad x(t_0) = x_0,$$

and for $M(0) = M_0$,

$$dM = [-\bar{\delta}M + (\bar{\beta}C + \bar{\delta}C) - \bar{\rho}F]dt + \sigma[-\bar{\delta}M + (\bar{\beta}C + \bar{\delta}C) - \bar{\rho}F]dw(t).$$

This completes the stochastic version of mathematical modeling of the marriage system of Kendall.

Illustration 4.1.9   (US government system [197]). We assume that all the assumptions, Articles I–V, the Bill of Rights, and the definitions and notations of Illustrations 4.2.9 [80] are given. Under these conditions, the deterministic mathematical model of the US government system is described by [80]

$$dx = (Ax + e)dt, \tag{4.1.56}$$

where $e = [e_1, e_2, e_3]^T$, $A = (a_{\ell j})_{3 \times 3}$, and

$$x_{\ell j} = \begin{cases} -\left(\alpha_{0\ell} + \sum\limits_{\substack{j \neq \ell}}^{3} \alpha_{j\ell} \right) & \text{for } j = \ell, \\ \alpha_{\ell j} & \text{for } j \neq \ell. \end{cases}$$

Source of random fluctuations.   The source of uncertainties is due to the environmental, national and international relations, and policy changes. By following the illustrations presented in this section, the random perturbations are through coefficient parameters of (4.1.56). Therefore, the deterministic model described by (4.1.56) reduces to the Itô–Doob-type stochastic differential equation

$$dx = (Ax + a)dt + (Bb + b)dw(t), \quad x(t_0) = x_0. \tag{4.1.57}$$

This completes the illustrations.

Exercises

1. *Reversible bacterial mutation* [166]. In the bacterial cell culture process, it has been observed that a variant of the initial form appears. The process of formation of a variant is called a mutation. A variant form of the initial form of

bacteria is termed a mutant. It is also known that certain variant cells are converted back to the original form by a process known as back-mutation/reversion. Let A and B be two types of bacterial cells that are genetically connected. B is formed by the mutation process from an initial cell type A. At time t, let N_1 and N_2 be the number of cells of types A and B, respectively. The dynamic interactions of these cells are described by $A \rightleftarrows B$; rates k_{11} and k_{22} are the fractional growth rates of A and B, respectively; k_{12} is the fractional mutation rate of the initial form of cell A to cell B; and k_{21} is the fractional back-mutation rate of variant cell B to the original cell. k_{ij} for all $1 \leq i, j \leq 2$ are considered to be positive.

(a) Derive the deterministic mathematical description of this dynamic mutation process.

(b) Is this mutation process free from random disturbances?

(c) If the answer to the question in (b) is no, then identify the parameters that can be influenced by the random disturbance. If possible, classify with justifications the nature of the random disturbance(s).

(d) Derive the stochastic mathematical description of this dynamic mutation process under a random environment.

2. **Pharmacokinetics: "prompt" injection** [188, 189]. Assume that all the conditions of Example 4.1.2 are satisfied.

(a) Derive Theorell's mathematical model for the drug distributions of intravenously administered "prompt" injection.

(b) Derive the stochastic mathematical model of this type of drug distribution problem.

(c) Compare the mathematical models in (a) and (b) with the models described in Example 4.1.2.

3. **Kendall's two-sex model** [66, 151]. In 1948, D. G. Kendall presented a two-sex deterministic model under the following assumptions:

(a) The growth of a single species is directly proportional to the size of the population.

(b) Let F and M be the number of females and the number of males at time t, and assume that $dF = [-\mu F + \frac{1}{2}\Lambda(F, M)]dt$ and $dM = [-\mu M + \frac{1}{2}\Lambda(F, M)]dt$, where $\Lambda(F, M)$ represents arbitrary interactions between the female and male populations. Rewrite Kendall's systems of differential equations if:

   (i) $\Lambda(F, M) = \rho F M$;

  (ii) $\Lambda(F, M) = 2\rho\sqrt{FM}$;

 (iii) $\Lambda(F, M) = 2\rho(F + M)$;

 (iv) give the interpretation for Kendall's interaction $\Lambda(F, M)$ in (i)–(iii);

  (v) if possible, write the corresponding stochastic mathematical models with justifications.

> **Hint:** Note that $d(F - M) = -\mu(F - M)dt$ and $F(0) - M(0)$ very small.

4. **Goodman's two-sex model** [48, 151]. Assume that all the conditions of Exercise 4.1.3 are satisfied.

   (a) Under Goodman's interaction function $\Lambda(F, M) = 2\rho \min\{F, M\}$, show Kendall's deterministic two-sex model reduces to $dF = (-\mu F + \rho M)dt$ and $dM = (-\mu + \rho)M\,dt$.

   (b) If possible, write the stochastic mathematical model corresponding to the model in (a).

5. Recall that dynamic modeling was based on the fact that the stopclock is perfect. In general, it is subject to random fluctuations. Let F be a cumulative distribution function (cdf) of a random variable T (time). For $x \in R^n$ be the state of a multidimensional compartmental system. Show that the mathematical model (4.1.37) reduces to

$$dc = (Ac + a)dF + (Bc + a)dw(t),$$

where A, c, and a are as defined. Moreover, give the interpretation of this model. **Hint:** An equivalent representation is $c(t) = c_0 + \int_{-\infty}^{t}(Ac(s) + a)dF(s)$. Under further assumptions on F, show that

$$c(t) = c_0 + \int_{t_0}^{t}(Ac(s) + a)dF(s) + \int_{t_0}^{t}(Bc(s) + a)dw(t).$$

6. Under assumptions of Exercises 5, reformulate Kendall's mathematical marriage model (Example 4.1.3).

7. Develop an Itô–Doob-type stochastic version of Richard Levin's population extinction model (Illustration 3.7.3) for an n-species subcommunity:

$$dN_i = \left[m_i N \sum_{i=j}^{n}(1 - N_j) - e_{ij}N_j \right]dt, \quad N_i(0) = N_{0i}.$$

Determine its linearized version under the steady-state assumption of the population.

8. Construct an Itô–Doob-type stochastic version of the epidemiological process (Illustration 3.1.4) of the following deterministic model:

$$\begin{cases} dS = b_1 + a_{11}S + a_{12}I + a_{13}R - b_{11}SI, & S(0) = S_0, \\ dI = a_{22}I + b_{12}SI, & I(0) = I_0, \\ dR = a_{32}I + a_{33}R, & R(0) = 0, \end{cases}$$

where S, I, and R stand for the version of susceptible, infective, and removal subpopulation, respectively, in a community; b_1 and b_{ij} are positive real numbers; for $j \neq 1$, a_{1j} and a_{32} are non-negative.

9. Develop an Itô–Doob-type stochastic version of the following chain reactions — the Rice–Herzfeld mechanism (Illustration 3.1.1) deterministic model:

$$\frac{d}{dt}[M] = -[M](k_1 + k_2[R_1]),$$

$$\frac{d}{dt}[R_1] = k_1[M] + k_3[R_2] - k_2[R_1][M] - k_{4a}[R_1]^2,$$

$$\frac{d}{dt}[R_2] = k_2[R_1][M] - k_3[R_2],$$

where M, $[R_1]$, and $[R_2]$ are as defined in Illustration 3.1.1. Furthermore, under the steady-state hypothesis, find the linearized version of the stochastic model.

10. Formulate an Itô–Doob-type stochastic version of Kendall's deterministic marriage model in Example 4.1.3 and Exercise 5 (Section 4.2, [80]) described by

$$dF = [-\delta_1 F + \beta_1 C + \delta_2 C - K(F, M)]dt,$$

$$dM = [-\delta_2 M + \beta_2 C + \delta_1 C - K(F, M)]dt,$$

$$dC = [-(\delta_1 + \delta_2)C + K(F, M)]dt,$$

where F, C, M, and $K(F, M) = aF + bM + cFM$, with a, b, and c being non-negative numbers; δ_1, δ_2, β_1, and β_2 are any positive real numbers. Moreover, determine a linearized version of Itô–Doob-type stochastic mathematical model.

4.2 Linear Homogeneous Systems

In this section, we extend the eigenvalue-type methods described in Sections 2.4 and 4.3 [80], to find a solution to the first-order linear homogeneous system of stochastic differential equations with constant coefficients. We imitate the procedure described in Section 4.3 of Chapter 4 (Volume 1) with regard to Itô–Doob-type stochastic systems of linear homogeneous differential equations with constant coefficient matrices. Moreover, considering the idea of two different timescales [the usual timescale "t" and the timescale induced by $w(t)$], the method of solving an Itô–Doob-type stochastic system of linear homogeneous differential equations reduces to the problem of solving a linear system of algebraic equations with constant coefficients. Furthermore, we note that all the presented discussions and results of this chapter include the discussions and results concerning systems of deterministic differential equations as a special case [80].

4.2.1 *General Problem*

Let us consider the following first-order linear homogeneous system of stochastic differential equations of the Itô–Doob type:

$$dx = Bx\, dw(t), \tag{4.2.1}$$

where x is an n-dimensional column vector (or an $n \times 1$ matrix), $B = (b_{ij})_{n \times n}$ is an $n \times n$ constant matrix whose entries b_{ij} are real numbers, and w is a scalar normalized Wiener process (Definition 1.2.12).

The goal is to present a procedure for finding a general solution to (4.2.1). We are also interested in solving an IVP associated with this system. In addition, we note the basic ideas and algorithms (theorems) that can be easily applied to:

(a) Time-varying systems.
(b) Nonhomogeneous systems, and nonlinear systems of differential equations.

For this purpose, we first need to define the concept of a solution process of (4.2.1).

Definition 4.2.1. Let $J = [a, b]$, $a, b \in R$, and let $J \subseteq R$ be an interval. A *solution* to the stochastic system of differential equations of the Itô–Doob-type (4.2.1) is an n-dimensional random column vector (or $n \times 1$ matrix) process/function x defined on J into R^n such that it satisfies system (4.2.1) on J in the sense of Itô–Doob stochastic calculus. In short, if we substitute $x(t) = (x_i(t))_{n \times 1}$ and its Itô–Doob differential $dx(t)$ into system (4.2.1), then the system of stochastic differential equations remains valid for all t in J.

Example 4.2.1. Given: $dx_1 = (x_1 + 2x_2)dw(t)$ and $dx_2 = (4x_1 + 3x_2)dw(t)$. Show that $x_1(t) = e^{-(\frac{1}{2}t + w(t))}[1 \ -1]^T$ and $x_2(t) = e^{(-\frac{25}{2}t + 5w(t))}[1 \ 2]^T$ are solutions to the given system of differential equations, where T stands for the transpose of a matrix.

Solution procedure. By using the product of matrices, the matrix representation of the given differential equations is $dx(t) = Bx \, dw(t)$, where B and x are 2×2 and 2×1 matrices, respectively,

$$B = \begin{bmatrix} 1 & 2 \\ 4 & 3 \end{bmatrix} \quad \text{and} \quad x = \begin{bmatrix} x_1 \\ x_2 \end{bmatrix}.$$

We apply the deterministic and stochastic differential calculus to $e^{-(\frac{1}{2}t + w(t))}$ $[1 \ -1]^T = x_1(t)$, and we compute the differential:

$$dx_1(t) = d(e^{-(\frac{1}{2}t + w(t))}[1 \ -1]^T)$$

(from the expression and the Itô–Doob differential)

$$= -e^{-(\frac{1}{2}t + w(t))}[1 \ -1]^T d\left(\frac{1}{2}t + w(t)\right)$$

$$+ e^{-(\frac{1}{2}t + w(t))}[1 \ -1]^T d\left(\frac{1}{2}t + w(t)\right)^2$$

(Theorem 1.3.4 and the chain rule)

$$= -e^{-(\frac{1}{2}t+w(t))}[1 \ \ -1]^T \left(\frac{1}{2}dt + dw(t)\right)$$

$$+ \frac{1}{2}(e^{-(\frac{1}{2}t+w(t))}[1 \ \ -1]^T dt \left[\text{Itô–Doob differential: } d\left(\frac{1}{2}t + w(t)\right)^2 = dt\right]$$

$$= -e^{-(\frac{1}{2}t+w(t))}[1 \ \ -1]^T dw(t) \quad \text{(by simplification)}.$$

We compute $Bx_1(t)dw(t)$ as follows:

$$Bx_1(t)dw(t) = \begin{bmatrix} 1 & 2 \\ 4 & 3 \end{bmatrix}\begin{bmatrix} e^{-(\frac{1}{2}t+w(t))} \\ -e^{-(\frac{1}{2}t+w(t))} \end{bmatrix} dw(t) \quad \text{(by substitution)}$$

$$= \begin{bmatrix} -e^{-(\frac{1}{2}t+w(t))} \\ e^{-(\frac{1}{2}t+w(t))} \end{bmatrix} dw(t) \quad \text{(by matrix multiplication)}$$

$$= -e^{-(\frac{1}{2}t+w(t))}[1 \ \ -1]^T dw(t)$$

$$= dx_1(t) \quad \text{(by substitution)}.$$

This shows that the given process $x_1(t) = e^{-(\frac{1}{2}t+w(t))}[1 \ \ -1]^T$ satisfies the given differential equation (Definition 4.2.1). Therefore, it is a solution to the given Itô–Doob system of stochastic differential equations. Similarly, by applying the above argument to $x_2(t) = e^{-(\frac{25}{2}t+5w(t))}[1 \ 2]^T$, we have

$$dx_2(t) = d(e^{(-\frac{25}{2}t+5w(t))}[1 \ 2]^T)dw(t) = 5e^{(-\frac{25}{2}t+5w(t))}[1 \ 2]^T dw(t),$$

$$Bx_2(t)dw(t) = \begin{bmatrix} 1 & 2 \\ 4 & 3 \end{bmatrix}\begin{bmatrix} e^{(-\frac{25}{2}t+5w(t))} \\ 2e^{(-\frac{25}{2}t+5w(t))} \end{bmatrix} dw(t) = 5\begin{bmatrix} e^{(-\frac{25}{2}t+5w(t))} \\ 2e^{(-\frac{25}{2}t+5w(t))} \end{bmatrix} dw(t).$$

By comparing the expressions of $dx_2(t)$ and $Bx_2(t)dw(t)$ in the context of the definition of the transpose of a matrix, we conclude that $x_2(t) = e^{(-\frac{25}{2}t+5w(t))}[1 \ 2]^T$ is a solution process of the given system of differential equations. This establishes the solution process of the example.

4.2.2 *Procedure for Finding a General Solution*

In the following, we imitate the procedure for solving deterministic systems of differential equations in Chapter 4 of Volume 1 with natural modifications [two timescales: t and $w(t)$]. We repeat a four-step procedure in the context of Itô–Doob stochastic calculus with natural modifications to determine a solution process of (4.2.1). The procedure is described below.

Step #1. Let us seek a solution to (4.2.1) of the following form:

$$x(t) \equiv x(t, w(t)) = \exp[\lambda t + \mu w(t)]c \qquad (4.2.2)$$

where $c = (c_1, c_2, \dots, c_n)^T$ is an n-dimensional unknown constant random vector (or an $n \times 1$ matrix) that is independent of the Wiener process w; λ and μ are unknown scalar quantities (real or complex numbers).

We note that if $c = 0$, i.e. $c_1 = c_2 = \cdots = c_n = 0$, then $x(t)$ in (4.2.2) is a trivial solution process of (4.2.1) (a zero solution, i.e. a zero random process/function on J).

We seek the nontrivial solution processes (nonzero solutions) of (4.2.1). For this purpose, we need to compute λ, μ, and c with $c \neq 0$. Applying the Itô–Doob differential formula (Theorem 1.3.1), we find the Itô–Doob stochastic differential of $x(t)$ in (4.2.2). We then substitute $x(t)$ and $dx(t)$ into (4.2.1) and get

$$dx(t) = \exp[\lambda t + \mu w(t)]c(\lambda \, dt + \mu \, dw(t)) + \frac{1}{2}\exp[\lambda t + \mu w(t)]c\mu^2 dt = Bx \, dw(t).$$

This implies that

$$\exp[\lambda t + \mu w(t)]c\left(\lambda + \frac{1}{2}\mu^2\right)dt + \exp[\lambda t + \mu w(t)]c\mu \, dw(t) = B\exp[\lambda t + \mu w(t)]c \, dw(t).$$

Hence,

$$\exp[\lambda t + \mu w(t)]\left[c\left(\lambda + \frac{1}{2}\mu^2\right)dt + c\mu \, dw(t)\right] = \exp[\lambda t + \mu w(t)][Bc \, dw(t)].$$

From this and the fact that $\exp[\lambda t + \mu w(t)] \neq 0$ on J, we have

$$c\left(\lambda + \frac{1}{2}\mu^2\right)dt + c\mu \, dw(t) = Bc \, dw(t).$$

Now, by comparing the coefficients of differentials dt and $dw(t)$ on both sides of the above equation [the method of undetermined coefficients: $\Delta t \neq b\Delta w(t)$, for any number b], we obtain

$$\left(\lambda + \frac{1}{2}\mu^2\right)c = 0, \quad \mu c = Bc.$$

Hence,

$$(B - \mu I)c = 0 \quad \text{and} \quad \left(\lambda + \frac{1}{2}\mu^2\right)c = 0, \qquad (4.2.3)$$

where I is an $n \times n$ identity matrix.

Step #2. As in the case of the deterministic system (Section 4.3.2 of Volume 1), the problem of finding a solution to the stochastic system of differential equations (4.2.1) reduces to the problem of solving the linear homogeneous system of algebraic equations (4.2.3) with an n-dimensional nonzero unknown constant random column vector (or an $n \times 1$ matrix) c, and unknown scalar quantities λ and μ (real or complex numbers).

The coupled systems of algebraic systems in (4.2.3) are in triangular form. This makes a solution computation process relatively simple. First, we solve the completely decoupled system containing unknown parameter μ [the first subsystem in (4.2.3)]. Then, we use this solution μ and the second subsystem in (4.2.3) to solve an unknown parameter λ.

Now, we apply Step #2 for Procedure 4.3.2 for finding a solution to the deterministic system of Section 4.3 [80]. An unknown parameter μ is determined by solving the following linear homogeneous system of algebraic equations (because $c \neq 0$):

$$\det(B - \mu I) = 0, \tag{4.2.4}$$

where "det" stands for the determinant of the matrix $(B - \mu I)$. Let $\mu_1, \ldots, \mu_k, \ldots, \mu_n$ be n roots of (4.2.4) (real or complex numbers) which are not necessarily all distinct.

For each k and $1 \leq k \leq n$, we substitute the above-determined roots $\mu = \mu_k$ into the second subsystem in (4.2.3), and obtain

$$\lambda + \frac{1}{2}\mu_k^2 = 0. \tag{4.2.5}$$

Now, we solve (4.2.5) for λ. Thus, for each k and $1 \leq k \leq n$, $\lambda = \lambda_k = -\frac{1}{2}\mu_k^2$.

We further remark that these roots are uniquely determined by the coefficient matrix B in (4.2.1). In short, they depend only on the entries of the matrix.

Step #3. For each $k = 1, 2, \ldots, n$, the characteristic root $\mu_k \equiv \mu_k(B)$ determined in Step #2, there exists (one can find) at least one nonzero eigenvector, $c_k \equiv c_k(B)$. This is computed by applying Step #3 of Procedure 4.3.2 [80] to solve the following linear homogeneous system of n algebraic equations:

$$(B - \mu_k I)c = 0, \tag{4.2.6}$$

for each $k = 1, 2, \ldots, n$. Of course, all the comments and conclusions regarding $c_k \equiv c_k(B)$ for each k, $k = 1, 2, \ldots, n$, can be outlined as in Section 4.3 [80].

Step #4. We note that by corresponding to n roots $\mu_1, \ldots, \mu_k, \ldots, \mu_n$ of (4.2.4) in Step #2, we find n roots $\lambda_1, \ldots, \lambda_k, \ldots \lambda_n$ of (4.2.5) and n eigenvectors

$c_1, \ldots, c_k, \ldots, c_n$ of the coefficient matrix B in (4.2.1). Thus, the procedure provides n solutions,

$$x_1(t) = \exp[\lambda_1 t + \mu_1 w(t)] c_1,$$

$$x_2(t) = \exp[\lambda_2 t + \mu_2 w(t)] c_2,$$

$$\cdots \qquad \cdots \cdots$$

$$x_k(t) = \exp[\lambda_k t + \mu_k w(t)] c_k,$$

$$\cdots \qquad \cdots \cdots$$

$$x_n(t) = \exp[\lambda_n t + \mu_n w(t)] c_n,$$

to the system of differential equations (4.2.1) corresponding to the n eigenvalues
[roots of (4.2.4) and (4.2.5)] and the corresponding n eigenvectors of B determined
by system (4.2.6).

Example 4.2.2. Given: $dx(t) = Bx\,dw(t)$, where $B = \begin{bmatrix} 1 & 3 & 0 \\ 7 & 1 & 5 \\ 0 & -4 & 1 \end{bmatrix}$. Find solutions
to the given system of Itô–Doob-type stochastic differential equations.

Solution process. This is the system of three ($n = 3$) first-order linear Itô–
Doob-type stochastic differential equations. To find the solutions to this system of
differential equations, we imitate Procedure 4.2.2. We seek a solution of the form
described in Step #1, i.e. $x(t) = \exp[\lambda t + \mu w(t)] c$. The prime goal is to find an
unknown nonzero three-dimensional vector $c = [c_1 \ c_2 \ c_3]^T$ and scalar numbers λ
and μ. For this purpose, we use (4.2.3), and we have

$$0 = (B - \lambda I)c = \left(\begin{bmatrix} 1 & 3 & 0 \\ 7 & 1 & 5 \\ 0 & -4 & 1 \end{bmatrix} - \mu \begin{bmatrix} 1 & 0 & 0 \\ 0 & 1 & 0 \\ 0 & 0 & 1 \end{bmatrix} \right) c \quad \text{and} \quad \left(\lambda + \frac{1}{2}\mu^2 \right) c = 0$$

$$= \begin{bmatrix} 1 - \mu & 3 & 0 \\ 7 & 1 - \mu & 5 \\ 0 & -4 & 1 - \mu \end{bmatrix} c \quad \text{and} \quad \left(\lambda + \frac{1}{2}\mu^2 \right) c = 0.$$

The first system of linear homogeneous algebraic equations can be rewritten as

$$\begin{bmatrix} 1 - \mu & 3 & 0 \\ 7 & 1 - \mu & 5 \\ 0 & -4 & 1 - \mu \end{bmatrix} c = \begin{bmatrix} (1 - \mu)c_1 + 3c_2 \\ 7c_1 + (1 - \mu)c_2 + 5c_3 \\ -4c_2 + (1 - \mu)c_3 \end{bmatrix} = 0,$$

which implies that

$$(1 - \mu)c_1 + 3c_2 = 0,$$

$$7c_1 + (1 - \mu)c_2 + 5c_3 = 0,$$

$$-4c_2 + (1 - \mu)c_3 = 0.$$

From Step #2 of Procedure 4.2.2, we obtain the characteristic equation (4.2.4) of the matrix B in the context of the given example as

$$0 = \det(B - \lambda I) = (1 - \mu)[(1 - \mu)^2 + 20] - 3[7(1 - \mu)] + 0(-28)$$

$$= (1 - \mu)^3 + 20(1 - \mu) - 21(1 - \mu)$$

$$= \mu(1 - \mu)(\mu - 2).$$

The eigenvalues are $\mu_1 = 0$, $\mu_2 = 1$ and $\mu_3 = 2$. Moreover, from (4.2.5), we have, $\lambda_1 = 0, \lambda_2 = -\frac{1}{2}, \lambda_3 = -2$. This completes Step #2. To find the eigenvectors corresponding to the eigenvalues, the corresponding systems of linear homogeneous algebraic equations are solved by following the argument (elementary row operations) used in Examples 1.3.15 and 1.3.16 (Chapter 1 of Volume 1):

(1) For $\mu_1 = 0$:

$$[B_1 \mid b] = \begin{bmatrix} 1 & 3 & 0 & 0 \\ 7 & 1 & 5 & 0 \\ 0 & -4 & 1 & 0 \end{bmatrix} \to \cdots \to \begin{bmatrix} 1 & 3 & 0 & 0 \\ 0 & 1 & -\frac{1}{4} & 0 \\ 0 & 0 & 0 & 0 \end{bmatrix}.$$

This implies that $c_1 + 3c_2 = 0$ and $c_2 - \frac{1}{4}c_3 = 0$. Hence, the corresponding solution set is $S_1 = \text{Span}([-3 \ 1 \ 4]^T)$.

(2) For $\mu_2 = 1$:

$$[B_1 \mid b] = \begin{bmatrix} 0 & 3 & 0 & 0 \\ 7 & 0 & 5 & 0 \\ 0 & -4 & 0 & 0 \end{bmatrix} \to \cdots \to \begin{bmatrix} 7 & 0 & 5 & 0 \\ 0 & 3 & 0 & 0 \\ 0 & -4 & 0 & 0 \end{bmatrix}.$$

This implies that $7c_1 + 5c_3 = 0$, $3c_2 = 0$ and $-4c_2 = 0$. Hence, the corresponding solution set is $\text{Span}([5 \ 0 \ -7]^T) = S_1$.

(3) For $\mu_3 = 2$:

$$[B_1 \mid b] = \begin{bmatrix} -1 & 3 & 0 & 0 \\ 7 & -1 & 5 & 0 \\ 0 & -4 & -1 & 0 \end{bmatrix} \to \cdots \to \begin{bmatrix} 1 & -3 & 0 & 0 \\ 0 & 1 & \frac{1}{4} & 0 \\ 0 & 0 & 0 & 0 \end{bmatrix}.$$

This implies that $c_1 - 3c_2 = 0$ and $c_2 + \frac{1}{4}c_3 = 0$. Hence, the corresponding solution set is $\text{Span}([3 \ 1 \ -4]^T) = S_1$.

From (4.2.3), the corresponding eigenvectors and eigenvalues of λ are

(1) $\lambda_1 = 0$, $[-3 \; 1 \; 4]^T$;
(2) $\lambda_2 = -\frac{1}{2}$, $[5 \; 0 \; -7]^T$;
(3) $\lambda_2 = -2$, $[3 \; 1 \; -4]^T$.

From Step #4 of Procedure 4.2.2, we have the following three solutions corresponding to the pairs of eigenvalues $\mu_1 = 0$ and $\lambda_1 = 0$, $\mu_2 = 1$ and $\lambda_2 = -\frac{1}{2}$, and $\mu_3 = 2$ and $\lambda_3 = -2$:

$$x_1(t) = e^{0(t+w(t))} \begin{bmatrix} -3 \\ 1 \\ 4 \end{bmatrix}, \quad x_2(t) = e^{(-\frac{1}{2}t+w(t))} \begin{bmatrix} 5 \\ 0 \\ -7 \end{bmatrix}, \quad x_3(t) = e^{2(-t+w(t))} \begin{bmatrix} 3 \\ 1 \\ -4 \end{bmatrix},$$

respectively. This completes the solution procedure.

Example 4.2.3. Given: $dx_1 = x_2 dw(t)$. Find a solution to this system of stochastic differential equations.

Solution procedure. We repeat the arguments used in Example 4.2.2 with an appropriate change of notations. In this case, $dx(t) = Bx \, dwt$, where B and x are 2×2 and 2×1 matrices, respectively. In fact,

$$B = \begin{bmatrix} 0 & 1 \\ 0 & 0 \end{bmatrix} \quad \text{and} \quad x = \begin{bmatrix} x_1 \\ x_2 \end{bmatrix}.$$

The characteristic equation (4.2.4) of the matrix B in the context of the given example is

$$0 = \det(B - \mu I) = (-\mu)(-\mu) = \mu^2.$$

The eigenvalues are $\mu_1 = 0$ and $\mu_2 = 0$. We note that $\mu_1 = 0$ is the double (repeated) root of the characteristic equation of B. From (4.2.3), the corresponding values of λ are $\lambda_1 = 0 = \lambda_2$.

The eigenvector corresponding to the eigenvalue $\mu_1 = 0 = \lambda_1$ is $c_1(B) = [1 \; 0]^T$. As per Step #4, we found one solution:

$$x_1(t) = e^{0t+w(t)} \begin{bmatrix} 1 \\ 0 \end{bmatrix} = \begin{bmatrix} 1 \\ 0 \end{bmatrix} \quad (\text{by } e^{0t+w(t)} = e^0 = 1).$$

This completes the solution procedure.

Theorem 4.2.1 (fundamental property). *Let $x_1(t), x_2(t), \ldots, x_k(t), \ldots, x_m(t)$ be m solutions to (4.2.1) on J. Then,*

$$x(t) = a_1 x_1(t) + a_2 x_2(t) + \cdots + a_k x_k(t) + \cdots + a_m x_m(t) = \sum_{k=1}^{m} a_k x_k(t) \quad (4.2.7)$$

is also a solution to (4.2.1) on J, where $a_1, a_2, \ldots, a_k, \ldots, a_m$ are arbitrary constants.

Proof. Let us find the differentials of both sides of the expressions in (4.2.7). This is achieved by using the rules of the differentials (Theorem 1.5.1, Observation 1.5.2 [80]), the concept of the solution to the system of Itô–Doob-type stochastic differential equations, and matrix algebra. Thus, we have

$$dx(t) = d[a_1 x_1(t) + a_2 x_2(t) + \cdots + a_k x_k(t) + \cdots + a_m x_m(t)] \qquad \text{[from (4.2.7)]}$$

$$= a_1 dx_1(t) + a_2 dx_2(t) + \cdots + a_k dx_k(t) + \cdots + a_m dx_m(t)$$

(Observation 1.5.2 [80])

$$= a_1 B x_1(t) dw + a_2 B x_2(t) dw + \cdots + a_k B x_k(t) dw + \cdots + a_m B x_m(t) dw$$

$$= B[a_1 x_1(t) + a_2 x_2(t) + \cdots + a_k x_k(t) + \cdots + a_m x_m(t)] dw(t) \qquad \text{[from (4.2.7)]}$$

$$= Bx(t) dw(t).$$

This shows that $x(t)$ is a solution to (4.2.1). This completes the proof of the theorem. $\qquad \square$

Example 4.2.4. Given: $dx_1 = (x_1 + 2x_2) dw(t)$ and $dx_1 = (4x_1 + 3x_2) dw(t)$. Show that:

(a) $x_1(t) = e^{(-\frac{1}{2}t + w(t))}[1 \ -1]^T$ and $x_2(t) = e^{(-\frac{25}{2}t + 5w(t))}[1 \ 2]^T$ are solutions to the given system of Itô–Doob-type stochastic differential equations, where T stands for the transpose of a matrix, and moreover

(b)

$$x(t) = a_1 x_1(t) + a_2 x_2(t)$$

$$= a_1 e^{-(\frac{1}{2}t + w(t))} \begin{bmatrix} 1 \\ -1 \end{bmatrix} + a_2 e^{(-\frac{25}{2}t + 5w(t))} \begin{bmatrix} 1 \\ 2 \end{bmatrix}$$

$$= \begin{bmatrix} a_1 e^{-(\frac{1}{2}t + w(t))} + a_2 e^{(-\frac{25}{2}t + 5w(t))} \\ -a_1 e^{-(\frac{1}{2}t + w(t))} + 2a_2 e^{(-\frac{25}{2}t + 5w(t))} \end{bmatrix} a$$

is the solution process of the given system of differential equations, where a_1 and a_2 are arbitrary real numbers.

Solution procedure. From Example 4.2.1, we conclude that the given processes $x_1(t)$ and $x_2(t)$ are solution processes of the given system of Itô–Doob-type stochastic differential equations. We need to show the validity of the mathematical statement in (b). For this purpose, we imitate the arguments used in the proof of

Theorem 4.2.1 and Example 4.2.1. We consider

$$dx(t) = d(a_1 x_1(t) + a_2 x_2(t))$$

$$= d\left(a_1 e^{-(\frac{1}{2}t+w(t))} \begin{bmatrix} 1 \\ -1 \end{bmatrix}\right) + d\left(a_2 e^{(-\frac{25}{2}t+5w(t))} \begin{bmatrix} 1 \\ 2 \end{bmatrix}\right) \quad \text{(by definition)}$$

$$= a_1 d\left(e^{-(\frac{1}{2}t+w(t))} \begin{bmatrix} 1 \\ -1 \end{bmatrix}\right) + a_2 d\left(e^{(-\frac{25}{2}t+5w(t))} \begin{bmatrix} 1 \\ 2 \end{bmatrix}\right)$$

$$= -a_1 e^{-(\frac{1}{2}t+w(t))} \begin{bmatrix} 1 \\ -1 \end{bmatrix} dw(t)$$

$$+ 5a_2 e^{(-\frac{25}{2}t+5w(t))} \begin{bmatrix} 1 \\ 2 \end{bmatrix} dw(t) \quad \text{(by Theorem 1.3.1)}$$

$$= \begin{bmatrix} -a_1 e^{-(\frac{1}{2}t+w(t))} + 5a_2 e^{(-\frac{25}{2}t+5w(t))} \\ a_1 e^{-(\frac{1}{2}t+w(t))} + 10a_2 e^{(-\frac{25}{2}t+5w(t))} \end{bmatrix} dw(t) \quad \text{(by Theorem 1.3.1S2 [80]).}$$

Now, we compute $Bx(t)dw(t)$ as follows:

$$Bx(t)dw(t) = \begin{bmatrix} 1 & 2 \\ 4 & 3 \end{bmatrix} \begin{bmatrix} a_1 e^{-(\frac{1}{2}t+w(t))} + a_2 e^{(-\frac{25}{2}t+5w(t))} \\ -a_1 e^{-(\frac{1}{2}t+w(t))} + 2a_2 e^{(-\frac{25}{2}t+5w(t))} \end{bmatrix} dw(t) \quad \text{(by substitution)}$$

$$= \begin{bmatrix} -a_1 e^{-(\frac{1}{2}t+w(t))} + 5a_2 e^{(-\frac{25}{2}t+5w(t))} \\ a_1 e^{-(\frac{1}{2}t+w(t))} + 10a_2 e^{(-\frac{25}{2}t+5w(t))} \end{bmatrix} dw(t) \quad \text{(by matrix multiplication)}.$$

By comparing these calculations, we conclude that $dx(t) = Bx(t)dw(t)$. From Definition 4.2.1, we infer that the linear combination $x(t)$ of the given solution processes $x_1(t)$ and $x_2(t)$ is also the solution to the given system of Itô–Doob-type stochastic differential equations. This completes the solution procedure.

Observation 4.2.1. In view of Theorem 4.2.1, an observation similar to Observation 4.3.2 of Chapter 4 [80] can be made with regard to the stochastic linear system (4.2.1). To avoid monotonicity, we minimize the repetitiveness of it.

4.2.3 *Initial Value Problem*

Let us formulate an IVP for the stochastic system of differential equations (4.2.1). We consider

$$dx = Bx\, dw(t), \quad x(t_0) = x_0. \tag{4.2.8}$$

Here x, B, and $w(t)$ are as defined in (4.2.1); t_0 is in J; and x_0 is an n-dimensional random vector (or an $n \times 1$ matrix) defined on the complete probability space $(\Omega, \mathfrak{F}, P)$, and it is independent of $w(t)$ for all t in J. (t_0, x_0) is called the *initial data* or *initial condition*. In short, the random vector x_0 in R^n is the value of the solution process at $t = t_0$ (the initial/given time); it is an initial/given state vector of the system defined on the probability space $(\Omega, \mathfrak{F}, P)$ into R^n, and it is independent of $w(t)$ for all t in J. The problem of finding a solution to (4.2.8) is called the *initial value problem* (IVP). Its solution is denoted by $x(t) = x(t, t_0, x_0)$ for $t \geq t_0$ and t in J.

Definition 4.2.2. A *solution process of the* IVP (4.2.8) is an n-dimensional random column vector (or an $n \times 1$ matrix) process/function x defined on J such that x and its Itô–Doob differential satisfy both the given system of differential equations (4.2.8) and the initial condition (t_0, x_0). In short, (i) $x(t)$ is the solution to (4.2.1) (Definition 4.2.1) and (ii) $x(t_0) = x(t_0, t_0, x_0) = x_0$.

Example 4.2.5. Given: $dx = B\, dw(t)$, $x(t_0) = x_0$, where $B = \begin{bmatrix} 0 & 1 \\ -1 & 0 \end{bmatrix}$. Show that

$$x(t) = \Phi(t, t_0)x_0 = e^{\frac{1}{2}(t-t_0)} \begin{bmatrix} \cos\theta(w(t)) & \sin\theta(w(t)) \\ -\sin\theta(w(t)) & \cos\theta(w(t)) \end{bmatrix} x_0$$

is the solution to the given IVP, where $\theta(w(t)) = (w(t) - w(t_0))$.

Solution procedure. We imitate the argument used in the solution process of Example 4.3.6 (Chapter 4 of Volume 1). For this purpose, we compute the following differential:

$$dx(t) = d(\Phi(t, t_0)x_0) \qquad \text{(by notation)}$$

$$= d\left(e^{\frac{1}{2}(t-t_0)} \begin{bmatrix} \cos\theta(w(t)) & \sin\theta(w(t)) \\ -\sin\theta(w(t)) & \cos\theta(w(t)) \end{bmatrix} x_0 \right) \qquad \text{(from the given expression)}$$

$$= \frac{1}{2} e^{\frac{1}{2}(t-t_0)} \begin{bmatrix} \cos\theta(w(t)) & \sin\theta(w(t)) \\ -\sin\theta(w(t)) & \cos\theta(w(t)) \end{bmatrix} x_0 dt$$

$$+ e^{\frac{1}{2}(t-t_0)} \begin{bmatrix} d\cos\theta(w(t)) & d\sin\theta(w(t)) \\ -d\sin\theta(w(t)) & d\cos\theta(w(t)) \end{bmatrix} x_0 \qquad \text{(by Theorem 1.3.1).}$$

We compute

$$d\cos\theta(w(t)) = -\sin\theta(w(t))d\theta(w(t)) - \frac{1}{2}\cos\theta(w(t))(d\theta(w(t)))^2$$

$$= -\sin\theta(w(t))dw(t) - \frac{1}{2}\cos\theta(w(t))dt$$

[from $\theta(w(t))$ and Theorem 1.3.1],

$$d\sin\theta(w(t)) = \cos\theta(w(t))d\theta(w(t)) - \frac{1}{2}\sin\theta(w(t))(d\theta(w(t)))^2$$

(by Theorem 1.3.1)

$$= \cos\theta(w(t))dw(t) - \frac{1}{2}\sin\theta(w(t))dt.$$

Substituting these expressions in the above and simplifying, we get

$$dx(t) = \frac{1}{2}e^{\frac{1}{2}(t-t_0)}\begin{bmatrix} \cos\theta(w(t)) & \sin\theta(w(t)) \\ -\sin\theta(w(t)) & \cos\theta(w(t)) \end{bmatrix}x_0 dt$$

$$+ e^{\frac{1}{2}(t-t_0)}\begin{bmatrix} -\frac{1}{2}\cos\theta(w(t)) & -\frac{1}{2}\sin\theta(w(t)) \\ \frac{1}{2}\sin\theta(w(t)) & -\frac{1}{2}\cos\theta(w(t)) \end{bmatrix}x_0 dt$$

$$+ e^{\frac{1}{2}(t-t_0)}\begin{bmatrix} -\sin\theta(w(t)) & \cos\theta(w(t)) \\ -\cos\theta(w(t)) & -\sin\theta(w(t)) \end{bmatrix}x_0 dw(t)$$

$$= e^{\frac{1}{2}(t-t_0)}\begin{bmatrix} -\sin\theta(w(t)) & \cos\theta(w(t)) \\ -\cos\theta(w(t)) & -\sin\theta(w(t)) \end{bmatrix}x_0 dw(t) \quad \text{(by simplification).}$$

On the other hand, we also compute $Bx(t)dw(t)$ as follows:

$$Bx(t)dw(t) = \begin{bmatrix} 0 & 1 \\ -1 & 0 \end{bmatrix}e^{\frac{1}{2}(t-t_0)}\begin{bmatrix} \cos\theta(w(t)) & \sin\theta(w(t)) \\ -\sin\theta(w(t)) & \cos\theta(w(t)) \end{bmatrix}x_0 dw(t)$$

$$= e^{\frac{1}{2}(t-t_0)}\begin{bmatrix} -\sin\theta(w(t)) & \cos\theta(w(t)) \\ -\cos\theta(w(t)) & -\sin\theta(w(t)) \end{bmatrix}x_0 dw(t).$$

By comparing these calculations, we conclude that $dx(t) = Bx(t)dw(t)$. We need to verify that $x(t_0) = x_0$. For this purpose, we substitute $t = t_0$ into the given solution process and simplify

$$x(t_0) = \Phi(t_0, t_0)x_0 \quad \text{(by substitution)}$$

$$= e^{\frac{1}{2}(t_0 - t_0)} \begin{bmatrix} \cos\theta(w(t_0)) & \sin\theta(w(t_0)) \\ -\sin\theta(w(t_0)) & \cos\theta(w(t_0)) \end{bmatrix} x_0$$

$$= \begin{bmatrix} \cos 0 & \sin 0 \\ -\sin 0 & \cos 0 \end{bmatrix} x_0 \quad [\text{by } e^0 = 1,\ \theta(w(t_0)) = 0 \text{ by simplifying}]$$

$$= \begin{bmatrix} 1 & 0 \\ 0 & 1 \end{bmatrix} x_0 \quad \text{(by } \cos 0 = 1,\ \sin 0 = 0,\ \text{identity matrix concept)}$$

$$= x_0$$

From Definition 4.2.2, we conclude that the given random function $x(t)$ is the solution to the given IVP. This completes the solution process.

In the following, we present a stochastic version of the deterministic procedure in Section 4.3.4 (Chapter 4 of Volume 1) for finding the solution process of systems of linear Itô–Doob-type stochastic differential equations with constant coefficients (4.2.8).

4.2.4 *Procedure for Solving the IVP*

Now, we briefly summarize the procedure for finding a solution to the IVP (4.2.8):

Step #1. From Theorem 4.2.1, it follows that a linear combination,

$$x(t) = a_1 x_1(t) + a_2 x_2(t) + \cdots + a_k x_k(t) + \cdots + a_m x_m(t), \qquad (4.2.9)$$

of m solutions $x_1(t), x_2(t), \ldots, x_k(t), \ldots x_m(t)$ to (4.2.1) is also a solution to (4.2.1), where a_k are arbitrary real numbers.

Step #2. In this step, we need to find an arbitrary constant in (4.2.9). For this purpose, we need to solve the following system of nonhomogeneous random algebraic equations:

$$x(t_0) = a_1 x_1(t_0) + a_2 x_2(t_0) + \cdots + a_k x_k(t_0) + \cdots + a_m x_m(t_0) = x_0. \qquad (4.2.10)$$

This system has a unique solution, provided that: (a) $m = n$ and (b) the determinant of the coefficient matrix of the system (4.2.10) is different from zero, i.e. $\det(x_{ik}(t_0)) \neq 0$, where $x_{ik}(t_0)$ is the ith component of the kth solution to (4.2.1), at $t = t_0$ for $k = 1, 2, \ldots, n$. Under these conditions, $a_1, a_2, \ldots, a_k, \ldots, a_m$ are uniquely determined by both x_0 and the Wiener process w.

Step #3. Under the conditions (a) and (b) in Step #2 and substituting the algebraic solution to (4.2.10) into (4.2.9), the solution to the IVP (4.2.8) is determined.

Observation 4.2.2. Again, all stipulations and comments about the deterministic version (Observation 4.3.3 [80]) of observation remain true for the stochastic systems. For the sake of completeness, we briefly summarize them:

(i) Under the conditions (a) $m = n$ and (b) $\det(x_{ik}(t_0)) \neq 0$, the solution to (4.2.1) in (4.2.9) can be rewritten as

$$x(t) = \Phi(t)a \equiv \Phi(t, w(t))a, \qquad (4.2.11)$$

where Φ is an $n \times n$ matrix process defined by $\Phi(t) = (x_{ik}(t))_{n \times n} \equiv (x_{ik}(t, w(t)))_{n \times n} = \Phi(t, w(t))$, whose columns $x_1(t), x_2(t), \ldots, x_k(t), \ldots x_n(t)$ are solutions of (4.2.1) defined in Theorem 4.2.1; an n-dimensional unknown constant column vector ($n \times 1$ matrix) $a = [a_1, a_2, \ldots, a_k, \ldots, a_n]^T$ is composed from arbitrary coefficients of the solutions $x_k(t)$ in (4.2.9) for $k = 1, 2, \ldots, n$.

(ii) The solution represented in (4.2.11) is called a *general solution* to (4.2.1). We remark that $\Phi(t) \equiv \Phi(t, w(t))$ defined in (4.2.11) depends on the eigenvalues and corresponding eigenvectors of the coefficient matrix B and the Wiener process w. However, $\Phi(t)$ is not uniquely determined by the eigenvectors corresponding to the eigenvalues of the matrix B.

(iii) We further remark that the observations in (i) and (ii) are also applicable to the IVP for the system of linear stochastic differential equations with a time-varying system (the matrix B is a function of the independent variable t in J).

(iv) In the case of the system (4.2.8) (time-invariant, i.e. B is a constant matrix), the validity of $\det(\Phi(t_0, w(t_0))) \neq 0$ is equivalent to the fact that $\det(\Phi(0)) \neq 0$ ($0 \neq \det(\Phi(t_0)) = \exp[\lambda_1 t_0 + \mu_1 w(t_0)] \exp[\lambda_2 t_0 + \mu_2 w(t_0)]] \cdots \exp[\lambda_n t_0 + \mu_n w(t_0)] \det(\Phi(0))$ by Theorem $1.4.1D_4$ [80]). This is in view of the representation of the solution (4.2.2) and $\exp[\lambda_k t_0 + \mu_k w(t_0)] \neq 0$ for any t_0 in R. In the light of this, it is enough to verify that the $n \times n$ matrix $\Phi(0) = [c_1, \ldots, c_k, \ldots, c_n]$ associated with n eigenvectors as the columns of the matrix is inevitable (Theorem 1.4.4 [80]). This implies that $\Phi(0)$ has inverse $\Phi^{-1}(t_0)$ (Observation 1.4.4 [80]). Hence, from (4.2.8) and (4.2.11) (Theorem 1.4.4 [80]), we have

$$a = \Phi^{-1}(t_0)x_0 \equiv \Phi^{-1}(t_0, w(t_0))x_0. \qquad (4.2.12)$$

This procedure for finding the general solution is summarized in Figure 4.2.1 flowchart.

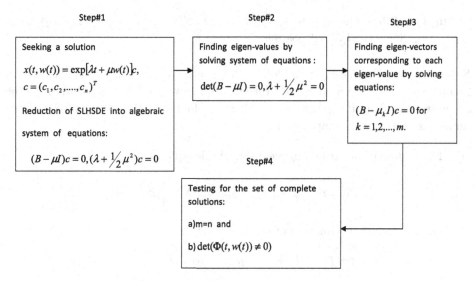

Fig. 4.2.1   Process of finding the general solution.

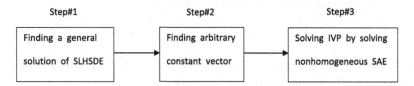

Fig. 4.2.2   Process of finding the solution of the IVP.

(v) The solution to the IVP (4.2.8) determined in Step #3 can be represented by substituting the expression (4.2.12) into (4.2.11). Hence, we have

$$x(t) = \Phi(t)\Phi^{-1}(t_0)x_0 \equiv \Phi(t, w(t))\Phi^{-1}(t_0, w(t_0))x_0$$

$$= \Phi(t, t_0)x_0 \equiv \Phi(t, w(t), t_0, w(t_0))x_0, \qquad (4.2.13)$$

where $\Phi(t, t_0) = \Phi(t)\Phi^{-1}(t_0) \equiv \Phi(t, w(t))\Phi^{-1}(t_0, w(t_0)) = \Phi(t, w(t), t_0, w(t_0))$. Thus, the solution $x(t, t_0, x_0) = x(t)$ to the IVP (4.2.8) is represented by $x(t, t_0, x_0) = x(t) = \Phi(t, t_0)x_0 \equiv \Phi(t, w(t))\Phi^{-1}(t_0, w(t_0))x_0 = \Phi(t, w(t), t_0, w(t_0))x_0 = x(t, w(t), t_0, w(t_0), x_0)$. This solution is referred to as a *particular solution* to (4.2.1).

(vi) The procedure for finding the solution to the IVP (4.2.8) is summarized in flowchart (Fig. 4.2.2):

(vii) From (iv), we further remark that $\Phi(0) = [c_1, \ldots, c_k, \ldots, c_n]$ is not uniquely determined. It depends on a particular choice of eigenvector corresponding to the eigenvalue of the matrix B. Hence, $\Phi(0)$ is not unique.

Example 4.2.6. Given: $dx_1 = (4x_1 - 2x_2)dw(t)$ and $dx_2 = (x_1 + x_2)dw(t)$, $x(t_0) = x_0$. Solve the IVP (if possible).

Solution procedure. First, we follow Procedure 4.2.2 and the argument used in Example 4.2.3. To use Step #1 of Procedure 4.2.4, we need to find solutions to the given problem. For that purpose, we follow the four-step Procedure 4.2.2, and we summarize simply the computations:

(i) The system of linear homogeneous algebraic equations:

$$0 = \det(B - \mu I)c \quad \text{and} \quad \left(\lambda + \frac{1}{2}\mu^2\right)c = 0.$$

(ii) The characteristic equation of the matrix B:

$$0 = \det(B - \mu I) = (4 - \mu)(1 - \mu) + 2 = (\mu - 2)(\mu - 3).$$

(iii) The eigenvalues $\mu_1 = 2$ and $\mu_2 = 3$ and the corresponding solutions of $\lambda : \lambda_1 = -2, \lambda_2 = -\frac{9}{2}$, and eigenvectors of the matrix B:

$$c_1(B) = [1\ 1]^T, \quad c_2(B) = [1\ 2]^T.$$

From Step #4 of Procedure 4.2.2, we have the following two solutions corresponding to the eigenvalues $\mu_1 = 2$ and $\mu_2 = 3$:

$$x_1(t) = e^{2(-t+w(t))} \begin{bmatrix} 1 \\ 1 \end{bmatrix} \quad \text{and} \quad x_2(t) = e^{(-\frac{9}{2}t+3w(t))} \begin{bmatrix} 1 \\ 2 \end{bmatrix}.$$

Applying Theorem 4.2.1 and using the argument in Example 4.2.4, we conclude that

$$x(t) = a_1 x_1(t) + a_2 x_2(t) = a_1 e^{2(-t+w(t))} \begin{bmatrix} 1 \\ 1 \end{bmatrix} + a_2 e^{(-\frac{9}{2}t+3w(t))} \begin{bmatrix} 1 \\ 2 \end{bmatrix}$$

is the solution process of the given system of differential equations. In fact, this solution expression can be rewritten as

$$x(t) = \Phi(t)a = \begin{bmatrix} e^{2(-t+w(t))} & e^{(-\frac{9}{2}t+3w(t))} \\ e^{2(-t+w(t))} & 2e^{(-\frac{9}{2}t+3w(t))} \end{bmatrix} a.$$

If $\det(\Phi(t_0)) \neq 0$, then $x(t)$ would be the general solution (Observation 4.2.2). To verify that $\det(\Phi(t_0)) \neq 0$, we apply the definition of the determinant to $\Phi(t_0)$ and obtain $\det(\Phi(t_0)) = e^{(-\frac{13}{2}t_0+5w(t_0))} \neq 0$ (by the property of the exponential function). Thus, we infer that the above linear combination of the solution is indeed the general solution to the given system of differential equations. Finally, we solve the given IVP. Following Step #3 of Procedure 4.2.4, we have

$$\Phi^{-1}(t_0) = \frac{1}{e^{\left(-\frac{13}{2}t_0 + 5w(t_0)\right)}} \begin{bmatrix} 2e^{\left(-\frac{9}{2}t_0 + 3w(t_0)\right)} & -e^{\left(-\frac{9}{2}t_0 + 3w(t_0)\right)} \\ -e^{2(-t_0 + w(t_0))} & e^{2(-t_0 + w(t_0))} \end{bmatrix}$$

$$= \begin{bmatrix} 2e^{-2(-t_0 + w(t_0))} & -e^{-2(-t_0 + w(t_0))} \\ -e^{-\left(-\frac{9}{2}t_0 + 3w(t_0)\right)} & e^{-\left(-\frac{9}{2}t_0 + 3w(t_0)\right)} \end{bmatrix} \quad \text{(Observation 1.4.4 [80])},$$

and

$$x(t, t_0, x_0) = x(t) = \Phi(t, t_0)x_0 \quad \text{(by substitution)}$$

$$= \begin{bmatrix} e^{2(-t + w(t))} & e^{\left(-\frac{9}{2}t + 3w(t)\right)} \\ e^{2(-t + w(t))} & 2e^{\left(-\frac{9}{2}t + 3w(t)\right)} \end{bmatrix} \begin{bmatrix} 2e^{-2(-t_0 + w(t_0))} & -e^{-2(-t_0 + w(t_0))} \\ -e^{-\left(-\frac{9}{2}t_0 + 3w(t_0)\right)} & e^{-\left(-\frac{9}{2}t_0 + 3w(t_0)\right)} \end{bmatrix} x_0$$

$$= \begin{bmatrix} 2e^{\theta(t - t_0)} - e^{\psi(t - t_0)} & -e^{\theta(t - t_0)} + e^{\psi(t - t_0)} \\ 2\left(e^{\theta(t - t_0)} - e^{\psi(t - t_0)}\right) & -e^{\theta(t - t_0)} + 2e^{\psi(t - t_0)} \end{bmatrix} \begin{bmatrix} x_{10} \\ x_{20} \end{bmatrix} \quad \text{(Definition 1.3.11 [80])},$$

where $\theta(t - t_0) \equiv -2(t - t_0) + 2(w(t) - w(t_0))$ and $-\frac{9}{2}(t - t_0) + 3(w(t) - w(t_0)) \equiv \psi(t - t_0)$. This completes the procedure for solving the IVP.

Observation 4.2.3

(i) We note that the Itô–Doob-type stochastic differential can be applied to the random matrix process. From the Itô–Doob differential of the random matrix process/function (Observation 1.5.2) and the matrix algebra (Theorem 1.3.1) [80], we have

$$d\Phi(t) = d(x_{ik}(t))_{n \times n} = d[x_1(t) \ x_2(t), \ldots, x_k(t), \ldots, x_n(t)] \quad \text{[from (4.2.11)]}$$

$$= (dx_{ik}(t))_{n \times n} = [dx_1(t) \ dx_2(t), \ldots, dx_k(t), \ldots, dx_n(t)] \quad \text{(by definition)}$$

$$= [Bx_1(t)dw \ Bx_2(t)dw, \ldots, Bx_k(t)dw, \ldots, Bx_n(t)dw]$$

$$\text{(by the solution concept)}$$

$$= B[x_1(t) \ x_2(t) \ldots, x_k(t), \ldots, x_n(t)]dw(t)$$

$$= B\Phi(t)dw(t) \quad \text{(by notation).} \quad (4.2.14)$$

This shows that the $\Phi(t)$ defined in (4.2.11) is also the solution to (4.2.1). If $\det(\Phi(t_0)) \neq 0$, then Φ is called a *general fundamental matrix solution* to (4.2.1) at $t = t_0$. Hence, $\Phi(t, t_0)$ is also the fundamental solution to (4.2.1) and it is a particular fundamental solution to (4.2.1). This is due to the fact that $\Phi(t, t_0) = \Phi(t)\Phi^{-1}(t_0)$, and hence $\Phi(t_0, t_0) = I$. $\Phi(t, t_0)$ is also referred to as the *state transition matrix* or the *normalized fundamental matrix solution* to (4.2.8) at $t = t_0$. Hence, $x(t, t_0, x_0) = x(t) = \Phi(t, t_0)x_0$ is referred to as the particular solution to (4.2.1).

(ii) We recall that the $\Phi(t)$ defined in (4.2.11) is not uniquely determined. Therefore, at this stage, we do not make any comment about the uniqueness of the solution $x(t, t_0, x_0) = x(t) = \Phi(t, t_0)x_0$ to (4.2.8). This will be addressed at a later stage.

Example 4.2.7. Given: $dx_1 = (x_1 + 2x_2)dw(t)$ and $dx_1 = (4x_1 + 3x_2)dw(t)$. Show that $[x_1(t)\ x_2(t)]$ is the general fundamental matrix solution process of the given system of Itô–Doob-type stochastic differential equations, where $x_1(t)$ and $x_2(t)$ are defined by $x_1(t) = e^{-(\frac{1}{2}t + w(t))}[1\ {-1}]^T$ and $x_2(t) = e^{(-\frac{25}{2}t + 5w(t))}[1\ 2]^T$, respectively, with T standing for the transpose of a matrix.

Solution procedure. From the solution process of Example 4.2.1, we conclude that the vector functions $x_1(t)$ and $x_2(t)$ are the solution process of the given system of differential equations $dx(t) = Bx\,dw(t)$, where B and x are 2×2 and 2×1 matrices, respectively. Following the argument used in Example 4.2.6, we set $\Phi(t)$ as

$$[x_1(t)\ x_2(t)]^T = \Phi(t) = \begin{bmatrix} e^{-(\frac{1}{2}t + w(t))} & e^{(-\frac{25}{2}t + 5w(t))} \\ -e^{-(\frac{1}{2}t + w(t))} & 2e^{(-\frac{25}{2}t + 5w(t))} \end{bmatrix}$$

with $\Phi(t_0) = \begin{bmatrix} e^{-(\frac{1}{2}t_0 + w(t_0))} & e^{(-\frac{25}{2}t_0 + 5w(t_0))} \\ -e^{-(\frac{1}{2}t_0 + w(t_0))} & 2e^{(-\frac{25}{2}t_0 + 5w(t_0))} \end{bmatrix}$. We compute $\det\Phi(t_0)$ as follows:

$$\det(\Phi(t_0)) = e^{-(\frac{1}{2}t_0 + w(t_0))}(2e^{(-\frac{25}{2}t_0 + 5w(t_0))})$$

$$+ (e^{-(\frac{1}{2}t_0 + w(t_0))})e^{(-\frac{25}{2}t_0 + 5w(t_0))} \quad \text{(by Definition 1.4.2 [80])}$$

$$= 3e^{(-13t_0 + 4w(t_0))} \quad \text{(by simplification)}$$

$$\neq 0 \quad \text{(by the property of the exponential function)}.$$

Therefore, by Observation 4.2.2, we conclude that the column vector functions of $\Phi(t)$ are linearly independent. Moreover, for $n = 2$, these solutions form a complete set of solutions to the given stochastic system of differential equations. We need to show that it is the general fundamental solution process of the given system of differential equations. For this purpose, we imitate the arguments used in Examples 4.2.4 and 4.2.5, and obtain

$$B\Phi(t)dw(t) = \begin{bmatrix} 1 & 2 \\ 4 & 3 \end{bmatrix} \begin{bmatrix} e^{-(\frac{1}{2}t + w(t))} & e^{(-\frac{25}{2}t + 5w(t))} \\ -e^{-(\frac{1}{2}t + w(t))} & 2e^{(-\frac{25}{2}t + 5w(t))} \end{bmatrix} dw(t) \quad \text{(by substitution)}$$

$$= \begin{bmatrix} -e^{-(\frac{1}{2}t + w(t))} & 5e^{(-\frac{25}{2}t + 5w(t))} \\ e^{-(\frac{1}{2}t + w(t))} & 10e^{(-\frac{25}{2}t + 5w(t))} \end{bmatrix} dw(t) \quad \text{(by simplification)},$$

and

$$d\Phi(t) = d\left(\begin{bmatrix} e^{-(\frac{1}{2}t+w(t))} & e^{(-\frac{25}{2}t+5w(t))} \\ -e^{-(\frac{1}{2}t+w(t))} & 2e^{(-\frac{25}{2}t+5w(t))} \end{bmatrix}\right) \qquad \text{(by the given expression)}$$

$$= \begin{bmatrix} de^{-(\frac{1}{2}t+w(t))} & de^{(-\frac{25}{2}t+5w(t))} \\ -de^{-(\frac{1}{2}t+w(t))} & 2de^{(-\frac{25}{2}t+5w(t))} \end{bmatrix} \qquad \text{(Observation 1.5.2 [80])}.$$

We compute

$$de^{-(\frac{1}{2}t+w(t))} = -e^{-(\frac{1}{2}t+w(t))}d\left(\frac{1}{2}t+w(t)\right) + \frac{1}{2}e^{-(\frac{1}{2}t+w(t))}\left(d\left(\frac{1}{2}t+w(t)\right)\right)^2$$

$$= -e^{-(\frac{1}{2}t+w(t))}d\left(\frac{1}{2}t+w(t)\right) + \frac{1}{2}e^{-(\frac{1}{2}t+w(t))}dt$$

$$= -e^{-(\frac{1}{2}t+w(t))}dw(t), \quad \left[\text{by } d\left(\frac{1}{2}t+w(t)\right)^2 = dt \text{ and Theorem 1.3.1}\right],$$

$$e^{(-\frac{25}{2}t+5w(t))} = e^{(-\frac{25}{2}t+5w(t))}d\left(-\frac{25}{2}t+5w(t)\right)$$

$$+ \frac{1}{2}e^{(-\frac{25}{2}t+5w(t))}\left(d\left(-\frac{25}{2}t+5w(t)\right)\right)^2 \qquad \text{(by Theorem 1.3.1)}$$

$$= e^{(-\frac{25}{2}t+5w(t))}d\left(-\frac{25}{2}t+5w(t)\right) + \frac{25}{2}e^{(-\frac{25}{2}t+5w(t))}dt$$

$$= 5e^{(-\frac{25}{2}t+5w(t))}dw(t) \quad \left[\text{by } d\left(\frac{1}{2}t+w(t)\right)^2 = dt\right.$$

and Taylor's formula].

Substituting these expressions and simplifying, we have

$$d\Phi(t) = \begin{bmatrix} -e^{-(\frac{1}{2}t+w(t))}dw(t) & 5e^{(-\frac{25}{2}t+5w(t))}dw(t) \\ e^{-(\frac{1}{2}t+w(t))}dw(t) & 10e^{(-\frac{25}{2}t+5w(t))}dw(t) \end{bmatrix}$$

$$= \begin{bmatrix} -e^{-(\frac{1}{2}t+w(t))} & 5e^{(-\frac{25}{2}t+5w(t))} \\ e^{-(\frac{1}{2}t+w(t))} & 10e^{(-\frac{25}{2}t+5w(t))} \end{bmatrix} dw(t).$$

Comparing the right-hand side expressions of $B\Phi(t)dw(t)$ with $d\Phi(t)$ and applying Definition 4.2.2, we conclude that $\Phi(t)$ is the *general fundamental matrix solution* to the given system of Itô–Doob-type stochastic differential equations at $t = t_0$.

Prior to the procedure for finding n eigenvectors and the corresponding n eigenvalues associated with (4.2.6), we present a stochastic calculus dependent test for verifying the validity of $\det(\Phi(t_0)) \neq 0$, where $\Phi(t) = (x_{ik}(t))_{n \times n} = [x_1(t), x_2(t), \ldots, x_k(t), \ldots, x_n(t)]$, and the column vectors: $x_1(t), x_2(t), \ldots, x_k(t), \ldots, x_n(t)$ are solutions of (4.2.1). This test is easy to verify and is computationally attractive. Moreover, it is applicable to both time-invariant and time-varying linear systems of Itô–Doob-type stochastic differential equations. To the best of our literature study, this result is new and it is a natural extension of the deterministic result.

Theorem 4.2.2   (stochastic version of Abel–Jacobi–Liouville formula [79, 80]). *Let $x_1(t), x_2(t), \ldots, x_k(t), \ldots, x_n(t)$ be any n solutions processes of (4.2.1) on J. Let $\Phi(t) = [x_1(t), x_2(t), \ldots, x_i(t), \ldots, x_n(t)] = [\Phi_1^T(t), \ldots, \Phi_i^T(t), \ldots, \Phi_n^T(t)]^T$ be the $n \times n$ matrix function defined on J, where $\Phi_i(t) = [x_{i1}(t), \ldots, x_{ik}(t), \ldots, x_{in}(t)]$ is the ith row of the matrix process $\Phi(t)$. Let $\mathrm{tr}(B) = \sum_{i=1}^n b_{ii}$ and $\mathbb{L}(B) = \frac{1}{2} \sum_{i=1}^n \sum_{k \neq i}^n (b_{ii} b_{kk} - b_{ik} b_{ki})$. Then,*

$$d \det(\Phi(t)) = \mathbb{L}(B) \det(\Phi(t)) dt + \mathrm{tr}(B) \det(\Phi(t)) dw(t), \tag{4.2.15}$$

$$\exp\left[-\int_{t_0}^t \left(\mathbb{L}(B) - \frac{1}{2}(\mathrm{tr}(B))^2 \right) ds - \int_{t_0}^t \mathrm{tr}(B) dw(s) \right] \det(\Phi(t)) = c, \tag{4.2.16}$$

where c is an arbitrary constant.

Proof.   From Observation 4.2.3, we conclude that $\Phi(t)$ is the matrix solution process of (4.2.1), i.e.

$$d\Phi(t) = (dx_{ik}(t))_{n \times n} = B\Phi(t) dw(t) = \left(\sum_{j=1}^n b_{ij} x_{jk}(t) dw(t) \right)_{n \times n}. \tag{4.2.17}$$

Applying Theorem 1.3.4 and Theorem 1.5.6 (Chapter 1 of Volume 1) to $\det(\Phi(t))$, we have

$$d \det(\Phi(t)) = \sum_{i=1}^n d\Phi_i(t) C_i^T(\Phi(t)) + \frac{1}{2!} \sum_{i=1}^n \sum_{k \neq i}^n d\Phi_i(t) \frac{\partial}{\partial \Phi_k} C_i^T(\Phi(t))(d\Phi_k(t))^T$$

$$= \sum_{i=1}^n W(\Phi_1(t), \ldots, d\Phi_i(t), \ldots, \Phi_n(t)) \text{ (by Observation 1.4.2 [80])}$$

$$+ \frac{1}{2!} \sum_{i=1}^n \sum_{k \neq i} W(\Phi_1(t), \ldots, d\Phi_i(t), \ldots, d\Phi_k(t), \Phi_n(t)). \tag{4.2.18}$$

Now, we compute the expressions for the terms on the right-hand side of (4.2.18). For each i, $1 \leq i \leq n$, first, we rewrite $W(\Phi_1(t), \ldots, d\Phi_i(t), \ldots, \Phi_n(t))$:

$$W(\Phi_1(t), \ldots, d\Phi_i(t), \ldots, \Phi_n(t))$$

$$= \begin{vmatrix} x_{11} & \cdots & x_{1k} & \cdots & x_{1n} \\ \cdots & \cdots & \cdots & \cdots & \cdots \\ dx_{i1} & \cdots & dx_{ik} & \cdots & dx_{in} \\ \cdots & \cdots & \cdots & \cdots & \cdots \\ x_{n1} & \cdots & x_{nk} & \cdots & x_{nn} \end{vmatrix}$$

$$
= \begin{vmatrix}
x_{11} & \cdots & x_{1k} & \cdots & x_{1n} \\
\cdots & \cdots & \cdots & \cdots & \cdots \\
\sum_{j=1}^{n} b_{ij} x_{j1} dw(t) & \cdots & \sum_{j=1}^{n} b_{ij} x_{jk} dw(t) & \cdots & \sum_{j=1}^{n} b_{ij} x_{jn} dw(t) \\
\cdots & \cdots & \cdots & \cdots & \cdots \\
x_{n1} & \cdots & x_{nk} & \cdots & x_{nn}
\end{vmatrix}.
$$

By the application of the property of the determinant (Theorem 1.4.1D$_6$, Chapter 1 [80]), we have

$$
W(\Phi_1(t), \ldots, d\Phi_i(t), \ldots, \Phi_n(t))
$$

$$
= \begin{vmatrix}
x_{11} & \cdots & x_{1k} & \cdots & x_{1n} \\
\cdots & \cdots & \cdots & \cdots & \cdots \\
b_{ii} x_{i1} dw(t) & \cdots & b_{ii} x_{ik} dw(t) & \cdots & b_{ii} x_{in} dw(t) \\
\cdots & \cdots & \cdots & \cdots & \cdots \\
x_{n1} & \cdots & x_{nk} & \cdots & x_{nn}
\end{vmatrix}.
$$

This together with Theorem 1.4.1D$_4$ (Chapter 1 of Volume 1) gives

$$
W(\Phi_1(t), \ldots, d\Phi_i(t), \ldots, \Phi_n(t)) = b_{ii} \det(\Phi(t)) dw(t),
$$

for each i, $1 \leq i \leq n$, and hence the expression for the first term in (4.2.18) is

$$
\sum_{i=1}^{n} W(\Phi_1(t), \ldots, d\Phi_i(t), \ldots, \Phi_n(t)) = \sum_{i=1}^{n} b_{ii} \det(\Phi(t)) dw(t)
$$

$$
= \operatorname{tr}(B) \det(\Phi(t)) dw(t). \qquad (4.2.19)
$$

Now, for each i, k, $k \neq i$, $i, k = 1, 2, \ldots, n$, we rewrite $W(\Phi_1(t), \ldots, d\Phi_i(t), \ldots, d\Phi_k(t), \ldots, \Phi_n(t))$ as follows:

$$
W(\Phi_1(t), \ldots, d\Phi_i(t), \ldots, d\Phi_k(t), \ldots, \Phi_n(t))
$$

$$
= \begin{vmatrix}
x_{11} & \cdots & x_{1i} & \cdots & x_{1k} & \cdots & x_{1n} \\
\cdots & \cdots & \cdots & \cdots & \cdots & \cdots & \cdots \\
dx_{i1} & \cdots & dx_{ii} & \cdots & dx_{ik} & \cdots & dx_{in} \\
\cdots & \cdots & \cdots & \cdots & \cdots & \cdots & \cdots \\
dx_{k1} & \cdots & dx_{ki} & \cdots & dx_{kk} & \cdots & dx_{kn} \\
\cdots & \cdots & \cdots & \cdots & \cdots & \cdots & \cdots \\
x_{n1} & \cdots & x_{ni} & \cdots & x_{nk} & \cdots & x_{nn}
\end{vmatrix}
$$

$$= (dw(t))^2 \begin{vmatrix} x_{11} & \cdots & x_{1i} & \cdots & x_{1k} & \cdots & x_{1n} \\ \cdots & \cdots & \cdots & \cdots & \cdots & \cdots & \cdots \\ \sum_{j=1}^{n} b_{ij}x_{j1} & \cdots & \sum_{j=1}^{n} b_{ij}x_{ji} & \cdots & \sum_{j=1}^{n} b_{ij}x_{jk} & \cdots & \sum_{j=1}^{n} b_{ij}x_{jn} \\ \cdots & \cdots & \cdots & \cdots & \cdots & \cdots & \cdots \\ \sum_{j=1}^{n} b_{kj}x_{j1} & \cdots & \sum_{j=1}^{n} b_{kj}x_{ji} & \cdots & \sum_{j=1}^{n} b_{kj}x_{jk} & \cdots & \sum_{j=1}^{n} b_{kj}x_{jn} \\ \cdots & \cdots & \cdots & \cdots & \cdots & \cdots & \cdots \\ x_{n1} & \cdots & x_{ni} & \cdots & x_{nk} & \cdots & x_{nn} \end{vmatrix}.$$

By repeated application of the determinant (Theorem 1.4.1D$_6$, Chapter 1 of Volume 1) and the usage of $E[(dw(t))^2] = dt$, we have

$$W(\Phi_1(t), \ldots, d\Phi_i(t), \ldots, d\Phi_k(t), \ldots, \Phi_n(t))$$

$$= dt \begin{vmatrix} x_{11} & \cdots & x_{1i} & \cdots & x_{1k} & \cdots & x_{1n} \\ \cdots & \cdots & \cdots & \cdots & \cdots & \cdots & \cdots \\ b_{ii}x_{i1} + b_{ik}x_{k1} & \cdots & b_{ii}x_{ii} + b_{ik}x_{ki} & \cdots & b_{ii}x_{ik} + b_{ik}x_{kk} & \cdots & b_{ii}x_{in} + b_{ik}x_{kn} \\ \cdots & \cdots & \cdots & \cdots & \cdots & \cdots & \cdots \\ b_{kk}x_{k1} + b_{ki}x_{i1} & \cdots & b_{kk}x_{ki} + b_{ki}x_{ii} & \cdots & b_{kk}x_{kk} + b_{ki}x_{ik} & \cdots & b_{kk}x_{kn} + b_{ki}x_{in} \\ \cdots & \cdots & \cdots & \cdots & \cdots & \cdots & \cdots \\ x_{n1} & \cdots & x_{ni} & \cdots & x_{nk} & \cdots & x_{nn} \end{vmatrix}.$$

Now, we apply the property of the determinant (Theorem 1.4.1D$_5$, Chapter 1 of Volume 1), and obtain

$$W(\Phi_1(t), \ldots, d\Phi_i(t), \ldots, d\Phi_k(t), \ldots, \Phi_n(t))$$

$$= dt \begin{vmatrix} x_{11} & \cdots & x_{1i} & \cdots & x_{1k} & \cdots & x_{1n} \\ \cdots & \cdots & \cdots & \cdots & \cdots & \cdots & \cdots \\ b_{ii}x_{i1} & \cdots & b_{ii}x_{ii} & \cdots & b_{ii}x_{ik} & \cdots & b_{ii}x_{in} \\ \cdots & \cdots & \cdots & \cdots & \cdots & \cdots & \cdots \\ b_{kk}x_{k1} + b_{ki}x_{i1} & \cdots & b_{kk}x_{ki} + b_{ki}x_{ii} & \cdots & b_{kk}x_{kk} + b_{ki}x_{ik} & \cdots & b_{kk}x_{kn} + b_{ki}x_{in} \\ \cdots & \cdots & \cdots & \cdots & \cdots & \cdots & \cdots \\ x_{n1} & \cdots & x_{ni} & \cdots & x_{nk} & \cdots & x_{nn} \end{vmatrix}$$

$$+ dt \begin{vmatrix} x_{11} & \cdots & x_{1i} & \cdots & x_{1k} & \cdots & x_{1n} \\ \cdots & \cdots & \cdots & \cdots & \cdots & \cdots & \cdots \\ b_{ik}x_{k1} & \cdots & b_{ik}x_{ki} & \cdots & b_{ik}x_{kk} & \cdots & b_{ik}x_{kn} \\ \cdots & \cdots & \cdots & \cdots & \cdots & \cdots & \cdots \\ b_{kk}x_{k1}+b_{ki}x_{i1} & \cdots & b_{kk}x_{ki}+b_{ki}x_{ii} & \cdots & b_{kk}x_{kk}+b_{ki}x_{ik} & \cdots & b_{kk}x_{kn}+b_{ki}x_{in} \\ \cdots & \cdots & \cdots & \cdots & \cdots & \cdots & \cdots \\ x_{n1} & \cdots & x_{ni} & \cdots & x_{nk} & \cdots & x_{nn} \end{vmatrix}.$$

Again, we repeat the application of properties of the determinant (Theorem 1.4.1D$_4$ and D$_5$ [80]), and we have

$$W(\Phi_1(t), \ldots, d\Phi_i(t), \ldots, d\Phi_k(t), \ldots, \Phi_n(t))$$

$$= dt \begin{vmatrix} x_{11} & \cdots & x_{1i} & \cdots & x_{1k} & \cdots & x_{1n} \\ \cdots & \cdots & \cdots & \cdots & \cdots & \cdots & \cdots \\ b_{ii}x_{i1} & \cdots & b_{ii}x_{ii} & \cdots & b_{ii}x_{ik} & \cdots & b_{ii}x_{in} \\ \cdots & \cdots & \cdots & \cdots & \cdots & \cdots & \cdots \\ b_{kk}x_{k1} & \cdots & b_{kk}x_{ki} & \cdots & b_{kk}x_{kk} & \cdots & b_{kk}x_{kn} \\ \cdots & \cdots & \cdots & \cdots & \cdots & \cdots & \cdots \\ x_{n1} & \cdots & x_{ni} & \cdots & x_{nk} & \cdots & x_{nn} \end{vmatrix}$$

$$+ dt \begin{vmatrix} x_{11} & \cdots & x_{1i} & \cdots & x_{1k} & \cdots & x_{1n} \\ \cdots & \cdots & \cdots & \cdots & \cdots & \cdots & \cdots \\ b_{ik}x_{k1} & \cdots & b_{ik}x_{ki} & \cdots & b_{ik}x_{kk} & \cdots & b_{ik}x_{kn} \\ \cdots & \cdots & \cdots & \cdots & \cdots & \cdots & \cdots \\ b_{ki}x_{i1} & \cdots & b_{ki}x_{ii} & \cdots & b_{ki}x_{ik} & \cdots & b_{ki}x_{in} \\ \cdots & \cdots & \cdots & \cdots & \cdots & \cdots & \cdots \\ x_{n1} & \cdots & x_{ni} & \cdots & x_{nk} & \cdots & x_{nn} \end{vmatrix}$$

$$+ dt \begin{vmatrix} x_{11} & \cdots & x_{1i} & \cdots & x_{1k} & \cdots & x_{1n} \\ \cdots & \cdots & \cdots & \cdots & \cdots & \cdots & \cdots \\ b_{ii}x_{i1} & \cdots & b_{ii}x_{ii} & \cdots & b_{ii}x_{ik} & \cdots & b_{ii}x_{in} \\ \cdots & \cdots & \cdots & \cdots & \cdots & \cdots & \cdots \\ b_{ki}x_{i1} & \cdots & b_{ki}x_{ii} & \cdots & b_{ki}x_{ik} & \cdots & b_{ki}x_{in} \\ \cdots & \cdots & \cdots & \cdots & \cdots & \cdots & \cdots \\ x_{n1} & \cdots & x_{ni} & \cdots & x_{nk} & \cdots & x_{nn} \end{vmatrix}$$

$$+\,dt\begin{vmatrix} x_{11} & \cdots & x_{1i} & \cdots & x_{1k} & \cdots & x_{1n} \\ \cdots & \cdots & \cdots & \cdots & \cdots & \cdots & \cdots \\ b_{ik}x_{k1} & \cdots & b_{ik}x_{ki} & \cdots & b_{ik}x_{kk} & \cdots & b_{ik}x_{kn} \\ \cdots & \cdots & \cdots & \cdots & \cdots & \cdots & \cdots \\ b_{kk}x_{k1} & \cdots & b_{kk}x_{ki} & \cdots & b_{kk}x_{kk} & \cdots & b_{kk}x_{kn} \\ \cdots & \cdots & \cdots & \cdots & \cdots & \cdots & \cdots \\ x_{n1} & \cdots & x_{ni} & \cdots & x_{nk} & \cdots & x_{nn} \end{vmatrix}$$

$$=\,b_{ii}b_{kk}dt\begin{vmatrix} x_{11} & \cdots & x_{1i} & \cdots & x_{1k} & \cdots & x_{1n} \\ \cdots & \cdots & \cdots & \cdots & \cdots & \cdots & \cdots \\ x_{i1} & \cdots & x_{ii} & \cdots & x_{ik} & x_{in} \\ \cdots & \cdots & \cdots & \cdots & \cdots & \cdots & \cdots \\ x_{k1} & \cdots & x_{ki} & \cdots & x_{kk} & x_{kn} \\ \cdots & \cdots & \cdots & \cdots & \cdots & \cdots & \cdots \\ x_{n1} & \cdots & x_{ni} & \cdots & x_{nk} & \cdots & x_{nn} \end{vmatrix}$$

$$+\,b_{ik}b_{ki}dt\begin{vmatrix} x_{11} & \cdots & x_{1i} & \cdots & x_{1k} & \cdots & x_{1n} \\ \cdots & \cdots & \cdots & \cdots & \cdots & \cdots & \cdots \\ x_{k1} & \cdots & x_{ki} & \cdots & x_{kk} & \cdots & x_{kn} \\ \cdots & \cdots & \cdots & \cdots & \cdots & \cdots & \cdots \\ x_{i1} & \cdots & x_{ii} & \cdots & x_{ik} & \cdots & x_{in} \\ \cdots & \cdots & \cdots & \cdots & \cdots & \cdots & \cdots \\ x_{n1} & \cdots & x_{ni} & \cdots & x_{nk} & \cdots & x_{nn} \end{vmatrix}$$

$$+\,b_{ii}b_{ki}dt\begin{vmatrix} x_{11} & \cdots & x_{1i} & \cdots & x_{1k} & \cdots & x_{1n} \\ \cdots & \cdots & \cdots & \cdots & \cdots & \cdots & \cdots \\ x_{i1} & \cdots & x_{ii} & \cdots & x_{ik} & \cdots & x_{in} \\ \cdots & \cdots & \cdots & \cdots & \cdots & \cdots & \cdots \\ x_{i1} & \cdots & x_{ii} & \cdots & x_{ik} & \cdots & x_{in} \\ \cdots & \cdots & \cdots & \cdots & \cdots & \cdots & \cdots \\ x_{n1} & \cdots & x_{ni} & \cdots & x_{nk} & \cdots & x_{nn} \end{vmatrix}$$

$$+\, b_{ik}b_{kk}dt \begin{vmatrix} x_{11} & \cdots & x_{1i} & \cdots & x_{1k} & \cdots & x_{1n} \\ \cdots & \cdots & \cdots & \cdots & \cdots & \cdots & \cdots \\ x_{k1} & \cdots & x_{ki} & \cdots & x_{kk} & \cdots & x_{kn} \\ \cdots & \cdots & \cdots & \cdots & \cdots & \cdots & \cdots \\ x_{k1} & \cdots & x_{ki} & \cdots & x_{kk} & \cdots & x_{kn} \\ \cdots & \cdots & \cdots & \cdots & \cdots & \cdots & \cdots \\ x_{n1} & \cdots & x_{ni} & \cdots & x_{nk} & \cdots & x_{nn} \end{vmatrix}. \tag{4.2.20}$$

Observing that the third and fourth terms of (4.2.20) are the determinants with the ith and kth rows being identical, and hence, applying the property of the determinant (Theorem 1.4.1D$_3$, Chapter 1 of Volume 1), (4.2.20) reduces to

$$W(\Phi_1(t),\ldots,d\Phi_i(t),\ldots,d\Phi_k(t),\ldots,\Phi_n(t))$$

$$= b_{ii}b_{kk}dt \begin{vmatrix} x_{11} & \cdots & x_{1i} & \cdots & x_{1k} & \cdots & x_{1n} \\ \cdots & \cdots & \cdots & \cdots & \cdots & \cdots & \cdots \\ x_{i1} & \cdots & x_{ii} & \cdots & x_{ik} & \cdots & x_{in} \\ \cdots & \cdots & \cdots & \cdots & \cdots & \cdots & \cdots \\ x_{k1} & \cdots & x_{ki} & \cdots & x_{kk} & \cdots & x_{kn} \\ \cdots & \cdots & \cdots & \cdots & \cdots & \cdots & \cdots \\ x_{n1} & \cdots & x_{ni} & \cdots & x_{nk} & \cdots & x_{nn} \end{vmatrix}$$

$$+\, b_{ik}b_{ki}dt \begin{vmatrix} x_{11} & \cdots & x_{1i} & \cdots & x_{1k} & \cdots & x_{1n} \\ \cdots & \cdots & \cdots & \cdots & \cdots & \cdots & \cdots \\ x_{k1} & \cdots & x_{ki} & \cdots & x_{kk} & \cdots & x_{kn} \\ \cdots & \cdots & \cdots & \cdots & \cdots & \cdots & \cdots \\ x_{i1} & \cdots & x_{ii} & \cdots & x_{ik} & \cdots & x_{in} \\ \cdots & \cdots & \cdots & \cdots & \cdots & \cdots & \cdots \\ x_{n1} & \cdots & x_{ni} & \cdots & x_{nk} & \cdots & x_{nn} \end{vmatrix}.$$

This, along with the property of the determinant (Theorem 1.4.1D$_2$) and the notation of the determinant of $\Phi(t)$, yields

$$W(\Phi_1(t),\ldots,d\Phi_i(t),\ldots,d\Phi_k(t),\ldots,\Phi_n(t))$$
$$= b_{ii}b_{kk}\det(\Phi(t))dt - b_{ik}b_{ki}\det(\Phi(t))dt$$
$$= (b_{ii}b_{kk} - b_{ik}b_{ki})\det(\Phi(t))dt, \tag{4.2.21}$$

for each i, k, $k \neq i$, $i, k = 1, 2, \ldots, n$, and hence the expression for the second term of (4.2.18) is

$$\frac{1}{2!} \sum_{i=1}^{n} \sum_{k \neq i} W(\Phi_1(t), \ldots, d\Phi_i(t), \ldots, d\Phi_k(t), \ldots, \Phi_n(t))$$

$$= \frac{1}{2} \sum_{i=1}^{n} \sum_{k \neq i} (b_{ii}b_{kk} - b_{ik}b_{ki}) \det(\Phi(t)) dt$$

$$= \mathbb{L}(B) \det(\Phi(t)) dt. \tag{4.2.22}$$

We substitute the expressions (4.2.19) and (4.2.22) into (4.2.18), and obtain Equation (4.2.15):

$$d \det(\Phi(t)) = \mathbb{L}(B) \det(\Phi(t)) dt + \mathrm{tr}(B) \det(\Phi(t)) dw(t).$$

This establishes the first-order linear scalar homogeneous Itô–Doob-type stochastic differential equation (4.2.15). By following the procedure of Chapter 2, one can obtain an expression for a general solution to (4.2.15),

$$\det(\Phi(t)) = \exp\left[\int_{t_0}^{t} \left(\mathbb{L}(B) - \frac{1}{2}(\mathrm{tr}(B))^2 \right) ds + \int_{t_0}^{t} \mathrm{tr}(B) dw(s) \right] c,$$

which is equivalent to the expression (4.2.16). For $t = t_0$, if $\det(\Phi(t_0))$ is known, then $\det(\Phi(t_0)) = c$. The c is determined by the initial data $(t_0, \det(\Phi(t_0)))$. Thus, the particular solution to (4.2.15) can be determined for any given value of Φ at $t = t_0$. $\qquad\square$

Example 4.2.8.   Let us consider Example 4.2.6. Show that $\det(\Phi(t))$ is expressed by

$$\det(\Phi(t)) = \exp\left[\int_{t_0}^{t} \left(\mathbb{L}(B) - \frac{1}{2}(\mathrm{tr}(B))^2 \right) ds + \int_{t_0}^{t} \mathrm{tr}(B) dw(s) \right] e^{(-\frac{13}{2}t_0 + 5w(t_0))},$$

where $B = \begin{bmatrix} 4 & -2 \\ 1 & 1 \end{bmatrix}$, $\mathrm{tr}(B) = b_{11} + b_{22} = 4 + 1 = 5$, and $= \frac{1}{2}[b_{11}b_{22} - b_{12}b_{21} + b_{22}b_{11} - b_{21}b_{12} = \mathbb{L}(B) = \frac{1}{2}[4 + 2 + 4 + 2] = 6$.

Solution procedure.   By following the argument used in the solution process and from Example 4.2.6, we have $\det(\Phi(t)) = e^{(-\frac{13}{2}t + 5w(t))}$. From this, we compute

$$d \det(\Phi(t)) = d(e^{(-\frac{13}{2}t + 5w(t))}) \quad \text{(from the given expression)}$$

$$= e^{(-\frac{13}{2}t + 5w(t))} d\left(-\frac{13}{2}t + 5w(t) \right)$$

$$+ \frac{1}{2} e^{(-\frac{13}{2}t+5w(t))} \left(d \left(-\frac{13}{2}t + 5w(t) \right) \right)^2$$

(by the differential property)

$$= e^{(-\frac{13}{2}t+5w(t))} \left(-\frac{13}{2}dt + 5dw(t) \right) + \frac{25}{2} e^{(-\frac{13}{2}t+5w(t))} dt$$

(by Theorem 1.3.1)

$$= 6e^{(-\frac{13}{2}t+5w(t))} dt + 5e^{(-\frac{13}{2}t+5w(t))} dw(t)$$

$$= 6 \det(\Phi(t))dt + 5 \det(\Phi(t))dw(t) \quad \text{[by substitution for } \det \Phi(t)]$$

$$= \mathbb{L}(B) \det(\Phi(t))dt + \text{tr}(B) \det(\Phi(t))dw(t) \quad \text{(by notation)}.$$

This shows that $d \det(\Phi(t))$ satisfies the differential equation (4.2.15). Moreover, by the method of finding a general solution to the linear scalar stochastic differential equation of Section 2.3 of Chapter 2, we have $\det(\Phi(t))$ $e^{(-\frac{13}{2}(t-t_0)+5(w(t)-w(t_0)))} = \det(\Phi) = \det(\Phi_0) \exp[\int_{t_0}^t (\mathbb{L}(B) - \frac{1}{2}(\text{tr}(B))^2)ds + \int_{t_0}^t \text{tr}(B)$ $dw(s)]$. From Example 4.2.6, we have $\det(\Phi_0) = e^{(-\frac{13}{2}t_0+5w(t_0))}$. Hence,

$$e^{(-\frac{13}{2}t_0+5w(t_0))} e^{(-\frac{13}{2}(t-t_0)+5(w(t)-w(t_0)))} = \det(\Phi(t)) = e^{(-\frac{13}{2}t+5w(t))}.$$

Observation 4.2.4. Again, an observation analogous to the deterministic version of Observation 4.3.5 [80] is reformulated. This re-enforces the fundamental ideas about finding the solution processes of (4.2.1) that are parallel to the method of solving deterministic systems differential equations.

(i) From (4.2.16), we can immediately conclude that $\det(\Phi(t)) \equiv \det(\Phi(t, w(t))) \neq 0$ on J if and only if $\det(\Phi(t_0)) \equiv \det(\Phi(t_0, w(t_0))) = c \neq 0$. Therefore, $\Phi(t)$ is a fundamental matrix solution to (4.2.1) if and only if $\det(\Phi(t_0)) = c \neq 0$.

(ii) This means that if $\det(\Phi(t^*)) = 0$ for some $t = t^*$ in J, then $\det(\Phi(t)) = 0$ for all t in J. This implies that the set of solutions is linearly dependent (Definition 1.4.5 [80], Chapter 1 of Volume 1).

(iii) Any set $\{x_1(t), x_2(t), \ldots, x_k(t), \ldots x_n(t)\}$ of n nonzero solutions to (4.2.1) is a *complete set of solutions* to (4.2.1) (the basis for the solution space) if $\det(\Phi(t^*)) \neq 0$ for some t^* in J. This idea can be used to construct the complete set of solutions to (4.2.8). For example, the initial states can be chosen from $\{e_1, e_2, \ldots, e_j, \ldots, e_n\} \subseteq R^n$, where $e_j = (\delta_{ij})_{n \times 1}$, $\delta_{ij} = 0$ for $i \neq j$, and $\delta_{ij} = 1$ for $i = j$.

(iv) We further remark that $\det(\Phi(t))$ is called the *Wronskian process* of the set $\{x_1(t), x_2(t), \ldots, x_k(t), \ldots, x_n(t)\}$ of solution processes of (4.2.1).

(v) $\det(\Phi(t)) \neq 0$ on J if and only if $\Phi(t)$ is inevitable on J (Theorem 1.4.4 [80]). At a later stage, we find the inverse of $\Phi(t)$, i.e. $\Phi^{-1}(t) \equiv \Phi^{-1}(t, w(t))$ on J.

(vi) We further note that by replacing $w(t)$ with t that the stochastic system of differential equations (4.2.1) and its solution representation (4.2.2) reduce to the deterministic system and its solution representation as a special case. Thus, the presented discussions in this chapter are directly applicable to systems of deterministic differential equations as a special case.

Example 4.2.9. Let us consider Example 4.2.2. Find:

(a) the Wronskian of the solution set $\{x_1(t), x_2(t), x_3(t)\}$, where

$$x_1(t) = e^{0(t+w(t))} \begin{bmatrix} -3 \\ 1 \\ 4 \end{bmatrix}, \quad x_2(t) = e^{(-\frac{1}{2}t+w(t))} \begin{bmatrix} 5 \\ 0 \\ -7 \end{bmatrix},$$

$$x_3(t) = e^{2(-t+w(t))} \begin{bmatrix} 3 \\ 1 \\ -4 \end{bmatrix},$$

and

(b) $\Phi^{-1}(t)$ (if possible).

Solution procedure. First, we observe that $n = 3$. From Example 4.2.2, $x_1(t)$, $x_2(t)$, and $x_3(t)$ are the solutions to the given system of Itô–Doob-type stochastic differential equations. From the definition of $\Phi(t)$ in (4.2.11) and Observation 4.2.2, we have

$$\Phi(t) = \begin{bmatrix} -3 & 5e^{(-\frac{1}{2}t+w(t))} & 3e^{2(-t+w(t))} \\ 1 & 0 & e^{2(-t+w(t))} \\ 4 & -7e^{(-\frac{1}{2}t+w(t))} & -4e^{2(-t+w(t))} \end{bmatrix}.$$

From Observation 4.2.4(iv), the Wronskian of the set $\{x_1(t), x_2(t), x_3(t)\}$ of solutions to the given system differential equations is $\det(\Phi(t))$. From the definition of determinant of the matrix (Definition 1.4.2 [80]), we obtain

$$\det(\Phi(t)) = (-3)(7e^{(-\frac{5}{2}t+3w(t))}) - 5e^{(-\frac{1}{2}t+w(t))}(-4e^{2(-t+w(t))} - 4e^{2(-t+w(t))})$$

$$+ 3e^{2(-t+w(t))}\left(-7e^{(-\frac{1}{2}t+w(t))}\right)$$

$$= -21e^{(-\frac{5}{2}t+3w(t))} + 40e^{(-\frac{5}{2}t+3w(t))} - 21e^{(-\frac{5}{2}t+3w(t))} \quad \text{(by simplification)}$$

$$= -2e^{(-\frac{5}{2}t+3w(t))} \quad \text{(by further simplification).}$$

This completes the solution process of Part (a). From this and Observation 4.2.4(iii), the given set of solutions is a complete set of solutions. In short, $\Phi(t)$ is invertible. To find $\Phi^{-1}(t)$, we repeat the procedure used in Examples 1.4.6 and 1.4.8 (Chapter 1 of Volume 1), and we get

$$\Phi^{-1}(t) = \frac{\operatorname{adj}\Phi(t)}{\det(\Phi(t))} = \frac{\begin{bmatrix} 7e^{(-\frac{5}{2}t+3w(t))} & -e^{(-\frac{5}{2}t+3w(t))} & 5e^{(-\frac{5}{2}t+w(t))} \\ 8e^{2(-t+w(t))} & 0 & 6e^{2(-t+w(t))} \\ -7e^{(-\frac{1}{2}t+w(t))} & -e^{(-\frac{1}{2}t+w(t))} & -5e^{2(-t+w(t))} \end{bmatrix}}{-2e^{(-\frac{5}{2}t+3w(t))}}$$

$$= \begin{bmatrix} -\frac{7}{2} & \frac{1}{2} & -\frac{5}{2} \\ -4e^{(\frac{1}{2}t-w(t))} & 0 & -3e^{(\frac{1}{2}t-w(t))} \\ \frac{7}{2}e^{2(t-w(t))} & \frac{1}{2}e^{2(t-w(t))} & \frac{5}{2}e^{2(t-w(t))} \end{bmatrix} \quad \text{(by simplification)}.$$

This completes an algebraic procedure for finding the inverse of $\Phi(t)$.

Example 4.2.10. Given: $dx(t) = Bx\,dw(t)$, $x(t_0) = x_0$, where B is a 3×3 matrix in Example 4.2.2.

(a) Find the state transition matrix.
(b) Solve the IVP.

Solution procedure. Again, we observe here that $n = 3$. Step #1 of Procedure 4.2.4 requires us to apply the four-step Procedure of 4.2.2 to find solutions to the given system of differential equations. This has been achieved in Example 4.2.2. Hence, we have

$$x_1(t) = e^{0(t+w(t))}\begin{bmatrix} -3 \\ 1 \\ 4 \end{bmatrix}, \quad x_2(t) = e^{(-\frac{1}{2}t+w(t))}\begin{bmatrix} 5 \\ 0 \\ -7 \end{bmatrix},$$

$$x_3(t) = e^{2(-t+w(t))}\begin{bmatrix} 3 \\ 1 \\ -4 \end{bmatrix}.$$

Repeating the argument used in Example 4.2.4, we can conclude that for arbitrary constants a_1, a_2, and a_3, $x(t) = a_1x_1(t) + a_2x_2(t) + a_3x_3(t)$ is also a solution process of the given system (see also Theorem 4.2.1). This completes Step #2 of Procedure 4.2.4. Step #3 of Procedure 4.2.4 can be completed by imitating the arguments used in Example 4.2.9. In this case,

$$\Phi^{-1}(t_0)x_0 = \begin{bmatrix} -\dfrac{7}{2} & \dfrac{1}{2} & -\dfrac{5}{2} \\ -4e^{(\frac{1}{2}t_0 - w(t_0))} & 0 & -3e^{(\frac{1}{2}t_0 - w(t_0))} \\ \dfrac{7}{2}e^{2(t_0 - w(t_0))} & \dfrac{1}{2}e^{2(t_0 - w(t_0))} & \dfrac{5}{2}e^{2(t_0 - w(t_0))} \end{bmatrix} \begin{bmatrix} x_{10} \\ x_{20} \\ x_{30} \end{bmatrix}.$$

The state transition matrix or the normalized fundamental matrix solution to the given system is:

$$\Phi(t)\Phi^{-1}(t_0) = \Phi(t, t_0)$$

$$= \begin{bmatrix} \phi_{11}(t - t_0, w(t) - w(t_0)) & \phi_{12}(t - t_0, w(t) - w(t_0)) & \phi_{13}(t - t_0, w(t) - w(t_0)) \\ \phi_{21}(t - t_0, w(t) - w(t_0)) & \phi_{22}(t - t_0, w(t) - w(t_0)) & \phi_{23}(t - t_0, w(t) - w(t_0)) \\ \phi_{31}(t - t_0, w(t) - w(t_0)) & \phi_{32}(t - t_0, w(t) - w(t_0)) & \phi_{33}(t - t_0, w(t) - w(t_0)) \end{bmatrix},$$

where

$$\phi_{11}(t - t_0, w(t) - w(t_0)) = \frac{21}{2} - 20e^{(-\frac{1}{2}(t-t_0) + (w(t) - w(t_0)))}$$
$$+ \frac{21}{2}e^{2(-(t-t_0) + (w(t) - w(t_0)))},$$

$$\phi_{12}(t - t_0, w(t) - w(t_0)) = -\frac{3}{2} + \frac{3}{2}e^{2(-(t-t_0) + (w(t) - w(t_0)))},$$

$$\phi_{13}(t - t_0, w(t) - w(t_0)) = \frac{15}{2} - 15e^{(-\frac{1}{2}(t-t_0) + (w(t) - w(t_0)))}$$
$$+ \frac{15}{2}e^{2(-(t-t_0) + (w(t) - w(t_0)))},$$

$$\phi_{21}(t - t_0, w(t) - w(t_0)) = -\frac{7}{2} + \frac{7}{2}e^{2(-(t-t_0) + (w(t) - w(t_0)))},$$

$$\phi_{22}(t - t_0, w(t) - w(t_0)) = \frac{1}{2} + \frac{1}{2}e^{2(-(t-t_0) + (w(t) - w(t_0)))},$$

$$\phi_{23}(t - t_0, w(t) - w(t_0)) = -\frac{5}{2} + \frac{5}{2}e^{2(-(t-t_0) + (w(t) - w(t_0)))},$$

$$\phi_{31}(t - t_0, w(t) - w(t_0)) = -14 + 28e^{(-\frac{1}{2}(t-t_0) + (w(t) - w(t_0)))}$$
$$- 14e^{2(-(t-t_0) + (w(t) - w(t_0)))},$$

$$\phi_{32}(t - t_0, w(t) - w(t_0)) = 2 - 2e^{2(-(t-t_0) + (w(t) - w(t_0)))},$$

$$\phi_{33}(t - t_0, w(t) - w(t_0)) = -10 + 21e^{(-\frac{1}{2}(t-t_0) + (w(t) - w(t_0)))}$$
$$- 10e^{2(-(t-t_0) + (w(t) - w(t_0)))}.$$

From Step #3 of Procedure 4.2.4, the solution to the given IVP is $x(t, t_0, x_0) = \Phi(t, t_0)x_0$, where $\Phi(t, t_0)$ is defined above, and $x_0 = [x_{10} \ x_{20} \ x_{30}]^T$. This completes the procedure of Part (b).

The following result extends the deterministic analytic method for finding an inverse of a fundamental matrix solution to (4.2.1). Moreover, this result will subsequently used. Again, this is a new result.

Theorem 4.2.3. *Let Φ be the fundamental matrix solution process of* (4.2.1). *Then, Φ is invertible and its inverse $\Phi^{-1}(t) \equiv \Phi^{-1}(t, w(t))$ satisfies the following stochastic matrix differential equation of the Itô–Doob type:*

$$d\Phi^{-1}(t) = \Phi^{-1}(t)B^2 dt - \Phi^{-1}(t)B \, dw(t). \tag{4.2.23}$$

Moreover,

$$d(\Phi^{-1})^T = (B^T)^2(\Phi^{-1}(t))^T dt - B^T(\Phi^{-1}(t))^T dw(t). \tag{4.2.24}$$

Proof. Let Φ be a fundamental matrix solution to (4.2.1). From Observation 4.2.4, it is invertible on J. Its inverse Φ^{-1} exists on J. By using this and Theorem 1.3.4 (product rule), we compute the Itô–Doob differential on both sides of the processes $\Phi\Phi^{-1} = I$, and we have

$$d(\Phi\Phi^{-1}) = d\Phi\Phi^{-1} + \Phi d\Phi^{-1} + d\Phi d\Phi^{-1} = dI = 0,$$

and hence

$$d\Phi\Phi^{-1} + \Phi d\Phi^{-1} + d\Phi d\Phi^{-1} = 0.$$

This implies that

$$\Phi(t)d\Phi^{-1}(t) = -d\Phi(t)\Phi^{-1}(t) - d\Phi(t)d\Phi^{-1}(t)$$

$$= -B\Phi(t)\Phi^{-1}(t)dw(t) - B\Phi(t)d\Phi^{-1}(t)dw(t) \quad \text{[from (4.2.14)]}.$$

Multiplying both sides of the above equation from the left by $\Phi^{-1}(t)$ and again using $\Phi\Phi^{-1} = I$, we get

$$d\Phi^{-1}(t) = -\Phi^{-1}(t)B \, dw(t) - \Phi^{-1}(t)B\Phi(t)d\Phi^{-1}(t)dw(t). \tag{4.2.25}$$

Using the unknown but the explicit nature of $d\Phi^{-1}(t)$ in (4.2.25) and the nature of the Itô–Doob differential calculus, (4.2.25) can be simplified:

$$d\Phi^{-1}(t) = -\Phi^{-1}(t)B \, dw(t)$$

$$- \Phi^{-1}(t)B\Phi(t)[-\Phi^{-1}(t)B \, dw(t) - \Phi^{-1}(t)B\Phi(t)d\Phi^{-1}(t)dw(t)]dw(t)$$

$$= -\Phi^{-1}(t)B \, dw(t) + [\Phi^{-1}(t)B^2 + \Phi^{-1}(t)B^2\Phi(t)d\Phi^{-1}(t)](dw(t))^2$$

$$\tag{4.2.26}$$

$$= -\Phi^{-1}(t)B \, dw(t) + \Phi^{-1}(t)B^2 dt \ [\text{by } B^2 d\Phi^{-1}(t))dw(t))^2 = o(dt)]$$

$$= \Phi^{-1}(t)B^2 dt - \Phi^{-1}(t)B \, dw(t) \quad \text{(by rearrangement)}. \tag{4.2.27}$$

This establishes the derivation of Equation (4.2.23). The proof of the expression (4.2.24) follows from the properties of the transpose, differential and inverse of matrices. This completes the proof of the theorem. □

Observation 4.2.5. We note that $(\Phi^{-1})^T(t)$ is a fundamental matrix solution to Itô–Doob-type linear system of stochastic differential equations:

$$dy = (B^T)^2 y\, dt - B^T y\, dw(t), \quad y(t_0) = x_0. \tag{4.2.28}$$

This statement is justified at a later stage of this section. System (4.2.28) is called the *adjoint to the system* in (4.2.8). System (4.2.28) is equivalent to the following system:

$$dy = yB^2 dt - yB\, dw(t), \quad y(t_0) = x_0^T. \tag{4.2.29}$$

We note that y in (4.2.29) is a row vector, and y in (4.2.28) is a column vector. In the light of this notational understanding, $\Phi^{-1}(t)$ is a fundamental matrix solution to (4.2.29). This can be justified by (4.2.23).

Example 4.2.11. Let us consider Example 4.2.10. Show that $\Phi^{-1}(t)$ is the fundamental matrix solution to $dy = yB^2 dt - yB\, dw(t)$, where B and $\Phi^{-1}(t)$ are as described in Examples 4.2.2 and 4.2.9, respectively.

Solution procedure. We remark that $\Phi^{-1}(t)$ is an invertible matrix process, and its inverse is $\Phi(t)$ (Example 4.2.9). Therefore, we only need to show that $\Phi^{-1}(t)$ satisfies (4.2.23). Now, we repeat the argument used in Example 4.2.7 to compute $\Phi^{-1}(t)B^2 dt$, $-\Phi^{-1}(t)B\, dw(t)$ and $d\Phi^{-1}(t)$:

$\Phi^{-1}(t)B^2 dt$

$$= \begin{bmatrix} -\dfrac{7}{2} & \dfrac{1}{2} & -\dfrac{5}{2} \\ -4e^{(\frac{1}{2}t-w(t))} & 0 & -3e^{(\frac{1}{2}t-w(t))} \\ \dfrac{7}{2}e^{2(t-w(t))} & \dfrac{1}{2}e^{2(t-w(t))} & \dfrac{5}{2}e^{2(t-w(t))} \end{bmatrix} \begin{bmatrix} 1 & 3 & 0 \\ 7 & 1 & 5 \\ 0 & -4 & 1 \end{bmatrix} \begin{bmatrix} 1 & 3 & 0 \\ 7 & 1 & 5 \\ 0 & -4 & 1 \end{bmatrix} dt$$

$$= \begin{bmatrix} -\dfrac{7}{2} & \dfrac{1}{2} & -\dfrac{5}{2} \\ -4e^{(\frac{1}{2}t-w(t))} & 0 & -3e^{(\frac{1}{2}t-w(t))} \\ \dfrac{7}{2}e^{2(t-w(t))} & \dfrac{1}{2}e^{2(t-w(t))} & \dfrac{5}{2}e^{2(t-w(t))} \end{bmatrix} \begin{bmatrix} 22 & 6 & 15 \\ 14 & 2 & 10 \\ -28 & -8 & -19 \end{bmatrix} dt$$

$$= \begin{bmatrix} 0 & 0 & 0 \\ -4e^{(\frac{1}{2}t-w(t))} & 0 & -3e^{(\frac{1}{2}t-w(t))} \\ 14e^{2(t-w(t))} & 2e^{2(t-w(t))} & 10e^{2(t-w(t))} \end{bmatrix} dt,$$

$$-\Phi^{-1}(t)B\,dw(t) = -\begin{bmatrix} -\dfrac{7}{2} & \dfrac{1}{2} & -\dfrac{5}{2} \\ -4e^{(\frac{1}{2}t-w(t))} & 0 & -3e^{(\frac{1}{2}t-w(t))} \\ \dfrac{7}{2}e^{2(t-w(t))} & \dfrac{1}{2}e^{2(t-w(t))} & \dfrac{5}{2}e^{2(t-w(t))} \end{bmatrix} \begin{bmatrix} 1 & 3 & 0 \\ 7 & 1 & 5 \\ 0 & -4 & 1 \end{bmatrix} dw(t)$$

$$= \begin{bmatrix} 0 & 0 & 0 \\ 4e^{(\frac{1}{2}t-w(t))} & 0 & 3e^{(\frac{1}{2}t-w(t))} \\ -7e^{2(t-w(t))} & -e^{2(t-w(t))} & -5e^{2(t-w(t))} \end{bmatrix} dw(t),$$

and

$$d\Phi^{-1}(t)$$

$$= d\left(\begin{bmatrix} -\dfrac{7}{2} & \dfrac{1}{2} & -\dfrac{5}{2} \\ -4e^{(\frac{1}{2}t-w(t))} & 0 & -3e^{(\frac{1}{2}t-w(t))} \\ \dfrac{7}{2}e^{2(t-w(t))} & \dfrac{1}{2}e^{2(t-w(t))} & \dfrac{5}{2}e^{2(t-w(t))} \end{bmatrix}\right)$$

$$= \begin{bmatrix} d\left(-\dfrac{7}{2}\right) & d\left(\dfrac{1}{2}\right) & d\left(-\dfrac{5}{2}\right) \\ d(-4e^{(\frac{1}{2}t-w(t))}) & d(0) & d(-3e^{(\frac{1}{2}t-w(t))}) \\ d\left(\dfrac{7}{2}e^{2(t-w(t))}\right) & d\left(\dfrac{1}{2}e^{2(t-w(t))}\right) & d\left(\dfrac{5}{2}e^{2(t-w(t))}\right) \end{bmatrix}$$

$$= \begin{bmatrix} 0 & 0 & 0 \\ -4e^{(\frac{1}{2}t-w(t))}(dt-dw(t)) & 0 & -3e^{(\frac{1}{2}t-w(t))}(dt-dw(t)) \\ 7e^{2(t-w(t))}(2dt-dw(t)) & e^{2(t-w(t))}(2dt-dw(t)) & 5e^{2(t-w(t))}(2dt-dw(t)) \end{bmatrix}$$

$$= \begin{bmatrix} 0 & 0 & 0 \\ -4e^{(\frac{1}{2}t-w(t))} & 0 & -3e^{(\frac{1}{2}t-w(t))} \\ 14e^{2(t-w(t))} & 2e^{2(t-w(t))} & 10e^{2(t-w(t))} \end{bmatrix} dt$$

$$+ \begin{bmatrix} 0 & 0 & 0 \\ 4e^{(\frac{1}{2}t-w(t))} & 0 & 3e^{(\frac{1}{2}t-w(t))} \\ -7e^{2(t-w(t))} & -e^{2(t-w(t))} & -5e^{2(t-w(t))} \end{bmatrix} dw(t).$$

The right-hand side expression of $d\Phi^{-1}(t)$ is exactly equal to the sum of the right-hand side expressions of $\Phi^{-1}(t)B^2dt$ and $-\Phi^{-1}(t)Bdw(t)$. Therefore, by the application of Definition 4.2.2, we conclude that $\Phi^{-1}(t)$ is the *general fundamental matrix solution* to the stochastic matrix differential equation (4.2.23) corresponding to the given system of Itô–Doob-type stochastic differential equations. This completes the solution procedure.

Exercises

1. Using the concept of the Wronskian, prove or disprove that the given functions are a linearly independent set of solutions to the system of stochastic differential equations:

(a) $dx = \begin{bmatrix} 1 & 1 \\ 4 & 1 \end{bmatrix} x\, dw(t)$, $x_1(t) = e^{-(\frac{1}{2}t + w(t))} \begin{bmatrix} -1 \\ 2 \end{bmatrix}$, $x_2(t) = e^{(-\frac{9}{2}t + w(t))} \begin{bmatrix} 1 \\ 2 \end{bmatrix}$;

(b) $dx = \begin{bmatrix} 3 & 0 & 1 \\ 9 & -1 & 2 \\ -9 & 4 & -1 \end{bmatrix} x\, dw(t)$, $x_1(t) = e^{(-\frac{9}{2}t + w(t))} \begin{bmatrix} 4 \\ 9 \\ 0 \end{bmatrix}$,

$$x_2(t) = e^{-t} \begin{bmatrix} \cos t \\ \sin t + 2\cos t \\ -4\cos t - \sin t \end{bmatrix}, x_3(t) = e^{-t} \begin{bmatrix} \sin t \\ 2\sin t - \cos t \\ \cos t - 4\sin t \end{bmatrix};$$

(c) $dx = \begin{bmatrix} 1 & 0 \\ -1 & 2 \end{bmatrix} x\, dw(t)$, $x_1(t) = e^{(-\frac{1}{2}t + w(t))} \begin{bmatrix} 1 \\ 1 \end{bmatrix}$, $x_2(t) = e^{2(-t + w(t))} \begin{bmatrix} 0 \\ 1 \end{bmatrix}$;

(d) $dx = \begin{bmatrix} 1 & 1 \\ 0 & 1 \end{bmatrix} x\, dw(t)$, $x_1(t) = e^{-(\frac{1}{2}t + w(t))} \begin{bmatrix} 1 \\ 0 \end{bmatrix}$, $x_2(t) = e^{-(\frac{1}{2}t + w(t))} \begin{bmatrix} w(t) \\ 1 \end{bmatrix}$;

(e) $dx = \begin{bmatrix} 1 & 0 \\ 1 & 1 \end{bmatrix} x\, dw(t)$, $x_1(t) = e^{(-\frac{1}{2}t + w(t))} \begin{bmatrix} 0 \\ 1 \end{bmatrix}$, $x_2(t) = e^{-(\frac{1}{2}t + w(t))} \begin{bmatrix} 1 \\ w(t) \end{bmatrix}$;

(f) $dx = \begin{bmatrix} 1 & 0 \\ 0 & -1 \end{bmatrix} x\, dw(t)$, $x_1(t) = e^{(-\frac{1}{2}t + w(t))} \begin{bmatrix} 1 \\ 0 \end{bmatrix}$, $x_2(t) = e^{-(\frac{1}{2}t + w(t))} \begin{bmatrix} 0 \\ 1 \end{bmatrix}$.

2. Find the Wronskian of the given functions in Exercise 1.
3. Using the results of Exercise 2, prove or disprove that the set of functions in Exercise 1 is a linearly independent set of solutions to the stochastic system of differential equations.
4. Which set of solutions in Exercise 1 forms the fundamental matrix solutions to the corresponding stochastic system of differential equations? Justify your answer.
5. Using the stochastic version of the Abel–Jacobi–Liouville formula, prove or disprove the answer to the question in Exercise 4.
6. Find the general solution to the differential equations in Exercise 1 (if possible).

7. Find the solutions to the IVPs (if possible) with regard to the systems in Exercises 1 with the initial condition:

(i) $(0, x_0) = (0, [1\ 2]^T)$;
(ii) $(1, x_0) = (1, [2\ 1]^T)$.

4.3 Procedure for Finding the Fundamental Matrix Solution

In this section, we present procedures for finding n linearly independent eigenvectors of the coefficient matrix B in (4.2.1) corresponding to the n eigenvalues determined by (4.2.4). Procedures for finding a complete set of linearly independent eigenvectors for stochastic systems are creatively developed. Moreover, they parallel the procedures for finding a complete set of eigenvectors for deterministic systems [80]. There are three cases. These cases depend on the nature of the eigenvalues of (4.2.4), namely:

(i) Distinct real numbers.
(ii) Distinct complex numbers.
(iii) Repeated, real, or complex numbers.

The procedures for finding n linearly independent eigenvectors of the coefficient matrix B and the corresponding n linearly independent solutions to (4.2.1) are briefly outlined in the following discussion.

Distinct Real Eigenvalues 4.3.1. Let $\mu_1, \ldots, \mu_k, \ldots, \mu_n$ be the real and distinct roots of (4.2.4) and corresponding to these real roots, let $\lambda_1, \ldots, \lambda_k, \ldots, \lambda_n$ be real roots of (4.2.5). In addition, $c_1, \ldots, c_k, \ldots, c_n$ be the eigenvectors corresponding to these eigenvalues determined by solving the systems of algebraic equations (4.2.6). Then, the set of solutions to (4.2.1) $\{x_1(t), x_2(t), \ldots, x_k(t), \ldots, x_n(t)\}$ is a complete set of solutions.

Discussion. In this case, it can be verified that the Wronskian process corresponding to the set $\{x_1(t), x_2(t), \ldots, x_k(t), \ldots, x_n(t)\}$ of solutions to (4.2.1) is different from zero. This means that $\det \Phi(t)) \neq 0$ on J. In fact, from Observation 4.2.4(iv), $\det(\Phi(0)) \neq 0$, where the $n \times n$ matrix $\Phi(0) = [c_1, \ldots, c_k, \ldots, c_n]$, whose n column vectors are eigenvectors corresponding to the matrix B in (4.2.1). Hence, $[c_1, \ldots, c_k, \ldots, c_n]$ is an invertible matrix, because the eigenvectors are linearly independent. Moreover, the set $\{c_1, \ldots, c_k, \ldots, c_n\}$ is a basis for R^n. By imitating the argument used in the deterministic case, one can validate these statements. For details, see Ref. 80. In summary, in this case, the general solution to (4.2.1) is as in (4.2.11), i.e. $x(t) = \Phi(t)a$. The solution to the IVP (4.2.8) is represented in (4.2.13), i.e. $x(t, t_0, x_0) = x(t) = \Phi(t, t_0)x_0$. This completes the outline of the procedure with regard to distinct real roots.

Illustration 4.3.1. Given: $dx(t) = Bx\,dw(t)$, $x(t_0) = x_0$, where $B = \begin{bmatrix} b_{11} & b_{12} \\ b_{21} & b_{22} \end{bmatrix}$,

with b_{ij} being any given real numbers satisfying the condition $d(B) = (\mathrm{tr}(B))^2 - 4\det(B) > 0$, where $\mathrm{tr}(B)$ is as defined in Theorem 4.2.2, and $\det(B)$ is the determinant (Definition 1.4.2 [80]). Find: (a) a general solution and (b) the solution to the given IVP.

Solution procedure. The argument used in the deterministic case is modified in the context of the Itô–Doob calculus and its effects in solving the systems of differential equations. Again, to solve an IVP (Steps #1–3 of Procedure 4.2.4), we need to find a general solution to the given system of differential equations [Steps #1–4 of Procedure 4.2.2 and Observation 4.2.3(i)–(ii)]. To find a general solution, we need to find a complete set of solutions (Steps #1–4 of Procedure 4.2.2 and Observation 4.2.1).

We seek a solution of the form $x(t) = \exp[\lambda t + \mu w(t)]c$. The main goal is to find an unknown nonzero two-dimensional vector $c = [c_1\ c_2]^T$ and scalar numbers λ and μ. For this purpose, we use (4.2.3), and we have

$$0 = (B - \mu I)c \quad \text{and} \quad \left(\lambda + \frac{1}{2}\mu^2\right)c = 0.$$

The characteristic equation (4.2.4) of the matrix B in the context of the given example is

$$0 = \det(B - \mu I) = \mu^2 - \mathrm{tr}(B)\mu + \det(B) = (\mu - \mu_1)(\mu - \mu_2),$$

and

$$\lambda_1 = -\frac{1}{2}\mu_1^2 \quad \text{and} \quad \lambda_2 = -\frac{1}{2}\mu_2^2,$$

where $\mu_1 = \dfrac{\mathrm{tr}(B) + \sqrt{(\mathrm{tr}(B))^2 - 4\det(B)}}{2}$ and $\mu_2 = \dfrac{\mathrm{tr}(B) - \sqrt{(\mathrm{tr}(B))^2 - 4\det(B)}}{2}$ are the roots of the quadratic polynomial equation $\mu^2 - \mathrm{tr}(B)\mu + \det(B) = 0$. They are also the eigenvalues of the coefficient matrix in the example. From the assumption $(\mathrm{tr}(B))^2 - 4\det(B) > 0$, we conclude that these are the distinct real roots of $0 = \det(B - \mu I)$.

Case 1. From $\det(B - \mu I) = 0 = (b_{11} - \mu)(b_{22} - \mu) - b_{21}b_{12}$, we note that "at least b_{12} and or $b_{21} = 0$ if and only if the eigenvalues μ_1 and μ_2 are diagonal elements of the matrix B." The corresponding eigenvectors and solutions follow:

(i) For $b_{21} = 0 = b_{12}$, $\mu_1 = b_{11}$ and $\lambda_1 = -\frac{1}{2}b_{11}^2 : x_1(t) = e^{-\frac{1}{2}b_{11}^2 t + b_{11}w(t)}\begin{bmatrix} 1 \\ 0 \end{bmatrix}$;

for $b_{21} = 0 = b_{12}$, $\mu_2 = b_{22}$ and $\lambda_2 = -\frac{1}{2}b_{22}^2 :$

$$x_2(t) = e^{-\frac{1}{2}b_{22}^2 t + b_{22}w(t)}\begin{bmatrix} 0 \\ 1 \end{bmatrix}.$$

(ii) For $b_{21} \neq 0$, $b_{12} = 0$, $\mu_1 = b_{11}$ and $\lambda_1 = -\frac{1}{2}b_{11}^2$:

$$x_1(t) = e^{-\frac{1}{2}b_{11}^2 t + b_{11} w(t)} \begin{bmatrix} b_{11} - b_{22} \\ b_{21} \end{bmatrix};$$

for $b_{21} \neq 0$, $b_{12} = 0$, $\mu_2 = b_{22}$ and $\lambda_2 = -\frac{1}{2}b_{22}^2$: $x_2(t) = e^{-\frac{1}{2}b_{22}^2 t + b_{22} w(t)} \begin{bmatrix} 0 \\ 1 \end{bmatrix}.$

(iii) For $b_{12} \neq 0$, $b_{21} = 0$, $\mu_1 = b_{11}$ and $\lambda_1 = -\frac{1}{2}b_{11}^2$: $x_1(t) = e^{-\frac{1}{2}b_{11}^2 t + b_{11} w(t)} \begin{bmatrix} 1 \\ 0 \end{bmatrix};$

for $b_{12} \neq 0$, $b_{21} = 0$, $\mu_2 = b_{22}$ and $\lambda_2 = -\frac{1}{2}b_{22}^2$:

$$x_2(t) = e^{-\frac{1}{2}b_{22}^2 t + b_{22} w(t)} \begin{bmatrix} b_{12} \\ b_{22} - b_{11} \end{bmatrix}.$$

Case 2. The negation of the statement in Case 1 is: "The off-diagonal elements are nonzero if and only if neither one of the eigenvalues μ_1 and μ_2 is a diagonal element of the matrix B." In this case, we can choose eigenvectors of the matrix B. Then the solutions corresponding to the eigenvalues μ_1 and μ_2 are

$$x_1(t) = e^{-\frac{1}{2}\mu_1^2 t + \mu_1 w(t)} \begin{bmatrix} b_{12} \\ \mu_1 - b_{11} \end{bmatrix} \quad \text{and} \quad x_2(t) = e^{-\frac{1}{2}\mu_2^2 t + \mu_2 w(t)} \begin{bmatrix} \mu_2 - b_{22} \\ b_{21} \end{bmatrix}.$$

The roots of the characteristic polynomials are distinct and real. Therefore, these solutions are linearly independent. Hence, the fundamental matrices formed from these solutions are:

Case 1

(i) $\Phi(t) \equiv \Phi(t, w(t)) = \begin{bmatrix} e^{-\frac{1}{2}b_{11}^2 t + b_{11} w(t)} & 0 \\ 0 & e^{-\frac{1}{2}b_{22}^2 t + b_{22} w(t)} \end{bmatrix}.$

(ii) $\Phi(t) \equiv \Phi(t, w(t)) = \begin{bmatrix} (b_{11} - b_{22})e^{-\frac{1}{2}b_{11}^2 t + b_{11} w(t)} & 0 \\ b_{21}e^{-\frac{1}{2}b_{11}^2 t + b_{11} w(t)} & e^{-\frac{1}{2}b_{22}^2 t + b_{22} w(t)} \end{bmatrix}.$

(iii) $\Phi(t) \equiv \Phi(t, w(t)) = \begin{bmatrix} e^{-\frac{1}{2}b_{11}^2 t + b_{11} w(t)} & b_{12}e^{-\frac{1}{2}b_{22}^2 t + b_{22} w(t)} \\ 0 & (b_{22} - b_{11})e^{-\frac{1}{2}b_{22}^2 t + b_{22} w(t)} \end{bmatrix}.$

Case 2

(i) $\Phi(t) \equiv \Phi(t, w(t)) = \begin{bmatrix} b_{12}e^{-\frac{1}{2}\mu_1^2 t + \mu_1 w(t)} & (\mu_2 - b_{22})e^{-\frac{1}{2}\mu_2^2 t + \mu_1 w(t)} \\ (\mu_1 - b_{11})e^{-\frac{1}{2}\mu_1^2 t + \mu_1 w(t)} & b_{21}e^{-\frac{1}{2}\mu_2^2 t + \mu_2 w(t)} \end{bmatrix}.$

Moreover, the general solutions (4.2.11) corresponding to the respective cases are represented by:

Case 1

(i) $x(t) = \Phi(t)a = \begin{bmatrix} e^{-\frac{1}{2}b_{11}^2 t + b_{11}w(t)} & 0 \\ 0 & e^{-\frac{1}{2}b_{22}^2 t + b_{22}w(t)} \end{bmatrix} a.$

(ii) $x(t) = \Phi(t)a = \begin{bmatrix} (b_{11} - b_{22})e^{-\frac{1}{2}b_{11}^2 t + b_{11}w(t)} & 0 \\ b_{21}e^{-\frac{1}{2}b_{11}^2 t + b_{11}w(t)} & e^{-\frac{1}{2}b_{22}^2 t + b_{22}w(t)} \end{bmatrix} a.$

(iii) $x(t) = \Phi(t)a = \begin{bmatrix} e^{-\frac{1}{2}b_{11}^2 t + b_{11}w(t)} & b_{12}e^{-\frac{1}{2}b_{22}^2 t + b_{22}w(t)} \\ 0 & (b_{22} - b_{11})e^{-\frac{1}{2}b_{22}^2 t + b_{22}w(t)} \end{bmatrix} a.$

Case 2

$$x(t) = \Phi(t)a = \begin{bmatrix} b_{12}e^{-\frac{1}{2}\mu_1^2 t + \mu_1 w(t)} & (\mu_2 - b_{22})e^{-\frac{1}{2}\mu_2^2 t + \mu_2 w(t)} \\ (\mu_1 - b_{11})e^{-\frac{1}{2}\mu_1^2 t + \mu_1 w(t)} & b_{21}e^{-\frac{1}{2}\mu_2^2 t + \mu_2 w(t)} \end{bmatrix} a.$$

where $a = [a_1\ a_2]^T$ is an arbitrary two-dimensional constant vector.

The solutions to the respective IVPs are determined by imitating Step #3 of Procedure 4.2.4, and hence they are:

Case 1

(i) $\Phi(t, w(t), t_0, w(t_0)) \equiv x(t) = \Phi(t, t_0)x_0 = \begin{bmatrix} e^{\nu(\Delta t, \Delta w(t))} & 0 \\ 0 & e^{\gamma(\Delta t, \Delta w(t))} \end{bmatrix} x_0.$

(ii) $\Phi(t, w(t), t_0, w(t_0)) \equiv x(t) = \Phi(t, t_0)x_0 =$

$$\begin{bmatrix} e^{\nu(\Delta t, \Delta w(t))} & 0 \\ \dfrac{b_{21}}{(b_{11} - b_{22})}(e^{\gamma(\Delta t, \Delta w(t))} - e^{\nu(\Delta t, \Delta w(t)) - a_{22}t_0}) & e^{\gamma(\Delta t, \Delta w(t))} \end{bmatrix} x_0.$$

(iii) $\Phi(t, w(t), t_0, w(t_0)) \equiv x(t) = \Phi(t, t_0)x_0 =$

$$\begin{bmatrix} e^{\nu(\Delta t, \Delta w(t))} & \dfrac{b_{12}}{(b_{22} - b_{11})}(e^{\nu(\Delta t, \Delta w(t))} - e^{\gamma(\Delta t, \Delta w(t)) - a_{11}t_0}) \\ 0 & e^{\gamma(\Delta t, \Delta w(t))} \end{bmatrix} x_0.$$

where $\nu(\Delta t, \Delta w(t)) = -\frac{1}{2}b_{11}^2(t - t_0) + b_{11}(w(t) - w(t_0))$, $\Delta t = (t - t_0)$, $\gamma(\Delta t, \Delta w(t)) = -\frac{1}{2}b_{22}^2(t - t_0) + b_{22}(w(t) - w(t_0))$, and $\Delta w(t) = w(t) - w(t_0)$.

Case 2

$$x(t) = \Phi(t)x_0 = \frac{2}{(d(B) + (b_{22} - b_{11})\sqrt{d(B)})} \begin{bmatrix} \phi_{11}(t - t_0) & \phi_{12}(t - t_0) \\ \phi_{21}(t - t_0) & \phi_{22}(t - t_0) \end{bmatrix} x_0,$$

where

$$\phi_{11}(t - t_0) = b_{21}b_{12}e^{\beta(\Delta t, \Delta w(t))} + (\mu_2 - b_{22})(b_{11} - \mu_1)e^{\alpha(\Delta t, \Delta w(t))},$$

$$\phi_{12}(t - t_0) = b_{12}(b_{22} - \mu_2)(e^{\beta(\Delta t, \Delta w(t))} - e^{\alpha(\Delta t, \Delta w(t))}),$$

$$\phi_{21}(t - t_0) = b_{21}(b_{11} - \mu_1)(e^{\alpha(\Delta t, \Delta w(t))} - e^{\beta(\Delta t, \Delta w(t))}),$$

$$\phi_{22}(t - t_0) = (\mu_1 - b_{11})(b_{22} - \mu_2)(e^{\beta(\Delta t, \Delta w(t))} + b_{21}b_{12}e^{\alpha(\Delta t, \Delta w(t))}),$$

$$\beta(\Delta t, \Delta w(t)) = -\frac{1}{2}\mu_1^2(t - t_0) + \mu_1(w(t) - w(t_0)),$$

$$\alpha(\Delta t, \Delta w(t)) = -\frac{1}{2}\mu_2^2(t - t_0) + \mu_2(w(t) - w(t_0)).$$

In the context of Case 2, we note that $2[b_{21}b_{12} + (\mu_1 - b_{11})(b_{22} - \mu_2)] = d(B) + (b_{22} - b_{11})\sqrt{d(B)} \neq 0$, since $d(B) = (b_{11} - b_{22})^2 + 4b_{21}b_{12}$. This completes the solution procedure.

Example 4.3.1. Given: $dx(t) = Bx\,dt$, $x(t_0) = x_0$ where $B = \begin{bmatrix} 4 & -3 \\ 2 & -1 \end{bmatrix}$ Find: (a) a general solution and (b) the solution to the given IVP.

Solution procedure. In this case, $d(B) = (4 + 1)^2 + 4(-6) = 1$. By repeating the arguments of Illustration 4.3.1 and Example 4.2.2, we have the following computations:

(i) solutions to the algebraic equations (4.2.3) with respect to the given matrix B:

$$0 = \det(B - \mu I) = (\mu - 2)(\mu - 1), \quad \mu_1 = 2, \quad \mu_2 = 1; \quad \lambda_1 = -\frac{1}{2}\mu_1^2;$$

$$\lambda_2 = -\frac{1}{2}\mu_2^2, \quad \lambda_1 = -2, \quad \lambda_2 = -\frac{1}{2}.$$

(ii) Eigenvectors [solving (4.2.6)] and the solutions (4.2.2), which correspond to the eigenvalues

$$x_1(t) = e^{2(-t+w(t))}\begin{bmatrix} -3 \\ -2 \end{bmatrix}, \quad x_2(t) = e^{-\frac{1}{2}t+w(t)}\begin{bmatrix} 2 \\ 2 \end{bmatrix}.$$

The roots of the characteristic polynomials are distinct and real. These solutions are linearly independent. Hence, the fundamental matrix solution formed by these solutions (Illustration 4.3.1, Case 2) is

$$\Phi(t) \equiv \Phi(t, w(t)) = \begin{bmatrix} -3e^{2(-t+w(t))} & 2e^{(-\frac{1}{2}t+w(t))} \\ -2e^{2(-t+w(t))} & 2e^{(-\frac{1}{2}t+w(t))} \end{bmatrix}.$$

Moreover, from (4.2.11), its general solution is described by

$$x(t) = \Phi(t, w(t)) = \begin{bmatrix} -3e^{2(-t+w(t))} & 2e^{(-\frac{1}{2}t+w(t))} \\ -2e^{2(-t+w(t))} & 2e^{(-\frac{1}{2}t+w(t))} \end{bmatrix} a,$$

where $a = [a_1 \ a_2]^T$ is an arbitrary two-dimensional constant vector. Thus, the solution to the IVP is determined by Step #3 of Procedure 4.2.4. It is

$$\Phi(t, w(t), t_0, w(t_0)) \equiv x(t) = \Phi(t, t_0)x = -\frac{2}{4} \begin{bmatrix} \phi_{11}(t - t_0) & \phi_{12}(t - t_0) \\ \phi_{21}(t - t_0) & \phi_{22}(t - t_0) \end{bmatrix} x_0,$$

where

$$\phi_{11}(t - t_0) = 4e^{-\frac{1}{2}(t-t_0)+(w(t)-w(t_0))} - 6e^{2(-(t-t_0)+(w(t)-w(t_0)))},$$

$$\phi_{12}(t - t_0) = -6(e^{-\frac{1}{2}(t-t_0)+(w(t)-w(t_0))} - e^{2(-(t-t_0)+(w(t)-w(t_0)))}),$$

$$\phi_{21}(t - t_0) = -4(e^{2(-(t-t_0)+(w(t)-w(t_0)))} - e^{-\frac{1}{2}(t-t_0)+(w(t)-w(t_0))}),$$

$$\phi_{22}(t - t_0) = -6e^{-\frac{1}{2}(t-t_0)+(w(t)-w(t_0))} + 4e^{2(-(t-t_0)+(w(t)-w(t_0)))}.$$

In the context of Case 2, we note that $d(B) + (b_{22} - b_{11})\sqrt{d(B)} = (1 - 5) \neq 0$, since $d(B) = (b_{11} - b_{22})^2 + 4b_{21}b_{12} = 25 - 24 = 1$. This completes the solution procedure.

Distinct Complex Eigenvalues 4.3.2. Let $\mu_1, \dots, \mu_k, \dots, \mu_n$ be the distinct and complex roots of (4.2.4) and, corresponding to these complex roots, let $\lambda_1, \dots, \lambda_k, \dots, \lambda_n$ be the complex roots of (4.2.5). In addition, $c_1, \dots, c_k, \dots, c_n$ be the eigenvectors corresponding to these eigenvalues determined by solving the systems of algebraic equations (4.2.6). Then, the set of solutions to (4.2.1) $\{x_1(t), x_2(t), \dots, x_k(t), \dots, x_n(t)\}$ is a complete set of solutions.

Discussion. We also note that real roots can be considered to be complex roots by the definition of the complex number. The rest of the discussion is similar to the linear deterministic systems of differential equations (Section 4.4, Chapter 4 of Volume 1). In fact, it is repeated for the ordered pairs (μ_k, λ_k) with certain modifications.

SDC 1. We recall that eigenvalues and corresponding eigenvectors of a matrix with real entries occur in complex conjugate pairs. The solutions corresponding to these complex eigenvalues and corresponding eigenvectors will also be complex-valued stochastic processes. Our goal is to find pairs of linearly independent real-valued solutions corresponding to the pairs of complex conjugate eigenvectors of the matrix B in (4.2.1). From (4.2.5), for each complex conjugate pair of eigenvalues $\mu = \mu_k = \alpha_k + i\beta_k$ and $\bar{\mu} = \bar{\mu}_k = \alpha_k - i\beta_k$ with $\operatorname{Im}\mu_k = \beta_k \neq 0$, we have

$$\lambda = \lambda_k = -\frac{1}{2}\mu_k^2 = -\frac{1}{2}(\alpha_k + i\beta_k)^2 = -\frac{1}{2}[(\alpha_k^2 - \beta_k^2) + i2\alpha_k\beta_k],$$

$$\bar{\lambda} = -\frac{1}{2}(\bar{\mu}_k)^2 = \bar{\lambda}_k = -\frac{1}{2}(\alpha_k - i\beta_k)^2 = -\frac{1}{2}[(\alpha_k^2 - \beta_k^2) - i2\alpha_k\beta_k],$$

and the corresponding complex conjugate eigenvectors $c_k = d_k + ie_k$ and $\bar{c}_k = d_k - ie_k$. Setting $\gamma_k = -\frac{1}{2}(\alpha_k^2 - \beta_k^2)$, $\nu_k = -\alpha_k\beta_k$, $\lambda_k = \gamma_k + i\nu_k$ and $\bar{\lambda} = \gamma_k - i\nu_k$, we write the corresponding solution representation (4.2.2) as:

$$x_k(t) = \exp[\lambda_k t + \mu_k w(t)]c_k \quad \text{[from (4.2.2)]}$$

$$= \exp[(\gamma_k + i\nu_k)t + (\alpha_k + i\beta_k)w(t)](d_k + ie_k) \quad \text{(by substitution)}$$

$$= \exp[\gamma_k t + \alpha_k w(t)]\exp[i(\nu_k t + \beta_k w(t))](d_k + ie_k)$$

$$\text{(by the law of exponents)}$$

$$= \exp[\gamma_k t + \alpha_k w(t)](\cos(\nu_k t + \beta_k w(t))$$

$$+ i\sin(\nu_k t + \beta_k w(t)))(d_k + ie_k) \quad \text{(by Euler's formula)}$$

$$= \exp[\gamma_k t + \alpha_k w(t)][[d_k\cos(\nu_k t + \beta_k w(t))$$

$$- e_k\sin(\nu_k t + \beta_k w(t))] + i[d_k\sin(\nu_k t + \beta_k w(t))$$

$$+ e_k\cos(\nu_k t + \beta_k w(t))]] \quad \text{(product of complex numbers).} \quad (4.3.1)$$

Similarly, we can write a complex conjugate solution corresponding to the complex conjugate eigenvalue and eigenvector as

$$\bar{x}_k(t) = \exp[\bar{\lambda}_k t + \bar{\mu}_k w(t)]\bar{c}_k \quad \text{[from (4.2.2)]}$$

$$= \exp[(\gamma_k - i\nu_k)t + (\alpha_k - i\beta_k)w(t)](d_k - ie_k) \quad \text{(by substitution)}$$

$$= \exp[\gamma_k t + \alpha_k w(t)]\exp[-i(\nu_k t + \beta_k w(t))](d_k - ie_k)$$

$$\text{(by the law of exponents)}$$

$$= \exp[\gamma_k t + \alpha_k w(t)](\cos(\nu_k t + \beta_k w(t))$$

$$- i\sin(\nu_k t + \beta_k w(t)))(d_k - ie_k) \quad \text{(by Euler's formula)}$$

$$= \exp[\gamma_k t + \alpha_k w(t)][[d_k\cos(\nu_k t + \beta_k w(t))$$

$$- e_k\sin(\nu_k t + \beta_k w(t))] - i[d_k\sin(\nu_k t + \beta_k w(t))$$

$$+ e_k\cos(\nu_k t + \beta_k w(t))]] \quad \text{(product of complex numbers).} \quad (4.3.2)$$

From (4.3.1) and (4.3.2), the sum and the difference of the complex conjugate pair of numbers/functions, we have

$$\operatorname{Re}x_k(t) = \frac{1}{2}x_k(t) + \frac{1}{2}\bar{x}_k(t) \quad \text{and} \quad \operatorname{Im}x_k(t) = \frac{1}{2}x_k(t) - \frac{1}{2}\bar{x}_k(t).$$

Moreover,

$$\operatorname{Re} x_k(t) = \exp[\gamma_k t + \alpha_k w(t)][d_k \cos(\nu_k t + \beta_k w(t)) - e_k \sin(\nu_k t + \beta_k w(t))], \quad (4.3.3)$$

$$\operatorname{Im} x_k(t) = \exp[\gamma_k t + \alpha_k w(t)][d_k \sin(\nu_k t + \beta_k w(t)) + e_k \cos(\nu_k t + \beta_k w(t)). \quad (4.3.4)$$

From the application of Theorem 4.2.1 to $\operatorname{Re} x_k(t)$ and $\operatorname{Im} x_k(t)$, it follows that $\operatorname{Re} x_k(t)$ and $\operatorname{Im} x_k(t)$ are solutions to (4.2.1). Furthermore, these are real-valued functions.

Illustration 4.3.2. Given: $dx(t) = Bx\, dw(t)$, $x(t_0) = x_0$, where $B = \begin{bmatrix} b_{11} & b_{12} \\ b_{21} & b_{22} \end{bmatrix}$, and b_{ij} are any given real numbers satisfying the condition $d(B) = (\operatorname{tr}(B))^2 - 4\det(B) < 0$, where $\operatorname{tr}(B)$ and $\det(B)$ are as defined in Illustration 4.3.1. Find: (a) the general solution and (b) the solution to the given IVP.

Solution procedure. Again, we note that $n = 2$. We imitate the argument used in Illustration 4.3.1 and Procedure 4.2.4 to find the solution to the IVP. The desired computations are:

$$0 = \det(B - \mu I) = \mu^2 - \operatorname{tr}(B)\mu + \det(B) = (\mu - \mu_1)(\mu - \mu_2),$$

$$\lambda_1 = -\frac{1}{2}\mu_1^2 \quad \text{and} \quad \lambda_2 = -\frac{1}{2}\mu_2^2,$$

where (μ_1, λ_1) and (μ_2, λ_2) are as described in Illustration 4.3.1. From the assumption $d(B) = (\operatorname{tr}(B))^2 - 4\det(B) < 0$ and SDC1, we conclude that μ_1 and μ_2 are complex conjugate roots of $0 = \det(B - \mu I)$. (For instance, $\mu_2 = \bar{\mu}_1$.) Moreover, the roots are distinct. We need to find eigenvectors corresponding to these eigenvalues μ_1 and μ_2. Again, from our discussion SDC1, it is enough to find the eigenvector corresponding to one of the eigenvalues (μ_1 or μ_2). Without loss of generality, we choose $\mu_1 = \alpha + i\beta$. With this, we imitate the arguments used in Illustration 4.3.1 in the context of Case 2. From SDC1, (4.3.1) and (4.3.3), we obtain eigenvectors as well as the real and imaginary parts of the solution corresponding to the eigenvalue $\mu_1 = \alpha + i\beta = \frac{\operatorname{tr}(B)}{2} + i\frac{\sqrt{-d(B)}}{2} = \frac{\operatorname{tr}(B) + i\sqrt{(\operatorname{tr}(B))^2 - 4\det(B)}}{2}$

$$\Phi(t) = e^{(\gamma t + \alpha w(t))} \begin{bmatrix} \phi_{11}(t, w(t)) & \phi_{12}(t, w(t)) \\ \phi_{21}(t, w(t)) & \phi_{22}(t, w(t)) \end{bmatrix},$$

where

$$\phi_{11}(t, w(t)) = b_{12} \cos(\nu t + \beta w(t)),$$

$$\phi_{12}(t, w(t)) = b_{12} \sin(\nu t + \beta w(t)),$$

$$\phi_{21}(t, w(t)) = (b_{11} - \alpha) \cos(\nu t + \beta w(t)) - \beta \sin(\nu t + \beta w(t)),$$

$$\phi_{22}(t, w(t)) = (b_{11} - \alpha) \sin(\nu t + \beta w(t)) - \beta \cos(\nu t + \beta w(t)),$$

$$\alpha = \frac{\text{tr}(B)}{2}, \quad \beta = \frac{\sqrt{-d(B)}}{2}, \quad \nu = -\alpha\beta = -\frac{\text{tr}(B)\sqrt{-d(B)}}{4},$$

$$\gamma = -\frac{1}{2}(\alpha^2 - \beta^2) = -\frac{(\text{tr}(B))^2 + d(B)}{8} = -\frac{(\text{tr}(B))^2 - 2\det(B)}{4}.$$

The general solution of the illustration (4.2.11) is given by

$$x(t) = \Phi(t)a = e^{(\gamma t + \alpha w(t))} \begin{bmatrix} \phi_{11}(t, w(t)) & \phi_{12}(t, w(t)) \\ \phi_{21}(t, w(t)) & \phi_{22}(t, w(t)) \end{bmatrix} a,$$

where $a = [a_1\ a_2]^T$ is an arbitrary two-dimensional constant vector. To solve the IVP (b), we imitate the presented arguments in Illustration 4.3.1. The details are left to the reader. This completes the solution process of the given illustration.

Example 4.3.2. Given: $dx(t) = Bx\,dw(t)$, $x(t_0) = x_0$, where $B = \begin{bmatrix} 2 & 1 \\ -4 & 2 \end{bmatrix}$. Find the solution to the given IVP.

Solution procedure. In this case, $d(B) = (2+2)^2 - 4(8) = -16$. By repeating the argument of Illustration 4.3.2, we arrive at

$$0 = \det(B - \mu I) = \mu^2 - 4\mu + 8 = (\mu - 2 - 2i)(\mu - 2 + 2i),$$

$$\lambda_1 = -\frac{1}{2}\mu_1^2 = -\frac{1}{2}(2 + 2i)^2 = -4i \quad \text{and} \quad \lambda_2 = -\frac{1}{2}(2 - 2i)^2 = 4i.$$

Here, $n = 2$, and the eigenvalues of the coefficient matrix are complex conjugate numbers $\mu_1 = 2 + 2i$ and $\mu_2 = 2 - 2i$. We set $\mu_1 = 2 + 2i = \alpha + i\beta$. This implies that $\alpha = 2$, $\beta = 2$. Hence, $\gamma = -\frac{1}{2}(\alpha^2 - \beta^2) = 0$, $\nu = -\alpha\beta = -4$. An eigenvector corresponding to $\mu_1 = 2 + 2i$ is $c_1 = d + ie$, where $d = [1,0]^T$ and $e = [0,2]^T$. Now, we repeat the argument used in Illustration 4.3.2, and we obtain the following linearly independent real solutions:

$$x_1(t) = e^{2w(t)} \begin{bmatrix} \cos(-4t + 2w(t)) \\ -2\sin(-4t + 2w(t)) \end{bmatrix},$$

$$x_2(t) = e^{2w(t)} \begin{bmatrix} \sin(-4t + 2w(t)) \\ 2\cos(-4t + 2w(t)) \end{bmatrix}.$$

Hence,

$$\Phi(t) = e^{2w(t)} \begin{bmatrix} \cos(-4t + 2w(t)) & \sin(-4t + \beta w(t)) \\ -2\sin(-4t + \beta w(t)) & 2\cos(-4t + 2w(t)) \end{bmatrix}.$$

We observe that $\det(\Phi(t)) = 2e^{2w(t)} \neq 0$. Therefore, $\Phi(t)$ is invertible (Theorem 1.4.4 [80]). Its general solution is given by

$$x(t) = \Phi(t)a = e^{2w(t)} \begin{bmatrix} \cos(-4t + 2w(t)) & \sin(-4t + 2w(t)) \\ -2\sin(-4t + 2w(t)) & 2\cos(-4t + 2w(t)) \end{bmatrix} a.$$

Now, we can retrace the procedure for finding a as per Steps #2 and #3 of Procedure 4.2.4, and obtain the solution to the given IVP

$$x(t) = \Phi(t)\Phi^{-1}(t_0)x_0$$

$$= e^{2(w(t)-w(t_0))} \begin{bmatrix} \cos\theta(\Delta t, \Delta w(t)) & \frac{1}{2}\sin\theta(\Delta t, \Delta w(t)) \\ -2\sin\theta(\Delta t), \Delta w(t) & \cos\theta(\Delta t), \Delta w(t) \end{bmatrix} x_0,$$

where $\theta(\Delta t, \Delta w(t)) = -4(t - t_0) + 2(w(t) - w(t_0))$. This completes the solution process of the given illustration.

Example 4.3.3 (predator–prey model). We consider a linearized version of the classical model of Lotka–Volterra with the one-predator–one-prey system described in Example 4.1.1:

$$dN = CN\,dw(t), \quad N(0) = N_0,$$

where

$$N = [N_1\ N_2]^T, \quad C = \begin{bmatrix} 0 & -\dfrac{\alpha_1\beta_0}{\beta_1} \\ \dfrac{\beta_1\alpha_0}{\alpha_1} & 0 \end{bmatrix}.$$

Solve the IVP.

Solution procedure. We imitate the procedure of Illustration 4.3.2. First, we note that the characteristic equation associated with the matrix C in the above-described linearized stochastic predator–prey mathematical model is $\det(C - \mu I) = 0 = \mu^2 + \alpha_0\beta_0$. We have $\mu_1 = \sqrt{\alpha_0\beta_0}i$, $\mu_2 = -\sqrt{\alpha_0\beta_0}i$, $\lambda_1 = -\frac{1}{2}\mu_1^2 = \frac{1}{2}\alpha_0\beta_0 = \lambda_2$. Here the characteristic roots are distinct complex numbers. Therefore, the procedure described in Illustration 4.3.2 is applicable. Thus, the solution to the system of stochastic linear homogeneous differential equations is represented by

$$x(t) = \Phi(t)x_0 = \Phi(t)\Phi^{-1}(t_0)x_0 = e^{\gamma(t-t_0)} \begin{bmatrix} \phi_{11}\Delta w(t) & \phi_{12}\Delta w(t) \\ \phi_{21}\Delta w(t) & \phi_{22}\Delta w(t) \end{bmatrix} x_0,$$

where $\phi_{11}\Delta w(t) = \cos\beta\Delta w(t)$, $\phi_{12}\Delta w(t) = -\frac{\alpha_1}{\beta_1}\sqrt{\frac{\beta_0}{\alpha_0}}\sin\beta\Delta w(t)$,

$$\phi_{21}\Delta w(t) = \frac{\beta_1}{\alpha_1}\sqrt{\frac{\alpha_0}{\beta_0}}\sin\beta\Delta w(t), \quad \phi_{22}\Delta w(t) = \cos\beta\Delta w(t),$$

$\alpha = 0$, $\beta = \frac{\sqrt{-d(B)}}{2} = \sqrt{\alpha_0 \beta_0}$, $\nu = -\alpha\beta = 0$, $\gamma = \frac{1}{2}\beta^2$, and $\Delta w(t) = w(t) - w(t_0)$. This completes the solution procedure.

Multiple Eigenvalues 4.3.3. Let $\mu_1, \ldots, \mu_k, \ldots, \mu_n$ be the roots of (4.2.4) and, corresponding to these roots, let $\lambda_1, \ldots, \lambda_k, \ldots, \lambda_n$ be the roots of (4.2.5). It is further assumed that there are m distinct roots and $m < n$. Determine the complete set of solutions to (4.2.1).

Solution procedure. It is known that there are m distinct roots and $m < n$. This implies that corresponding to m distinct roots of (4.2.4) and (4.2.5), there are at least m linearly independent eigenvectors. Moreover, there is at least one root of the characteristic system of equations (4.2.4) of multiplicity $p > 1$. The conclusions and definitions of the deterministic system of the differential equations with multiple eigenvalues [80] are presented under some modifications. In particular, an algorithm for finding defective eigenvectors of B and solutions to (4.2.1) needs modifications. In the following, we outline this algorithm.

SME 1. We recall that there is at least one eigenvector corresponding to each eigenvalue. Moreover, there are two types of repeated or multiple eigenvalues of a matrix with real entries: (a) complete and (b) not complete. With reference to the defective pair $(\mu, \lambda) = (\mu_i, \lambda_i)$ of the eigenvalue of the multiplicity p and $p - q$ unrecovered solutions corresponding to $(\mu, \lambda) = (\mu_i, \lambda_i)$, we note that the maximum length of the longest chain is $r = 1 + (p - q)$. For more details, see Section 4.4 of Volume 1, where q stands for the number of ordinary eigenvectors relative to the pair of eigenvalues $(\mu, \lambda) = (\mu_i, \lambda_i)$.

Algorithm For finding defective eigenvectors and solutions. Let $c_i(\ell)$ be an ordinary eigenvector corresponding to the eigenvalue $\mu = \mu_i$ of the matrix B for $1 \leq \ell \leq q$. In addition, $\{c_i(\ell), c_i(\ell, k - 1), \ldots, c(\ell, 2), c_i(\ell, 1)\}$ be a length k chain of generalized eigenvectors with respect to the ordinary eigenvector $c_i(\ell)$. We seek a solution to (4.2.1) of the form

$$x(t) = \exp\left[-\frac{\mu^2}{2}t + \mu w(t)\right] P_{k-1}\left(w(t)I - \left(\frac{1}{2}(B - \mu I) + \mu I\right)t\right), \qquad (4.3.5)$$

where $\mu = \mu_i$; $P_{k-1}\left(w(t)I - \left(\frac{1}{2}(B - \mu I) + \mu I\right)t\right) \equiv P_{k-1}(w(t), t)$ and

$$P_{k-1}(w(t), t)$$

$$= \frac{[w(t)I - (\frac{1}{2}(B - \mu I) + \mu I)t]^{k-1}}{(k-1)!} c_i(\ell)$$

$$+ \cdots + \left[w(t)I - \left(\frac{1}{2}(B - \mu I) + \mu I\right)t\right]c_i(\ell, 2) + c_i(\ell, 1),$$

and $c_i(\ell, 1)$ is a nonzero solution to

$$(B - \mu I)^k c = 0, \qquad (4.3.6)$$

for $1 \leq k \leq 1 + (p - q)$; and $c_i(\ell, k-1), \ldots, c_i(\ell, 2), c_i(\ell, 1)$ are unknown vectors to be determined. We note that

$$
\begin{aligned}
&\frac{\partial}{\partial t} P_{k-1}(w(t), t) \\
&= -\frac{\left[w(t)I - \left(\frac{1}{2}(B - \mu I) + \mu I \right) t \right]^{k-2}}{(k-2)!} \left(\frac{1}{2}(B - \mu I) + \mu I \right) c_i(\ell) \\
&\quad - \cdots - \left(\frac{1}{2}(B - \mu I) + \mu I \right) c_i(\ell, 2),
\end{aligned}
$$

$$
\begin{aligned}
\frac{\partial}{\partial w} P_{k-1}(w(t), t) &= \frac{\left[w(t)I - \left(\frac{1}{2}(B - \mu I) + \mu I \right) t \right]^{k-2}}{(k-2)!} c_i(\ell) \\
&\quad + \cdots + \left[w(t)I - \left(\frac{1}{2}(B - \mu I) + \mu I \right) t \right] c_i(\ell, 3) + c_i(\ell, 2),
\end{aligned}
$$

$$
\begin{aligned}
&\frac{\partial^2}{\partial w^2} P_{k-1}(w(t), t) \\
&= \frac{\left[w(t)I - \left(\frac{1}{2}(B - \mu I) + \mu I \right) t \right]^{k-3}}{(k-3)!} c_i(\ell) \\
&\quad + \cdots + \left[w(t)I - \left(\frac{1}{2}(B - \mu I) + \mu I \right) t \right] c_i(\ell, 4) + c_i(\ell, 3). \quad (4.3.7)
\end{aligned}
$$

Moreover, from the expressions of $\frac{\partial}{\partial t} P_{k-1}$ and $\frac{\partial}{\partial w} P_{k-1}$, we have

$$
\begin{aligned}
\frac{\partial}{\partial t} P_{k-1}(w(t), t) &= -\mu \frac{\partial}{\partial w} P_{k-1}(w(t), t) \\
&\quad - \frac{1}{2} \left[\frac{\left[w(t)I - \left(\frac{1}{2}(B - \mu I) + \mu I \right) t \right]^{k-2}}{(k-2)!} (B - \mu I) c_i(\ell) \right. \\
&\quad \left. + \cdots + (B - \mu I) c_i(\ell, 2) \right]. \quad\quad (4.3.8)
\end{aligned}
$$

We compute the Itô–Doob differential on both sides of (4.3.5), and we have

$$
\begin{aligned}
dx(t) &= d \left(\exp\left[-\frac{\mu^2}{2} t + \mu w(t) \right] P_{k-1}(w(t), t) \right) \quad \text{[from (4.3.5)]} \\
&= \mu \exp\left[-\frac{\mu^2}{2} t + \mu w(t) \right] P_{k-1}(w(t), t) dw(t)
\end{aligned}
$$

$$+ \exp\left[-\frac{\mu^2}{2}t + \mu w(t)\right]\left[\frac{\partial}{\partial t}P_{k-1}(w(t), t)dt + \frac{\partial}{\partial w}P_{k-1}(w(t), t)dw(t)\right.$$

$$\left. +\frac{1}{2}\frac{\partial^2}{\partial w^2}P_{k-1}(w(t), t)dt\right]$$

$$+ \exp\left[-\frac{\mu^2}{2}t + \mu w(t)\right]\mu\frac{\partial}{\partial w}P_{k-1}(w(t), t)dt$$

(by Theorems 1.3.1 and 1.3.4)

$$= \exp\left[-\frac{\mu^2}{2}t + \mu w(t)\right]\left[\mu P_{k-1}(w(t), t) + \frac{\partial}{\partial w}P_{k-1}(w(t), t)\right]dw(t)$$

$$+ \exp\left[-\frac{\mu^2}{2}t + \mu w(t)\right]\left[\frac{\partial}{\partial t}P_{k-1}(w(t), t) + \mu\frac{\partial}{\partial w}P_{k-1}(w(t), t)\right.$$

$$\left. +\frac{1}{2}\frac{\partial^2}{\partial w^2}P_{k-1}(w(t), t)\right]dt \quad \text{(by regrouping terms)}.$$

Under the assumption that the expression (4.3.5) is a solution to (4.2.1), we have

$$dx(t) = Bx(t)dw(t)$$

$$= B\exp\left[-\frac{\mu^2}{2}t + \mu w(t)\right]P_{k-1}(w(t), t)dw(t) \quad \text{(by Theorem 4.2.1)}.$$

Hence, by comparing the two expressions of $dx(t)$, we obtain

$$B\exp\left[-\frac{\mu^2}{2}t + \mu w(t)\right]P_{k-1}(w(t), t)dw(t)$$

$$= \exp\left[-\frac{\mu^2}{2}t + \mu w(t)\right]\left[\mu P_{k-1}(w(t), t) + \frac{\partial}{\partial w}P_{k-1}(w(t), t)\right]dw(t)$$

$$+ \exp\left[-\frac{\mu^2}{2}t + \mu w(t)\right]\left[\frac{\partial}{\partial t}P_{k-1}(w(t), t) + \mu\frac{\partial}{\partial w}P_{k-1}(w(t), t)\right.$$

$$\left. +\frac{1}{2}\frac{\partial^2}{\partial w^2}P_{k-1}(w(t), t)\right]dt. \tag{4.3.9}$$

Now, we compare the coefficients of the terms $dw(t)$ and dt in (4.3.9) (an application of the method of undetermined coefficients), and we have

$$BP_{k-1}(w(t), t) = \mu P_{k-1}(w(t), t)) + \frac{\partial}{\partial w}P_{k-1}(w(t), t), \tag{4.3.10}$$

and

$$\frac{\partial}{\partial t}P_{k-1}(w(t), t) + \mu\frac{\partial}{\partial w}P_{k-1}(w(t), t) + \frac{1}{2}\frac{\partial^2}{\partial w^2}P_{k-1}(w(t), t) = 0, \tag{4.3.11}$$

respectively. Now, again, for $1 \le j \le k-1$, by recalling the notation and definition of $P_{k-1}(w(t), t)$ in (4.3.5) $[P_{k-1}(w(t)I - (\frac{1}{2}(B - \mu I) + \mu I)t) \equiv P_{k-1}(w(t), t)$ and comparing the two sides of the coefficients of the terms $[w(t)I - (\frac{1}{2}(B - \mu I) + \mu I)t)]^j$ in (4.3.10), we obtain the following algorithm

$$Bc_i(\ell, 1) = \mu c_i(\ell, 1) + c_i(\ell, 2),$$
$$Bc_i(\ell, 2) = \mu c_i(\ell, 2) + c_i(\ell, 3),$$

$$\cdots \cdots \qquad \cdots \cdots$$

$$Bc_i(\ell, k - 1) = \mu c_i(\ell, k - 1) + c_i(\ell),$$
$$Bc_i(\ell) = \mu c_i(\ell), \qquad (4.3.12)$$

which implies that

$$(B - \mu I)c_i(\ell, 1) = c_i(\ell, 2),$$
$$(B - \mu I)c_i(\ell, 2) = c_i(\ell, 3),$$

$$\cdots \cdots \qquad \cdots \cdots$$

$$(B - \mu I)c_i(\ell, k - 1) = c_i(\ell),$$
$$(B - \mu I)c_i(\ell) = 0. \qquad (4.3.13)$$

We need to verify that the coefficients of terms of the $P_{k-1}(w(t), t)$, which satisfy (4.3.13) must also satisfy (4.3.11). In fact, (4.3.11) is the *consistency condition* for the validity of the algorithm in (4.3.13). For this purpose, we consider the left-hand expression in (4.3.11) and show that it is equal to zero in the context of (4.3.13). We start with

$$\frac{\partial}{\partial t} P_{k-1}(w(t), t) + \mu \frac{\partial}{\partial w} P_{k-1}(w(t), t) + \frac{1}{2} \frac{\partial^2}{\partial w^2} P_{k-1}(w(t), t)$$

$$= \mu \frac{\partial}{\partial w} P_{k-1}(w(t), t) + \frac{1}{2} \frac{\partial^2}{\partial w^2} P_{k-1}(w(t), t) - \mu \frac{\partial}{\partial w} P_{k-1}(w(t), t)$$

$$- \frac{1}{2} \left[\frac{[w(t) - (\frac{1}{2}(B - \mu I) + \mu)t]^{k-2}}{(k - 2)!} (B - \mu I)c_i(\ell) \right.$$

$$\left. + \cdots + (B - \mu I)c_i(\ell, 2) \right] \quad \text{[from (4.3.8)]}$$

$$= \frac{1}{2} \frac{\partial^2}{\partial w^2} P_{k-1} \left(w(t)I - \left(\frac{1}{2}(B - \mu I) + \mu I \right) t \right)$$

$$- \frac{1}{2} \left[\frac{[w(t) - (\frac{1}{2}(B - \mu I) + \mu)t]^{k-2}}{(k - 2)!} (B - \mu I)c_i(\ell) \right.$$

$$\left. + \cdots + (B - \mu I)c_i(\ell, 2) \right] \quad \text{(by simplification)}$$

$$= \frac{1}{2} \left[\frac{[w(t) - (\frac{1}{2}(B - \mu I) + \mu)t]^{k-3}}{(k-3)!} c_i(\ell) \right.$$

$$+ \cdots + \left[w(t)I - \left(\frac{1}{2}(B - \mu I) + \mu I \right) t \right] c_i(\ell, 4) + c_i(\ell, 3)$$

$$- \frac{[w(t) - (\frac{1}{2}(B - \mu I) + \mu)t]^{k-2}}{(k-2)!}(B - \mu I)c_i(\ell)$$

$$\left. - \cdots - (B - \mu I)c_i(\ell, 2) \right] \quad \text{[from (4.3.7)]}$$

$$= \frac{1}{2} \left[\frac{[w(t) - (\frac{1}{2}(B - \mu I) + \mu I)t]^{k-3}}{(k-3)!} c_i(\ell) \right.$$

$$+ \cdots + \left[w(t)I - \left(\frac{1}{2}(B - \mu I)t + \mu I \right) t \right] c_i(\ell, 4) + c_i(\ell, 3)$$

$$- \frac{\left[w(t)I - (\frac{1}{2}(B - \mu I) + \mu I) t \right]^{k-3}}{(k-3)!}(B - \mu I)c_i(\ell, k-1)$$

$$- \cdots - \left[w(t)I - \left(\frac{1}{2}(B - \mu I)t + \mu I \right) t \right] (B - \mu I)c_i(\ell, 3)$$

$$\left. - (B - \mu I)c_i(\ell, 2) \right] \quad \text{[from (4.3.13)]}$$

$$= \frac{1}{2} \left[\frac{[w(t) - (\frac{1}{2}(B - \mu I) + \mu I)]^{k-3}}{(k-3)!}[c_i(\ell) - (B - \mu I)c_i(\ell, k-1)] \right.$$

$$+ \cdots + \left[w(t)I - \left(\frac{1}{2}(B - \mu I)t + \mu I \right) t \right] [c_i(\ell, 4) - (B - \mu I)c_i(\ell, 3)]$$

$$\left. + [c_i(\ell, 3) - (B - \mu I)c_i(\ell, 2] \right] \quad \text{(by rearranging)}$$

$$= 0 \quad \text{[by using (4.3.13)]}.$$

This shows that the algorithm (4.3.13) satisfies the consistency condition (4.3.11). Thus, (4.3.13) provides the computational algorithm for finding the unknown coefficient vectors $c_i(\ell, k-1), \ldots, c_i(\ell, 2), c_i(\ell, 1)$ of $x(t)$ defined in (4.3.5). By following the argument used in the procedure for deterministic systems of differential equations (Section 4.4 of Volume 1), we conclude that $c_i(\ell, k-1), \ldots, c_i(\ell, 1)$ are generalized eigenvectors associated with $\mu = \mu_i$. In particular, for $q = 1 = \ell, r = p = k$, the set $\{c_i, c_i(\ell, k-1), \ldots, c_i(\ell, 2), c_i(\ell, 1)\}$ is referred to as *a length k chain of generalized eigenvectors based on the eigenvector c_i*. The corresponding k linearly independent solutions are

$$x_i^1 = \exp\left[-\frac{\mu_i^2}{2} + \mu_i w(t)\right] c_i,$$

$$x_i^2 = \exp\left[-\frac{\mu_i^2}{2} + \mu_i w(t)\right] P_1(w(t), t),$$

$$x_i^3 = \exp\left[-\frac{\mu_i^2}{2} + \mu_i w(t)\right] P_2(w(t), t),$$

$$\cdots \quad \cdots \quad \cdots \quad \cdots$$

$$x_i^k = \exp\left[-\frac{\mu_i^2}{2} + \mu_i w(t)\right] P_{k-1}(w(t), t), \tag{4.3.14}$$

where

$$P_{j-1}(w(t), t)$$

$$= \frac{[w(t)I - (\frac{1}{2}(B - \mu I) + \mu I)t]^{j-1}}{(j-1)!} c_i(\ell)$$

$$+ \cdots + \left[w(t)I - \left(\frac{1}{2}(B - \mu I) + \mu I\right)t\right] c_i(\ell, 2) + c_i(\ell, 1),$$

for $j = 1, 2, \ldots, k-1$, and the coefficients are determined by (4.3.13).

Example 4.3.4. Given: $dx = Bxw(t)$, $x(t_0) = x_0$, where $B = \begin{bmatrix} 0 & 0 \\ 0 & 0 \end{bmatrix}$. Solve the IVP.

Solution procedure. We follow the reasoning and the procedure described in Illustration 4.3.1. $\mu_1 = 0$ is the eigenvalue of multiplicity 2. By solving $(B - \mu_1 I)c = 0$, the eigenvector and the solution corresponding to the eigenvalue $\mu_1 = 0$ $(\lambda_1 = 0)$ are given by

$$0c_{11} + 0c_{12} = 0,$$

$$0c_{11} + 0c_{12} = 0.$$

In this case, the solution space is $R^2 = \text{Span}(\{[1\ 0]^T, [0\ 1]^T\})$, and hence the vectors $[1\ 0]^T, [0\ 1]^T$ form a complete set of linearly independent eigenvectors of a 2×2 zero matrix. Moreover, the complete set of linearly independent solutions of the example is

$$x_1(t) = e^{0t} \begin{bmatrix} 1 \\ 0 \end{bmatrix} = \begin{bmatrix} 1 \\ 0 \end{bmatrix} \quad \text{and} \quad x_2(t) = e^{0t} \begin{bmatrix} 0 \\ 1 \end{bmatrix} = \begin{bmatrix} 0 \\ 1 \end{bmatrix}.$$

Its fundamental matrix solution and the general solutions are

$$\Phi(t) = \begin{bmatrix} 1 & 0 \\ 0 & 1 \end{bmatrix} \quad \text{and} \quad x(t) = \Phi(t)a,$$

respectively. Thus by employing the arguments and procedure used in Example 4.2.6, the solution to the given IVP is

$$x(t, t_0, x_0) = \begin{bmatrix} 1 & 0 \\ 0 & 1 \end{bmatrix} \begin{bmatrix} x_{10} \\ x_{20} \end{bmatrix} = \begin{bmatrix} x_{10} \\ x_{20} \end{bmatrix}.$$

This completes the solution process of the given example.

Illustration 4.3.3. Given: $dx(t) = Bx\,dw(t)$, $x(t_0) = x_0$, where $B = \begin{bmatrix} b_{11} & b_{12} \\ b_{21} & b_{22} \end{bmatrix}$ and b_{ij} are any given real numbers satisfying the condition $d(B) = (\text{tr}(B))^2 - 4\det(B) = 0$, where $\text{tr}(B)$ and $\det(B)$ are as defined in Illustration 4.3.1. Find: (a) a general solution and (b) the solution to the given IVP.

Solution procedure. Again, we note that $n = 2$. We imitate the arguments used in Illustration 4.3.1 and Procedure 4.2.4 to find the solution to the IVP. We have the following computational results:

$$0 = \det(B - \mu I) = \mu^2 - \text{tr}(B)\mu + \det(B) = (\mu - \mu_1)(\mu - \mu_2),$$

and

$$\lambda_1 = -\frac{1}{2}\mu_1^2, \quad \lambda_2 = -\frac{1}{2}\mu_2^2,$$

where (μ_1, λ_1) and (μ_2, λ_2) are as described in Illustration 4.3.1. μ_1 and μ_2 are roots of the quadratic polynomial equation $\mu^2 - \text{tr}(B)\mu + \det(B) = 0$ with real coefficients. From the assumption $d(B) = (\text{tr}(B))^2 - 4\det(B) = 0$, we conclude that μ_1 and μ_2 are real and equal roots of $0 = \det(B - \mu I)$. In fact, its repeated root is $\mu_1 = \frac{\text{tr}(B)}{2} = \pm\sqrt{\det(B)}$. Thus, the quadratic polynomial equation has a root of multiplicity 2. By repeating the argument used in Illustration 4.3.1 and Procedure 4.2.2, we find an eigenvector and solution corresponding to this eigenvalue μ_1. Let c_1 be an eigenvector corresponding to $\mu_1 = \pm\sqrt{\det(B)}$, and let the corresponding solution to the given system be $x_1(t) = e^{(-\frac{(\text{tr}(B))^2}{4}t + \frac{\text{tr}(B)}{2}w(t))}c_1$.

To find the complete set of eigenvectors, we need to follow the procedure that is presented with regard to the multiplicity of eigenvalues of the coefficient matrix B. First, we recall the algorithm SME1 (4.3.5), (4.3.13), and (4.3.14) for finding the complete set of solutions concerning the multiplicity of eigenvalues. In this illustration, $\mu = \frac{\text{tr}(B)}{2} = \pm\sqrt{\det(B)}$ is of multiplicity 2, $n = 2$, $p = 2$, $m = 1$. Therefore, $n - m = 1$, and the remaining linearly independent eigenvector needs to be determined. $p - q = 2 - 1 = 1$ is the deficiency.

To find the remaining linearly independent solution, we need to imitate the algorithm for finding the defective eigenvector and the corresponding solution.

We seek the solution:

$$x(t) = e^{(-\frac{1}{2}\mu_1 t + \mu_1 w(t))} \left(\left[w(t)I - \left(\frac{1}{2}(B - \mu_1 I) + \mu_1 I \right) t \right] c_1 + c_1(1,1) \right),$$

where $c_1(1,1)$ is a generalized eigenvector corresponding to the ordinary eigenvector c_1. By following the algorithm SME1, we obtain

$$(B - \mu_1 I)c_1(1,1)) = c_1,$$

$$(B - \mu_1 I)c_1 = 0.$$

Knowing c_1, we can solve the first system of linear nonhomogeneous algebraic equations. Its solution is a generalized eigenvector $c_1(1,1)$ of the coefficient matrix B. Thus, the corresponding solution is

$$x_2(t) = e^{(-\frac{1}{2}\mu_1 t + \mu_1 w(t))} \left(\left[w(t)I - \left(\frac{1}{2}(B - \mu_1 I) + \mu_1 I \right) t \right] c_1 + c_1(1,1) \right).$$

In this case, the general solution to the given system is $x(t) = \Phi(t)a$, where

$$\Phi(t) = e^{(-\frac{1}{2}\mu_1 t + \mu_1 w(t))} \left[c_1 \left[w(t)I - \left(\frac{1}{2}(B - \mu_1 I) + \mu_1 I \right) t \right] c_1 + c_1(1,1) \right].$$

Finally, the solution to the IVP can be computed by following the arguments used in Illustrations 4.3.1 and 4.3.2. This completes the solution process of the given illustration.

Example 4.3.5. Given: $dx = Bxw(t)$, $x(t_0) = x_0$, where $B = \begin{bmatrix} 0 & 1 \\ 0 & 0 \end{bmatrix}$. Solve the IVP.

Solution procedure. We follow the argument and the procedure described in Illustration 4.3.1. From Example 4.2.3, we already have the eigenvalues of the given matrix and a corresponding solution. In this example, $\mu_1 = 0$ and

$$x_1(t) = e^{0t} \begin{bmatrix} 1 \\ 0 \end{bmatrix} = \begin{bmatrix} 1 \\ 0 \end{bmatrix}.$$

Again, recalling Example 4.2.3 and Illustration 4.3.3, $n = 2$, $p = 2$, $m = 1$. Therefore, $n - m = 1$, and the remaining linearly independent eigenvector needs to be determined. $p - q = 2 - 1 = 1$ is the deficiency. To find the remaining linearly independent solution, we need to imitate the algorithm for finding the defective eigenvector corresponding to $\mu_1 = 0$ with eigenvector $c_1 = [1 \ 0]^T$. We need to find $c_1(1,1)$ in the following system of algebraic equations:

$$(B - \mu_1 I)c_1(1,1)) = c_1,$$

$$(B - \mu_1 I)c_1 = 0.$$

We solve this system of equations, and we have the generalized eigenvector

$$c_1(1,1) = \begin{bmatrix} 0 \\ 1 \end{bmatrix}.$$

Thus, from Illustration 4.3.3, the second solution is given by

$$x_2(t) = e^{(-\frac{1}{2}\mu_1 t + \mu_1 w(t))} \left(\left[w(t)I - \left(\frac{1}{2}(B - \mu_1 I) + \mu_1 I \right) t \right] c_1 + c_1(1,1) \right)$$

$$= e^{0t} \left(\left[\begin{matrix} w(t) & 0 \\ 0 & w(t) \end{matrix} \right] - \begin{bmatrix} 0 & \frac{1}{2} \\ 0 & 0 \end{bmatrix} t \right) c_1 + c_1(1,1)$$

$$= \begin{bmatrix} w(t) & -\frac{1}{2}t \\ 0 & w(t) \end{bmatrix} c_1 + c_1(1,1)$$

$$= \begin{bmatrix} w(t) & -\frac{1}{2}t \\ 0 & w(t) \end{bmatrix} \begin{bmatrix} 1 \\ 0 \end{bmatrix} + \begin{bmatrix} 0 \\ 1 \end{bmatrix}$$

$$= \begin{bmatrix} w(t) \\ 0 \end{bmatrix} + \begin{bmatrix} 0 \\ 1 \end{bmatrix} = \begin{bmatrix} w(t) \\ 1 \end{bmatrix}.$$

The general solution to the given system is $x(t) = \Phi(t)a$, where

$$x(t) = \Phi(t)a = \begin{bmatrix} 1 & w(t) \\ 0 & 1 \end{bmatrix} a.$$

By employing the arguments and the procedure used in Example 4.2.6, the solution to the given IVP is

$$x(t, t_0, x_0) = \begin{bmatrix} 1 & w(t) - w(t_0) \\ 0 & 1 \end{bmatrix} \begin{bmatrix} x_{10} \\ x_{20} \end{bmatrix} = \begin{bmatrix} x_{10} + (w(t) - w(t_0))x_{20} \\ x_{20} \end{bmatrix}.$$

This completes the solution process of the given example.

Example 4.3.6. Given: $dx = Bxw(t)$, $x(t_0) = x_0$, where $B = \begin{bmatrix} 1 & 1 & 0 & 0 \\ 0 & 1 & 1 & 0 \\ 0 & 0 & 1 & 0 \\ 0 & 6 & 0 & 1 \end{bmatrix}.$

Solve the IVP.

Solution procedure. Here, $n = 4$. We follow the argument and the procedure described in Illustrations 4.3.1 and 4.3.3. Again, the characteristic equation (4.2.4) of the matrix B in the context of the given example is

$$0 = \det(B - \mu I) = (1 - \mu)^4 = (\mu - \mu_1)(\mu - \mu_2)(\mu - \mu_3)(\mu - \mu_4),$$

with $\mu_1 = \mu_2 = \mu_3 = \mu_4 = 1$ and

$$\lambda_1 = \lambda_2 = \lambda_3 = \lambda_4 = -\frac{1}{2}\mu_1^2 = -\frac{1}{2}\mu_2^2 = -\frac{1}{2}\mu_3^2 = -\frac{1}{2}\mu_4^2 = -\frac{1}{2}.$$

By solving $(B - \mu_1 I)c = 0$, the eigenvector and the solution corresponding to the eigenvalue $\mu_1 = 1$ are

$$0c_1 + c_2 + 0c_3 + 0c_4 = 0,$$
$$0c_1 + 0c_2 + c_3 + 0c_4 = 0,$$
$$0c_1 + 0c_2 + 0c_3 + 0c_4 = 0,$$
$$0c_1 + 6c_2 + 0c_3 + 0c_4 = 0.$$

By inspection, this implies that $c_2 = 0 = c_3$. Therefore, the solution space is

$$S = \{c \in R^4 : c = [c_1, 0, 0, c_4]^T\} = \text{Span}([1, 0, 0, 0]^T, [0, 0, 0, 1]^T).$$

We note that $\mu_1 = 1$ is the root of multiplicity 4, i.e. $p = 4$. $p - q = 2$ is the deficiency in the number of the eigenvector. To find the remaining linearly independent eigenvectors, we pick a very general ordinary eigenvector with respect to $\mu_1 = 1$. For example, we choose an arbitrary nonzero vector $c_1 = [k_1, 0, 0, k_2]^T$ in $\text{Span}([1, 0, 0, 0]^T, [0, 0, 0, 1]^T) = S$. Let us check the rank or order of this ordinary eigenvector instead of the algorithm SME (4.3.5). For this purpose, from (4.3.13), let us consider $c(1, 2) = [u_1, u_2, u_3, u_4]^T$ as a generalized eigenvector of rank 2. Then,

$$(B - \mu_1 I)c(1,2) = \begin{bmatrix} 0 & 1 & 0 & 0 \\ 0 & 0 & 1 & 0 \\ 0 & 0 & 0 & 0 \\ 0 & 6 & 0 & 0 \end{bmatrix} c(1,2) = \begin{bmatrix} u_2 \\ u_3 \\ 0 \\ 6u_2 \end{bmatrix} = \begin{bmatrix} k_1 \\ 0 \\ 0 \\ k_2 \end{bmatrix},$$

$c(1, 2)$ is of order 2, and it is a chain if and only if $u_2 = k_1$, $u_3 = 0$, $6u_2 = 6k_1 = k_2$, and $0u_1 + 0u_2 + 0u_3 + 0u_4 = 0$. Therefore, u_1 and u_4 are arbitrary variables (free variables). Here, we choose $u_1 = k_3$ and $u_4 = k_4$. Thus, for every choice of k_1, k_3, and k_4, we have a chain of length 2 as follows: $c(1, 2) = [k_3, k_1, 0, k_4]^T$ and $c_1 = [k_1, 0, 0, 6k_1]^T$. We look for a generalized eigenvector of order 3. We repeat the discussion with respect to $c(1, 2)$ by considering $c(1, 1) = [v_1, v_2, v_3, v_4]^T$, and have

$$(B - \mu_1 I)c(1, 1) = \begin{bmatrix} 0 & 1 & 0 & 0 \\ 0 & 0 & 1 & 0 \\ 0 & 0 & 0 & 0 \\ 0 & 6 & 0 & 0 \end{bmatrix} c(1, 1) = \begin{bmatrix} v_2 \\ v_3 \\ 0 \\ 6v_2 \end{bmatrix} = \begin{bmatrix} k_3 \\ k_1 \\ 0 \\ k_4 \end{bmatrix}.$$

This yields $v_2 = k_3$, $v_3 = k_1$, $6v_2 = 6k_3 = k_4$, and $0v_1 + 0v_2 + 0v_3 + 0v_4 = 0$. Therefore, v_1 and v_4 are arbitrary variables (free variables). Here, we choose $v_1 = k_5$ and $v_4 = k_6$. Thus, for every choice of k_1, k_3, k_5, and k_6, we have a chain of length 3 as follows: $c(1, 1) = [k_5, k_3, k_1, k_6]^T$, $c(1, 2) = [k_3, k_1, 0, 6k_3]^T$, and $c_1 = [k_1, 0, 0, 6k_1]^T$. For the choice of $k_1 = k_3 = k_5 = k_6 = 1$. We further note that

$$P_1(t, w(t)) = \left(\left[w(t)I - \left(\frac{1}{2}(B - I) + I \right) t \right] c(1, 2) + c(1, 1) \right)$$

$$= \left(\left[w(t)I - \left(\frac{1}{2}(B + I) \right) t \right] c(1, 2) + c(1, 1) \right)$$

$$= \begin{bmatrix} w(t) - t & -\frac{1}{2}t & 0 & 0 \\ 0 & w(t) - t & -\frac{1}{2}t & 0 \\ 0 & 0 & w(t) - t & 0 \\ 0 & -3t & 0 & w(t) - t \end{bmatrix} \begin{bmatrix} 1 \\ 1 \\ 0 \\ 6 \end{bmatrix} + \begin{bmatrix} 1 \\ 1 \\ 1 \\ 1 \end{bmatrix}$$

$$= \begin{bmatrix} w(t) - \frac{3}{2}t \\ w(t) - t \\ 0 \\ 6w(t) - 9t \end{bmatrix} + \begin{bmatrix} 1 \\ 1 \\ 1 \\ 1 \end{bmatrix} = \begin{bmatrix} 1 + w(t) - \frac{3}{2}t \\ 1 + w(t) - t \\ 1 \\ 1 + 6w(t) - 9t \end{bmatrix} = \begin{bmatrix} (w(t) - t) - \frac{1}{2}t + 1 \\ (w(t) - t) + 1 \\ 1 \\ 6(w(t) - t) - 3t + 1 \end{bmatrix},$$

$$P_2(t, w(t)) = \left(\left[w(t)I - \left(\frac{1}{2}(B - I) + I \right) t \right]^2 c_1 \right.$$

$$+ \left. \left[w(t)I - \left(\frac{1}{2}(B - I) + I \right) t \right] c(1, 2) + c(1, 1) \right)$$

$$= \left[w(t)I - \left(\frac{1}{2}(B - I) + I \right) t \right]^2 c_1 + \begin{bmatrix} 1 + w(t) - \frac{3}{2}t \\ 1 + w(t) - t \\ 1 \\ 1 + 6w(t) - 9t \end{bmatrix}$$

$$
= \begin{bmatrix} (w(t)-t)^2 & -t(w(t)-t) & \frac{1}{4}t^2 & 0 \\ 0 & (w(t)-t)^2 & -t(w(t)-t) & 0 \\ 0 & 0 & (w(t)-t)^2 & 0 \\ 0 & -6t(w(t)-t) & \frac{3}{2}t^2 & 6(w(t)-t)^2 \end{bmatrix} \begin{bmatrix} 1 \\ 0 \\ 0 \\ 0 \\ 6 \end{bmatrix}
$$

$$
+ \begin{bmatrix} 1+w(t)-\frac{3}{2}t \\ 1+w(t)-t \\ 1 \\ 1+6w(t)-9t \end{bmatrix}
$$

$$
= \begin{bmatrix} (w(t)-t)^2 \\ 0 \\ 0 \\ 6(w(t)-t)^2 \end{bmatrix} + \begin{bmatrix} 1+w(t)-t-\frac{3}{2}t \\ 1+w(t)-t \\ 1 \\ 1+6w(t)-9t \end{bmatrix} = \begin{bmatrix} (w(t)-t)^2+1+w(t)-\frac{3}{2}t \\ 1+w(t)-t \\ 1 \\ 6(w(t)-t)^2+1+6w(t)-9t \end{bmatrix}
$$

$$
= \begin{bmatrix} (w(t)-t)^2+(w(t)-t)-\frac{1}{2}t+1 \\ (w(t)-t)+1 \\ 1 \\ 6(w(t)-t)^2+6(w(t)-t)-3t+1 \end{bmatrix}.
$$

By using (4.3.14), the linearly independent solutions are

$$
x_1(t) = e^{(-\frac{1}{2}t+w(t))} \begin{bmatrix} 1 \\ 0 \\ 0 \\ 0 \end{bmatrix}, \quad x_3(t) = e^{(-\frac{1}{2}t+w(t))} \begin{bmatrix} w(t)-t-\frac{1}{2}t+1 \\ (w(t)-t)+1 \\ 1 \\ 6(w(t)-t)-3t+1 \end{bmatrix},
$$

$$
x_2(t) = e^{(-\frac{1}{2}t+w(t))} \begin{bmatrix} 1 \\ 0 \\ 0 \\ 6 \end{bmatrix}, \quad x_4(t) = e^{(-\frac{1}{2}t+w(t))} \begin{bmatrix} (w(t)-t)^2+(w(t)-t)-\frac{1}{2}t+1 \\ (w(t)-t)+1 \\ 1 \\ 6(w(t)-t)^2+6(w(t)-t)-3t+1 \end{bmatrix}.
$$

$\Phi(t) = [x_1(t), x_2(t), x_3(t), x_4(t)]$, $\det(\Phi(t)) = e^{(-\frac{1}{2}t + w(t))}$ $(w(t) - t)^2 \neq 0$. $\Phi(t)$ is invertible. The general solution to the given problem is provided by $x(t) = \Phi(t)a$. The rest of the problem can be completed by following Procedure 4.2.4. The details are left to the reader.

Observation 4.3.1. This observation is a direct extension of the deterministic Observation 4.4.1 (Chapter 4 of Volume 1) to stochastic systems of Itô–Doob differential equations. From Theorem 1.5.8(e) (Chapter 1 of Volume 1), we note that $\Phi(t) \equiv \Phi(t, w(t)) = \exp\left[-\frac{B^2}{2}t + Bw(t)\right]$ is a matrix solution process of (4.2.1). In this case, for any n-dimensional random vector c in R^n which is independent of Wiener process $w(t)$, $x(t) = \Phi(t)c = \exp\left[-\frac{B^2}{2}t + Bw(t)\right]c$ is a solution to (4.2.1). In fact, $x(t) = \exp\left[-\frac{B^2}{2}t + Bw(t)\right]c$ can be calculated explicitly in terms of the matrix B. By noting that $-\frac{1}{2}B^2 = -\left[\frac{1}{2}(B - \mu I)^2 + \mu(B - \mu I) + \frac{1}{2}\mu^2 I\right]$, for any type of eigenvalue μ with any type of eigenvectors associated with eigenvalue μ, we have

$$x(t) = \exp\left[-\frac{1}{2}B^2 t + Bw(t)\right]c$$

$$= \exp\left[-\left(\frac{1}{2}(B - \mu I)^2 + \mu(B - \mu I) + \frac{1}{2}\mu^2 I\right)t + \mu I w(t) + (B - \mu I)w(t)\right]c$$

$$= \exp\left[-\frac{1}{2}\mu^2 I t + \mu I w(t)\right]$$

$$\times \exp\left[-\left(\frac{1}{2}(B - \mu I)^2 + \mu(B - \mu I)\right)t + (B - \mu I)w(t)\right]c$$

$$= \exp\left[-\frac{1}{2}\mu^2 t + \mu w(t)\right] I \exp\left[\left(w(t)I - \left(\frac{1}{2}(B - \mu I) + \mu I\right)t\right)(B - \mu I)\right]c$$

$$= \exp\left[-\frac{\mu^2}{2}t + \mu w(t)\right] \exp\left[\left(w(t)I - \left(\frac{1}{2}(B - \mu I) + \mu I\right)t\right)(B - \mu I)\right]c$$

$$= \exp\left[-\frac{\mu^2}{2}t + \mu w(t)\right]\left[I + \left[\left(w(t)I - \left(\frac{1}{2}(B - \mu I) + \mu I\right)t\right)(B - \mu I)\right]\right.$$

$$+ \frac{[(w(t)I - (\frac{1}{2}(B - \mu I) + \mu I)t)(B - \mu I)]^2}{2!}$$

$$+ \cdots + \frac{[(w(t)I - (\frac{1}{2}(B - \mu I) + \mu I)t)(B - \mu I)]^r}{r!}$$

$$\left. + \cdots + \frac{[(w(t)I - (\frac{1}{2}(B - \mu I) + \mu I)t)(B - \mu I)]^j}{j!} + \cdots\right]c$$

$$= \exp\left[-\frac{\mu^2}{2}t + \mu w(t)\right]\left[I + \left[\left(w(t)I - \left(\frac{1}{2}(B - \mu I) + \mu I\right)t\right)(B - \mu I)\right]\right.$$

$$+ \frac{[(w(t)I - (\frac{1}{2}(B - \mu I) + \mu I)t)(B - \mu I)]^2}{2!}$$

$$+ \cdots + \frac{[(w(t)I - (\frac{1}{2}(B - \mu I) + \mu I)t)(B - \mu I)]^j}{j!}$$

$$+ \cdots + \left.\frac{[(w(t)I - (\frac{1}{2}(B - \mu I) + \mu I)t)(B - \mu I)]^{r-1}}{r - 1!} + \cdots\right]c$$

$$[\text{from } (B - \mu I)^r = 0]$$

$$= \exp\left[-\frac{\mu^2}{2}t + \mu w(t)\right]\left[Ic + \left[\left(w(t)I - \left(\frac{1}{2}(B - \mu I) + \mu I\right)t\right)(B - \mu I)\right]c\right.$$

$$+ \frac{[(w(t)I - (\frac{1}{2}(B - \mu I) + \mu I)t)(B - \mu I)]^2}{2!}c$$

$$+ \cdots + \frac{[(w(t)I - (\frac{1}{2}(B - \mu I) + \mu I)t)(B - \mu I)]^j}{j!}c$$

$$+ \cdots + \left.\frac{[(w(t)I - (\frac{1}{2}(B - \mu I) + \mu I)t)(B - \mu I)]^{r-1}}{r - 1!}c\right]$$

$$= \exp\left[-\frac{\mu^2}{2}t + \mu w(t)\right]\left[Ic + \left[w(t)I - \left(\frac{1}{2}(B - \mu I) + \mu I\right)t\right](B - \mu I)c\right.$$

$$+ \frac{[(w(t)I - (\frac{1}{2}(B - \mu I) + \mu I)t)(B - \mu I)]^2}{2!}c$$

$$+ \cdots + \frac{[(w(t)I - (\frac{1}{2}(B - \mu I) + \mu I)t)(B - \mu I)]^j}{j!}c$$

$$+ \cdots + \left.\frac{[(w(t)I - (\frac{1}{2}(B - \mu I) + \mu I)t)(B - \mu I)]^{r-1}}{(r - 1)!}c\right]$$

$$= \exp\left[-\frac{\mu^2}{2}t + \mu w(t)\right]\left[Ic + \left[w(t)I - \left(\frac{1}{2}(B - \mu I) + \mu I\right)t\right](B - \mu I)c\right.$$

$$+ \frac{[w(t)I - (\frac{1}{2}(B - \mu I) + \mu I)t]^2}{2!}(B - \mu I)^2 c$$

$$+ \cdots + \frac{[w(t)I - (\frac{1}{2}(B - \mu I) + \mu I)t]^j}{j!}(B - \mu I)^j c$$

$$+ \cdots + \left.\frac{[(w(t)I - (\frac{1}{2}(B - \mu I) + \mu I)t]^{r-1}}{r - 1!}(B - \mu I)^{r-1}c\right] \quad \text{(by rewriting)}$$

$$= \exp\left[-\frac{\mu^2}{2}t + \mu w(t)\right]\left[c_1 + \left[w(t)I - \left(\frac{1}{2}(B - \mu I) + \mu I\right)t\right]c_2\right.$$

$$+ \frac{[(w(t)I - (\frac{1}{2}(B - \mu I) + \mu I)t)]^2}{2!} c_3$$

$$+ \cdots + \frac{[(w(t)I - (\frac{1}{2}(B - \mu I) + \mu I)t)]^j}{j!} c_{j+1}$$

$$+ \cdots + \frac{[(w(t)I - (\frac{1}{2}(B - \mu I) + \mu I)t)]^{r-1}}{r-1!} c_r \Bigg].$$

In view of Cayley–Hamilton theorem (Observation 4.3.1 [80]) in matrix algebra, and by setting the following:

$$c = c_1,$$

$$(B - \mu I)c_1 = c_2,$$

$$(B - \mu I)^2 c = (B - \mu I)c_2 = c_3,$$

$$\dots \dots \quad \dots$$

$$(B - \mu I)^{j-1} c = (B - \mu I)c_{j-1} = c_j,$$

$$\dots \dots \quad \dots$$

$$(B - \mu I)^{r-1} c = (B - \mu I)c_{r-1} = c_r,$$

$$(B - \mu I)^r c = (B - \mu I)c_r = 0.$$

The above calculation is valid for every eigenvalue of B with or without its multiplicity. Hence, it determines all n linearly independent solutions to (4.2.1). This shows that $\Phi(t) = \Phi(t, w(t)) = \exp\left[-\frac{B^2}{2}t + Bw(t)\right]$ is indeed a fundamental solution to (4.2.1). We further note that the calculation provides another algebraic (linear algebra) method of finding a fundamental matrix solution (4.2.1). In addition, it gives a basis for seeking a solution form as described in (4.3.5). The following flowchart describes the procedure for finding a fundamental matrix solution:

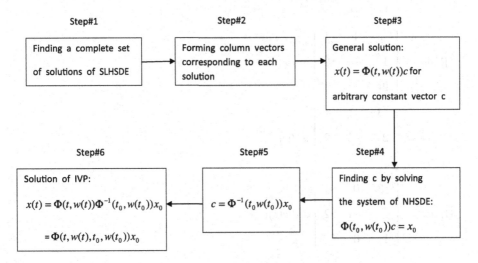

Fig. 4.3.1   Process of solving the IVP.

Exercises

1. Find the general solution to each system of differential equations:

(a) $dx = \begin{bmatrix} -1 & 1 \\ 4 & -1 \end{bmatrix} x \, dw(t);$

(b) $dx = \begin{bmatrix} -1 & 1 \\ 1 & -1 \end{bmatrix} x \, dw(t);$

(c) $dx = \begin{bmatrix} 1 & -1 \\ -1 & 1 \end{bmatrix} x \, dw(t);$

(d) $dx = \begin{bmatrix} 1 & -2 \\ -2 & 1 \end{bmatrix} x \, dw(t);$

(e) $dx = \begin{bmatrix} 1 & -1 \\ 1 & -1 \end{bmatrix} x \, dw(t);$

(f) $dx = \begin{bmatrix} -1 & -1 \\ 1 & -1 \end{bmatrix} x \, dw(t);$

(g) $dx = \begin{bmatrix} -1 & 0 \\ 1 & -1 \end{bmatrix} x \, dw(t);$

(h) $dx = \begin{bmatrix} 1 & -1 \\ 1 & 1 \end{bmatrix} x \, dw(t);$

(i) $dx = \begin{bmatrix} -1 & -1 \\ 1 & -1 \end{bmatrix} x \, dw(t);$

(j) $dx = \begin{bmatrix} -3 & 1 & 1 \\ 1 & -4 & 2 \\ 2 & 2 & -5 \end{bmatrix} x \, dw(t);$

(k) $dx = \begin{bmatrix} -3 & -1 & 1 \\ 2 & -4 & 2 \\ 1 & -2 & -5 \end{bmatrix} x \, dw(t);$

(l) $dx = \begin{bmatrix} -3 & 0 & 1 \\ 2 & -4 & 2 \\ 1 & 0 & -5 \end{bmatrix} x \, dw(t);$

(m) $dx = \begin{bmatrix} -3 & 1 & 1 \\ -2 & -4 & -2 \\ 1 & 2 & -5 \end{bmatrix} x \, dw(t);$

(n) $dx = \begin{bmatrix} 3 & 1 \\ 1 & 4 \end{bmatrix} x \, dw(t).$

2. Solve the IVPs (a)–(i) in Exercise 1 with $x(t_0) = x_0$:

   (a) $x_0 = [1 \ 1]^T$;
   (b) $x_0 = [2 \ 1]^T$;
   (c) $x_0 = [1 \ 2]^T$;
   (d) $x_0 = [2 \ 2]^T$;
   (e) any x_0.

3. Solve the IVPs (j)–(m) in Exercise 1 with $x(t_0) = x_0$:

   (a) $x_0 = [1 \ 1 \ 1]^T$;
   (b) $x_0 = [2 \ 2 \ 1]^T$;
   (c) $x_0 = [1 \ 2 \ 1]^T$;
   (d) any x_0.

4. Determine (if possible) the closed-form solutions of the linearized version of Itô–Doob-type stochastic Richard Levin's population extinction (Exercise 7), epidemiological process (Exercise 8), chain reactions — Rice–Herzfeld mechanism (Exercise 9), and Kendall's deterministic marriage (Exercise 10) models in Section 4.1.

4.4 General Homogeneous Systems

In this section, we utilize the eigenvalue-type method developed in Section 4.2 to compute a solution to a general stochastic system of Itô–Doob-type first-order linear homogeneous differential equations with constant coefficients. This computational technique is feasible for a certain class of Itô–Doob-type first-order systems of differential equations. Otherwise, it is not computationally feasible. However, we see that, it is possible to find a solution conceptually.

4.4.1 *General Problem*

Let us consider the following general first-order linear homogeneous system of Itô–Doob-type stochastic differential equations with constant coefficients:

$$dx = Ax \, dt + Bx \, dw(t), \qquad (4.4.1)$$

where B is as defined in (4.2.1), $A = (a_{ij})_{n \times n}$ is an $n \times n$ constant matrix whose entries a_{ij} are real numbers $(a_{ij} \in R)$, and w is a scalar normalized Wiener process (Definition 1.2.12).

The goal is to present a procedure for finding a closed-form solution process of (4.4.1). In addition, we are interested in solving an IVP associated with this problem. We note that the concept of the solution process of (4.4.1) can be defined in conjunction with Definition 4.2.1. The details are left to the reader.

4.4.2 *Procedure for Finding a General Solution*

We extend the usage of the eigenvalue-type method developed in Sections 2.3 and 4.2 for finding a closed solution to a general first-order linear homogeneous system of Itô–Doob-type stochastic differential equations with constant coefficients. Moreover, it parallels Procedure 4.5.2 (Volume 1) for finding a closed-form solution to the deterministic perturbed system of differential equations. We note that system (4.4.1) can be called an Itô–Doob-type stochastic perturbed system of differential equations with respect to either a deterministic or a stochastic unperturbed system of differential equations.

Step #1. (Decomposition). We decompose (4.4.1) into its two timescales, namely the deterministic time t and the stochastic timescale characterized by the Wiener process $w(t)$:

$$dx = Ax\, dt \qquad \text{(deterministic unperturbed/perturbation)}, \qquad (4.4.2)$$

$$dx = Bx\, dw(t) \quad \text{(stochastic perturbation/unperturbed)}, \qquad (4.4.3)$$

respectively. By using the notations and definitions of Section 2.3 and 4.3, our goal of finding a general solution of (4.4.1) is decomposed into the following subgoals:

(a) To find the fundamental matrix solution processes $\Phi_d(t)$ and $\Phi_s(t)$ of (4.4.2) and (4.4.3), respectively.
(b) To create a candidate for the fundamental matrix solution process of (4.4.1) as

$$\Phi(t) \equiv \Phi_d(t)\Phi_s(t) \equiv \Phi_d(t)\Phi_s(t, w(t)). \qquad (4.4.4)$$

(c) To test the validity of the fundamental matrix solution $\Phi(t)$.

In the following, a procedure for fulfilling the subgoals (a)–(c) is outlined.

Step #2. (Computation). We imitate the eigenvalue-type method developed in Sections 4.2 and 4.3 [Observations 4.2.2–4.2.4(vi)] to find the fundamental matrix solution processes of both (4.4.2) and (4.4.3). Following the notations and definitions of Sections 2.3, 4.2, and 4.3, the fundamental matrix solution processes of (4.4.2) and (4.4.3) are denoted by $\Phi_d(t)$ and $\Phi_s(t) \equiv \Phi_s(t, w(t))$, respectively. This fulfills the subgoal (a) in Step #1.

Step #3. (Validation). From Observations 4.2.2–4.2.4(vi), we recall that $\Phi_d(t)$ and $\Phi_s(t)$ satisfy (4.4.2) and (4.4.3), i.e.

$$\Phi_d(t) = A\Phi_d(t)dt, \tag{4.4.5}$$

$$\Phi_s(t) = B\Phi_s(t)dw(t), \tag{4.4.6}$$

respectively. Now, we utilize Theorem 1.3.4, and we compute

$$d\Phi = d\Phi_d(t)\Phi_s(t) + \Phi_d(t)d\Phi_s(t) + d\Phi_d(t)d\Phi_s(t) \quad \text{[from (4.4.5) and (4.4.6)]}$$

$$= A\Phi_d(t)dt\Phi_s(t) + \Phi_d(t)B\Phi_s(t)dw(t) \quad \text{(by Substitution)}$$

$$\quad + A\Phi_d(t)dt\, B\Phi_s(t)dw(t)$$

$$= A\Phi_d(t)\Phi_s(t)dt + \Phi_d(t)B\Phi_s(t)dw(t) \quad \text{(by Itô–Doob calculus)}.$$

Thus,

$$d\Phi(t) = A\Phi_d(t)\Phi_s(t)dt + \Phi_d(t)B\Phi_s(t)dw(t). \tag{4.4.7}$$

If we assume that $\Phi_d(t)$ is a normalized fundamental matrix solution to (4.4.2), and the matrices A and B satisfy the commutative law for the multiplication, i.e.

$$AB = BA, \tag{4.4.8}$$

then, from (4.4.4), (4.4.7) reduces to

$$d\Phi(t) = A\Phi_d(t)\Phi_s(t)dt + B\Phi_d(t)\Phi_s(t)dw(t) \quad \text{(by Theorem 1.5.8(d) [80])} \tag{4.4.9}$$

$$= A\Phi(t)dt + B\Phi(t)dw(t) \quad \text{[from (4.4.4)]}. \tag{4.4.10}$$

This shows that the $\Phi(t)$ defined in (4.4.4) is indeed a matrix solution process of (4.4.1). Moreover, from Theorems 1.4.2 and 1.4.4 (Chapter 1 of Volume 1), $\Phi(t)$ is a nonsingular matrix because $\Phi_d(t)$ and $\Phi_s(t)$ are nonsingular matrices. Hence, $\Phi(t)$ is the fundamental matrix solution to (4.4.1). This establishes the validity of both of the subgoals (b) and (c), whenever (4.4.8) is satisfied. The question concerns what we can say about $\Phi(t)$ if

$$AB \neq BA. \tag{4.4.11}$$

This case requires special attention and will be considered at a later stage.

Step #4. (Conclusion). From Observation 4.2.2, in particular (4.2.11), for arbitrary n-dimensional column vector a, we conclude that $x(t) = \Phi(t)a$ is a general solution to (4.4.1). This completes the procedure for finding a general solution to (4.4.1), provided that the matrices A and B satisfy (4.4.8).

The procedure for finding the general solution is summarized in the following flowchart:

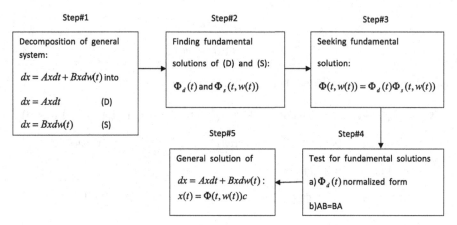

Fig. 4.4.1   Process of finding the general solution.

Observation 4.4.1

(i) In the preceding procedure, if $\Phi_s(t)$ is the normalized fundamental matrix of (4.4.3) and $AB = BA$, then one can also replace (4.4.4) by

$$\Phi(t) = \Phi_s(t)\Phi_d(t), \qquad (4.4.12)$$

and show that $\Phi(t)$ is the fundamental matrix solution to (4.4.1).

(ii) From Observations 4.2.4(vi) and 4.3.1, we note that $\Phi_d(t) = \exp[At]$ and $\Phi_s(t) = \exp\left[-\frac{B^2}{2}t + Bw(t)\right]$. We also note that these processes are the normalized fundamental matrices of (4.4.2) and (4.4.3), respectively. Under the condition $AB = BA$, the general solution to (4.4.1) is also represented by

$$x(t) = \Phi(t)a = \exp[At]\exp\left[-\frac{B^2}{2}t + Bw(t)\right]a$$

$$= \exp\left[\left(A - \frac{B^2}{2}\right)t + Bw(t)\right]a$$

[by Theorem 1.5.8(d), Volume 1]. $\qquad (4.4.13)$

Hence,

$$\Phi(t) = \exp\left[\left(A - \frac{B^2}{2}\right)t + Bw(t)\right]. \qquad (4.4.14)$$

We note that the fundamental solution process of (4.4.1) is given *explicitly or in closed form* in terms of its coefficient matrices A and B. In general, this is not feasible.

(iii) The above-described procedure is based on direct imitation of the procedures described in Sections 2.3, 4.2, and 4.3. It is applicable to the coefficient matrices in (4.4.1) which:

    (a) Satisfy the commutative law for the matrix multiplication (4.4.8).

    (b) One the fundamental matrix solutions is the normalized form (Example 4.5.1 [80]).

A direct proof can be given by using Theorem 4.2.2.

Illustration 4.4.1. Given: $dx = Ax\, dt + Bx\, dw(t)$, where

$$A = \begin{bmatrix} a_{11} & a_{12} \\ a_{21} & a_{22} \end{bmatrix} \quad \text{and} \quad B = \begin{bmatrix} b_{11} & b_{12} \\ b_{21} & b_{22} \end{bmatrix}.$$

Find the general solution to the stochastic system of differential equations.

Solution procedure. From Observation 4.2.4(vi), and Illustrations 4.3.1–4.3.3, we find the normalized fundamental matrix solution $\Phi_d(t)$ of (4.4.2) and the fundamental matrix solution $\Phi_s(t)$ (4.4.3), respectively. From Step #3 of Procedure 4.4.2, the matrix $\Phi(t)$ defined by $\Phi(t) = \Phi_d(t)\Phi_s(t)$ is the fundamental matrix solution to the given illustration if the coefficient matrices A and B satisfy the condition (4.4.8), i.e. $AB = BA$. $AB = BA$ if and only if either

(i) $\frac{a_{12}}{b_{12}} = \frac{a_{21}}{b_{21}} = \frac{a_{11}-a_{22}}{b_{11}-b_{22}} = r$ or

(ii) $\frac{b_{12}}{a_{12}} = \frac{b_{21}}{a_{22}} = \frac{b_{11}-b_{22}}{a_{11}-a_{22}} = r,$

for some real number r. Under this condition, from Step #4, the general solution of the illustration can be determined by $x(t) = \Phi(t)a$. This completes the procedure for finding the general solution. Further details are left to the reader.

Example 4.4.1. Given: $dx = Ax\, dt + Bx\, dw(t)$, where $A = \begin{bmatrix} 3 & 1 \\ 1 & 4 \end{bmatrix}$ and $B = \begin{bmatrix} 2 & 1 \\ 1 & 3 \end{bmatrix}$. Find: (a) the fundamental solution and (b) the general solutions to the given system of differential equations.

Solution Procedure. Goal of the example: To find the general solution to the given differential equation.

    Using Step #1, the decomposition of the given system of stochastic differential equations is:

$$dx = Ax\, dt \quad \text{and} \quad dx = Bx\, dw(t).$$

Now, following the ideas given in Step #2 of Procedure 4.4.2 [Section 4.2 and Observation 4.2.4(vi)], we see that the eigenvalues and corresponding eigenvectors of the matrix A are $\lambda_1^d = \frac{7+\sqrt{5}}{2}$, $c_1^d = \begin{bmatrix} 1 & \frac{1+\sqrt{5}}{2} \end{bmatrix}^T$ and $\lambda_2^d = \frac{7-\sqrt{5}}{2}$, $c_2^d = \begin{bmatrix} -\frac{1+\sqrt{5}}{2} & 1 \end{bmatrix}^T$. Similarly, the eigenvalues and corresponding eigenvectors of the matrix B are $\mu_1 = \frac{5+\sqrt{5}}{2}$, $\lambda_1^s = -\frac{1}{2}\mu_1^2 = -\frac{15+5\sqrt{5}}{4}$, $c_1^s = \begin{bmatrix} 1 & \frac{1+\sqrt{5}}{2} \end{bmatrix}^T$ and $\mu_2 = \frac{5-\sqrt{5}}{2}$, $\lambda_2^s = -\frac{1}{2}\mu_2^2$ $= -\frac{15-5\sqrt{5}}{4}$, $c_2^s = \begin{bmatrix} -\frac{1+\sqrt{5}}{2} & 1 \end{bmatrix}^T$. The normalized fundamental matrix $\Phi_d(t,0)$ and the fundamental matrix $\Phi_s(t)$ solution processes of the deterministic and stochastic components of the given problem are:

$$
\Phi_d(t,0) = \frac{4}{4+(1+\sqrt{5})^2}
\begin{bmatrix}
e^{\lambda_1^d t} & -\dfrac{1+\sqrt{5}}{2}e^{\lambda_2^d t} \\[2ex]
\dfrac{1+\sqrt{5}}{2}e^{\lambda_1^d t} & e^{\lambda_2^d t}
\end{bmatrix}
\begin{bmatrix}
1 & \dfrac{1+\sqrt{5}}{2} \\[2ex]
-\dfrac{1+\sqrt{5}}{2} & 1
\end{bmatrix}
$$

$$
= \frac{4}{4+(1+\sqrt{5})^2}
\begin{bmatrix}
e^{\lambda_1^d t} + \dfrac{(1+\sqrt{5})^2}{4}e^{\lambda_2^d t} & \dfrac{1+\sqrt{5}}{2}(e^{\lambda_1^d t} - e^{\lambda_2^d t}) \\[2ex]
\dfrac{1+\sqrt{5}}{2}(e^{\lambda_1^d t} - e^{\lambda_2^d t}) & \dfrac{(1+\sqrt{5})^2}{4}e^{\lambda_1^d t} + e^{\lambda_2^d t}
\end{bmatrix}
$$

$$
= \frac{1}{4+(1+\sqrt{5})^2}
\begin{bmatrix}
4e^{\lambda_1^d t} + (1+\sqrt{5})^2 e^{\lambda_2^d t} & 2(1+\sqrt{5})(e^{\lambda_1^d t} - e^{\lambda_2^d t}) \\[2ex]
2(1+\sqrt{5})(e^{\lambda_1^d t} - e^{\lambda_2^d t}) & (1+\sqrt{5})^2 e^{\lambda_1^d t} + 4e^{\lambda_2^d t}
\end{bmatrix},
$$

and

$$
\Phi_s(t) =
\begin{bmatrix}
e^{\lambda_1^s t + \mu_1 w(t)} & -\dfrac{1+\sqrt{5}}{2}e^{\lambda_2^s t + \mu_2 w(t)} \\[2ex]
\dfrac{1+\sqrt{5}}{2}e^{\lambda_1^s t + \mu_1 w(t)} & e^{\lambda_2^s t + \mu_2 w(t)}
\end{bmatrix}.
$$

This completes Step #2 of Procedure 4.4.2. As per (4.4.4), we define

$$
\Phi(t) = \Phi_d(t)\Phi_s(t) =
\begin{bmatrix}
\phi_{11}(t, w(t)) & \phi_{12}(t, w(t)) \\
\phi_{21}(t, w(t)) & \phi_{22}(t, w(t))
\end{bmatrix},
$$

where

$$
\phi_{11}(t, w(t)) = \frac{4e^{\lambda_1^d t} + (1+\sqrt{5})^2 e^{\lambda_2^d t} + (1+\sqrt{5})^2(e^{\lambda_1^d t} - e^{\lambda_2^d t})}{4+(1+\sqrt{5})^2} e^{\lambda_1^s t + \mu_1 w(t)}
$$

$$
= \frac{(4+(1+\sqrt{5})^2)e^{\lambda_1^d t}}{4+(1+\sqrt{5})^2} e^{\lambda_1^s t + \mu_1 w(t)} \quad \text{(by simplification)}
$$

$$
= e^{(\lambda_2^d + \lambda_1^s)t + \mu_1 w(t)} \quad \text{(by simplification)},
$$

$$
\phi_{12}(t, w(t)) = \frac{-(1+\sqrt{5})[4e^{\lambda_1^d t} + (1+\sqrt{5})^2 e^{\lambda_2^d t}] + 4(1+\sqrt{5})(e^{\lambda_1^d t} - e^{\lambda_2^d t})}{4+(1+\sqrt{5})^2}
$$

$$
\times \frac{1}{2}e^{\lambda_2^s t + \mu_2 w(t)}
$$

$$
= \frac{-(1+\sqrt{5})[(1+\sqrt{5})^2 + 4]}{4+(1+\sqrt{5})^2} \frac{1}{2}e^{(\lambda_2^d + \lambda_2^s)t + \mu_2 w(t)} \quad \text{(by simplification)}
$$

$$
= -\frac{1}{2}(1+\sqrt{5})e^{(\lambda_2^d + \lambda_2^s)t + \mu_2 w(t)} \quad \text{(by simplification)},
$$

$$\phi_{21}(t, w(t)) = \frac{(1+\sqrt{5})[4(e^{\lambda_1^d t} - e^{\lambda_2^d t}) + (1+\sqrt{5})^2 e^{\lambda_1^d t} + 4e^{\lambda_2^d t}]}{4 + (1+\sqrt{5})^2} \frac{1}{2} e^{\lambda_1^s t + \mu_1 w(t)}$$

$$= \frac{(1+\sqrt{5})[4 + (1+\sqrt{5})^2]}{4 + (1+\sqrt{5})^2} \frac{1}{2} e^{(\lambda_1^d + \lambda_1^s)t + \mu_1 w(t)} \quad \text{(by simplification)}$$

$$= \frac{1}{2}(1+\sqrt{5}) e^{(\lambda_1^d + \lambda_1^s)t + \mu_1 w(t)} \quad \text{(by simplification)},$$

$$\phi_{22}(t, w(t)) = \frac{-(1+\sqrt{5})^2 (e^{\lambda_1^d t} - e^{\lambda_2^d t}) + (1+\sqrt{5})^2 e^{\lambda_1^d t} + 4e^{\lambda_2^d t}}{4 + (1+\sqrt{5})^2} e^{\lambda_2^s t + \mu_2 w(t)}$$

$$= \frac{[4 + (1+\sqrt{5})^2]}{4 + (1+\sqrt{5})^2} e^{(\lambda_2^d + \lambda_2^s)t + \mu_2 w(t)} \quad \text{(by simplification)}$$

$$= e^{(\lambda_2^d + \lambda_2^s)t + \mu_2 w(t)} \quad \text{(by simplification)}.$$

We observe that the coefficient matrices A and B satisfy the commutative law for the multiplication, i.e.

$$AB = \begin{bmatrix} 3 & 1 \\ 1 & 4 \end{bmatrix} \begin{bmatrix} 2 & 1 \\ 1 & 3 \end{bmatrix} = \begin{bmatrix} 7 & 6 \\ 6 & 13 \end{bmatrix} = BA.$$

Therefore, from Step #3 of Procedure 4.4.2, we conclude that $\Phi(t)$ is the fundamental matrix solution process of the given system of stochastic differential equations. From Step #4 of Procedure 4.4.2, we have

$$x(t) = \Phi(t)a.$$

This is the desired general solution to the given system of stochastic differential equations. This completes the solution process of the example.

Example 4.4.2. Given: $dx = Ax\,dt + Bx\,dw(t)$, where $A = \begin{bmatrix} 2 & 1 \\ 1 & 3 \end{bmatrix}$ and $B = \begin{bmatrix} 3 & 1 \\ 1 & 4 \end{bmatrix}$. Find: (a) the fundamental solution and (b) the general solutions to the given system of differential equations.

Solution procedure. To avoid repetition, we just follow the mechanical argument used in the solution process of Example 4.4.1. In this case, the eigenvalues and corresponding eigenvectors of the matrix A are $\lambda_1^d = \frac{5+\sqrt{5}}{2}$, $c_1^d = \begin{bmatrix} 1 & \frac{1+\sqrt{5}}{2} \end{bmatrix}^T$ and $\lambda_2^d = \frac{5-\sqrt{5}}{2}$, $c_2^d = \begin{bmatrix} -\frac{1+\sqrt{5}}{2} & 1 \end{bmatrix}^T$. Similarly, the eigenvalues and corresponding eigenvectors of the matrix B are $\mu_1 = \frac{7+\sqrt{5}}{2}$ and $\lambda_1^s = -\frac{1}{2}\mu_1^2 = -\frac{27+7\sqrt{5}}{4}$ with $c_1^s = \begin{bmatrix} 1 & \frac{1+\sqrt{5}}{2} \end{bmatrix}^T$ and $\mu_2 = \frac{7-\sqrt{5}}{2}$ and $\lambda_2^s = -\frac{1}{2}\mu_2^2 = -\frac{27-7\sqrt{5}}{4}$ with

$c_2^s = \left[-\frac{1+\sqrt{5}}{2} \ 1 \right]^T$. The remaining argument is the direct imitation of the argument used in Example 4.4.1. Thus, we have

$$\Phi(t) = \Phi_d(t, 0)\Phi_s(t) \quad \text{and} \quad x(t) = \Phi(t)a,$$

where the description of $\Phi(t)$ is exactly similar to the description of $\Phi(t)$ in Example 4.4.1. The details are left to the reader.

In the following, we present a theoretical algorithm that provides a systematic approach to find a general solution process of (4.4.1). This procedure does not require the condition (4.4.8) and the condition for one of the fundamental matrices to be the normalized form. Furthermore, the result signifies that the deterministic system (4.4.2) is an unperturbed system of the stochastic perturbed system (4.4.1). In this case, the stochastic perturbed system (4.4.1) is reduced to a stochastic system with a time-varying deterministic coefficient matrix function.

Theorem 4.4.1 (reduced stochastic system with time-varying coefficients). *Let us assume that* (H_1) $\Phi_d(t)$ *is a fundamental matrix solution to* (4.4.2); *and* (H_2) $x(t) = \Phi_d(t)c(t)$ *be a solution to* (4.4.1), *where* $c(t)$ *is an unknown n-dimensional smooth vector function. Then,* $c(t)$ *is a solution process of the following system of linear Itô–Doob-type stochastic differential equations with the time-varying deterministic coefficient matrix process* $\Phi_d^{-1}(t)B\Phi_d(t)$:

$$dc = \Phi_d^{-1}(t)B\Phi_d(t)c\,dw(t). \tag{4.4.15}$$

Proof. From Observation 4.2.4(vi), we assert that $\Phi_d(t)$ is invertible, and hence $\Phi_d^{-1}(t)$ exists on J. Moreover, from Theorem 4.3.3 [80], we have

$$d\Phi_d^{-1}(t) = -\Phi_d^{-1}(t)A\,dt. \tag{4.4.16}$$

Let $x(t)$ be a solution process of (4.4.1). We assume that $x(t) = \Phi_d(t)c(t)$ can be rewritten as

$$c(t) = \Phi_d^{-1}(t)x(t). \tag{4.4.17}$$

Under the assumptions of the theorem and using the product rule (Theorem 1.3.4), we compute the following Itô–Doob differential:

$$\begin{aligned}
dc(t) &= d\Phi_d^{-1}(t)x(t) + \Phi_d^{-1}(t)dx(t) + d\Phi_d^{-1}(t)dx(t) \quad \text{(by Theorem 1.3.4)} \\
&= -\Phi_d^{-1}(t)Ax(t)dt + \Phi_d^{-1}(t)[Ax(t)dt + Bx(t)dw(t)] \\
&\quad + d\Phi_d^{-1}(t)[Ax(t)dt + Bx(t)dw(t)] \quad \text{[by Definition 4.2.1 and (4.4.16)]} \\
&= -\Phi_d^{-1}(t)Ax(t)dt + \Phi_d^{-1}(t)Ax(t)dt + \Phi_d^{-1}(t)Bx(t)dw(t) \\
&\quad - \Phi_d^{-1}(t)A^2x(t)dt\,dt - \Phi_d^{-1}(t)AB\,x(t)dt\,dw(t) \\
&\quad \text{(by regrouping and simplifying)}
\end{aligned}$$

$$= \Phi_d^{-1}(t)Bx(t)dw(t) \quad \text{(by Itô–Doob-type calculus and simplifications)}$$

$$= \Phi_d^{-1}(t)B\Phi_d(t)c(t)dw(t) \quad [\text{by substitution } x(t) = \Phi_d(t)c(t)].$$

This shows that the unknown vector process $c(t)$ satisfies the system of Itô–Doob-type linear stochastic differential equations (4.4.15) with the time-varying deterministic coefficient matrix process $\Phi_d^{-1}(t)B\Phi_d(t)$. This completes the proof of the theorem. $\qquad\square$

The following result is an alternate form of Theorem 4.4.1. It is obtained by interchanging the roles of the pairs $(\Phi_d(t), A)$ and $(\Phi_d(t), B)$. Moreover, the result signifies that the stochastic system (4.4.3) is an unperturbed system of the stochastic perturbed system (4.4.1). In this case, the stochastic perturbed system (4.4.1) is reduced to a deterministic system with the time-varying stochastic coefficient matrix process.

Corollary 4.4.1 (reduced deterministic system with stochastic process-varying coefficients). *Let us assume that (H_1) $\Phi_s(t)$ is a fundamental matrix solution to (4.4.3); and (H_2) $x(t) = \Phi_s(t)c(t)$ be a solution to (4.4.1), where $c(t)$ is an unknown n-dimensional smooth vector function. Then, $c(t)$ is a solution process of the following system of linear deterministic differential equations with stochastic process varying coefficient matrix $\Phi_s^{-1}(t)A\Phi_s(t)$:*

$$dc = \Phi_s^{-1}(t)A\Phi_s(t)c\,dt. \tag{4.4.18}$$

Proof. The proof of the corollary can be constructed parallel to the proof of Theorem 4.4.1. Here, we utilize Observation 4.2.4 and Theorem 4.2.3. The details are left as exercise to the reader. $\qquad\square$

Now, we present a theoretical algorithm that determines a general solution to (4.4.1). This result does not require the assumption (4.4.8) and the condition for one of the fundamental matrices to be the normalized form.

Theorem 4.4.2. *Let $\Phi_d(t)$ be a fundamental matrix solution to (4.4.2) and let $\Phi_T(t) \equiv \Phi_T(t, w(t))$ be a fundamental matrix solution to (4.4.15). Then, the fundamental matrix solution and the general solution to (4.4.1) are represented by (i) $\Phi(t) = \Phi_d(t)\Phi_T(t)$ and (ii) $x(t) = \Phi_d(t)\Phi_T(t)a$, where a is an arbitrary constant vector.*

Proof. $\Phi_d(t)$ and $\Phi_T(t)$ are as described in the theorem. We define $\Phi(t) = \Phi_d(t)\Phi_T(t)$. We know that $\Phi(t)$ is nonsingular (Theorem 1.4.2 [80]). We compute

$$d\Phi(t) = d\Phi_d(t)\Phi_T(t) + \Phi_d(t)d\Phi_T(t) + d\Phi_d(t)d\Phi_T(t) \quad \text{(by Theorem 1.3.4)}$$

$$= A\Phi_d(t)\Phi_T(t)dt + \Phi_d(t)\Phi_d^{-1}(t)B\Phi_d(t)\Phi_T(t)dw(t)$$

$$+ A\Phi_d(t)\Phi_T(t)dt + \Phi_d^{-1}(t)B\Phi_d(t)\Phi_T(t)dw(t) \quad \text{[from (4.4.5) and (4.4.15)]}$$

$$= A\Phi(t)dt + B\Phi(t)dw(t) \qquad \text{(by simplification)}.$$

This proves Part (i). Part (ii) follows from the definition of the fundamental solution. Furthermore, the solution process of the system of perturbed differential equations $dx = Ax\,dt + Bx\,dw(t)$ is the composition of the solution process of the unperturbed component of the differential equation and the solution process of the transformed system with a time-varying deterministic coefficient matrix function. This establishes the conclusion of the theorem. In addition, we do not need any assumptions about the matrices A, B, $\Phi_d(t)$ and $\Phi_T(t)$. $\qquad\square$

The following corollary is parallel to Theorem 4.4.2. It is formulated by interchanging the roles of $(\Phi_d(t), A)$ and $(\Phi_s(t), B)$.

Corollary 4.4.2. *Let $\Phi_s(t)$ be a fundamental matrix solution to (4.4.3) and let $\Phi_T(t)$ be a fundamental matrix solution to (4.4.18). Then, the fundamental matrix solution and the general solution to (4.4.1) are represented by (a) $\Phi(t) = \Phi_s(t)\Phi_T(t)$ and (b) $x(t) = \Phi_s(t)\Phi_T(t)a$, where a is an arbitrary constant vector.*

Proof. The proof of the corollary can be constructed parallel to the proof of Theorem 4.4.2. Here, we utilize Observation 4.2.4 and Theorem 4.2.3. The details are left as an exercise to the reader. $\qquad\square$

Example 4.4.3. Given: $dx = Ax\,dt + Bx\,dw(t)$, where $A = \begin{bmatrix} 2 & 5 \\ 1 & 3 \end{bmatrix}$ and $B = \begin{bmatrix} -1 & 0 \\ 1 & -2 \end{bmatrix}$. Find: (a) the fundamental solution and (b) the general solutions to the given system of differential equations.

Solution procedure. By imitating the argument used in Theorem 4.4.1 and Illustration 4.3.1 (Illustration 4.4.1 [80]), we have

$$\Phi_d(t) = \begin{bmatrix} 5e^{\frac{5+\sqrt{21}}{2}t} & -\dfrac{1+\sqrt{21}}{2}e^{\frac{5-\sqrt{21}}{2}t} \\[2mm] \dfrac{1+\sqrt{21}}{2}e^{\frac{5+\sqrt{21}}{2}t} & e^{\frac{5-\sqrt{21}}{2}t} \end{bmatrix}.$$

$\det(\Phi_d(t)) = \frac{20+(1+\sqrt{21})^2}{4}e^{5t}$ and its inverse is

$$\Phi_d^{-1}(t) = \frac{4}{20 + (1+\sqrt{21})^2} \begin{bmatrix} e^{-\frac{5+\sqrt{21}}{2}t} & \dfrac{1+\sqrt{21}}{2}e^{-\frac{5+\sqrt{21}}{2}t} \\[2mm] -\dfrac{1+\sqrt{21}}{2}e^{-\frac{5-\sqrt{21}}{2}t} & 5e^{-\frac{5-\sqrt{21}}{2}t} \end{bmatrix}.$$

The solution process of the system of original Itô–Doob-type stochastic differential equations $dx = Ax\,dt + Bx\,dw(t)$ is determined by using the fundamental solution process of the transformed system of Itô–Doob-type stochastic differential equations with the deterministic coefficient matrix function (4.4.15) (Theorem 4.4.2):

$$\Phi_d^{-1}(t)B\Phi_d(t)$$

$$= \Phi_d^{-1}(t) \begin{bmatrix} -1 & 0 \\ 1 & -2 \end{bmatrix} \begin{bmatrix} 5e^{\frac{5+\sqrt{21}}{2}t} & -\dfrac{1+\sqrt{21}}{2}e^{\frac{5-\sqrt{21}}{2}t} \\ \dfrac{1+\sqrt{21}}{2}e^{\frac{5+\sqrt{21}}{2}t} & e^{\frac{5-\sqrt{21}}{2}t} \end{bmatrix}$$

$$= \Phi_d^{-1}(t) \begin{bmatrix} -5e^{\frac{5+\sqrt{21}}{2}t} & \dfrac{1+\sqrt{21}}{2}e^{\frac{5-\sqrt{21}}{2}t} \\ (4-\sqrt{21})e^{\frac{5+\sqrt{21}}{2}t} & -\dfrac{5+\sqrt{21}}{2}e^{\frac{5-\sqrt{21}}{2}t} \end{bmatrix}.$$

From this, we write

$$\frac{20+(1+\sqrt{21})^2}{4}\Phi_d^{-1}(t)B\Phi_d(t)$$

$$= \begin{bmatrix} e^{-\frac{5+\sqrt{21}}{2}t} & \dfrac{1+\sqrt{21}}{2}e^{-\frac{5+\sqrt{21}}{2}t} \\ -\dfrac{1+\sqrt{21}}{2}e^{-\frac{5-\sqrt{21}}{2}t} & 5e^{-\frac{5-\sqrt{21}}{2}t} \end{bmatrix}$$

$$\times \begin{bmatrix} -5e^{\frac{5+\sqrt{21}}{2}t} & \dfrac{1+\sqrt{21}}{2}e^{\frac{5-\sqrt{21}}{2}t} \\ (4-\sqrt{21})e^{\frac{5+\sqrt{21}}{2}t} & -\dfrac{5+\sqrt{21}}{2}e^{\frac{5-\sqrt{21}}{2}t} \end{bmatrix}$$

$$= \begin{bmatrix} -5+\dfrac{1+\sqrt{21}}{2}(4-\sqrt{21}) & \dfrac{1+\sqrt{21}}{2}\left[1-\dfrac{5+\sqrt{21}}{2}\right]e^{-\sqrt{21}t} \\ 5\left[\dfrac{1+\sqrt{21}}{2}+(4-\sqrt{21})\right]e^{\sqrt{21}t} & -\dfrac{(1+\sqrt{21})(1+\sqrt{21})}{4}-\dfrac{5(5+\sqrt{21})}{2} \end{bmatrix}$$

$$= \begin{bmatrix} \dfrac{3(-9+\sqrt{21})}{2} & \dfrac{-(24+4\sqrt{21})}{4}e^{-\sqrt{21}t} \\ \dfrac{5(9-\sqrt{21})}{2}e^{\sqrt{21}t} & -\dfrac{72+12\sqrt{21}}{4} \end{bmatrix}.$$

Thus,

$$\Phi_d^{-1}(t)B\Phi_d(t) = \frac{4}{20+(1+\sqrt{21})^2} \begin{bmatrix} \dfrac{3(-9+\sqrt{21})}{2} & \dfrac{-(24+4\sqrt{21})}{4}e^{-\sqrt{21}t} \\ \dfrac{5(9-\sqrt{21})}{2}e^{-\sqrt{21}t} & -\dfrac{72+12\sqrt{21}}{4} \end{bmatrix}.$$

The closed-form fundamental matrix solution to systems of differential equations of the time-varying coefficient matrix is not feasible. However, using the conceptual formulation, its existence is definitely feasible. The details about the fundamental solutions process, the general solution process, and the solution to the IVP are presented in Section 4.6.

Example 4.4.4. Given: $dx = Ax\,dt + Bx\,dw(t)$, where $A = \begin{bmatrix} 2 & 5 \\ 1 & 3 \end{bmatrix}$ and $B = \begin{bmatrix} -\frac{1}{2} & 0 \\ 1 & -2 \end{bmatrix}$. Find: (a) the fundamental solution and (b) the general solutions to the given differential equation.

Solution procedure. By imitating the argument used in Theorem 4.4.1 (Corollary 4.4.1) and Illustration 4.3.1, we have

$$\Phi_s(t) = \begin{bmatrix} e^{-(\frac{1}{2}t+w(t))} & 0 \\ e^{-(\frac{1}{2}t+w(t))} & e^{-2(t+w(t))} \end{bmatrix}.$$

$\det(\Phi_s(t)) = e^{-(\frac{5}{2}t+3w(t))}$ and its inverse is

$$\Phi_s^{-1}(t) = \begin{bmatrix} e^{(\frac{1}{2}t+w(t))} & 0 \\ -e^{2(t+w(t))} & e^{2(t+w(t))} \end{bmatrix}.$$

The solution process of the system of original Itô–Doob-type stochastic differential equations $dx = Ax\,dt + Bx\,dw(t)$ is determined by using the fundamental solution process of the transformed system of deterministic differential equations with the random coefficient matrix function (4.4.18) (Corollary 4.4.1):

$$\Phi_s^{-1}(t)A\Phi_s(t)$$

$$= \begin{bmatrix} e^{(\frac{1}{2}t+w(t))} & 0 \\ -e^{2(t+w(t))} & e^{2(t+w(t))} \end{bmatrix} \begin{bmatrix} 2 & 5 \\ 1 & 3 \end{bmatrix} \begin{bmatrix} e^{-(\frac{1}{2}t+w(t))} & 0 \\ e^{-(\frac{1}{2}t+w(t))} & e^{-2(t+w(t))} \end{bmatrix}$$

$$= \begin{bmatrix} e^{(\frac{1}{2}t+w(t))} & 0 \\ -e^{2(t+w(t))} & e^{2(t+w(t))} \end{bmatrix} \begin{bmatrix} 7e^{-(\frac{1}{2}t+w(t))} & 5e^{-2(t+w(t))} \\ 4e^{-(\frac{1}{2}t+w(t))} & 3e^{-2(t+w(t))} \end{bmatrix}$$

$$= \begin{bmatrix} 7 & 5e^{-(\frac{3}{2}t+w(t))} \\ -3e^{(\frac{3}{2}t+w(t))} & -2 \end{bmatrix}.$$

The rest of the argument is similar to the argument used in Example 4.4.3. This completes the solution procedure of the example.

Observation 4.4.2

(i) Under the assumptions of $AB = BA$ and the normalized fundamental matrices (Example 4.5.1 [80]) $\Phi_s(t)$ and $\Phi_d(t)$, we have $\Phi_s^{-1}(t)A\Phi_s(t) = A$ and $\Phi_d^{-1}(t)B\Phi_d(t) = B$. Hence, the system of linear differential equations with time-varying coefficients in (4.4.15) and (4.4.18) reduces to the system of linear differential equations with constant coefficients (4.4.3) and (4.4.2), respectively. In this case, the general solutions to (4.4.15) and (4.4.18) are the general solutions to (4.4.3) and (4.4.2), respectively. Moreover, they are determined by the fundamental solutions $\Phi_d(t)$ to (4.4.2) and $\Phi_s(t)$ to (4.4.3), respectively. Therefore, applying Theorem 4.4.2 and Corollary 4.4.2, the fundamental solution to (4.4.1) is $\Phi_d(t)\Phi_s(t) = \Phi_s(t)\Phi_d(t)$, whenever one of the matrices is in the normalized form. Hence, $x(t)$ defined in (H_2) of Theorem 4.4.1 and Theorem 4.4.2 (or Corollary 4.4.1 and Corollary 4.4.2) reduces to

$$
\begin{aligned}
x(t) &= \Phi_d(t)c(t) \\
&= \Phi_d(t)\Phi_s(t)a \text{ [by substitution } c(t) = \Phi_d(t)a] \\
&= \Phi_d(t)a \text{ [from (4.4.4)]} \\
&= \exp\left[\left(A - \frac{B^2}{2}\right)t + Bw(t)\right]a \text{ [from (4.4.14)].}
\end{aligned}
\tag{4.4.19}
$$

This is indeed the general solution process that was directly determined by using the developed procedures in Sections 2.3 and 4.2.

(ii) From the above observation (i), the problem of finding a general solution process of (4.4.1) with $AB \neq BA$ reduces to the problem of finding a general fundamental matrix solution to a system of linear differential equations with time-varying coefficient matrices of either (4.4.15) or (4.4.18). Unfortunately, in general, there is no computational procedure that would enable us to find a closed-form computational representation of the fundamental matrix solution to such systems of linear differential equations with time-varying coefficients in terms of the coefficient matrices. In literature, this issue has been addressed in the conceptual framework [8, 43, 71, 98, 141, 182]. This topic is presented in the last section of this chapter.

(iii) From the above observations (i) and (ii) of Observation 4.2.3, we further note that finding the general solution process of (4.4.1) depends on finding the fundamental solution to (4.4.1). Finding the fundamental solution to (4.4.1) depends on finding the fundamental solutions to (4.4.2) and (4.4.15) or (4.4.3) and (4.4.18). The task of justifying the existence of the fundamental solution to systems of linear differential equations with time-varying coefficients is the topic of discussion in the final section of this chapter.

(iv) We further recall a newly introduced study of a system of deterministic differential equations of the type (Section 4.5 of Volume 1)

$$dx = Ax\,dt + Bx\,dt.$$

It is called the deterministic perturbed system. If A and B satisfy the condition (4.4.8), then the above procedure provides a closed-form solution to this deterministic perturbed system as a special case.

4.4.3 *Initial Value Problem*

Let us formulate an IVP for the general first-order linear system of Itô–Doob-type homogeneous stochastic differential equations with constant coefficients. We consider

$$dx = Ax\,dt + Bx\,dw(t), \quad x(t_0) = x_0. \tag{4.4.20}$$

Here x, A, B, and $w(t)$ are as defined in (4.4.1). We note that the concept of the solution of the *initial value problem* (IVP) (4.4.20) can be defined in conjunction with Definition 4.2.2. Again, its solution is represented by $x(t) = x(t, t_0, x_0)$, for $t \geq t_0$ and t in J.

4.4.4 *Procedure for solving the IVP*

In the following, we apply the procedure for finding a solution process of the IVP (4.2.8) to the IVP (4.4.20) with $AB = BA$. We briefly summarize the procedure for finding a solution to the IVP (4.4.20):

Step #1. From the procedure for finding a general solution to (4.4.1) and Observation 4.4.1, we have

$$x(t) = \Phi(t)c = \exp\left[\left(A - \frac{B^2}{2}\right)t + Bw(t)\right]c, \tag{4.4.21}$$

where $c \in R^n$, $t \in J = [a, b] \subseteq R$, and $\Phi(t)$ is as defined in (4.4.14). Here, c is an n-dimensional arbitrary constant vector.

Step #2. In this step, we need to find an arbitrary constant c in (4.4.21). For this purpose, we utilize the given initial data (t_0, x_0), and solve the system of algebraic equations

$$x(t_0) = \Phi(t_0)c,$$

$$x_0 = \exp\left[\left(A - \frac{B^2}{2}\right)t_0 + Bw(t_0)\right]c, \tag{4.4.22}$$

for any given t_0 in J. Since $\det(\exp\left[\left(A - \frac{B^2}{2}\right)t_0 + Bw(t_0)\right] \neq 0$, the algebraic equation (4.4.22) has a unique solution, and it is given by

$$c = \Phi^{-1}(t_0)x_0 = \exp\left[-\left(A - \frac{B^2}{2}\right)t_0 + Bw(t_0)\right]x_0. \qquad (4.4.23)$$

Step #3. The solution to the IVP (4.4.20) is determined by substituting the expression (4.4.23) into (4.4.21). Hence, we have

$$x(t) = \Phi(t)c \qquad \text{[from (4.4.13)]}$$

$$= \Phi(t)\Phi^{-1}(t_0)x_0 \quad \text{[by substitution from (4.4.23)]}$$

$$= \exp\left[-\left(A - \frac{B^2}{2}\right)(t - t_0) + B(w(t) - w(t_0))\right]x_0 \quad \text{(by laws of exponents)}$$

$$= \Phi(t, t_0)x_0 \qquad \text{(by notation)}, \qquad (4.4.24)$$

where

$$\Phi(t)\Phi^{-1}(t_0) = \Phi(t, t_0) = \exp\left[\left(A - \frac{B^2}{2}\right)(t - t_0) + B(w(t) - w(t_0))\right]. \quad (4.4.25)$$

Thus, the solution process $x(t) = x(t, t_0, x_0)$ of the IVP (4.4.20) is represented by

$$x(t) = x(t, t_0, x_0)$$

$$= \Phi(t, t_0)x_0 \quad \text{[from (4.4.25)]}$$

$$= \exp\left[\left(A - \frac{B^2}{2}\right)(t - t_0) + B(w(t) - w(t_0))\right]x_0 \quad \text{[from (4.4.25)]}. \quad (4.4.26)$$

The solution process in (4.4.26) is referred to as a *particular solution* to (4.4.1). $\Phi(t, t_0)$ is called the *normalized fundamental solution* to (4.4.1) since $\Phi(t_0, t_0) = I$.

The procedure for finding the solution to the IVP is summarized in the following flowchart:

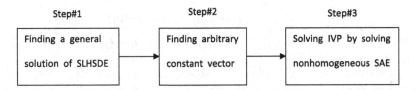

Step#1	Step#2	Step#3
Finding a general solution of SLHSDE	Finding arbitrary constant vector	Solving IVP by solving nonhomogeneous SAE

Fig. 4.4.2   Process of finding the solution of IVP.

Illustration 4.4.2. Given: $dx = Ax\,dt + Bx\,dw(t)$, $x(t_0) = x_0, x_0 \in R^2$, where $A = \begin{bmatrix} a_{11} & a_{12} \\ a_{21} & a_{22} \end{bmatrix}$ and $B = \begin{bmatrix} b_{11} & b_{12} \\ b_{21} & b_{22} \end{bmatrix}$. Find the solution to the IVP.

Solution procedure. By repeating the procedure outlined in Illustration 4.4.1, we arrive at the general solution $x(t) = \Phi(t)a$ of the illustration. From this, we imitate the argument used in Illustrations 4.3.1–4.3.3, and obtain the solution to the given IVP. The details are left as an exercise. This completes the procedure for finding the general solution.

Example 4.4.5. Given: $dx = Ax\,dt + Bx\,dw(t)$, $x(t_0) = x_0, x_0 \in R^2$, where

$$A = \begin{bmatrix} 5 & -\sqrt{5} \\ \sqrt{5} & 1 \end{bmatrix} \quad \text{and} \quad B = \begin{bmatrix} 4\sigma & -\sqrt{5}\sigma \\ \sqrt{5}\sigma & 0 \end{bmatrix}, \text{ with } \sigma \neq 0. \text{ Find the solution to}$$

the IVP.

Solution procedure. From Step #1 of Procedure 4.4.4, we need to find the general solution to this IVP. First, we find the fundamental matrix solutions to both the deterministic and the stochastic components of the given system of differential equations. In this case, the eigenvalues of the matrices A and B are $3 \pm i$ and $2\sigma \pm \sigma i$, respectively. Based on the procedures [Section 4.3 and Observation 4.2.4(vi)] for finding the eigenvector and solutions of differential equations with distinct complex eigenvalues, it is sufficient to find one eigenvector corresponding to each distinct complex eigenvalue. In this case, the eigenvector and solutions corresponding to eigenvalue $\lambda_1^d = 3 + i$ of A are

$$c_i^d = \begin{bmatrix} 2+i \\ \sqrt{5} \end{bmatrix}, \quad x_1^d(t) = e^{3t} \begin{bmatrix} 2\cos t - \sin t \\ \sqrt{5}\cos t \end{bmatrix}, \quad \text{and} \quad x_2^d(t) = e^{3t} \begin{bmatrix} \cos t + 2\sin t \\ \sqrt{5}\sin t \end{bmatrix}.$$

Similarly, the eigenvector and solution corresponding to eigenvalue $\mu_1 = \alpha_1 + \beta_1 i = 2\sigma + \sigma i$ of B with $\lambda_1^s = -\frac{1}{2}\mu_1^2 = -\frac{1}{2}(3\sigma^2 + 4\sigma^2 i) = -\frac{3}{2}\sigma^2 - 2\sigma^2 i = \gamma_1 + \nu_1 i$ [for notations (4.3.3)] are

$$c_i^s = \begin{bmatrix} 2+i \\ \sqrt{5} \end{bmatrix}, \quad x_1^s(t) = e^{\gamma_1 t + \alpha_1 w(t)} \begin{bmatrix} 2\cos(\nu_1 t + \beta_1 w(t)) - \sin(\nu_1 t + \beta_1 w(t)) \\ \sqrt{5}\cos(\nu_1 t + \beta_1 w(t)) \end{bmatrix},$$

$$x_2^s(t) = e^{\gamma_1 t + \alpha_1 w(t)} \begin{bmatrix} 2\sin(\nu_1 t + \beta_1 w(t)) + \cos(\nu_1 t + \beta_1 w(t)) \\ \sqrt{5}\sin(\nu_1 t + \beta_1 w(t)) \end{bmatrix},$$

where $\gamma_1 t + \alpha_1 w(t) = -\frac{3}{2}\sigma^2 t + 2\sigma w(t)$ and $\nu_1 t + \beta_1 w(t) = -2\sigma^2 t + \sigma w(t)$. For any nonzero real numbers a and b and from $a\sin\theta \pm b\cos\theta$, we recall that

$$a\sin\theta \pm b\cos\theta = \sqrt{a^2+b^2}\left[\frac{a}{\sqrt{a^2+b^2}}\sin\theta \pm \frac{b}{\sqrt{a^2+b^2}}\cos\theta\right].$$

Via trigonometric expressions and setting $\cos\psi = \frac{a}{\sqrt{a^2+b^2}}$ and $\sin\psi = \frac{b}{\sqrt{a^2+b^2}}$, we have

$$a\sin\theta \pm b\cos\theta = \sqrt{a^2+b^2}[\sin\theta\cos\psi \pm \sin\psi\cos\theta] = \sqrt{a^2+b^2}\sin(\theta \pm \psi).$$

Similarly, by defining $\cos\psi = \frac{c}{\sqrt{c^2+d^2}}$ and $\sin\psi = \frac{d}{\sqrt{c^2+d^2}}$, we also have

$$c\cos\theta \pm d\sin\theta = \sqrt{c^2+d^2}[\cos\theta\cos\psi \pm \sin\theta\sin\psi] = \sqrt{c^2+d^2}\cos(\theta \mp \psi).$$

From these preliminary trigonometric identities and setting $\psi = \arctan\left(\frac{1}{2}\right)$, we write some of the components of the above solution process:

$$2\cos t - \sin t = \sqrt{5}\cos(t+\psi), \quad 2\sin t + \cos t = \sqrt{5}\sin(t+\psi),$$

$$2\cos(\nu_1 t + \beta_1 w(t)) - \sin(\nu_1 t + \beta_1 w(t)) = \sqrt{5}\cos(\nu_1 t + \beta_1 w(t) + \psi),$$

$$2\sin(\nu_1 t + \beta_1 w(t)) + \cos(\nu_1 t + \beta_1 w(t)) = \sqrt{5}\sin(\nu_1 t + \beta_1 w(t) + \psi).$$

With this mathematical compactification and from the above discussion, we write the fundamental matrices of the deterministic and stochastic components of differential equations as:

$$\Phi_d(t) = e^{3t}\begin{bmatrix} \sqrt{5}\cos(t+\psi) & \sqrt{5}\sin(t+\psi) \\ \sqrt{5}\cos t & \sqrt{5}\sin t \end{bmatrix},$$

$$\Phi_s(t) = e^{\gamma_1 t + \alpha_1 w(t)}\begin{bmatrix} \sqrt{5}\cos(\nu_1 t + \beta_1 w(t) + \psi) & \sqrt{5}\sin(\nu_1 t + \beta_1 w(t) + \psi) \\ \sqrt{5}\cos(\nu_1 t + \beta_1 w(t)) & \sqrt{5}\sin(\nu_1 t + \beta_1 w(t)) \end{bmatrix}.$$

To find the fundamental matrix solution to the overall system of stochastic differential equations for the given example, we need to know at least one of the normalized fundamental matrix solutions with respect to $\Phi_d(t)$ and $\Phi_s(t)$. To get this, we find this with regards to $\Phi_d(t)$. For this purpose, we use Observation 4.2.3(iv) and Step #3 of Procedure 4.2.4 with $t_0 = 0$. We compute $\Phi_d(0), \Phi_d^{-1}(0)$ and the normalized fundamental matrix solution with respect to $\Phi_d(t)$ as

$$\Phi_d(0) = \begin{bmatrix} \sqrt{5}\cos\psi & \sqrt{5}\sin\psi \\ \sqrt{5} & 0 \end{bmatrix} = \begin{bmatrix} 2 & 1 \\ \sqrt{5} & 0 \end{bmatrix}, \quad \Phi_d^{-1}(0) = \begin{bmatrix} 0 & \frac{1}{\sqrt{5}} \\ 1 & \frac{-2}{\sqrt{5}} \end{bmatrix},$$

$$\Phi_d(t,0) = \Phi_d(t)\Phi_d^{-1}(0) = e^{3t}\begin{bmatrix} \sqrt{5}\cos(t+\psi) & \sqrt{5}\sin(t+\psi) \\ \sqrt{5}\cos t & \sqrt{5}\sin t \end{bmatrix}\begin{bmatrix} 0 & \frac{1}{\sqrt{5}} \\ 1 & \frac{-2}{\sqrt{5}} \end{bmatrix}.$$

$$= e^{3t}\begin{bmatrix} \sqrt{5}\sin(t+\psi) & -\sqrt{5}\sin t \\ \sqrt{5}\sin t & -\sqrt{5}\sin(t-\psi) \end{bmatrix}.$$

Thus, from (4.4.4), the fundamental matrix solution to the overall system of stochastic differential equations in the given example is

$$\Phi(t) = \Phi_d(t,0)\Phi_s(t)$$

$$= e^{3t+\alpha(t,w(t))}\begin{bmatrix} \phi_{11}(t,w(t),\psi) & \phi_{12}(t,w(t),\psi) \\ \phi_{21}(t,w(t),\psi) & \phi_{22}(t,w(t),\psi) \end{bmatrix},$$

where $\phi \equiv \phi(t, w(t)) \equiv -2\sigma^2 t + \sigma w(t)$ and $\alpha(t, w(t)) \equiv -\frac{3}{2}\sigma^2 t + 2\sigma w(t)$,

$$\phi_{11}(t, w(t), \psi) = 5[\sin(t + \psi)\cos(\phi + \psi) - \sin t \cos \phi],$$
$$\phi_{12}(t, w(t), \psi) = 5[\sin(t + \psi)\sin(\phi + \psi) - \sin t \sin \phi],$$
$$\phi_{21}(t, w(t), \psi) = 5[\sin t \cos(\phi + \psi) - \sin(t - \psi)\cos \phi],$$
$$\phi_{22}(t, w(t), \psi) = 5[\sin t \sin(\phi + \psi) - \sin(t - \psi)\sin \phi].$$

We further confirm our calculations by the following observations. First, we note that

$$\sin(t + \psi)\sin(t - \psi)$$
$$= \frac{1}{2}[\cos 2\psi - \cos 2t] \quad \left(\text{by } \sin\theta\sin\phi = \frac{1}{2}[\cos(\theta - \phi) - \cos(\theta + \phi)]\right)$$
$$= \frac{1}{2}[1 - 2\sin^2\psi - (1 - 2\sin^2 t)] \quad (\text{by } \cos 2\theta = 1 - 2\sin^2 t)$$
$$= [\sin^2 t - \sin^2 \psi] \quad (\text{by simplification}).$$

Now, we compute

$$\det(\Phi_d(t, 0))$$
$$= -5e^{6t}[\sin(t + \psi)\sin(t - \psi) - \sin^2 t]$$
$$= -5e^{6t}[\sin^2 t - \sin^2\psi - \sin^2 t] \quad (\text{by } \sin(t + \psi)\sin(t - \psi) = [\sin^2 t - \sin^2 \psi])$$
$$= 5e^{6t}\sin^2\psi = e^{6t} \quad \left(\text{by } \sin\psi = \frac{1}{\sqrt{5}}\right).$$

It satisfies the conclusion of Theorem 4.2.2, in particular (4.2.16). Similarly,

$$\det(\Phi_s(t))e^{-2\alpha(t,w(t))} \quad (\text{by Theorem 1.4.1D}_4 \text{ [80]})$$
$$= 5([\sin(\phi(t, w(t)))\cos(\phi(t, w(t)) + \psi) - \sin(\phi(t, w(t)) + \psi)\cos(\phi(t, w(t)))]$$
$$= 5[\sin\phi(t, w(t)) - \psi - \phi(t, w(t))]e^{2\alpha(t,w(t))} = 5\sin(-\psi) = -\sqrt{5},$$

and hence $\det(\Phi_s(t, 0)) = -\sqrt{5}e^{2\alpha(t,w(t))}$, which satisfies the conclusion of Theorem 4.2.2, in particular (4.2.16). We further compute

$$\det(\Phi(t))e^{-2(3t+\alpha(t,w(t)))}$$
$$= 25[(\sin(t + \psi)\cos(\phi + \psi) - \sin t \cos \phi)(\sin t \sin(\phi + \psi) - \sin(t - \psi)\sin \phi)$$
$$- (\sin(t + \psi)\sin(\phi + \psi) - \sin t \sin \phi)(\sin t \cos(\phi + \psi) - \sin(t - \psi)\cos \phi)]$$

$$= 25[[\sin(t+\psi)\cos(\phi+\psi)\sin t\sin(\phi+\psi)+\sin t\cos\phi\sin(t-\psi)\sin\phi$$
$$-\sin(t+\psi)\cos(\phi+\psi)\sin(t-\psi)\sin\phi-\sin t\cos\phi\sin t\sin(\phi+\psi)]$$
$$-[\sin(t+\psi)\sin(\phi+\psi)\sin t\cos(\phi+\psi)+\sin t\sin\phi\sin(t-\psi)\cos\phi)$$
$$-\sin(t+\psi)\sin(\phi+\psi)\sin(t-\psi)\cos\phi-\sin t\cos(\phi+\psi)\sin t\sin\phi]]$$
$$= 25[[\sin(t+\psi)\sin(t-\psi)[\sin(\phi+\psi)\cos\phi-\cos(\phi+\psi)\sin\phi]]$$
$$+\sin^2 t[\cos(\phi+\psi)\sin\phi-\cos\phi\sin(\phi+\psi)]] \quad \text{(by simplification)}$$
$$= 25[\sin(t+\psi)\sin(t-\psi)\sin\psi-\sin^2 t\sin\psi]$$
$$= 25\sin\psi[\sin(t+\psi)\sin(t-\psi)-\sin^2 t] \quad \text{(by further simplification)}.$$

This is based on the fact that

$$\sin(\phi+\psi)\sin(t-\psi)$$

$$= \frac{1}{2}[\cos 2\psi-\cos 2t] \quad \left(\text{by } \sin\theta\sin\phi=\frac{1}{2}[\cos(\theta-\phi)-\cos(\theta+\phi)]\right)$$

$$= \frac{1}{2}[1-2\sin^2\psi-(1-2\sin^2 t)] \quad \text{(by } \cos 2\theta=1-2\sin^2 t)$$

$$= [\sin^2 t-\sin^2\psi] \quad \text{(by simplification)},$$

and we have

$$\det(\Phi(t))e^{-2(3t+\alpha(t,w(t)))} = 25\sin\psi[\sin(t+\psi)\sin(t-\psi)-\sin^2 t]$$

$$= 25\sin\psi[-\sin^2\psi+\sin^2 t-\sin^2 t] \quad \text{(by simplification)}$$

$$= -25\sin^3\psi = \frac{-25}{5\sqrt{5}} = -\sqrt{5}.$$

This implies that $\det(\Phi(t)) = -\sqrt{5}e^{2(3t+\alpha(t,w(t)))} = \det(\Phi_d(t))\det(\Phi_s(t))$ (Theorem 1.4.2, Chapter 1 of Volume 1). We further note that the coefficient matrices in this example commute, i.e. $AB = BA$. Therefore, with these considerations, the remaining argument is a direct imitation of the argument used in Example 4.4.1. In fact, the fundamental matrix and the general solutions to the given problems are defined by

$$\Phi(t) = \Phi_d(t)\Phi_s(t), \quad x(t) = \Phi(t)a,$$

in view of the fact that $\det(\Phi(t)) = \det(\Phi_d(t))\det(\Phi_s(t)) = -\sqrt{5}e^{6t+2\alpha(t,w(t))t} \neq 0$. Thus, the solution to the IVP is given by $x(t) = \Phi(t)\Phi^{-1}(t_0)x_0$. This completes the solution process.

Observation 4.4.3. Note that the form of a general fundamental solution expression in (4.4.14) for the stochastic differential equation (4.4.1) is an extension of the form of the corresponding general fundamental solution expressions in Observations 4.3.1 and 4.2.4(vi) of the linear deterministic and stochastic systems of differential equations (4.4.2) and (4.4.3), respectively. In fact, (i) for the choice of $B \equiv 0$, i.e. in the absence of random perturbations/effects, it includes a fundamental solution expression in Observation 4.3.1 as a special case of the fundamental solution expression in (4.4.14). For the choice of $A \equiv 0$, i.e. in the absence of deterministic dynamic effects, it includes a fundamental solution expression in Observation 4.3.1 as a special case of the fundamental solution expression in (4.4.14).

Exercises

1. Let us denote by $a, b, c \ldots$ arbitrary but given real numbers. For the following given pairs of matrices (A, B), identify the pair matrices that satisfy the commutative property of matrix multiplication:

(a) $A = \begin{bmatrix} 5 & \sqrt{5} \\ -\sqrt{5} & 1 \end{bmatrix}$, $B = \begin{bmatrix} 8 & -\sqrt{5} \\ -2\sqrt{5} & 0 \end{bmatrix}$;

(b) $A = \begin{bmatrix} 2 & 0 \\ 0 & 1 \end{bmatrix}$, $B = \begin{bmatrix} a & 0 \\ 0 & a \end{bmatrix}$;

(c) $A = \begin{bmatrix} 3 & 2 \\ -1 & 2 \end{bmatrix}$, $B = \begin{bmatrix} 1.5 & 1 \\ -0.5 & 1 \end{bmatrix}$;

(d) $A = \begin{bmatrix} 1 & 0 \\ 0 & -1 \end{bmatrix}$, $B = \begin{bmatrix} \sqrt{3} & 0 \\ 0 & 2 \end{bmatrix}$;

(e) $A = \begin{bmatrix} 4 & 1 \\ -4 & 8 \end{bmatrix}$, $B = \begin{bmatrix} 1 & 3 \\ 3 & 2 \end{bmatrix}$;

(f) $A = \begin{bmatrix} 1 & 0 \\ 0 & 1 \end{bmatrix}$, $B = \begin{bmatrix} 3 & 0 \\ 0 & 2 \end{bmatrix}$;

(g) $A = \begin{bmatrix} 3 & 2 \\ 1 & 4 \end{bmatrix}$, $B = \begin{bmatrix} 1 & 1 \\ 1 & 2 \end{bmatrix}$;

(h) $A = \begin{bmatrix} a & 0 \\ 0 & b \end{bmatrix}$, $B = \begin{bmatrix} \sqrt{a} & 0 \\ 0 & \sqrt{b} \end{bmatrix}$;

(i) $A = \begin{bmatrix} -1.1 & 1 \\ -1 & 0.1 \end{bmatrix}$, $B = \begin{bmatrix} 0.1 & -1 \\ 1 & 1.1 \end{bmatrix}$;

(j) $A = \begin{bmatrix} a & 0 \\ 0 & b \end{bmatrix}$, $B = \begin{bmatrix} b\sqrt{a} & 0 \\ 0 & a\sqrt{b} \end{bmatrix}$;

(k) $A = \begin{bmatrix} 3 & 2 \\ -1 & 4 \end{bmatrix}$, $B = \begin{bmatrix} 1 & 1 \\ 1 & 2 \end{bmatrix}$;

(l) $A = \begin{bmatrix} 0 & 1 \\ 0 & 0 \end{bmatrix}$, $B = \begin{bmatrix} 0 & 0 \\ 1 & 0 \end{bmatrix}$.

2. Given: $A = \begin{bmatrix} \alpha & \alpha\delta \\ \delta & \beta \end{bmatrix}$, $B = \begin{bmatrix} \gamma + b(\alpha - \beta) & ab\delta \\ b\delta & \gamma \end{bmatrix}$, where a, b, α, β, δ, and γ are arbitrary real numbers such that a, b, and $\alpha - \beta$ are nonzero. Verify that $AB = BA$.

3. Find the fundamental matrix solutions to the following system of differential equations: (D) $dx = Ax\,dt$ and (S) $dx = Bx\,dw(t)$, where the matrices A and B are as given in Exercise 1.

4. Find the fundamental matrix solutions to the following system of differential equations: (D) $dx = Bx\,dt$ and (S) $dx = Ax\,dw(t)$, where the matrices A and B are as given in Exercise 1.

5. Find the fundamental matrix solutions to the following system of differential equations: $dx = Ax\,dt + Bx\,dw(t)$, where the matrices A and B are as given in Exercise 1.

6. Find the general solutions to the following system of differential equations: $dx = Ax\,dt + Bx\,dw(t)$, where the matrices A and B are as given in Exercise 1.

7. Find the general solution to each system of differential equations:

(a) $dx = Ax\,dt + Bx\,dw(t)$;

(b) $dx = Bx\,dt + Ax\,dw(t)$, where $A = \begin{bmatrix} -1 & 1 \\ 1 & -1 \end{bmatrix}$, $B = \begin{bmatrix} 1 & -2 \\ -2 & 1 \end{bmatrix}$;

(c) $dx = Ax\,dt + Bx\,dw(t)$;

(d) $dx = Bx\,dt + Ax\,dw(t)$, where $A = \begin{bmatrix} -1 & 0 \\ 1 & -1 \end{bmatrix}$, $B = \begin{bmatrix} 1 & -1 \\ 1 & 1 \end{bmatrix}$;

(e) $dx = Ax\,dt + Bx\,dw(t)$;

(f) $dx = Bx\,dt + Ax\,dw(t)$, where $A = \begin{bmatrix} -1 & -1 \\ 1 & -1 \end{bmatrix}$, $B = \begin{bmatrix} 3 & 1 \\ 1 & 4 \end{bmatrix}$.

8. Solve the IVPs in Exercises 5 and 6 with $x(t_0) = x_0$:

(a) $x_0 = [1\ 1]^T$;
(b) $x_0 = [2\ 1]^T$;
(c) $x_0 = [1\ 2]^T$;
(d) $x_0 = [2\ 2]^T$;
(e) any x_0.

9. Prove Corollary 4.4.1.

10. Verify the conclusion of Corollary 4.4.2.

11. Show that:
    (a) $\Phi(t) = \Phi_s(t)\Phi_T(t)$ is the fundamental matrix solution and the general solution to (4.4.1);
    (b) $x(t) = \Phi_s(t)\Phi_T(t)a$ is the general solution to (4.4.1), where a is an arbitrary constant vector, $\Phi_s(t)$, and $\Phi_T(t)$ are as defined in Corollary 4.4.2.

12. For $i = 1, 2, \ldots, k$, let A, B_i be $n \times n$ matrices with real entries, and let $w_i(t)$ be the Wiener process (Theorem 1.2.2) defined on a complete probability space $(\Omega, \mathfrak{F}, P)$. Further, assume that for $i \neq j$, $i, j = 1, 2, \ldots, k$, $w_i(t)$ and $w_j(t)$ are mutually independent on J and satisfy $E[dw_i(t)dw_j(t)] = 0$. Show that:

    (a) If for $i = 1, 2, \ldots, k$, A and B_i mutually commute, then the process defined by

    $$x(t, w_1(t), \ldots w_k(t)) = \exp\left[\left(A - \frac{1}{2}\sum_{i=1}^{k} B_i^2\right)t + \sum_{i=1}^{k} B_i w_i(t)\right]c$$

    is the solution process of the following Itô–Doob stochastic differential equation:

    $$dx = Ax\,dt + \sum_{i=1}^{k} B_i x\,dw_i(t). \tag{$*$}$$

    (b) In the absence of the mutual commutativity property of coefficient matrices, we have

    $$dc = \Phi_d^{-1}(t)\left(\sum_{i=1}^{k} B_i\right)\Phi_d(t)c\,dw(t), \tag{$**$}$$

    $$dc = \prod_{i=1}^{k} \Phi_{si}^{-1}(t)A\prod_{i=1}^{k}\Phi_{si}(t)c\,dt, \tag{$***$}$$

    where $\Phi_{si}(t)$ the fundamental matrix solution process of

    $$dx_i = B_i x_i dw_i(t), \quad i = 1, 2, \ldots, k. \tag{$****$}$$

13. Find the general solution to $dx = Ax\,dt + B_1 x\,dw_1(t) + B_2 x\,dw_2(t)$, where
    $$A = \begin{bmatrix} 2 & 5 \\ 1 & 3 \end{bmatrix}, B_1 = \begin{bmatrix} -1 & 0 \\ 1 & -2 \end{bmatrix}, \text{ and } B_2 = \begin{bmatrix} 2 & 5 \\ 1 & 3 \end{bmatrix}.$$

14. Let F be a continuous cumulative distribution function (cdf) of a random time variable T, and let A, B, B_1, and B_2 be 2×2 matrices. Moreover, $w(t)$ equates to the earlier defined Wiener process. Apply the procedure outlined in Sections 4.2 and 4.4 to find the general solution process of the following differential equations:

    (a) $dx = Ax\,dt + Bx\,dF(t)$;

(b) $dx = Ax\, dt + B_1 x\, dF(t) + B_2 x\, dw_2(t)$;

(c) Solve (a) and (b) for given matrices in Exercise 8 with $B = B_1$ in (a).

4.5 Linear Nonhomogeneous Systems

In this section, we utilize the *method of variation of constants parameters* to compute a particular solution to first-order linear nonhomogeneous systems of stochastic differential equations of the Itô–Doob-type. We recall that this method is efficient and useful for solving both deterministic and stochastic first-order linear nonhomogeneous systems of differential equations. It is applicable to both time-invariant and time-varying linear and nonlinear systems of both deterministic and stochastic differential equations.

4.5.1 *General Problem*

Let us consider the following first-order linear nonhomogeneous system of stochastic differential equations of the Itô–Doob-type:

$$dx = [Ax + p(t)]dt + [Bx + q(t)]dw(t), \qquad (4.5.1)$$

where x, A, and B are as defined in (4.4.1) and satisfy the condition (4.4.8); p and q are n-dimensional continuous functions defined on an interval $J \subseteq R$ into R^n; and w is a scalar normalized Wiener process (Definition 1.2.12).

The goal is to present a procedure for finding a solution process of (4.5.1). In addition, we are interested in solving an IVP associated with (4.5.1). The concept of the solution process defined in Definition 4.2.1 is directly applicable to (4.5.1). The details are left to the reader.

4.5.2 *Procedure for Finding a General Solution*

To find a general solution to (4.5.1), we imitate the conceptual ideas of finding a general solution to linear nonhomogeneous algebraic equations described in Theorem 1.4.7 of Volume 1. In particular, we recast Procedure 2.4.2 for finding a solution process of first-order linear homogeneous scalar stochastic differential equations corresponding to (2.4.1) in Section 2.4. The basic ideas are:

(i) To find a general solution process x_c of the first-order linear homogeneous system of stochastic differential equations corresponding to (4.5.1).

(ii) To find a particular solution process x_p of (4.5.1).

(iii) To validate the random function $x = x_c + x_p$ to be a general solution to (4.5.1).

Here, we reintroduce the method of variation of constants parameters to Itô–Doob systems of stochastic differential equations to find a particular solution process x_p of (4.5.1). This technique is a very powerful, systematic, and an elegant approach to find a particular solution to (4.5.1).

Step #1. First, find a fundamental solution to first-order linear homogeneous system of stochastic differential equations corresponding to (4.5.1), i.e.

$$dx = Ax\,dt + Bx\,dw(t). \tag{4.5.2}$$

We note that this stochastic system of Itô–Doob differential equations equals (4.4.1). Therefore, we imitate Steps #1–4 described in Procedure 4.4.2 for finding a general solution process of (4.4.1), and determine a fundamental solution to (4.5.2) (Observation 4.4.1):

$$\Phi(t) = \Phi_d(t)\Phi_s(t) \quad \text{[from (4.4.4)]}$$

$$= \exp\left[\left(A - \frac{B^2}{2}\right)t + Bw(t)\right] \quad \text{[from (4.4.14)]}. \tag{4.5.3}$$

Step #2 (method of variation of constants parameters). From Step #1, Observation 4.2.4, and Theorem 1.3.3 I1 (Volume 1), we have

$$\Phi^{-1}(t) = \Phi_s^{-1}(t)\Phi_d^{-1}(t). \tag{4.5.4}$$

Moreover, from Observation 4.2.4(vi) and Theorem 4.2.3, we have

$$d\Phi_d^{-1}(t) = -\Phi_d^{-1}(t)A\,dt, \tag{4.5.5}$$

$$d\Phi_s^{-1}(t) = \Phi_s^{-1}(t)B^2dt - \Phi_s^{-1}(t)B\,dw(t). \tag{4.5.6}$$

Now, applying the Itô–Doob product formula (Theorem 1.3.4) to $\Phi^{-1}(t)$ and using $AB = BA$ in (4.4.8), we get

$$d\Phi^{-1}(t)$$

$$= d\Phi_s^{-1}(t)\Phi_d^{-1}(t) + \Phi_s^{-1}(t)d\Phi_d^{-1}(t) + d\Phi_s^{-1}(t)d\Phi_d^{-1}(t) \quad \text{(by Theorem 1.3.4)}$$

$$= [\Phi_s^{-1}(t)B^2dt - \Phi_s^{-1}(t)B\,dw(t)]\Phi_d^{-1}(t) - \Phi_s^{-1}(t)\Phi_d^{-1}(t)A\,dt$$

$$\quad - [\Phi_s^{-1}(t)B^2dt - \Phi_s^{-1}(t)B\,dw(t)]\Phi_d^{-1}(t)A\,dt \quad \text{[from (4.5.5) and (4.5.6)]}$$

$$= [\Phi_s^{-1}(t)B^2\Phi_d^{-1}(t) - \Phi_s^{-1}(t)\Phi_d^{-1}(t)A]dt - \Phi_s^{-1}(t)B\,dw(t)\Phi_d^{-1}(t)$$

(by substitution)

$$= [\Phi_s^{-1}(t)\Phi_d^{-1}(t)B^2 - \Phi_s^{-1}(t)\Phi_d^{-1}(t)A]dt - \Phi_s^{-1}(t)\Phi_d^{-1}(t)B\,dw(t) \quad \text{[from (4.4.8)]}$$

$$= \Phi^{-1}(t)(B^2 - A)dt - \Phi^{-1}(t)B\,dw(t) \quad \text{[from (4.5.4)]}. \tag{4.5.7}$$

With this discussion, we are ready to determine a particular solution to (4.5.1). A particular solution process x_p of (4.5.1) is a random function that satisfies the system of stochastic differential equations (4.5.1). Using the ideas of Section 2.4, we define a random function:

$$x_p(t) = \Phi(t)c(t), \quad c(a) = c_a, \tag{4.5.8}$$

where $\Phi(t)$ is as determined in Step #1, and $c(t)$ is an unknown random function with $c(a) = c_a$. Our goal of finding a particular solution process $x_p(t)$ of (4.5.1) is equivalent to finding an unknown random process $c(t)$ of (4.5.8) passing through c_a at $t = a$. For this purpose, we assume that $x_p(t)$ is a solution process of (4.5.1). By using this assumption and applying the Itô–Doob product formula (Theorem 1.3.4) to

$$c(t) = \Phi^{-1}(t)x_p(t), \tag{4.5.9}$$

we obtain

$$
\begin{aligned}
dc(t) &= \Phi^{-1}(t)dx_p(t) + d\Phi^{-1}(t)x_p(t) + d\Phi^{-1}(t)dx_p(t) \quad \text{(by definition of } x_p) \\
&= \Phi^{-1}(t)[[Ax_p(t) + p(t)]dt + [Bx_p(t) + q(t)]dw(t)] \\
&\quad + d\Phi^{-1}(t)x_p(t) + d\Phi^{-1}(t)dx_p(t) \quad \text{(by Theorem 1.3.4)} \\
&= \Phi^{-1}(t)[[Ax_p(t) + p(t)]dt + [Bx_p(t) + q(t)]dw(t)] \\
&\quad + [\Phi^{-1}(t)(B^2 - A)dt - \Phi^{-1}(t)B\,dw(t)]x_p(t) \\
&\quad + [\Phi^{-1}(t)(B^2 - A)dt - \Phi^{-1}(t)B\,dw(t)][[Ax_p(t) + p(t)]dt \\
&\quad + [Bx_p(t) + q(t)]dw(t)] \quad \text{[from (4.5.7), definition and substitution of } x_p] \\
&= \Phi^{-1}(t)[[Ax_p(t) + p(t)]dt + [Bx_p(t) + q(t)]dw(t)] \\
&\quad + \Phi^{-1}(t)[(B^2 - A)x_p(t)dt - Bx_p(t)dw(t)] \\
&\quad - \Phi^{-1}(t)[B^2x_p(t) + Bq(t)](dw(t))^2 \quad \text{(by grouping)} \\
&= \Phi^{-1}(t)[[Ax_p(t) + (B^2 - A)x_p(t) + p(t)]dt \\
&\quad + [Bx_p(t) + q(t) - Bx_p(t)]dw(t) \\
&\quad - [B^2x_p(t) + Bq(t)]dt] \quad \text{[by } (dw(t))^2 = dt \text{ and simplifications]} \\
&= \Phi^{-1}(t)[[p(t) - Bq(t)]dt + q(t)dw(t)] \quad \text{(by simplifications).} \tag{4.5.10}
\end{aligned}
$$

We recall that the problem of finding a particular solution $x_p(t)$ defined in (4.5.8) reduces to the problem of finding the unknown process $c(t)$ of (4.5.8) passing through c_a at $t = a$. This problem further reduces to a problem of solving the following IVP:

$$dc(t) = \Phi^{-1}(t)[p(t) - Bq(t)]dt + \Phi^{-1}(t)q(t)dw(t), \quad c(a) = c_a. \tag{4.5.11}$$

Employing the method of solving the IVP described in Section 2.2, the solution to the IVP (4.5.11) is given by

$$c(t) = c_a + \int_a^t \Phi^{-1}(s)[p(s) - Bq(s)]ds + \int_a^t \Phi^{-1}(s)q(s)dw(s). \qquad (4.5.12)$$

We substitute the expression for $c(t)$ in (4.5.12) into (4.5.8), and we obtain

$$x_p(t) = \Phi(t)\left[c_a + \int_a^t \Phi^{-1}(s)[p(s) - B(s)q(s)]ds + \int_a^t \Phi^{-1}(s)q(s)dw(s)\right]$$

$$= \Phi(t)c_a + \int_a^t \Phi(t)\Phi^{-1}(s)[p(s) - B(s)q(s)]ds + \int_a^t \Phi(t)\Phi^{-1}(s)q(s)dw(s).$$

$$\qquad (4.5.13)$$

Step #3. From Step #1 and Observation 4.2.2, for any arbitrary constant C, the random function defined by $x_c(t) = \Phi(t)C$ is the general solution to (4.5.2). By the conceptual algebraic ideas described in Theorem 1.4.7 of Volume 1, we define a random function as $x(t) = x_c(t) + x_p(t)$, where $x_p(t)$ is as determined in Step #2. Now, we claim that this process $x(t) = x_c(t) + x_p(t)$ is a general solution to (4.5.1). For this purpose, we consider

$$x(t) = x_c(t) + x_p(t)$$

$$= \Phi(t)C + \Phi(t)c_a + \int_a^t \Phi(t)\Phi^{-1}(s)[p(s) - Bq(s)]ds$$

$$+ \int_a^t \Phi(t)\Phi^{-1}(s)q(s)dw(s) \quad \text{(by substitution)}$$

$$= \Phi(t)(C + c_a) + \int_a^t \Phi(t)\Phi^{-1}(s)[p(s) - Bq(s)]ds$$

$$+ \int_a^t \Phi(t)\Phi^{-1}(s)q(s)dw(s) \quad \text{(by Theorem 1.3.1 [80])}$$

$$= \Phi(t)c + \int_a^t \Phi(t)\Phi^{-1}(s)[p(s) - Bq(s)]ds$$

$$+ \int_a^t \Phi(t)\Phi^{-1}(s)q(s)dw(s) \quad \text{(by rewriting)}$$

$$= \Phi(t)\left[c + \int_a^t \Phi^{-1}(s)[p(s) - Bq(s)]ds\right.$$

$$\left. + \int_a^t \Phi^{-1}(s)q(s)dw(s)\right] \qquad \text{(by recasting)}, \qquad (4.5.14)$$

where $c = C + c_a$ is an arbitrary constant due to the fact that C and c_a are arbitrary constants. Hence, $\Phi(t)c$ is also the general solution to (4.5.2). We show that $x(t)$ is a general solution to (4.5.1).

$dx(t)$

$$= d\Phi(t)\left[c + \int_a^t \Phi^{-1}(s)[p(s) - Bq(s)]ds + \int_a^t \Phi^{-1}(s)q(s)dw(s)\right]$$

$$+ \Phi(t)d\left[c + \int_a^t \Phi^{-1}(s)[p(s) - Bq(s)]ds + \int_a^t \Phi^{-1}(s)q(s)dw(s)\right]$$

$$+ d\Phi(t)d\left[c + \int_a^t \Phi^{-1}(s)[p(s) - Bq(s)]ds + \int_a^t \Phi^{-1}(s)q(s)dw(s)\right]$$

(by Theorem 1.3.4)

$$= d\Phi(t)\left[c + \int_a^t \Phi^{-1}(s)[p(s) - Bq(s)]ds + \int_a^t \Phi^{-1}(s)q(s)dw(s)\right]$$

$$+ \Phi(t)d\left[\int_a^t \Phi^{-1}(s)[p(s) - Bq(s)]ds + \int_a^t \Phi^{-1}(s)q(s)dw(s)\right]$$

$$+ d\Phi(t)d\left[\int_a^t \Phi^{-1}(s)[p(s) - Bq(s)]ds + \int_a^t \Phi^{-1}(s)q(s)dw(s)\right]$$

(by Theorem 1.3.1 [80])

$$= d\Phi(t)\left[c + \int_a^t \Phi^{-1}(s)[p(s) - Bq(s)]ds + \int_a^t \Phi^{-1}(s)q(s)dw(s)\right]$$

$$+ \Phi(t)d\left[\int_a^t \Phi^{-1}(s)[p(s) - Bq(s)]ds\right] + \Phi(t)d\left[\int_a^t \Phi^{-1}(s)q(s)dw(s)\right]$$

$$+ d\Phi(t)d\left[\int_a^t \Phi^{-1}(s)[p(s) - Bq(s)]ds\right]$$

$$+ d\Phi(t)d\left[\int_a^t \Phi^{-1}(s)q(s)dw(s)\right] \quad \text{(by Theorem 1.3.1 [80])}. \quad (4.5.15)$$

We know that:

(i) $d\Phi(t) = A\Phi(t)dt + B\Phi(t)dw(t)$ [from (4.4.10)];
(ii) from the fundamental theorem of elementary integral calculus [127, 168] and Itô–Doob calculus, we have

$$d\left[\int_a^t \Phi^{-1}(s)[p(s) - \sigma(s)q(s)]ds\right] = \Phi^{-1}(t)[p(t) - Bq(t)]dt$$

(by the fundamental theorem);

(iii) $d[\int_a^t \Phi^{-1}(s)q(s)dw(s)] = \Phi^{-1}(t)q(t)dw(t)$ (by Itô–Doob calculus).

From (i)–(iii) and (4.5.13), (4.5.15) reduces to

$dx(t)$

$$
= [A\Phi(t)dt + B\Phi(t)dw(t)]\left[c + \int_a^t \Phi^{-1}(s)[p(s) - Bq(s)]ds + \int_a^t \Phi^{-1}(s)q(s)dw(s)\right]
$$

$$
+ \Phi(t)\Phi^{-1}(t)[p(t) - Bq(t)]dt + \Phi(t)\Phi^{-1}(t)q(t)dw(t)
$$

$$
+ [A\Phi(t)dt + B\Phi(t)dw(t)]\Phi^{-1}(t)[p(t) - Bq(t)]dt
$$

$$
+ [A\Phi(t)dt + B\Phi(t)dw(t)]\Phi^{-1}(t)q(t)dw(t) \quad \text{(by substitution)}
$$

$$
= A\Phi(t)\left[c + \int_a^t \Phi^{-1}(s)[p(s) - Bq(s)]ds + \int_a^t \Phi^{-1}(s)q(s)dw(s)\right]dt
$$

$$
+ B(t)\Phi(t)\left[c + \int_a^t \Phi^{-1}(s)[p(s) - Bq(s)]ds + \int_a^t \Phi^{-1}(s)q(s)dw(s)\right]dw(t)
$$

$$
+ \Phi(t)\Phi^{-1}(t)[p(t) - Bq(t)]dt + \Phi(t)\Phi^{-1}(t)q(t)dw(t)
$$

$$
+ B\Phi(t)dw(t)\Phi^{-1}(t)q(t)dw(t) + A\Phi(t)dt\Phi^{-1}(t)[p(t) - Bq(t)](dt)
$$

$$
+ B\Phi(t)dw(t)\Phi^{-1}(t)[p(t) - Bq(t)]dt + A\Phi(t)dt\Phi^{-1}(t)q(t)dw(t)
$$

  (by grouping)

$$
= A\Phi(t)\left[c + \int_a^t \Phi^{-1}(s)[p(s) - Bq(s)]ds + \int_a^t \Phi^{-1}(s)q(s)dw(s)\right]dt
$$

$$
+ B\Phi(t)\left[c + \int_a^t \Phi^{-1}(s)[p(s) - Bq(s)]ds + \int_a^t \Phi^{-1}(s)q(s)dw(s)\right]dw(t)
$$

$$
+ \Phi(t)\Phi^{-1}(t)[p(t) - Bq(t)]dt + \Phi(t)\Phi^{-1}(t)q(t)dw(t)
$$

$$
+ B\Phi(t)\Phi^{-1}(t)q(t)(dw(t))^2 \quad \text{(by simplifications)}
$$

$$
= A\Phi(t)\left[c + \int_a^t \Phi^{-1}(s)[p(s) - Bq(s)]ds + \int_a^t \Phi^{-1}(s)q(s)dw(s)\right]dt
$$

$$
+ B\Phi(t)\left[c + \int_a^t \Phi^{-1}(s)[p(s) - Bq(s)]ds + \int_a^t \Phi^{-1}(s)q(s)dw(s)\right]dw(t)
$$

$$
+ p(t)dt + q(t)dw(t) + Bq(t)(dw(t))^2 - Bq(t)dt \quad \text{(by further simplifications)}
$$

$$
= A\Phi(t)\left[c\int_a^t \Phi^{-1}(s)[p(s) - Bq(s)]ds + \int_a^t \Phi^{-1}(s)q(s)dw(s)\right]dt
$$

$$
+ B\Phi(t)\left[c + \int_a^t \Phi^{-1}(s)[p(s) - Bq(s)]ds + \int_a^t \Phi^{-1}(s)q(s)dw(s)\right]dw(t)
$$

$$
+ p(t)dt + q(t)dw(t) \quad [(dw(t))^2 \approx dt \text{ by further simplifications}]
$$

$$= A[x_c(t) + x_p(t)]dt + B[x_c(t) + x_p(t)]dw(t)$$

$$+ p(t)dt + q(t)dw(t) \quad \text{[by substitution of } c, x_c(t) \text{ and } x_p(t) \text{ in (4.5.14)]}$$

$$= Ax(t)dt + Bx(t)dw(t) + p(t)dt$$

$$+ q(t)dw(t) \quad \text{[from the definition of } x(t) = x_c(t) + x_p(t) \text{ in (4.5.14)]}$$

$$= [Ax(t) + p(t)]dt + [Bx(t) + q(t)]dw(t) \quad \text{(by regrouping)}. \tag{4.5.16}$$

This shows that $x(t) = x_c(t) + x_p(t)$ is indeed a solution to (4.5.1), and supports the claim that $x(t) = x_c(t) + x_p(t)$ is a solution to (4.5.1). Moreover, from (4.4.14), the general solution to (4.5.1) is represented by

$$x(t) = \Phi(t)\left[c + \int_a^t \Phi^{-1}(s)[p(s) - Bq(s)]ds + \int_a^t \Phi^{-1}(s)q(s)dw(s)\right]$$

$$= \exp\left[\left(A - \frac{1}{2}B^2\right)(t - a) + B(w(t) - w(a))\right]c$$

$$+ \int_a^t \exp\left[\left(A - \frac{1}{2}B^2\right)(t - s) + B(w(t) - w(s))\right][p(s) - Bq(s)]ds$$

$$+ \int_a^t \exp\left[\left(A - \frac{1}{2}B^2\right)(t - s) + B(w(t) - w(s))\right]q(s)dw(s). \tag{4.5.17}$$

Observation 4.5.1

(i) From (4.4.13) in Observation 4.4.1, we note that $x_c(t) = \Phi(t)c$ is the general solution to (4.5.2), where $\Phi(t)$ is the fundamental solution to (4.5.2), and c is an arbitrary constant vector. $x_c(t) = \Phi(t)c$ is also referred to as the *complementary solution* to (4.5.1). However, to find a particular solutions to (4.5.1) in Step #2, we assumed that $x_p(t) = \Phi(t)c(t)$ is a particular solution to (4.5.1), where $c(t)$ is an unknown random process with t as the parameter. The structures of the complementary and the particular solutions are the same. In the case of the particular solution, we treated the constant c in the complementary solution as an unknown function of the independent variable t as the parameter. Because of this, the method of finding a particular solution to (4.5.1) is called the *method of variation of constants parameters*.

(ii) Instead of using a general fundamental solution in (4.4.13), one can use a normalized fundamental solution $\Phi(t, t_0)$ corresponding to (4.5.2), and compute a particular solution to (4.5.1).

(iii) We further observe that the method of variation of parameters determines the general solution to (4.5.1). For instance, if $c_a = 0$ in (4.5.8), then the particular solution in (4.5.13) reduces to

$$x_p(t) = \Phi(t)\left[\int_a^t \Phi^{-1}(s)[p(s) - B(s)q(s)]ds + \int_a^t \Phi^{-1}(s)q(s)dw(s)\right]$$

$$= \int_a^t \Phi(t)\Phi^{-1}(s)[p(s) - B(s)q(s)]ds$$

$$+ \int_a^t \Phi(t)\Phi^{-1}(s)q(s)dw(s) \quad \text{(by simplification)}$$

$$= \int_a^t \Phi(t,s)[p(s) - Bq(s)]ds$$

$$+ \int_a^t \Phi(t,s)q(s)dw(s) \quad \text{[from notation and (4.4.25)],} \qquad (4.5.18)$$

and if c_a in (4.5.8) is any arbitrary constant c, then the particular solution in (4.5.13) is indeed a general solution to (4.5.1), i.e.

$$x(t) = \Phi(t)\left[c + \int_a^t \Phi^{-1}(s)[p(s) - B(s)q(s)]ds + \int_a^t \Phi^{-1}(s)q(s)dw(s)\right]$$

$$= \Phi(t)c + \int_a^t \Phi(t,s)[p(s) - B(s)q(s)]ds + \int_a^t \Phi(t,s)q(s)dw(s). \quad (4.5.19)$$

This solution process is equivalent to the general solution process described in (4.5.17).

The process of finding the general solution is summarized in the following flowchart:

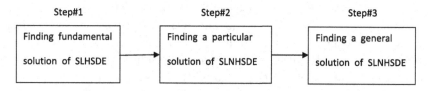

Fig. 4.5.1   Process of finding the general solution.

Illustration 4.5.1.   Given: $dx = [Ax + p(t)]dt + [Bx + q(t)]dw(t)$, where $A = \begin{bmatrix} a_{11} & a_{12} \\ a_{21} & a_{22} \end{bmatrix}$, $B = \begin{bmatrix} b_{11} & b_{12} \\ b_{21} & b_{22} \end{bmatrix}$ and b_{ij} and a_{ij} are any given real numbers satisfying the condition:

(a) $d(A) = (\text{tr}(A))^2 - 4\det(A) > 0$;
(b) $AB = BA$;
(c) p and q are two-dimensional continuous functions defined on an interval $J \subseteq R$ into R^2, and w is a scalar normalized Wiener process.

Find:

(i) a general solution to the given differential equation, particularly, for:
(ii) $B = 0$ and $q(t) \equiv 0$;
(iii) $B = 0$ and $p(t) \equiv 0$;
(iv) $A = 0$ and $q(t) \equiv 0$;
(v) $A = 0$ and $p(t) \equiv 0$.

Solution procedure. The goal of the illustration is to find a general solution to the given system of differential equations. For this purpose, we recall Procedure 4.5.2. We need to find a fundamental matrix solution corresponding to the overall linear system and then find solutions of special cases of the given problem in the illustration. First, from the solution processes of Illustrations 4.3.1 and 4.4.1, assumptions (i) $d(A) = (\text{tr}(A))^2 - 4\det(A) > 0$ and (ii) $AB = BA$, we note that $\frac{a_{12}}{b_{12}} = \frac{a_{21}}{b_{21}} = \frac{a_{11}-a_{22}}{b_{11}-b_{21}} = r$ (or $\frac{b_{12}}{a_{12}} = \frac{b_{21}}{a_{22}} = \frac{b_{11}-b_{22}}{a_{11}-a_{21}} = r$). This implies that $a_{12} = rb_{12}$, $a_{21} = rb_{21}$ and $a_{11} - a_{22} = r(b_{11} - b_{22})$ [or $b_{12} = ra_{12}$, $b_{21} = ra_{21}$ and $b_{11} - b_{22} = r(a_{11} - a_{22})$]. Hence, $d(A) = (\text{tr}(A))^2 - 4\det(A) = (a_{11} - a_{22})^2 + 4a_{12}a_{21} = r^2(b_{11} - b_{22})^2 + r^2 4b_{12}b_{21} = r^2[(b_{11} - b_{22})^2 + 4b_{12}b_{21}] = r^2[(\text{tr}(B))^2 - 4\det(B)] = r^2 d(B)$ (or $d(A) = r^2 d(B)$). This shows that $d(B)$ and $d(A)$ have the same sign, i.e. if $d(A) > 0$ $(d(A) < 0)$, then $d(B) > 0$ $(d(B) < 0)$. Moreover, $\lambda_1 - a_{11} = \frac{\text{tr}(A)+\sqrt{d(A)}-2a_{11}}{2} = \frac{a_{22}-a_{11}+\sqrt{d(A)}}{2} = \frac{r(b_{22}-b_{11})+|r|\sqrt{d(B)}}{2} = r\frac{(b_{22}-b_{11})+\text{sgn}(r)\sqrt{d(B)}}{2} = r(\mu_1 - b_{11})$, for $r > 0$. A similar comment can be made for $r < 0$.

As per Step #1 of Procedure 4.5.2, first, we need a fundamental solution process of the corresponding system of first-order linear homogeneous stochastic differential equations $dx = Ax\,dt + Bx\,dw(t)$. By imitating the solution procedure of Illustration 4.3.1, the fundamental matrix solutions of deterministic and stochastic components of the above system of first-order linear homogeneous stochastic differential equations are:

Case 1

(i) $\Phi_d(t) = \begin{bmatrix} e^{a_{11}t} & 0 \\ 0 & e^{a_{22}t} \end{bmatrix}$;

(ii) $\Phi_d(t) = \begin{bmatrix} (a_{11} - a_{22})e^{a_{11}t} & 0 \\ a_{21}e^{a_{11}t} & e^{a_{22}t} \end{bmatrix}$;

(iii) $\Phi_d(t) = \begin{bmatrix} e^{a_{11}t} & a_{12}e^{a_{22}t} \\ 0 & (a_{22} - a_{11})e^{a_{22}t} \end{bmatrix}$.

Case 2 $\quad \Phi_d(t) = \begin{bmatrix} a_{12}e^{\lambda_1 t} & (\lambda_2 - a_{22})e^{\lambda_2 t} \\ (\lambda_1 - a_{11})e^{\lambda_1 t} & a_{21}e^{\lambda_2 t} \end{bmatrix}$.

Case 1

(i) $\Phi_s(t) \equiv \Phi(t, w(t)) = \begin{bmatrix} e^{-\frac{1}{2}b_{11}^2 t + b_{11} w(t)} & 0 \\ 0 & e^{-\frac{1}{2}b_{22}^2 t + b_{22} w(t)} \end{bmatrix}$;

(ii) $\Phi_s(t) \equiv \Phi_s(t, w(t)) = \begin{bmatrix} (b_{11} - b_{22}) e^{-\frac{1}{2}b_{11}^2 t + b_{11} w(t)} & 0 \\ b_{21} e^{-\frac{1}{2}b_{11}^2 t + b_{11} w(t)} & e^{-\frac{1}{2}b_{22}^2 t + b_{22} w(t)} \end{bmatrix}$;

(iii) $\Phi_s(t) \equiv \Phi_s(t, w(t)) = \begin{bmatrix} e^{-\frac{1}{2}b_{11}^2 t + b_{11} w(t)} & b_{12} e^{-\frac{1}{2}b_{22}^2 t + b_{22} w(t)} \\ 0 & (b_{22} - b_{11}) e^{-\frac{1}{2}b_{22}^2 t + b_{22} w(t)} \end{bmatrix}$.

Case 2

$$\Phi_s(t) \equiv \Phi_s(t, w(t)) = \begin{bmatrix} b_{12} e^{-\frac{1}{2}\mu_1^2 t + \mu_1 w(t)} & (\mu_2 - b_{22}) e^{-\frac{1}{2}\mu_2^2 t + \mu_2 w(t)} \\ (\mu_1 - b_{11}) e^{-\frac{1}{2}\mu_1^2 t + \mu_1 w(t)} & b_{21} e^{-\frac{1}{2}\mu_2^2 t + \mu_2 w(t)} \end{bmatrix}.$$

Moreover, we need one of the fundamental matrix solutions to be normalized (Step #3 of Procedure 4.4.2 and Observation 4.4.1). Therefore, without loss of generality, we find the normalized fundamental matrix solution corresponding to the deterministic part $\Phi_d(t)$. For this purpose, we use Observation 4.2.3(iv) and Step #3 of Procedure 4.2.4 with $t_0 = 0$. We compute $\Phi_d(0)$, $\Phi_d^{-1}(0)$ and the normalized fundamental matrix solution $\Phi_d(t, 0)$ with respect to each case of $\Phi_d(t)$ as

Case 1

(i) $\Phi_d(0) = \begin{bmatrix} e^{a_{11}0} & 0 \\ 0 & e^{a_{22}0} \end{bmatrix} = \begin{bmatrix} 1 & 0 \\ 0 & 1 \end{bmatrix}$, $\Phi_d^{-1}(0) = \begin{bmatrix} 1 & 0 \\ 0 & 1 \end{bmatrix}$, $\Phi_d(t, 0) = \begin{bmatrix} e^{a_{11}t} & 0 \\ 0 & e^{a_{22}t} \end{bmatrix}$;

(ii) $\Phi_d(0) = \begin{bmatrix} (a_{11} - a_{22}) e^{a_{11}0} & 0 \\ a_{21} e^{a_{11}0} & e^{a_{22}0} \end{bmatrix} = \begin{bmatrix} (a_{11} - a_{22}) & 0 \\ a_{21} & 1 \end{bmatrix}$,

$\Phi_d^{-1}(0) = \begin{bmatrix} \dfrac{1}{a_{11} - a_{22}} & 0 \\ \dfrac{-a_{21}}{a_{11} - a_{22}} & 1 \end{bmatrix}$, $\Phi_d(t, 0) = \begin{bmatrix} e^{a_{11}t} & 0 \\ \dfrac{a_{21}(e^{a_{22}t} - e^{a_{11}t})}{a_{11} - a_{22}} & e^{a_{22}t} \end{bmatrix}$;

(iii) $\Phi_d(0) = \begin{bmatrix} e^{a_{11}0} & a_{12} e^{a_{22}0} \\ 0 & (a_{22} - a_{11}) e^{a_{22}0} \end{bmatrix} = \begin{bmatrix} 1 & a_{12} \\ 0 & a_{22} - a_{11} \end{bmatrix}$

$\Phi_d^{-1}(0) = \begin{bmatrix} 1 & \dfrac{-a_{21}}{a_{22} - a_{11}} \\ 0 & \dfrac{1}{a_{22} - a_{11}} \end{bmatrix}$, $\Phi_d(t, 0) = \begin{bmatrix} e^{a_{11}t} & \dfrac{a_{12}(e^{a_{11}t} - e^{a_{22}t})}{a_{22} - a_{11}} \\ 0 & e^{a_{22}t} \end{bmatrix}$.

Case 2

$$\Phi_d(0) = \begin{bmatrix} a_{12}e^{\lambda_1 0} & (\lambda_2 - a_{22})e^{\lambda_2 0} \\ (\lambda_1 - a_{11})e^{\lambda_1 0} & a_{21}e^{\lambda_2 0} \end{bmatrix} = \begin{bmatrix} a_{12} & \lambda_2 - a_{22} \\ \lambda_1 - a_{11} & a_{21} \end{bmatrix},$$

$$\Phi_d^{-1}(0) = \frac{2}{(d(A) + (a_{22} - a_{11})\sqrt{d(A)}} \begin{bmatrix} a_{21} & a_{22} - \lambda_2 \\ a_{11} - \lambda_1 & a_{12} \end{bmatrix},$$

$$\Phi_d(t,0) = \begin{bmatrix} \phi_{11}(t) & \phi_{12}(t) \\ \phi_{21}(t) & \phi_{22}(t) \end{bmatrix},$$

where

$$\phi_{11}(t) = \frac{2[a_{21}a_{12}e^{\lambda_1 t} + (\lambda_2 - a_{22})(a_{11} - \lambda_1)e^{\lambda_2 t}]}{d(A) + (a_{22} - a_{11})\sqrt{d(A)}},$$

$$\phi_{12}(t) = \frac{2[a_{12}(a_{22} - \lambda_2)(e^{\lambda_1 t} - e^{\lambda_2 t})]}{d(A) + (a_{22} - a_{11})\sqrt{d(A)}},$$

$$\phi_{21}(t) = \frac{2[a_{21}(a_{11} - \lambda_1)(e^{\lambda_2 t} - e^{\lambda_1 t})]}{d(A) + (a_{22} - a_{11})\sqrt{d(A)}},$$

$$\phi_{22}(t) = \frac{2[(\lambda_1 - a_{11})(a_{22} - \lambda_2)e^{\lambda_1 t} + a_{21}a_{12}e^{\lambda_2 t}]}{d(A) + (a_{22} - a_{11})\sqrt{d(A)}}.$$

The computations of $\Phi_s(0)$, $\Phi_s^{-1}(0)$ and the normalized fundamental matrix solution $\Phi_s(t, w(t), 0)$ with respect to $\Phi_s(t)$ can be reproduced analogously. The details are left as an exercise.

Now, from Step #2 of Procedure 4.4.2, suitable candidates for the fundamental matrix solutions corresponding to the linear homogeneous system of differential equations are:

Case 1

(i) $\Phi(t, w(t), 0) = \Phi_d(t,0)\Phi_s(t, w(t), 0) = \begin{bmatrix} e^{a_{11}t} & 0 \\ 0 & e^{a_{22}t} \end{bmatrix} \begin{bmatrix} e^{\beta_{11}(w(t))} & 0 \\ 0 & e^{\beta_{22}(w(t))} \end{bmatrix}$

$$= \begin{bmatrix} e^{a_{11}t + \beta_{11}(w(t))} & 0 \\ 0 & e^{a_{22}t + \beta_{22}(w(t))} \end{bmatrix}.$$

(ii) For

$\Phi(t, w(t), 0) = \Phi_d(t,0)\Phi_s(t, w(t), 0)$

$$= \begin{bmatrix} e^{a_{11}t} & 0 \\ \dfrac{a_{21}(e^{a_{22}t} - e^{a_{11}t})}{a_{11} - a_{22}} & e^{a_{22}t} \end{bmatrix} \begin{bmatrix} e^{\beta_{11}(w(t))} & 0 \\ \dfrac{b_{21}(e^{\beta_{22}(w(t))} - e^{\beta_{11}(w(t))})}{b_{11} - b_{22}} & e^{\beta_{22}(w(t))} \end{bmatrix}$$

$$= \begin{bmatrix} e^{a_{11}t + \beta_{11}(w(t))} & 0 \\ \dfrac{a_{21}e^{\beta_{11}(w(t))}(e^{a_{22}t} - e^{a_{11}t})}{a_{11} - a_{22}} + \dfrac{b_{21}e^{a_{22}t}(e^{\beta_{22}(w(t))} - e^{\beta_{11}(w(t))})}{b_{11} - b_{22}} & e^{a_{22}t + \beta_{22}(w(t))} \end{bmatrix}.$$

(iii) $\Phi(t, w(t), 0) = \Phi_d(t, 0)\Phi_s(t, w(t), 0)$

$$= \begin{bmatrix} e^{a_{11}t} & \dfrac{a_{12}(e^{a_{11}t} - e^{a_{22}t})}{a_{22} - a_{11}} \\ 0 & e^{a_{22}t} \end{bmatrix} \begin{bmatrix} e^{\beta_{11}(w(t))} & \dfrac{b_{12}(e^{\beta_{11}(w(t))} - e^{\beta_{22}(w(t))})}{b_{22} - b_{11}} \\ 0 & e^{\beta_{22}(w(t))} \end{bmatrix}$$

$$= \begin{bmatrix} e^{a_{11}t + \beta_{11}(w(t))} & \dfrac{a_{12}e^{\beta_{22}(w(t))}(e^{a_{11}t} - e^{a_{22}t})}{a_{22} - a_{11}} + \dfrac{b_{12}e^{a_{11}t}(e^{\beta_{11}(w(t))} - e^{\beta_{22}(w(t))})}{b_{22} - b_{11}} \\ 0 & e^{a_{22}t + \beta_{22}(w(t))} \end{bmatrix},$$

where $\beta_{11}(w(t)) = -\frac{1}{2}b_{11}^2 t + b_{11}w(t)$ and $\beta_{22}(w(t)) = -\frac{1}{2}b_{22}^2 t + b_{22}w(t)$

Case 2

$$\Phi(t, w(t)) = \Phi_d(t, 0)\Phi_s(t, w(t), 0)$$

$$= \begin{bmatrix} \phi_{11}(t) & \phi_{12}(t) \\ \phi_{21}(t) & \phi_{22}(t) \end{bmatrix} \begin{bmatrix} \varpi_{11}(w(t)) & \varpi_{12}(w(t)) \\ \varpi_{21}(w(t)) & \varpi_{22}(w(t)) \end{bmatrix}$$

$$= \begin{bmatrix} \varphi_{11}(t, w(t)) & \varphi_{12}(t, w(t)) \\ \varphi_{21}(t, w(t)) & \varphi_{22}(t, w(t)) \end{bmatrix},$$

where

$$\varpi_{11}(w(t)) = \frac{2[b_{21}b_{12}e^{\mu_{11}(w(t))} + (\mu_2 - b_{22})(b_{11} - \mu_1)e^{\mu_{22}(w(t))}]}{d(B) + (b_{22} - b_{11})\sqrt{d(B)}},$$

$$\varpi_{12}(w(t)) = \frac{2[b_{12}(b_{22} - \mu_2)(e^{\mu_{11}(w(t))} - e^{\mu_{22}(w(t))})]}{d(B) + (b_{22} - b_{11})\sqrt{d(B)}},$$

$$\varpi_{21}(w(t)) = \frac{2[b_{21}(b_{11} - \mu_1)(e^{\mu_{22}(w(t))} - e^{\mu_{11}(w(t))})]}{d(B) + (b_{22} - b_{11})\sqrt{d(B)}},$$

$$\varpi_{22}(w(t)) = \frac{2[(\mu_1 - b_{11})(b_{22} - \mu_2)e^{\mu_{11}(w(t))} + b_{21}b_{12}e^{\mu_{22}(w(t))}]}{d(B) + (b_{22} - b_{11})\sqrt{d(B)}},$$

$$\mu_{11}(w(t)) = -\frac{1}{2}\mu_1^2 t + \mu_1 w(t) \quad \text{and} \quad \mu_{22}(w(t)) = -\frac{1}{2}\mu_2^2 t + \mu_2 w(t),$$

$$\varphi_{11}(t, w)) = \phi_{11}(t)\varpi_{11}(w(t)) + \phi_{12}(t)\varpi_{21}(w(t)),$$

$$\varphi_{12}(t, w)) = \phi_{11}(t)\varpi_{12}(w(t)) + \phi_{12}(t)\varpi_{22}(w(t)),$$

$$\varphi_{21}(t, w)) = \phi_{21}(t)\varpi_{11}(w(t)) + \phi_{22}(t)\varpi_{21}(w(t)),$$

$$\varphi_{22}(t, w)) = \phi_{21}(t)\varpi_{12}(w(t)) + \phi_{22}(t)\varpi_{22}(w(t)).$$

For $r > 0$,

$\phi_{11}(t)\varpi_{11}(w(t))$

$$= \frac{2[a_{21}a_{12}e^{\lambda_1 t} + (\lambda_2 - a_{22})(a_{11} - \lambda_1)e^{\lambda_2 t}]}{d(A) + (a_{22} - a_{11})\sqrt{d(A)}}$$

$$\times \frac{2[b_{21}b_{12}e^{\mu_{11}(w(t))} + (\mu_2 - b_{22})(b_{11} - \mu_1)e^{\mu_{22}(w(t))}]}{d(B) + (b_{22} - b_{11})\sqrt{d(B)}}$$

$$= \frac{r^2 4[b_{21}b_{12}e^{\lambda_1 t} + (\mu_2 - b_{22})(b_{11} - \mu_1)e^{\lambda_2 t}]}{r^2(d(B) + (b_{22} - b_{11})\sqrt{d(B)})}$$

$$\times \frac{b_{21}b_{12}e^{\mu_{11}(w(t))} + (\mu_2 - b_{22})(b_{11} - \mu_1)e^{\mu_{22}(w(t))}}{d(B) + (b_{22} - b_{11})\sqrt{d(B)}}$$

$$= \frac{4[b_{21}b_{12}e^{\lambda_1 t} + (\mu_2 - b_{22})(b_{11} - \mu_1)e^{\lambda_2 t}]}{(d(B) + (b_{22} - b_{11})\sqrt{d(B)})^2}$$

$$\times \frac{b_{21}b_{12}e^{\mu_{11}(w(t))} + (\mu_2 - b_{22})(b_{11} - \mu_1)e^{\mu_{22}(w(t))}}{(d(B) + (b_{22} - b_{11})\sqrt{d(B)})^2}$$

$$= \frac{4[b_{21}b_{12}(\mu_2 - b_{22})(b_{11} - \mu_1)e^{\lambda_1 t + \mu_{22}(w(t))} + b_{21}b_{12}(\mu_2 - b_{22})(b_{11} - \mu_1)e^{\lambda_2 t + \mu_{11}(w(t))}]}{(d(B) + (b_{22} - b_{11})\sqrt{d(B)})^2}$$

$$+ \frac{4[(b_{21}b_{12})^2 e^{\lambda_1 t + \mu_{11}(w(t))} + [(\mu_2 - b_{22})(b_{11} - \mu_1)]^2 e^{\lambda_2 t + \mu_{22}(w(t))}]}{(d(B) + (b_{22} - b_{11})\sqrt{d(B)})^2},$$

$\phi_{12}(t)\varpi_{21}(w(t))$

$$= \frac{2[a_{12}(a_{22} - \lambda_2)(e^{\lambda_1 t} - e^{\lambda_2 t})]}{d(A) + (a_{22} - a_{11})\sqrt{d(A)}}$$

$$\times \frac{2[b_{21}(b_{11} - \mu_1)(e^{\mu_{22}(w(t))} - e^{\mu_{11}(w(t))})]}{d(B) + (b_{22} - b_{11})\sqrt{d(B)}}$$

$$= \frac{4[b_{12}(b_{22} - \mu_2)(e^{\lambda_1 t} - e^{\lambda_2 t})][b_{21}(b_{11} - \mu_1)(e^{\mu_{22}(w(t))} - e^{\mu_{11}(w(t))})]}{(d(B) + (b_{22} - b_{11})\sqrt{d(B)})^2}$$

$$= \frac{4[b_{21}b_{12}(b_{22} - \mu_2)(b_{11} - \mu_1)(e^{\lambda_1 t} - e^{\lambda_2 t})e^{\mu_{22}(w(t))}]}{(d(B) + (b_{22} - b_{11})\sqrt{d(B)})^2}$$

$$- \frac{b_{21}b_{12}(b_{22} - \mu_2)(b_{11} - \mu_1)(e^{\lambda_1 t} - e^{\lambda_2 t})e^{\mu_{11}(w(t))}}{(d(B) + (b_{22} - b_{11})\sqrt{d(B)})^2}$$

$$= \frac{4[-b_{21}b_{12}(\mu_2 - b_{22})(b_{11} - \mu_1)e^{\lambda_1 t + \mu_{22}(w(t))}]}{(d(B) + (b_{22} - b_{11})\sqrt{d(B)})^2}$$

$$- \frac{b_{21}b_{12}(\mu_2 - b_{22})(b_{11} - \mu_1)e^{\lambda_2 t + \mu_{11}(w(t))}}{(d(B) + (b_{22} - b_{11})\sqrt{d(B)})^2}$$

$$+ \frac{4[b_{21}b_{12}(\mu_2 - b_{22})(b_{11} - \mu_1)e^{\lambda_1 t + \mu_{11}(w(t))}]}{(d(B) + (b_{22} - b_{11})\sqrt{d(B)})^2}$$

$$+ \frac{b_{21}b_{12}(\mu_2 - b_{22})(b_{11} - \mu_1)e^{\lambda_2 t + \mu_{22}(w(t))}}{(d(B) + (b_{22} - b_{11})\sqrt{d(B)})^2}.$$

Hence,

$$\varphi_{11}(t, w(t)) = \phi_{11}(t)\varpi_{11}(w(t)) + \phi_{12}(t)\varpi_{21}(w(t))$$

$$= \frac{4[(b_{21}b_{12})^2 e^{\lambda_1 t + \mu_{11}(w(t))} + [(\mu_2 - b_{22})(b_{11} - \mu_1)]^2 e^{\lambda_2 t + \mu_{22}(w(t))}]}{(d(B) + (b_{22} - b_{11})\sqrt{d(B)})^2}$$

$$+ \frac{4[b_{21}b_{12}(\mu_2 - b_{22})(b_{11} - \mu_1)e^{\lambda_2 t + \mu_{22}(w(t))}]}{(d(B) + (b_{22} - b_{11})\sqrt{d(B)})^2}$$

$$+ \frac{b_{21}b_{12}(\mu_2 - b_{22})(b_{11} - \mu_1)e^{\lambda_1 t + \mu_{11}(w(t))}}{(d(B) + (b_{22} - b_{11})\sqrt{d(B)})^2}$$

$$= \frac{4[(b_{21}b_{12})^2 e^{\lambda_1 t + \mu_{11}(w(t))} + b_{21}b_{12}(\mu_2 - b_{22})(b_{11} - \mu_1)e^{\lambda_1 t + \mu_{11}(w(t))}]}{(d(B) + (b_{22} - b_{11})\sqrt{d(B)})^2}$$

$$+ \frac{4[b_{21}b_{12}(\mu_2 - b_{22})(b_{11} - \mu_1)e^{\lambda_2 t + \mu_{22}(w(t))}]}{(d(B) + (b_{22} - b_{11})\sqrt{d(B)})^2}$$

$$+ \frac{[(\mu_2 - b_{22})(b_{11} - \mu_1)]^2 e^{\lambda_2 t + \mu_{22}(w(t))}}{(d(B) + (b_{22} - b_{11})\sqrt{d(B)})^2}$$

$$= \frac{4[(b_{21}b_{12})^2 + b_{21}b_{12}(\mu_2 - b_{22})(b_{11} - \mu_1)e^{\lambda_1 t + \mu_{11}(w(t))}]}{(d(B) + (b_{22} - b_{11})\sqrt{d(B)})^2}$$

$$+ \frac{4[b_{21}b_{12}(\mu_2 - b_{22})(b_{11} - \mu_1) + [(\mu_2 - b_{22})(b_{11} - \mu_1)]^2]e^{\lambda_2 t + \mu_{22}(w(t))}}{(d(B) + (b_{22} - b_{11})\sqrt{d(B)})^2},$$

$$\phi_{11}(t)\varpi_{12}(w(t))$$

$$= \frac{2[a_{21}a_{12}e^{\lambda_1 t} + (\lambda_2 - a_{22})(a_{11} - \lambda_1)e^{\lambda_2 t}]}{d(A) + (a_{22} - a_{11})\sqrt{d(A)}}$$

$$\times \frac{2[b_{12}(b_{22} - \mu_2)(e^{\mu_{11}(w(t))} - e^{\mu_{22}(w(t))})]}{d(B) + (b_{22} - b_{11})\sqrt{d(B)}}$$

$$= \frac{4[b_{21}b_{12}e^{\lambda_1 t} + (\mu_2 - b_{22})(b_{11} - \mu_1)e^{\lambda_2 t}]}{d(B) + (b_{22} - b_{11})\sqrt{d(B)}}$$

$$\times \frac{b_{12}(b_{22} - \mu_2)(e^{\mu_{11}(w(t))} - e^{\mu_{22}(w(t))})}{d(B) + (b_{22} - b_{11})\sqrt{d(B)}}$$

$$= \frac{4[b_{21}b_{12}^2(b_{22} - \mu_2)e^{\lambda_1 t + \mu_{11}(w(t))} - b_{12}(\mu_2 - b_{22})^2(b_{11} - \mu_1)e^{\lambda_2 t + \mu_{22}(w(t))}]}{(d(B) + (b_{22} - b_{11})\sqrt{d(B)})^2}$$

$$+ \frac{4[b_{12}(\mu_2 - b_{22})^2(b_{11} - \mu_1)e^{\lambda_2 t + \mu_{11}(w(t))} - b_{21}b_{12}^2(b_{22} - \mu_2)e^{\lambda_1 t + \mu_{22}(w(t))}]}{(d(B) + (b_{22} - b_{11})\sqrt{d(B)})^2},$$

$$\phi_{12}(t)\varpi_{22}(w(t))$$

$$= \frac{2[a_{12}(a_{22} - \lambda_2)(e^{\lambda_1 t} - e^{\lambda_2 t})]}{d(A) + (a_{22} - a_{11})\sqrt{d(A)}} \frac{2[(\mu_1 - b_{11})(b_{22} - \mu_2)(e^{\mu_{11}(w(t))} + b_{21}b_{12}e^{\mu_{22}(w(t))})]}{d(B) + (b_{22} - b_{11})\sqrt{d(B)}}$$

$$= \frac{4[b_{21}(b_{22} - \mu_2)(e^{\lambda_1 t} - e^{\lambda_2 t})]}{d(B) + (b_{22} - b_{11})\sqrt{d(B)}} \frac{(\mu_1 - b_{11})(b_{22} - \mu_2)e^{\mu_{11}(w(t))} + b_{21}b_{12}e^{\mu_{22}(w(t))}}{d(B) + (b_{22} - b_{11})\sqrt{d(B)}}$$

$$= \frac{4[b_{12}(\mu_1 - b_{11})(b_{22} - \mu_2)^2 e^{\lambda_1 t + \mu_{11}(w(t))} - b_{21}b_{12}^2(b_{22} - \mu_2)e^{\lambda_2 t + \mu_{22}(w(t))}]}{(d(B) + (b_{22} - b_{11})\sqrt{d(B)})^2}$$

$$+ \frac{b_{21}b_{12}^2(b_{22} - \mu_2)e^{\lambda_1 t + \mu_{22}(w(t))} - b_{12}(\mu_1 - b_{11})(b_{22} - \mu_2)^2 e^{\lambda_2 t + \mu_{11}(w(t))}}{(d(B) + (b_{22} - b_{11})\sqrt{d(B)})^2}.$$

Hence,

$$\varphi_{12}(t, w(t)) = \phi_{11}(t)\varpi_{12}(w(t)) + \phi_{12}(t)\varpi_{22}(w(t))$$

$$= \frac{4[b_{21}b_{12}^2(b_{22} - \mu_2)e^{\lambda_1 t + \mu_{11}(w(t))} - b_{12}(\mu_2 - b_{22})^2(b_{11} - \mu_1)e^{\lambda_2 t + \mu_{22}(w(t))}]}{(d(B) + (b_{22} - b_{11})\sqrt{d(B)})^2}$$

$$+ \frac{4[b_{12}(\mu_1 - b_{11})(b_{22} - \mu_2)^2 e^{\lambda_1 t + \mu_{11}(w(t))} - b_{21}b_{12}^2(b_{22} - \mu_2)e^{\lambda_2 t + \mu_{22}(w(t))}]}{(d(B) + (b_{22} - b_{11})\sqrt{d(B)})^2}$$

$$= \frac{4[b_{21}b_{12}^2(b_{22} - \mu_2)e^{\lambda_1 t + \mu_{11}(w(t))} + b_{12}(b_{11} - \mu_1)(b_{22} - \mu_2)^2 e^{\lambda_1 t + \mu_{11}(w(t))}]}{(d(B) + (b_{22} - b_{11})\sqrt{d(B)})^2}$$

$$- \frac{4[b_{12}(b_{22} - \mu_2)^2(b_{11} - \mu_1)e^{\lambda_2 t + \mu_{22}(w(t))} + b_{12}b_{12}^2(b_{22} - \mu_2)e^{\lambda_2 t + \mu_{22}(w(t))}]}{(d(B) + (b_{22} - b_{11})\sqrt{d(B)})^2}$$

$$= \frac{4b_{12}(b_{22} - \mu_2)[b_{21}b_{12} + (b_{11} - \mu_1)(b_{22} - \mu_2)]e^{\lambda_1 t + \mu_{11}(w(t))}}{(d(B) + (b_{22} - b_{11})\sqrt{d(B)})^2}$$

$$- \frac{4b_{12}(b_{22} - \mu_2)[(b_{11} - \mu_1)(b_{22} - \mu_2) + b_{21}b_{12}]e^{\lambda_2 t + \mu_{22}(w(t))}}{(d(B) + (b_{22} - b_{11})\sqrt{d(B)})^2}$$

$$= \frac{4b_{12}(b_{22} - \mu_2)[b_{21}b_{12} + (b_{11} - \mu_1)(b_{22} - \mu_2)][e^{\lambda_1 t + \mu_{11}(w(t))} - e^{\lambda_2 t + \mu_{22}(w(t))}]}{(d(B) + (b_{22} - b_{11})\sqrt{d(B)})^2}.$$

We observe that by interchanging the subscripts of all the parameters λ_1, λ_2, μ_1, μ_2, b_{ij} in the numerators of $\phi_{11}(t, w(t))$ and $\phi_{12}(t, w(t))$, the expressions for $\phi_{22}(t, w(t))$ and $\phi_{21}(t, w(t))$ can be obtained. In fact, by interchanging $11 \leftrightarrow 22$, $2 \leftrightarrow 1$, and $12 \leftrightarrow 21$, we have

$$\varphi_{22}(t, w(t)) = \phi_{21}(t)\varpi_{12}(w(t)) + \phi_{22}(t)\varpi_{22}(w(t))$$

$$= \frac{4[(b_{21}b_{12})^2 + b_{21}b_{12}(\mu_1 - b_{11})(b_{22} - \mu_2)]e^{\lambda_2 t + \mu_{22}(w(t))}}{(d(B) + (b_{22} - b_{11})\sqrt{d(B)})^2}$$

$$+ \frac{4[b_{21}b_{12}(\mu_1 - b_{11})(b_{22} - \mu_2) + [(\mu_1 - b_{11}(b_{22} - \mu_2)]^2]e^{\lambda_1 t + \mu_{11}(w(t))}}{(d(B) + (b_{22} - b_{11})\sqrt{d(B)})^2},$$

$$\varphi_{21}(t, w(t)) = \phi_{21}(t)\varpi_{11}(w(t)) + \phi_{22}(t)\varpi_{21}(w(t))$$

$$= \frac{4b_{21}(b_{11} - \mu_1)[b_{21}b_{12} + (b_{22} - \mu_2)(b_{11} - \mu_1)][e^{\lambda_2 t + \mu_{22}(w(t))} - e^{\lambda_1 t + \mu_{11}(w(t))}]}{(d(B) + (b_{22} - b_{11})\sqrt{d(B)})^2}.$$

This is the complete representation of the normalized fundamental matrix solution process of the corresponding first-order linear system of differential equations.

Now, by imitating Steps #2 and #3 of Procedure 4.5.2 as well as Example 2.4.1, one can easily find a particular solution. Hence, the general solution of the given illustration depends on the two cases. However, the form of the general solution is given by (4.5.19):

$$x(t) = \Phi(t)c + \int_a^t \Phi(t,s)[p(s) - B(s)q(s)]ds + \int_a^t \Phi(t,s)(t,s)q(s)dw(s),$$

where the fundamental matrix $\Phi(t) \equiv \Phi(t, w(t))$ is as determined above. This completes the solution process of (a). From Example 4.4.2, the normalized fundamental solution processes corresponding to problems (b), (c), and (d), (e) are $\Phi_d(t,0)$ and $\Phi_s(t, w(t), 0)$, respectively. With this observation, the general solution processes of (b)–(e) reduce to

$$x(t) = \Phi_d(t,0)c + \int_a^t \Phi_d(t,s)p(s)ds,$$

$$x(t) = \Phi_d(t,0)c + \int_a^t \Phi_d(t,s)q(s)dw(s),$$

$$x(t) = \Phi_s(t, w(t), 0)c + \int_a^t \Phi_s(t, w(s), s)p(s)ds,$$

and

$$x(t) = \Phi_s(t, w(t), 0)c - \int_a^t \Phi_s(t, w(s), s)Bq(s)ds + \int_a^t \Phi_s(t, w(s), s)q(s)dw(s),$$

respectively.

Example 4.5.1. Given: $dx = [Ax + p(t)]dt + [Bx + q(t)]dw(t)$, where $A = \begin{bmatrix} 2\alpha & \alpha \\ \alpha & \alpha \end{bmatrix}$, $B = \begin{bmatrix} 2\sigma & \sigma \\ \sigma & \sigma \end{bmatrix}$, σ and α are constants, p, and q and are two-dimensional continuous functions defined on an interval $J \subseteq R$ into R^2. Find:

(a) a general solution to the given differential equation, particularly, for:
(b) $\sigma = 0$ and $q(t) \equiv 0$;
(c) $\sigma = 0$ and $p(t) \equiv 0$;
(d) $\alpha = 0$ and $q(t) \equiv 0$;
(e) $\alpha = 0$ and $p(t) \equiv 0$.

Solution procedure. Following Procedure 4.5.2, we find a general solution of the given example. We need to find the fundamental matrix solution, and then a particular solution to the given problem. First, we imitate the solution procedures of Illustrations 4.3.1 and 4.5.1. In this case, $\lambda_1^d = \frac{3+\sqrt{5}}{2}\alpha$ with $c_1^d = \begin{bmatrix} 1 & \frac{-1+\sqrt{5}}{2} \end{bmatrix}^T$ and $\lambda_2^d = \frac{3-\sqrt{5}}{2}\alpha$ with $c_2^d = \begin{bmatrix} \frac{1-\sqrt{5}}{2} & 1 \end{bmatrix}^T$. Similarly, the eigenvalues and corresponding eigenvectors of the matrix B are $\mu_1 = \frac{3+\sqrt{5}}{2}\sigma$ and $\lambda_1^s = -\frac{1}{2}\mu_1^2 = -\frac{7+3\sqrt{5}}{4}\sigma^2$ with $c_1^s = \left[1 \quad \frac{-1+\sqrt{5}}{2}\right]^T$ and $\mu_2 = \frac{3-\sqrt{5}}{2}\sigma$ and $\lambda_2^s = -\frac{1}{2}\mu_2^2 = -\frac{7-3\sqrt{5}}{4}\sigma^2$ with $c_2^s = \left[\frac{1-\sqrt{5}}{2} \quad 1\right]^T$. The normalized fundamental matrix solution processes $\Phi_d(t,0)$ and $\Phi_s(t,0)$ of the deterministic and stochastic components of the given problem are:

$$\Phi_d(t,0) = \frac{4}{\alpha^2[4+(1-\sqrt{5})^2]} \begin{bmatrix} \alpha e^{\lambda_1^d t} & \frac{1-\sqrt{5}}{2}\alpha e^{\lambda_2^d t} \\ \frac{-1+\sqrt{5}}{2}\alpha e^{\lambda_1^d t} & \alpha e^{\lambda_2^d t} \end{bmatrix}$$

$$\times \begin{bmatrix} \alpha & -\frac{1-\sqrt{5}}{2}\alpha \\ \frac{1-\sqrt{5}}{2}\alpha & \alpha \end{bmatrix}$$

$$= \begin{bmatrix} \dfrac{4e^{\lambda_1^d t} + (1-\sqrt{5})^2 e^{\lambda_2^d t}}{4+(1-\sqrt{5})^2} & \dfrac{2(1-\sqrt{5})(e^{\lambda_2^d t} - e^{\lambda_1^d t})}{4+(1-\sqrt{5})^2} \\ \dfrac{-2(1-\sqrt{5})(e^{\lambda_1^d t} - e^{\lambda_2^d t})}{4+(1-\sqrt{5})^2} & \dfrac{(1-\sqrt{5})^2 e^{\lambda_1^d t} + 4e^{\lambda_2^d t}}{4+(1-\sqrt{5})^2} \end{bmatrix},$$

$$\Phi_s(t,0) = \begin{bmatrix} \dfrac{4e^{\beta_{11}(w(t))} + (1-\sqrt{5})^2 e^{\beta_{22}(w(t))}}{4+(1-\sqrt{5})^2} & \dfrac{2(1-\sqrt{5})(e^{\beta_{22}(w(t))} - e^{\beta_{11}(w(t))})}{4+(1-\sqrt{5})^2} \\ \dfrac{-2(1-\sqrt{5})(e^{\beta_{11}(w(t))} - e^{\beta_{22}(w(t))})}{4+(1-\sqrt{5})^2} & \dfrac{(1-\sqrt{5})^2 e^{\beta_{11}(w(t))} + 4e^{\beta_{22}(w(t))}}{4+(1-\sqrt{5})^2} \end{bmatrix},$$

where $\beta_{11}(w(t)) = -\frac{7+3\sqrt{5}}{4}\sigma^2 t + \frac{3+\sqrt{5}}{2}\sigma w(t)$ and $-\frac{7-3\sqrt{5}}{4}\sigma^2 t + \frac{3-\sqrt{5}}{2}\sigma w(t) = \beta_{22}(w(t))$.

From Step #1 of Procedure 4.5.2, the fundamental matrix solution to the corresponding homogeneous system $dx = Ax\,dt + Bx\,dw(t)$ is given by

$$\Phi(t,w(t)) = \Phi_d(t,0)\Phi_s(t,w(t),0) = \begin{bmatrix} \varphi_{11}(t,w(t)) & \varphi_{12}(t,w(t)) \\ \varphi_{21}(t,w(t)) & \varphi_{22}(t,w(t)) \end{bmatrix},$$

$\varphi_{11}(t,w(t))$

$$= \frac{4e^{\lambda_1^d t} + (1-\sqrt{5})^2 e^{\lambda_2^d t}}{4+(1-\sqrt{5})^2} \frac{4e^{\beta_{11}(w(t))} + (1-\sqrt{5})^2 e^{\beta_{22}(w(t))}}{4+(1-\sqrt{5})^2}$$

$$+ \frac{2(1-\sqrt{5})(e^{\lambda_2^d t} - e^{\lambda_1^d t})}{4+(1-\sqrt{5})^2} \frac{-2(1-\sqrt{5})(e^{\beta_{11}(w(t))} - e^{\beta_{22}(w(t))})}{4+(1-\sqrt{5})^2}$$

$$= \frac{16e^{\lambda_1^d t + \beta_{11}(w(t))} + (1 - \sqrt{5})^4 e^{\lambda_2^d t + \beta_{22}(w(t))}}{[4 + (1 - \sqrt{5})^2]^2}$$

$$+ \frac{4(1 - \sqrt{5})^2 [e^{\lambda_2^d t + \beta_{11}(w(t))} + e^{\lambda_1^d t + \beta_{22}(w(t))}]}{[4 + (1 - \sqrt{5})^2]^2}$$

$$+ \frac{4(1 - \sqrt{5})^2 [e^{\lambda_2^d t + \beta_{22}(w(t))} - e^{\lambda_2^d t + \beta_{11}(w(t))} + e^{\lambda_1^d t + \beta_{11}(w(t))} - e^{\lambda_1^d t + \beta_{22}(w(t))}]}{[4 + (1 - \sqrt{5})^2]^2}$$

$$= \frac{16e^{\lambda_1^d t + \beta_{11}(w(t))} + (1 - \sqrt{5})^4 e^{\lambda_2^d t + \beta_{22}(w(t))}}{[4 + (1 - \sqrt{5})^2]^2}$$

$$+ \frac{4(1 - \sqrt{5})^2 [e^{\lambda_2^d t + \beta_{22}(w(t))} + e^{\lambda_1^d t + \beta_{11}(w(t))}]}{[4 + (1 - \sqrt{5})^2]^2}$$

$$= \frac{[16 + 4(1 - \sqrt{5})^2]e^{\lambda_1^d t + \beta_{11}(w(t))} + [4(1 - \sqrt{5})^2 + (1 - \sqrt{5})^4]e^{\lambda_2^d t + \beta_{22}(w(t))}}{[4 + (1 - \sqrt{5})^2]^2}$$

$$= \frac{4\exp[\lambda_1^d t + \beta_{11}(w(t))] + (1 - \sqrt{5})^2 \exp[\lambda_2^d t + \beta_{22}(w(t))]}{4 + (1 - \sqrt{5})^2},$$

$$\varphi_{12}(t, w(t)) = \frac{4e^{\lambda_1^d t} + (1 - \sqrt{5})^2 e^{\lambda_2^d t}}{4 + (1 - \sqrt{5})^2} \frac{2(1 - \sqrt{5})(e^{\beta_{22}(w(t))} - e^{\beta_{11}(w(t))})}{4 + (1 - \sqrt{5})^2}$$

$$+ \frac{2(1 - \sqrt{5})(e^{\lambda_2^d t} - e^{\lambda_1^d t})}{4 + (1 - \sqrt{5})^2} \frac{(1 - \sqrt{5})^2 e^{\beta_{11}(w(t))} + 4e^{\beta_{22}(w(t))}}{4 + (1 - \sqrt{5})^2}$$

$$= \frac{8(1 - \sqrt{5})e^{\lambda_1^d t + \beta_{22}(w(t))} + 2(1 - \sqrt{5})^3 e^{\lambda_2^d t + \beta_{22}(w(t))}}{[4 + (1 - \sqrt{5})^2]^2}$$

$$- \frac{8(1 - \sqrt{5})e^{\lambda_1^d t + \beta_{11}(w(t))} + 2(1 - \sqrt{5})^3 e^{\lambda_2^d t + \beta_{11}(w(t))}}{[4 + (1 - \sqrt{5})^2]^2}$$

$$+ \frac{2(1 - \sqrt{5})^3 e^{\lambda_2^d t + \beta_{11}(w(t))} + 8(1 - \sqrt{5})e^{\lambda_2^d t + \beta_{22}(w(t))}}{[4 + (1 - \sqrt{5})^2]^2}$$

$$- \frac{2(1 - \sqrt{5})^3 e^{\lambda_1^d t + \beta_{11}(w(t))} + 8(1 - \sqrt{5})e^{\lambda_1^d t + \beta_{22}(w(t))}}{[4 + (1 - \sqrt{5})^2]^2}$$

$$= \frac{2(1 - \sqrt{5})[(1 - \sqrt{5})^2 + 4][e^{\lambda_2^d t + \beta_{22}(w(t))} - e^{\lambda_1^d t + \beta_{11}(w(t))}]}{[4 + (1 - \sqrt{5})^2]^2}$$

$$= \frac{2(1 - \sqrt{5})[e^{\lambda_2^d t + \beta_{22}(w(t))} - e^{\lambda_1^d t + \beta_{11}(w(t))}]}{4 + (1 - \sqrt{5})^2},$$

$$\varphi_{21}(t, w(t)) = \frac{-2(1-\sqrt{5})(e^{\lambda_1^d t} - e^{\lambda_2^d t})}{4 + (1-\sqrt{5})^2} \frac{4e^{\beta_{11}(w(t))} + (1-\sqrt{5})^2 e^{\beta_{22}(w(t))}}{4 + (1-\sqrt{5})^2}$$

$$+ \frac{(1-\sqrt{5})^2 e^{\lambda_1^d t} + 4e^{\lambda_2^d t}}{4 + (1-\sqrt{5})^2} \frac{-2(1-\sqrt{5})(e^{\beta_{11}(w(t))} - e^{\beta_{22}(w(t))})}{4 + (1-\sqrt{5})^2}$$

$$= \frac{-8(1-\sqrt{5})e^{\lambda_1^d t + \beta_{11}(w(t))} + 2(1-\sqrt{5})^3 e^{\lambda_2^d t + \beta_{22}(w(t))}}{[4 + (1-\sqrt{5})^2]^2}$$

$$+ \frac{8(1-\sqrt{5})e^{\lambda_2^d t + \beta_{11}(w(t))} - 2(1-\sqrt{5})^3 e^{\lambda_1^d t + \beta_{22}(w(t))}}{[4 + (1-\sqrt{5})^2]^2}$$

$$+ \frac{2(1-\sqrt{5})^3 e^{\lambda_1^d t + \beta_{22}(w(t))} + 8(1-\sqrt{5})e^{\lambda_2^d t + \beta_{22}(w(t))}}{[4 + (1-\sqrt{5})^2]^2}$$

$$- \frac{2(1-\sqrt{5})^3 e^{\lambda_1^d t + \beta_{11}(w(t))} + 8(1-\sqrt{5})e^{\lambda_2^d t + \beta_{11}(w(t))}}{[4 + (1-\sqrt{5})^2]^2}$$

$$= \frac{2(1-\sqrt{5})[(1-\sqrt{5})^2 + 4][e^{\lambda_2^d t + \beta_{22}(w(t))} - e^{\lambda_1^d t + \beta_{11}(w(t))}]}{4 + (1-\sqrt{5})^2}$$

$$= \frac{-2(1-\sqrt{5})[e^{\lambda_1^d t + \beta_{11}(w(t))} - e^{\lambda_2^d t + \beta_{22}(w(t))}]}{4 + (1-\sqrt{5})^2},$$

$$\varphi_{22}(t, w(t))$$

$$= \frac{-2(1-\sqrt{5})(e^{\lambda_1^d t} - e^{\lambda_2^d t})}{4 + (1-\sqrt{5})^2} \frac{2(1-\sqrt{5})(e^{\beta_{22}(w(t))} - e^{\beta_{11}(w(t))})}{4 + (1-\sqrt{5})^2}$$

$$+ \frac{(1-\sqrt{5})^2 e^{\lambda_1^d t} + 4e^{\lambda_2^d t}}{4 + (1-\sqrt{5})^2} \frac{(1-\sqrt{5})^2 e^{\beta_{11}(w(t))} + 4e^{\beta_{22}(w(t))}}{4 + (1-\sqrt{5})^2}$$

$$= \frac{-4(1-\sqrt{5})^2 e^{\lambda_1^d t + \beta_{22}(w(t))} + 4(1-\sqrt{5})^2 e^{\lambda_2^d t + \beta_{22}(w(t))}}{[4 + (1-\sqrt{5})^2]^2}$$

$$+ \frac{4(1-\sqrt{5})^2 e^{\lambda_1^d t + \beta_{11}(w(t))} - 4(1-\sqrt{5})^2 e^{\lambda_2^d t + \beta_{11}(w(t))}}{[4 + (1-\sqrt{5})^2]^2}$$

$$+ \frac{(1-\sqrt{5})^4 e^{\lambda_1^d t + \beta_{11}(w(t))} + 4(1-\sqrt{5})^2 e^{\lambda_1^d t + \beta_{22}(w(t))}}{[4 + (1-\sqrt{5})^2]^2}$$

$$+ \frac{4(1-\sqrt{5})^2 e^{\lambda_2^d t + \beta_{11}(w(t))} + 16 e^{\lambda_2^d t + \beta_{22}(w(t))}}{[4 + (1-\sqrt{5})^2]^2}$$

$$= \frac{[16 + 4(1-\sqrt{5})^2]e^{\lambda_2^d t + \beta_{22}(w(t))} + [4(1-\sqrt{5})^2 + (1-\sqrt{5})^4]e^{\lambda_1^d t + \beta_{11}(w(t))}}{[4 + (1-\sqrt{5})^2]^2}$$

$$= \frac{4\exp[\lambda_2^d t + \beta_{22}(w(t))] + (1-\sqrt{5})^2 \exp[\lambda_1^d t + \beta_{11}(w(t))]}{4 + (1-\sqrt{5})^2}.$$

This is the complete representation of the normalized fundamental matrix solution process of the first-order linear system of differential equations. Following the argument used in Illustration 4.5.1, one can find the general solution to the given system. This completes the solution process of (a). Following the argument used in Illustration 4.5.1, the general solution processes of (b), (c), (d), and (d) reduce to

$$x(t) = \Phi_d(t, 0)c + \int_a^t \Phi_d(t, s)p(s)ds,$$

$$x(t) = \Phi_d(t, 0)c + \int_a^t \Phi_d(t, s)q(s)dw(s),$$

$$x(t) = \Phi_s(t, w(t), 0)c + \int_a^t \Phi_s(t, s)p(s)ds,$$

$$x(t) = \Phi_s(t, w(t), 0)c - \int_a^t \Phi_s(t, w(s), s)Bq(s)ds$$

$$+ \int_a^t \Phi_s(t, w(s), s)q(s)dw(s),$$

respectively, where $\Phi_d(t, 0)$ and $\Phi_s(t, w(t), 0)$ are the normalized solution processes of the deterministic $(dx = Ax\, dt)$ and stochastic parts $[dx = Bx\, dw(t)]$ of the given system of differential equations. This completes the solution process of the example.

Illustration 4.5.2. Given: $dx = [Ax + p(t)]dt + [Bx + q(t)]dw(t)$, where

$$A = \begin{bmatrix} a_{11} & a_{12} \\ a_{21} & a_{22} \end{bmatrix}, \quad B = \begin{bmatrix} b_{11} & b_{12} \\ b_{21} & b_{22} \end{bmatrix},$$

using b_{ij} and a_{ij} being any given real numbers satisfying the condition:

(a) $d(A) = (\operatorname{tr}(A))^2 - 4\det(A) < 0$;
(b) $AB = BA$;
(c) p and q are two-dimensional continuous functions defined on an interval $J \subseteq R$ into R^n, and w is a scalar normalized Wiener process.

Find:

(i) a general solution of given differential equation, in particular, for
(ii) $B = 0$ and $q(t) \equiv 0$;
(iii) $B = 0$ and $p(t) \equiv 0$;
(iv) $A = 0$ and $q(t) \equiv 0$;
(v) $A = 0$ and $p(t) \equiv 0$.

Solution procedure. Again, we note that $n = 2$. We imitate the arguments used in Illustrations 4.3.2, 4.4.1, and 4.5.1 and Procedure 4.5.2 to find the general solution to the given system of differential equations. We further recall that $d(A) = r^2 d(B)$.

In this case, we have complex conjugate eigenvalues. We repeat the procedures described in Observation 4.2.4(vi) and Illustrations 4.3.2, and 4.5.1, and get

$$\Phi_d(t) = \begin{bmatrix} a_{12}e^{\alpha^d t}\cos\beta^d t & a_{12}e^{\alpha^d t}\sin\beta^d t \\ e^{\alpha^d t}[(\alpha^d - a_{11})\cos\beta^d t - \beta^d \sin\beta^d t] & e^{\alpha^d t}[(\alpha^d - a_{11})\sin\beta^d t + \beta^d \cos\beta^d t] \end{bmatrix},$$

$$\Phi_d(0) = \begin{bmatrix} a_{12} & 0 \\ \alpha^d - a_{11} & \beta^d \end{bmatrix}, \quad \Phi_d^{-1}(0) = \begin{bmatrix} \dfrac{1}{a_{12}} & 0 \\ -\dfrac{\alpha^d - a_{11}}{a_{12}\beta^d} & \dfrac{1}{\beta^d} \end{bmatrix},$$

$$\Phi_d(t,0) = \begin{bmatrix} e^{\alpha^d t}\dfrac{\beta^d \cos\beta^d t - (\alpha^d - a_{11})\sin\beta^d t}{\beta^d} & a_{12}e^{\alpha^d t}\dfrac{\sin\beta^d t}{\beta^d} \\ -e^{\alpha^d t}\dfrac{[(\alpha^d - a_{11})^2 + \beta^{d^2}]\sin\beta^d t}{a_{12}\beta^d} & e^{\alpha^d t}\dfrac{(\alpha^d - a_{11})\sin\beta^d t + \beta^d \cos\beta^d t}{\beta^d} \end{bmatrix},$$

where $\lambda^d = \alpha^d + \beta^d i$, and

$$\Phi_s(t, w(t)) = e^{(\gamma t + \alpha w(t))} \begin{bmatrix} \phi_{11}(t, w(t)) & \phi_{12}(t, w(t)) \\ \phi_{21}(t, w(t)) & \phi_{22}(t, w(t)) \end{bmatrix},$$

$$\Phi_s(0) = \begin{bmatrix} b_{12} & 0 \\ b_{11} - \alpha & \beta^d \end{bmatrix}, \quad \Phi_s^{-1}(0) = \begin{bmatrix} \dfrac{1}{b_{12}} & 0 \\ -\dfrac{b_{11} - \alpha}{b_{12}\beta} & \dfrac{1}{\beta^d} \end{bmatrix},$$

and hence

$$\Phi_s(t, w(t), 0) = e^{(\gamma t + \alpha w(t))} \begin{bmatrix} \dfrac{\beta\phi_{11}(t, w(t)) - (b_{11} - a)\phi_{12}(t, w(t))}{\beta b_{12}} & \dfrac{\phi_{12}(t, w(t))}{\beta} \\ \dfrac{\beta\phi_{21}(t, w(t)) - (b_{11} - a)\phi_{22}(t, w(t))}{\beta b_{12}} & \dfrac{\phi_{22}(t, w(t))}{\beta} \end{bmatrix},$$

where

$$\phi_{11}(t, w(t)) = b_{12}\cos(\nu t + \beta w(t)),$$

$$\phi_{12}(t, w(t)) = b_{12}\sin(\nu t + \beta w(t)),$$

$$\phi_{21}(t, w(t)) = (b_{11} - \alpha)\cos(\nu t + \beta w(t)) - \beta\sin(\nu t + \beta w(t)),$$

$$\phi_{22}(t, w(t)) = (b_{11} - \alpha)\sin(\nu t + \beta w(t)) + \beta\cos(\nu t + \beta w(t)),$$

$\mu = \alpha + \beta i$, $\alpha = \dfrac{\text{tr}(B)}{2}$, $\beta = \dfrac{\sqrt{-d(B)}}{2}$, $\nu = -\alpha\beta = -\dfrac{\text{tr}(B)\sqrt{-d(B)}}{4}$, $\gamma = -\frac{1}{2}(\alpha^2 - \beta^2) = -\dfrac{(\text{tr}(B))^2 + d(B)}{8} = -\dfrac{(\text{tr}(B))^2 - 2\det(B)}{4}$.

From Step #1 of Procedure 4.5.2, the fundamental matrix solution to the corresponding homogeneous system $dx = Ax\, dt + Bx\, dw(t)$ is given by $\Phi(t, w(t)) =$

$\Phi_d(t,0)\Phi_s(t,w(t),0)$, where $\Phi_d(t,0)$ is as determined above $\Phi_s(t,w(t),0)$. We note that $\Phi(t,w(t))$ is a normalized fundamental matrix solution to the corresponding homogeneous system of differential equations of the given system of linear nonhomogeneous differential equations. To complete the remaining parts of the problem, the procedure of Illustration 4.5.2 can be imitated in the context of the given problem. The details are left to the reader.

Example 4.5.2. Given: $dx = [Ax + p(t)]dt + [Bx + q(t)]dw(t)$, where $A = \begin{bmatrix} 3\alpha & -\sqrt{5}\alpha \\ \sqrt{5}\alpha & -\alpha \end{bmatrix}$, $B = \begin{bmatrix} 4\sigma & -2\sqrt{5}\sigma \\ 2\sqrt{5}\sigma & -4\sigma \end{bmatrix}$, σ and α are constants, p and q are two-dimensional continuous functions defined on an interval $J \subseteq R$ into R^2. Find:

(a) a general solution to the given differential equation, particularly for:
(b) $\sigma = 0$ and $q(t) \equiv 0$;
(c) $\sigma = 0$ and $p(t) \equiv 0$;
(d) $\alpha = 0$ and $q(t) \equiv 0$;
(e) $\alpha = 0$ and $p(t) \equiv 0$.

Solution procedure. By following the solution procedure of Illustrations 4.5.1 and 4.5.2 and Example 4.5.1, we conclude that for $n = 2$, the eigenvalues of the coefficient matrices A and B are complex conjugate numbers $(1 \pm i)\alpha$ and $\pm 2\sigma i$, respectively. Moreover, $\lambda_1^d = (1+i)\alpha$ with $c_1^d = [-\sqrt{5}\alpha - (2-i)\alpha]^T$ and $\lambda_2^d = (1-i)\alpha$ with $c_1^d = [(2+i)\alpha \ \sqrt{5}\alpha]^T$. Similarly, the eigenvalues and corresponding eigenvectors of the matrix B are $\mu_1 = 2\sigma i$ and $\lambda_1^s = -\frac{1}{2}\mu_1^2 = 2\sigma^2$ with $c_1^s = [-\sqrt{5}\sigma - (2-i)\sigma]^T$, and $\mu_2 = -2\sigma i$ and $\lambda_2^s = -\frac{1}{2}\mu_2^2 = 2\sigma^2$ with $c_2^s = [(2+i)\sigma \ \sqrt{5}\sigma]^T$. Here, by following the notations of Illustrations 4.3.2 and 4.5.2, we have $\gamma = 2\sigma^2$ and $\nu = 0$. The normalized fundamental matrix solution processes $\Phi_d(t,0)$ and $\Phi_s(t,0)$ of the deterministic and stochastic components of the given problem are:

$$\Phi_d(t,0) = -\frac{e^{\alpha t}}{\alpha^2\sqrt{5}} \begin{bmatrix} -\alpha\sqrt{5}\cos\alpha t & -\alpha\sqrt{5}\sin\alpha t \\ -\alpha(2\cos\alpha t + \sin\alpha t) & \alpha(\cos\alpha t - 2\sin\alpha t) \end{bmatrix} \begin{bmatrix} \alpha & 0 \\ 2\alpha & -\alpha\sqrt{5} \end{bmatrix}$$

$$= \frac{e^{\alpha t}}{\sqrt{5}} \begin{bmatrix} \sqrt{5}\cos\alpha t + 2\sqrt{5}\sin\alpha t & -5\sin\alpha t \\ 2\cos\alpha t + \sin\alpha t - 2\cos\alpha t + 4\sin\alpha t & \sqrt{5}\cos\alpha t - 2\sqrt{5}\sin\alpha t \end{bmatrix}$$

$$= e^{\alpha t} \begin{bmatrix} \cos\alpha t + 2\sin\alpha t & -\sqrt{5}\sin\alpha t \\ \sqrt{5}\sin\alpha t & \cos\alpha t - 2\sin\alpha t \end{bmatrix}$$

$$= e^{\alpha t} \begin{bmatrix} \sqrt{5}\cos(\alpha t - \theta) & -\sqrt{5}\sin\alpha t \\ \sqrt{5}\sin\alpha t & \sqrt{5}\cos(\alpha t + \theta) \end{bmatrix}, \quad \theta = \arctan 2,$$

and by repeating the above computations, we get

$$\Phi_s(t, w(t), 0)$$

$$= e^{2\sigma^2 t} \begin{bmatrix} \cos 2\sigma w(t) + 2\sin 2\sigma w(t) & -\sqrt{5}\sin 2\sigma w(t) \\ \sqrt{5}\sin 2\sigma w(t) & \cos 2\sigma w(t) - 2\sin 2\sigma w(t) \end{bmatrix}$$

$$= e^{2\sigma^2 t} \begin{bmatrix} \sqrt{5}\cos(2\sigma w(t) - \theta) & -\sqrt{5}\sin 2\sigma w(t) \\ \sqrt{5}\sin 2\sigma w(t) & \sqrt{5}\cos(2\sigma w(t) + \theta) \end{bmatrix}, \quad \theta = \arctan 2.$$

The fundamental matrix solution to the corresponding homogeneous system is given by

$$\Phi(t, w(t), 0) = \Phi_d(t, 0)\Phi_s(t, w(t), 0)$$

$$= e^{(\alpha + 2\sigma^2)t} \begin{bmatrix} \cos(2\sigma w(t) + \alpha t) + 2\sin(2\sigma w(t) + \alpha t) & -\sqrt{5}\sin(2\sigma w(t) + \alpha t) \\ \sqrt{5}\sin(2\sigma w(t) + \alpha t) & \cos(2\sigma w(t) + \alpha t) - 2\sin(2\sigma w(t) + \alpha t) \end{bmatrix}$$

$$= e^{(\alpha + 2\sigma^2)t} \begin{bmatrix} \sqrt{5}\cos((2\sigma w(t) + \alpha t) - \theta) & -\sqrt{5}\sin(2\sigma w(t) + \alpha t) \\ \sqrt{5}\sin(2\sigma w(t) + \alpha t) & \sqrt{5}\cos((2\sigma w(t) + \alpha t) + \theta) \end{bmatrix}.$$

We recall that $\Phi_d(t, 0)$, $\Phi_s(t, w(t), 0)$ and $\Phi(t, w(t), 0)$ are the normalized fundamental matrix solutions corresponding to the deterministic part $(dx = Ax\, dt)$, the stochastic part $[dx = Bx\, dw(t)]$, and the overall homogeneous system $[dx = Ax\, dt + Bx\, dw(t)]$ of the given nonhomogeneous system, respectively. By employing these normalized fundamental matrix solutions and following the argument used in Illustrations 4.5.1 and 4.5.2, the solution process of the example can be completed. The details are left to the reader.

Illustration 4.5.3. Given: $dx = [Ax + p(t)]dt + [Bx + q(t)]dw(t)$, where $A = \begin{bmatrix} a_{11} & a_{12} \\ a_{21} & a_{22} \end{bmatrix}$, $B = \begin{bmatrix} b_{11} & b_{12} \\ b_{21} & b_{22} \end{bmatrix}$, with b_{ij} and a_{ij} being any given real numbers satisfying the condition:

(a) $d(A) = (\text{tr}(A))^2 - 4\det(A) = 0$;
(b) $AB = BA$;
(c) p and q are two-dimensional continuous functions defined on an interval $J \subseteq R$ into R^2, and w is a scalar normalized Wiener process.

Find:

(i) a general solution to the given differential equation, particularly for:
(ii) $B = 0$ and $q(t) \equiv 0$;
(iii) $B = 0$ and $p(t) \equiv 0$;
(iv) $A = 0$ and $q(t) \equiv 0$;
(v) $A = 0$ and $p(t) \equiv 0$.

Solution procedure. Again, we note that $n = 2$. We imitate the arguments used in Illustrations 4.3.3, 4.4.1, and 4.5.1, and Procedure 4.5.2 to find the general solution to the given system. Under the condition (b), we further note that $d(B) = (\mathrm{tr}(B))^2 - 4\det(B) = 0$, and

$$\lambda_1 - a_{11} = \frac{\mathrm{tr}(A) - 2a_{11}}{2} = \frac{a_{22} - a_{11}}{2} = r\frac{b_{22} - b_{11}}{2}$$

$$= r\frac{\mathrm{tr}(B) - 2b_{11}}{2} = r(\mu_1 - b_{11}) = \pm|r|\sqrt{\det(B)}.$$

Thus, this quadratic polynomial equation has a root of multiplicity 2. Now, we repeat the procedures outlined in Observation 4.2.4(vi) and Illustrations 4.3.3 and 4.5.1, and obtain the general solution process with regard to the given system of differential equations as well as its particular cases (ii)–(v). The details are left to the reader as an exercise. This completes the solution process of the given illustration.

4.5.3 *Initial Value Problem*

Let us formulate an IVP for first-order linear nonhomogeneous systems of stochastic differential equations of the Itô–Doob-type (3.5.1). We consider

$$dx = [Ax + p(t)]dt + [Bx + q(t)]dw(t), \quad x(t_0) = x_0. \tag{4.5.20}$$

4.5.4 *Procedure for Solving the IVP*

In the following, we apply the procedure for finding a solution process of a first-order linear nonhomogeneous system of stochastic differential equations of the Itô–Doob-type (4.5.1) to the IVP (4.5.20). We briefly summarize the procedure:

Step #1. Following the procedure for finding a general solution to (4.5.1), we have

$$x(t) = \Phi(t)\left[c + \int_a^t \Phi^{-1}(s)[p(s) - Bq(s)]ds + \int_a^t \Phi^{-1}(s)q(s)dw(s)\right]$$

$$= \Phi(t)c + \int_a^t \Phi(t,s)[p(s) - Bq(s)]ds + \int_a^t \Phi(t,s)q(s)dw(s), \tag{4.5.21}$$

where $a \in R$, $t \in J = [a, b] \subseteq R$, $\Phi(t)$ is as defined in (4.4.14), and $\Phi(t, s) = \Phi(t)\Phi^{-1}(s)$. Here, c is an arbitrary n-dimensional constant vector.

Step #2. In this step, we need to find an arbitrary constant c in (4.5.21). For this purpose, we utilize the given initial given data (t_0, x_0), and solve the following system of algebraic equations:

$$x(t_0) = \Phi(t_0)\left[c + \int_a^{t_0} \Phi^{-1}(s)[p(s) - Bq(s)]ds + \int_a^{t_0} \Phi^{-1}(s)q(s)dw(s)\right] = x_0,$$

$$\tag{4.5.22}$$

for any given t_0 in J. Since $\exp\left[(A - \frac{1}{2}B^2)(t_0 - a) + B(w(t_0) - w(a))\right] \neq 0$ for any A and B, the system of algebraic equations (4.5.22) has a unique solution. This solution is given by

$$c = \Phi^{-1}(t_0)x_0 - \left[\int_a^{t_0} \Phi^{-1}(s)[p(s) - Bq(s)]ds + \int_a^{t_0} \Phi^{-1}(s)q(s)dw(s)\right]. \quad (4.5.23)$$

Step #3. The solution to the IVP (4.5.20) is determined by substituting the expression (4.5.23) into (4.5.21). Hence, we have

$$x(t) = \Phi(t)c + \int_a^t \Phi(t,s)[p(s) - Bq(s)]ds + \int_a^t \Phi(t,s)q(s)dw(s)$$

$$= \Phi(t)\left[\Phi^{-1}(t_0)x_0 - \int_a^{t_0} \Phi^{-1}(s)[p(s) - Bq(s)]ds - \int_a^{t_0} \Phi^{-1}(s)q(s)dw(s)\right]$$

$$+ \int_a^t \Phi(t,s)[p(s) - Bq(s)]ds + \int_a^t \Phi(t,s)q(s)dw(s) \quad \text{[from (4.5.23)]}$$

$$= \Phi(t,t_0)x_0 - \int_a^{t_0} \Phi(t)\Phi^{-1}(s)[p(s) - Bq(s)]ds - \int_a^{t_0} \Phi(t)\phi^{-1}(s)q(s)dw(s)$$

$$+ \int_a^t \Phi(t,s)[p(s) - Bq(s)]ds + \int_a^t \Phi(t,s)q(s)dw(s)$$

$$= \Phi(t,t_0)x_0 - \int_a^{t_0} \Phi(t,s)[p(s) - Bq(s)]ds - \int_a^{t_0} \Phi(t,s)q(s)dw(s)$$

$$+ \int_a^t \Phi(t,s)[p(s) - Bq(s)]ds + \int_a^t \Phi(t,s)q(s)dw(s) \quad \text{[from (4.4.25)]}$$

$$= \Phi(t,t_0)x_0 + \int_{t_0}^t \Phi(t,s)[p(s) - Bq(s)]ds$$

$$+ \int_{t_0}^t \Phi(t,s)q(s)dw(s) \quad \text{(by grouping)}, \quad (4.5.24)$$

where $\Phi(t,t_0)$ is as defined in (4.4.25).

Thus, from (4.4.25) and (4.4.26), the solution process $x(t) = x(t,t_0,x_0)$ of the IVP (4.5.20) in (4.5.24) is also represented by

$$x(t) = x(t,t_0,x_0)$$

$$= \exp\left[\left(A - \frac{B^2}{2}\right)(t - t_0) + B(w(t) - w(t_0))\right]x_0 \quad \text{[from (4.4.14)]}$$

$$+ \int_{t_0}^t \exp\left[\left(A - \frac{B^2}{2}\right)(t - s) + B(w(t) - w(s))\right][p(s) - Bq(s)]ds$$

$$+ \int_{t_0}^t \exp\left[\left(A - \frac{B^2}{2}\right)(t - s) + B(w(t) - w(s))\right]q(s)dw(s). \quad (4.5.25)$$

This solution is referred to as a *particular solution* to (4.5.1).

The procedure for finding the solution to the IVP is summarized in the following flowchart:

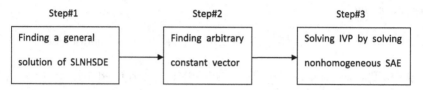

Fig. 4.5.2   Process of finding solution of IVP.

Illustration 4.5.4   (diabetes mellitus model [13, 54, 77]). Carbohydrates, fats, and proteins are the three major components of food that are essential for the maintenance of the body. The digestive products are monosaccharides, particularly glucose. Glucose is easily absorbed in blood. After one eats a meal, the normal concentration of the glucose in the blood ranges from 90 to 140 mg%. The glucose is about 40% of the energy source. Moreover, it is the only nutrient that can be utilized by the brain, retina, and germinal epithelium for energy. The blood glucose is controlled by hormones, namely insulin, glucagon, epinephrine, glucocorticoid, thyroxin, etc.

TDM1 (glucose).   The increase in the concentration of blood glucose causes increased glucose absorption in the liver. This in turn causes glucogen formation from glucose and its storage in the liver. Conversely, the drop in the blood glucose concentration reverses the above process (glycogenolysis). Moreover, in the case of low carbohydrates in the diet, the release of the glucose supply is due to gluconeogenesis. Besides the rate of intake of carbohydrates in the diet, the glucose concentration in plasma and extracellular is governed by the following mechanisms:

(a) The ability of the liver to buffer the blood glucose concentration.
(b) An automatic feedback process due to the insulin secretion by the beta cells in the islets of Langerhans (one of the tissues of the pancreas).
(c) An automatic feedback process of glucose from the liver (glycogenolysis-glucagon secreted by alpha cells in the pancreas and epinephrine released by adrenal medullae; gluconeogenesis-glucocorticoid secreted by adrenal cortex, namely cortisol).
(d) Glucose utilization by body tissue.

TDM2 (insulin).   Insulin is secreted by the beta cells in the pancreas. After a person eats some food containing carbohydrates, the beta cells in the pancreas secrete insulin. The insulin facilitates glucose absorption in blood. It is referred to as the carrier of energy to the body. Under the normal physiological conditions, an increase in the glucose concentration of blood stimulates the pancreas to produce insulin at a faster rate. On the other hand, a drop in the concentration lowers the insulin production by the islets of Langerhans (one of the tissues of the pancreas).

Moreover, an increase in the concentration of insulin causes the glucose to pass through the membrane of the cells in the body, thus increasing the absorption of glucose from the blood. In summary, the effects of insulin are:

(a) Enhanced rate of glucose metabolism
(b) Decreased blood glucose concentration
(c) Increased glycogen storage in tissue.

TDM3 (diabetes mellitus). The effects of lack of insulin are:

(a) Decreased utilization of glucose in the blood, resulting in an increase in the glucose concentration in the blood.
(b) Significant increase in the mobility of fats from the fat storage areas, causing abnormalities and lipid buildup.
(c) Diminished deposition of protein in the tissues of the body. Among the factors causing diabetes mellitus are:

   (i) The excess amount of adenohypophysis (adenohypophysis is composed of three major cells — chromophobias, acidophils, and basophils) may cause the beta cells to burn out.
   (ii) A hereditary disease for most of the persons.

TDM4. The diffusion of glucose through cell membrane depends on the metabolic rate inside the cell, the influence of ionization on the cell membrane (Illustration 2.4.5 — cell growth problem), as well as the homogeneity of the liquid in the surrounding medium and the membrane permeability of the cell. As a result of this, it is natural to consider the influence of random fluctuations. The randomness can be described by the Brownian motion process.

By following the deterministic model of Bolie [13], we formulate its stochastic version. For detecting the diabetes, a large dose of insulin is given to the person. Let x_1 and x_2 be the extracellular glucose (G) and insulin (I) concentrations at a time t, and let Δx_1 and Δx_2 be the change of glucose and insulin concentrations over an interval of time Δt, respectively. Then, we have

$$\begin{aligned} \Delta x_1 = {} & \text{(Net change in G)} = \text{(Change due to the external sources of G)} \\ & + \text{(Change due to G)} \\ & + \text{(Change due to the above normal size of insulin I)} \\ = {} & e_1 \Delta t - k_{11} x_1 \Delta t - k_{12} x_2 \Delta t \ \text{(by TDM1 and TDM2)} \\ = {} & (e_1 - k_{11} x_1 - k_{12} x_2) \Delta t, \end{aligned}$$

and

$$\begin{aligned} \Delta x_2 = {} & \text{(Net change in I)} = \text{(Change due to the external sources of I)} \\ & + \text{(Change due to I)} \\ & + \text{(Change due to the above normal size of insulin G)} \end{aligned}$$

$$= e_2 \Delta t + k_{21} x_1 \Delta t - k_{22} x_2 \Delta t \text{ (by TDM1 and TDM2)}$$

$$= (e_2 + k_{21} x_1 - k_{22} x_2) \Delta t,$$

where $e_1 > 0$ and $e_2 > 0$ characterize the constant input of glucose and insulin, respectively; $k_{11} > 0$ and $k_{22} > 0$ characterize the density-dependent effects of glucose and insulin, respectively; $k_{12} > 0$ characterizes the effects of the insulin concentration on the rate of glucose, and the negative sign of $k_{12} > 0$ reflects the presence of insulin in the glucose metabolism in body cells and/or to build up the glucagon in the liver and body cells (TDM2); and $k_{21} > 0$ characterizes the effects of the presence of glucose on the insulin production in the pancreas (TDM1).

From TDM4, the above formulation under the assumption of single random disturbance, the mathematical model is described by

$$\Delta x_1 = (e_1 - k_{11} x_1 - k_{12} x_2) \Delta t + \sigma (e_1 - k_{11} x_1 - k_{12} x_2) \Delta w(t),$$

$$\Delta x_2 = (e_2 + k_{21} x_1 - k_{22} x_2) \Delta t + \sigma (e_2 + k_{21} x_1 - k_{22} x_2) \Delta w(t),$$

which implies that

$$dx_1 = (e_1 - k_{11} x_1 - k_{12} x_2) dt + \sigma (e_1 - k_{11} x_1 - k_{12} x_2) dw(t),$$

$$dx_2 = (e_2 + k_{21} x_1 - k_{22} x_2) dt + \sigma (e_2 + k_{21} x_1 - k_{22} x_2) dw(t),$$

Hence,

$$dx = (Ax + p)dt + (Bx + q)dw(t), \quad x(t_0) = x_0, \tag{4.5.26}$$

where $x = [x_1 \ x_2]^T$, $p = [e_1 \ e_2]^T$, $q = [\sigma e_1 \ \sigma e_2]^T$,

$$A = \begin{bmatrix} -k_{11} & -k_{12} \\ k_{21} & -k_{22} \end{bmatrix} \quad \text{and} \quad B = \begin{bmatrix} -\sigma k_{11} & -\sigma k_{12} \\ \sigma k_{21} & -\sigma k_{22} \end{bmatrix}.$$

By following the procedure for solving the IVP associated with the system of stochastic differential equations (4.5.26), the general and particular solution processes of this system are given by

$$x(t) = \Phi(t, w(t))c + \int_a^t (\Phi(t, w(t), s, w(s)))(p - Bq)ds$$

$$+ \int_a^t \Phi(t, w(t), s, w(s)) q \, dw(s),$$

$$x(t) = \Phi(t, w(t)) x_0 + \int_{t_0}^t (\Phi(t, w(t), s, w(s)))(p - Bq)ds$$

$$+ \int_{t_0}^t \Phi(t, w(t), s, w(s)) \, dw \, q(s),$$

respectively, where the elements of $\Phi(t, w(t))$ are defined by

$$\Phi(t, w(t)) = \frac{k_{12}k_{21}}{k_{12}k_{21} - (k_{11} - k_{22})^2} e^{\alpha_0(t-t_0)}(\phi_{ij}(t, w(t), \psi))_{2\times 2},$$

$$\phi_{11}(t, w(t), \psi) = \cos \varphi_d(t, \psi) \cos \varphi_s(t, \psi) + \sin \varphi_d(t) \sin \varphi_s(t),$$

$$\phi_{12}(t, w(t), \psi) = -\sqrt{\frac{k_{12}}{k_{21}}} [\cos \varphi_d(t, \psi) \sin \varphi_s(t) + \sin \varphi_d(t) \cos \varphi_s(t, \psi)]$$

$$\phi_{21}(t, w(t), \psi) = -\sqrt{\frac{k_{21}}{k_{12}}} [\sin \varphi_d(t) \cos \varphi_s(t, \psi) + \cos \varphi_d(t, \psi) \sin \varphi_s(t)],$$

$$\phi_{22}(t, w(t), \psi) = \cos \varphi_d(t, \psi) \cos \varphi_s(t, \psi) + \sin \varphi_d(t) \sin \varphi_s(t),$$

$\psi = \arctan\left[\frac{a_{11}-\alpha}{\beta}\right]$, $\varphi_d(t) = \beta(t-t_0)$, $\varphi_s(t) = \gamma(t-t_0) + \beta(w(t) - w(t_0))$, $\varphi_d(t, \psi) = \varphi_d(t) - \psi$, $\varphi_s(t, \psi) = \varphi_s(t, w(t)) - \psi$, $\alpha = \frac{\text{tr}(A)}{2} = -\frac{k_{11}+k_{22}}{2}$, $\beta = \frac{\sqrt{-d(A)}}{2}$, $d(A) = (\text{tr}(A))^2 - 4\det(A)$, and $\alpha_0(t-t_0) = \left(\alpha - \frac{\sigma^2}{2}(\alpha^2 - \beta^2)\right)(t-t_0) = -\frac{1}{4}[(k_{11}(2+\sigma^2 k_{11}) + k_{22}(2 + \sigma^2 k_{22}) - 2\sigma^2 k_{12}k_{21}].$

Exercises

1. Find the general solution to each system of differential equations:

   (a) $dx = (Ax + p)dt + (Bx + q)dw(t)$;
   (b) $dx = (B + q)x\, dt + (Ax + p)dw(t)$,

   where $A = \begin{bmatrix} -1 & 1 \\ 1 & -1 \end{bmatrix}$, $B = \begin{bmatrix} 1 & -2 \\ -2 & 1 \end{bmatrix}$, $p = \begin{bmatrix} 1 \\ 1 \end{bmatrix}$, $q = \begin{bmatrix} 1 \\ 2 \end{bmatrix}$;

   (c) $dx = (Ax + p)dt + (Bx + q)dw(t)$;
   (d) $dx = (B + q)x\, dt + (Ax + p)dw(t)$,

   where $A = \begin{bmatrix} -1 & 0 \\ 1 & -1 \end{bmatrix}$, $B = \begin{bmatrix} 1 & -1 \\ 1 & 1 \end{bmatrix}$, $p = \begin{bmatrix} 1 \\ 0 \end{bmatrix}$, $q = \begin{bmatrix} 0 \\ 2 \end{bmatrix}$;

   (e) $dx = (Ax + p)dt + (Bx + q)dw(t)$;
   (f) $dx = (B + q)x\, dt + (Ax + p)dw(t)$,

   where $A = \begin{bmatrix} -1 & -1 \\ 1 & -1 \end{bmatrix}$, $B = \begin{bmatrix} 3 & 1 \\ 1 & 4 \end{bmatrix}$, $p = \begin{bmatrix} 2 \\ 1 \end{bmatrix}$, $q = \begin{bmatrix} 1 \\ 2 \end{bmatrix}$.

2. Solve the IVPs (a)–(f) in Exercise 1 with:

   (a) $x_0 = [1 \quad 1]^T$;
   (b) $x_0 = [2 \quad 1]^T$;
   (c) $x_0 = [1 \quad 2]^T$;
   (d) $x_0 = [2 \quad 2]^T$;
   (e) any x_0.

3. Assume that all hypotheses of Theorem 4.4.1 are satisfied. Let $x(t)$ be a solution process of (4.5.1) defined by the transformation $x(t) = \Phi_d(t)c(t)$, where $c(t)$ is an unknown smooth process. Show that $c(t)$ satisfies the following stochastic system of differential equations:

$$dc = \Phi_d^{-1}(t)p(t)\Phi_d(t)dt + \Phi_d^{-1}(t)[B\Phi_d(t)c + q(t)]dw(t).$$

4. Let us assume that all the conditions of Corollary 4.4.1 remain valid. Further assume that $x(t)$ is a solution process of (4.5.1) defined by the transformation $x(t) = \Phi_s(t)c(t)$, where $c(t)$ is an unknown smooth process. Prove or disprove that $c(t)$ satisfies the following stochastic system of differential equations:

$$dc = \Phi_s^{-1}(t)[A\Phi_s(t)c + p(t) - Bq(t)]dt + \Phi_s^{-1}q(t)dw(t).$$

5. Complete the solution process of Illustrations 4.5.2 and 4.5.3.

6. For $i = 1, 2, \ldots, k$, let A, B_i be $n \times n$ matrices with real entries, and and let $w_i(t)$ be a Wiener process (Theorem 1.2.2) defined on a complete probability space $(\Omega, \mathfrak{F}, P)$. Further assume that for $i \neq j$, $i, j = 1, 2 \ldots, k$, $w_i(t)$ and $w_j(t)$ are mutually independent on J and satisfy $E[dw_i(t)dw_j(t)] = 0$. If for $i = 1, 2, \ldots, k$, A and B_i mutually commute, find the solution process of the following system of stochastic differential equations:

$$dx = [Ax + p(t)]dt + \left[\sum_{i=1}^{k} B_i x + q(t)\right] dw(t).$$

7. Find the general solution to

$$dx = (Ax + p)dt + (B_1 x + q_1)dw_1(t) + (B_2 x + q_2)dw_2(t),$$

where $A = \begin{bmatrix} 2 & 5 \\ 1 & 3 \end{bmatrix}$, $B_1 = \begin{bmatrix} -1 & 0 \\ 1 & -2 \end{bmatrix}$, $B_2 = \begin{bmatrix} 2 & 5 \\ 1 & 3 \end{bmatrix}$, $p = \begin{bmatrix} 1 \\ 1 \end{bmatrix}$, $q_1 = \begin{bmatrix} 1 \\ 2 \end{bmatrix}$, $q_2 = \begin{bmatrix} 0 \\ 1 \end{bmatrix}$.

8. Let F be a continuous cumulative distribution function (cdf) of a random time variable T, and let A, B, B_1, and B_2 be 2×2 matrices. Moreover, the Wiener process $w(t)$ is as defined earlier. Apply the procedure outlined in Sections 4.2, 4.4, and 4.5 to find the general solution process of the following differential equations:

(a) $dx = (Ax + p)dt + (Bx + q)dF(t)$;
(b) $dx = Ax\, dt + (B_1 x + q_1)dF(t) + (B_2 x + q_2)dw(t)$.

9. Solve (a) and (b) for given matrices in Exercise 7 with $B = B_1$ in (a).

4.6 Fundamental Conceptual Algorithms and Analysis

In Sections 4.2–4.5, we presented the computational procedures/methods for finding solution processes of linear systems of Itô–Doob-type stochastic differential equations with constant coefficients. Again, the set of questions similar to the set of questions in Section 2.5 can be raised for linear systems of Itô–Doob-type stochastic differential equations. The presented techniques provide the basis for addressing/raising several important questions in the modeling of dynamic processes in the biological, chemical, engineering, medical, physical, and social sciences. Moreover, the presented results of Section 2.5 and the deterministic systems of differential equations of Chapter 4 of Volume 1 are extended to systems of nonhomogeneous Itô–Doob type stochastic differential equations with time-varying coefficients. To avoid repetition, we provide the answers to the three most basic questions (#2, #3, and #10 in Section 2.5) with regard to linear systems of stochastic differential equations of the Itô–Doob type. These questions generate the three most important problems in the dynamic modeling and the theory of systems of differential equations, namely:

(i) Existence problem.
(ii) Uniqueness problem.
(iii) Fundamental properties of solution processes.

The presented procedures for finding solutions to IVPs in Sections 4.2–4.5 justify the existence of solutions to IVPs for linear first-order systems of Itô–Doob-type stochastic differential equations. We also note that these computational procedures were applicable to time-invariant coefficient matrices. In fact, the problem of finding solutions to such systems reduce to the problem of solving linear time-varying systems (Theorems 4.4.1 and 4.4.2 and Corollaries 4.4.1 and 4.4.2). The computational procedures are very limited in exhibiting the existence of the solution process. As a result of this, we need to find a theoretical procedure that assures the existence of the solution process of such systems. This can be summarized in the following result. The existence result and the results presented thereafter play an important role, theoretically and practically, in the study of:

(a) Time-varying and time-invariant (nonstationary and stationary) nonlinear systems.
(b) Nonlinear systems of stochastic differential equations of the Itô–Doob type.

Theorem 4.6.1 (existence and uniqueness theorem). *Let us consider the initial value problem*

$$dy = [A(t)y + p(t)]dt + [B(t)y + q(t)]dw(t), \quad y(t_0) = y_0, \qquad (4.6.1)$$

where dx stands for the stochastic differential of the Itô–Doob-type; A and B are $n \times n$ continuous matrix functions defined on an interval $J = [a, b]$ $(J \subseteq R)$; p and q are n-dimensional continuous vector functions defined on an interval $J = [a, b]$

into R^n; w is a scalar normalized Wiener process; initial data/conditions $(t_0, y_0) \in J \times R^n$; and y_0 is an R^n-valued random vector defined on a complete probability space $(\Omega, \mathfrak{F}, P)$, and is independent of $w(t)$ for all t in J and $E[\|y_0\|^2] < \infty$.

Then, the IVP (4.6.1) has a unique solution process $y(t, t_0, y_0) = y(t)$ through any given initial data/conditions (t_0, y_0) for $t \in [t_0, b] = I \subseteq J$. Moreover,

$$y(t) = y_0 + \int_{t_0}^t [A(s)y(s) + p(s)]ds + \int_{t_0}^t [B(s)y(s) + q(s)]dw(s). \qquad (4.6.2)$$

Proof. For $x, y \in R^n$, the distance between points associated with x and y is defined by

$$\|x - y\| = \sqrt{(x_1 - y_1)^2 + (x_2 - y_2)^2 + \cdots + (x_i - y_i)^2 + \cdots + (x_n - y_n)^2}.$$

We observe that the drift $C(t, y) = A(t)y + p(t)$ and the diffusion $F(t, y) = B(t)y + q(t)$ coefficients satisfy the following conditions:

$$\|C(t, y)\| + \|F(t, y)\| \leq K\|y\| + N \quad \text{(linear growth condition)}, \qquad (4.6.3)$$

and

$$\|C(t, x) - C(t, y)\| = \|F(t, x) - F(t, y)\| \leq L\|x - y\| \quad \text{(Lipschitz condition)},$$
$$(4.6.4)$$

for all $(t, x), (t, y) \in I \times R^n$ ($I = [t_0, b] \subseteq J \subseteq R$). We further note that these functions are continuous on $I \times R^n$.

Now, we define the following approximation procedure:

$$y_n(t) = \begin{cases} y_0 & \text{for } t_0 \leq t \leq b, \\ y_0 + \int_{t_0}^t [A(s)y_{n-1}(s) + p(s)]ds + \int_{t_0}^t [B(s)y_{n-1}(s) + q(s)]dw(s), \end{cases}$$
$$(4.6.5)$$

for $n \geq 1$. We recall that the first- and second-integral terms on the right-hand side of (4.6.5) are in the sense of Riemann–Cauchy [127, 168] and Itô–Doob calculus [7, 8, 32, 43, 56, 98], respectively. Using the relations (4.6.3) and (4.6.4), the continuity of drift and diffusion coefficients on $[t_0, b] \times R^n$, probabilistic and advanced real analysis techniques (the concept of uniform convergence, Borel–Cantelli lemma, Weierstrass's convergence test, etc.), one can show that the sequence of random functions $\{y_n\}_{n=0}^{\infty}$ defined by (4.6.5) converges uniformly to a random function y defined on I. Moreover, the uniform limit random function y satisfies the IVP (4.6.1). This establishes the basic scheme about the existence of the solution process of the given IVP. Any further details require the technical understanding of both probabilistic and advanced stochastic analysis [7, 8, 32, 98]. The validity of (4.6.2) follows from (4.6.5) and the uniform convergence $\{y_n\}_{n=0}^{\infty}$ on I.

In view of the Lipschitz condition in (4.6.4), the uniqueness of the solution process of the IVP (4.6.1) can be established by following the standard technical argument [8, 32, 43, 56, 98]. The details are left to the reader to satisfy his or her desire to undertake the challenges of the future study of stochastic differential equations. □

Corollary 4.6.1. *Let us consider the following first-order linear homogeneous system of stochastic differential equations corresponding to (4.6.1):*

$$dx = A(t)x\,dt + B(t)x\,dw(t), \quad x(t_0) = x_0, \tag{4.6.6}$$

where the matrix functions A and B, the Wiener process w, and the initial data (t_0, x_0) satisfy the conditions of Theorem 4.6.1. Then the IVP (4.6.6) has a unique solution process $x(t, t_0, x_0) = x(t)$ through any given initial data/conditions (t_0, x_0) for $t \in I$, $t_0 \in J$. Moreover,

$$x(t) = x_0 + \int_{t_0}^t A(s)x(s)ds + \int_{t_0}^t B(s)x(s)dw(s). \tag{4.6.7}$$

Observation 4.6.1

(i) We note that the IVP (4.5.20) is a special case of the IVP (4.6.1). Because of this, the IVP (4.5.20) has two representations or expressions: (a) computationally explicit (4.5.25) and (b) conceptually implicit (4.6.2).

(ii) The uniqueness part of the theorem (Theorem 4.6.1) conceptualizes the solution process $y(t) = y(t, t_0, y_0)$ of the IVP (4.6.1) as a function of three variables (t, t_0, y_0). From the definition of the function, for given (t, t_0, y_0), the output $y(t, t_0, y_0)$ [the value of the solution process at (t, t_0, y_0)] is uniquely determined under the conditions of the uniqueness theorem. This conceptualization plays a very important role in the advanced study of differential equations [7, 8, 32, 43, 56, 98]. Moreover, this notation $y(t, t_0, y_0)$ is just a natural symbol for the solution process of (4.6.1).

(iii) We also note that the IVPs (4.2.8), and (4.4.20) are special cases of the IVP (4.6.6). Furthermore, Parts (i) and (ii) of the remark are also valid with regard to the IVP (4.6.6).

Based on our understanding and the knowledge of the "Basis and Scope of Conceptual Algorithms" of Chapters 2 and 4 (Volume 1), we need the basic information on the fundamental solution process of the corresponding linear systems of nonhomogeneous differential equations. Therefore, prior to the study of fundamental properties of (4.6.1), we need to establish the basic properties of the solution process of the linear homogeneous time-varying coefficients of system of stochastic differential equations (4.6.6).

Theorem 4.6.2 (fundamental property). *Let the hypotheses of Corollary* 4.6.1 *be satisfied. In addition,* $x_1(t), x_2(t), \ldots, x_k(t), \ldots, x_m(t)$ *be m solutions to* (4.6.6) *on I with the corresponding initial vectors* $x_{10}, x_{20}, \ldots, x_{k0}, \ldots, x_{m0}$ *in* R^n. *Then,*

$$x(t) = a_1 x_1(t) + a_2 x_2(t) + \cdots + a_k x_k(t) + \cdots + a_m x_m(t) = \sum_{k=1}^{m} a_k x_k(t) \quad (4.6.8)$$

is also a solution to (4.6.6) *on I, where* $a_1, a_2, \ldots, a_k, \ldots, a_m$ *are arbitrary real constants.*

Proof. The proof of the theorem can be imitated in conjunction with the proof of Theorem 4.2.1. The details of the proof are left as an exercise to the reader. $\square$

Remark 4.6.1.

(i) From Corollary 4.6.1, we note that if $x(t_0) = 0$, then $x(t, t_0, 0) = 0$.

(ii) We remark that an observation similar to Observation 4.2.1 can be reformulated with regard to the time-varying system (4.6.6). The verification of this remark is left as an exercise to the reader.

(iii) For $m = n$ and by following the arguments used in Observation 4.2.2 in the context of Observation 4.2.4(vi), we can rewrite the expression of $x(t)$ in (4.6.8) as:

$$x(t) = \Phi(t)a, \quad (4.6.9)$$

where Φ is an $n \times n$ matrix defined by $\Phi(t) = (x_{ik}(t))_{n \times n} \equiv (x_{ik}(t, w(t)))_{n \times n} = \Phi(t, w(t))$, whose columns are the solutions $x_1(t), x_2(t), \ldots, x_k(t), \ldots, x_m(t)$ defined in Theorem 4.6.1 with the corresponding initial vectors $x_{10}, x_{20}, \ldots, x_{k0}, \ldots, x_{n0}$ in R^n; an n-dimensional unknown constant random column vector (or an $n \times 1$ matrix) $a = [a_1, a_2, \ldots, a_k, \ldots, a_n]^T$ is formed from arbitrary coefficients of the solutions $x_k(t)$ for $k = 1, 2, \ldots, n$ in (4.6.8). The solution represented in (4.6.9) is called a *general solution* process of (4.6.6). We remark that $\Phi(t) \equiv \Phi(t, w(t))$, defined in (4.6.9) depends on the n arbitrary given initial vectors $x_{10}, x_{20}, \ldots, x_{k0}, \ldots, x_{n0}$ in R^n as well as the Wiener process $w(t)$. Hence, $\Phi(t)$ is not uniquely determined by an arbitrary n but by given initial vectors at a given initial time $t = t_0$. This suggests that the observation similar to Observation 4.2.2 in the context of Observation 4.2.4(vi) can be rewritten by replacing a set of eigenvectors $\{c_1, \ldots, c_k, \ldots, c_n\}$ with the set of initial vectors $\{x_{10}, x_{20}, \ldots, x_{k0}, \ldots, x_{n0}\}$, which corresponds to n solutions to (4.6.6). Furthermore, Span($\{x_{10}, x_{20}, \ldots, x_{k0}, \ldots, x_{n0}\}$) $= R^n$ if and only if $0 \neq \det(\Phi(t_0))$ $[\Phi(t_0) = (x_{ik}(t_0, w(t_0)))_{n \times n} = (x_{ik0})_{n \times n}]$ (Theorem 1.4.6 of Volume 1), and by the uniqueness (Corollary 4.6.1) $\det(\Phi(t)) \neq 0$ on I. The details are left as an exercise.

(iv) Following the argument used in Observation 4.2.3 in the context of Observation 4.2.4(vi), we remark that

$$d\Phi(t) = A(t)\Phi(t)dt + B(t)\Phi(t)dw(t), \qquad (4.6.10)$$

where $\Phi(t)$ is as defined in (4.6.9). Moreover, the notions of general and normalized fundamental matrix solutions to (4.6.6) at $t = t_0$ can be introduced in a natural way. In fact, one can use the initial vectors in (4.6.6) as $x_{k0} = e_k = (\delta_{ik})_{n \times 1}$, where $\delta_{ik} = 0$ for $i \neq k$, $\delta_{ik} = 1$ for $i = k$, and $k = 1, 2, \ldots, n$. Hence, the solution to the IVP (4.6.6) $x(t, t_0, x_0)$ can be represented as

$$x(t, t_0, x_0) = x(t) = \Phi(t)\Phi^{-1}(t_0)x_0 = \Phi(t, t_0)x_0. \qquad (4.6.11)$$

The detailed imitation of Observation 4.2.3 is left to the reader as an exercise.

In the following, we present a result similar to Theorem 4.2.2. This result provides an alternative analytic test (a stochastic generalized version of Abel's formula) rather than the algebraic test described in Remark 4.6.1(iii) [Span($\{x_{10}, x_{20}, \ldots, x_{k0}, \ldots, x_{n0}\}$) $= R^n$] for the fundamental matrix solution. It extends the deterministic results [80] in a systematic way.

Theorem 4.6.3 (generalized stochastic version of the Abel–Jacobi–Liouville theorem [79, 80]). *Let the hypotheses of Corollary 4.6.1 be satisfied. Let $x_1(t)$, $x_2(t), \ldots, x_k(t), \ldots, x_n(t)$ be any n solutions processes of (4.6.6) on I. Let $\Phi(t) = [x_1(t), x_2(t), \ldots, x_i(t), \ldots, x_n(t)] = [\Phi_1^T(t), \ldots, \Phi_i^T(t), \ldots, \Phi_n^T(t)]^T$ be the $n \times n$ matrix process defined on I, where $\Phi_i(t) = [x_{i1}, x_{i2}, \ldots, x_{ik}, \ldots, x_{in}]$ is the ith row of the matrix process $\Phi(t)$. Let*

$$\mathbb{L}(B(t)) = \frac{1}{2}\sum_{i=1}^{n}\sum_{k \neq i}^{n}(b_{ii}(t)b_{kk}(t) - b_{ik}(t)b_{ki}(t)),$$

$$\text{tr}(A(t)) = \sum_{i=1}^{n} a_{ii}(t),$$

$$\text{tr}(B(t)) = \sum_{i=1}^{n} b_{ii}(t).$$

Then,

$$d\det(\Phi(t)) = [\text{tr}(A(t)) + \mathbb{L}(B(t))]\det(\Phi(t))dt + \text{tr}(B(t))\det(\Phi(t))dw(t), \quad (4.6.12)$$

$$\exp\left[-\int_{t_0}^{t}\left[\text{tr}(A(s)) + \mathbb{L}(B(s)) - \frac{1}{2}(\text{tr}(B(s)))^2\right]ds - \int_{t_0}^{t}\text{tr}(B(s))dw(s)\right]\det(\Phi(t))$$

$$= c(constant). \qquad (4.6.13)$$

Moreover, $\det(\Phi(t)) \neq 0$ [a general fundamental matrix solution process of (4.6.6)] if and only if $c \neq 0$.

Proof. From (4.6.10), we conclude that $\Phi(t)$ is the matrix solution process of (4.6.6), i.e.

$$d\Phi(t) = (dx_{ik}(t))_{n \times n} = A(t)\Phi(t)dt + B(t)\Phi(t)dw(t)$$

$$= \left(\sum_{j=1}^{n} a_{ij}(t)x_{jk}(t)dt + \sum_{j=1}^{n} b_{ij}(t)x_{jk}(t)dw(t) \right)_{n \times n}. \qquad (4.6.14)$$

Applying Theorems 1.5.6 of Volume 1 and 1.3.4 to $\det(\Phi(t))$ and following the proof of Theorem 4.2.2, we arrive at

$$W(\Phi_1(t), \ldots, d\Phi_i(t), \ldots, \Phi_n(t))$$

$$= \begin{vmatrix} x_{11} & \cdots & \cdots & x_{1n} \\ a_{ii}(t)x_{i1}(t)dt + b_{ii}(t)x_{i1}dw(t) & \cdots & \cdots & a_{ii}(t)x_{i1}(t)dt + b_{ii}(t)x_{in}dw(t) \\ \cdots & \cdots & \cdots & \cdots \\ x_{n1} & \cdots & \cdots & x_{nn} \end{vmatrix}.$$

This together with Theorem 1.4.1D$_4$, D$_5$ and D$_6$ (Volume 1) yields

$$W(\Phi_1(t), \ldots, d\Phi_i(t), \ldots, \Phi_n(t)) = a_{ii}(t)\det(\Phi(t))dt + b_{ii}(t)\det(\Phi(t))dw(t).$$

The rest of the proof can be constructed by employing Itô–Doob stochastic calculus and the argument used in Theorem 4.2.2. The details are left as an exercise to the reader. $\qquad \square$

Remark 4.6.2. We remark that an observation similar to Observation 4.2.4 remains true with regard to the time-varying system (4.6.6). The verification of this remark is left as an exercise to the reader.

In the following, we present a new result that is similar to Theorem 4.2.3 with regard to the time-varying system (4.6.6). The following result extends the deterministic result [80].

Theorem 4.6.4. *Let the hypotheses of Corollary 4.6.1 be satisfied. In addition, Φ be the fundamental matrix solution process of (4.6.6). Then, Φ is invertible and its inverse Φ^{-1} satisfies the following stochastic matrix differential equation of the Itô–Doob-type:*

$$d\Phi^{-1}(t) = \Phi^{-1}(t)[-A(t) + B^2(t)]dt - \Phi^{-1}(t)B(t)dw(t). \qquad (4.6.15)$$

Moreover,

$$d(\Phi^{-1})^T = [-A^T(t) + (B^T(t))^2](\Phi^{-1}(t))^T dt - B^T(t)(\Phi^{-1}(t))^T dw(t). \qquad (4.6.16)$$

Proof. The proof of the theorem can be constructed by following the proofs of Theorems 4.2.3 and (4.5.7). The details are left to the reader. $\qquad \square$

Remark 4.6.3

(i) We remark that an observation similar to Observation 4.2.5 remains true with regard to the time-varying system (4.6.6). The verification of this remark is left as an exercise to the reader. In fact, $(\Phi^{-1})^T$ is a fundamental matrix solution to

$$dz = [-A^T(t) + (B^T(t))^2]z\,dt - B^T(t)z\,dw(t). \tag{4.6.17}$$

System (4.6.17) is called the *adjoint* to system (4.6.6). Moreover, it is equivalent to the system

$$dz = z[-A(t) + (B^2(t)]dt - zB(t)dw(t); \tag{4.6.18}$$

where z in (4.6.18) is a row vector, and z in (4.6.17) is a column vector. In the light of this notional understanding, $\Phi^{-1}(t)$ is a fundamental matrix solution to (4.6.18). This can be justified from (4.6.15).

(ii) We further remark that (4.6.18) provides another simple calculus approach to find the inverse of a general fundamental solution to (4.6.6).

Our experience in finding a solution to the IVPs (4.2.8), (4.4.20), and (4.6.6) is based on the construction of a general fundamental matrix solution [Observation 4.2.2 and Remark 4.6.1(iii)]. The construction of a general fundamental matrix solution process depends on the choice of eigenvectors/initial states of solutions. Therefore, the general fundamental matrix solution process is not uniquely determined. The following result provides the relationship between two fundamental matrix solution processes.

Lemma 4.6.1. *Let Φ_1 be a fundamental matrix solution process of (4.6.6). Φ_2 is any other general fundamental matrix solution process of (4.6.6) if and only if $\Phi_2(t, w(t)) = \Phi_1(t, w(t))C$ on I for some nonsingular constant random matrix C.*

Proof. We assume that $\Phi_1(t, w(t))$ is a given general fundamental matrix solution process of (4.6.6), and $\Phi_2(t, w(t))$ is any other general fundamental matrix solution process of (4.6.6). First, we prove that $\Phi_2 = \Phi_1 C$ on I for some nonsingular constant random matrix C. For this purpose, we define $\Psi = \Phi_1^{-1}\Phi_2$ on I. This is defined in view of Theorems 4.6.3 and 4.6.4 (Observation 4.2.4). Our goal reduces to showing that Ψ is a nonsingular constant random matrix. For this purpose, we compute the Itô–Doob stochastic differential of Ψ:

$$\begin{aligned}
d\Psi &= d(\Phi_1^{-1}\Phi_2) &&\text{(by definition of } \Psi) \\
&= d(\Phi_1^{-1})\Phi_2 + \Phi_1^{-1}d\Phi_2 + d(\Phi_1^{-1})d\Phi_2 &&\text{(by Theorem 1.3.4)} \\
&= [\Phi_1^{-1}[-A(t) + B^2(t)]dt - \Phi_1^{-1}B(t)dw(t)]\Phi_2 \\
&\quad + \Phi_1^{-1}[A(t)\Phi_2 dt + B(t)\Phi_2 dw(t)]
\end{aligned}$$

$$+ [\Phi_1^{-1}[-A(t) + B^2(t)]dt - \Phi_1^{-1}B(t)dw(t)][A(t)\Phi_2 dt$$

$$+ B(t)\Phi_2 dw(t)] \quad \text{[by Theorem 4.6.4 and from (4.6.10)]}$$

$$= \Phi_1^{-1}[-A(t) + B^2(t)]\Phi_2 dt - \Phi_1^{-1}B(t)dw(t)\Phi_2 + \Phi_1^{-1}A(t)\Phi_2 dt$$

$$+ \Phi_1^{-1}B(t)\Phi_2 dw(t) + \Phi_1^{-1}[-A(t) + B^2(t)]B(t)\Phi_2 dw(t)dt$$

$$- \Phi_1^{-1}B^2(t)\Phi_2 dt \quad \text{(by simplification and Itô–Doob calculus)}$$

$$= 0 \qquad\qquad \text{(final simplification)}. \tag{4.6.19}$$

From (4.6.19), we conclude that $\Psi(t) = C$ is constant on I due to the fact that the Itô–Doob differential of the constant stochastic process on an interval I for $t \geq t_0$ is zero. By substitution and simplification, we conclude that $\Phi_2 = \Phi_1 C$ on I. From Theorems 1.4.4 [80] and 4.6.3, $\det(\Phi_2) = \det(\Phi_1)\det(C)$ (Theorem 1.4.2 [80]), $\det(\Phi_2) \neq 0 \neq \det(\Phi_1)$. Hence, $0 \neq \det(C)$. Therefore, C is a nonsingular constant random matrix. This completes the proof of the "if" part.

To prove the "only if" part, we start with $\Phi_2 = \Phi_1 C$, where C is a nonsingular constant matrix, and show that Φ_2 is the general fundamental solution process of (4.6.6). We find the differential (Itô–Doob sense) of both sides of the expressions, and obtain

$$d\Phi_2 = d\Phi_1 C \quad \text{(by Theorem 1.3.1)}$$

$$= [A(t)\Phi_1 dt + B(t)\Phi_1 dw(t)]C \quad \text{[by substitution from (4.6.10)]}$$

$$= A(t)\Phi_1 C dt + B(t)\Phi_1 C dw(t) \quad \text{(Theorem 1.3.1M2, M3 [80])}$$

$$= A(t)\Phi_2 dt + B(t)\Phi_2 dw(t) \quad \text{(by substitution for $\Phi_1 C$)}.$$

Hence, $\Phi_1 C$ is a solution to (4.6.6). Since $\det(\Phi_2) = \det(\Phi_1)\det(C) \neq 0$ for $C \neq 0$. Therefore, we conclude that $\Phi_1 C$ is the general fundamental solution process of (4.6.6). This completes the proof of the theorem. $\qquad\square$

Observation 4.6.2

(i) Obviously, two different linear homogeneous systems differential equations cannot have the same general fundamental solution process. In fact, from (4.6.10), we note that $A(t)dt + B(t)dw(t) = d\Phi(t)\Phi^{-1}(t)$. This is true in view of Remark 4.6.1(iii) and (iv).

(ii) Let us assume that Φ_1 and Φ_2 are two given general fundamental solution processes of (4.6.6). In addition, let $t_0 \in J$ be given. From Lemma 4.6.1, we have $\Phi_2(t) = \Phi_1(t)C$ on I. For $t_0 \in J$, we have $\Phi_2(t_0) = \Phi_1(t_0)C$. This implies that $C = \Phi_1^{-1}(t_0)\Phi_2(t_0)$. From this, we obtain $\Phi_2(t) = \Phi_1(t)\Phi_1^{-1}(t_0)\Phi_2(t_0)$. This implies that

$$\Phi_2(t, w(t))\Phi_2^{-1}(t_0, w(t_0)) = \Phi_1(t, w(t))\Phi_1^{-1}(t_0, w(t_0)) \quad \text{(by Theorem 4.6.4)}.$$

From this and using the notation of the normalized fundamental solution process of (4.6.6), we have

$$\Phi(t, t_0) \equiv \Phi_2(t, w(t), t_0) = \Phi_1(t, w(t), t_0). \qquad (4.6.20)$$

This shows that the normalized fundamental solution process of (4.6.6) is uniquely determined by A, B, w, and t_0. The normalized fundamental solution process $\Phi(t, t_0)$ of (4.6.6), defined in (4.6.20), is referred to as the *normalized fundamental solution process of* (4.6.6) *at* $t = t_0$.

In the following, we present a few algebraic properties of the normalized fundamental solution process of (4.6.6). The proof of the result is based on the observed properties of a general fundamental solution process, Observation 4.2.1, Theorem 4.2.3, and differential/integral calculus.

Lemma 4.6.2. *Let* $\Phi(t, w(t), t_0) \equiv \Phi(t, t_0)$ *and* $\Psi(t, w(t), t_0) \equiv \Psi(t, t_0)$ *be the normalized fundamental matrix solution processes of* (4.6.6) *and* (4.6.18) *at* $t = t_0$, *respectively. In addition, let* $\Phi(t, t_1)$ *be the normalized fundamental matrix solution of* (4.6.6) *at* $t = t_1$. *Then, for all* t_0, t_1, s, *and* t *in* I:

(a) $\Psi(t, t_0)\Phi(t, t_0) = I_{n \times n} = \Phi(t_0, t)\Phi(t, t_0),$ *for* $t_0 \leq t$;   (4.6.21)

where $I_{n \times n}$ *is an* $n \times n$ *identity matrix, and*

$$\Psi(t, t_0) = \Phi^{-1}(t, t_0) = \Phi(t_0, t), \qquad \text{\textit{for} } t_0 \leq t; \quad (4.6.22)$$

(b) $\Phi_1(t, t_1) = \Phi(t, t_0)\Phi_1(t_0, t_1),$ *for* $t_0 \leq t$;   (4.6.23)

(c) $\Phi(t, t_0) = \Phi(t, s)\Phi(s, t_0),$ *for* $t_0 \leq t$;   (4.6.24)

(d) $\Phi(t, s) = \Phi(t, t_0)\Psi(s, t_0) = \Phi(t, t_0)\Phi(t_0, s),$ *for* $t_0 \leq t$;   (4.6.25)

(e) $\partial_s \Phi(t, s) = \Phi(t, s)[-A(s) + B^2(s)]ds - \Phi(t, s)B(s)dw(s),$ (4.6.26)

where $\partial_s \Phi(t, s)$ *is the Itô–Doob stochastic partial differential of* $\Phi(t, s)$ *with respect to* s *for fixed* t.

Proof. To prove (a), using the Itô–Doob stochastic differential, we compute

$d(\Psi(t, t_0)\Phi(t, t_0))$

$\quad = d\Psi(t, t_0)\Phi(t, t_0) + \Psi(t, t_0)d\Phi(t, t_0) + d\Psi(t, t_0)d\Phi(t, t_0)$ (by Theorem 1.3.4)

$\quad = \Psi(t, t_0)([-A(t) + B^2(t)]dt - B(t)dw(t))\Phi(t, t_0)$

$\qquad + \Psi(t, t_0)(A(t)\Phi(t, t_0)dt + B(t)\Phi(t, t_0)dw(t))$ [from (4.6.15) and (4.6.10)]

$\qquad + (-\Psi(t, t_0)B(t)dw(t))B(t)\Phi(t, t_0)dw(t)$

$\qquad$ (by neglecting the higher-order terms)

$\quad = 0$ (final simplification).

This establishes the fact that $\Psi(t, t_0)\Phi(t, t_0) = C$, where C is an arbitrary constant random matrix. This is due the fact that the Itô–Doob differential of a constant stochastic process on an interval I for $t \geq t_0$ is zero. Thus, the matrix process $\Psi(t, t_0)\Phi(t, t_0)$ is a constant random matrix on I. Moreover, $\Psi(t, t_0)$ and $\Phi(t, t_0)$ are the normalized solution processes (4.6.18) and (4.6.6) at $t = t_0$, respectively. Therefore,

$$\Psi(t, t_0)\Phi(t, t_0) = C = \Psi(t_0, t_0)\Phi(t_0, t_0) = I_{n \times n}. \qquad (4.6.27)$$

This shows that $\Psi(t, t_0)$ is the algebraic inverse of $\Phi(t, t_0)$, and it is denoted by $\Phi(t_0, t)$. This statement is equivalent to other notations in (4.6.22). In view of these notations and (4.6.27), we have

$$\Psi(t, t_0)\Phi(t, t_0) = I_{n \times n} = \Phi(t_0, t)\Phi(t, t_0), \quad \text{for } t_0 \leq t.$$

This completes the proof of (4.6.21).

To prove (b), from Lemma 4.6.1 and Observation 4.6.2(ii), we have $\Phi_1(t, t_1) = \Phi(t, t_0)C$. By using the assumption of the lemma and $t = t_1$, we get $C = \Phi_1(t_0, t_1)$. After substitution, we obtain

$$\Phi_1(t, t_1) = \Phi(t, t_0)\Phi_1(t_0, t_1), \quad \text{for } t_0 \leq t.$$

This establishes the relation (4.6.23).

To prove (c), for $t_0 \leq t$ and any s in I, we utilize elementary algebraic properties and the structure of $\Phi(t, t_0)$ described in (4.6.11), and obtain

$$
\begin{aligned}
\Phi(t, t_0) &= \Phi(t)\Phi^{-1}(t_0) && \text{[by definition (4.6.11)]} \\
&= \Phi(t)\Phi^{-1}(s)\Phi(s)\Phi^{-1}(t_0) \text{ (by } \Phi^{-1}(s) && \text{(Theorem 4.6.3)} \\
&= \Phi(t, s)\Phi(s, t_0) && \text{[by (4.6.21)].}
\end{aligned}
$$

This completes the proof of (4.6.24).

For the proof of (d), from (4.6.24) and solving for $\Phi(t, s)$, we have

$$
\begin{aligned}
\Phi(t, s) &= \Phi(t, t_0)\Phi^{-1}(s, t_0) && \text{[by (4.6.24)]} \\
&= \Phi(t, t_0)\Psi(s, t_0) && \text{[by notation (4.6.22)]} \\
&= \Phi(t, t_0)\Phi(t_0, s) && \text{for } t_0 \leq t \quad \text{[by notation (4.6.22)].}
\end{aligned}
$$

This proves (d).

For the proof of (e), we apply the Itô–Doob differential (Theorem 1.3.4) on both sides with respect to s for fixed (t, t_0) to the expression (4.6.25), and obtain

$$
\begin{aligned}
\partial_s \Phi(t, s) &= \Phi(t, t_0)[\Psi(s, t_0)[-A(s) + B^2(s)]ds - \Psi(s, t_0)B(s)dw(s)] \quad \text{[by (4.6.15)]} \\
&= ([-\Phi(t, t_0)\Psi(s, t_0)A(s) + \Phi(t, t_0)\Psi(s, t_0)B^2(s)]ds \\
&\quad - \Phi(t, t_0)\Psi(s, t_0)B(s)dw(s)) \quad \text{(by Theorem 1.3.1M2 [80])}
\end{aligned}
$$

$$= ([-\Phi(t,s)A(s) + \Phi(t,s)B^2(s)]ds - \Phi(t,s)B(s)dw(s)) \quad \text{[by (4.6.25)]}$$

$$= \Phi(t,s)[-A(s) + B^2(s)]ds$$

$$- \Phi(t,s)B(s)dw(s)) \quad \text{(by Theorem 1.3.1M2 [80])}.$$

This establishes the result in (e). This completes the proof of the lemma. $\square$

Observation 4.6.3

(i) We further emphasize that the proof of Lemma 4.6.2 depends heavily on Theorems 4.6.3 and 4.6.4. It is motivated by Theorems 4.2.2 and 4.2.3 as well as the closed-form representations of $\Phi(t, t_0)$ in Sections 4.2–4.4.

(ii) An alternative proof of these expressions in Lemma 4.6.2 can be given by utilizing the conceptual approach. In fact, the proofs of Parts (a), (c), and (e) will be illustrated at a later stage of this section.

In the following, we present a few more results that shed light on the algebraic properties of the solution processes of the IVPs (4.6.6) and (4.6.18). Based on Observation 4.6.1, the IVPs (4.6.6) and (4.6.18) include the IVPs (4.2.8), (4.4.20) and (4.2.29) as special cases. These properties are utilized for studying fundamental properties of the solution process of the IVP (4.6.1). Moreover, these results provide an auxiliary conceptual tool and insight for undertaking the study of nonlinear and nonstationary scalar as well as systems of differential equations.

Lemma 4.6.3 (principle of superposition). *Let* $x_1(t, t_0, x_0) = x_1(t)$ *and* $x_2(t, t_0, x_0) = x_2(t)$ *be solutions to the IVP* (4.6.6) *through* (t_0, x_0) *for* $t \geq t_0$, *and* $t \in I$, $t_0 \in J$. *In addition,* α *and* β *be any two scalar real numbers. Then,*

$$\alpha x_1(t, t_0, x_0) + \beta x_2(t, t_0, x_0) \tag{4.6.28}$$

is a solution to (4.6.6) *through the initial data* $(t_0, \alpha x_0 + \beta x_0)$ *for* $t \geq t_0$, *and* t, t_0 *in* I.

Proof. The proof of the lemma can be reconstructed by imitating the argument used in the proof of Lemma 2.5.3. To minimize repetition, we leave the details to the reader as an exercise. $\square$

Observation 4.6.4

(i) From Illustration 1.4.4, Observation 1.4.5 of Volume 1, and the conclusion of Lemma 4.6.3, we infer that an arbitrary linear combination of solutions to (4.6.6) is also a solution to the IVP (4.6.6).

(ii) In fact, from Theorem 1.4.5 [80] and (i), we infer that the span $S(\{x_1, x_2, \ldots, x_k, \ldots, x_n\})$ of the solution processes $x_1, x_2, \ldots, x_k, \ldots, x_n$ of (4.6.6) is a vector subspace of $C[[t_0, b], R[\Omega, R^n]]$.

(iii) From Definition 1.4.6 of Volume 1), the dimension of the subspace $S(\{x_1, x_2, \ldots, x_k, \ldots, x_n\})$ in (ii) is at most n. This is due to the fact that the general solution to (4.6.6) depends on n arbitrary constants. In fact, it is a linear combination of n random functions.

The following result shows that the normalized fundamental matrix solution $\Phi(t, t_0)$ to (4.6.6) possesses certain basic algebraic properties as a function of each fixed (t, t_0) for $t \geq t_0$ and $t \in I$, $t_0 \in J$.

Lemma 4.6.4. Let $x(t) = x(t, t_0, x_0)$ be a solution process of the IVP (4.6.6) through (t_0, x_0) for $t \geq t_0$, and $t \in I$, $t_0 \in J$. For fixed (t, t_0), the transformation/mapping $\Phi(t, t_0)$ [in (4.6.11)] is defined on R^n into R^n by $\Phi(t, t_0) x_0 = x(t, t_0, x_0)$ for x_0 in R^n. Then, for every x_0, x_1, x_2 in R and α in R,

$$\Phi(t, t_0)(x_1 + x_2) = \Phi(t, t_0)x_1 + \Phi(t, t_0)x_2, \tag{4.6.29}$$

$$\Phi(t, t_0)(\alpha x_0) = \alpha \Phi(t, t_0)x_0. \tag{4.6.30}$$

Proof. Again, the proof of the lemma can be reconstructed by following the argument used in the proof of Lemma 2.5.4. The details are left to the reader as an exercise. $\qquad \square$

Observation 4.6.5

(i) We observe that the mapping/transformation that satisfies the properties described in (4.6.29) and (4.6.30) is called a *linear transformation/mapping*. For fixed (t, t_0), the normalized fundamental solution $\Phi(t, t_0)$ to (4.6.6) at $t = t_0$ is a linear transformation defined on R^n into itself. In fact, $\Phi(t, t_0)x_0$ is a first-degree polynomial process in n variables.

(ii) The second part of the observation is similar to the second part of Observation 2.5.6, which can be easily reformulated.

In the following, we present a relationship between the solution processes of the linear homogeneous system of differential equations (4.6.6) and its adjoint system of differential equations (4.6.18).

Lemma 4.6.5. Let $x(t) = \Phi(t, t_0)x_0$ and $z(t) = x_0^T \Psi(t, t_0)$ be the solution processes of the IVPs (4.6.6) and (4.6.18), respectively, through (t_0, x_0), where $\Phi(t, t_0)$ and $\Psi(t, t_0)$ are the normalized fundamental solution processes of (4.6.6) and (4.6.18) for $t \geq t_0$, and $t \in I$, $t_0 \in J$. Then,

$$z(t)x(t) = c, \tag{4.6.31}$$

where c is a constant random variable. Moreover, $c = x_0^T x_0$ and

$$\Psi(t, t_0)\Phi(t, t_0) = I_{n \times n}, \tag{4.6.32}$$

for all $t \geq t_0$ and $t \in I$, $t_0 \in J$.

Proof. By imitating the proof of Lemma 4.6.2(a), we compute

$d(z(t)x(t))$

$$= dz(t)x(t) + z(t)dx(t) + dy\,dx \qquad \text{(by Theorem 1.3.4)}$$

$$= (z(t)[-A(t) + B^2(t)]dt - z(t)B(t)dw(t))x(t)$$

$$+ (z(t)[-A(t) + B^2(t)]dt - z(t)B(t)dw(t))(A(t)x(t)dt + B(t)x(t)dw(t))$$

$$+ z(t)(A(t)x(t)dt + B(t)x(t)dw(t)) \quad \text{[from (4.6.6) and (4.6.18)]}$$

$$= z(t)[-A(t) + B^2(t)]x(t)dt - z(t)B(t)x(t)dw(t)$$

$$+ z(t)A(t)x(t)dt + z(t)B(t)x(t)dw(t)$$

$$- z(t)B^2(t)x(t)dt \quad \text{(by Theorem 1.3.1M2 [80] and Theorem 1.3.4)}$$

$$= 0 \qquad \text{(by final simplification)}.$$

This establishes the fact that $z(t)x(t) = c$, where c is a constant random variable. This is due to the fact that the Itô–Doob differential of a constant stochastic process is zero. This completes the proof of (4.6.31). The proof of $c = x_0^T x_0$ follows from the facts that $z(t_0)x(t_0) = c = x_0^T x_0$, $\Phi(t_0, t_0) = I_{n \times n} = \Psi(t_0, t_0)$ and $z(t)x(t) = x_0^T x_0 = x_0\Psi(t, t_0)\Phi(t, t_0)x_0$, for any x_0 in R^n. The details are left to the reader. Finally, from $x_0\Psi(t, t_0)\Phi(t, t_0)x_0 = x_0^T x_0$, we have $\Psi(t, t_0)\Phi(t, t_0) = I_{n \times n}$. This completes the proof of (4.6.32). Thus, the proof of the lemma is complete. $\square$

Observation 4.6.6

(i) The conclusions of Lemma 4.6.5 provide a relationship between the normalized fundamental matrix solution processes of (4.6.6) and (4.6.18). Moreover, Lemma 4.6.5 provides an alternative conceptual proof of Lemma 4.6.2(a).

(ii) In addition, Lemma 4.6.5, conceptually confirms that the solution process provided by Theorem 4.6.4 is indeed the algebraic inverse of a fundamental matrix solution to (4.6.6). Of course, Theorem 4.6.4 provides an analytic method for finding an inverse of any fundamental matrix solution process of (4.6.6).

In the following, we provide an analytic (conceptual) alternative proof of Lemma 4.6.2(c). Its proof was based on the elementary-algebraic approach. This result provides a very general argument for establishing the validity of the relation (4.6.24). Moreover, the presented argument provides a conceptual reasoning for undertaking the study of more complex properties of solutions to more general differential equations.

Lemma 4.6.6. *Let $\Phi(t, t_0)$ be a normalized fundamental matrix solution to (4.6.6) at $t = t_0$, and let $x(t, t_0, x_0)$ be the solution process of the IVP (4.6.6) through*

(t_0, x_0). *Then, for all* $t \geq t_0$, *and* $t \in I$, $t_0 \in J$,

(a) $x(t, s, x(s, t_0, x_0)) = x(t, t_0, x_0)$,   (4.6.33)
(b) $\Phi(t, s)\Phi(s, t_0) = \Phi(t, t_0)$.   (4.6.34)

Proof. The proof of the lemma can be reproduced by imitating the proof of Lemma 2.5.6 through replacing R by R^n in the context of (4.6.6). The details are omitted.   ☐

Prior to presenting several additional basic results concerning (4.6.1), we need to present a final result with respect to the system (4.6.6). This result deals with the differentiability of the solution process of (4.6.6) with respect to all three variables (t, t_0, x_0) in the sense of Itô–Doob calculus. The byproduct of the result is an analytic (conceptual) alternative proof of Lemma 4.6.2(e) that was based on the elementary-algebraic approach.

Lemma 4.6.7. *Let* $x(t, t_0, x_0)$ *be the solution process of the IVP* (4.6.6) *through* (t_0, x_0). *Then, for all* $\geq t_0$, *and* $t \in I$, $t_0 \in J$,

(a) $\frac{\partial}{\partial x_0} x(t) = \frac{\partial}{\partial x_0} x(t, t_0, x_0)$ *exists and it satisfies the matrix differential equation*

$$dX = A(t)X \, dt + B(t)X \, dw(t), \quad \frac{\partial}{\partial x_0} x(t_0) = I_{n \times n}, \quad (4.6.35)$$

where $\frac{\partial}{\partial x_0} x(t, t_0, x_0)$ *stands for a deterministic partial derivative of the solution process* $x(t, t_0, x_0)$ *of* (3.6.6) *with respect to* x_0 *at* (t, t_0, x_0) *for fixed* (t, t_0);

(b) $\frac{\partial}{\partial t_0} x(t) = \frac{\partial}{\partial t_0} x(t, t_0, x_0)$ *exists (in the generalized sense, this is beyond the level of the course) and it satisfies the differential equation*

$$dx = A(t)x \, dt + B(t)x \, dw(t), \quad \frac{\partial}{\partial t_0} x(t_0, t_0, x_0) = z_0, \quad (4.6.36)$$

where $\partial_{t_0} x(t, t_0, x_0) = \frac{\partial}{\partial t_0} x(t, t_0, x_0) dt_0$ *stands for the Itô–Doob-type stochastic partial differential of the solution process* $x(t, t_0, x_0)$ *of* (4.6.6), *with respect to* t_0 *at* (t, t_0, x_0) *for fixed* (t, x_0); *an initial differential* $\partial x(t_0) = \partial_{t_0} x(t_0, t_0, x_0)$ $\left(\frac{dx_0(t_0)}{dt_0} = \frac{\partial}{\partial t_0} x(t_0) = z_0 \right)$ *satisfies the system of stochastic differential equations*

$$\frac{\partial}{\partial t_0} x(t_0, t_0, x_0) dt_0 \equiv \partial x(t_0)$$

$$= [-A(t_0) + B^2(t_0)] x_0 dt_0 - B(t_0) x_0 dw(t_0) \quad (4.6.37)$$

or

$$\frac{\partial}{\partial t_0} x(t_0, t_0, x_0) = z_0 = [-A(t_0) + B^2(t_0)] x_0 - B(t_0) x_0 \frac{d}{dt_0} w(t_0),$$

where $\frac{d}{dt_0}w(t_0)$ is the white noise [*Observation* 1.2.11(v)] *process of the Wiener process.*

Proof. From Lemma 4.6.4, the solution $x(t, t_0, x_0) = \Phi(t, t_0)x_0$ to the IVP (4.6.6) through (t_0, x_0) is a linear function of x_0 defined on R^n into R^n. By following the argument used in the proof of Lemma 2.5.8, one can establish the validity of the statement (a). In fact, one can conclude that

$$\frac{\partial}{\partial x_0}x(t) = \frac{\partial}{\partial x_0}x(t, t_0, x_0) = \Phi(t, t_0). \tag{4.6.38}$$

Moreover, $\frac{\partial}{\partial x_0}x(t, t_0, x_0)$ is a continuous function in x_0. This is due to the fact that $\frac{\partial}{\partial x_0}x(t, t_0, x_0) = \Phi(t, t_0)$ is independent of x_0 (i.e. constant) for each fixed (t, t_0), for $t \geq t_0$, and $t \in I$, $t_0 \in J$. From (4.6.38), it is clear that $\frac{\partial}{\partial x_0}x(t) = \frac{\partial}{\partial x_0}x(t, t_0, x_0)$ is the normalized fundamental solution process of (4.6.6).

To prove the statement (b), for fixed (t, t_0), again from the definition of the solution $x(t, t_0, x_0) = \Phi(t, t_0)x_0$ (4.6.11), we apply Lemma 4.6.2(e) to $x(t, t_0, x_0) = \Phi(t, t_0)x_0$ in (4.6.11), and we get

$$\partial_{t_0}x(t, t_0, x_0) = \partial_{t_0}\Phi(t, t_0)x_0 \qquad \text{[from (4.6.11)]}$$

$$= \Phi(t, t_0)[-A(t_0) + B^2(t_0)]dt_0$$

$$\quad - \Phi(t, t_0)B(t_0)dw(t_0)x_0 \quad \text{[from (4.6.26)]}$$

$$= \Phi(t, t_0)\partial x(t_0) \qquad \text{[from (4.6.37)]}, \tag{4.6.39}$$

where $\partial_{t_0}x(t, t_0, x_0)$ denotes the Itô–Doob partial differential of the solution process $x(t, t_0, x_0)$ of (4.6.6) with respect to t_0 at (t, t_0, x_0); $\partial x(t_0)$ as defined in the lemma and satisfies the system of differential equations (4.6.37).

From (4.6.39), it is clear that $\partial_{t_0}x(t, t_0, x_0)$ satisfies the IVP (4.6.36). $\partial x(t_0) = \partial_{t_0}x(t_0, t_0, x_0)$ satisfies the system of differential equations (4.6.37). Moreover, from notation of $\partial_{t_0}x(t, t_0, x_0)$, we have $\frac{\partial}{\partial t_0}x(t, t_0, x_0) = \Phi(t, t_0)$. This completes the proof of the theorem. $\qquad\square$

Observation 4.6.7

(i) From the proof of Lemma 4.6.7 we observe that $\frac{\partial}{\partial x_0}x(t, t_0, x_0) = \Phi(t, t_0)$ is independent of x_0, i.e. $\frac{\partial}{\partial x_0}x(t, t_0, x_0) = \Phi(t, t_0)$ [Jacobian matrix of $x(t, t_0, x_0)$ for fixed (t, t_0)]. It is obvious that $\frac{\partial}{\partial x_0}x(t, t_0, x_0)$ is sample continuous process in the (t, t_0, x_0). Moreover, its second partial derivative of $x(t, t_0, x_0)$ is $\frac{\partial^2}{\partial x_0^2}x(t, t_0, x_0)$ [Hessian matrix of $x(t, t_0, x_0)$ for fixed (t, t_0)], and $\frac{\partial^2}{\partial x_0^2}x(t, t_0, x_0) = 0$.

(ii) Using (i) and Lemma 4.6.7, we introduce the Itô–Doob partial differentials of the solution process $x(t, t_0, x_0)$ of (4.6.6) with respect to x_0 and t_0 at

(t, t_0, x_0) as:

$$\partial_{x_0} x(t, t_0, x_0) = \frac{\partial}{\partial x_0} x(t, t_0, x_0) dx_0 = \Phi(t, t_0) dx_0, \qquad (4.6.40)$$

$$\partial_{t_0} x(t, t_0, x_0) = \frac{\partial}{\partial t_0} x(t, t_0, x_0) \partial x(t_0) = \Phi(t, t_0) \partial x(t_0), \qquad (4.6.41)$$

where dx_0 is the Itô–Doob differential of $x(t, t_0, x_0)$ at t_0, i.e.

$$dx(t_0) = A(t_0) x(t_0) dt_0 + B(t_0) x(t_0) dw(t_0), \qquad (4.6.42)$$

and $\frac{\partial}{\partial t_0} x(t_0, t_0, x_0) dt_0 = \partial x(t_0)$ is the partial differential of $x(t, t_0, x_0)$ with respect to t_0 at $t = t_0$ for fixed t_0. Note the difference between dx_0 and $\partial x(t_0)$.

(iii) From the notation of (4.6.40) and (4.6.41), the Itô–Doob mixed partial differentials of the solution process $x(t, t_0, x_0)$ of (4.6.6) are

$$\partial_{t_0 x_0}^2 x(t, t_0, x_0) = \partial_{t_0} (\partial_{x_0} x(t, t_0, x_0))$$

(by the definition of the partial differential)

$$= \partial_{t_0} \left(\frac{\partial}{\partial x_0} x(t, t_0, x_0) dx_0 \right) \quad \text{[from (4.6.40)]}$$

$$= \Phi(t, t_0)[[-A(t_0) + B^2(t_0)] dt_0 - \Phi(t, t_0) B(t_0) dw(t_0)] dx_0$$

$$\text{[from (4.6.26)]}$$

$$= -\Phi(t, t_0) B^2(t_0) x(t_0)] dt_0 \quad \text{[from Itô–Doob calculus and (4.6.42)]}.$$

$$(4.6.43)$$

On the other hand,

$$\partial_{x_0 t_0}^2 x(t, t_0, x_0) = \partial_{x_0} (\partial_{t_0} x(t, t_0, x_0))$$

(by the definition of the partial differential)

$$= \partial_{x_0} (\Phi(t, t_0) \partial x(t_0) \quad \text{[from (4.6.41)]}$$

$$= \Phi(t, t_0) \partial_{x_0} (\partial x(t_0)) \quad \text{(from Observation 1.5.2 [80])}$$

$$= \Phi(t, t_0)[-A(t_0) + B^2(t_0)] dt_0 - \Phi(t, t_0) B(t_0) dw(t_0)] \partial_{x_0} x_0$$

$$\text{[from (4.6.39)]}$$

$$= \Phi(t, t_0)[-A(t_0) + B^2(t_0)] dt_0 - \Phi(t, t_0) B(t_0) dw(t_0)] dx_0 \quad \text{(by notation)}$$

$$= -\Phi(t, t_0) B^2(t_0) x(t_0)] dt_0 \quad \text{[from Itô–Doob calculus and (4.6.42)]}.$$

$$(4.6.44)$$

(iv) We further observe that Lemma 4.6.7(b) provides a conceptual proof for Lemma 4.6.2(e). In fact, from (4.6.37) and (4.6.39), we have

$$\frac{\partial}{\partial x_0}(\partial_{t_0} x(t, t_0, x_0))$$

$$= \frac{\partial}{\partial x_0}(\Phi(t, t_0)\partial x(t_0)) \quad \text{[from (4.6.39)]}$$

$$= \frac{\partial}{\partial x_0}(\Phi(t, t_0)[-A(t_0) + B^2(t_0)]dt_0 - \Phi(t, t_0)B(t_0)dw(t_0)]x_0)$$

[from (4.6.37)]

$$= \Phi(t, t_0)\left([-A(t_0) + B^2(t_0)]dt_0 - \Phi(t, t_0)B(t_0)dw(t_0)]\frac{\partial}{\partial x_0}x_0\right)$$

(Observation 1.5.2 [80])

$$= \Phi(t, t_0)([-A(t_0) + B^2(t_0)]dt_0 - \Phi(t, t_0)B(t_0)dw(t_0))\left(\frac{\partial}{\partial x_0}x_0 = I_{n\times n}\right).$$

$$(4.6.45)$$

On the other hand,

$$\partial_{t_0}\left(\frac{\partial}{\partial x_0}x(t, t_0, x_0)\right) = \partial_{t_0}(\Phi(t, t_0)) \quad \text{[from (4.6.38)]}$$

$$= \partial_{t_0}\Phi(t, t_0) \quad \text{(by notation).} \quad (4.6.46)$$

From (iii), (4.6.45) and (4.6.46), the validity of (4.6.26) follows immediately.

Hence, we conclude that its mixed second derivative exists (in the generalized sense),

$$\frac{\partial^2}{\partial x_0 \partial t_0}x(t, t_0, x_0) = \frac{\partial^2}{\partial t_0 \partial x_0}x(t, t_0, x_0),$$

and it $\left[\frac{\partial^2}{\partial t_0 \partial x_0}x(t, t_0, x_0)\right]$ is the normalized fundamental solution process of the adjoint differential equation (4.6.18). In the following, we present a result that provides a relationship between the solution process of (4.6.1) and (4.6.6). This result gives another alternative solution representation of (4.6.1) in the form of an integral equation. It also provides an analytic tool for investigating the behavior of the solution process of (4.6.1), knowing the behavior of (4.6.6). This idea will be highlighted in Chapter 6.

Theorem 4.6.5 (method of variation of parameters). *Let the assumption of Theorem 4.6.1 be satisfied. In addition, let $y(t) = y(t, t_0, y_0)$ and $x(t) = x(t, t_0, y_0)$ be the solution processes of (4.6.1) and (4.6.6), respectively, through the same initial* *data* (t_0, y_0), *for all* $t \geq t_0$, *and* $t \in I$, $t_0 \in J$. *Then,*

$$y(t) = x(t) + \int_{t_0}^{t} \Phi(t, s)[p(s) - B(s)q(s)]ds + \int_{t_0}^{t} \Phi(t, s)q(s)dw(s), \qquad (4.6.47)$$

for $t \in I$, $t_0 \in J$.

Proof. Let $y(t) = y(t, t_0, y_0)$ and $x(t) = x(t, t_0, y_0)$ be given solution processes of (4.6.1) and (4.6.6), respectively, through the same initial data (t_0, y_0) for all $t \geq t_0$, and $t \in I$, $t_0 \in J$. From (4.6.11) and Theorem 4.6.3, we recall that $x(t) = \Phi(t, t_0)y_0$, where $\Phi(t, t_0)$ is the normalized fundamental matrix solution process of (4.6.6). Knowing the normalized fundamental matrix solution $\Phi(t, t_0)$ to (4.6.6), we imitate Procedure 4.5.4 for finding the solution process of the IVP (4.6.1), and we arrive at (4.5.24), i.e.

$$y(t) = \Phi(t, t_0)y_0 + \int_{t_0}^{t} \Phi(t, s)[p(s) - B(s)q(s)]ds + \int_{t_0}^{t} \Phi(t, s)q(s)dw(s). \quad (4.6.48)$$

This together with (4.6.11) implies that

$$y(t) = x(t) + \int_{t_0}^{t} \Phi(t, s)[p(s) - B(s)q(s)]ds + \int_{t_0}^{t} \Phi(t, s)q(s)dw(s),$$

for $t \in I$, $t_0 \in J$. This completes the proof of the theorem. $\qquad\square$

A very important byproduct of the uniqueness theorem is that the knowledge of the solution process at a future time depends only on the knowledge of the initial data, and it is independent of its knowledge between the initial time and the future time. This is demonstrated by Lemma 2.5.7 in the context of the solution process of the IVP (2.5.1). The following result further extends this idea to the IVP (4.6.1).

Lemma 4.6.8. *Let the hypotheses of Theorem 4.6.1 be satisfied. For $t_0 \leq s \leq t$, let $y(t) = (t, t_0, y_0)$ and $y(t, s, y(s)) = y(t, s, y(s, t_0, y_0))$ be the solution processes of (4.6.1) through (t_0, y_0) and $(s, y(s))$, respectively. Then,*

$$y(t) = y(t, t_0, y_0) = y(t, s, y(s, t_0, y_0)), \quad \text{for } t_0 \leq s \leq t. \qquad (4.6.49)$$

Proof. For $t_0 \leq s \leq t$, let $y(t, t_0, y_0)$ and $y(t, s, y(s))$ be solutions to (4.6.1) through (t_0, y_0) and $(s, y(s)) = (s, y(s, t_0, y_0))$, respectively. From the representation of $y(t) = y(t, t_0, y_0)$ in (4.6.47), Lemma 4.6.2(c) or Lemma 4.6.6, and imitating the proof of Lemma 2.5.7, one can easily establish the conclusion of the lemma. The details are left to the reader as an exercise. $\qquad\square$

Observation 4.6.8. Utilizing Lemmas 2.5.7 and 4.6.8, the uniqueness part of Theorem 4.6.1 as well as the implicit (4.6.2) and the explicit (4.6.47) representations of the solution process (4.6.1), one can reformulate an observation similar to Observation 2.5.8. The details are left to the reader as an exercise.

The following result establishes the continuity of the solution process $y(t, t_0, y_0)$ with respect to (t, t_0, y_0). We can state the result without its proof. The proof is left as an exercise to the reader.

Theorem 4.6.6 (continuous dependence of the solution on initial conditions). *Let the hypotheses of Theorem 4.6.1 be satisfied. Then, the solution process* $y(t) = y(t, t_0, y_0)$ *is continuous with respect to* (t, t_0, y_0) *for* $t \geq t_0$, *and* $t \in I$, $t_0 \in J$. *In particular, it is continuous with respect to the initial conditions/data.*

In fact, in the following, we present a result that exhibits the differentiability of the solution process of (4.6.1) with respect to all three variables (t, t_0, y_0) in the sense of Itô–Doob.

Theorem 4.6.7 (differentiability of solutions with respect to initial conditions). *Let the hypotheses of Theorem 4.6.1 be satisfied. In addition, let* $y(t, t_0, y_0)$ *be the solution process of the IVP* (4.6.1). *Then, for all* $t \geq t_0$, *and* $t \in I$, $t_0 \in J$:

(i) $\frac{\partial}{\partial y_0} y(t) = \frac{\partial}{\partial y_0} y(t, t_0, y_0)$ *exists, and it satisfies the IVP* (4.6.35);
(ii) $\frac{\partial}{\partial t_0} y(t) = \frac{\partial}{\partial t_0} y(t, t_0, y_0)$ *exists, and it satisfies the IVP* (4.6.36) *with* $(t_0, \partial y(t_0))$,

where $\partial_{t_0} y(t, t_0, y_0) = \frac{\partial}{\partial t_0} y(t, t_0, y_0) dt_0$ *stands for the Itô–Doob-type stochastic partial differential of solution process* $y(t, t_0, y_0)$ *of* (4.6.1) *with respect to* t_0 *at* (t, t_0, y_0) *for fixed* (t, y_0); *its initial state* $\partial y(t_0) = \partial_{t_0} y(t_0, t_0, y_0)$ $\left[\frac{dy_0(t_0)}{dt_0} = \frac{\partial}{\partial t_0} y(t_0) = z_0 \right]$ *satisfies the system of stochastic differential equations*

$$\frac{\partial}{\partial t_0} y(t_0, t_0, y_0) dt_0 \equiv \partial y(t_0)$$

$$= [-A(t_0) + B^2(t_0)] y_0 - p(t_0) + B(t_0) q(t_0)] dt_0$$

$$- [B(t_0) y_0 + q(t_0)] dw(t_0) \tag{4.6.50}$$

or

$$\frac{\partial}{\partial t_0} y(t_0 t_0, x_0) = z_0 = [-A(t_0) + B^2(t_0)] y_0 - p(t_0) + B(t_0) q(t_0)] dt_0$$

$$- [B(t_0) y_0 + q(t_0)] \frac{d}{dt_0} w(t_0),$$

where $\frac{d}{dt_0} w(t_0)$ *is the white noise (Observation 1.2.11(v)) process of the Wiener process.*

Proof. The proof of the theorem can be constructed by imitating the proof of Theorem 2.5.4. The details are left as an exercise to the reader. □

Observation 4.6.9. We note that the proof of Theorem 4.6.7 can be reconstructed by utilizing the conceptual aspect of Lemma 4.6.7. For the sake of simplicity, we chose to prove it by utilizing algebraic properties coupled with the solution representation (4.6.2) of (4.6.1).

In the following, we present a very important result that connects the solution process of (4.6.1) with the corresponding homogeneous system of differential equations (4.6.6). The Alekseev-type approach [80, 98] for the method of variation of constant parameters is used to prove the following result. Furthermore, the result offers a conceptual tool and insight for undertaking the study of both nonstationary and nonlinear systems of differential equations. In addition, it provides an alternative method for investigating the qualitative properties of more complex differential equations.

Theorem 4.6.8 (variation of constant formula — Alekseev type). *Let the hypotheses of Theorem 4.6.1 be satisfied. In addition, $y(t) = y(t, t_0, y_0)$ and $x(t) = x(t, t_0, y_0)$ be the solution processes of the IVPs (4.6.1) and (4.6.6) through the same initial data (t_0, y_0) on I, respectively. Then,*

$$y(t, t_0, y_0) = x(t, t_0, y_0) + \int_{t_0}^{t} \Phi(t, s)[p(s) - B(s)q(s)]ds + \int_{t_0}^{t} \Phi(t, s)q(s)dw(s),$$

$$(4.6.51)$$

for $t \in I$, $t_0 \in J$.

Proof. From the assumption of the theorem, for $t_0 \leq s \leq t$, we define a solution to the IVP (4.6.6) through $(s, y(s))$, as:

$$x(t, s, y(s, t_0, y_0)) = x(t, s, y(s)),\qquad (4.6.52)$$

where $y(s) = y(s, t_0, y_0)$ is a solution to the IVP (4.6.1) through (t_0, y_0). From (4.6.52) and the definition of the IVP, we observe that for $s = t_0$,

$$x(t, t_0, y(t, t_0, y_0)) = x(t, t_0, y_0).\qquad (4.6.53)$$

From the application of Lemma 4.6.7, the Itô–Doob differential formula (Theorem 1.3.4), the chain rule, and Observation 4.6.7 with respect to s for $t_0 \leq s \leq t$, we have

$$d_s x(t, s, y(s))$$

$$= \partial_{t_0} x(t, s, y(s)) + \partial_{x_0} x(t, s, y(s))$$

$$\quad + \frac{1}{2} \partial^2_{x_0 x_0} x(t, s, y(s)) + \partial^2_{t_0 x_0} x(t, s, y(s)) \quad \text{(by Theorem 1.3.4)}$$

$$= \frac{\partial}{\partial t_0} x(t, t_0, x_0) \partial y(s) + \frac{\partial}{\partial x_0} x(t, s, y(s)) dy(s)$$

$$\quad + \partial_{t_0} \left(\frac{\partial}{\partial x_0} x(t, s, y(s)) dy(s) \right) \qquad \text{[from (4.6.40)–(4.6.43)]}$$

$$= \Phi(t, s)\partial y(s) + \Phi(t, s)dy(s)$$

$$\quad + \partial_s \Phi(t, s)dy(s) \qquad \text{[from (4.6.40)–(4.6.43)]}$$

$$= \Phi(t,s)([-A(s) + B^2(s)]y(s)ds - B(s)y(s)dw(s))$$

$$+ \Phi(t,s)([A(s)y(s) + p(s)]ds + [B(s)y(s) + q(s)]dw(s))$$

$$+ \Phi(t,s)([-A(s) + B^2(s)]ds - B(s)dw(s)([A(s)y(s) + p(s)]ds$$

$$+ [B(s)y(s) + q(s)]dw(s) \quad \text{[from Lemma 4.6.7 [(4.6.37)] and (4.6.42)]}$$

$$= \Phi(t,s)[-A(s) + B^2(s)]y(s)ds - \Phi(t,s)B(s)y(s)dw(s)$$

$$+ \Phi(t,s)[A(s)y(s) + p(s)]ds + \Phi(t,s)[B(s)y(s) + q(s)]dw(s)$$

$$- \Phi(t,s)[(B^2(s)y(s) + B(s)q(s)]ds \quad \text{(by Itô–Doob calculus)}$$

$$= \Phi(t,s)[p(s) - B(s)q(s)]ds + \Phi(t,s)q(s)dw(s) \quad \text{(by simplifying)}. \quad (4.6.54)$$

Integrating (4.6.54) on both sides with respect to s from t_0 to t, and using Lemma 4.6.6 and the notations and the uniqueness part of the solution process of (4.6.6) (Corollary 4.6.1), we get

$$x(t,t,y(t,t_0,y_0)) - x(t,t_0,y(t_0)) = y(t,t_0,y_0) - x(t,t_0,y_0)$$

$$= \int_{t_0}^{t} \Phi(t,s)[p(s) - B(s)q(s)]ds + \int_{t_0}^{t} \Phi(t,s)q(s)dw(s).$$

This is written as

$$y(t,t_0,y_0) = x(t,t_0,y_0) + \int_{t_0}^{t} \Phi(t,s)[p(s) - B(s)q(s)]ds + \int_{t_0}^{t} \Phi(t,s)q(s)dw(s).$$

This completes the proof of the theorem. Moreover, from (4.6.11), we have

$$y(t,t_0,y_0) = \Phi(t,t_0)y_0 + \int_{t_0}^{t} \Phi(t,s)[p(s) - B(s)q(s)]ds + \int_{t_0}^{t} \Phi(t,s)q(s)dw(s),$$

which is identical to (4.6.48). □

Observation 4.6.10

(i) From the conclusion of Theorem 4.6.5, it is obvious that Theorem 4.6.8 provides another alternative conceptual approach to the closed-form representation of the solution to (4.6.1). In fact, one can compare the solution expressions on the right-hand sides of (4.6.47) and (4.6.51). These expressions are exactly the same.

(ii) We further observe that the solution representation of (4.6.1) in either (4.6.47) or (4.6.51) was derived from the conceptual knowledge of the corresponding homogeneous differential equation (4.6.6). In particular, the smoothness (conceptual) properties of the solution process with respect to its initial data/conditions outlined in Observation 4.6.7 were utilized.

(iii) For the validity of Theorem 4.6.5, we merely needed the existence of the solution process of (4.6.1).

(iv) All presented results include the results of Section 4.7 [80] as special cases.

Exercises

1. Complete the proof of Theorem 4.6.2.
2. Verify the assertions made in Remark 4.6.1.
3. Complete the proof of Theorem 4.6.3.
4. Verify the assertions made in Remark 4.6.2.
5. Complete the proof of Theorem 4.6.4.
6. Verify the assertions in Remark 4.6.3.
7. Complete the proof of Lemma 4.6.3.
8. Complete the proof of Lemma 4.6.4.
9. Given: $x\Psi(t,t_0)\Phi(t,t_0)x = xx$ for any $x \neq 0$. Show that $\Psi(t,t_0)\Phi(t,t_0) = 1$.
10. Construct the proof of Lemma 4.6.6.
11. Complete the proof of Lemma 4.6.7.
12. Complete the proof of Theorem 4.6.6.
13. Complete the proof of Theorem 4.6.7.
14. Using the integral representation of the solution process of (4.6.1) in (4.6.6), show that the conclusion of Lemma 4.6.7 remains true.

4.7 Notes and Comments

The development of the material in this chapter is new. The development of the stochastic modeling procedure in Section 4.1 is a modified form of the classical modeling processes [2, 57, 98, 131, 132, 158, 174, 195]. It is a parallel extension of single state system dynamic processes of Section 2.1 to multivariate processes, and also an extension of deterministic multivariate processes of Section 4.2 [80] to stochastic multivariate processes. It is based on an elementary descriptive statistical approach which includes the classical approach in a unified way. The examples, exercises, and illustrations are based on the second author's class notes over a period of more than 40 years. The initiation of an eigenvalue and eigenvector-type approach to both univariate and multivariate processes is the most suitable approach to the introduction of stochastic modeling, methods, and motivation for an advanced study by any undergraduate and graduate students with a minimal mathematical background. The eigenvalue and eigenvector-type approach developed in Section 2.3 is systematically connected and extended to include the existing deterministic [80] approach to solving systems of linear homogeneous differential equations in Section 4.2. The fundamental results and concepts, namely stochastic versions of the principle of superposition, the Abel–Jacobi–Liouville theorem and its generalizations, are adapted from Refs. 77 and 79. The procedure for finding the fundamental matrix process of linear systems of stochastic differential equations with constant coefficient in Section 4.3 is also adapted from [77, 79]. Methods of finding solutions to general first-order linear homogeneous stochastic differential equations in Section 4.4 are based on Refs. 77 and 79. The methods described in Section 4.5 are new and based on the material in Sections 4.2–4.4. The developed material in Section 4.6 is based on the second author's research. Several numerical and applied examples are based on interdisciplinary research [1, 10–14, 23–26, 30, 33, 36, 41, 48, 49, 54, 55, 58, 59, 62, 64, 66–69, 96, 113, 115–119, 126, 128–130, 143, 151, 154–161, 170–172, 175, 179–189, 191–195, 197]. The objective of this is to generate curiosity, motivate, and to challenge the undergraduate to learn more than what will be gained from this course.

Chapter 5

Higher-Order Differential Equations

Introduction

The presentation of the material in this chapter follows the same framework as for Chapters 2 and 4. Section 5.1 begins with mathematical models. Higher-order linear Itô–Doob-type differential equations are presented in Section 5.2. Examples are given to illustrate the closed-form solutions to higher-order Itô–Doob-type stochastic differential equations. The structure of the stochastic companion system and the results developed for the general linear systems of stochastic differential equations are used to develop the techniques for solving higher-order stochastic differential equations with constant coefficients, in Section 5.3. By employing the stochastic companion system of differential equations, a variation-of-constants formula is developed, and is utilized to solve the higher-order linear nonhomogeneous stochastic differential equations in Section 5.4. Section 5.5 deals with then Itô–Doob-type stochastic Laplace transform. Finally, the stochastic Laplace transform technique is applied for finding the closed-form solutions to IVPs in Section 5.6. Several examples are given to illustrate the method.

5.1 Mathematical Modeling

The material in this section is a straightforward stochastic version of the material in Section 5.1 of Volume 1. It is a special but important case of Chapter 4. In fact, the development of the mathematical model is embedded in the modeling process of Section 4.1. We will also present a few illustrations regarding the mathematical model building process under a randomly varying environment.

Illustration 5.1.1 (stochastic version of the spring problem [55, 80, 82]). This illustration is a prototype for several mechanical systems when bodies are attached to the ends of springs. The motion of the particle (Illustration 5.2.1 of Volume 1) under the environmental random perturbations is disturbed by the influence of the surrounding environment. This causes resistance to particle's motion. It is assumed

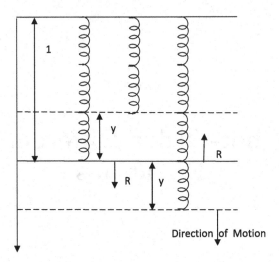

Fig. 5.1.1   Spring mass dynamic.

that the resistance force is directly proportional to the speed of the particle. In view of this, the parameters associated with resistance and restoring forces [coefficients of resistance or friction (μ) and spring stiffness (k)] are subject to random fluctuations described by the stationary Gaussian process with independent increments. As a result of this, assumptions SHM2–SHM5 [80] are modified to incorporate the resistance and the random environmental perturbations. Imitating the development of the deterministic mathematical model of simple harmonic motion (SHM) (Illustration 5.2.1 of Volume 1) and applying Modified Brownian Motion Model 4.1.2, we have

$$dy^{(1)} + (a_1 y^{(1)} + a_0 y)dt + (b_1 y^{(1)} + b_0 y)dw(t) = 0, \tag{5.1.1}$$

where $a_0 = \frac{k}{m}$ (k and m are as defined in Illustration 5.2.1 of Volume 1), and w is a Wiener process.

Illustration 5.1.2 (RLC circuit under random perturbations [55, 80, 82]). We recall the basic components and ideas of the circuit network (Illustration 4.1.4 and Exercise 2.1.6). The illustration is a stochastic version of Illustration 5.2.2 of Volume 1. In the RLC circuit (Figure 5.1.2), it is assumed that the randomness is due to the presence of register and capacitor, and it is described by the stationary Gaussian process with independent increments.

By following the development of the dynamic processes in Section 4.1 and the deterministic model (Volume 1), we have

$$dQ^{(1)} + \left[\frac{R}{L}Q^{(1)} + \frac{1}{LC}Q - \frac{1}{L}E(t)\right]dt + [\sigma_1 Q^{(1)}$$
$$+ \sigma_0 Q - \Lambda(w(t))]dw(t) = 0, \tag{5.1.2}$$

where $Q(t)$, $E(t)$, R, L, and C are as defined in Illustration 5.2.2 [80] w is a Wiener process, and Λ is a smooth process with a zero mean.

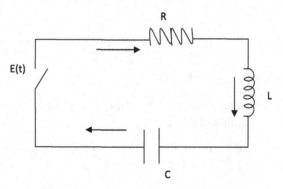

Fig. 5.1.2   RLC circuit.

Illustration 5.1.3 (stochastic dynamic of creatinine clearance [26, 80]). The deterministic Illustration 5.2.4 [80] of the intravenously injected creatinine clearance model is based on assumptions ICC1–ICC6. The major drawback of the deterministic model is the assumption (ICC1) that the creatinine diffusion through the membrane is deterministic. The sources of randomness are described in Illustrations 2.4.4–2.4.6 as well as in Exercises 4 and 5 (Section 2.1). In addition, the creatinine clearance (via the kidneys) depends not only on the concentration of creatinine diffusion in the plasma but also the speed of the creatinine diffusion in the plasma. Under this consideration and following the development of the model-building process of Example 4.1.2 and Illustration 5.2.4 [80], the dynamic of creatinine clearance is described by the following stochastic differential equations

$$dm = [-\kappa c_1 - \alpha(c_1 - c_2)]dt, \tag{5.1.3}$$

$$de = \kappa c_1\, dt + [\sigma_1 c_1^{(1)} + \sigma_0 c_1]dw(t), \tag{5.1.4}$$

and

$$I = c_1 V_1 + c_2 V_2 + e(t), \tag{5.1.5}$$

where m, c_1, c_2, α, κ, I, V_1, and V_2 are as defined in Illustration 5.2.4 [80]. Imitating the arguments used in Illustration 5.1.4 of Volume 1, (5.1.3)–(5.1.5) are reduced to

$$V_1 dc_1 = [-\kappa c_1 - \alpha(c_1 - c_2)]dt \quad \text{(from the definitions of } c\text{'s)}, \tag{5.1.6}$$

$$e(t) = \int_0^t \kappa c_1(s)ds + \int_0^t [\sigma_1 c_1^{(1)}(s)$$
$$+ \sigma_0 c_1(s)]dw(s) \quad \text{(from integration)}, \tag{5.1.7}$$

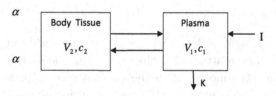

Fig. 5.1.3   Creatinine clearance dynamic process.

and

$$c_2 = \frac{I - (c_1 V_1 + \int_0^t \kappa c_1(s) ds + \int_0^t [\sigma_1 c_1^{(1)}(s) + \sigma_0 c_1(s)] dw(s))}{V_2}$$

[from (5.1.5) and (5.1.7)].   (5.1.8)

Substituting the expression for c_2 in (5.1.8) into (5.1.6) and then finding the differential and algebraic simplifications, we get

$$dc^{(1)} + \left[\left(\frac{\kappa + \alpha}{V_1} + \frac{\alpha}{V_2} \right) c_1^{(1)} + \frac{\alpha \kappa}{V_1 V_2} c_1 \right] dt + \frac{\alpha}{V_1 V_2} \left[\sigma_1 c_1^{(1)} + \sigma_0 c_1 \right] dw(t) = 0.$$

(5.1.9)

This is second-order linear Itô–Doob-type stochastic differential equation with constant coefficients, and it describes the dynamic of creatinine clearance under random perturbations.

Exercises

1. **Stochastic angular motion of the cupula** (Groen, Emond, and Jonnkee [80, 191]). The deterministic version of the angular motion of the cupula (Exercise 5.2(1) [80]) is extended to its stochastic version. The cupula conditions may generate roughness, and affect the normal deflection process, thus causing a random uncertainty in the stiffness coefficient (α). Moreover, the motion of the cupula is in a fluid, and this motion generates a force due to the fluid which can be under the influence of random perturbations. This force is called the drag force [55]. The drag force is assumed to be $F_d = -\beta \frac{dx}{dt}$. This is similar to the frictional force acting on the body attached to the spring. Let m be the moment of inertia instead of mass, since we are dealing with the angular moment. Following the argument used in Illustration 5.2.1, the dynamic of the cupula under random perturbations is described by

$$m \, dx^{(1)} + [\beta x^{(1)} + \alpha x] dt + [\sigma_1 x^{(1)} + \sigma_0 x] dw(t) = 0.$$

2. Using Illustration 5.1.2, show that an alternate mathematical model of the RLC circuit is described by

$$dI^{(1)} + \left[\frac{R}{L} I^{(1)} + \frac{1}{LC} I - \frac{1}{L} \frac{d}{dt} E(t) \right] dt + [\sigma_1 I^{(1)} + \sigma_0 I - \Lambda(w(t))] dw(t) = 0.$$

In particular, if E is a constant function, then

$$dI^{(1)} + \left[\frac{R}{L} I^{(1)} + \frac{1}{LC} I \right] dt + [\sigma_1 I^{(1)} + \sigma_0 I - \Lambda(w(t))] dw(t) = 0.$$

3. Let us consider a light rod of length ℓ pivoted smoothly at one end, with a particle with a mass m attached at the other end. The particle is called the bob of the pendulum. The motion of the bob is in a vertical plane. The medium of the motion of the bob is not smooth, and it is subject to random perturbations.

Let θ be the angle made by the rod with the vertical line at a time t. The bob is free to move in a vertical plane under the gravitational force. Show that the motion of the bob in a vertical plane is described by

$$\ell \, d\theta^{(1)} + mg \sin \theta \, dt + [b_1 \theta^{(1)} + b_0 \sin \theta] dw(t) = 0.$$

5.2 Linear Homogeneous Equations

In this section, we examine the eigenvalue-type method described in Sections 2.3, 4.2 and 4.4 to find solutions to linear higher-order Itô–Doob-type homogeneous differential equations with constant coefficients.

5.2.1 *General Problem*

Let us consider the following higher-order linear homogeneous Itô–Doob-type stochastic differential equation:

$$dy^{(n-1)} + L^d(d)y + L^s(d)y$$
$$\equiv dy^{(n-1)} + [a_{n-1}y^{(n-1)} + a_{n-2}y^{(n-2)} + \cdots + a_1 y' + a_0 y] dt$$
$$+ [b_{n-1}y^{(n-1)} + b_{n-2}y^{(n-2)} + \cdots + b_1 y' + b_0 y] dw(t) = 0, \qquad (5.2.1)$$

where $a_{n-1}, a_{n-2}, \ldots, a_1, a_0, b_{n-1}, b_{n-2}, \ldots, b_1$, and b_0 are any given real numbers; $y^{(k)}$ stands for the kth sample path derivative of y for $k = 1, 2, \ldots, n-1$; n is a positive integer and $n \geq 2$; and L^d and L^s are linear ordinary differential operators with constant coefficients defined by

$$L^d(d) = a_{n-1}d\frac{d^{(n-2)}}{dt^{n-2}} + \cdots + \cdots + a_1 d^{(1)} + a_0 I, \qquad (5.2.2)$$

$$L^s(d) = b_{n-1}d\frac{d^{(n-2)}}{dt^{n-2}} + \cdots + \cdots + b_1 d^{(1)} + b_0 I, \qquad (5.2.3)$$

we note that $dy^{(k-1)} = \frac{d^k}{dt^k} y \, dt$ for $k = 1, 2, \ldots, k \ldots, n$ with $d^{(0)}y = y \, dt (d^{(0)} = I)$; and w is a normalized scalar Wiener process (Definition 1.2.12/Theorem 1.2.2).

In this subsection, based on our background and experience, we explore the feasibility of a procedure for finding a general solution to (5.2.1). As usual, we are also interested in solving an IVP associated with these type of equations. This provides a tool for solving real-world problems. An attempt is made to extend the existing methods of solving higher-order deterministic differential equations to higher-order stochastic differential equations of the Itô–Doob type in a natural way.

Definition 5.2.1. Let $J = [a, b]$, for $a, b \in R$, and hence $J \subseteq R$ be an interval. A solution to an nth-order linear differential equation of the type (5.2.1) is an $n-1$ times continuously sample path differentiable process y defined on J into R such that it satisfies (5.2.1) on J in the sense of Itô–Doob calculus. In short, the usual Itô–Doob-type differential $dy^{(n-1)}$ of the $(n-1)$th derivative of y, its lower-order sample derivatives $y^{(n-1)}, \ldots, y^{(1)}$, and y satisfy the given differential equation (5.2.1).

Example 5.2.1. Let us consider

$$dy^{(1)} + L^d(d)y + L^s(d)y \equiv dy^{(1)} + y\,dt - [\cos ty^{(1)} + \sin ty]dw(t) = 0.$$

Stochastic processes $y_1(t)$ and $y_2(t)$ are

$$y_1(t) = C_1 \cos t, \quad y_2(t) = \left[p(t) + \cos t \int^t p(s)dw(s) \right] C_2,$$

where C_1 and C_2 are arbitrary constants with $C_2 \neq 0$; and $p(t) = \sin te(t)$, with $e(t) = \exp\left[-\frac{1}{2}\int^t \cos^2 s\,ds - \int^t \cos s\,dw(s) \right]$. Show that $y_1(t)$ and $y_2(t)$ are solutions to the given the differential equation.

Solution procedure. First, we note that the given stochastic differential equation is second-order. We apply Definition 5.2.1 to show that the given processes $y_1(t)$ and $y_2(t)$ are the solution to the given differential equation. Using elementary calculus [127], we substitute $y_1(t) = C_1 \cos t$, $y_1^{(1)}(t) = -C_1 \sin t$, and $dy_1^{(1)}(t) = -C_1 \cos t\,dt$ into the stochastic differential equation. In this case, $y_1(t)$ and its derivatives satisfy the given differential equation.

We apply Theorems 1.3.1, 1.3.3 and 1.3.4 to $y_2(t)$, and determine $y_2^{(1)}(t)$ and $dy_2^{(1)}(t)$ (if possible):

$$dy_2(t) = d\left[p(t) + \cos t \int^t p(s)dw(s) \right] C_2 \quad \text{(from Theorem 1.3.3)}$$

$$= C_2 \left[d(\sin te(t)) + d\left(\cos t \int^t \sin se(s)dw(s) \right) \right]$$

$$\text{(from Theorems 1.3.1 and 1.3.4)}$$

$$= C_2 \left[\left[\cos te(t) - \sin t \int^t \sin se(s)dw(s) \right] dt \right.$$

$$+ \sin te(t) \left(-\frac{1}{2} \cos^2 t\,dt - \cos t\,dw(t) + \frac{1}{2} \cos^2 t\,dt \right)$$

$$\left. + \cos td\left(\int^t \sin se(s)dw(s) \right) \right] \quad \text{(by simplifying)}$$

$$= C_2 \left[\left[\cos te(t) - \sin t \int^t \sin se(s)dw(s) \right] dt \right.$$

$$\left. - \sin t \cos te(t)dw(s) + \sin t \cos te(t)dw(t) \right] \quad \text{(by simplifying)}$$

$$= C_2 \left[\cos te(t) - \sin t \int^t \sin se(s)dw(s) \right] dt \quad \text{(by algebraic simplification)}.$$

From the definition of the deterministic differential, we have

$$y_2^{(1)}(t) = \left[\cos te(t) - \sin t \int^t \sin se(s)dw(s)\right] C_2.$$

Now, by imitating the above argument, we compute $dy_2^{(1)}(t)$ as follows:

$$dy_2^{(1)}(t) = dC_2 \left[\cos te(t) - \sin t \int^t \sin se(s)dw(s)\right] \text{ (from Theorems 1.3.1 and 1.3.4)}$$

$$= C_2 \left[\left[-\sin te(t) - \cos t \int^t \sin se(s)dw(s)\right] dt - \sin^2 te(t)dw(t)\right.$$

$$\left. + \cos te(t)\left(-\frac{1}{2}\cos^2 t\, dt - \cos t\, dw(t) + \frac{1}{2}\cos^2 t\, dt\right)\right]$$

$$= C_2 \left[\left[-\sin te(t) - \cos t \int^t \sin se(s)dw(s)\right] dt\right.$$

$$\left. -(\sin^2 t + \cos^2 t)e(t)dw(t)\right] \qquad \text{(by regrouping and simplifying)}$$

$$= -y_2(t)dt - C_2(\sin^2 t + \cos^2 t)e(t)dw(t) \quad \text{(by substitution).}$$

We observe that

$$\cos t y_2^{(1)}(t) + \sin t y_2(t)$$

$$= \cos t \left[\cos te(t) - \sin t \int^t p(s)dw(s)\right] C_2$$

$$+ \sin t \left[\sin te(t) + \cos t \int^t p(s)dw(s)\right] C_2 \quad \text{(by substitution)}$$

$$= \left[\cos^2 te(t) - \cos t \sin t \int^t p(s)dw(s)\right.$$

$$\left. + \sin^2 te(t) + \cos t \sin t \int^t p(s)dw(s)\right] C_2 \quad \text{(by simplifying)}$$

$$= C_2(\sin^2 t + \cos^2 t)e(t) \qquad \text{(by simplifying).}$$

Substituting $\cos t y_2^{(1)}(t) + \sin t y_2(t)$ for $C_2(\sin^2 t + \cos^2 t)e(t)$ in the expression of $dy_2^{(1)}(t)$, we have

$$dy_2^{(1)}(t) = -y_2(t)dt - \left(\cos t y_2^{(1)}(t) + \sin t y_2(t)\right) dw(t),$$

which implies that

$$dy_2^{(1)}(t) + y_2(t)dt + \left[\cos t y_2^{(1)}(t) + \sin t y_2(t)\right] dw(t) = 0.$$

This shows that the given process $y_2(t)$ is the solution process of the stochastic differential equation.

Example 5.2.2. Let $y_1(t)$ and $y_2(t)$ be stochastic processes defined as

$$y_1(t) = \left[e(t) + b_0 \exp[-t] \int^t p(s)dw(s)\right] C_1,$$

$$y_2(t) = C_2 \exp[-t],$$

where $e(t) = \exp\left[\left(2 - \frac{1}{2}b_0^2\right)t - b_0 w(t)\right]$; $p(t) = \exp[t]e(t)$; b_0, C_1 and C_2 are arbitrary constants, and both C_1 and b_0 are nonzero real numbers. Verify that $y_1(t)$ and $y_2(t)$ are solution processes of the following second-order stochastic differential equation:

$$dy^{(1)}(t) + L^d(d)y + L^s(d)y$$
$$\equiv dy^{(1)} - (y^{(1)} + 2y)dt + b_0[y^{(1)} + y]dw(t) = 0.$$

Solution procedure. We imitate the argument used in Example 5.2.1 to verify that $y_1(t)$ and $y_2(t)$ are solution processes of the given stochastic differential equation. Again, using elementary calculus, we substitute $y_2(t) = C_2 \exp[-t]$, $-C_2 \exp[-t] = y_1^{(1)}(t)$, and $y_2^{(2)}(t) = C_2 \exp[-t]$ into the stochastic differential equation. In this case, $y_2(t)$ and its derivatives satisfy the given differential equation. We conclude that $y_2(t)$ is the solution process of the given differential equation.

Now, for the verification of the solution process of $y_1(t)$, we apply Theorems 1.3.1, 1.3.3 and 1.3.4 to $y_1(t)$, and determine $y_1^{(1)}(t)$ and $dy_1^{(1)}(t)$ (if possible):

$$dy_1(t) = d\left[e(t) + b_0 \exp[-t] \int^t p(s)dw(s)\right] C_1 \quad \text{(from Theorem 1.3.3)}$$

$$= C_1\left[de(t) + d\left(b_0 \exp[-t] \int^t p(s)dw(s)\right)\right]$$

$$\text{(from Theorems 1.3.1 and 1.3.4)}$$

$$= C_1\left[-b_0 \exp[-t] \int^t p(s)dw(s)dt\right.$$

$$+ e(t)\left(\left(2 - \frac{1}{2}b_0^2\right)dt - b_0 dw(t) + \frac{1}{2}b_0^2 dt\right)$$

$$\left. + b_0 \exp[-t]d\left(\int^t p(s)dw(s)\right)\right] \quad \text{(by simplifying)}$$

$$= C_1\left[\left[-b_0 \exp[-t] \int^t p(s)dw(s)\right]dt\right.$$

$$\left. + 2e(t)dt - b_0 e(t)dw(t) + b_0 e(t)dw(t)\right] \quad \text{(by simplifying)}$$

$$= C_1\left[-b_0 \exp[-t] \int^t p(s)dw(s) + 2e(t)\right]dt$$

$$\text{(by algebraic simplification).}$$

From the definition of the deterministic differential [127], we have

$$y_1^{(1)}(t) = C_1 \left[-b_0 \exp[-t] \int^t p(s)dw(s) + 2e(t) \right].$$

Now, by imitating the argument uses in Example 5.2.1, we compute the following:

$$dy_1^{(1)}(t) = dC_1 \left[-b_0 \exp[-t] \int^t p(s)dw(s) + 2e(t) \right] \quad \text{(from Theorems 1.3.1 and 1.3.4)}$$

$$= C_1 \left[b_0 \exp[-t] \int^t p(s)dw(s)dt - b_0 e(t)dw(t) \right.$$

$$+ 2e(t) \left[\left(2 - \frac{1}{2}b_0^2 \right) dt - b_0 dw(t) + \frac{1}{2}b_0^2 dt \right] \right]$$

$$= C_1 \left[\left[b_0 \exp[-t] \int^t e(s)dw(s) + 4e(t) \right] dt \right.$$

$$\left. - 3b_0 e(t)dw(t) \right] \quad \text{(by regrouping and simplifying)}$$

Based on the above computations, we observe that

$$y_1^{(1)}(t) + 2y_1(t) = C_1 \left[-b_0 \exp[-t] \int^t p(s)dw(s) + 2e(t) \right.$$

$$\left. + 2e(t) + 2b_0 \exp[-t] \int^t p(s)dw(s) \right] \quad \text{(by substitution)}$$

$$= C_1 \left[4e(t) + b_0 \exp[-t] \int^t p(s)dw(s) \right] \quad \text{(by simplifying)},$$

$$-b_0(y_1^{(1)}(t) + y_1(t)) = -b_0 \left[C_1 \left[-b_0 \exp[-t] \int^t p(s)dw(s) + 2e(t) \right] \right.$$

$$\left. + C_1 \left[b_0 \exp[-t] \int^t p(s)dw(s) + e(t) \right] \right] \quad \text{(by grouping)}$$

$$= -b_0 C_1 \left[e(t) + b_0 \exp[-t] \int^t p(s)dw(s) \right.$$

$$\left. - b_0 \exp[-t] \int^t p(s)dw(s) + 2e(t) \right] \quad \text{(by simplifying)}.$$

$$= -3b_0 C_1 e(t).$$

Substituting $y_1^{(1)}(t) + 2y_1(t)$ and $-b_0(y_1^{(1)}(t) + y_1(t))$ in the expression of $dy_1^{(1)}(t)$, we have

$$dy_1^{(1)}(t) = (y_1^{(1)}(t) + 2y_1(t))dt - b_0(y_1^{(1)}(t) + y_1(t))dw(t).$$

This implies that

$$dy_1^{(1)}(t) - (y_1^{(1)}(t) + 2y_1(t))dt + b_0(y_1^{(1)}(t) + y_1(t))dw(t) = 0.$$

This shows that the given process $y_1(t)$ is the solution process of the stochastic differential equation.

In the following, we examine the usage of the eigenvalue-type method to find the general and particular solutions to higher-order linear stochastic differential equations.

5.2.2 *Feasibility of Finding a General Solution*

Let us imitate the procedure initiated in Section 2.3.2. Following Step #1 of Procedure 2.3.2, we decompose (5.2.1) into its deterministic and stochastic parts:

$$dy^{(n-1)} + L^d(d)y \equiv dy^{(n-1)} + [a_{n-1}y^{(n-1)}$$

$$+ a_{n-2}y^{(n-2)} + \cdots + a_1 y^{(1)} + a_0 y]dt = 0, \qquad (5.2.4)$$

$$dy^{(n-1)} + L^s(d)y \equiv dy^{(n-1)} + [b_{n-1}y^{(n-1)}$$

$$+ b_{n-2}y^{(n-2)} + \cdots + b_1 y^{(1)} + b_0 y]dw(t) = 0, \qquad (5.2.5)$$

respectively. As the part in Step #1 of Procedure 2.3.2, by using the procedure developed in Section 5.3 of Chapter 5 [80], we can fulfill the goal of finding the solution y^d of the deterministic part (5.2.4). To test the extent of the limitation of finding the solution y^s of the stochastic part (5.2.5), we consider the following example.

Example 5.2.3. Let us consider $dy^{(1)} - (y^{(1)} + 2y)dt + [b_1 y^{(1)} + b_0 y]dw(t) = 0.$

Solution process. Our goal is to apply the eigenvalue-type method developed in Section 2.3.2 to find the general solution to the given differential equation. The decomposition of the stochastic differential equation follows:

(D) $\quad dy^{(1)} - (y^{(1)} + 2y)dt = 0 \quad$ and $\quad$ (S) $\quad dy^{(1)} + [b_1 y^{(1)} + b_0 y]dw(t) = 0.$

The set of linearly independent solutions of the Part (D) is $y_1(t) = \exp[-t]$ and $y_2(t) = \exp[2t]$. Thus, the general solution of deterministic part of (5.2.4) in the context of the example is given by

$$y(t) = c_1 \exp[-t] + c_2 \exp[2t].$$

To determine a complete set of solutions of the stochastic Part (S), we seek a solution in the usual form,

$$y^s(t) \equiv y^s(t, w(t)) = \exp\left[\int_a^t \lambda_1^s(s)ds + \int_a^t \lambda_2^s(s)dw(s)\right]c^s, \quad \text{for } t \in J,$$

as discussed in Section 2.3.2. The sample paths of the process $y^s(t)$ are not differentiable. Therefore, we cannot continue this procedure further.

In short, at this stage, using the eigenvalue approach parallel to the developed approach [procedures in Section 4.2 and Section 5.3, Chapter 5, Volume 1] is not directly applicable for higher-order stochastic differential equations. This poses a challenge to the problem-solver. This problem is further examined in the next section.

In the following section, we extend the concept of a companion system associated with the deterministic higher-order linear differential equation with constant coefficients [80] to the Itô–Doob higher-order stochastic differential equations (5.2.1). We further continue to examine the feasibility of finding the closed-form solution processes of this type of stochastic differential equations.

5.3 Companion Systems

In this section, we introduce the representation of the nth-order linear nonhomogeneous Itô–Doob-type stochastic differential equation as a system of linear nonhomogeneous differential equations. This representation paves the way for employing the procedures and conceptual analysis developed in Chapter 4 as well as the deterministic version (Chapters 4 and 5 of Volume 1) to study the higher-order linear Itô–Doob-type stochastic differential equations in a systematic way. Moreover, an attempt is made to utilize the particular structure of the coefficient matrix of the linear system of differential equations corresponding to the higher-order linear homogeneous differential equations to systematically simplify the procedures for finding the general solution to higher-order differential equations. In short, this section deals with an alternative way of examining the feasibility of finding solution processes of higher-order linear homogeneous and nonhomogeneous Itô–Doob-type stochastic differential equations with constant coefficients.

Let us consider the following Itô–Doob-type nth-order nonhomogeneous stochastic differential equation:

$$dy^{(n-1)} + \left[a_{n-1}y^{(n-1)} + a_{n-2}y^{(n-2)} + \cdots + a_1 y^{(1)} + a_0 y - g(t)\right] dt$$

$$+ \left[b_{n-1}y^{(n-1)} + b_{n-2}y^{(n-2)} + \cdots + b_1 y^{(1)} + b_0 y - h(t, w(t))\right] dw(t) = 0,$$

$$(5.3.1)$$

where $a_{n-1}, a_{n-2}, \ldots, a_1, a_0, b_{n-1}, b_{n-2}, \ldots, b_1$, and b_0 are given real numbers; $g(t)$ is a continuous function defined on an interval $J = [a, b]$ into R; h is a sample path continuous stochastic function defined on $J \times R$ into R; and the process w is as defined before.

We imitate the procedure for transforming higher-order linear deterministic differential equations [80].

Theorem 5.3.1. *Every nth-order Itô–Doob-type stochastic differential equation of the type* (5.3.1) *is equivalent to the following first-order system of stochastic differential equations of the Itô–Doob type:*

$$dx = [Ax + p(t)]dt + [Bx + q(t)]dw(t), \tag{5.3.2}$$

where A and B are $n \times n$ constant matrices; x, p, and q are $n \times 1$ matrix functions;

$$A = \begin{bmatrix} 0 & 1 & \cdots & 0 & 0 \\ 0 & 0 & \cdots & 0 & 0 \\ \cdots & \cdots & \cdots & \cdots & \cdots \\ 0 & 0 & \cdots & 0 & 1 \\ -a_0 & -a_1 & \cdots & -a_{n-2} & -a_{n-1} \end{bmatrix}, \tag{5.3.3}$$

$$B = \begin{bmatrix} 0 & 0 & \cdots & 0 & 0 \\ 0 & 0 & \cdots & 0 & 0 \\ \cdots & \cdots & \cdots & \cdots & \cdots \\ 0 & 0 & \cdots & 0 & 0 \\ -b_0 & -b_1 & \cdots & -b_{n-2} & -b_{n-1} \end{bmatrix}, \tag{5.3.4}$$

$x^T = [x_1 \ x_2 \ \cdots \ x_{n-1} \ x_n]$, $[p(t)]^T = [0, 0, \ldots, g(t)]$, *and* $[q(t)]^T = [0, 0, \ldots, h(t, w(t))]^T$ *and T stands for a transpose of the matrix.*

Proof. Let an nth-order Itô–Doob-type stochastic differential equation of the type (5.3.1) be given. Let us introduce a transformation as

$$y = x_1, \quad y^{(1)} = x_2, \quad y^{(2)} = x_3, \ldots, y^{(i-1)} = x_i, \ldots, y^{(n-2)} = x_{n-1},$$

$$y^{(n-1)} = x_n. \tag{5.3.5}$$

Using this transformation and the concept of the differential, we rewrite (5.3.1) as

$$dy = dx_1 = x_2 dt,$$

$$dy^{(1)} = dx_2 = x_3 dt,$$

$$dy^{(2)} = dx_3 = x_4 dt,$$

$$\cdots \qquad \cdots \qquad \cdots \qquad \cdots$$

$$dy^{(i-1)} = dx_i = x_{i+1} dt,$$

$$\cdots \qquad \cdots \qquad \cdots \qquad \cdots$$

$$dy^{(n-2)} = dx_{n-1} = x_n dt,$$

$$dy^{(n-1)} = dx_n = -[a_0 x_1 + a_1 x_2 + \cdots + a_{n-2} x_{n-1} + a_{n-1} x_n - g(t)] \, dt$$

$$- [b_0 x_1 + b_1 x_2 + \cdots + b_{n-2} x_{n-1} + b_{n-1} x_n - h(t, w(t))] \, dw(t). \tag{5.3.6}$$

From (5.3.6), notations for the derivative of a matrix function, and matrix algebra, the first-order system of differential equations (5.3.6) can be rewritten as (5.3.2).

Conversely, by reversing the above argument, any system of first-order stochastic differential equations of the Itô–Doob type (5.3.2) can be rewritten as (5.3.1). The details are left to the reader. □

Definition 5.3.1. Matrices of the form defined in (5.3.3) and (5.3.4) are called companion matrices associated with the nth-order Itô–Doob-type stochastic differential equation of the type (5.3.1). They are denoted by A_c and B_c.

Definition 5.3.2. The system (5.3.2) is called the Itô–Doob-type stochastic companion system of differential equations for the nth-order Itô–Doob-type nonhomogeneous stochastic differential equation (5.3.1).

The following result illustrates the importance of the companion system in a natural way.

Theorem 5.3.2. *Let $y(t)$ and $x(t)$ be any solution processes of (5.3.1) and (5.3.2), respectively. Then,*

(i) $[x_y(t)]^T = [y(t), y^{(1)}(t), \ldots, y^{(i-1)}(t), \ldots, y^{(n-1)}(t)]$ *is the solution process of (5.3.2),*

(ii) $dx_1^{(n-1)} = -(a_0 x_1 + a_1 x_2 + \cdots + a_{n-2} x_{n-1} + a_{n-1} x_n - g(t))dt - (b_0 x_1 + b_1 x_2 + \cdots + b_{n-2} x_{n-1} + b_{n-1} x_n - h(t, w(t)))dw(t)$,
where $x_1(t)$ is the first component of the solution process $x(t)$ of (5.3.2).

Proof. To prove (i), let us assume that $y(t)$ is any solution process of (5.3.1). By using the argument used in the proof of Theorem 5.3.1, we arrive at (5.3.6). From (5.3.6) and using the matrix algebra and calculus, we conclude that the process

$$[x_y(t)]^T = [y(t), y'(t), \ldots, y^{(i-1)}(t), \ldots, y^{(n-1)}(t)]$$
$$= [x_1(t), x_2(t), \ldots, x_i(t), \ldots, x_n(t)]$$

is also the solution process of (5.3.2). This completes the proof of (i).

To prove (ii), we assume that $x(t)$ is any solution to (5.3.2). This implies that $x(t)$ and $dx(t)$ satisfy the companion system. Again, from (5.3.6), we have

$$dx_n = -(a_0 x_1 + a_1 x_2 + \cdots + a_{n-2} x_{n-1} + a_{n-1} x_n - g(t))dt$$
$$- (b_0 x_1 + b_1 x_2 + \cdots + b_{n-2} x_{n-1} + b_{n-1} x_n - h(t, w(t)))dw(t).$$

This together with (5.3.5) yields

$$dx_1^{(n-1)} = -\left[a_0 x_1 + a_1 x_1' + \cdots + a_{n-2} x_1^{(n-2)} + a_{n-1} x_1^{(n-1)} - g(t) \right] dt$$
$$- \left[b_0 x_1 + b_1 x_1' + \cdots + b_{n-2} x_1^{(n-2)} + b_{n-1} x_1^{(n-1)} - h(t) \right] dw(t).$$

This establishes the proof of (ii). □

Theorem 5.3.2 provides a natural basis for defining the general solution to (5.3.1).

Definition 5.3.3. The general solution to the nth-order Itô–Doob-type stochastic differential equation (5.3.1) is defined to be the first component of the general solution to the corresponding Itô–Doob-type stochastic companion system of differential equations (5.3.2).

Theorems 5.3.1 and 5.3.2 are useful for establishing the existence and uniqueness of solution results of (5.3.1). We simply state the results without any further details. The "principles of superposition" for the higher-order differential equations can also be constructed.

Theorem 5.3.3 (existence and uniqueness). *Let us consider the IVP*

$$dy^{(n-1)} + \left[a_{n-1}y^{(n-1)} + a_{n-2}y^{(n-2)} + \cdots + a_1y^{(1)} + a_0y - g(t) \right] dt$$
$$+ \left[b_{n-1}y^{(n-1)} + b_{n-2}y^{(n-2)} + \cdots + b_1y^{(1)} + b_0y - h(t, w(t)) \right] dw(t) = 0,$$
$$(5.3.7)$$

$$y(t_0) = y_{10}, y^{(1)}(t_0) = y_{20}, y^{(2)}(t_0) = y_{30}, \ldots, y^{(i-1)}(t_0) = y_{i0}, \ldots, y^{(n-1)}(t_0) = y_{n0}.$$

It has a unique solution.

Theorem 5.3.4 (principle of superposition). *If $y_1(t)$ and $y_2(t)$ are any solutions to*

$$dy^{(n-1)} + \left[a_{n-1}y^{(n-1)} + a_{n-2}y^{(n-2)} + \cdots + a_1y^{(1)} + a_0y - g_1(t) \right] dt$$
$$+ \left[b_{n-1}y^{(n-1)} + b_{n-2}y^{(n-2)} + \cdots + b_1y^{(1)} + b_0y - h_1(t, w(t)) \right] dw(t) = 0,$$

and

$$dy^{(n-1)} + \left[a_{n-1}y^{(n-1)} + a_{n-2}y^{(n-2)} + \cdots + a_1y^{(1)} + a_0y - g_2(t) \right] dt$$
$$+ \left[b_{n-1}y^{(n-1)} + b_{n-2}y^{(n-2)} + \cdots + b_1y^{(1)} + b_0y - h_2(t, w(t)) \right] dw(t) = 0,$$

respectively, then $y(t) = c_1 y_1(t) + c_2 y_2(t)$ is the solution to

$$dy^{(n-1)} + \left[a_{n-1}y^{(n-1)} + a_{n-2}y^{(n-2)} + \cdots + a_1y^{(1)} + a_0y \right] dt$$
$$+ \left[b_{n-1}y^{(n-1)} + b_{n-2}y^{(n-2)} + \cdots + b_1y^{(1)} + b_0y \right] dw(t)$$
$$= [c_1 g_1(t) + c_2 g_2(t)] dt + [c_1 h_1(t, w(t)) + c_2 h_2(t, w(t))]. \qquad (5.3.8)$$

Definition 5.3.4. The algebraic equation

$$\Lambda(\lambda) = \lambda^n + a_{n-1}\lambda^{n-1} + \cdots + a_1\lambda + a_0 = 0 \tag{5.3.9}$$

is said to be an auxiliary equation associated with the deterministic part (5.2.4) of the corresponding homogeneous higher-order stochastic differential equation (5.2.1), i.e.

$$dy^{(n-1)} + \left[a_{n-1}y^{(n-1)} + a_{n-2}y^{n-2} + \cdots + a_1 y^{(1)} + a_0 y \right] dt = 0.$$

Observation 5.3.1.

(i) The linear homogeneous system of stochastic differential equations corresponding to (5.3.2) is

$$dx = A_c x\, dt + B_c x\, dw(t), \tag{5.3.10}$$

where A_c and B_c are $n \times n$ deterministic and stochastic companion constant matrices defined in (5.3.3) and (5.3.4), respectively; and x is an $n \times 1$ matrix function.

(ii) The system of Itô–Doob-type stochastic differential equations (5.3.10) is also a companion system of the nth-order homogeneous Itô–Doob-type stochastic differential equation of the type (5.2.1).

(iii) The IVP (5.3.7) with respect to the stochastic companion system of differential equations (5.3.10) is described by

$$dx = A_c x\, dt + B_c x\, dw(t), \quad x(t_0) = x_0,$$

where $x_0 = [y_{10}, y_{20}, \ldots, y_{i0}, \ldots, y_{n0}]^T = [y(t_0), y^{(1)}(t_0), \ldots, y^{(i-1)}(t_0), \ldots, y^{(n-1)}(t_0)]^T$.

(iv) The stochastic companion system (5.3.10) is a special case of the general first-order linear homogeneous system of Itô–Doob-type stochastic differential equations (4.4.1). Moreover, we note that $A_c B_c \neq B_c A_c$, i.e. A_c and B_c do not commute.

The remainder of the section is devoted to exploring the feasibility of finding a closed-form solution to a class of a system of Itô–Doob-type stochastic differential equations (5.3.10).

5.3.1 *Procedure for Finding the General Solution*

We apply the eigenvalue-type method developed in Section 4.4 to compute a solution to the companion system of first-order linear homogeneous differential equations with constant coefficients (5.3.10). Here, the homogeneous companion system (5.3.10) can be treated as a perturbed homogeneous companion system of either deterministic or stochastic perturbation.

Step #1. We decompose (5.3.10) into its unperturbed and perturbation parts

$$dx = A_c x\, dt \text{ (deterministic unperturbed/perturbation system)}, \quad (5.3.11)$$

and

$$dx = B_c x\, dw(t)\text{(stochastic perturbation/unperturbed system)}, \quad (5.3.12)$$

respectively. The goals are parallel to those in Step #1 of Procedure 4.4.2:

(a) to find the fundamental matrix solution processes $\Phi_D(t)$ and $\Phi_S(t)$ of (5.3.11) and (5.3.12), respectively;

(b) to create a candidate for the fundamental matrix solution process of (5.3.10):

$$\Phi_{cs}(t) = \Phi_D(t)\Phi_S(t); \quad (5.3.13)$$

(c) to test the validity of the fundamental matrix solution.

In the following, a procedure for fulfilling the stated subgoals in (a)–(c) is outlined.

Step #2. Steps #2 and #3 of Procedure 4.4.2 are repeated and the fundamental matrices $\Phi_D(t)$ and $\Phi_S(t)$ of (5.3.11) and (5.3.12) are determined. Thus, we have

$$d\Phi_U(t) = A_c\Phi_D(t)dt, \quad (5.3.14)$$

$$d\Phi_S(t) = B_c\Phi_S(t)dw(t), \quad (5.3.15)$$

and

$$
\begin{aligned}
d\Phi_{cs}(t) &= d\Phi_D(t)\Phi_S(t) + \Phi_D(t)d\Phi_S(t) && \text{(from Theorem 1.3.4)}\\
&= A_c\Phi_D(t)dt\Phi_S(t) + \Phi_D(t)B_c\Phi_S(t)dw(t) && \text{[from (5.3.14) and (5.3.15)]}\\
&= A_c\Phi_{cs}(t)dt + \Phi_D(t)B_c\Phi_S dw(t) && \text{(by simplifications). (5.3.16)}
\end{aligned}
$$

We note that the stochastic companion matrices A_c and B_c do not commute [Observation 5.3.1(iv)], i.e. $A_c B_c \neq B_c A_{cc}$. In fact,

$$
A_c B_c =
\begin{bmatrix}
0 & 1 & \cdots & 0 & 0\\
0 & 0 & \cdots & 0 & 0\\
\cdots & \cdots & \cdots & \cdots & \cdots\\
0 & 0 & \cdots & 0 & 1\\
-a_0 & -a_1 & \cdots & -a_{n-2} & -a_{n-1}
\end{bmatrix}
\begin{bmatrix}
0 & 0 & \cdots & 0 & 0\\
0 & 0 & \cdots & 0 & 0\\
\cdots & \cdots & \cdots & \cdots & \cdots\\
0 & 0 & \cdots & 0 & 0\\
-b_0 & -b_1 & \cdots & -b_{n-2} & -b_{n-1}
\end{bmatrix}
$$

$$
=
\begin{bmatrix}
0 & 0 & \cdots & 0 & 0\\
0 & 0 & \cdots & 0 & 0\\
\cdots & \cdots & \cdots & \cdots & \cdots\\
-b_0 & -b_1 & \cdots & -b_{n-2} & -b_{n-1}\\
b_0 a_{n-1} & b_1 a_{n-1} & \cdots & b_{n-2} a_{n-1} & b_{n-1} a_{n-1}
\end{bmatrix},
$$

$$
B_c A_c = \begin{bmatrix} 0 & 0 & \cdots & 0 & 0 \\ 0 & 0 & \cdots & 0 & 0 \\ \cdots & \cdots & \cdots & \cdots & \cdots \\ 0 & 0 & \cdots & 0 & 0 \\ -b_0 & -b_1 & \cdots & -b_{n-2} & -b_{n-1} \end{bmatrix} \begin{bmatrix} 0 & 1 & \cdots & 0 & 0 \\ 0 & 0 & \cdots & 0 & 0 \\ \cdots & \cdots & \cdots & \cdots & \cdots \\ 0 & 0 & \cdots & 0 & 1 \\ -a_0 & -a_1 & \cdots & -a_{n-2} & -a_{n-1} \end{bmatrix}
$$

$$
= \begin{bmatrix} 0 & 0 & \cdots & 0 & 0 \\ 0 & 0 & \cdots & 0 & 0 \\ \cdots & \cdots & \cdots & \cdots & \cdots \\ 0 & 0 & \cdots & 0 & 0 \\ a_0 b_{n-1} & -b_0 + a_1 b_{n-1} & \cdots & -b_{n-3} + a_{n-2} b_{n-1} & -b_{n-2} + a_{n-1} b_{n-1} \end{bmatrix}.
$$

From a comparison of the expressions $A_c B_c$ and $B_c A_c$, it is clear that $A_c B_c = B_c A_c$ if and only if $B_c = 0$. This statement is equivalent to the deterministic companion system. Therefore, the candidate $\Phi_{cs}(t)$ for the fundamental matrix for the stochastic companion system (5.3.10) cannot be validated. Moreover, we cannot proceed beyond (5.3.16). Moreover, this is consistent with the stochastic (Chapter 4) and deterministic (Chapter 4 [80]) linear systems of differential equations with constant coefficient matrices.

Observation 5.3.2.

(i) Let us formulate a statement parallel to Theorem 4.4.1 in the context of the stochastic companion system (5.3.10): let $\Phi_D(t)$ be a fundamental matrix solution to (5.3.11). In addition, let $x(t) = \Phi_D(t)c(t)$ be a solution to (5.3.10), with $c(t)$ being an n-dimensional unknown vector function. Then, $c(t)$ is a solution process of the following stochastic system of linear differential equations with time-varying coefficient matrix $\Phi_D^{-1}(t)B_c \Phi_D(t)$:

$$
dc = \Phi_D^{-1}(t)B_c \Phi_D(t)c\, dw(t). \tag{5.3.17}
$$

(ii) To examine the the structure of the coefficient matrix $\Phi_D^{-1}(t)B_c \Phi_D(t)$, we use the structure of the stochastic companion system (A_c and B_c) and the structure of the fundamental matrix $[\Phi_D(t)]$ associated with the deterministic companion unperturbed system (5.3.11). For this purpose, we apply Theorem 4.4.1 to the stochastic companion system (5.3.10) and obtain

$$
B_c \Phi_D(t) = \begin{bmatrix} B_{c1} \\ \cdots \\ B_{ci} \\ \cdots \\ B_{cn} \end{bmatrix} \begin{bmatrix} \Phi_D^1(t) & \cdots & \Phi_D^j(t) & \cdots & \Phi_D^n(t) \end{bmatrix} \quad \text{(Observation 1.3.1 [80])}
$$

$$
= \begin{bmatrix} B_{c1}\Phi_D^1(t) & \cdots & B_{c1}\Phi_D^j(t) & \cdots & B_{c1}\Phi_D^n(t) \\ \cdots & & \cdots & & \cdots \\ B_{ci}\Phi_D^1(t) & \cdots & B_{ci}\Phi_D^j(t) & \cdots & B_{ci}\Phi_D^n(t) \\ \cdots & & \cdots & & \cdots \\ B_{cn}\Phi_D^1(t) & \cdots & B_{cn}\Phi_D^j(t) & \cdots & B_{cn}\Phi_D^n(t) \end{bmatrix}
$$

$$
= \begin{bmatrix} 0 & \cdots & 0 & \cdots & 0 \\ \cdots & \cdots & \cdots & \cdots & \\ 0 & \cdots & 0 & \cdots & 0 \\ \cdots & \cdots & \cdots & \cdots & \\ B_{cn}\Phi_D^1(t) & \cdots & B_{cn}\Phi_D^j(t) & \cdots & B_{cn}\Phi_D^n(t) \end{bmatrix}. \tag{5.3.18}
$$

Denoting $\Phi_D(t) = (\psi_{ij}(t))_{n\times n}$, $\Phi_D^{-1}(t) = (\phi_{ij}(t))_{n\times n}$ and using (5.3.18), we compute $\Phi_D^{-1}(t)B_c\Phi_D(t)$ as

$$
\Phi_D^{-1}(t)B_c\Phi_D(t) = \begin{bmatrix} \phi_{1n}B_{cn}\Phi_D^1(t) & \cdots & \phi_{1n}B_{cn}\Phi_D^j(t) & \cdots & \phi_{1n}B_{cn}\Phi_D^n(t) \\ \cdots & & \cdots & & \cdots \\ \phi_{in}B_{cn}\Phi_D^1(t) & \cdots & \phi_{in}B_{cn}\Phi_D^j(t) & \cdots & \phi_{in}B_{cn}\Phi_D^n(t) \\ \cdots & & \cdots & & \cdots \\ \phi_{nn}B_{cn}\Phi_D^1(t) & \cdots & \phi_{nn}B_{cn}\Phi_D^j(t) & \cdots & \phi_{nn}B_{cn}\Phi_D^n(t) \end{bmatrix}.
$$
$$\tag{5.3.19}$$

In the following, we present a result. This result provides a tool for finding a closed-form solution process of classes of higher-order linear stochastic differential equations with constant coefficients.

Lemma 5.3.1. *All column vectors except any one column of the matrix* $\Phi_D^{-1}(t)B_c\Phi_D(t)$ *are zeroes if and only if*

$$
B_{cn}\Phi_D^j(t) = 0 \text{ for all } j \neq k, \text{ for any given } k \in I(1,n). \tag{5.3.20}
$$

Proof. The validity of the necessary condition follows from

$$
\Phi_D^{-1}(t)B_c\Phi_D(t) = \begin{bmatrix} 0 & \cdots & \phi_{1n}(t)B_{cn}\Phi_D^k(t) & \cdots & 0 \\ \cdots & \cdots & \cdots & & \cdots \\ 0 & \cdots & \phi_{kn}(t)B_{cn}\Phi_D^k(t) & \cdots & 0 \\ \cdots & \cdots & \cdots & & \cdots \\ 0 & \cdots & \phi_{nn}(t)B_{cn}\Phi_D^k(t) & \cdots & 0 \end{bmatrix}, \tag{5.3.21}
$$

for any given column $k \in I(1,n)$. For the sufficient condition, we note that (5.3.20) is true for all $j \neq k$ and any given $k \in I(1,n)$. Under this assumption, (5.3.19) reduces to (5.3.21). This establishes the lemma. □

Observation 5.3.3. We recall that rate coefficient matrix $\Phi_D^{-1}(t)B_c\Phi_D(t)$ in (5.3.17) is a time-varying matrix. The eigenvector of the deterministic companion matrix A_c in (5.3.11) has the form $c(\lambda) = [1, \lambda, \ldots, \lambda^i, \ldots, \lambda^{n-1}]^T$, corresponding to each of its eigenvalue λ (Observation 5.4.1 [80]). Hence, the algebraic condition (5.3.20) $[B_{cn}\Phi_D^j(t) = 0]$ implies that $B_{cn} = [-b_0, \ldots, -b_i, \ldots, -b_n]$ is perpendicular to the jth column vector of $\Phi_D(t)$. Thus, B_{cn} is perpendicular to $c(\lambda_j) = [1, \lambda_j, \ldots, \lambda_j^i, \ldots, \lambda_j^{n-1}]^T$. This is also equivalent to the statement that λ_j is a characteristic root of

$$b_{n-1}\lambda_j^{n-1} + b_{n-2}\lambda_j^{n-2} + \cdots + b_1\lambda_j^1 + b_0\lambda_j = 0.$$

This is equivalent to the fact that for $j \neq k$, $j, k \in I(1, n)$, $y_j(t) = \psi_{1j}(t)$ are solution processes of the following $(n-1)$th order linear deterministic differential equation with constant coefficients, $b_{n-1}, b_{n-2}, \ldots, b_1, b_0$:

$$L(y) \equiv b_{n-1}y^{(n-1)} + b_{n-2}y^{(n-2)} + \cdots + b_1y^{(1)} + b_0y = 0. \tag{5.3.22}$$

Furthermore, the condition (5.3.20) signifies that at most one of the components of the dynamic process, namely $y, y^{(1)}, \ldots, y^{(k)}, \ldots, y^{(n-1)}$, is subject to random perturbations. This is the cost for finding closed-form solutions. In addition, the coefficients of $y, y^{(1)}, \ldots, y^{(k)}, \ldots, y^{(n-1)}$ are influenced by the random perturbations.

Now, we present a very general theoretical algorithm that provides a closed-form solution process of (5.2.1).

Theorem 5.3.5. *Let the hypotheses of Lemma 5.3.1 be satisfied. Then,*

(i) *the higher-order stochastic differential equation (5.2.1) is solvable;*

(ii) *n solutions to (5.2.1) are represented by*

$$\begin{cases} y_j(t) = \psi_{1j}(t), & \text{for } j \neq k, \ j, k \in I(1, n), \\ y_k(t) = \displaystyle\sum_{j \neq k}^{n} \psi_{1j}(t) \int_a^t \phi_{jn}(s)B_{cn}\Phi_D^k(s) \\ \qquad \times \exp[\nu(s, w(s))]dw(s) + \psi_{1k}(t)\exp[\nu(t, w(t))], & \text{for } j = k, \end{cases} \tag{5.3.23}$$

where $\nu(t, w(t)) = -\frac{1}{2}\int_a^t(\phi_{kn}(s)B_{cn}\Phi_D^k(s))^2 ds + \int_a^t \phi_{kn}(s)B_{cn}\Phi_D^k(s)dw(s)$;

(iii) *a closed-form general solution to (5.2.1) is*

$$y(t) = \sum_{j=1}^{n} c_j y_j(t),$$

where c_j are arbitrary constants with $c_k \neq 0$.

Proof. From (5.3.17), (5.3.25) and (5.3.26) we have

$$\begin{cases} dc_j = \phi_{jn} B_{cn} \Phi_D^k(t) c_k dw(t), & \text{for } j \neq k, \ j,k \in I(1,n), \\ dc_k = \phi_{kn} B_{cn} \Phi_D^k(t) c_k dw(t), & \text{for some } k \in I(1,n). \end{cases} \tag{5.3.24}$$

Using Procedure 2.3.2, the closed-form solution component c_k of (5.3.17) is

$$c_k(t) = c_{k0} \exp[\nu(t, w(t))]$$
$$= c_{k0} \varphi_{kk}(t) \quad \text{(by notation)}, \tag{5.3.25}$$

where c_{k0} is an arbitrary constant.

Substituting $c_k(t)$ in (5.3.25) into the first equation in (5.3.24) and applying Procedure 2.2.2, we get

$$c_j(t) = c_{j0} + c_{k0} \int_a^t \phi_{jn}(s) B_{cn} \Phi_D^k(s) \exp[\nu(s, w(s))] dw(s)$$
$$= c_{j0} + \varphi_{jk}(t) \quad \text{(by notation)} \tag{5.3.26}$$

where c_{j0} are arbitrary constants, for $j \neq k$, $j,k \in I(1,n)$.

From (5.3.24), (5.3.25) and (5.3.26) the fundamental solution process $\Phi_T(t) \equiv \Phi_T(t, w(t))$ of the transformed system (5.3.17) in the context of (5.3.21) is

$$\Phi_T(t) = \begin{bmatrix} 1 & \cdots & \varphi_{1k}(t) & \cdots & 0 \\ \cdots\cdots & \cdots & \cdots\cdots \\ 0 & \cdots & \varphi_{kk}(t) & \cdots & 0 \\ \cdots\cdots & \cdots & \cdots\cdots \\ 0 & \cdots & \varphi_{nk}(t) & \cdots & 1 \end{bmatrix}, \tag{5.3.27}$$

where $\Phi_T(t) = (\varphi_{il}(t))_{n \times n}$ and

$$\varphi_{il}(t) = \begin{cases} 1, & \text{for } i = l \text{ and } l \neq k, \\ \exp[\nu(t, w(t))], & \text{for } i = l = k, \\ \int_a^t \phi_{in}(s) B_{cn} \Phi_D^k(s) \exp[\nu(s, w(s))] dw(s), & \text{for } i \neq l = k, \\ 0, & \text{otherwise (i.e. } i,l \neq k). \end{cases} \tag{5.3.28}$$

From (5.3.10), Observation 5.3.2, (5.3.20), (5.3.27), (5.3.28), and the application of Theorem 4.4.2, we conclude that

$$\Phi_{cs}(t) = \Phi_D(s)\Phi_T(t) = \begin{bmatrix} \psi_{11}(t) & \cdots & \sum_{i=1}^n \psi_{1i}(t)\varphi_{ik}(t) & \cdots & \psi_{1n}(t) \\ \cdots & \cdots & \cdots & \cdots & \cdots \\ \psi_{k1}(t) & \cdots & \sum_{i=1}^n \psi_{ki}(t)\varphi_{ik}(t) & \cdots & \psi_{kn}(t) \\ \cdots & \cdots & \cdots & \cdots & \cdots \\ \psi_{n1}(t) & \cdots & \sum_{i=1}^n \psi_{ni}(t)\varphi_{ik}(t) & \cdots & \psi_{nn}(t) \end{bmatrix} \tag{5.3.29}$$

is the fundamental matrix solution to (5.3.10). Therefore, applying Theorem 5.3.2, we conclude that the general solution to (5.2.1) is

$$y(t) = \sum_{j \neq k}^{n} c_j \psi_{1j}(t) + c_k \sum_{i=1}^{n} \psi_{1i}(t)\varphi_{ik}(t) = \sum_{j=1}^{n} c_j y_j(t), \qquad (5.3.30)$$

where c_j are arbitrary constants; for $j \neq k$, $y_j(t) = \psi_{1j}(t)$ and $y_k(t) = \sum_{i=1}^{n} \psi_{1i}(t)\varphi_{ik}$.

This establishes (i)–(iii), and completes the proof of the theorem. $\qquad\Box$

Next, we exhibit the application of Theorem 5.3.5 for finding a closed-form solution to a higher-order stochastic differential equation with constant coefficients.

Example 5.3.1. Let us consider the following third-order linear homogeneous stochastic differential equation with constant coefficients:

$$dy^{(2)} + [y^{(2)} + y^{(1)} + y]dt + [y^{(2)} + y]dw(t) = 0.$$

Prove or disprove the feasibility of a closed-form solution to the stochastic differential equation. If it is feasible, then find its general solution.

Solution procedure. First, we present a systematic procedure for justifying the feasibility of the solution:

Step #1. (companion form representation). Following the proof of Theorem 5.3.1, the companion form of the given stochastic differential equation is

$$dx = A_c x\, dt + B_c x\, dw(t),$$

where $x \in R^3$; and A_c and B_c are 3×3 deterministic and stochastic companion constant matrices in (5.3.3) and (5.3.4) relative to the given differential equation, and they are

$$A_c = \begin{bmatrix} 0 & 1 & 0 \\ 0 & 0 & 1 \\ -1 & -1 & -1 \end{bmatrix}, \quad \text{and} \quad B_c = \begin{bmatrix} 0 & 0 & 0 \\ 0 & 0 & 0 \\ -1 & 0 & -1 \end{bmatrix}.$$

Step #2. (finding the fundamental solution to the deterministic part of the given stochastic differential equation). By imitating Procedure 5.3.1, we have

$$\Phi_D(t) = (\psi_{ij}(t))_{3\times3} = \begin{bmatrix} \psi_{11}(t) & \psi_{12}(t) & \psi_{13}(t) \\ \psi_{21}(t) & \psi_{21}(t) & \psi_{23}(t) \\ \psi_{21}(t) & \psi_{32}(t) & \psi_{33}(t) \end{bmatrix} \quad \text{[by notation (5.3.18)]}$$

$$= \begin{bmatrix} \cos t & \sin t & \exp[-t] \\ -\sin t & \cos t & -\exp[-t] \\ -\cos t & -\sin t & \exp[-t] \end{bmatrix},$$

and the inverse of $\Phi_D(t)$ is

$$\Phi_D^{-1}(t) = (\phi_{ij}(t))_{3\times 3} = \begin{bmatrix} \phi_{11}(t) & \phi_{12}(t) & \phi_{13}(t) \\ \phi_{21}(t) & \phi_{21}(t) & \phi_{23}(t) \\ \phi_{21}(t) & \phi_{32}(t) & \phi_{33}(t) \end{bmatrix} \quad \text{[by notation (5.3.18)]}$$

$$= \frac{1}{2} \begin{bmatrix} \cos t - \sin t & -2\sin t & -(\cos t + \sin t) \\ \cos t + \sin t & 2\cos t & \cos t - \sin t \\ \exp[t] & 0 & \exp[t] \end{bmatrix}.$$

Step #3. [*transformed stochastic system* (5.3.17)]. The stochastic transformed system relative to the given differential equation is given by

$$dc = \Phi_D^{-1}(t)B_c\Phi_D(t)c\,dw(t) = \begin{bmatrix} 0 & 0 & (\cos t + \sin t)\exp[-t] \\ 0 & 0 & -(\cos t - \sin t)\exp[-t] \\ 0 & 0 & -1 \end{bmatrix} c\,dw(t)$$

Thus, the time-varying coefficient rate matrix (5.3.17), which corresponds to the given stochastic differential equation, satisfies the hypothesis of Lemma 5.3.1. Hence, the given stochastic differential equation satisfies the conditions of Theorem 5.3.5. Therefore, by the application of Theorem 5.3.5, the conclusion (i) of Theorem 5.3.5 assures the feasibility of a closed-form solution. Moreover, the conclusion (iii) of Theorem 5.3.5 provides the closed-form representation of the general solution to the given differential equation. Thus, we have

$$y(t) = c_1 \cos t + c_2 \sin t + c_3 \left[\exp\left[-\frac{3}{2}t - w(t) \right] \right.$$

$$+ \cos t \int_a^t (\cos s + \sin s)\exp\left[-\frac{3}{2}s - w(s) \right] dw(s)$$

$$\left. - \sin t \int_a^t (\cos s - \sin s)\exp\left[-\frac{3}{2}s - w(s) \right] dw(s) \right].$$

In the following, we illustrate the usefulness of Theorem 5.3.5 in the context of second-order linear stochastic differential equations with constant coefficients. This is achieved in the framework of three cases: (i) real and distinct roots, (ii) complex roots, and (iii) repeated roots of the auxiliary equation (5.3.9) associated with the deterministic part of the second-order homogeneous stochastic differential equation.

Illustration 5.3.1 [real and distinct roots of the auxiliary equation (5.3.9)]. Let us consider

$$\begin{cases} dy^{(1)} + (a_1 y^{(1)} + a_0 y)dt + [b_1 y^{(1)} + b_0 y]dw(t) = 0, \\ y(t_0) = y_0, \quad y^{(1)}(t_0) = y_0^{(1)}, \end{cases} \quad (5.3.31)$$

where a_1, a_0, b_1, and b_0 are any given real numbers; and w is a normalized Wiener process. We assume that the auxiliary equation (5.3.9) associated with the deterministic part of the given second-order homogeneous stochastic differential equation satisfies the condition $a_1^2 - 4a_0 > 0$. Determine:

(i) The classes of solvable differential equations (5.3.31).
(ii) The closed-form solution of the representative of each class.
(iii) Solve the corresponding IVP.

Solution procedure. We outline the solution process of the illustration:

Step #1. (*companion form representation*). Following the proof of Theorem 5.3.1, the companion form of (5.3.31) is

$$dx = A_c x \, dt + B_c x \, dw(t), \tag{5.3.32}$$

where $x \in R^2$; and A_c and B_c are 2×2 constant deterministic and stochastic companion matrices in (5.3.3) and (5.3.4) relative to (5.3.31), and they are

$$A_c = \begin{bmatrix} 0 & 1 \\ -a_0 & -a_1 \end{bmatrix} \quad \text{and} \quad B_c = \begin{bmatrix} 0 & 0 \\ -b_0 & -b_1 \end{bmatrix}. \tag{5.3.33}$$

Step #2. [*finding the fundamental solution to the deterministic part of* (5.3.32)]. Imitating Procedure 5.3.1, we have

$$\Phi_D(t) = (\psi_{ij}(t))_{2\times2} = \begin{bmatrix} \psi_{11}(t) & \psi_{12}(t) \\ \psi_{21}(t) & \psi_{22}(t) \end{bmatrix} \quad \text{[by notation (5.3.18)]}$$

$$= \begin{bmatrix} \exp[\lambda_1 t] \begin{bmatrix} 1 \\ \lambda_1 \end{bmatrix} & \exp[\lambda_2 t] \begin{bmatrix} 1 \\ \lambda_2 \end{bmatrix} \end{bmatrix} \quad \text{(by Observation 5.3.3)}, \tag{5.3.34}$$

and the inverse of $\Phi_D(t)$ is

$$\Phi_D^{-1}(t) = (\phi_{ij}(t))_{2\times2} = \begin{bmatrix} \phi_{11}(t) & \phi_{12}(t) \\ \phi_{21}(t) & \phi_{22}(t) \end{bmatrix} \quad \text{[by notation (5.3.18)]}$$

$$= \frac{1}{\lambda_2 - \lambda_1} \begin{bmatrix} \lambda_2 \exp[-\lambda_1 t] & -\exp[-\lambda_1 t] \\ -\lambda_1 \exp[-\lambda_2 t] & \exp[-\lambda_2 t] \end{bmatrix} \quad \text{(by Observation 1.4.4 [80])}. \tag{5.3.35}$$

Step #3. [*transformed stochastic system* (5.3.17)]. The transformed stochastic system relative to (5.3.31) is given by

$$dc = \Phi_D^{-1}(t) B_c \Phi_D(t) c \, dw(t)$$

$$= \begin{bmatrix} \frac{b_0 + b_1 \lambda_1}{\lambda_2 - \lambda_1} & \frac{b_0 + b_1 \lambda_2}{\lambda_2 - \lambda_1} \exp[(\lambda_2 - \lambda_1)t] \\ -\frac{b_0 + b_1 \lambda_1}{\lambda_2 - \lambda_1} \exp[(\lambda_1 - \lambda_2)t] & -\frac{b_0 + b_1 \lambda_2}{\lambda_2 - \lambda_1} \end{bmatrix} c \, dw(t). \tag{5.3.36}$$

Step #4. (*determination of classes*). In this case, there are two nontrivial classes of differential equations (5.3.31) with distinct real roots ($CDRR$). They are denoted by

$$CDRR_1 = \{\text{SDE (5.3.31)} : b_0 + b_1\lambda_1 = 0\},$$

and

$$CDRR_2 = \{\text{SDE (5.3.31)} : b_0 + b_1\lambda_2 = 0\}.$$

Moreover, from Observation 5.3.3, these classes are redescribed as

$$CDRR_1 = \{\text{SDE (5.3.31)} : \exp[\lambda_1 t] \text{ is the solution to both (5.3.11) and (5.3.22)}\},$$

and

$$CDRR_2 = \{\text{SDE (5.3.31)} : \exp[\lambda_2 t] \text{ is the solution to both (5.3.11) and (5.3.22)}\}.$$

Step #5. (*reduced transformed systems*). The transformed systems (5.3.36) with respect to the determined classes $CDRR_1$ and $CDRR_2$ in Step #4 are

$$dc = \begin{bmatrix} 0 & \frac{b_0+b_1\lambda_2}{\lambda_2-\lambda_1}\exp[(\lambda_2-\lambda_1)t] \\ 0 & -\frac{b_0+b_1\lambda_2}{\lambda_2-\lambda_1} \end{bmatrix} c\,dw(t), \tag{5.3.37}$$

and

$$dc = \begin{bmatrix} \frac{b_0+b_1\lambda_1}{\lambda_2-\lambda_1} & 0 \\ -\frac{b_0+b_1\lambda_1}{\lambda_2-\lambda_1}\exp[(\lambda_1-\lambda_2)t] & 0 \end{bmatrix} c\,dw(t). \tag{5.3.38}$$

Step #6. [*general solution to* (5.3.31)]. From the application of Theorem 5.3.5, the general solutions corresponding to the classes $CDRR_1$ and $CDRR_2$, i.e. the reduced systems (5.3.37) and (5.3.38), are

$$y(t) = c_1 \exp[\lambda_1 t] + c_2 \exp\left[\left(\lambda_2 - \frac{1}{2}\nu_2^2\right)t + \nu_2 w(t)\right]$$

$$+ c_2 \frac{b_0 + b_1\lambda_2}{\lambda_2 - \lambda_1}\exp[\lambda_1 t]\int_a^t \exp\left[\left(\lambda_2 - \lambda_1 - \frac{1}{2}\nu_2^2\right)s + \nu_2 w(s)\right]dw(s), \tag{5.3.39}$$

and

$$y(t) = c_2 \exp[\lambda_2 t] + c_1 \exp\left[\left(\lambda_1 - \frac{1}{2}\nu_1^2\right)t + \nu_1 w(t)\right]$$

$$- c_1 \frac{b_0 + b_1\lambda_1}{\lambda_2 - \lambda_1}\exp[\lambda_2 t]\int_a^t \exp\left[\left(\lambda_1 - \lambda_2 - \frac{1}{2}\nu_1^2\right)s + \nu_1 w(s)\right]dw(s), \tag{5.3.40}$$

where $\nu_1 = \frac{b_0+b_1\lambda_1}{\lambda_2-\lambda_1}$ and $\nu_2 = -\frac{b_0+b_1\lambda_2}{\lambda_2-\lambda_1}$, respectively.

Step #7. [*solution of the IVP* (5.3.31)]. Using the initial data $(y(t_0) = y_0, y^{(1)}(t_0) = y_0^{(1)})$, we solve arbitrary constants c_1 and c_2 in general solutions with regard to the classes. The details are left to the reader.

Example 5.3.2. We consider the second-order stochastic differential equation

$$dy^{(1)} + (y^{(1)} - 2y)dt + [3y^{(1)} + 6y]dw(t).$$

Determine:

(a) The class of solvable differential equations.

(b) The closed-form solution to the representative of each class.

Solution process. First, we note that $a_1^2 - 4a_0 = 1 + 8 = 9 > 0$. Therefore, we can apply the steps outlined in Illustration 5.3.1. From Step #1, we have the deterministic and stochastic companion constant matrices

$$A_c = \begin{bmatrix} 0 & 1 \\ 2 & -1 \end{bmatrix}, B_c = \begin{bmatrix} 0 & 0 \\ -6 & -3 \end{bmatrix}.$$

From Step #2, the fundamental solution $\Phi_D(t)$ of the deterministic part and its inverse are

$$\Phi_D(t) = \begin{bmatrix} \exp[-2t] & \exp[t] \\ -2\exp[-2t] & \exp[t] \end{bmatrix} \quad \text{and} \quad \Phi_D^{-1}(t) = \frac{1}{3}\begin{bmatrix} \exp[2t] & -\exp[2t] \\ 2\exp[-t] & \exp[-t] \end{bmatrix}.$$

From Step #3, the transformed stochastic system for the given stochastic differential equation is

$$dc = \Phi_D^{-1}(t)B_c\Phi_D(t)c\,dw(t)$$

$$= \begin{bmatrix} b_1\frac{\frac{b_0}{b_1}+\lambda_1}{\lambda_2-\lambda_1} & b_1\frac{\frac{b_0}{b_1}+\lambda_2}{\lambda_2-\lambda_1}\exp[\lambda_2-\lambda_1 t] \\ -b_1\frac{\frac{b_0}{b_1}+\lambda_1}{\lambda_2-\lambda_1}\exp[(\lambda_1-\lambda_2)t] & -b_1\frac{\frac{b_0}{b_1}+\lambda_2}{\lambda_2-\lambda_1} \end{bmatrix} c\,dw(t)$$

$$= \begin{bmatrix} 0 & -3\exp[3t] \\ 0 & 3 \end{bmatrix} c\,dw(t).$$

Using Step #4, we conclude that the given stochastic differential equation belongs to the class $CDRR_1$ determined by $b_0 + b_1\lambda_1 = 0$. This completes the solution process of (a).

By imitating Steps #5 and #6, we have the following closed-form general solution process of the given differential equation:

$$y(t) = c_1\exp[-2t] + c_2\exp\left[\left(1 - \frac{9}{2}\right)t + 3w(t)\right]$$

$$- c_2 3\exp[-2t]\int_a^t \exp\left[\left(3 - \frac{9}{2}\right)s + 3w(s)\right]ds$$

$$= c_1\exp[-2t] + c_2\left(\exp\left[-\frac{7}{2}t + 3w(t)\right]\right.$$

$$\left. - 3\exp[-2t]\int_a^t \exp\left[-\frac{3}{2}s + 3w(s)\right]dw(s)\right).$$

This completes the solution process of the differential equation.

In the following, we examine the case of repeated roots for a second-order linear stochastic differential equation with constant coefficients.

Illustration 5.3.2 [repeated roots of the auxiliary equation (5.3.9)]. Let us consider

$$\begin{cases} dy^{(1)} + (a_1 y^{(1)} + a_0 y)dt + [b_1 y^{(1)} + b_0 y]dw(t) = 0, \\ y(t_0) = y_0, \quad y^{(1)}(t_0) = y_0^{(1)}, \end{cases} \tag{5.3.41}$$

where a_1, a_0, b_1, and b_0 are any given real numbers; and w is a normalized Wiener process. We assume that the auxiliary equation (5.3.9) associated with the deterministic part of the given second-order homogeneous stochastic differential equation satisfies the condition $a_1^2 - 4a_0 = 0$. Determine:

(i) The classes of solvable differential equations (5.3.41).
(ii) The closed-form solution of the representative of each class.
(iii) Solve the corresponding IVP.

Solution procedure. We imitate the solution process of Illustration 5.3.1. In fact, Step #1 is exactly as described in (5.3.32) and (5.3.33).

Step #2. [*finding the fundamental solution of the deterministic part of* (5.3.41)]. Imitating Procedure 5.3.1 and Step #2 in Illustration 5.3.1, we have

$$\Phi_D(t) = \left[\exp[\lambda_1 t] \begin{bmatrix} 1 \\ \lambda_1 \end{bmatrix} \exp[\lambda_1 t] \begin{bmatrix} t \\ (\lambda_1 t + 1) \end{bmatrix} \right], \tag{5.3.42}$$

and the inverse of $\Phi_D(t)$ is

$$\Phi_D^{-1}(t) = (\phi_{ij}(t))_{2\times 2} = \exp[-\lambda_1 t] \begin{bmatrix} (\lambda_1 t + 1) & -t \\ -\lambda_1 & 1 \end{bmatrix}. \tag{5.3.43}$$

Step #3. [*transformed stochastic system* (5.3.17)]. The transformed stochastic system relative to (5.3.41) is given by

$$dc = \Phi_D^{-1}(t) B_c \Phi_D(t) c \, dw(t)$$

$$= \begin{bmatrix} t(b_0 + b_1 \lambda_1) & b_1 t + t^2(b_0 + b_1 \lambda_1) \\ -(b_0 + b_1 \lambda_1) & -b_1 - t(b_0 + b_1 \lambda_1) \end{bmatrix} c \, dw(t). \tag{5.3.44}$$

Step #4. (*determination of classes*). The classes with respect to the repeated root λ_1 are determined by

$$\mathcal{CDRR}_1 = \{\text{SDE } (5.3.41) : b_0 + b_1 \lambda_1 = 0\} \text{ and } \mathcal{CDRR}_2 = \{\text{SDE } (5.3.41) : b_1 = 0\}.$$

Moreover, from Observation 5.3.3, these classes are redescribed as

$$\mathcal{CDRR}_1 = \{\text{SDE } (5.3.41) : \exp[\lambda_1 t] \text{ is a solution to both } (5.3.11) \text{ and } (5.3.22)\},$$

and

$$CDRR_2 = \{\text{SDE } (5.3.39) : \exp[\lambda_1 t] \text{ is a solution to } (5.3.11) \text{ only}\}.$$

Step #5. (*reduced transformed systems*). The transformed systems (5.3.44) with respect to the determined classes $CDRR_1$ and $CDRR_2$ in Step #4 are

$$dc = \begin{bmatrix} 0 & b_1 t \\ 0 & -b_1 \end{bmatrix} c \, dw(t), \tag{5.3.45}$$

and

$$dc = b_0 \begin{bmatrix} t & t^2 \\ -1 & -t \end{bmatrix} c \, dw(t). \tag{5.3.46}$$

Step #6. [*general solution to* (5.3.41)]. We note that Theorem 5.3.5 is applicable to only the class $CDRR_1$ [reduced system (5.3.45)]. Thus, the closed-form general solution process of (5.3.41) corresponding to the class $CDRR_1$ is given by

$$y(t) = c_1 \exp[\lambda_1 t] + c_2 \left(t \exp \left[\left(\lambda_1 - \frac{1}{2} \nu_2^2 \right) t + \nu_2 w(t) \right] \right.$$

$$+ b_1 \exp[\lambda_1 t] \int_a^t s \exp \left[-\frac{1}{2} \nu_2^2 s + \nu_2 w(s) \right] dw(s) \Bigg), \tag{5.3.47}$$

where $\nu_2 = -b_1$ and $\lambda_1 = -\frac{1}{2} a_1$.

For obvious reasons, we emphasize that Theorem 5.3.5 is not applicable to the class $CDRR_2$ [reduced system (5.3.46)]. Moreover, the coefficient rate matrix relative to the class $CDRR_2$ is a time-varying matrix, and its structure exhibits nontrivial coupled interactions with the components of state vector c. Because of this, one is not able to find the closed-form solution to this type of system. Thus, it is not feasible to find the closed-form solution to (5.3.41) corresponding to the class $CDRR_2$.

By following Step #7, the solution to the IVP (5.3.41) with the initial data $(y(t_0) = y_0, \; y^{(1)}(t_0) = y_0^{(1)})$ can be solved by the arbitrary constants c_1 and c_2 in the general solutions to (5.3.41) with regard to the class $CDRR_1$. The details are left to the reader.

Example 5.3.3. We consider the second-order stochastic differential equation

$$dy^{(1)} + (2y^{(1)} + y)dt + [y^{(1)} + y]dw(t) = 0.$$

Determine:

(a) The class of solvable differential equations.
(b) The closed-form solution of the representative of each class.

Solution process. First, we note that $a_1^2 - 4a_0 = 4 - 4 = 0$. Therefore, we can apply the steps outlined in Illustration 5.3.2. From Step #1, we have the deterministic and stochastic companion constant matrices

$$A_c = \begin{bmatrix} 0 & 1 \\ -1 & -2 \end{bmatrix}, \quad B_c = \begin{bmatrix} 0 & 0 \\ -1 & -1 \end{bmatrix}.$$

From Step #2, the fundamental solution $\Phi_D(t)$ of the deterministic part and its inverse are

$$\Phi_D(t) = \exp[-t] \begin{bmatrix} 1 & t \\ -1 & 1-t \end{bmatrix} \quad \text{and} \quad \Phi_D^{-1}(t) = \exp[t] \begin{bmatrix} 1-t & -t \\ 1 & 1 \end{bmatrix}.$$

From Step #3, the transformed stochastic system for the given stochastic differential equation is given by

$$dc = \Phi_D^{-1}(t) B_c \Phi_D(t) c \, dw(t)$$

$$= \begin{bmatrix} t(1-1) & t+t^2(1-1) \\ -(1-1) & -1+t(1-1) \end{bmatrix} c \, dw(t) = \begin{bmatrix} 0 & t \\ 0 & -1 \end{bmatrix} c \, dw(t).$$

Using Step #4, we conclude that the given stochastic differential equation belongs to the class $\mathcal{CDRR}_1$ determined by $b_0 + b_1 \lambda_1 = 0$ with $\lambda_1 = -1$. This completes the solution process of (a).

By imitating Steps #5 and #6, we have the following closed-form general solution process of the given differential equation:

$$y(t) = c_1 \exp[-t] + c_2 t \exp\left[\left(-1 - \frac{1}{2}\right)t - w(t)\right]$$

$$+ c_2 \exp[-t] \int_a^t s \exp\left[-\frac{1}{2}s - w(s)\right] dw(s)$$

$$= c_1 \exp[-t] + c_2 \left(t \exp\left[-\frac{3}{2}t - w(t)\right]\right.$$

$$\left. + \exp[-t] \int_a^t s \exp\left[-\frac{1}{2}s - w(s)\right] dw(s)\right).$$

This completes the solution process of the differential equation.

Finally, we consider the case of nontrivial complex roots for a second-order linear stochastic differential equation with constant coefficients.

Illustration 5.3.3   [complex roots of the auxiliary equation (5.3.9)]. Let us consider

$$\begin{cases} dy^{(1)} + (a_1 y^{(1)} + a_0 y)dt + [b_1 y^{(1)} + b_0 y]dw(t) = 0, \\ y(t_0) = y_0, \quad y^{(1)}(t_0) = y_0^{(1)}, \end{cases} \qquad (5.3.48)$$

where a_1, a_0, b_1, and b_0 are any given real numbers, and w is a normalized Wiener process. We assume that the auxiliary equation (5.3.9) associated with the deterministic part of the given second-order homogeneous stochastic differential equation satisfies the condition $a_1^2 - 4a_0 < 0$. Determine:

(i) The classes of solvable differential equations (5.3.48).
(ii) The closed-form solution of the representative of each class.
(iii) Solve the corresponding IVP.

Solution procedure. We imitate the solution process of Illustration 5.3.1. In fact, Step #1 is exactly as described in (5.3.32) and (5.3.33).

Step #2. [*finding the fundamental solution of the deterministic part of* (5.3.48)]. Imitating Procedure 5.3.1, Step #2 in Illustration 5.3.1 and $a_1^2 - 4a_0 < 0$, the auxiliary equation (5.3.9) has complex roots ($\lambda_1 = \alpha + i\beta$ and $\lambda - 2 = \alpha - i\beta$, with $\beta \neq 0$). Thus, for $j = 1, 2$, $b_0 + b_1\lambda_j = 0$ is impossible, unless $b_0 = b_1 = 0$. From this and Observation 5.3.3, we conclude that the time-varying coefficient rate matrix of the transformed system (5.3.17) corresponding to (5.3.48) cannot be reduced to (5.3.21). Therefore, in the case of this illustration, one cannot find a closed-form solution to (5.3.48). In fact, in this case, the transformed system is $dc = \Phi_D^{-1}(t)B_c\Phi_D(t)c\,dw(t)$, where

$$\Phi_D^{-1}(t)B_c\Phi_D(t)$$
$$= \begin{bmatrix} \sin\beta t(b_0\cos\beta t + rb_1\cos(\theta + \beta t) & \sin\beta t(b_0\cos\beta t + rb_1\sin(\theta + \beta t) \\ -\cos\beta t(b_0\cos\beta t - rb_1\cos(\theta + \beta t) & -\cos\beta t(b_0\sin\beta t + rb_1\sin(\theta + \beta t) \end{bmatrix},$$

where $r = \sqrt{\alpha^2 + \beta^2}$ and $\theta = \arctan\left(\frac{\beta}{\alpha}\right)$ for $\alpha \neq 0$ or $\frac{\pi}{2}$.

Observation 5.3.4. Another reason that it is difficult to find a closed-form solution to (5.3.48) is that it is impossible to find the closed-form solution of the reduced transformed system (5.3.46) in the context of repeated roots (Illustration 5.3.2). For justification, we refer to Example 5.2.1.

Exercises

1. Under the hypotheses of Theorem 5.3.5, find the closed-form solution representation for the IVP associated with (5.3.10) (Observation 5.3.1).
2. Solve the IVPs (5.3.31) and (5.3.41) (Illustrations 5.3.1 and 5.3.2).
3. Prove or disprove that the following Itô–Doob-type stochastic differential equations are solvable.

   (a) $dy^{(1)} + (y^{(1)} - 2y)dt + \sigma[y^{(1)} - y]dw(t) = 0$, for any $\sigma \neq 0$;
   (b) $dy^{(1)} + (y^{(1)} - 2y)dt + \sigma[y^{(1)} + 2y]dw(t) = 0$, for any $\sigma \neq 0$;
   (c) $dy^{(1)} + (y^{(1)} - 2y)dt + [y^{(1)} + y]dw(t) = 0$;
   (d) $dy^{(1)} - (y^{(1)} - 2y)dt + \sigma[y^{(1)} - 2y]dw(t) = 0$, for any $\sigma \neq 0$;

(e) $dy^{(1)} - (y^{(1)} - 2y)dt + \sigma[y^{(1)} - 2y]dw(t) = 0;$

(f) $dy^{(2)} - [y^{(2)} - y^{(1)} + y]dt + \sigma[y^{(2)} + y]dw(t) = 0,$ for any $\sigma \neq 0;$

(g) $dy^{(3)} + [y^{(2)} - 4y^{(1)} - 4y]dt + \sigma[y^{(2)} + 3y^{(1)} + 2y]dw(t) = 0,$ for any $\sigma \neq 0;$

(h) $dy^{(3)} + [y^{(2)} - 4y^{(1)} - 4y]dt + \sigma[y^{(2)} - y^{(1)} + 2y]dw(t) = 0,$ for any $\sigma \neq 0;$

(i) $dy^{(3)} + [y^{(2)} - 4y^{(1)} - 4y]dt + \sigma[y^{(2)} - 4y]dw(t) = 0,$ for any $\sigma \neq 0;$

(j) $dy^{(3)} - y\,dt + \sigma[y^{(2)} + y]dw(t) = 0,$ for any $\sigma \neq 0;$

(k) $dy^{(1)} - (4y^{(1)} - 4y)dt + \sigma[y^{(1)} - 2y]dw(t) = 0,$ for any $\sigma \neq 0;$

(l) $dy^{(2)} - [3y^{(2)} - 3y^{(1)} + y]dt + \sigma[y^{(2)} - 2y^{(1)} + y]dw(t) = 0,$ for any $\sigma \neq 0;$

(m) $dy^{(4)} + y^{(3)}dt + \sigma[y^{(4)} + y^{(2)}]dw(t) = 0,$ for any $\sigma \neq 0;$

(n) $dy^{(3)} + y^{(1)}dt + \sigma[y^{(3)} + y]dw(t) = 0,$ for any b_0 and $b_1.$

4. If the stochastic differential equations in Exercise 3 are solvable, find:

 (a) Their fundamental solutions.

 (b) The closed-form general solutions.

5. Let us assume that $y(t_0) = y_0$, $y^{(1)}(t_0) = y_0^{(1)}$, $y^{(2)}(t_0) = y_0^{(2)}$, $y^{(3)}(t_0) = y_0^{(3)}$, and $y^{(4)}(t_0) = y_0^{(4)}$ are given arbitrary initial states of the system at the initial time $t = t_0$. By appropriately choosing the given initial data, solve for the IVPs corresponding to the dynamic equations in Exercise 4.

6. Applying Corollary 4.4.1, find the transformed deterministic system of differential equations with respect to the stochastic companion system (5.3.10).

5.4 Linear Nonhomogeneous Equations

In this section, we utilize the *method of variation of constants parameters* to compute a particular solution to higher-order linear nonhomogeneous stochastic differential equations of the Itô–Doob-type (5.3.1). The significance of this method for solving the deterministic and stochastic first-order linear nonhomogeneous systems of differential is repeatedly exhibited. This method has been applied to both time-varying linear (Section 4.6) and nonlinear systems of deterministic [80] and stochastic differential equations.

5.4.1 *General Problem*

Let us consider the following first-order linear nonhomogeneous systems of stochastic differential equations of the Itô–Doob-type (5.3.2) associated with (5.3.1):

$$dx = [A_c x + p(t)]dt + [B_c x + q(t)]dw(t), \tag{5.4.1}$$

where x, A_c, and B_c are as defined in (5.3.10); p and q are n-dimensional continuous functions defined on an interval $J \subseteq R$ into R^n in (5.3.3) and (5.3.4); and w is a scalar normalized Wiener process (Definition 1.2.12).

 A procedure for finding a solution process of (5.4.1) is exactly parallel to the procedure developed in Section 4.5. In addition, we are also interested in solving an IVP associated with (5.4.1). The concept of the solution process defined in Definition 5.2.1 is directly applicable to (5.4.1). The details are left to the reader.

5.4.2 *Procedure for Finding the General Solution*

To find a general solution to (5.4.1), we imitate the conceptual ideas of finding a general solution to the linear nonhomogeneous algebraic equations described in Theorem 1.4.7 [80]. In particular, we recast Procedure 4.5.2 for finding a solution process of the first order linear homogeneous system of stochastic differential equations corresponding to (4.5.1) in Section 4.5. The basic ideas are:

(i) To find a general solution process x_c of first-order linear homogeneous companion systems of stochastic differential equations corresponding to (5.2.1).

(ii) To find a particular solution process x_p of (5.3.2).

(iii) To validate the random function $x = x_c + x_p$ to be a general solution to (5.4.1).

Here, we reintroduce the method of variation of constants parameters to Itô–Doob systems of stochastic differential equations to find a particular solution process x_p of (5.4.1).

Step #1. First, find a fundamental solution to first-order linear homogeneous companion system of stochastic differential equations corresponding to (5.4.1), i.e.

$$dx = A_c x \, dt + B_c x \, dw(t). \tag{5.4.2}$$

We note that this stochastic system of Itô–Doob differential equations is exactly the same as (5.3.10). Therefore, we imitate Steps #1 and #2, described in Procedure 5.3.1 for finding a general solution process of (5.3.10), and determine a fundamental solution to (5.4.2) (Observation 4.4.1):

$$\Phi_{cs}(t) = \Phi_D(t)\Phi_S(t) \quad \text{[from (5.3.13)]}. \tag{5.4.3}$$

Step #2 (method of variation of constants parameters). From Observation 5.3.2, we reduce the system (5.3.17), and then further reduce the time-varying coefficient rate matrix $\Phi_D^{-1}(t)B_c\Phi_D(t)$ into the form (5.3.21). Finally, applying Theorem 5.3.5, the fundamental matrix solution $\Phi_{cs}(t)$ (5.3.29) is determined as

$$\Phi_{cs}(t) = \Phi_D(s)\Phi_T(t) = \begin{bmatrix} \psi_{11}(t) & \cdots & \sum_{i=1}^{n}\psi_{1i}(t)\varphi_{ik}(t) & \cdots & \psi_{1n}(t) \\ \cdots & \cdots & \cdots & \cdots & \cdots \\ \psi_{k1}(t) & \cdots & \sum_{i=1}^{n}\psi_{ki}(t)\varphi_{ik}(t) & \cdots & \psi_{kn}(t) \\ \cdots & \cdots & \cdots & \cdots & \cdots \\ \psi_{n1}(t) & \cdots & \sum_{i=1}^{n}\psi_{ni}(t)\varphi_{ik}(t) & \cdots & \psi_{nn}(t) \end{bmatrix}. \tag{5.4.4}$$

The remainder of the procedure is exactly parallel to the procedure developed in Section 4.5. In fact, we assume that a particular solution $x_p(t)$ to (5.4.1) has the following representation:

$$x_p(t) = \Phi_{cs}(t)c(t), \quad c(a) = c_a, \tag{5.4.5}$$

where $c(t)$ is an unknown function.

By imitating the argument used in Procedure 4.5.1, we get

$$c(t) = c_a + \int_a^t \Phi_{cs}^{-1}(s)[p(s) - Bq(s)]ds + \int_a^t \Phi_{cs}^{-1}(s)q(s)dw(s). \tag{5.4.6}$$

We substitute the expression for $c(t)$ in (5.4.6) into (5.4.5), and we obtain

$$x_p(t) = \Phi_{cs}(t)\left[c_a + \int_a^t \Phi_{cs}^{-1}(s)[p(s) - B(s)q(s)]ds + \int_a^t \Phi_{cs}^{-1}q(s)dw(s)\right]$$

$$= \Phi_{cs}(t)c_a + \int_a^t \Phi_{cs}(t)\Phi_{cs}^{-1}(s)[p(s) - B(s)q(s)]ds$$

$$+ \int_a^t \Phi_{cs}(t)\Phi_{cs}^{-1}(s)q(s)dw(s). \tag{5.4.7}$$

From this, using the relation $\Phi_{cs}(t, s) = \Phi_{cs}(t)\Phi_{cs}^{-1}(s)$, Observation 4.5.1 and given c_a is an arbitrary constant vector, we have the following representation for the general solution to (5.4.1):

$$x(t) = \Phi(t)c + \left[\int_a^t \Phi_{cs}(t, s)[p(s) - Bq(s)]ds + \int_a^t \Phi_{cs}(t, s)q(s)dw(s)\right]. \tag{5.4.8}$$

Step #3. From Theorem 5.3.2, the general solution to (5.3.1) is the first component of $x(t)$ in (5.4.8).

Observation 5.4.1. We note that the IVP corresponding to (5.4.1) is exactly the same as the IVP (5.3.7). The procedure for solving the closed-form solution to the IVP associated with (5.4.1) is exactly the same as Procedure 4.5.4. The details are left as an exercise to the reader.

Exercises

1. Under the hypotheses of Theorem 5.3.5 and imitating Procedure 4.5.4, find the closed-form solution representation of the IVP associated with (5.4.1).

2. Prove or disprove that the following Itô–Doob-type stochastic differential equations are solvable:

   (a) $dy^{(1)} + [y^{(1)} - 2y + \sin t]dt + \sigma[y^{(1)} - y + \sin w(t)]dw(t) = 0$, for any $\sigma \neq 0$;

   (b) $dy^{(1)} + [y^{(1)} - 2y + \sin t]dt + \sigma[y^{(1)} + 2y + \sin w(t)]dw(t) = 0$, for any $\sigma \neq 0$;

   (c) $dy^{(1)} + [y^{(1)} - 2y + \sin t]dt + [y^{(1)} + y + \sin w(t)]dw(t) = 0$;

   (d) $dy^{(1)} - [y^{(1)} - 2y + \sin t]dt + \sigma[y^{(1)} - 2y + \sin w(t)]dw(t) = 0$, for any $\sigma \neq 0$;

   (e) $dy^{(1)} - [y^{(1)} - 2y + t]dt + \sigma[y^{(1)} - 2y + w(t)]dw(t) = 0$;

(f) $dy^{(2)} - [y^{(2)} - y^{(1)} + y + t^2]dt + \sigma[y^{(2)} + y + w^2(t)]dw(t) = 0$, for any $\sigma \neq 0$;

(g) $dy^{(3)} + [y^{(2)} - 4y^{(1)} - 4y + 1]dt + \sigma[y^{(2)} + 3y^{(1)} + 2y]dw(t) = 0$, for any $\sigma \neq 0$;

(h) $dy^{(3)} + [y^{(2)} - 4y^{(1)} - 4y]dt + \sigma[y^{(2)} - y^{(1)} + 2y + 1]dw(t) = 0$, for any $\sigma \neq 0$;

(i) $dy^{(3)} + [y^{(2)} - 4y^{(1)} - 4y + 2t]dt + \sigma[y^{(2)} - 4y + 3]dw(t) = 0$, for any $\sigma \neq 0$;

(j) $dy^{(3)} + [-y + \exp[t]]dt + \sigma[y^{(2)} + y + \exp[w(t))]]dw(t) = 0$, for any $\sigma \neq 0$;

(k) $dy^{(1)} - [4y^{(1)} - 4y + \ln t]dt + \sigma[y^{(1)} - 2y + \cos w(t)]dw(t) = 0$, for any $\sigma \neq 0$;

(l) $dy^{(2)} - [3y^{(2)} - 3y^{(1)} + y]dt + \sigma[y^{(2)} - 2y^{(1)} + y + 1]dw(t) = 0$, for any $\sigma \neq 0$;

(m) $dy^{(4)} + [y^{(3)} + \exp[t]]dt + \sigma[y^{(4)} + y^{(2)} + \exp[-w(t)]]dw(t) = 0$, for any $\sigma \neq 0$;

(n) $dy^{(4)} + [y + \sin t]dt + [b_1 y^{(1)} + b_0 y + w(t)]dw(t) = 0$, for any b_0 and b_1.

3. If the stochastic differential equations in Exercise 2 are solvalable, Find:

   (a) the fundamental solutions;
   (b) the closed-form general solutions.

4. Let us assume that $y(t_0) = y_0$, $y^{(1)}(t_0) = y^{(1)}$, $y^{(2)}(t_0) = y_0^{(2)}$, $y^{(3)}(t_0) = y_0^{(3)}$, and $y^{(4)}(t_0) = y_0^{(4)}$ are given arbitrary initial states of the system at the initial time $t = t_0$. By appropriately choosing the given initial data, solve the initial value problems corresponding to the dynamic equations in Exercise 3.

5.5 The Laplace Transform

In this section, we present the concept of the Laplace transform and its applications to solve the higher-order linear nonhomogeneous differential equations with constant coefficients.

Definition 5.5.1. Let f be a real-valued function of two variables $(t, w(t))$ defined for all real numbers $t \geq 0$, with $w(t)$ being the Wiener process. The Laplace transform of f in the sense of the Cauchy–Riemann integral is defined by

$$F(s) = \mathcal{L}(f)(s) = \int_0^\infty e^{-st} f(t, w(t))dt = \lim_{T \to \infty} \left[\int_0^T e^{-st} f(t, w(t))dt \right], \quad (5.5.1)$$

for all values of s for which this improper integral exists. It is denoted by $F(s) = \mathcal{L}(f)(s)$. Moreover, the Laplace transform of f in the sense of the Itô–Doob integral, denoted by $F^w(s) = \mathcal{L}^w(f)(s)$, is defined by

$$F^w(s) = \mathcal{L}^w(f)(s) = \int_0^\infty e^{-st} f(t, w(t))dw(t) = \lim_{T \to \infty} \left[\int_0^T e^{-st} f(t, w(t))dw(t) \right], \quad (5.5.2)$$

for all values of s for which this improper integral exists.

Example 5.5.1. Find the Laplace transform of $f(t, w(t)) = c$ in the sense of the Itô–Doob integral, a constant function, for $c \neq 0$.

Solution process. Applying the Itô–Doob differential formula Theorem 1.3.1 to $ce^{-st}w(t)$, we obtain

$$d(ce^{-st}w(t)) = -cse^{-st}w(t)dt + ce^{-st}dw(t).$$

Now, applying the Itô–Doob improper integrals in Definition 5.5.1, we have

$$c \lim_{T\to\infty} [e^{-st}w(t)]_0^T = -cs \lim_{T\to\infty} \left[\int_0^T e^{-st}w(t)dt \right] + c \lim_{T\to\infty} \left[\int_0^T e^{-st}dw(t) \right].$$

Along with the properties of the Wiener process, this gives

$$c \lim_{T\to\infty} \left[\int_0^T e^{-st}dw(t) \right] = cs \lim_{T\to\infty} \left[\int_0^T e^{-st}w(t)dt \right].$$

Hence,

$$\mathcal{L}^w(c)(s) = s\mathcal{L}(cw)(s).$$

Thus, for $s > 0$ and $c \neq 0$,

$$\mathcal{L}^w(c)(s) = s\mathcal{L}(cw)(s) \quad \text{if and only if} \quad \mathcal{L}(w)(s) = \frac{\mathcal{L}^w(1)(s)}{s}. \tag{5.5.3}$$

Example 5.5.2. Find the Laplace transform of $f(t, w(t)) = w(t)$ in the sense of the Itô–Doob integral.

Solution process. Again, applying the Itô–Doob differential formula Theorem 1.3.1 to $e^{-st}w^2(t)$, we obtain

$$d(e^{-st}w^2(t)) = -se^{-st}w^2(t)dt + 2e^{-st}w(t)dw(t) + e^{-st}dt.$$

Using the Itô–Doob improper integrals (Definition 5.5.1), we have

$$\lim_{T\to\infty} [e^{-st}w^2(t)]_0^T = -s \lim_{T\to\infty} \left[\int_0^T e^{-st}w^2(t)dt \right]$$

$$+ 2 \lim_{T\to\infty} \left[\int_0^T e^{-st}w(t)dw(t) \right] + \lim_{T\to\infty} \left[\int_0^T e^{-st}dt \right].$$

By following the argument used in Example 5.5.1, we get

$$2 \lim_{T\to\infty} \left[\int_0^T e^{-st}w(t)dw(t) \right] = s \lim_{T\to\infty} \left[\int_0^T e^{-st}w^2(t)dt \right] - \lim_{T\to\infty} \left[\int_0^T e^{-st}dt \right],$$

and hence

$$2\mathcal{L}^w(w)(s) = s\mathcal{L}(w^2)(s) - \frac{1}{s}.$$

Thus, for $s > 0$,

$$\frac{2\mathcal{L}^w(w)(s)}{s} = \mathcal{L}(w^2)(s) - \frac{1}{s^2} \quad \text{if and only if} \quad \mathcal{L}(w^2)(s) = \frac{2\mathcal{L}^w(w)(s)}{s} + \frac{1}{s^2}. \quad (5.5.4)$$

To use the Laplace transform, one needs to know under what condition(s) the Laplace transform is defined. For this purpose, we present the following to define a class of functions.

Definition 5.5.2. Let $L[f]$ be a class of smooth functions (random) that satisfy the exponential growth condition

$$|f(t, w(t))| \leq K e^{Mt + \Lambda w(t)}, \quad s > M, \quad (5.5.5)$$

for some $M > 0$, $\Lambda > 0$, and $K \geq 0$.

From Definition 5.5.1 and the properties of both deterministic and Itô–Doob integrals (Section 1.3), it is clear that the Laplace transform obeys the following property.

Theorem 5.5.1. *Let $f_1, f_2 \in L[f]$, and let c_1 and c_2 be arbitrary given constants. Then,*

$$\mathcal{L}(c_1 f_1 + c_2 f_2)(s) = c_1 \mathcal{L}(f_1)(s) + c_2 \mathcal{L}(f_2)(s), \quad (5.5.6)$$

and

$$\mathcal{L}^w(c_1 f_1 + c_2 f_2)(s) = c_1 \mathcal{L}^w(f_1)(s) + c_2 \mathcal{L}^w(f_2)(s). \quad (5.5.7)$$

Example 5.5.3. Find the Laplace transform of $f(t, w(t)) = e^{(a - \frac{\sigma^2}{2})t + \sigma w(t)}$ in the sense of the Itô–Doob integral for $\sigma \neq 0$ and $\sigma, a \in R$.

Solution process. Applying the Itô–Doob differential formula Theorem 1.3.1 to $e^{-(s - a + \frac{\sigma^2}{2})t + \sigma w(t)}$, we obtain

$$d(e^{-(s - a + \frac{\sigma^2}{2})t + \sigma w(t)}) = -\left(s - a + \frac{\sigma^2}{2}\right) e^{-(s - a + \frac{\sigma^2}{2})t + \sigma w(t)} dt$$

$$+ \sigma e^{-(s - a + \frac{\sigma^2}{2})t + \sigma w(t)} dw(t) + \frac{1}{2}\sigma^2 e^{-(s - a + \frac{\sigma^2}{2})t + \sigma w(t)} dt.$$

Now, using the strong law of large numbers and the Itô–Doob improper integral (Definition 5.5.1), we have

$$\lim_{T \to \infty} [e^{-(s - a + \frac{\sigma^2}{2})t + \sigma w(t)}]_0^T = \lim_{T \to \infty} \left[\sigma \int_0^T e^{-(s - a + \frac{\sigma^2}{2})t + \sigma w(t)} dw(t) \right]$$

$$- (s - a) \lim_{T \to \infty} \left[\int_0^T e^{-(s - a + \frac{\sigma^2}{2})t + \sigma w(t)} dt \right],$$

which yields

$$(s-a)\mathcal{L}(e^{(a-\frac{\sigma^2}{2})t+\sigma w(t)})(s) - 1 = \mathcal{L}^w(\sigma e^{(a-\frac{\sigma^2}{2})t+\sigma w(t)})(s).$$

Thus, for $s > a - \frac{\sigma^2}{2}$,

$$\mathcal{L}(e^{(a-\frac{\sigma^2}{2})t+\sigma w(t)})(s) = \frac{1+\sigma\mathcal{L}^w(e^{(a-\frac{\sigma^2}{2})t+\sigma w(t)})(s)}{s-a}$$

$$= \frac{1}{s-a} + \frac{\sigma\mathcal{L}^w(e^{(a-\frac{\sigma^2}{2})t+\sigma w(t)})(s)}{s-a}. \qquad (5.5.8)$$

Example 5.5.4. Find the Laplace transform of $f(t, w(t)) = e^{at+\sigma w(t)}$ in the sense of the Itô–Doob integral for $\sigma \neq 0$ and $\sigma, a \in R$.

Solution process. From Example 5.5.3, we observe that for $s > a + \frac{\sigma^2}{2}$

$$\mathcal{L}(e^{at+\sigma w(t)})(s) = \frac{1+\sigma\mathcal{L}^w(e^{at+\sigma w(t)})(s)}{s-a-\frac{\sigma^2}{2}} = \frac{1}{s-a-\frac{\sigma^2}{2}} + \frac{\sigma\mathcal{L}^w(e^{at+\sigma w(t)})(s)}{s-a-\frac{\sigma^2}{2}}.$$
$$(5.5.9)$$

Moreover, $\sigma = 0$, and (5.5.8) or (5.5.9) reduces to the deterministic version of the Laplace transform

$$\mathcal{L}(e^{at})(s) = \frac{1}{s-a} \qquad (5.5.10)$$

as a special case.

Example 5.5.5. Find the Laplace transform of $f(t, w(t)) = \cos(at+\sigma w(t))$ in the sense of the Itô–Doob integral for $\sigma \neq 0$ and $\sigma, a \in R$.

Solution process. The Itô–Doob differential of $e^{-st}\sin(at+\sigma w(t))$ (Theorem 1.3.1) is

$$d(e^{-st}\sin(at+\sigma w(t))) = -\left(s+\frac{1}{2}\sigma^2\right)e^{-st}\sin(at+\sigma w(t))dt$$
$$+ ae^{-st}\cos(at+\sigma w(t))dt + \sigma e^{-st}\cos(at+\sigma w(t))dw(t),$$

which implies that

$$\lim_{T\to\infty}(e^{-sT}\sin(aT+\sigma w(T))) - e^{-s0}\sin(a0+\sigma w(0))$$

$$= -\left(s+\frac{1}{2}\sigma^2\right)\lim_{T\to\infty}\int_0^T e^{-st}\sin(at+\sigma w(t))dt$$

$$+ a\lim_{T\to\infty}\int_0^T e^{-st}\cos(at+\sigma w(t))dt + \sigma\lim_{T\to\infty}\int_0^T e^{-st}\cos(at+\sigma w(t))dw(t).$$

Hence,

$$\left(s + \frac{1}{2}\sigma^2\right) \mathcal{L}(\sin(at + \sigma w(t)))(s) - a\mathcal{L}(\cos(at + \sigma w(t)))(s)$$

$$= \sigma \mathcal{L}^w(\cos(at + \sigma w(t)))(s), \qquad (5.5.11)$$

which is equivalent to

$$\mathcal{L}(\sin(at + \sigma w(t)))(s) = \frac{a\mathcal{L}(\cos(at + \sigma w(t)))(s) + \sigma \mathcal{L}^w(\cos(at + \sigma w(t)))(s)}{s + \frac{1}{2}\sigma^2}.$$

$$(5.5.12)$$

Example 5.5.6. Find the Laplace transform of $f(t, w(t)) = \sin(at + \sigma w(t))$ in the sense of the Itô–Doob integral for $\sigma \neq 0$ and $\sigma, a \in R$.

Solution process. By following the earlier solution procedure, we have

$$d(e^{-st}\cos(at + \sigma w(t)))$$

$$= -\left(s + \frac{1}{2}\sigma^2\right)e^{-st}\cos(at + \sigma w(t))dt - ae^{-st}\sin(at + \sigma w(t))dt$$

$$- \sigma e^{-st}\sin(at + \sigma w(t))dw(t).$$

This implies that

$$\sigma \mathcal{L}^w(\sin(at + \sigma w(t)))(s) = 1 - \left(s + \frac{1}{2}\sigma^2\right)\mathcal{L}\cos(at + \sigma w(t)))(s)$$

$$- a\mathcal{L}(\sin(at + \sigma w(t)))(s), \qquad (5.5.13)$$

which is equivalent to

$$\mathcal{L}(\cos(at + \sigma w(t)))(s) = \frac{1 - a\mathcal{L}(\sin(at + \sigma w(t)))(s) - \sigma \mathcal{L}^w(\sin(at + \sigma w(t)))(s)}{s + \frac{1}{2}\sigma^2}.$$

$$(5.5.14)$$

Observation 5.5.1.

(i) From (5.5.11) and (5.5.13), one can easily obtain

$$\mathcal{L}(\cos(at + \sigma w(t)))(s)$$

$$= \frac{\left(s + \frac{1}{2}\sigma^2\right) - a\sigma\mathcal{L}^w(\cos(at + \sigma w(t)))(s) - \sigma\left(s + \frac{1}{2}\sigma^2\right)\mathcal{L}^w(\sin(at + \sigma w(t)))(s)}{\left(s + \frac{1}{2}\sigma^2\right)^2 + a^2},$$

$$(5.5.15)$$

$$\mathcal{L}(\sin(at + \sigma w(t)))(s)$$

$$= \frac{a - a\sigma\mathcal{L}^w(\sin(at + \sigma w(t)))(s) + \sigma\left(s + \frac{1}{2}\sigma^2\right)\mathcal{L}^w(\cos(at + \sigma w(t)))(s)}{\left(s + \frac{1}{2}\sigma^2\right)^2 + a^2}.$$

$$(5.5.16)$$

(ii) For $a = 0$, (5.5.15) and (5.5.16) reduce to

$$\mathcal{L}(\cos(\sigma w(t)))(s) = \frac{1 - \sigma \mathcal{L}^w(\sin(\sigma w(t)))(s)}{s + \frac{1}{2}\sigma^2}, \tag{5.5.17}$$

$$\mathcal{L}(\sin(\sigma w(t)))(s) = \frac{\sigma \mathcal{L}^w(\cos(\sigma w(t)))(s)}{s + \frac{1}{2}\sigma^2}, \tag{5.5.18}$$

respectively.

(iii) Furthermore, for $\sigma = 0$, (5.5.15) and (5.5.16) reduce to the deterministic version of the Laplace transform, as a special case, i.e.

$$\mathcal{L}(\cos(at))(s) = \frac{s}{s^2 + a^2}, \tag{5.5.19}$$

$$\mathcal{L}(\sin(at))(s) = \frac{a}{s^2 + a^2}. \tag{5.5.20}$$

For easy reference, we will present a known result [80] concerning the Laplace transform of a derivative of the deterministic function.

Theorem 5.5.2 (Laplace transform of a derivative). *Let us suppose that f has $n - 1$ continuous derivatives on $[0, \infty)$, and for each i, $0 \le i \le n - 1$, let $f^{(i)} \in L[f]$. Further assume that $f^{(n)}$ is piecewise continuous in every subinterval $0 \le t < b$. Then, $f^{(n)} \in L[f]$ and*

$$\mathcal{L}(f^{(n)})(s) = s^n \mathcal{L}(f)(s) - s^{n-1}f(0) - s^{n-2}f'(0) - \cdots - f^{(n-1)}(0). \tag{5.5.21}$$

Now, we present a result concerning the Laplace transform of an indefinite integral of Itô–Doob type.

Theorem 5.5.3. (Laplace transform of an indefinite integral). *Let us assume that $f, \frac{\partial f}{\partial w} \in L[f]$. In addition, let I be Itô–Doob indefinite integrals of f. Then, $I \in L[f]$,*

$$\mathcal{L}(I)(s) = \mathcal{L}\left(\int_0^t f(u, w(u))dw(u)\right)(s) = \frac{\mathcal{L}^w(f)(s)}{s}. \tag{5.5.22}$$

Proof. From the smoothness (for example, piecewise continuity) of the function f and the properties of the indefinite integral of the Itô–Doob indefinite integral $I \in L[f]$. Moreover, to prove (5.5.22), we compute the Itô–Doob differential of $e^{-st}I(t)$ as

$$d(e^{-st}I(t)) = -se^{-st}I(t)dt + e^{-st}f(t, w(t))dw(t) \quad \text{(by Theorem 1.3.1).} \tag{5.5.23}$$

This implies

$$\lim_{T \to \infty} [e^{-st}I(t)|_0^T] = -s \lim_{T \to \infty}\left[\int_0^T e^{-st}I(t)dt\right] + \lim_{T \to \infty}\left[\int_0^T e^{-st}f(t, w(t))dw(t)\right].$$

Hence,

$$\mathcal{L}(I)(s) = \frac{1}{s} \int_0^\infty e^{-st} f(t, w(t)) dw(t) = \frac{\mathcal{L}^w(f)(s)}{s}.$$

This completes the proof of the theorem. □

Example 5.5.7. Find the Laplace transform of $(bw(t) + ct)e^{(a-\frac{\sigma^2}{a})t+\sigma w(t)}$ in the sense of the Itô–Doob integral for σ, a, b and $c \in R$.

Solution process. Again, we apply the Itô–Doob differential formula (Theorem 1.3.1) to $(bw(t) + ct)e^{(a-\frac{\sigma^2}{2})t+\sigma w(t)}$, and we obtain

$$d((bw(t) + ct)e^{(a-\frac{\sigma^2}{2})t+\sigma w(t)})$$

$$= \left(a - \frac{\sigma^2}{2}\right)(bw(t) + ct)e^{(a-\frac{\sigma^2}{2})t+\sigma w(t)} dt + ce^{(a-\frac{\sigma^2}{2})t+\sigma w(t)} dt$$

$$+ [\sigma(bw(t) + ct) + b]e^{(a-\frac{\sigma^2}{2})t+\sigma w(t)} dw(t)$$

$$+ \frac{1}{2}[\sigma[\sigma(bw(t) + ct) + b] + b\sigma]e^{(a-\frac{\sigma^2}{2})t+\sigma w(t)} dt,$$

which implies that

$$(bw(t) + ct)e^{(a-s-\frac{\sigma^2}{2})t+\sigma w(t)}$$

$$= e^{-st} \int_0^t [a(bw(u) + cu) + c + b\sigma]e^{(a-\frac{\sigma^2}{2})u+\sigma w(u)} du$$

$$+ e^{-st} \int_0^t [\sigma(bw(u) + cu) + b]e^{(a-\frac{\sigma^2}{2})u+\sigma w(u)} dw(u).$$

We note that all the terms of the above expression are continuous functions and belong to the class $L[f]$, defined in Definition 5.5.2. By applying the Laplace transform and Theorem 5.5.3, we get

$$\mathcal{L}((bw(t) + ct)e^{(a-\frac{\sigma^2}{2})t+\sigma w(t)})(s)$$

$$= \frac{a\mathcal{L}((bw(t) + ct)e^{(a-\frac{\sigma^2}{2})t+\sigma w(t)})(s)}{s} + \frac{(c + b\sigma)\mathcal{L}(e^{(a-\frac{\sigma^2}{2})t+\sigma w(t)})(s)}{s}$$

$$+ \frac{\sigma\mathcal{L}^w((bw(t) + ct)e^{(a-\frac{\sigma^2}{2})t+\sigma w(t)})(s)}{s} + \frac{b\mathcal{L}^w(e^{(a-\frac{\sigma^2}{2})t+\sigma w(t)})(s)}{s}.$$

By simplifying the above expression, we have

$$(s - a)\mathcal{L}((bw(t) + ct)e^{(a-\frac{\sigma^2}{2})t+\sigma w(t)})(s) - (c + b\sigma)\mathcal{L}(e^{(a-\frac{\sigma^2}{2})t+\sigma w(t)})(s)$$

$$= \sigma\mathcal{L}^w((bw(t) + ct)e^{(a-\frac{\sigma^2}{2})t+\sigma w(t)})(s) + b\mathcal{L}^w(e^{(a-\frac{\sigma^2}{2})t+\sigma w(t)})(s). \quad (5.5.24)$$

From this, $\sigma \neq 0$ and Example 5.5.3,

$$\sigma^2 \mathcal{L}^w((bw(t) + ct)e^{(a-\frac{\sigma^2}{2})t+\sigma w(t)})(s)$$

$$= (s-a)\sigma\mathcal{L}((bw(t) + ct)e^{(a-\frac{\sigma^2}{2})t+\sigma w(t)})(s) - \sigma(c+b\sigma)\mathcal{L}(e^{(a-\frac{\sigma^2}{2})t+\sigma w(t)})(s)$$

$$\quad - (s-a)b\mathcal{L}(e^{(a-\frac{\sigma^2}{2})t+\sigma w(t)})(s) + b \text{ [from (5.5.8)]}$$

$$= (s-a)[\sigma\mathcal{L}((bw(t) + ct)e^{(a-\frac{\sigma^2}{2})t+\sigma w(t)})(s) - b\mathcal{L}(e^{(a-\frac{\sigma^2}{2})t+\sigma w(t)})(s)] + b$$

$$\quad - \sigma(c+b\sigma)\mathcal{L}(e^{(a-\frac{\sigma^2}{2})t+\sigma w(t)})(s).$$

This is equivalent to

$$\sigma\mathcal{L}((bw(t) + ct)e^{(a-\frac{\sigma^2}{2})t+\sigma w(t)})(s) - b\mathcal{L}(e^{(a-\frac{\sigma^2}{2})t+\sigma w(t)})(s)$$

$$= \frac{\sigma^2 \mathcal{L}^w((bw(t) + ct)e^{(a-\frac{\sigma^2}{2})t+\sigma w(t)})(s)}{s-a} - \frac{b}{s-a}$$

$$\quad + \frac{\sigma(c+b\sigma)\mathcal{L}(e^{(a-\frac{\sigma^2}{2})t+\sigma w(t)})(s)}{s-a}. \tag{5.5.25}$$

Observation 5.5.2.

(i) For $\sigma = 0 = c$, $b = 1$ and a, (5.5.24) reduces to

$$(s-a)\mathcal{L}(w(t)e^{at})(s) = \mathcal{L}^w(e^{at})(s) \quad \text{if and only if}$$

$$\mathcal{L}(w(t)e^{at})(s) = \frac{a\mathcal{L}(w(t)e^{at})(s)}{s} + \frac{\mathcal{L}^w(e^{at})(s)}{s}. \tag{5.5.26}$$

(ii) For $b = 0$, $a, \sigma \neq 0$, and $c = 1$, (5.5.24) reduces to

$$\mathcal{L}(te^{(a-\frac{\sigma^2}{2})t+\sigma w(t)})(s) = \frac{\sigma\mathcal{L}^w(te^{(a-\frac{\sigma^2}{2})t+\sigma w(t)})(s) + \mathcal{L}(e^{(a-\frac{\sigma^2}{2})t+\sigma w(t)})(s)}{s-a} \tag{5.5.27}$$

if and only if

$$\mathcal{L}(te^{(a-\frac{\sigma^2}{2})t+\sigma w(t)})(s) = \frac{\mathcal{L}((at+1)e^{(a-\frac{\sigma^2}{2})t+\sigma w(t)})(s)}{s}$$

$$\quad + \frac{\sigma\mathcal{L}^w(te^{(a-\frac{\sigma^2}{2})t+\sigma w(t)})(s)}{s}. \tag{5.5.28}$$

(iii) For σ, a, b, $c \in R$ from Example 5.5.4, (5.5.29) reduces to

$$\sigma\mathcal{L}((bw(t) + ct)e^{at+\sigma w(t)})(s) - b\mathcal{L}(e^{at+\sigma w(t)})(s)$$

$$= \frac{\sigma^2\mathcal{L}^w((bw(t) + ct)e^{at+\sigma w(t)})(s)}{s - a - \frac{\sigma^2}{2}} - \frac{b}{s - a - \frac{\sigma^2}{2}}$$

$$+ \frac{\sigma(c + b\sigma)\mathcal{L}(e^{\sigma w(t)}(s))}{s - a - \frac{\sigma^2}{2}}. \tag{5.5.29}$$

(iv) For $b = 0 = a$, $\sigma \neq 0$ and $c = 1$, (5.5.29) reduces to

$$\mathcal{L}(te^{\sigma w(t)})(s) = \frac{\sigma\mathcal{L}^w(te^{\sigma w(t)})(s)}{s - \frac{\sigma^2}{2}} + \frac{\mathcal{L}(e^{\sigma w(t)})(s)}{s - \frac{\sigma^2}{2}} \text{ if and only if} \tag{5.5.30}$$

$$\mathcal{L}(te^{\sigma w(t)})(s) = \frac{\sigma\mathcal{L}^w(te^{\sigma w(t)})(s)}{s} + \frac{\mathcal{L}\left(\left(1 + \frac{\sigma^2}{2}t\right)e^{\sigma w(t)}\right)(s)}{s}. \tag{5.5.31}$$

Definition 5.5.3. A real-valued function $\kappa_{R_+} \equiv \kappa$ is called a characteristic function with respect to a set $[0, \infty)$, if

$$\kappa_{R_+}(t) \equiv \kappa(t) = \begin{cases} 0, & \text{if } t < 0, \\ 1, & \text{if } t \geq 0, \end{cases} \quad \kappa(t - c) = \begin{cases} 0, & \text{if } t < c, \\ 1, & \text{if } t \geq c, \end{cases} \tag{5.5.32}$$

for any $c \in R$.

Theorem 5.5.4. *If $\mathcal{L}^w(g)(s)$ and $0 \leq c, g(t)$ is defined on $(-c, \infty)$. Then,*

$$\mathcal{L}^w(g(t - c + c\kappa(t - c), w(t - c + c\kappa(t - c)))\kappa(t - c))(s) = e^{-sc}\mathcal{L}^w(g)(s). \tag{5.5.33}$$

Proof. From Definition 5.5.1, the conclusion of the theorem remains valid. □

Example 5.5.8. Given: $g(t, w(t)) = \begin{cases} 3, & \text{if } 0 < t < 1, \\ w(t), & \text{if } t \geq 1. \end{cases}$

Find $\mathcal{L}(g)(s)$.

Solution process. First, we rewrite $g(t, w(t))$ in terms of the unit step function,

$$g(t, w(t)) = 3 - 3\kappa(t - 1) + (w(t - 1 + \kappa(t - 1))\kappa(t - 1).$$

This is in the form of $g(t - 1, w(t - 1))$. Hence, we can apply Theorem 5.5.4, and we have

$$\mathcal{L}(g(t, w(t)))(s) = \frac{3}{s} - \frac{3}{s}e^{-s} + e^{-s}\mathcal{L}(w(t)) = \frac{3}{s} - \frac{3}{s}e^{-s} + \frac{e^{-s}\mathcal{L}^w(1)}{s}.$$

Now, we introduce the concept of the convolution integral of two functions. Moreover, we obtain an expression for the Laplace transform for the convolution integral of two functions.

Definition 5.5.4. Let f and g be piecewise-continuous functions defined on $t \geq 0$. The Cauchy–Riemann and Itô–Doob convolution integrals of f and g are defined by

$$(f * g)(t) = \int_0^t g(t - u)f(u)dw(u). \tag{5.5.34}$$

Theorem 5.5.5 (Laplace transform of the convolution integral). *Let us assume that $f, g \in L[f]$. Then, $(f * g) \in L[f]$,*

$$\mathfrak{L}(f * g)(s) = \mathfrak{L}\left(\int_0^t g(t - u)f(u, w(u))du\right)(s) = \mathfrak{L}(g)(s)\mathfrak{L}^w(f)(s). \tag{5.5.35}$$

Proof. One can easily verify the correctness of the conclusion. The details are left to the reader. From Definition 5.5.2, one can conclude that $(f * g) \in L[f]$, and from Definitions 5.5.1 and 5.5.4, one has

$$\mathfrak{L}(f * g)(s) = \int_0^\infty e^{-s}\left[\int_0^t g(t - u)f(u, w(u))dw(u)\right]dt$$

$$= \int_0^\infty \int_0^t e^{-st}g(t - u)f(u, w(u))dw(u)dt \text{ (by Fubini's theorem)}$$

$$= \int_0^\infty \int_u^\infty e^{-st}g(t - u)f(u, w(u))dt\, dw(u)$$

$$= \int_0^\infty \int_0^\infty e^{-st}g(t - u)\kappa(t - u)f(u, w(u))dt\, dw(u)$$

$$[\text{by the definition of } \kappa(t - u)]$$

$$= \int_0^\infty e^{-su}\mathfrak{L}(g)(s)f(u, w(u))dw(u)$$

$$= \mathfrak{L}(g)(s)\int_0^\infty e^{-su}f(u, w(u))dw(u)$$

$$= \mathfrak{L}(g)(s)\mathfrak{L}^w(f)(s).$$

This completes the proof of the theorem. □

Example 5.5.9. Solve the given equation $g(t) = w(t) + \int_0^t \sin(t - u)g(u)du$.

Solution process. Let $\mathfrak{L}(g)(s)$. Then, $\mathfrak{L}(\sin t)(s) = \frac{1}{1+s^2}$. Now, applying Theorem 5.5.5, we have

$$\mathfrak{L}(g)(s) = \mathfrak{L}(f * g)(s) = \mathfrak{L}(w(t)) + \mathfrak{L}(g)(s)\mathfrak{L}(f)(s) = \frac{\mathfrak{L}^w(1)}{s} + \frac{\mathfrak{L}(g)(s)}{1 + s^2}.$$

By solving for $\mathcal{L}(g)(s)$, we obtain

$$\mathcal{L}(g)(s) = \frac{(1+s^2)\mathcal{L}^w(1)}{s^3} = \frac{1+s^2}{s^2}\frac{\mathcal{L}^w(1)}{s} = \left(1 + \frac{1}{s^2}\right)\frac{\mathcal{L}^w(1)}{s}.$$

The inverse Laplace transform of this is given by

$$g(t) = \mathcal{L}^{-1}\left[\left(1 + \frac{1}{s^2}\right)\frac{\mathcal{L}^w(1)}{s}\right]$$

$$= \int_0^t (1 + (t-u))\left(\int_0^u dw(v)\right) du$$

$$= \int_0^t (1 + (t-u))w(u)du.$$

Table 5.1. A short table of Laplace transforms.

$f(t)$	$\mathcal{L}(f(t))(s)$	$f(t, w(t))$	$\mathcal{L}^w(f(t))(s)$
c	$\frac{c}{s}, s > 0$	c	$\mathcal{L}(cw)(s) = \frac{c\mathcal{L}^w(1)(s)}{s}$
e^{at}	$\frac{1}{s-a}, s > a$	$e^{at+\sigma w(t)}$	$\mathcal{L}(\sigma e^{at+\sigma w(t)})(s) = \frac{1+\sigma\mathcal{L}^w(e^{at+\sigma w(t)})(s)}{s-a-\frac{1}{2}\sigma^2}$
$\sin at$	$\frac{a}{s^2+a^2}$	$\sin \sigma w(t)$	$\mathcal{L}(\sin(\sigma w(t)))(s) = \frac{\sigma\mathcal{L}^w(\cos(\sigma w(t)))(s)}{s+\frac{1}{2}\sigma^2}$
$\cos at$	$\frac{s}{s^2+a^2}$	$\cos \sigma w(t)$	$\mathcal{L}(\cos \sigma w(t))(s) = \frac{1-\sigma\mathcal{L}^w(\sin \sigma w(t))(s)}{s+\frac{1}{2}\sigma^2}$
t^n	$\frac{n!}{s^{n+1}}, s > 0$	$w^n(t)$	$\mathcal{L}(w^n)(s) = \frac{n\mathcal{L}^w(w^{n-1})(s)}{s} + \frac{n(n-1)\mathcal{L}(w^{n-2})(s)}{2s}$
$t^n e^{at}$	$\frac{n!}{(s-a)^{n+1}}, s > 0$	$w(t)e^{at}$	$\mathcal{L}(w(t)e^{at})(s) = \frac{a\mathcal{L}(w(t)e^{at})(s)}{s} + \frac{\mathcal{L}^w(e^{at})(s)}{s}$
		$te^{\sigma w(t)}$	$\mathcal{L}(te^{\sigma w(t)})(s) = \frac{\sigma\mathcal{L}(te^{\sigma w(t)})(s)}{s-\frac{\sigma^2}{2}} + \frac{\mathcal{L}(e^{\sigma w(t)})(s)}{s-\frac{\sigma^2}{2}}$
$t\sin at$	$\frac{2as}{(s^2+a^2)^2}$		
$t\cos at$	$\frac{s^2-a^2}{(s^2+a^2)^2}$		
$e^{-at}\sin bt$	$\frac{b}{(s+a)^2+b^2}$		
$e^{-at}\cos bt$	$\frac{s+a}{(s+a)^2+b^2}$		

Exercises

1. Find the Laplace transform of the following functions:
   (a) $f(t, w(t)) = \cos^2 \sigma w(t) - \sin^2 \sigma w(t)$;
   (b) $f(t, w(t)) = \sin \sigma w(t) \cos^2 \sigma w(t)$;
   (c) $f(t, w(t)) = \cos h(bt + \sigma w(t))$;
   (d) $f(t, w(t)) = \sin h(bt + \sigma w(t))$;
   (e) $f(t, w(t)) = \sin(at + \sigma w(t)) \cos(at + \sigma w(t))$;
   (f) $f(t, w(t)) = (at + \sigma w(t)) \cos(bt + \sigma w(t))$;

(g) $f(t, w(t)) = (at + \sigma w(t)) \sin(bt + \sigma w(t))$;

(h) $f(t, w(t)) = \sin(bt + \sigma w(t)) e^{\left(a - \frac{\sigma^2}{2}\right)t + \sigma w(t)}$;

(i) $f(t, w(t)) = \cos(bt + \sigma w(t)) e^{\left(a - \frac{\sigma^2}{2}\right)t + \sigma w(t)}$.

2. Prove or Disprove.

(a) $\mathcal{L}(\cos(at + \sigma w(t)))(s)$

$$= \frac{\left(s + \frac{1}{2}\sigma^2\right) - a\sigma\mathcal{L}^w(\cos(at + \sigma w(t)))(s) - \sigma\left(s + \frac{1}{2}\sigma^2\right)\mathcal{L}^w(\sin(at + \sigma w(t)))(s)}{\left(s + \frac{1}{2}\sigma^2\right)^2 + a^2};$$

(b) $\mathcal{L}(\sin(at + \sigma w(t)))(s) = \dfrac{a - a\sigma\mathcal{L}^w(\sin(at + \sigma w(t)))(s) + \sigma\left(s + \frac{1}{2}\sigma^2\right)\mathcal{L}^w(\cos(at + \sigma w(t)))(s)}{\left(s + \frac{1}{2}\sigma^2\right)^2 + a^2}$;

(c) $\mathcal{L}(w(t)e^{at})(s) = \dfrac{\mathcal{L}^w(e^{at})}{s - a}$ if and only if $\dfrac{\mathcal{L}^w(e^{at})}{s} = \dfrac{(s - a)\mathcal{L}(w(t)e^{at})(s)}{s}$;

(d) $\dfrac{\sigma\mathcal{L}^w(te^{\sigma w(t)})(s)}{s} = \mathcal{L}(te^{\sigma w(t)})(s) - \dfrac{\mathcal{L}\left(\left(1 + \frac{\sigma^2}{2}t\right)e^{\sigma w(t)}\right)(s)}{s}$.

3. Find $\mathcal{L}(f(t, w(t)))(s)$ if $f(t, w(t))$ is given by:

(a) $f(t, w(t)) = \begin{cases} 3, & \text{if } 0 < t < 2, \\ w^2(t), & \text{if } t \geq 2, \end{cases}$

(b) $f(t, w(t)) = \begin{cases} 0, & \text{if } 0 < t < 1, \\ \sin \sigma w(t), & \text{if } 1 \leq t < 2, \\ t, & \text{if } 2 \leq t, \end{cases}$

(c) $f(t, w(t)) = \begin{cases} 3, & \text{if } 0 < t < 2, \\ 1, & \text{if } t \geq 2. \end{cases}$

4. Find the inverse Laplace transform of the following:

(a) $F(s) = \frac{2}{s^2 + 4s + 8}$;

(b) $G(s) = \frac{s^3 + 1}{(s^2 + 1)s^5}$;

(c) $H(s) = \frac{s^2 + 1}{(s^2 + a^2)(s^3 + b^3)}$;

(d) $G(s) = \frac{s + 2}{s^2 + 2s + 8}$;

(e) $M(s) = \frac{s}{s^2 + 4} - \frac{1}{s^2} - \left(\frac{s}{s^2 + 4} - \frac{1}{s^2}\right)e^{-s}$.

5. Solve the given integral equation:

(a) $g(t) = e^{\sigma w(t)} + \int_0^t \cos(t - u)g(u)du$;

(b) $x(t) = w(t) + \int_0^t \exp[-2(t - u)]x(u)du$;

(c) $y(t) = t + \int_0^t \exp[(t - u)]y(u)dw(u)$;

(d) $h(t) = t^2 + \int_0^t \sin(t - u)h(u)dw(u)$.

5.6 Applications of the Laplace Transform

The Laplace transform will be used to solve the IVPs. It transforms a linear differential equation with constant coefficients into an algebraic equation. The techniques for solving the algebraic equations are easier than the methods of solving the IVPs.

Example 5.6.1. Use the Laplace transform to solve the IVP

$$dy' + y\,dt = \sigma dw(t), \quad y(0) = 0, \quad y'(0) = 1, \quad \text{for } \sigma \neq 0.$$

Solution procedure. We note that the Itô–Doob differential equation is equivalent to the integral equation

$$y'(t) = y'(0) - \int_0^t y(u)du + \sigma \int_0^t dw(u).$$

Now, we apply the Laplace transform to both sides, and obtain

$$\mathcal{L}(y'(t)) = \mathcal{L}\left(e^{-st}\left[y'(0) - \int_0^t y(u)du + \sigma \int_0^t dw(u) \right] \right) \qquad \text{(by Definition 5.5.1)}$$

$$= \mathcal{L}(e^{-st}y'(0)) - \mathcal{L}\left(\int_0^t y(u)du \right) + \sigma\mathcal{L}\left(\int_0^t dw(u) \right) \quad \text{(by Theorem 5.5.1)}$$

$$= \frac{1}{s} - \frac{\mathcal{L}(y(t))}{s} + \frac{\sigma\mathcal{L}^w(1)}{s} \quad \text{(by Theorem 5.5.3 and the initial data).}$$

Moreover, from Theorem 5.5.2, we have $\mathcal{L}(y'(t)) = s\mathcal{L}(y(t)) - y(0)$. From this and using the initial conditions, the above expression reduces to

$$s\mathcal{L}(y(t)) = \frac{1}{s} - \frac{\mathcal{L}(y(t))}{s} + \frac{\sigma\mathcal{L}^w(1)}{s}.$$

Now, we solve for $\mathcal{L}(y(t))$

$$\mathcal{L}(y(t)) = \frac{s}{1+s^2}\left(\frac{1}{s} + \frac{\sigma\mathcal{L}^w(1)}{s} \right) = \frac{1}{1+s^2} + \frac{s}{1+s^2}\frac{\sigma\mathcal{L}^w(1)}{s}.$$

By applying the inverse Laplace transform to both sides, we get

$$y(t) = \mathcal{L}^{-1}\left(\frac{1}{1+s^2} + \frac{s}{1+s^2}\frac{\sigma\mathcal{L}^w(1)}{s} \right)$$

$$= \mathcal{L}^{-1}\left(\frac{1}{1+s^2} \right) + \mathcal{L}^{-1}\left(\frac{s}{1+s^2}\frac{\sigma\mathcal{L}^w(1)}{s} \right) \qquad (\mathcal{L}^{-1} \text{ is a linear operator})$$

$$= \sin t + \sigma \int_0^t \cos(t-u)w(u)du \quad \text{(from Table 5.1 and Theorem 5.5.5).}$$

Thus, the solution to the IVP is given by

$$y(t) = \sin t + \sigma \int_0^t \cos(t-u)w(u)du.$$

This completes the solution process of the given problem.

Example 5.6.2 (Langevin equation [125, 139]). Use the Laplace transform to solve the IVP

$$dy' + \beta y' \, dt = \sigma \, dw(t), \quad y(0) = y_0, \quad y'(0) = v_0, \quad \text{for } \sigma \neq 0 \text{ and } \beta > 0.$$

Solution process. We note that the Itô–Doob differential equation is equivalent to the integral equation

$$y'(t) = y'(0) - \int_0^t y'(u)du + \sigma \int_0^t dw(u).$$

Now, we apply the Laplace transform to both sides, and obtain

$$\mathcal{L}(y'(t)) = \mathcal{L}\left(e^{-st}\left[y'(0) - \beta \int_0^t y'(u)du + \sigma \int_0^t dw(u)\right]\right) \qquad \text{(by Definition 5.5.1)}$$

$$= \mathcal{L}(e^{-st}y'(0)) - \beta\mathcal{L}\left(\int_0^t y'(u)du\right) + \sigma\mathcal{L}\left(\int_0^t dw(u)\right) \quad \text{(by Theorem 5.5.1)}$$

$$= \frac{v_0}{s} - \frac{\beta\mathcal{L}(y'(t))}{s} + \frac{\sigma\mathcal{L}^w(1)}{s} \quad \text{(by Theorem 5.5.3 and the initial data).}$$

Moreover, from Theorem 5.5.2, we have $\mathcal{L}(y'(t)) = s\mathcal{L}(y(t)) - y(0)$. From this and using the initial conditions, the above expression reduces to

$$s\mathcal{L}(y(t)) - y_0 = \frac{v_0}{s} - \frac{s\beta\mathcal{L}(y(t)) - \beta y_0}{s} + \frac{\sigma\mathcal{L}^w(1)}{s}.$$

Now, we solve for $\mathcal{L}(y(t))$

$$\mathcal{L}(y(t)) = \frac{v_0}{\beta s + s^2} + \frac{(\beta + s)y_0}{\beta s + s^2} + \frac{\sigma\mathcal{L}^w(1)}{\beta s + s^2}$$

$$= \frac{v_0}{s(\beta + s)} + \frac{y_0}{s} + \frac{\sigma\mathcal{L}^w(1)}{s(\beta + s)} \quad \text{(by simplification)}$$

$$= \frac{v_0}{\beta}\left[\frac{1}{s} - \frac{1}{\beta + s}\right] + \frac{y_0}{s} + \frac{1}{\beta + s}\frac{\sigma\mathcal{L}^w(1)}{s}$$

(by the method of partial fractions).

By applying the inverse Laplace transform to both sides, we get

$$y(t) = \mathcal{L}^{-1}\left(\frac{v_0}{\beta}\left[\frac{1}{s} - \frac{1}{\beta + s}\right] + \frac{y_0}{s} + \frac{1}{\beta + s}\frac{\sigma\mathcal{L}^w(1)}{s}\right)$$

$$= y_0 + \frac{v_0}{\beta}(1 - e^{-\beta t}) + \sigma\int_0^t e^{-\beta(t-u)}w(u)du$$

(from Table 5.1 and Theorem 5.5.5).

Thus, the solution to the IVP is given by

$$y(t) = y_0 + \frac{v_0}{\beta}(1 - e^{-\beta t}) + \sigma \int_0^t e^{-\beta(t-u)} w(u) du.$$

From this solution, one can determine the mean, covariance, and variance of the solution process. In fact, the mean of $y(t)$ is

$$E[y(t)] = E[y_0] + \frac{E[v_0]}{\beta}(1 - e^{-\beta t}).$$

The covariance and the variance of the solution process can be determined accordingly. The details are left to the reader. This completes the solution process of the given problem.

Example 5.6.3 (Chandrasekhar's equation [21]). Use the Laplace transform to solve the IVP

$$dy' + (\beta y' + \nu^2 y)dt = \sigma dw(t), \quad y(0) = y_0, \quad y'(0) = v_0, \quad \text{for } \sigma \neq 0 \text{ and } \beta > 0.$$

Solution process. We note that the Itô–Doob differential equation is equivalent to the integral equation

$$y'(t) = y'(0) - \beta \int_0^t y'(u) du - \nu^2 \int_0^t y(u) du + \sigma \int_0^t dw(u).$$

Now, we apply the Laplace transform to both sides

$$\mathcal{L}(y'(t))$$

$$= \mathcal{L}\left(e^{-st}\left[y'(0) - \beta \int_0^t y'(u) du - \nu^2 \int_0^t y(u) du\right.\right.$$

$$\left.\left. + \sigma \int_0^t dw(u)\right]\right) \qquad \text{(by Definition 5.5.1)}$$

$$= \mathcal{L}(e^{-st} y'(0)) - \beta \mathcal{L}\left(\int_0^t y'(u) du\right) - \nu^2 \mathcal{L}\left(\int_0^t y(u) du\right)$$

$$+ \sigma \mathcal{L}\left(\int_0^t dw(u)\right) \qquad \text{(by Theorem 5.5.1)}$$

$$= \frac{v_0}{s} - \frac{\beta \mathcal{L}(y'(t))}{s} - \frac{\nu^2 \mathcal{L}(y(t))}{s} + \frac{\sigma \mathcal{L}^w(1)}{s}$$

$$\text{(by Theorem 5.5.3 and the initial data).}$$

Moreover, from Theorem 5.5.2, we have $\mathcal{L}(y'(t)) = s\mathcal{L}(y(t)) - y(0)$, and imitating the argument used in Example 5.6.2, we obtain

$$s\mathcal{L}(y(t)) - y_0 = \frac{v_0 - \beta(s\mathcal{L}(y(t)) - y(0)) - \nu^2 \mathcal{L}(y(t)) + \sigma \mathcal{L}^w(1)}{s}$$

$$= \frac{v_0 - (\beta s + \nu^2)\mathcal{L}(y(t)) + \beta y(0) + \sigma \mathcal{L}^2(1)}{s}.$$

From an algebraic manipulation, we have

$$\mathcal{L}(y(t)) = \frac{v_0 + \beta y(0) + \sigma \mathcal{L}^w(1)}{\nu^2 + \beta s + s^2} + \frac{sy(0)}{\nu^2 + \beta s + s^2}$$

$$= \frac{v_0}{\left(s + \frac{1}{2}\beta\right)^2 + \frac{4\nu^2 - \beta^2}{4}} + \frac{(\beta + s)y_0}{\left(s + \frac{1}{2}\beta\right)^2 + \frac{4\nu^2 - \beta^2}{4}}$$

$$+ \frac{s}{\left(s + \frac{1}{2}\beta\right)^2 + \frac{4\nu^2 - \beta^2}{4}} \frac{\sigma \mathcal{L}^2(1)}{s}.$$

By applying the inverse Laplace transform to both sides, we get

$$\dot{y}(t) = \mathcal{L}^{-1}\left(\frac{v_0}{\left(s + \frac{1}{2}\beta\right)^2 + \frac{4\nu^2 - \beta^2}{4}} + \frac{(\beta + s)y_0}{\left(s + \frac{1}{2}\beta\right)^2 + \frac{4\nu^2 - \beta^2}{4}}\right.$$

$$\left. + \frac{s}{\left(s + \frac{1}{2}\beta\right)^2 + \frac{4\nu^2 - \beta^2}{4}} \frac{\sigma \mathcal{L}^2(1)}{s}\right),$$

and hence

$$y(t) = \begin{cases} e^{-\frac{\beta}{2}t}\left[\frac{2v_0 + \beta y(0)}{b} + \sin\frac{1}{2}bt + y_0 \cos\frac{1}{2}bt\right] \\ \quad + \sigma \int_0^t \left[\cos\frac{1}{2}b(t-u) - \frac{\beta}{b}\sin\frac{1}{2}b(t-u)\right]w(u)du, \\ \qquad\qquad \text{if } b^2 = (4\nu^2 - \beta^2) > 0; \\[2mm] 2v_0 t e^{-\frac{1}{2}\beta t} + y_0 e^{-\frac{1}{2}\beta t} \\ \quad + \sigma \int_0^t \left[e^{-\frac{1}{2}\beta(t-u)} - \beta(t-u)e^{-\frac{1}{2}\beta(u-u)}\right]w(u)du, \quad \text{if } (4\nu^2 - \beta^2) = 0; \\[2mm] \frac{v_0\left[e^{-\frac{1}{2}(\beta-b)t} - e^{-\frac{1}{2}(\beta+b)t}\right]}{4b} + \frac{y_0\left[e^{-\frac{1}{2}(\beta-b)t} - e^{-\frac{1}{2}(\beta+b)t}\right]}{2} \\ \quad + \sigma \int_0^t \left[\frac{\left[e^{-\frac{1}{2}(\beta-b)(t-u)} + e^{-\frac{1}{2}(\beta+b)(t-u)}\right]}{2}\right. \\ \qquad \left. - \frac{\left[e^{-\frac{1}{2}(\beta-b)(t-u)} - e^{-\frac{1}{2}(\beta+b)(t-u)}\right]}{4b}\right]w(u)\,du, \quad \text{if } (4\nu^2 - \beta^2) < 0. \end{cases}$$

Depending on the nature of magnitudes of ν^2 and β^2 and the sign of $4\nu^2 - \beta^2$, the representation of the solution process of Chandrasekhar's equation is derived. From this solution process, one determine the mean, covariance, and variance of the solution process. The details are left to the reader. This completes the solution process of the given problem.

Exercises

1. Use the Laplace transform to solve the following IVPs:

   (a) $dx^{(1)} + \alpha^2 x\, dt = 2\sin bw(t)dw(t)$, $x(0) = x_0$, $x^{(1)}(0) = v_0$;

   (b) $dx^{(2)} + [3\alpha x^{(2)} + 3\alpha x^{(2)} + a^3 x]dt = \sigma\, dw(t)$, $x(0) = 1$, $x^{(1)}(0) = 0$, $x^{(2)}(0) = 1$;

   (c) $dx^{(1)} + [6x^{(1)} + 9x]dt - w(t)te^{-3t}dw(t) = 0$, $x(0) = 1$, $x^{(1)}(0) = 0$;

   (d) $dx^{(1)} + [6x^{(1)} + 5x]dt - te^{\sigma w(t)}dw(t) = 0$, $x(0) = 1$, $x^{(1)}(0) = 0$.

2. Use the Laplace transform to solve the following IVPs:

   (a) $dx = Ax\, dw(t)$, where $A = \begin{bmatrix} 0 & -\beta \\ \beta & 0 \end{bmatrix}$, $x(0) = \begin{bmatrix} 1 & 1 \end{bmatrix}$;

   (b) $dx = Ax\, dw(t)$, where $A = \begin{bmatrix} -3 & 2 \\ 1 & -2 \end{bmatrix}$, $x(0) = \begin{bmatrix} 1 & 1 \end{bmatrix}$;

   (c) $dx = Ax\, dw(t)$, where $A = \begin{bmatrix} -1 & 0 \\ \beta & -2 \end{bmatrix}$, $x(0) = \begin{bmatrix} 2 & 1 \end{bmatrix}$.

5.7 Notes and Comments

The development of the material in this chapter is original. It is adapted from Refs. 70 and 145. The development of the stochastic modeling procedure involving higher-order rates of state of a system is demonstrated by a few applied examples in Section 5.1. The examples, exercises, and illustrations are based on the second author's academic experience. In Section 5.2, examples are given to justify that the linear higher-order Itô–Doob stochastic differential equations with constant coefficients possess closed-form solutions. Section 5.3 deals with the formulation of an Itô–Doob stochastic companion system of differential equations corresponding to linear higher-order nonhomogeneous stochastic differential equations. For further details, see Ref. 145. In addition, again, using the method of variation of parameters, the closed-form solution of the linear nonhomogeneous stochastic differential is outlined in Section 5.4. Section 5.5 introduces the classical Laplace transform into a certain class of Wiener-process-varying functions. It is adapted from Ref. 70. Finally, in Section 5.6, the stochastic Laplace transform technique is applied to find the closed-form solution processes of certain linear higher-order stochastic differential equations, particularly, Langevin [125] and Chandrasekhar [21] types of stochastic differential equations as well as linear stochastic integral equations. The development of the methods for finding solutions of the higher-order stochastic differential is still in its infancy. For additional recent developments, see Refs. 144 and 145.

Chapter 6

Topics in Differential Equations

Introduction

A brief review of Itô–Doob-type stochastic methods, analysis and new trends in the modeling of nonlinear and nonstationary dynamic processes are presented via concepts, basic algorithms, and examples. In a classical modeling approach, the evolution of states of dynamic systems can be considered to be a dynamic flow evolving on timescales. The goal is to find some information (qualitative/quantitative) about dynamic flows to use in planning and/or improving the quality of the performance of an underlying dynamic process. Information about the process can be gathered by either knowing the closed/exact expressions or the qualitative/quantitative properties of the solution process. Until recently [80] and for various reasons, the development of the area of methods for finding closed-form solutions of dynamic processes operating on single [classical timescale $T_1(t) = t$] or multiple timescales [two or more, for example, $T_1(t) = t$ and $T_2(t) = w(t)$ — Wiener process generated due to the randomly varying parameter] has not been updated to the level of developments in the various aspects and directions of dynamic processes described by differential equations. In general, a closed/exact form representation of a time evolution solution flow described by a nonlinear or linear nonstationary interconnected systems operating on a single timescale $[T_1(t) = t]$ is not always feasible. During the past 80 years, there have been a few analytic nonlinear techniques not requiring the closed-form solution of dynamic processes. The development of this chapter's material parallels its deterministic version (Chapter 6 of Volume 1). In Section 6.1, we briefly describe the fundamental conceptual results similar to the results outlined in Sections 2.5 and 4.6. Short outlines of two of the best-known nonlinear analytic methods, namely (i) nonlinear variation of constants parameters and (ii) the comparison method coupled with differential inequalities, are presented. Respectively, Sections 6.2 and 6.3 deal with a method of variation of constants/parameters and its extensions for the nonlinear systems of stochastic differential equations. A comparison method and its extensions in the context of differential inequalities are highlighted in Sections 6.4–6.6. Section 6.7 introduces linear stochastic hybrid systems through drug prescription and administration problems. The ideas and basic results about hereditary systems are presented in Section 6.8. An example is given to illustrate the hereditary effects. Certain qualitative properties of solution processes of stochastic differential equations are addressed in Section 6.9. In Section 6.10, several examples are given to exhibit the effects of randomly varying internal or external fluctuations on the various types of critical points associated with two-dimensional dynamic vector fields. Finally, in Section 6.11, a few illustrations and examples are extracted to illustrate the effects of random disturbances. An attempt is made to exhibit the limitations on finding the closed-form mean and variance of the solution processes. Moreover, it is shown that the solution to the differential equations and the mean of the solutions are not easily comparable. Finally, the effects of random perturbations are classified as: (i) passive, (ii) conditionally passive and active, (iii) partially active, and (iv) active. The purpose is to justify and to partially resolve the problem of "stochastic versus deterministic" ("uncertainty versus certainty") in nonlinear and nonstationary dynamic processes.

6.1 Fundamental Conceptual Algorithms and Analysis

In a classical modeling approach, the evolution of the states of the dynamic systems can be considered to be dynamic flow evolving on multiple timescales. The goal is to find an information (qualitative/quantitative) about dynamic flows. In general, a closed/exact form representation of a time evolution solution flow described by a nonlinear/linear nonstationary interconnected system of differential equations is not always feasible.

Let us consider a simple mathematical description of a nonlinear dynamic phenomenon under random environmental perturbations described by a system of nonlinear nonstationary Itô–Doob-type stochastic differential equations:

$$dx = f(t,x)dt + \sigma(t,x)dw(t), \quad x(t_0) = x_0, \tag{6.1.1}$$

where x, $f \in C[J \times R^n, R^n]$, $\sigma \in C[J \times R^n, R^{nm}]$, and $\sigma = [\sigma^1 \sigma^2 \cdots \sigma^j \cdots \sigma^m]$ is the column matrix representation of σ; where $C[J \times R^n, R^k]$ stands for the class of continuous functions defined on $J \times R^n$ into R^k for a positive integer k, and $J = [t_0, t_0+a)$ for some real number $a > 0$; x_0 is an n-dimensional random vector defined on a complete probability space (Ω, F, P); F_t is an increasing family of sub-σ-algebra of F; $w(t) = (w_1(t), w_2(t), \ldots, w_m(t))^T$ is an m-dimensional normalized Wiener process with independent increments; x_0 and $w(t)$ are mutually independent for each $t \geq t_0$.

In the following, we present a few fundamental mathematical problems. These problems play a very important role in the modeling of dynamic phenomena under the influence of random fluctuations.

Problem 6.1.1 (existence of a solution process [7, 8, 32, 43, 56, 71, 98, 141, 182, 195]). The first and the most basic test for validating a proposed mathematical dynamic model of linear or nonlinear dynamic phenomena evolving under stochastic uncertainties is the existence of a time-evolving dynamic flow satisfying the IVP (6.1.1). This is the most basic mathematical problem in the area of dynamic modeling of nonlinear processes in the biological, chemical, engineering, medical, physical and social sciences. Otherwise, the mathematical model described by (6.1.1) is vacuous.

It is well known [8, 56, 60, 98] that in addition to the above-stated standard assumptions on f, σ and the Wiener process, for some positive constants M and N, f and σ fulfill the following growth condition for $(t, x) \in J \times R^n$:

$$\|f(t,x)\| + \|\sigma(t,x)\| \leq N + M\|x\| \quad \text{(linear growth condition)}, \quad (6.1.2)$$

and $E[\|x_0\|^4] \leq c$ for some constant $c > 0$, and the IVP (6.1.1) has a solution. We also note that we may not be able to find a solution to (6.1.1) in closed form. In general, we do not need to know the exact or closed-form solution. It is enough to know the qualitative properties of the solution process of (6.1.1). We further emphasize that other than conceptual information, there is no need to know f, σ, and x_0, completely and explicitly.

Problem 6.1.2 (uniqueness of the solution process [7, 8, 32, 43, 56, 98, 141, 182, 195]). In many chemical, biological engineering, physical, and social systems, the uniqueness of the solution to (6.1.1) is a very important problem. Otherwise, for the users of this type of dynamic models, it creates ambiguities in the decision-making process. In addition to the conditions specified on f, σ, and x_0 in (6.1.1) and to insure the uniqueness of the solution process of (6.1.1), f and σ need to satisfy certain conditions, such as the Lipschitz condition:

$$\|f(t,x) - f(t,y)\| + \|\sigma(t,x) - \sigma(t,x)\| \leq L\|x - y\| \quad \text{(Lipschitz condition)}, \quad (6.1.3)$$

where L is some positive constant. We note that the Lipschitz condition (6.1.3) is global. The significance of the uniqueness of the solution process is the fact that the future knowledge of the dynamic state of the process is uniquely determined by the initial knowledge (initial data) of the state.

In the absence of the uniqueness assumption, the IVP (6.1.1) can have infinitely many solutions. This is illustrated by the following simple example:

Example 6.1.1. We consider a very simple IVP

$$dx = 2x^{1/2}dt + \nu x\, dw(t), \quad x(0) = 0. \quad (6.1.4)$$

Solution procedure. We note that $f(t,x) = 2x^{1/2}$ and $\sigma(t,x) = \nu x$. We consider (6.1.4) as a perturbed differential equation of the unperturbed stochastic differential equation $(dy = \nu y\, dw(t)$; its solution is $y(t) = \exp[-\frac{1}{2}\nu^2 t + \nu w(t)])$; we seek a solution to (6.1.4) as $x(t) = y(t)c(t)$. Employing Corollary 4.4.1 (the transformed deterministic differential equation is $dc = p(t)c^{1/2}dt$, where $\exp[\frac{1}{4}\nu^2 t - \frac{1}{2}\nu w(t)] = p(t)$), and applying the method of variable separable to this deterministic differential equation with stochastic process varying coefficients (Section 3.5, Chapter 3 of Volume 1), the given IVP is solved in a closed form. Its solution is

$$x(t) = \left(\int_0^t \exp\left[-\frac{1}{4}\nu^2(t-s) + \frac{1}{2}\nu(w(t) - w(s)) \right] ds \right)^2,$$

and a second solution is $x(t) \equiv 0$. In fact, we have

$$x(t) = \begin{cases} 0, & \text{if } 0 \le t \le c \text{ if } 0 \le t \le c, \\ \left(\int_c^t \exp\left[-\frac{1}{4}\nu^2(t-s) + \frac{1}{2}\nu(w(t) - w(s)) \right] ds \right)^2, & \text{if } t \ge c, \end{cases}$$

for any $0 \le c$. This shows that it has infinitely many solutions to the IVP (6.1.4). We remark that $f(t, x) = 2x^{1/2}$ is differentiable on $(0, \infty)$. However, its derivative is not bounded in the neighborhood of "0."

Observation 6.1.1. We note that the respective growth and Lipschitz conditions (6.1.2) and (6.1.3) are global. However, if (6.1.2) and (6.1.3) are satisfied in a neighborhood of a given point (t_0, x_0), then they are called local growth and Lipschitz conditions. Furthermore, a sufficient condition for the local Lipschitz condition on f and σ is the boundedness of the derivative of both f and σ with respect to x in the neighborhood of a given point (t_0, x_0). The proof follows by the application of Lemma 6.2.1 of Volume 1.

Problem 6.1.3 (dependence of the solution process on system parameters [43, 98]). In the process of developing the mathematical model for dynamic processes, rate functions and the initial data are subject to errors. These errors arise due to the selection process of the system parameters, generated by the experimental conditions/assumptions, including the initial data (t_0, x_0). Because of this, a solution flow depends on these system parameters. In light of this, the two basic questions are: To what extent do these parametric changes influence the errors in solution flows with respect to the nominal system? Are they harmless (in some sense)? The answer to these question addresses the issue of the robustness of dynamic modeling. We try to discuss the basic properties of the solution flow as a function of parameters in dynamic processes. To address this problem, we need to introduce system parameters, explicitly, into the mathematical description (6.1.1). For precision, we modify the dynamic system description (6.1.1) as

$$dx = f(t, x, \lambda)dt + \sigma(t, x, \lambda)dw(t), \quad x(t_0, \lambda) = x_0(\lambda), \qquad (6.1.5)$$

and the corresponding nominal system,

$$dx = f(t, x, \lambda_0)dt + \sigma(t, x, \lambda_0)dw(t), \quad x(t_0) = x_0, \qquad (6.1.6)$$

where $(t_0, x_0, \lambda_0) \in J \times R^n \times \Lambda$, with Λ being an open system parameter λ set in R^m.

Next, we present a very simple result that exhibits the continuous dependence of the solution process of (6.1.5) with respect to $(t_0, x_0, \lambda_0) \in R \times R^n \times \Lambda$.

Theorem 6.1.1. *Assume that:*

(H$_{6.1a}$) *The m column vectors of σ and $f \in C[J \times R^n \times \Lambda, R^n]$.*
(H$_{6.1b}$) *There exist positive numbers M and N such that*

$$\|f(t, x, \lambda)\| + \|\sigma(t, x, \lambda)\| \leq N + M\|x\|,$$

for $(t, x, \lambda) \in J \times R^n \times \Lambda$.
(H$_{6.1c}$) *There exists a positive number L such that*

$$\|f(t, x, \lambda) - f(t, y, \lambda)\| + \|\sigma(t, x, \lambda) - \sigma(t, y, \lambda)\| \leq L\|x - y\|,$$

for $(t, x, \lambda), (t, y, \lambda) \in J \times R^n \times \Lambda$.
(H$_{6.1d}$) *$x(t_0, \lambda) = x_0(\lambda)$ is independent of $w(t)$ and*

$$\lim_{\lambda \to \lambda_0} E[\|x(t_0, \lambda) - x(t_0, \lambda_0)\|^2] = 0.$$

(H$_{6.1e}$) *$E[\|x(t_0, \lambda)\|^4] \leq c_0$ for some constant $c_0 > 0$.*
(H$_{6.1f}$) *$\epsilon > 0$, $\rho > 0$,*

$$\lim_{\lambda \to \lambda_0} P\left[\sup_{\|x\| < \rho} (\|f(t, x, \lambda) - f(t, x, \lambda_0)\| + \|\sigma(t, x, \lambda) - \sigma(t, x, \lambda_0)\| > \epsilon \right] = 0.$$

Then, the IVPs (6.1.5) and (6.1.6) admit unique solution processes $x(t, t_1, x_0(\lambda), \lambda)$ and $x(t, t_0, x_0, \lambda_0)$ through $(t_1, x_0(\lambda))$ and $(t_0, x_0(\lambda_0))$, respectively. Moreover, for a given $\epsilon > 0$, there exists a $\delta(\epsilon) > 0$ such that

$$(E[\|x(t, t_1, x_0(\lambda), \lambda) - x(t, t_0, x_0, \lambda_0)\|^2])^{1/2} < \epsilon, \quad t \in J, \tag{6.1.7}$$

whenever

$$|t_1 - t_0| + E[\|x_0(\lambda) - x_0\|^2]^{1/2} + \|\lambda - \lambda_0\| < \delta(\epsilon).$$

Proof. Under hypotheses (H$_{6.1a}$)–(H$_{6.1e}$), the IVPs (6.1.5) and (6.1.6) provide unique solution processes through $(t_1, x_0, (\lambda))$ and $(t_0, x_0, (\lambda_0))$, respectively. The respective solution processes are represented by $x(t, t_1, x_0(\lambda), \lambda)$ and $x(t, t_0, x_0, \lambda_0)$. We set $x(t, \lambda) = x(t, t_1, x_0(\lambda), \lambda)$ and $x(t, \lambda_0) = x(t, t_0, x_0, \lambda_0)$. To prove the continuity of the solution with respect to $(t_0, x_0(\lambda_0), \lambda_0)$, we apply Itô–Doob calculus to

$\|x(t, t_1, x_0(\lambda), \lambda) - x(t, t_0, x_0, \lambda_0)\|^2$ (Theorem 1.3.2), and we obtain

$$d(\|x(t, \lambda) - x(t, \lambda_0)\|^2|)$$

$$= 2(x(t, \lambda) - x(t, \lambda_0))(f(t, x(t, \lambda), \lambda) - f(t, x(t, \lambda_0), \lambda_0))dt$$

$$+ \operatorname{tr}((\sigma(t, x(t, \lambda), \lambda) - \sigma(t, x(t, \lambda_0), \lambda_0))(\sigma(t, x(t, \lambda), \lambda) - \sigma(t, x(t, \lambda_0), \lambda_0))^T)dt$$

$$+ 2(x(t, \lambda) - x(t, \lambda_0))(\sigma(t, x(t, \lambda), \lambda) - \sigma(t, x(t, \lambda_0), \lambda_0))dw(t). \tag{6.1.8}$$

Without loss of generality, we assume that $t_1 > t_0$. From this, we have $\|x_0(\lambda) - x(t_1, \lambda_0)\|^2$.

Now, using the properties of the Itô–Doob stochastic integral, (6.1.8) reduces to

$$E[\|x(t, t_1, x_0(\lambda), \lambda) - x(t, t_0, x_0, \lambda_0)\|^2]$$

$$= 2E\left[\int_{t_1}^t (x(s, \lambda) - x(s, \lambda_0))(f(s, x(s, \lambda), \lambda) - f(s, x(s, \lambda_0), \lambda_0))ds\right]$$

$$+ E[\|x_0(\lambda) - x(t_1, t_0)\|^2] + \int_{t_1}^t E[\|\sigma(s, x(s, \lambda), \lambda) - \sigma(s, x(s, \lambda_0), \lambda_0))\|^2]ds.$$

$$\tag{6.1.9}$$

We need to estimate the expression (6.1.9). For this purpose, we consider the following:

$$2(x(s, \lambda) - x(s, \lambda_0))(f(s, x(s, \lambda), \lambda) - f(s, x(s, \lambda_0), \lambda_0))$$

$$\leq 2\|x(s, \lambda) - x(s, \lambda_0)\|\|f(s, x(s, \lambda), \lambda) - f(s, x(s, \lambda_0), \lambda_0)\|. \tag{6.1.10}$$

From ($H_{6.1c}$), we estimate the factor in the right-hand-side as:

$$\|f(s, x(s, \lambda), \lambda) - f(s, x(s, \lambda_0), \lambda_0)\|$$

$$= \|f(s, x(s, \lambda), \lambda) - f(s, x(s, \lambda_0), \lambda) + f(s, x(s, \lambda_0), \lambda) - f(s, x(s, \lambda_0), \lambda_0)\|$$

$$\leq \|f(s, x(s, \lambda), \lambda) - f(s, x(s, \lambda_0), \lambda)\|$$

$$+ \|f(s, x(s, \lambda_0), \lambda) - f(s, x(s, \lambda_0), \lambda_0)\|$$

$$\text{(properties of the norm, Theorem 1.5.7 [80])}$$

$$\leq L\|x(s, \lambda) - x(s, \lambda_0)\|$$

$$+ \|f(s, x(s, \lambda_0), \lambda) - f(s, x(s, \lambda_0), \lambda_0)\| \quad [\text{by } (H_{6.1c})]. \tag{6.1.11}$$

Using (6.1.11), (6.1.10) reduces to

$$2(x(s, \lambda) - x(s, \lambda_0))(f(s, x(s, \lambda), \lambda) - f(s, x(s, \lambda_0), \lambda_0))$$

$$\leq 2\|x(s, \lambda) - x(s, \lambda_0)\|[L\|x(s, \lambda) - x(s, \lambda_0)\|$$

$$+ \|f(s, x(s, \lambda_0), \lambda) - f(s, x(s, \lambda_0), \lambda_0)\|]$$

$$\leq 2L\|x(s,\lambda) - x(s,\lambda_0)\|^2$$
$$+ 2\|x(s,\lambda) - x(s,\lambda_0)\|\|f(s,x(s,\lambda_0),\lambda) - f(s,x(s,\lambda_0),\lambda_0)\|(2ab \leq a^2 + b^2)$$
$$\leq (2L+1)\|x(s,\lambda) - x(s,\lambda_0)\|^2$$
$$+ \|f(s,x(s,\lambda_0),\lambda) - f(s,x(s,\lambda_0),\lambda_0)\|^2 \quad \text{(by regrouping).} \tag{6.1.12}$$

Similarly, from $(H_{6.1c})$, we estimate the following as

$$\|\sigma(s,x(s,\lambda),\lambda) - \sigma(s,x(s,\lambda_0),\lambda_0)\|$$
$$= \|\sigma(s,x(s,\lambda),\lambda) - \sigma(s,x(s,\lambda_0),\lambda) + \sigma(s,x(s,\lambda_0),\lambda) - \sigma(s,x(s,\lambda_0),\lambda_0)\|$$
$$\leq \|\sigma(s,x(s,\lambda),\lambda) - \sigma(s,x(s,\lambda_0),\lambda)\|$$
$$+ \|\sigma(s,x(s,\lambda_0),\lambda) - \sigma(s,x(s,\lambda_0),\lambda_0)\|$$
$$\text{(properties of the norm, Theorem 1.5.7).}$$

We square both sides of the above inequality, and obtain

$$\|\sigma(s,x(s,\lambda),\lambda) - \sigma(s,x(s,\lambda_0),\lambda_0)\|^2$$
$$\leq (\|\sigma(s,x(s,\lambda),\lambda) - \sigma(s,x(s,\lambda_0),\lambda)\| + \|\sigma(s,x(s,\lambda_0),\lambda) - \sigma(s,x(s,\lambda_0),\lambda)\|)^2$$
$$\leq \|\sigma(s,x(s,\lambda),\lambda) - \sigma(s,x(s,\lambda_0),\lambda)\|^2$$
$$+ 2\|\sigma(s,x(s,\lambda),\lambda) - \sigma(s,x(s,\lambda_0),\lambda)\|\|\sigma(s,x(s,\lambda_0),\lambda) - \sigma(s,x(s,\lambda_0),\lambda_0)\|$$
$$+ \|\sigma(s,x(s,\lambda_0),\lambda) - \sigma(s,x(s,\lambda_0),\lambda_0)\|^2 \quad \text{(using: } (a+b)^2 = a^2 + 2ab + b^2\text{)}$$
$$\leq 2\|\sigma(s,x(s,\lambda),\lambda) - \sigma(s,x(s,\lambda_0),\lambda)\|^2$$
$$+ 2\|\sigma(s,x(s,\lambda_0),\lambda) - \sigma(s,x(s,\lambda_0),\lambda_0)\|^2 \quad \text{(using: } 2ab \leq a^2 + b^2\text{)}$$
$$\leq 2L^2\|x(s,\lambda) - x(s,\lambda_0)\|^2$$
$$+ 2\|\sigma(s,x(s,\lambda_0),\lambda) - \sigma(s,x(s,\lambda_0),\lambda_0)\|^2 \quad \text{[by } (H_{6.1c})\text{].} \tag{6.1.13}$$

Let us denote $p(t)$ and $m(t)$ by

$$p(t) = 2[\|f(s,x(s,\lambda_0),\lambda) - f(s,x(s,\lambda_0),\lambda_0)\|^2$$
$$+ \|\sigma(s,x(s,\lambda_0),\lambda) - \sigma(s,x(s,\lambda_0),\lambda_0)\|^2], \tag{6.1.14}$$
$$m(t) = E[\|x(t,t_1,x_0(\lambda),\lambda) - x(t,t_0,x_0,\lambda_0)\|^2], \tag{6.1.15}$$

with

$$m(t_1) = E[\|x_0(\lambda) - x(t_1,\lambda_0)\|^2. \tag{6.1.16}$$

From (6.1.12)–(6.1.16) and properties of the first moment, (6.1.9) is reduced to

$$E[\|x(t,t_1,x_0(\lambda),\lambda) - x(t,t_0,x_0,\lambda_0)\|^2 \leq E[\|x_0(\lambda) - x(t_1,t_0)\|^2]$$
$$+ (2L + 2L^2 + 1)\int_{t_1}^{t} [E[\|x(s,\lambda) - x(s,\lambda_0)\|^2] + E[p(s)]]\,ds. \tag{6.1.17}$$

From (6.1.17), setting $\alpha = 1 + 2L(1 + L)$, and applying Lemma 6.5.1 of Volume 1, we have

$$E[\|x(t, t_1, x_0(\lambda), \lambda) - x(t, t_0, x_0, \lambda_0)\|^2]$$

$$\leq E[\|x_0(\lambda) - x(t_1, t_0)\|^2] \exp[\alpha(t - t_1)] + \int_{t_1}^t E[p(s)] \exp[\alpha(t - s)]ds, \quad t \geq t_1.$$

$$(6.1.18)$$

From ($H_{6.1f}$), (6.1.14), (6.1.7), (6.1.18), the concept of the solution process, the Lebesgue convergence theorem [169], and imitating the argument used in deriving the estimates in (6.1.13), we obtain

$$m(t_1) = E[\|x_0(\lambda) - x_0(\lambda_0) + x_0(\lambda_0) - x(t_1, t_0, x_0, \lambda_0)\|^2]$$

$$\leq E[(\|x_0(\lambda) - x_0(\lambda_0)\| + \|x_0(\lambda_0) - x(t_1, t_0, x_0(\lambda_0), \lambda_0)\|)^2]$$

$$\leq 2E[\|x_0(\lambda) - x_0(\lambda_0)\|^2] + 2E[\|x(t_0, t_0, x_0(\lambda_0), \lambda_0) - x(t_1, t_0, x_0(\lambda_0), \lambda_0)\|^2],$$

for $t_1, t \in [t_0, t_0 + a]$, and for given $\epsilon > 0$, there exists a $\delta(\epsilon) > 0$, so that

$$E[\|x(t, t_1, x_0(\lambda), \lambda) - x(t, t_0, x_0(\lambda_0), \lambda_0)\|^2] < \epsilon^2,$$

whenever

$$|t_1 - t_0| + E[\|x_0(\lambda) - x_0(\lambda_0)\|^2] + \|\lambda - \lambda_0\| < \delta(\epsilon).$$

This completes the proof of the lemma. $\qquad\qquad\square$

The following result provides the differentiability of the solution process of (6.1.1) with respect to the initial data.

Theorem 6.1.2. *Assume that σ, f, and x_0 in (6.1.1) satisfy the hypotheses ($H_{6.1a}$), ($H_{6.1b}$), ($H_{6.1d}$), and ($H_{6.1e}$). Furthermore, σ and f are continuously differentiable with respect to x for fixed t. Let $x(x, t_0, x_0)$ be the solution process of (6.1.1) existing for $t \geq t_0$. Then,*

$$\Phi(t, t_0, x_0) = \frac{\partial}{\partial x} x(t, t_0, x_0) \qquad (6.1.19)$$

exists and is the fundamental matrix solution to

$$dy = H(t, t_0, x_0)y \, dt + \Gamma(t, t_0, x_0)y \, dw(t) \qquad (6.1.20)$$

such that $\Phi(t_0, t_0, x_0)$ is $n \times n$ identity matrix, where $n \times n$ matrices $f_x(t, x)$ and $\sigma_x^j(t, x)$ are continuous in (t, x) for $j = 1, 2, \ldots, m$; $\sigma_x(t, x)$ is $n \times nm$ matrix $\sigma_x(t, x) = [\sigma_x^1(t, x)\sigma_x^2(t, x) \cdots \sigma_x^j(t, x) \cdots \sigma_x^m(t, x)]$; $H(t, t_0, x_0) = f_x(t, x(t, t_0, x_0))$ and $\Gamma(t, t_0, x_0) = \sigma_x(t, x(t, t_0, x_0))$.

Proof. Under the assumption on σ, f and the application of Lemma 6.2.1 of Volume 1, we conclude that the σ and f satisfy the local Lipschitz condition. Therefore, from Theorem 6.1.1, it is obvious that $x(t) = x(t, t_0, x_0)$ is the unique solution to the IVP (6.1.1). Moreover, the solution is continuous with respect to the initial data (t_0, x_0). In the following, we show that (6.1.19) is valid. For this purpose, we consider the following. For small λ, $x(t, \lambda) = x(t, t_0, x_0 + \lambda e_k)$ and $x(t) = x(t, t_0, x_0)$ are solution processes of (6.1.1) through $(t_0, x_0 + \lambda e_k)$ and (t_0, x_0), respectively, where $e_k = (0, 0, \ldots, 1, \ldots, 0)^T$, whose kth component is 1. From Theorem 6.1.1, it clear that

$$\lim_{\lambda \to 0} x(t, \lambda) = x(t) \quad \text{uniformly on } J. \tag{6.1.21}$$

Set

$$\Delta x(t, \lambda) = x(t, \lambda) - x(t), \quad \Delta x_\lambda(t) = \frac{\Delta x(t, \lambda)}{\lambda}, \quad \Delta x_\lambda(t_0, \lambda) = e_k, \lambda \neq 0 \tag{6.1.22}$$

where $x = x(t)$ and $x(t, \lambda)$. Along with the application of Lemma 6.2.1 [80], this yields

$$f(t, x(t, \lambda)) - f(t, x(t)) = \int_0^1 \left[\frac{\partial}{\partial x} f(t, x(t) + \theta \Delta x(t, \lambda)) \right] \Delta x(t, \lambda) d\theta, \tag{6.1.23}$$

and

$$\sigma(t, x(t, \lambda)) - \sigma(t, x(t)) = \int_0^1 \left[\frac{\partial}{\partial x} \sigma(t, x(t) + \theta \Delta x(t, \lambda)) \right] \Delta x(t, \lambda) d\theta$$

$$= \sum_{j=1}^m \int_0^1 \left[\frac{\partial}{\partial x} \sigma^j(t, x(t) + \theta \Delta x(t, \lambda)) \right] \Delta x(t, \lambda) d\theta. \tag{6.1.24}$$

Let us set

$$\frac{\partial}{\partial x} f(t, x(t), \Delta x(t, \lambda), \lambda) = \int_0^1 \left[\frac{\partial}{\partial x} f(t, x(t) + \theta \Delta x(t, \lambda)) \right] d\theta, \tag{6.1.25}$$

and

$$\frac{\partial}{\partial x} \sigma(t, x(t), \Delta x(t, \lambda), \lambda) = \sum_{j=1}^m \int_0^1 \left[\frac{\partial}{\partial x} \sigma^j(t, x(t) + \theta \Delta x(t, \lambda)) \right] d\theta. \tag{6.1.26}$$

We note that the integrals in (6.1.25) and (6.1.26) are the Cauchy–Riemann integrals. From the hypotheses of the theorem, $n \times n$ matrices $f_x(t, x, \lambda)$ and $\sigma_x^j(t, x, \lambda)$ are continuous in (t, x, λ) for $j = 1, 2, \ldots, m$. Furthermore, from (6.1.25) and (6.1.26), we have

$$\lim_{\lambda \to 0} \frac{\partial}{\partial x} f(t, x(t), \Delta x(t, \lambda), \lambda) = \frac{\partial}{\partial x} f(t, x(t)), \tag{6.1.27}$$

and

$$\lim_{\lambda \to 0} \frac{\partial}{\partial x} \sigma(t, x(t), \Delta x(t, \lambda), \lambda) = \frac{\partial}{\partial x} \sigma(t, x(t)) \tag{6.1.28}$$

uniformly in $\theta \in [0, 1]$.

From (6.1.22)–(6.1.26), we have

$$d\Delta x_\lambda(t) = \frac{\partial}{\partial x} f(t, x(t), \Delta x(t, \lambda), \lambda) \Delta x_\lambda(t) dt$$

$$+ \frac{\partial}{\partial x} \sigma(t, x(t), \Delta x_\lambda(t, \lambda), \lambda) \Delta x_\lambda(t) dw(t), \quad \Delta x_\lambda(t_0) = e_k. \quad (6.1.29)$$

In view of (6.1.27) and (6.1.28), system (6.1.20) can be the nominal system corresponding to (6.1.29) with initial data $y(t_0) = e_k$, i.e.

$$dy = H(t, t_0, x_0)y\, dt + \Gamma(t, t_0, x_0)y\, dw(t), \quad y(t_0) = e_k. \quad (6.1.30)$$

It is obvious that the IVP (6.1.29) satisfies all the hypotheses of Theorem 6.1.1 and hence by its application, we have

$$\lim_{\lambda \to 0} \Delta x_\lambda(t) = y(t) \quad \text{uniformly on } J, \quad (6.1.31)$$

where $y(t)$ is the solution process of (6.1.30). Because of (6.1.21), we note that the limit of $\Delta x_\lambda(t)$ in (6.1.29) is equivalent to the derivative $\frac{\partial}{\partial x_k} x(t, t_0, x_0)$. Hence, $\frac{\partial}{\partial x_k} x(t, t_0, x_0)$ exists and is the solution process of (6.1.29). This is true for each $k \in I(1, n)$. Thus, $\frac{\partial}{\partial x} x(t, t_0, x_0)$ is the fundamental matrix solution process of (6.1.30) and satisfies the matrix differential equation (6.1.20). Moreover, $\frac{\partial}{\partial x} x(t, t_0, x_0)$ is denoted by $\Phi(t, t_0, x_0)$, and $\Phi(t_0, t_0, x_0)$ is $n \times n$ identity matrix. This completes the proof of the theorem. $\qquad\square$

Corollary 6.1.1. *Let us assume that all hypotheses of Theorem 6.1.2 are satisfied. Further assume that $\frac{\partial^2}{\partial x^2}\sigma$ and $\frac{\partial^2}{\partial x^2}f$ are continuously differentiable with respect to x for fixed t. Let $x(t, t_0, x_0)$ be the solution process of (6.1.1) existing for $t \geq t_0$. Then,*

$$\frac{\partial}{\partial x_0} \Phi(t, t_0, x_0) = \frac{\partial^2}{\partial x_0 \partial x_0} x(t, t_0, x_0) \quad (6.1.32)$$

exists and is the solution to the following stochastic matrix differential equations:

$$dY = [H(t, t_0, x_0)Y + P(t)]dt + [\Gamma(t, t_0, x_0)Y + Q(t)]dw(t), \quad y(t_0) = 0, \quad (6.1.33)$$

with

$$P(t) = \frac{\partial^2}{\partial x \partial x} f(t, x(t))\Phi(t, t_0, x_0)\Phi(t, t_0, x_0),$$

$$Q(t) = \frac{\partial^2}{\partial x \partial x} \sigma(t, x(t))\Phi(t, t_0, x_0)\Phi(t, t_0, x_0),$$

where $\frac{\partial^2}{\partial x \partial x} f(t, x(t))\Phi(t, t_0, x_0)$ and $\frac{\partial^2}{\partial x \partial x}\sigma(t, x(t))\Phi(t, t_0, x_0)$ are Hessian matrices of appropriate size whose entries are n-dimensional vectors; $P(t)$ and $Q(t)$ are $n \times n$ stochastic matrix processes.

Proof. The proof is left as an exercise to the reader. $\qquad\square$

Example 6.1.2. Let us consider a scalar nonlinear stochastic differential equation:

$$dI = \alpha I(\bar{N} - I)dt + \beta I\, dw(t), \quad I(t_0) = I_0. \tag{6.1.34}$$

(a) Is the solution to the IVP differentiable with respect to initial state?
(b) Find $\frac{\partial}{\partial I_0} I(t, t_0, I_0)$ and $\frac{\partial^2}{\partial I_0^2} I(t, t_0, I_0)$.

Solution process. We note that $f(t, I) = \alpha I(\bar{N} - I)$ and $\sigma(t, I) = \beta I$ are twice continuously differentiable with respect to I. In fact, $\beta = \frac{\partial}{\partial I} \sigma(t, I)$, $\frac{\partial}{\partial I} f(t, I) = \alpha \bar{N} - 2I$ and $\frac{\partial^2}{\partial I^2} f(t, I) = -2$ and $\frac{\partial^2}{\partial I^2} \sigma(t, I) = 0$. The closed-form solution to (6.1.34) is

$$I(t, t_0, I_0) = \left[\Phi(t, t_0) I_0^{-1} + \alpha \int_{t_0}^{t} \Phi(t, r)dr \right]^{-1}, \tag{6.1.35}$$

where $\Phi(t, t_0) = \exp[-(\alpha \bar{N} - \frac{1}{2}\beta^2)(t - t_0) - \beta(w(t) - w(t_0))]$ (Example 3.7.3). It is clear that the nontrivial solution to (6.1.34) is continuously differentiable with respect to (t, t_0, I_0). Moreover, in this case, the expressions (6.1.19) and (6.1.32) reduce to

$$\frac{\partial}{\partial I_0} I(t, t_0, I_0) = \frac{\Phi(t, t_0)}{(\Phi(t, t_0) + \alpha I_0 \int_{t_0}^{t} \Phi(t, r)dr)^2}, \tag{6.1.36}$$

$$\frac{\partial^2}{\partial I_0 \partial I_0} I(t, t_0, I_0) = \frac{-2\alpha \Phi(t, t_0) \int_{t_0}^{t} \Phi(t, r)dr}{(\Phi(t, t_0) + \alpha I_0 \int_{t_0}^{t} \Phi(t, r)dr)^3}. \tag{6.1.37}$$

Example 6.1.3. Consider a scalar nonlinear autonomous differential equation:

$$dx = \left[-a(t)\frac{1}{2}x^3 + b(t)x \right] dt + \sigma(t)x\, dw(t), \quad x(t_0) = x_0, \tag{6.1.38}$$

where a, b, and σ are continuous functions defined on R into R.

(a) Find the closed-from general solution, and the solution to the IVP.
(b) Is it differentiable with respect to x_0?
(c) Find $\frac{\partial}{\partial x_0} x(t, t_0, x_0)$ and $\frac{\partial^2}{\partial x_0 \partial x_0} x(t, t_0, x_0)$ (if possible).

Solution process. We note that this IVP can be solved in closed from by two methods:

(i) The energy function method applied to the Bernoulli-type stochastic differential equation (Section 3.7 of Chapter 3).
(ii) The method of variation of parameters.

Here, we use the approach (ii). We treat (3.1.38) as the stochastic perturbed differential equation of the following stochastic unperturbed and deterministic perturbation equations as

$$dy = b(t)y\,dt + \sigma(t)y\,dw(t) \qquad \text{(unperturbed)}, \tag{6.1.39}$$

$$dx = f(t,x)dt = -a(t)\frac{1}{2}x^3 dt \quad \text{(deterministic perturbation)}. \tag{6.1.40}$$

Using the method of finding the solution process of Section 2.3, the general solution process of (6.1.39) is

$$y(t) = \Phi(t)c = \exp\left[\int_a^t \left[b(s) - \frac{1}{2}\sigma^2(s)\right]ds + \int_a^t \sigma(s)dw(s)\right]c. \tag{6.1.41}$$

Now, considering a constant c to be a smooth unknown function, we assume that

$$x(t) = \Phi(t)c = \exp\left[\int_a^t \left[b(s) - \frac{1}{2}\sigma^2(s)\right]ds + \int_a^t \sigma(s)dw(s)\right]c(t) \tag{6.1.42}$$

is a solution process of (6.1.38). Imitating the proof of Theorem 4.4.1, we have

$$\begin{aligned}
dc &= d\Phi^{-1}(t)x(t) + \Phi^{-1}(t)dx(t) + d\Phi^{-1}(t)dx(t) \quad \text{(by Theorem 1.3.4)}\\
&= \Phi^{-1}(t)[(-b(t) + \sigma^2(t))dt - \sigma(t)dw(t)]x(t)\\
&\quad + \Phi^{-1}(t)\left[\left[-a(t)\frac{1}{2}x^3(t) + b(t)x(t)\right]dt + \sigma(t)x(t)dw(t)\right]\\
&\quad - \Phi^{-1}(t)\sigma(t)dw(t)\sigma(t)x(t)dw(t) \qquad\qquad \text{(by substitution)}\\
&= -\Phi^{-1}(t)a(t)\frac{1}{2}x^3(t) \qquad\qquad\qquad \text{(by simplifying)}\\
&= -\frac{1}{2}a(t)\Phi^{-1}(t)\Phi^3(t)c^3(t) \quad \text{(by substitution } x(t) = \Phi(t)c(t))\\
&= -\frac{1}{2}p(t)c^3 dt, \tag{6.1.43}
\end{aligned}$$

where the stochastic process p is defined by

$$p(t) \equiv p(t, w(t)) = \Phi^{-1}(t)a(t)\Phi^3(t) = a(t)\Phi^2(t, w(t)). \tag{6.1.44}$$

This is a transformed variable separable deterministic differential equation with a stochastic time-varying rate function. We apply the energy function method developed in Section 3.6, Chapter 3 of Volume 1. The closed-form solution to the transformed deterministic differential equation (6.1.43) is

$$c(t) = \frac{1}{\sqrt{c + \int^t p(s)ds}}, \tag{6.1.45}$$

where c is an arbitrary constant. From (6.1.42) and (6.1.45), the general solution to (6.1.38) is

$$x(t) = \Phi(t)c(t) = \frac{\Phi(t)}{\sqrt{c + \int^t p(s)ds}}. \tag{6.1.46}$$

For given $x(t_0) = x_0$, solve the algebraic equation

$$x_0 = \frac{\Phi(t_0)}{\sqrt{c + \int^{t_0} p(s)ds}}. \tag{6.1.47}$$

Using the solution c to (6.1.47), algebraic manipulations, and (6.1.44), here is the closed-form solution to the IVP (6.1.38)

$$x(t, t_0, x_0) = \frac{\Phi(t, t_0)|x_0|}{\sqrt{1 + x_0^2 \Phi^{-2}(t_0) \int_{t_0}^t p(s)ds}} = \frac{\Phi(t, t_0)|x_0|}{\sqrt{1 + x_0^2 \int_{t_0}^t a(s)\Phi^2(s, t_0)ds}}. \tag{6.1.48}$$

This completes the solution process of Part (a). The answer to the question in Part (b) is "Yes" except at $x_0 = 0$. In this case, by the uniqueness of the solution process of (6.1.38), $x(t, t_0, x_0) \equiv 0$. This process is always differentiable with respect to x_0. This completes Part (b). The solution expression (6.1.48) is a rational (algebraic) function of x_0. Therefore, Part (c) is left as an exercise to the reader. This completes the solution process of the example.

Illustration 6.1.1 (population dynamics).   We discuss a simple mathematical model in population dynamics described in Illustration 6.2.1 [80] under a randomly varying environment. Here, we assume all the assumptions in Illustration 6.2.1 except to PDM4. We introduce the modified assumption PDM4. It is obvious that abiotic environmental factors are not constants. They are subject to random perturbations, as described in Illustration 3.1.3 (SEP6). This mathematical model involves competing species. The assumptions relative to the density-dependent and crowding effects reflect the fact that the growth of any species in a restricted environment must eventually be limited by a shortage of resources.

Under the above assumptions, a mathematical model of the n-species community is described by

$$dN_i = N_i F_i(t, N)dt + \sigma_i(t)N_i dw(t), \quad N_i(t_0) = N_{i0}, \tag{6.1.49}$$

where N_i is the population density of the i-species in the model ecosystem for $i \in I(1, n)$; for each $i \in I(1, n)$, $F_i \in C[J \times D, R]$, and it describes the per capita growth rate of the i-species. Moreover, the sign of $\frac{\partial}{\partial N_j}F_i$ describes the interactions between ith and the jth-species in the community. For examples, competition $(-, -)$, commensalism $(+, +)$, and predator $(+, -)$, i.e. for $i \neq j$, $\frac{\partial}{\partial N_j}F_i$ and $\frac{\partial}{\partial N_i}F_j$

are positive and negative ($\frac{\partial}{\partial N_j}F_i > 0$ and $\frac{\partial}{\partial N_i}F_j < 0$), respectively. For $i = j$, $\frac{\partial}{\partial N_j}F_i$ describes the interspecific effects within the species. In particular,

$$F_i(t, N) = a_i(t) - b_{ii}(t)N_i - \sum_{j=1}^{n} b_{ij}(t)N_j, \quad j \neq i, \tag{6.1.50}$$

where $i, j \in I(1, n)$, a_i, σ_i, and b_{ij} can be considered to be system parameters; w is a normalized Wiener process. It is obvious that $F_i(t, N)$ in (6.1.50) are continuous with respect to a_i, σ_i, and b_{ij}, uniformly on a closed and bounded ball. It is also obvious that in the absence of cross-interactions between the species in the community, the mathematical model of isolated members of the community obey the Verhulst–Pearl logistic equation:

$$dN_i = N_i(a_i - b_{ii}N_i)dt + \sigma_i N_i dw(t), \quad N_i(t_0) = N_{i0}, \tag{6.1.51}$$

for each $i \in I(1, n)$. In the competition model, b_{ii} describes the inhibiting effects of each species. There are several examples in literature that show the close with agreement the growth of an actual laboratory population.

Exercises

1. Verify that the theoretical curve associated with the solution to the differential equation (6.1.51) is

$$N_i(t, t_0, N_{i0}) = N_{i0}\left(\Phi_i(t, t_0) + b_{ii}N_{i0}\int_{t_0}^{t}\Phi_i(t, s)ds\right)^{-1},$$

where $\Phi_i(t, t_0) = \exp\left[\left(-a_i + \frac{1}{2}\sigma_i^2\right)(t - t_0) - \sigma_i(w(t) - w(t_0))\right]$.

2. Let N_i be the solution process of (6.1.51). Find $\frac{\partial}{\partial N_0}N_i(t, t_0, N_{i0})$ and $\frac{\partial^2}{\partial N_0 \partial N_0}N_i(t, t_0, N_{i0})$ (if possible).

3. Solve the following nonlinear differential equations:

$$dx = -x^{2n+1}dt + \sigma x\, dw(t), \quad x(t_0) = x_0,$$

for $n \in I(1, \infty) = \{1, 2, 3, \ldots\}$. Moreover, determine $\frac{\partial}{\partial x_0}x(t, t_0, x_0)$ and $\frac{\partial^2}{\partial x_0 \partial x_0}x(t, t_0, x_0)$.

4. If possible, solve the differential equation

$$dx = (-a(t)x^{2n+1} + b(t)x)dt + \sigma(t)x\, dw(t), \quad x(t_0) = x_0,$$

where a and b are continuous functions defined on R into R and $n \in I(1, \infty)$. Moreover, find $\frac{\partial}{\partial x_0}x(t, t_0, x_0)$ and $\frac{\partial^2}{\partial x_0 \partial x_0}x(t, t_0, x_0)$.

5. Let us consider the following system of linear time-varying differential equations

$$dx = f(t, x)dt + \sum_{j=1}^{m}\Lambda_j x\, dw_j(t), \quad x(t_0) = x_0,$$

where Λ_j are pairwise-commutative $n \times n$ matrices and f is an n-dimensional continuously differentiable vector function in x.

(a) Reduce the stochastic differential to the nonlinear system of deterministic differential equation with a random function varying rate function. Moreover, show that $\frac{\partial}{\partial x_0} x(t, t_0, x_0)$ exists.

(b) Under what additional conditions on f and Λ_j does $\frac{\partial^2}{\partial x_0 \partial x_0} x(t, t_0, x_0)$ exist?

6. (a) Solve the following IVPs:

   (i) $dx = 4x^{3/4} dt + \nu x \, dw(t)$, $x(0) = 0$;

   (ii) $dx = 4x^{3/4} dt + \nu x \, dw(t)$, $x(1) = 1$.

(b) Is $x(t) \equiv 0$ the solution to (i)? Justify.

(c) Is $x(t) = \begin{cases} 0, & \text{if } 0 \leq t \leq c, \quad \text{if } 0 \leq t \leq c, \\ \left(\int_c^t \exp\left[-\frac{1}{8}\nu^2(t-s) + \frac{1}{4}\nu(w(t) - w(s)) \right] ds \right)^4, & \text{if } t \geq c, \end{cases}$

the solution to (i)? Justify.

(d) Are the solutions to the IVPs in (i) and (ii) uniquely determined? Justify.

6.2 Method of Variation of Parameters

In this section, one of the commonly used methods for finding information on the dynamic processes in sciences and engineering is the method of nonlinear variation of constants/parameters [80]. By having knowledge about the existence of the solution process and utilizing this method, one can find information about the behavior of the dynamic flow.

We also recall (Chapters 2, 4, and 5) that the method of variation of parameters has played a very powerful role in finding the solution representation of differential equations. The idea is very simple. One decomposes a complex system of differential equations into two parts in such a way that the differential equation corresponding to the simpler part is either easily solvable in a closed form or analytically analyzable, but the original complex system of differential equations is neither easily solvable in a closed form nor analytically analyzable. The method of variation of parameters provides a formula for a solution to the original complex system (a perturbed system generally not easily solvable in a closed form) in terms of the solution process of the differential equation (an unperturbed system) corresponding to the simpler part of the rate function. The following theorem is an extension of Lagrange [80].

Following the discussion in Section 4.4, the system of stochastic differential equations (6.1.1) is considered to be a perturbed system of either a deterministic or a stochastic unperturbed/perturbation system. The following result generalizes Theorem 4.4.1.

Theorem 6.2.1. *Let us consider the following system of stochastic differential equations (6.1.1) as a perturbed system with a deterministic unperturbed system:*

$$dx = f(t, x)dt + \sigma(t, x)dw(t), \quad x(t_0) = x_0, \tag{6.2.1}$$

$$dy = f(t, y)dt, \quad y(t_0) = x_0, \tag{6.2.2}$$

where

(H$_{6.2.1}$)(i) *m column vectors of σ and $f \in C[J \times R^n, R^n]$, $x, x_0 \in R^n$, and $J = [t_0, t_0 + a]$;*

(H$_{6.2.1}$)(ii) *the IVP (6.2.2) has a unique solution $y(t, t_0, x_0)$ existing for $t \geq t_0$;*

(H$_{6.2.1}$)(iii) *$\frac{\partial^2}{\partial x_0 \partial x_0} y(t, t_0, x_0)$ exists and is continuous in x_0 for fixed (t_0, x_0);*

(H$_{6.2.1}$)(iv) *the inverse of $\frac{\partial}{\partial x_0} y(t, t_0, x_0) = \Phi(t, t_0, x_0)$ exists and is continuous in (t_0, x_0) for fixed t;*

(H$_{6.2.1}$)(v) *the Wiener process w and x_0 are independent.*

Then, any solution $x(t, t_0, x_0)$ of the IVP (6.2.1) satisfies the relation

$$x(t, t_0, x_0) = y(t, t_0, z(t)), \tag{6.2.3}$$

where $z(t)$ is a solution to the IVP

$$dz = F(t, y(t, t_0, z))dt + \Sigma(t, y(t, t_0, z))dw(t), \quad z(t_0) = x_0, \tag{6.2.4}$$

where

$$\Sigma(t, y(t, t_0, z)) = \Phi^{-1}(t, t_0, z)\sigma(t, y(t, t_0, z)) \tag{6.2.5}$$

$$F(t, y(t, t_0, z)) = -\frac{1}{2}\Phi^{-1}(t, t_0, z)\, \mathrm{tr}\left(\frac{\partial^2}{\partial x_0 \partial x_0} y(t, t_0, z(t))\Psi(t, x(t), z)\right), \tag{6.2.6}$$

$$\Psi(t, x(t), z) = \Phi^{-1}(t, t_0, z)\sigma(t, y(t, t_0, z))\sigma^T(t, y(t, t_0, z))\Phi^{-1T}(t, t_0, z). \tag{6.2.7}$$

Proof. From the hypotheses of the theorem, the IVPs (6.2.1) and (6.2.2) have solution processes $x(t) = x(t, t_0, x_0)$ and $y(t) = y(t, t_0, x_0)$ through (t_0, x_0), respectively. The elementary method of variation of parameters requires the determination of a function $z(t)$ so that

$$x(t, t_0, x_0) = y(t, t_0, z(t)), \quad z(t_0) = x_0 \tag{6.2.8}$$

is a solution process of (6.2.1). From Theorems 6.1.1 and 1.3.2, we have

$$dx(t, t_0, x_0) = dy(t, t_0, z(t))$$

$$= \partial_t y(t, t_0, z(t)) + \frac{\partial}{\partial x_0} y(t, t_0, z(t))dz(t)$$

$$+ \frac{1}{2}\, \mathrm{tr}\left(\frac{\partial^2}{\partial x_0 \partial x_0} y(t, t_0, z(t))dz(t)(dz(t))^T\right) \quad \text{(by Theorem 1.3.2).}$$

$$\tag{6.2.9}$$

From (6.2.5)–(6.2.7) and the assumptions (iii) and (iv), we have

$$f(t, x(t))dt + \sigma(t, x(t))dw(t)$$

$$= dy(t, t_0, z(t)) + \frac{\partial}{\partial x_0} y(t, t_0, z(t))dz + \frac{1}{2}\,\mathrm{tr}\left(\frac{\partial^2}{\partial x_0 \partial x_0} y(t, t_0, z(t))dz(dz)^T\right)$$

$$= f(t, y(t))dt + \Phi(t, t_0, z(t))dz + \frac{1}{2}\,\mathrm{tr}\left(\frac{\partial^2}{\partial x_0 \partial x_0} y(t, t_0, z(t))dz(dz)^T\right)$$

$$= f(t, x(t))dt + \Phi(t, t_0, z(t))dz + \frac{1}{2}\,\mathrm{tr}\left(\frac{\partial^2}{\partial x_0 \partial x_0} y(t, t_0, z(t))dz(dz)^T\right).$$

From this, stochastic calculus, and assumption (iv), we have

$$dz = F(t, y(t, t_0, z))dt + \Sigma(t, y(t, t_0, z))dw(t).$$

This implies that

$$z(t) = x_0 + \int_{t_0}^t F(s, y(s, t_0, z(s)))ds + \int_{t_0}^t \Sigma(s, y(s, t_0, z(s)))dw(t). \qquad (6.2.10)$$

This completes the proof of the theorem. $\qquad\square$

Another version of the variation-of-constants formula for the solution process of (6.2.1) is presented in the following corollary.

Corollary 6.2.1. *Under the hypotheses of Theorem 6.2.1, the following relation is also valid:*

$$x(t, t_0, x_0) = y(t, t_0, x_0) + \int_{t_0}^t \Phi(t, t_0, z(s))(s)dz(s)$$

$$+ \int_{t_0}^t \frac{1}{2}\,\mathrm{tr}\left(\frac{\partial^2}{\partial x_0 \partial x_0} y(t, t_0, z(s))\Psi(s, y(s, t_0, z(s)))\right)ds, \qquad (6.2.11)$$

where $z(t)$ is any solution to (6.2.4).

Proof. For $t_0 \le s \le t$, from (6.2.4), we obtain

$$dy(t, t_0, z(s)) = \frac{\partial}{\partial x_0} y(t, t_0, z(s))dz(s)$$

$$+ \frac{1}{2}\frac{\partial^2}{\partial x_0 \partial x_0} y(t, t_0, z(s))dz(s)(dz(s))^T \quad \text{(by Theorem 1.3.2)}$$

$$= \Phi(t, t_0, z(s))\left[\Sigma(s, y(s, t_0, z(s)))dw(s) + F(s, y(s, t_0, z(s)))ds\right]$$

$$+ \frac{1}{2}\,\mathrm{tr}\left(\frac{\partial^2}{\partial x_0 \partial x_0} y(t, t_0, z(s))\Psi(s, y(s, t_0, z(s)))\right)ds. \qquad (6.2.12)$$

From (6.2.5), (6.2.6), and integrating (6.2.12), we have

$$y(t, t_0, z(t)) = y(t, t_0, x_0) + \int_{t_0}^{t} \Phi(t, t_0, z(s)) F(s, y(s, t_0, z(s))) ds$$

$$+ \int_{t_0}^{t} \Phi(t, t_0, z(s)) \Sigma(s, y(s, t_0, z(s))) dw(s)$$

$$+ \int_{t_0}^{t} \frac{1}{2} \operatorname{tr} \left(\frac{\partial^2}{\partial x_0 \partial x_0} y(t, t_0, z(s)) \Psi(s, y(s, t_0, z(s))) \right) ds.$$

Along with (6.2.8), this yields desired expression (6.2.11):

$$x(t, t_0, x_0) = y(t, t_0, x_0) + \int_{t_0}^{t} \Phi(t, t_0, z(s)) F(s, y(s, t_0, z(s))) ds$$

$$+ \int_{t_0}^{t} \Phi(t, t_0, z(s)) \Sigma(s, y(s, t_0, z(s))) dw(s)$$

$$+ \int_{t_0}^{t} \frac{1}{2} \operatorname{tr} \left(\frac{\partial^2}{\partial x_0 \partial x_0} y(t, t_0, z(s)) \Psi(s, y(s, t_0, z(s))) \right) ds$$

$$= y(t, t_0, x_0) + \int_{t_0}^{t} \Phi(t, t_0, z(s)) F(s, x(s)) ds$$

$$+ \int_{t_0}^{t} \Phi(t, t_0, z(s)) \Sigma(s, x(s)) dw(s)$$

$$+ \int_{t_0}^{t} \frac{1}{2} \operatorname{tr} \left(\frac{\partial^2}{\partial x_0 \partial x_0} y(t, t_0, z(s)) \Psi(s, x(s)) \right) ds. \qquad (6.2.13)$$

In the following, we present a variation-of-constants formula which is analogous to the known result due to Alekseev [80, 98]. Moreover, we shall show that the obtained nonlinear variation-of-constants formula is an alternative to the formula (6.2.11). $\qquad \square$

Theorem 6.2.2. *Assume that f and σ satisfy the hypotheses of Theorems 6.1.1 and Corollary 6.1.1. Then, any solution $x(t) = x(t, t_0, x_0)$ to (6.2.1) satisfies (6.2.3) and (6.2.11). Moreover, we have the stochastic version of the Alekseev-type formula:*

$$x(t, t_0, x_0) = y(t, t_0, x_0) + \int_{t_0}^{t} \Phi(t, s, x(s)) \sigma(s, x(s)) dw(s)$$

$$+ \int_{t_0}^{t} \frac{1}{2} \operatorname{tr} \left(\frac{\partial^2}{\partial x_0 \partial x_0} y(t, s, x(s)) \sigma(s, x(s)) \sigma(s, x(s))^T \right) ds, \quad (6.2.14)$$

where $y(t, t_0, x_0)$ is the solution to (6.2.2).

Proof. From Corollary 6.1.1, $y(t) = y(t, t_0, x_0)$ is a unique solution to (6.2.2), $\frac{\partial}{\partial x_0} y(t, t_0, x_0) = \Phi(t, t_0, x_0)$, and $\Phi(t, t_0, x_0)$ is the fundamental matrix solution to

(6.1.19). Moreover, $\Phi^{-1}(t, t_0, x_0)$ exists and is continuous in (t, t_0, x_0). Thus, f, σ, $\Phi(t, t_0, x_0)$, and $\Phi^{-1}(t, t_0, x_0)$ satisfy the hypotheses (ii)–(iv) of Theorem 6.2.1. Therefore, by the application of Theorem 6.2.1, any solution $x(t) = x(t, t_0, x_0)$ to the IVP (6.1.1) satisfies formulas (6.2.3) and (6.2.11). To show that $x(t, t_0, x_0)$ satisfies (6.2.14), for $t_0 \leq s \leq t$, we consider $x(s) = x(s, t_0, x_0)$ and $y(t, s, x(s))$ as solution processes of (6.2.1) and (6.2.2) through (t_0, x_0) and $(s, x(s))$, respectively. Then,

$$dy(t, s, x(s)) = \frac{\partial}{\partial x_0} y(t, s, x(s)) dx(s) + \frac{1}{2} \frac{\partial^2}{\partial x_0 \partial x_0} y(t, s, x(s)) dx(s)(dx(s))^T$$

$$+ \frac{\partial}{\partial t_0} y(t, s, x(s)) ds \qquad \text{(by Theorem 1.3.2)}$$

$$= \Phi(t, s, x(s))[f(s, x(s) ds + \sigma(s, x(s)) dw(s)]$$

$$+ \frac{1}{2} \operatorname{tr}\left(\frac{\partial}{\partial x_0} y(t, s, x(s)) \sigma(s, x(s))(\sigma(s, x(s))^T \right) ds$$

$$- \Phi(t, s, x(s)) f(s, x(s) ds \qquad \text{(by Theorem 6.2.1, Volume 1)}.$$

Along with (6.18) [80], (6.1.19), and (6.2.1), this yields

$$dy(t, s, x(s)) = \Phi(t, s, x(s)) \sigma(s, x(s)) dw(s)$$

$$+ \frac{1}{2} \operatorname{tr}\left(\frac{\partial}{\partial x_0} y(t, s, x(s)) \sigma(s, x(s))(\sigma(s, x(s)))^T \right) ds. \qquad (6.2.15)$$

Upon integration, (6.2.15) reduces to

$$y(t, t, x(t)) = y(t, t_0, x_0) + \int_{t_0}^t \Phi(t, s, x(s)) \sigma(s, x(s)) dw(s)$$

$$+ \frac{1}{2} \operatorname{tr}\left(\frac{\partial}{\partial x_0} y(t, s, x(s)) \sigma(s, x(s))(\sigma(s, x(s)))^T \right) ds. \qquad (6.2.16)$$

By the uniqueness of the solution process of (6.2.2), we have

$$x(t, t_0, x_0) = y(t, t_0, x_0) + \int_{t_0}^t \Phi(t, s, x(s)) \sigma(s, x(s)) dw(s)$$

$$+ \int_{t_0}^t \frac{1}{2} \operatorname{tr}\left(\frac{\partial}{\partial x_0} y(t, s, x(s)) \sigma(s, x(s))(\sigma(s, x(s)))^T \right) ds.$$

This completes the proof of the theorem. $\qquad \square$

Observation 6.2.1

(i) We recall that Theorem 6.2.1 is an extension of Theorem 4.4.1 to the nonlinear and nonstationary stochastic system (6.1.1).

(ii) Corollary 6.2.1 provides an analytic solution representation of (6.1.1) with regard to the solution representation of the deterministic unperturbed system (6.2.2) in the context of the transformed system (6.2.4). Theorem 6.2.2 provides the solution representation of the perturbed system (6.1.1) with respect to (6.2.2) in the context of itself. This relationship plays an important role in the study of qualitative properties of (6.1.1) in the context of (6.2.2).

(iii) We further note that $t_0 \leq s \leq t$, and from (6.2.8) and the uniqueness of the solution to (6.2.2), we have

$$y(t, s, x(s)) = y(t, s, y(s, t_0, z(s))) = y(t, t_0, z(s)), \qquad (6.2.17)$$

where $y(t, s, x(s))$ and $x(s) = y(s, t_0, z(s))$ are as defined above, and $z(s)$ is a solution process of (6.2.4) through (t_0, x_0).

The following result is an alternate form of Theorem 6.2.1. It is obtained by interchanging the role of the deterministic unperturbed system (6.2.2) and the stochastic unperturbed system of the stochastic perturbed system (6.1.1). It is an extension of Corollary 4.4.1 to nonlinear and nonstationary stochastic systems.

Theorem 6.2.3. *Let us consider the stochastic differential equations (6.2.1) as a perturbed system with a stochastic unperturbed system:*

$$dy = \sigma(t, y)dw(t), \quad y(t_0) = x_0, \qquad (6.2.18)$$

where

($H_{6.2.2}$)(i) *m column vectors of σ and $f \in C[J \times R^n, R^n]$, $x, x_0 \in R^n$, and $J = [t_0, t_0 + a]$;*

($H_{6.2.2}$)(ii) *the IVP (6.2.18) has a unique solution $y(t, t_0, x_0)$ existing for $t \geq t_0$;*

($H_{6.2.2}$)(iii) *$\frac{\partial^2}{\partial x_0 \partial x_0} y(t, t_0, x_0)$ exists and is continuous in x_0 for fixed (t_0, x_0);*

($H_{6.2.2}$)(iv) *the inverse of $\frac{\partial}{\partial x_0} y(t, t_0, x_0) = \Theta(t, w(t), t_0, w(t_0), x_0)$ exists and is continuous in (t_0, x_0) for fixed t;*

($H_{6.2.2}$)(v) *the Wiener process w and x_0 are independent.*

Then, any solution $x(t, t_0, x_0)$ to the IVP (6.2.1) satisfies the relation

$$x(t, t_0, x_0) = y(t, t_0, z(t)), \qquad (6.2.19)$$

where $z(t)$ is a solution to the IVP

$$dz = F(t, y(t, t_0, z))dt, \quad z(t_0) = x_0, \qquad (6.2.20)$$

and

$$F(t, y(t, t_0, z)) = \Theta^{-1}(t, w(t), t_0, w(t_0), z)f(t, y(t, t_0, z)). \qquad (6.2.21)$$

Proof. From the hypotheses of the theorem, the IVPs (6.1.1) and (6.2.18) have the solution processes $x(t) = x(t, t_0, x_0)$ and $y(t) = y(t, t_0, x_0)$ through (t_0, x_0), respectively. The elementary method of variation of parameters requires the determination of a function $z(t)$ so that

$$x(t, t_0, x_0) = y(t, t_0, z(t)), \quad z(t_0) = x_0 \tag{6.2.22}$$

is a solution process of (6.2.1). From Theorems 6.1.1 and 1.3.2, we have

$$dx(t, t_0, x_0) = dy(t, t_0, z(t))$$

$$= \partial_t y(t, t_0, z(t)) + \frac{\partial}{\partial x_0} y(t, t_0, z(t)) dz(t)$$

$$+ \frac{1}{2} \operatorname{tr} \left(\frac{\partial^2}{\partial x_0 \partial x_0} y(t, t_0, z(t)) \dot{dz}(t)(dz(t))^T \right) \quad \text{(by Theorem 1.3.2).}$$

$$\tag{6.2.23}$$

From (6.2.1), (6.2.18), (6.2.23) and the assumptions (iii) and (iv), we have

$$f(t, x(t)) dt + \sigma(t, x(t)) dw(t)$$

$$= dy(t, t_0, z(t)) + \frac{\partial}{\partial x_0} y(t, t_0, z(t)) dz + \frac{1}{2} \operatorname{tr} \left(\frac{\partial^2}{\partial x_0 \partial x_0} y(t, t_0, z(t)) dz(dz)^T \right)$$

$$= \sigma(t, y(t)) dw(t) + \Theta(t, t_0, z(t)) dz + \frac{1}{2} \operatorname{tr} \left(\frac{\partial^2}{\partial x_0 \partial x_0} y(t, t_0, z(t)) dz(dz)^T \right).$$

From (6.2.21), the method of undetermined coefficients and the assumption (iv), we have

$$dz = F(t, y(t, t_0, z)) dt.$$

This implies that

$$z(t) = x_0 + \int_{t_0}^t F(s, y(s, t_0, z(s))) ds. \tag{6.2.24}$$

This completes the proof of the theorem. $\qquad\qquad\square$

Observation 6.2.2. In the framework of Theorem 6.2.3, a corollary and a theorem similar to Corollary 6.2.1 and Theorem 6.2.2 can be reformulated. The details are under investigation. In the following, we present a very general mathematical illustration to exhibit the observation.

Illustration 6.2.1. For simplicity, we consider the scalar stochastic perturbed and unperturbed differential equations

$$dx = [f(t)x + p(t, x)] dt + [\sigma(t)x + q(t, x)] dw(t), \quad x(t_0) = x_0, \tag{6.2.25}$$

and

$$dy = f(t)y\,dt + \sigma(t)y\,dw(t), \quad y(t_0) = x_0, \tag{6.2.26}$$

respectively, where f and σ are any continous functions defined on J into R and p along with q are any smooth functions defined on $J \times R$ into R that insure the existence of the solution process of (6.2.25).

Solution process. Let $x(t) = x(t, t_0, x_0)$ and $y(t) = y(t, t_0, x_0)$ be solution processes of (6.2.25) and (6.2.26) through (t_0, x_0), respectively. From Lemma 2.5.8 and Observation 2.5.9, we have $y(t) = y(t, t_0, x_0) = \Phi(t, t_0)x_0$, $\frac{\partial}{\partial x_0}y(t, t_0, x_0) = \Phi(t, t_0)$, $\frac{\partial^2}{\partial x_0 \partial x_0}y(t, t_0, x_0) = 0$, and the partial differential of $y(t, t_0, x_0)$ with respect to t_0 is

$$\partial_{t_0}x(t_0) = [-f(t_0) + \sigma^2(t_0)]x_0 dt_0 - \sigma(t_0)x_0 dw(t_0). \tag{6.2.27}$$

The differential equations (6.2.25) and (6.2.26) satisfy all the assumptions of Theorem 6.2.1 and 6.2.2, Corollary 6.2.1. Thus, by the applications of these respective results (Theorem 6.2.1, Corollary 6.2.1, and Theorem 6.2.2) to (6.2.25) with its unperturbed stochastic differential equation (6.2.26), we have the following respective results:

(i) $\begin{cases} dz = \Phi^{-1}(t, t_0)[p(t, y(t, t_0, z))dt + p(t, y(t, t_0, z))]dw(t), \quad z(t_0) = x_0, \\ x(t) \equiv x(t, t_0, x_0) = y(t, t_0, z(t)), \end{cases}$

$$\tag{6.2.28}$$

where $z(t)$ is a solution to the IVP (6.2.28);

(ii) $$x(t, t_0, x_0) = y(t, t_0, x_0) + \int_{t_0}^{t} \Phi(t, t_0)dz(s), \tag{6.2.29}$$

(iii) $$x(t, t_0, x_0) = y(t, t_0, x_0) + \int_{t_0}^{t} \Phi(t, s)[p(s, x(s)) - \sigma(s)q(s, x(s))]ds$$

$$+ \int_{t_0}^{t} \Phi(t, s)q(s, x(s))dw(s). \tag{6.2.30}$$

We note that Theorems 2.5.5 and 4.6.8 are special cases of the result (iii).

Next, we present a few of examples to illustrate the presented results.

Example 6.2.1. We consider a scalar nonlinear deterministic unperturbed differential equation and its stochastic perturbed differential equations:

$$dy = -\frac{1}{2}y^3 dt, \quad y(t_0) = x_0, \tag{6.2.31}$$

and

$$dx = -\frac{1}{2}y^3 dt + \sigma(t, x)dw(t), \quad x(t_0) = x_0. \tag{6.2.32}$$

Solution process. From Example 6.2.2 of Volume 1, we have $y(t, t_0, x_0) =$ signum$(x_0)x_0[1 + x_0^2(t - t_0)]^{-1/2}$, $\frac{\partial}{\partial x_0}(t, t_0, x_0) =$ signum$(x_0)[1 + x_0^2(t - t_0)]^{-3/2}$ and $\frac{\partial^2}{\partial x_0 \partial_0} y(t, t_0, x_0) = -3x_0(t - t_0)[1 + x_0^2(t - t_0)]^{-5/2}$. Applying Theorem 6.2.2, we have the solution representation in the context of the unperturbed deterministic system:

$$x(t, t_0, x_0) = y(t, t_0, x_0) + \int_{t_0}^{t} \text{signum}(x(s))\sigma(s, x(s))[1 + x^2(s)(t - s)]^{-3/2} dw(s)$$

$$- \frac{1}{2} \int_{t_0}^{t} 3x(s)(t - s)[1 + x^2(s)(t - s)]^{-5/2}\sigma^2(s, x(s)) ds. \qquad (6.2.33)$$

Example 6.2.2. Let us consider Example 6.1.2 as an unperturbed stochastic differential equation for the following nonlinear stochastic perturbed differential equation:

$$dI = [\alpha I(\bar{N} - I) + g(t, I)]dt + [\beta I + \sigma(t, I)]dw(t), \quad I(t_0) = I_0, \qquad (6.2.34)$$

where g and σ are functions smooth enough to have a solution to the IVP (6.2.34).

Solution process. From the stochastic unperturbed (6.1.34) and perturbed (6.2.34) differential equations, it is clear that the stochastic perturbation is also a stochastic differential equation:

$$dI = g(t, I)dt + \sigma(t, I)dw(t), \quad I(t_0) = I_0. \qquad (6.2.35)$$

From Example 6.1.2, we recall the expressions (6.1.36) and (6.1.37) as

$$\frac{\partial}{\partial I_0} I(t, t_0, I_0) = \frac{\Phi(t, t_0)}{(\Phi(t, t_0) + \alpha I_0 \int_{t_0}^{t} \Phi(t, r)dr)^2},$$

$$\frac{\partial^2}{\partial I_0^2} I(t, t_0, I_0) = \frac{-2\alpha\Phi(t, t_0) \int_{t_0}^{t} \Phi(t, r)dr}{(\Phi(t, t_0) + \alpha I_0 \int_{t_0}^{t} \Phi(t, r)dr)^3}.$$

By denoting solutions $x(t, t_0, x_0)$ and $y(t, t_0, x_0)$ as the solution processes of (6.2.34) and (6.1.34), respectively, and imitating the proof of Theorem 6.2.3, we have

$$dx(t, t_0, x_0) = dy(t, t_0, z(t))$$

$$= \partial_t y(t, t_0, z(t)) + \frac{\partial}{\partial x_0} y(t, t_0, z(t))dz(t)$$

$$+ \frac{1}{2} \text{tr} \left(\frac{\partial^2}{\partial x_0 \partial x_0} y(t, t_0, z(t))dz(t)(dz(t))^T \right) \quad \text{(by Theorem 1.3.2).}$$

$$(6.2.36)$$

Substituting for $dx(t, t_0, x_0)$, $dy(t, t_0, z_0)$, using (6.2.8) and simplifying, we obtain

$$[ax(\bar{N} - x) + g(t, x)]dt + [\beta x + \sigma(t, x)]dw(t)$$

$$= dy(t, t_0, z(t)) + \frac{\partial}{\partial x_0}y(t, t_0, z(t))dz + \frac{1}{2}\text{tr}\left(\frac{\partial^2}{\partial x_0 \partial x_0}y(t, t_0, z(t))dz(dz)^T\right)$$

$$= ay(\bar{N} - y)dt + \beta y\, dw(t) + \Theta(t, t_0, z(t))dz$$

$$+ \frac{1}{2}\text{tr}\left(\frac{\partial^2}{\partial x_0 \partial x_0}y(t, t_0, z(t))dz(dz)^T\right)$$

$$= ax(\bar{N} - x)dt + \beta x\, dw(t) + \Theta(t, t_0, z(t))dz$$

$$+ \frac{1}{2}\text{tr}\left(\frac{\partial^2}{\partial x_0 \partial x_0}y(t, t_0, z(t))dz(dz)^T\right).$$

This implies that

$$g(t, x)dt + \sigma(t, x)dw(t) = \Theta(t, t_0, z(t))dz + \frac{1}{2}\text{tr}\left(\frac{\partial^2}{\partial x_0 \partial x_0}y(t, t_0, z(t))dz(dz)^T\right),$$

and hence

$$dz = \Theta^{-1}(t, t_0, z(t))\left[g(t, y(t, t_0, z(t)))dt + \sigma(t, y(t, t_0, z(t)))dw(t)\right.$$

$$\left. - \frac{1}{2}\text{tr}\left(\frac{\partial^2}{\partial x_0 \partial x_0}y(t, t_0, z(t))dz(dz)^T\right)\right].$$

Now, using the argument in Theorem 6.2.1, we conclude that $z(t)$ satisfies the transformed stochastic differential equation (6.2.4) with the following drift and diffusion coefficients:

$$F(t, I(t, t_0, z))$$

$$= \Theta^{-1}(t, t_0, z)\left[g(t, I(t, t_0, z(t))) - \frac{1}{2}\text{tr}\left[\frac{\partial^2}{\partial x_0 \partial x_0}I(t, t_0, z(t))\Psi(t, x(t), z)\right]\right]$$

$$= \frac{(\Phi(t, t_0) + \alpha z \int_{t_0}^{t}\Phi(t, r)dr)^2}{\Phi(t, t_0)}$$

$$\times \left[g(t, I(t, t_0, z(t))) + \frac{\alpha\Phi(t, t_0)\int_{t_0}^{t}\Phi(t, r)dr}{(\Phi(t, t_0) + \alpha z \int_{t_0}^{t}\Phi(t, r)dr)^3}\Psi(t, x(t), z)\right]$$

$$= \frac{(\Phi(t, t_0) + \alpha z \int_{t_0}^{t}\Phi(t, r)dr)^2}{\Phi(t, t_0)}$$

$$\times \left[g(t, I(t, t_0, z(t))) + \frac{\alpha\int_{t_0}^{t}\Phi(t, r)dr(\Phi(t, t_0) + \alpha z \int_{t_0}^{t}\Phi(t, r)dr)}{\Phi(t, t_0)}\right.$$

$$\left. \sigma^2(t, I(t, t_0, z))\right], \tag{6.2.37}$$

$$\Sigma(t, I(t, t_0, z)) = \frac{(\Phi(t, t_0) + \alpha z \int_{t_0}^{t} \Phi(t, r)dr)^2}{\Phi(t, t_0)} \sigma(t, I(t, t_0, z)), \tag{6.2.38}$$

$$\Psi(t, x(t), z) = \Theta^{-1}(t, t_0, z)\sigma(t, I(t, t_0, z))\sigma^T(t, I(t, t_0, z))\Theta^{-1T}(t, t_0, z)$$

$$= \frac{(\Phi(t, t_0) + \alpha z \int_{t_0}^{t} \Phi(t, r)dr)^4}{\Phi^2(t, t_0)} \sigma^2(t, I(t, t_0, z)). \tag{6.2.39}$$

Example 6.2.3. Again, we consider a simple scalar differential equation,

$$dy = \frac{y}{1+t}dt, \quad y(t_0) = x_0, \tag{6.2.40}$$

and its corresponding stochastic perturbed equation,

$$dx = \frac{x}{1+t}dt + \sigma(t, x)dw(t), \quad x(t_0) = x_0, \tag{6.2.41}$$

where σ is as defined in Example 6.2.2.

Solution process. Here, $y(t, t_0, x_0) = \frac{1+t}{1+t_0}x_0$, $\Phi(t, t_0, x_0) = \frac{1+t}{1+t_0}$ and $0 = \frac{\partial^2}{\partial x_0 \partial x_0} y(t, t_0, x_0)$. Imitating the proof of Theorem 6.2.2, we arrive at

$$x(t) = y(t) + \int_{t_0}^{t} \Phi(t, s)\sigma(s, x(s))dw(s)$$

$$= \frac{1+t}{1+t_0}x_0 + \int_{t_0}^{t} \frac{(1+t)}{(1+s)}\sigma(s, x(s))dw(s). \tag{6.2.42}$$

Exercises

1. By treating the system (6.1.49) in the context of (6.1.50) as a perturbed system of (6.1.51), find the variation-of-constants formula (6.2.14).

2. It is assumed that the solution process of the following stochastic differential equations exists on $J = [t_0, \infty)$. For the following pairs of perturbed and unperturbed nonlinear stochastic/deterministic differential equations, find the variation-of-constants formulas (6.2.11) and (6.2.14) for the perturbed system (P) with respect to the unperturbed differential equation (U):

   (a) $\begin{cases} dx = -\frac{1}{2n}x^{2n+1}dt + \sigma(t, x)dw(t) & x(t_0) = x_0 \quad (P), \\ dy = -\frac{1}{2n}y^{2n+1}dt & y(t_0) = x_0 \quad (U), \end{cases}$

   where $\sigma \in C[J \times R, R]$, and $n \in I(1, \infty)$.

   (b) $\begin{cases} dx = -\frac{a(t)}{2n}x^{2n+1}dt + \sigma(t, x)dw(t) & x(t_0) = x_0 \quad (P), \\ dy = -\frac{a(t)}{2n}y^{2n+1}dt & y(t_0) = x_0 \quad (U), \end{cases}$

   respectively, where $\sigma \in C[J \times R, R]$, $a \in C[J, R]$ is a continuous function, and $n \in I(1, \infty)$.

(c) $\begin{cases} dx = A(t)x\,dt + \sigma(t,x)dw(t) & x(t_0) = x_0 \quad (P), \\ dy = A(t)y\,dt & y(t_0) = x_0 \quad (U), \end{cases}$

respectively, where $A(t)$ is an $n \times n$ continuous matrix function and $\sigma \in C[J \times R^n, R^n]$.

(d) $\begin{cases} dx = Ax\,dt + [Bx + \sigma(t,x)]dw(t) & x(t_0) = x_0 \quad (P), \\ dy = Ay\,dt & y(t_0) = x_0 \quad (U), \end{cases}$

respectively, where A and B are $n \times n$ matrices, and function $\sigma \in C[J \times R^n, R^n]$.

3. Under the hypotheses of Theorem 6.2.3, develop the results parallel to Corollary 6.2.1 and Theorem 6.2.2.

4. It is assumed that the solution process of the following stochastic differential equations exists on $J = [t_0, \infty)$. For the following pairs of perturbed and unperturbed nonlinear stochastic/deterministic differential equations, find the variation-of-constants formulas (6.2.29) and (6.2.30) for the perturbed system (P) with respect to the unperturbed differential equation (U):

(a) $\begin{cases} dx = -\frac{1}{2n}x^{2n+1}dt + \sigma(t,x)dw(t) & x(t_0) = x_0 \quad (P), \\ dy = f(t)y\,dt + \sigma(t)y\,dw(t) & y(t_0) = x_0 \quad (U), \end{cases}$

where $\sigma \in C[J \times R, R]$ and $n \in I(1, \infty)$.

(b) $\begin{cases} dx = -\frac{a(t)}{2n}x^{2n+1}dt + \sigma(t,x)dw(t) & x(t_0) = x_0 \quad (P), \\ dy = f(t)y\,dt + \sigma(t)y\,dw(t) & y(t_0) = x_0 \quad (U), \end{cases}$

respectively, where $\sigma \in C[J \times R, R]$, $a \in C[J, R]$ is a continuous function, and $n \in I(1, \infty)$.

(c) $\begin{cases} dx = A(t)x\,dt + \sigma(t,x)dw(t) & x(t_0) = x_0 \quad (P), \\ dy = A(t)y\,dt + Bx\,dw(t) & y(t_0) = x_0 \quad (U), \end{cases}$

respectively, where $A(t)$ is an $n \times n$ continuous matrix function and $\sigma \in C[J \times R^n, R^n]$.

(d) $\begin{cases} dx = Ax\,dt + [Bx + \sigma(t,x)]dw(t) & x(t_0) = x_0 \quad (P), \\ dy = Ay\,dt + Bx\,dw(t) & y(t_0) = x_0 \quad (U), \end{cases}$

respectively, where A and B are $n \times n$ matrices, and function $\sigma \in C[J \times R^n, R^n]$.

6.3 Method of Generalized Variation of Parameters

By following the classical modeling approach, the evolution of states of a dynamic system can be considered to be a dynamic flow described by (6.1.1). The goal is to find the information (qualitative or quantitative) on the dynamic flow. In general, a closed-form representation of the flow [solution process (6.1.1)] of a nonlinear and nonstationary dynamic system is not possible. Historically, a well-known technique is to measure a dynamic flow by means of a suitable auxiliary measurement device, and then to use this measured dynamic flow to determine the desired information about the original dynamic solution flow. In the following, an energy/Lyapunov-like function is used as a measurement device, and the relationship between the solution processes of a complex system is established in terms of the solution process of an auxiliary dynamic system. This result is called the generalized variation-of-constants formula.

Let us consider the following Itô–Doob-type stochastic auxiliary system of differential equations as a perturbed system,

$$dv = a(t, v)dt + \kappa(t, v)dw(t), \quad z(t_0) = z_0, \tag{6.3.1}$$

and its deterministic auxiliary system (unperturbed),

$$dz = a(t, z)dt, \quad z(t_0) = z_0, \tag{6.3.2}$$

where column vectors of κ and $a \in C[J \times R^n, R^n]$ and are smooth enough to assure the existence of a solution process; in addition, a is twice continuously differentiable with respect to z; the solution $z(t, t_0, z_0)$ to (6.3.2) exists and is unique for $t \geq t_0$; $z(t, t_0, z_0)$ is twice continuously differentiable with respect to the initial state z_0; and further assume that its second derivative $\frac{\partial^2}{\partial z_0 \partial z_0} z(t, s, z_0)$ is locally Lipschitzian in z_0 for each (t, s).

Theorem 6.3.1. *Let us assume that:*

($H_{6.3.1}$)(i) *$z(t, t_0, z_0)$ is the solution process of the deterministic auxiliary system of differential equations (6.3.2) existing for $t \geq t_0$.*

($H_{6.3.1}$)(ii) *$V \in C[R_+ \times R^n, R^N]$ and its partial derivatives V_t, $\frac{\partial}{\partial x} V$ and $\frac{\partial^2}{\partial x \partial x} V$ exist and are continuous on $R_+ \times R^N$.*

($H_{6.3.1}$)(iii) *Let $x(t)$ are solution processes of (6.1.1).*

Then,

$$V(t, x(t)) = V(t, z(t)) + \int_{t_0}^t LV(s, x(s))ds$$

$$+ \int_{t_0}^t V_x(s, z(t, s, x(s)))\Phi(t, s, x(s))\sigma(s, x(s))dw(s), \tag{6.3.3}$$

where

$$dV(s, x(s)) = LV(s, x(s))ds + V_x(s, z(t, s, x(s)))\Phi(t, s, x(s))\sigma(s, x(s))dw(s)$$

$$LV(s, x(s)) = \frac{\partial}{\partial s} V(s, z(t, s, x(s)))$$

$$+ V_x(s, z(t, s, x(s)))\Phi(t, s, x(s))[f(s, x(s)) - a(s, x(s))]$$

$$+ \frac{1}{2} V_x(s, z(t, s, x(s))) \operatorname{tr} \left(\frac{\partial^2}{\partial z_0 \partial z_0} z(t, s, x(s))b(s, x(s)) \right)$$

$$+ \frac{1}{2} \operatorname{tr} \left(V_{xx}(s, z(t, s, x(s)))c(t, s, x(s)) \right) \tag{6.3.4}$$

$$b(s, x(s)) = \sigma(s, x(s))(\sigma(s, x(s))^T,$$

and

$$c(t, s, x(s)) = \Phi(t, s, x(s))b(s, x(s))\Phi^T(t, s, x(s)).$$

Proof. Let $x(t)$ and $z(t)$ be the solution processes of (6.1.1) and (6.3.2) through (t_0, x_0) and (t_0, z_0), respectively. For $t_0 \leq s \leq t$, we apply the Itô–Doob differential rule (Theorem 1.3.2) to $z(t, s, x(s))$ and $V(s, z(t, s, x(s)))$ with respect to s for fixed t, and obtain

$$d_s z(t, s, x(s)) = \frac{\partial}{\partial t_0} z(t, s, x(s))ds + \frac{\partial}{\partial z_0} z(t, s, x(s))dx(s)$$

$$+ \frac{1}{2} \operatorname{tr}\left(\frac{\partial^2}{\partial z_0 \partial z_0} z(t, s, x(s))dx(s)(dx(s))^T\right),$$

and

$$dV(s, z(t, s, x(s))) = \frac{\partial}{\partial s} V(s, z(t, s, x(s)))ds + \frac{\partial}{\partial x} V(s, z(t, s, x(s)))d_s z(t, s, x(s))$$

$$+ \frac{1}{2} \operatorname{tr}\left(\frac{\partial^2}{\partial x \partial x} V(s, z(t, s, x(s)))d_s z(t, s, x(s))(d_s z(t, s, x(s)))^T\right)$$

$$= V_s(s, z(t, s, x(s)))ds + V_x(s, z(t, s, x(s)))\left[\frac{\partial}{\partial z_0} z(t, s, x(s)))dx(s)\right.$$

$$+ \frac{\partial}{\partial t_0} z(t, s, x(s))ds + \frac{1}{2}\operatorname{tr}\left(\frac{\partial^2}{\partial z_0 \partial z_0} z(t, s, x(s))dx(s)(dx(s))^T\right)\bigg]$$

$$+ \frac{1}{2}\operatorname{tr}(V_{xx}(s, y(t, s, x(s)))\Phi(t, s, x(s))dx(s)(\Phi(t, s, x(s))dx(s))^T)$$

$$= V_s(s, z(t, s, x(s)))ds$$

$$+ V_x(s, z(t, s, x(s)))\left[\Phi(t, s, x(s))[[f(s, x(s)) - a(s, x(s))]ds + \sigma(s, x(s))dw(s)]\right.$$

$$+ \frac{1}{2}\operatorname{tr}\left(\frac{\partial^2}{\partial z_0 \partial z_0} z(t, s, x(s))\sigma(s, x(s))(\sigma(s, x(s)))^T\right)ds\bigg]$$

$$+ \frac{1}{2}\operatorname{tr}(V_{xx}(s, z(t, s, x(s)))\Phi(t, s, x(s))\sigma(s, x(s))(\Phi(t, s, x(s))\sigma(s, x(s)))^T)ds.$$

From this and following the proof of the Theorem 6.4.1 [80], we have

$$dV(s, z(t, s, x(s)))$$

$$= V_s(s, z(t, s, x(s)))ds$$

$$+ V_x(s, z(t, s, x(s)))\left[\Phi(t, s, x(s))[[f(s, x(s)) - a(s, x(s))]ds + \sigma(s, x(s))dw(s)]\right.$$

$$+ \frac{1}{2}\operatorname{tr}\left(\frac{\partial^2}{\partial z_0 \partial z_0} z(t, s, x(s))\sigma(s, x(s))(\sigma(s, x(s)))^T\right)ds\bigg]$$

$$+ \frac{1}{2}\operatorname{tr}(V_{xx}(s, z(t, s, x(s)))\Phi(t, s, x(s))\sigma(s, x(s))(\Phi(t, s, x(s))\sigma(s, x(s)))^T)ds$$

$$= LV(s, y(t, s, x(s)))ds$$

$$+ \frac{\partial}{\partial x} V(s, z(t, s, x(s))) \Phi(t, s, x(s)) \sigma(s, x(s)) dw(s). \tag{6.3.5}$$

From (6.2.17) and integrating both sides of (6.3.5) with respect to s from t_0 to t, we get

$$V(t, z(t, t, x(t))) - V(t, z(t)) = \int_{t_0}^t dV(s, x(s)). \tag{6.3.6}$$

From (6.3.6) and the uniqueness of the solution to (6.3.2), the validity of (6.3.3) follows, immediately. This completes the proof of the theorem. □

Observation 6.3.1. We note that if $V(t, x) = x$ and $a = f$ in (6.3.2), then the generalized variation-of-constants formula (6.3.3) reduces to the well-known result (6.2.14) [80, 98].

Example 6.3.1. If $V(t, x) = \frac{1}{2} \|x\|^2$, then $V_t(t, x) \equiv 0$ and $\frac{\partial}{\partial x} V(t, x) = x^T$. Moreover the choice of auxiliary function a in (6.3.2) is f in (6.2.2) $(a = f)$. In this case, the generalized formula (6.3.3) reduces to

$$\|x(t)\|^2 = \|y(t)\|^2 + \int_{t_0}^t y^T(t, s, x(s)) \Phi(t, s, x(s)) \sigma(s, x(s)) dw(s)$$

$$+ \sum_{i,j=1}^n \left[\int_{t_0}^t \left[y^T(t, s, x(s)) \, \mathrm{tr} \left(\frac{\partial^2}{\partial x_{0i} \partial x_{0j}} y(t, s, x(s)) b(s, x(s)) \right) \right. \right.$$

$$+ \frac{1}{2} \sum_{k=1}^n \int_{t_0}^t \frac{\partial}{\partial x_{0i}} y_k(t, s, x(s)) \frac{\partial}{\partial x_{0j}} y_k(t, s, x(s)) b_{ij}(s, x(s)) \right] ds \right], \tag{6.3.7}$$

where $b = (b_{ij})_{n \times n}$ matrix defined in (6.3.4).

Illustration 6.3.1. We pick $V(t, x) = \frac{1}{2} |x|^2$. Let us consider Illustration 6.2.1. Treating (6.2.26) as an auxiliary differential equation and imitating the discussion in Illustration 6.2.1, we have the following expression:

(a) $$|x(t)|^2 = |y(t)|^2 + \int_{t_0}^t \Phi^2(t, s) x(s) [p(s, x(s)) - \sigma(s) x(s) q(s, x(s))$$

$$+ q^2(s, x(s))] ds + \int_{t_0}^t \Phi^2(t, s) x(s) q(s, x(s)) dw(s). \tag{6.3.8}$$

(b) We consider Example 6.2.3. Here, again treating (6.2.40) as an auxiliary differential equation, the expression for the solution process (6.2.41) is

$$|x(t)|^2 = |y(t)|^2 + \int_{t_0}^t \Phi^2(t, s) x(s) \sigma(s, x(s)) dw(s), \tag{6.3.9}$$

respectively.

The following result provides a deviation of the solution process of the perturbed system (6.2.1) with respect to the solution process of the unperturbed system (6.2.2).

Theorem 6.3.2. *Assume that all the hypotheses of Theorem 6.3.1 hold. Furthermore, $y(t) = y(t, t_0, y_0)$ is the solution process of (6.2.2) through (t_0, y_0). Then,*

$$V(t, x(t) - y(t)) - V(t, z(t, t_0, x_0) - z(t, t_0, y_0))$$

$$= \int_{t_0}^{t} dV(s, x(s) - y(s)) ds. \tag{6.3.10}$$

Proof. By following the proof of Theorem 6.3.1, we have the relation

$$dV(s, z(t, s, x(s)) - z(t, s, y(s)))$$

$$= [V_s(s, \Delta z(t, s))$$

$$+ V_x(s, \Delta z(t, s))[[\Phi(t, s, x(s))\Delta x(f, a) - \Phi(t, s, y(s))\Delta y(f, a)]$$

$$+ \frac{1}{2} \operatorname{tr}\left(\frac{\partial^2}{\partial z_0 \partial z_0} \Delta z(t, s) b(s, x(s)))\right)$$

$$+ \frac{1}{2} \operatorname{tr}(V_{xx}(s, \Delta z(t, s))\Phi(t, s, x(s))\sigma(s, x(s))(\Phi(t, s, x(s))\sigma(s, x(s)))^T)]ds$$

$$+ V_x(s, \Delta z(t, s))\Delta \Phi(\sigma, x, y)dw(s)$$

$$= LV(s, \Delta z(t, s))ds + V_x(s, \Delta z(t, s))\Phi(t, s, x(s))\sigma(s, x(s))dw(s),$$

where $\Delta z(t, s) = z(t, s, x(s)) - z(t, s, y(s))$, $\Delta x(f, a) = f(s, x(s)) - a(s, x(s))$, and $\Delta y(f, a) = f(s, y(s)) - a(s, y(s))$. $\qquad\square$

Exercises

1. By treating the system (6.1.49) in the context of (6.1.50) as a perturbed system of (6.1.51), find the generalized variation-of-constants formula (6.3.3) in the context of a quadratic energy function of the type $V(t, N) = (N_1^2, N_2^2, \ldots, N_i^2, \ldots, N_n^2)^T$.

2. It is assumed that the solution process of the following stochastic differential equations exist on $J = [t_0, \infty)$. For the following pairs of stochastic (S) and auxiliary (A) differential equations determine the generalized variation-of-constants formula (6.3.3) with respect to the energy functions:

   (a) $V(t, x) = |x|^{2n}$,
   (b) $V(t, x) = |x|^{2m}$,

   where $m \in I[2, \infty)$.

   a_1 $\begin{cases} dx = -\frac{1}{2n}x^{2n+1}dt + \sigma(t, x)dw(t) & x(t_0) = x_0 \quad \text{(S)}, \\ dz = a(t)z\, dt & z(t_0) = z_0 \quad \text{(A)}, \end{cases}$

where $\sigma \in C[J \times R, R]$ and $n \in I(1, \infty)$.

a_2
$$\begin{cases} dx = -\frac{1}{2n}x^{2n+1}dt + \sigma(t,x)dw(t) & x(t_0) = x_0 \quad \text{(S)}, \\ dz = -\frac{1}{2n}z^{2n+1}dt & z(t_0) = z_0 \quad \text{(A)}, \end{cases}$$
where $\sigma \in C[J \times R, R]$ and $n \in (1, \infty)$.

b_1
$$\begin{cases} dx = [A(t) + B(t)]x\,dt + \sigma(t,x)dw(t) & x(t_0) = x_0 \quad \text{(S)}, \\ dz = A(t)z\,dt & z(t_0) = z_0 \quad \text{(A)}, \end{cases}$$
respectively, where A and B are $n \times n$ continuous matrix functions, and $\sigma \in C[J \times R^n, R^n]$.

b_2
$$\begin{cases} dx = [A(t) + B(t)]x\,dt + \sigma(t,x)dw(t), & x(t_0) = x_0 \quad \text{(S)}, \\ dy = [A(t) + B(t)]z\,dt & z(t_0) = z_0 \quad \text{(A)}, \end{cases}$$
respectively, where A and B are $n \times n$ continuous matrix functions, and $\sigma \in C[J \times R^n, R^n]$.

c_1
$$\begin{cases} dx = Ax\,dt + [Bx + \sigma(t,x)]dw(t), & x(t_0) = x_0 \quad \text{(S)}, \\ dy = Az\,dt & z(t_0) = z_0 \quad \text{(A)}, \end{cases}$$
respectively, where A and B are $n \times n$ constant matrices, and function $\sigma \in C[J \times R^n, R^n]$.

c_2
$$\begin{cases} dx = Ax\,dt + [Bx + \sigma(t,x)]dw(t), & x(t_0) = x_0 \quad \text{(S)}, \\ dy = Bz\,dw(t) & z(t_0) = z_0 \quad \text{(A)}, \end{cases}$$
respectively, where A and B are $n \times n$ constant matrices, and function $\sigma \in C[J \times R^n, R^n]$.

c_3
$$\begin{cases} dx = Ax\,dt + Bx + \sigma(t,x)dw(t), & x(t_0) = x_0 \quad \text{(S)}, \\ dy = Az\,dt + Bz\,dw(t) & z(t_0) = z_0 \quad \text{(A)}, \end{cases}$$
respectively, where A and B are $n \times n$ constant matrices, and function $\sigma \in C[J \times R^n, R^n]$.

6.4 Differential Inequalities and Comparison Theorem

In general, a closed/exact form representation of time evolution flow described by a nonlinear or linear nonstationary interconnected system is not always feasible. Having the knowledge about the existence and in the absence of a closed/exact form representation of dynamic flow, the time evolution of flow or functional of flow, satisfying a scalar dynamic inequality generated by silent or characteristic features of the dynamic system, is estimated by the corresponding comparison flow dynamic. In particular, it is well known [80, 98, 112, 178] that an arbitrary measured dynamic flow satisfying differential inequality is estimated by the extremal solution of the corresponding comparison system of differential equations. This technique is referred to as the method of comparison. This idea is analogous to the "comparison test" for the convergence or divergence of series in calculus [127, 168]. In this section, the above-stated technique is presented in its simplified form. For this purpose, let m be an arbitrary measured dynamic flow.

Prior to the presentation of a basic differential inequality result and the existence of more than one solution to a given differential equation (Example 6.1.1), we need to introduce the concepts of maximal and minimal solutions to the IVP.

We consider the following simpler system of stochastic differential equations with initial conditions:

$$du = g(t, u)dt + G(t, u)dw(t), \quad u(t_0) = u_0, \tag{6.4.1}$$

where G is an $N \times m$ matrix function; m column vectors of G and $g \in C[E, R^N]$, with E being an open (t, u) set in R^{1+N}; and w is an m-dimensional Wiener process as defined in (6.1.1). This type of differential equation is called a system of stochastic comparison differential equations.

Definition 6.4.1. A solution $r(t)$ to the comparison system of differential equations (6.4.1) defined on $[t_0, t_0 + a)$ is said to be a maximal solution process of (6.4.1) if every solution $u(t)$ to (6.4.1) existing on $[t_0, t_0 + a)$ satisfies the inequality

$$u(t) \leq r(t), \quad \text{for } t \in [t_0, t_0 + a). \tag{6.4.2}$$

A solution $\rho(t)$ to the differential equation (6.4.1) defined on $[t_0, t_0 + a)$ is said to be a minimal solution to (6.4.1) if any solution $u(t)$ to (6.4.1) satisfies the inequality

$$\rho(t) \leq u(t), \quad \text{for } t \in [t_0, t_0 + a). \tag{6.4.3}$$

Example 6.4.1. We consider the following very simple IVP:

$$du = 2u^{1/2}dt + \nu u \, dw(t), \quad u(0) = 0.$$

Solution procedure. From Example 6.1.1, we remark that the solution flow of this IVP is

$$u(t) = \begin{cases} 0, & \text{if } 0 \leq t \leq c, \\ \left(\int_c^t \exp\left[-\frac{1}{4}\nu^2(t - s) + \frac{1}{2}\nu(w(t) - w(s)) \right] ds \right)^2, & \text{if } c \leq t. \end{cases}$$

We further note that the maximal and minimal solutions processes are $r(t) = \left(\int_c^t \exp[-\frac{1}{4}\nu^2(t - s) + \frac{1}{2}\nu(w(t) - w(s))]ds \right)^2$ and $\rho(t) = 0$, respectively, for $t \geq 0$.

The concepts of the maximal and minimal solutions to differential equations are the backbone of the study of nonlinear nonstationary dynamic systems. Therefore, it is essential to be assured of the existence of extremal solutions. However, we do not undertake this task at this level. In our discussion, we will be considering some very simple comparison differential equations whose extremum solutions are easily guaranteed.

Theorem 6.4.1 (comparison theorem). *Let us suppose that the drift and diffusion coefficients g and G in (6.4.1) satisfy the hypotheses $(H_{6.1a})$, $(H_{6.1b})$, $(H_{6.1d})$, and $(H_{6.1e})$. Let us further assume that a domain of g and G is $[t_0, t_0 + a) \times R^N$.*

Furthermore, g and G satisfy the following conditions:

(H$_{6.4.1}$)(i) $g(t, u)$ *is quasimonotone nondecreasing in u for each t;*

(H$_{6.4.1}$)(ii) $r(t) = r(t, t_0, u_0)$ *is the maximal solution to (6.4.1) existing on $[t_0, t_0 + a)$;*

(H$_{6.4.1}$)(iii) $\sum_{j=1}^{m} |G_{ij}(t, u) - G_{ij}(t, v)| \leq h_i(|u - v|)$ *on $(t, u), (t, v) \in E, i, h_i \in [R_+, R_+], h_i(0) = 0, h_i$ is nondecreasing and $\int_{0+} \frac{ds}{h_i(s)} = \infty$;*

(H$_{6.4.1}$)(iv) $m, \alpha \in C[[t_0, t_0 + a), R^N], (t, m(t)) \in E$, *and m is a solution to*

$$\begin{cases} dm = \alpha(t)dt + G(t, m)dw(t), & m(t_0) = m_0, \\ \alpha(t) \leq g(t, m(t)). \end{cases} \tag{6.4.4}$$

Then,

(a) $\quad\begin{cases} dm = \alpha(t)dt + G(t, m)dw(t), & m(t_0) = m_0 \\ dm(t) \leq g(t, m(t))dt + G(t, m(t))dw(t), & for\ t \in [t_0, t_0 + a), \end{cases} \tag{6.4.5}$

where $\Delta t = dt\ (\Delta t > 0)$;

(b) $\quad\quad\quad m(t) \leq r(t, t_0, u_0), \quad for\ t \in [t_0, t_0 + a), \tag{6.4.6}$

whenever

$$m(t_0) \leq u_0. \tag{6.4.7}$$

Proof. The proof is left to the reader [98]. $\quad\quad\quad\quad\quad\quad\quad\quad\quad\square$

The following result provides a lower estimate for the solution process of (6.1.1) via the energy/Lyapunov function as a measure of the original dynamic flow.

Corollary 6.4.1 (comparison theorem). *Assume that all hypotheses of Theorem 6.4.1 are satisfied, except that the maximal solution $r(t)$ is replaced by the minimal solution $\rho(t)$ and the inequality (6.4.4) is reversed. Then,*

$$\rho(t) \leq m(t), \quad for\ t \in [t_0, t_0 + a). \tag{6.4.8}$$

Illustration 6.4.1. Let $m, \lambda \in C[t_0, \infty), R]$ and $\lambda \geq 0$. In addition, let us consider the following system of random differential inequalities:

$$\begin{cases} dm(t) = \lambda(t)\phi(t)dt + \sigma(t)m(t)dw(t), & for\ t \geq t_0, \\ \phi(t) \leq m(t), & for\ t \geq t_0, \\ m(t_0) \leq r_0, \end{cases} \tag{6.4.9}$$

where $\Delta t = dt\ (\Delta t > 0)$. Show that

$$dm(t) \leq \lambda(t)m(t)dt + \sigma(t)m(t)dw(t), \quad\quad\quad for\ t \geq t_0, \tag{6.4.10}$$

$$m(t) \leq r_0 \exp\left[\int_{t_0}^{t}\left(f(s) - \frac{1}{2}\sigma^2(s)\right) + \int_{t_0}^{t}\sigma(s)dw(s)\right], \quad for\ t \geq t_0. \tag{6.4.11}$$

Solution process. From (6.4.9), the non-negativity of λ, and the differential $dt(\Delta t > 0)$, we have

$$\lambda(t)\phi(t) \leq \lambda(t)m(t), \tag{6.4.12}$$

and hence the stochastic differential inequality

$$dm(t) \leq \lambda(t)m(t)dt + \sigma(t)m(t)dw(t), \quad \text{for } t \geq t_0.$$

This establishes the stochastic differential inequality (6.4.10). In this case, the following linear stochastic differential equation is the comparison equation:

$$du(t) = \lambda(t)u\,dt + \sigma(t)u\,dw(t), \quad \text{for } t \geq t_0, \quad u(t_0) = r_0. \tag{6.4.13}$$

It has a unique solution process. Therefore, the solution to the IVP (6.4.13) is the maximal solution, $r(t, t_0, r_0)$. Its closed-form representation (Chapter 2) is

$$r(t, t_0, r_0) = r_0 \exp\left[\int_{t_0}^{t}\left(f(s) - \frac{1}{2}\sigma^2(s)\right) + \int_{t_0}^{t}\sigma(s)dw(s)\right]. \tag{6.4.14}$$

We note that

$$|G(t, u) - G(t, v)| = |\sigma(t)u - \sigma(t)v| \leq H|u - v| \text{ on } [t_0, t_0 + a) \times R, \tag{6.4.15}$$

for some positive constant H.

From (6.4.9), (6.4.10), (6.4.12)–(6.4.15) and applying Theorem 6.4.1, we have the desired inequality, i.e.

$$m(t) \leq r_0 \exp\left[\int_{t_0}^{t}\left(f(s) - \frac{1}{2}\sigma^2(s)\right) + \int_{t_0}^{t}\sigma(s)dw(s)\right], \quad \text{for } t \geq t_0,$$

and hence

$$m(t) \leq r(t, t_0, r_0), \quad \text{for } t \geq t_0,$$

where $r(t, t_0, r_0)$ is the maximal solution to the stochastic comparison differential equation (6.4.13).

Exercises

1. Let $E = (t_0, t_0 + a) \times (c, d) \subseteq R^2$ and $g \in C[E, R]$. Assume that α and β are solution processes of

$$\begin{cases} d\alpha = f_1(t, \alpha)dt + \sigma(t, \alpha)dw(t), & \alpha(t_0) = \alpha_0, \\ d\beta = f_2(t, \beta)dt + \sigma(t, \beta)dw(t), & \beta(t_0) = \beta_0, \end{cases}$$

where w is a normalized Wiener process; $\Delta t = dt$ ($\Delta t > 0$). Further, assume that $f_1(t, x) < f_2(t, x)$, $|\sigma(t, x) - \sigma(t, y)| \leq h(|x - y|)$ for $(t, x), (t, y) \in E$ and h satisfies condition $H_{(6.4.1)}$(iii) of Theorem 6.4.1 and $\alpha(t_0) < \beta(t_0)$. Then,

(a) $\begin{cases} d\alpha = f_1(t,\alpha)dt + \sigma(t,\alpha)dw(t), & \alpha(t_0) = \alpha_0, \\ d\beta > f_1(t,\beta)dt + \sigma(t,\beta)dw(t), & \beta(t_0) = \beta_0; \end{cases}$

(b) $\alpha(t) < \beta(t)$, $t \in (t_0, t_0 + a)$.

2. Let $m, \phi, \nu, \lambda \in C[[t_0, t_0 + a), R]$, with λ being a non-negative function on R. Further, assume that ϕ, ν, λ, and ϕ satisfy the following inequality:

$$\begin{cases} dm(t) = [\lambda(t)\phi(t) + \nu(t)]dt + \sigma(t)m(t)dw(t), & \text{for } t \geq t_0, \\ \phi(t) \leq m(t), & \text{for } t \geq t_0, \\ m(t_0) \leq r_0, \end{cases}$$

where $\Delta t = dt$ $(\Delta t > 0)$. Show that for:

(a) $dm(t) \leq [\lambda(t)m(t) + \nu(t)]dt + \sigma(t)m(t)dw(t)$, for $t \geq t_0$;

(b) the stochastic comparison differential equation

$$du = [\lambda(t)u + \nu(t)]dt + \sigma(t)u\,dw(t), \quad \text{for } t \geq t_0, \quad u(t_0) = r_0;$$

(c) $\phi(t) \leq \Phi(t, t_0)r_0 + \int_{t_0}^t \Phi(t, s)v(s)ds$ where $\Phi(t, s) = \exp\left[\int_s^t [\lambda(\theta - \frac{1}{2}\sigma^2(\theta)] \, d\theta + \int_s^t \sigma(\theta)dw(\theta)\right]ds$.

3. Let $m, \phi, \varphi, \mu \in C[[t_0, t_0 + a), R]$, with μ being a non-negative function on R. Further, assume that ϕ, φ, and λ satisfy the following inequality:

$$\begin{cases} dm(t) = [\mu(t)\phi(t) + \varphi(t)]dt + \sigma(t)m(t)dw(t), & \text{for } t \geq t_0, \\ \phi(t) \geq m(t), & \text{for } t \geq t_0, \\ m(t_0) \geq r_0, \end{cases}$$

where $\Delta t = dt$ $(\Delta t > 0)$. Show that $\Delta t = dt$ $(\Delta t > 0)$:

(a) $dm(t) \geq [\mu(t)m(t) + \varphi(t)]dt + \sigma(t)m(t)dw(t)$, for $t \geq t_0$;

(b) the stochastic comparison differential equation

$$du = [\mu(t)u + \varphi(t)]dt + \sigma(t)u\,dw(t), \quad \text{for } t \geq t_0, \quad u(t_0) = r_0;$$

(c) $\phi(t) \geq \Phi(t, t_0)r_0 + \int_{t_0}^t \Phi(t, s)\varphi(s)$ where $\Phi(t, s) = \exp\left[\int_s^t [\mu(\theta - \frac{1}{2}\sigma^2(\theta)d\theta] + \int_s^t \sigma(\theta)dw(\theta)\right]ds$.

4. Let $\phi, m \in C[[t_0, t_0 + a), R]$, and let

$$\begin{cases} dm(t) = \phi(t)dt + \sigma m(t)dw(t), & \text{for } t \geq t_0, \\ \phi(t) \leq m(t), & \text{for } t \geq t_0, \\ m(t_0) = r_0 = c. \end{cases}$$

Show that:

(a) $\phi(t) \leq c\exp\left[(f - \frac{1}{2}\sigma^2)(t - t_0) + \sigma(w(t) - w(t_0))\right]$, for $t \geq t_0$;

(b) for $2f < \sigma^2$, $\lim_{t \to \infty} \phi(t) \leq 0$;

(c) if $0 \leq \phi(t)$ on $[t_0, t_0 + a)$, what can you say about (a) and (b)?

6.5 Comparison Theorems

It is well known that Lyapunov's second method has played a very significant role in the qualitative and quantitative analysis of nonlinear nonstationary systems of dynamic systems in the biological, engineering, physical, and social sciences. The concept of the Lyapunov/energy function is considered to be a suitable auxiliary measurement device for measuring a dynamic flow determined by (6.1.1). Let us consider a Lyapunov-like/energy function (Section 3.2 of Chapter 3) that satisfies:

(H$_{6.5.1}$)(a) $V \in C[R_+ \times R^n, R^N]$, and its partial derivatives, V_t, $\frac{\partial}{\partial x}V$, and $\frac{\partial^2}{\partial x \partial x}V$ exist and are continuous on $R_+ \times R^n$;

(H$_{6.5.1}$)(b) V_t, V_x, and V_{xx} satisfy the relation

$$dV(t, x) = LV(t, x)dt + V_x(t, x)\sigma(t, x)dw(s), \qquad (6.5.1)$$

where the operator L (defined in Theorem 1.3.2) is with respect to (6.1.1) as

$$LV(t, x) = V_t(t, x) + V_x(t, x)f(t, x) + \frac{1}{2}\text{tr}(V_{xx}(t, x)b(t, x)), \quad (6.5.2)$$

$$b(t, x) = \sigma(t, x)\sigma^T(t, x)$$

We present a very basic comparison theorem in the framework of the Lyapunov-like function.

Theorem 6.5.1. *Let $V \in C[R_+ \times R^n, R^N]$, and let it satisfy the hypothesis (H$_{6.5.1}$). In addition, let us assume that:*

(H$_{6.5.2}$) (i) *the Itô–Doob-type differential $dV(t, x)$, defined in (6.5.1), satisfies the inequality*

$$\begin{cases} LV(t, x) \le g(t, V(t, x)), \\ V_x(t, x)\sigma(t, x) = G(t, V(t, x)), \end{cases} \qquad (6.5.3)$$

where the functions g and G satisfy the conditions of Theorem 6.4.1;

(H$_{6.5.2}$) (ii) *$r(t) = r(t, t_0, u_0)$ is the maximal solution to the system of differential equations*

$$du = g(t, u)dt + G(t, u)dw(t), \quad u(t_0) = u_0, \qquad (6.5.4)$$

existing to the right of t_0;

(H$_{6.5.2}$) (iii) *$x(t) = x(t, t_0, x_0)$ is any solution to (6.1.1) existing for $t \ge t_0$ such that*

$$V(t_0, x_0) \le u_0. \qquad (6.5.5)$$

Then,

$$V(t, x(t)) \le r(t), \quad t \ge t_0. \qquad (6.5.6)$$

Proof. Let $x(t)$ be a solution process of (6.1.1). We set

$$m(t) = V(t, x(t)), \quad m(t_0) = V(t_0, x_0). \tag{6.5.7}$$

For sufficiently small $\Delta t > 0$, we have

$$
\begin{aligned}
m(t + \Delta t) - m(t) &= V(t + \Delta t, x(t + \Delta t)) - V(t, x(t)) \\
&= V(t + \Delta t, x(t) + \Delta t f(t, x(t)) + \sigma(t, x(t))\Delta w(t)) - V(t, x(t)) \\
&\quad + V(t + \Delta t, x(t + \Delta t)) \\
&\quad - V(t + \Delta t, x(t) + \Delta t f(t, x(t)) + \sigma(t, x(t))\Delta w(t)).
\end{aligned}
$$

From this, (6.5.3), the smoothness of V, and following the formal argument used in the proof of Theorem 1.3.2, we have

$$
\begin{aligned}
dm(t) = dV(t, x(t)) &= LV(t, x(t))dt + V_x(t, x(t))\sigma(t, x(t))dw(s) \\
&= LV(t, x(t))dt + G(t, V(t, x))dw(s), \tag{6.5.8}
\end{aligned}
$$

where the differential operator L and the function G are as defined in (6.5.2) and (6.5.3), respectively. This together with (6.5.3), (6.5.8), and for $\Delta t > 0$ yields the inequality

$$dV(t, x(t)) \leq g(t, V(t, x))dt + G(t, V(t, x))dw(t), \tag{6.5.9}$$

where $\Delta t = dt$ is a positive differential of the function t. From (6.5.7) and (6.5.8), (6.5.9) reduces to

$$dm(t) \leq g(t, m(t))dt + G(t, m(t))dw(t), \quad t \geq t_0,$$

and

$$m(t_0) \leq u_0. \tag{6.5.10}$$

From (6.5.5), (6.5.8)–(6.5.10) and under the assumptions of the theorem, all the assumptions of Theorem 6.4.1 are satisfied. Hence, by the application of Theorem 6.4.1, we have

$$m(t) \leq r(t), \quad t \geq t_0.$$

Thus,

$$V(t, x(t)) \leq r(t), \quad t \geq t_0.$$

This completes the proof of the theorem. $\qquad \square$

Some common special cases of Theorem 6.5.1 in literature are exhibited by the following corollary:

Corollary 6.5.1

(a) *For $g(t, u) \equiv 0$ and $G(t, u) \equiv 0$, the inequality (6.5.9) reduces to $dV(t, x) \leq 0$. The stochastic comparison differential equation is $du = 0$. Then, the function*

$V(t, x(t))$ is nonincreasing in t, and the relation (6.5.6) reduces to

$$V(t, x(t)) \leq V(t_0, x_0), \quad t \geq t_0.$$

(b) *For $g(t, u) \equiv \lambda(t)u$ and $G(t, u) = [G_1(t)u, \ldots, G_j(t)u, \ldots, G_m(t)u]$, the comparison stochastic differential equation is $du = \lambda(t)u\,dt + u\sum_{j=1}^{m} G_j(t)w_j(t)$. The inequality (6.5.9) reduces to*

$$dV(t, x)dt \leq \lambda(t)V(t, x)dt + V(t, x)\sum_{j=1}^{m} G_j(t)w_j(t). \tag{6.5.11}$$

Then, the relation (6.5.6) reduces to

$$V(t, x(t)) \leq V(t_0, x_0)\left[\exp\int_{t_0}^{t}\left(\lambda(s) - \frac{1}{2}\sum_{j=1}^{m} G_j^2(s)\right)ds + \sum_{j=1}^{m}\int_{t_0}^{t} G_j(s)w_j(s)\right],$$

$$t \geq t_0.$$

The following variant of Theorem 6.5.1 is more useful in several applications. The following comparison result is in the framework of deterministic differential inequalities and comparison theorems.

Theorem 6.5.2. *Assume that hypotheses all Theorem 6.5.1 of except for the representation $V_x(t, x)\sigma(t, x) = G(t, V(t, x))$ in (6.5.3) are valid. Further assume that:*

$(H_{6.5.2})(iv)$ *g in (6.5.9) is concave in u for fixed t;*
$(H_{6.5.2})(v)$ *the expected value of $E[V(t, x(t))]$ exists;*
$(H_{6.5.2})(vi)$ *$E[V(t_0, x_0)] \leq u_0$.*

Then,

$$E[V(t, x(t))] \leq r(t), \quad for \ t \geq t_0, \tag{6.5.12}$$

where $r(t) = r(t, t_0, u_0)$ is the maximal solution to the system of nonlinear deterministic comparison differential equations

$$du = g(t, u)dt, \quad u(t_0) = u_0, \tag{6.5.13}$$

where g satisfies all the assumptions of Theorem 6.4.1.

Proof. The proof of this is parallel to the proof of Theorem 6.5.1 and its deterministic version (Chapter 6, Volume I). The details are left as an exercise. □

Observation 6.5.1. Corollary 6.5.1 and several other versions of comparison results play a very important role in the study of highly complex systems of stochastic differential equations. The use of the deterministic differential inequalities and comparison theorems are attractive and computationally easy to verify. In particular, the relation $V_x(t, x)\sigma(t, x) = G(t, V(t, x))$ in (6.5.3) is very restrictive. However, in the absence of this relation, the comparison results of the type in Theorem 6.5.2 have played a significant role in the theory of stochastic systems of differential equations [98, 112].

Illustration 6.5.1. Let us consider the following system of linear time-varying differential equations:

$$dx = A(t, x)x\, dt + B(t, x)x\, dw(t), \quad x(t_0) = x_0, \qquad (6.5.14)$$

where $n \times n$ matrix functions A and B are defined and continuous on $R_+ \times R^n$. Further assume that

$$\frac{1}{2}x^T[A^T(t, x) + A(t, x)]x \le \lambda(t)\|x\|^2, \qquad (6.5.15)$$

$$\frac{1}{2}^T[B^T(t, x) + B(t, x)]x = \sigma(t)\|x\|^2, \qquad (6.5.16)$$

and

$$\text{tr}(B(t, x)xx^T B^T(t, x)) \le \gamma(t)\|x\|^2. \qquad (6.5.17)$$

Show that

$$\|x(t)\|^2 \le \|x_0\|^2 \exp\left[\int_{t_0}^t \left(f(s) - \frac{1}{2}\sigma^2(s)\right) + \int_{t_0}^t \sigma(s)dw(s)\right], \quad \text{for } t \ge t_0,$$
$$(6.5.18)$$

where the norm of x is an n-dimensional Euclidean norm and $f = \lambda + \gamma$.

Solution process. Let $V(t, x) = \frac{1}{2}\|x(t)\|^2 = \frac{1}{2}x^T x$. From Theorem 1.3.2, we have

$$LV(t, x(t)) = \frac{1}{2}x^T[A^T(t, x) + A(t, x)]x + \text{tr}(B(t, x)xx^T B^T(t, x)), \qquad (6.5.19)$$

and

$$V_x(t, x)\sigma(t, x) = \frac{1}{2}x^T[B^T(t, x) + B(t, x)]x. \qquad (6.5.20)$$

From (6.5.14)–(6.5.17), (6.5.19), (6.5.20), we have

$$\begin{cases} dV(t, x) = LV(t, x)dt + G(t, V(t, x))dw(t), \\ dV(t, x) \le g(t, V(t, x)) + G(t, V(t, x))dw(t), \end{cases} \qquad (6.5.21)$$

where

$$g(t, V(t, x)) = (\lambda(t) + \gamma(t))V(t, x) \quad \text{and} \quad G(t, V(t, x)) = \sigma(t)V(t, x). \quad (6.5.22)$$

The stochastic comparison differential equation (6.5.4) reduces to

$$du(t) = (\lambda(t) + \gamma(t))u\, dt + \sigma(t)u\, dw(t), \quad \text{for } t \ge t_0, \quad u(t_0) = u_0. \qquad (6.5.23)$$

Let $r(t, t_0, u_0)$ be the maximal, and let $x(t) = x(t, t_0, x_0)$ be any solution processes of (6.5.23) and (6.5.14), respectively, satisfying the relation

$$V(t_0, x_0) = \frac{1}{2}\|x_0\|^2 \le u_0. \tag{6.5.24}$$

From (6.5.21), (6.5.23), (6.5.24) and the application of comparison Theorem 6.5.1, we have the desired estimate on the solution to (6.5.14)

$$\|x(t)\|^2 \le \|x_0\|^2 \exp\left[\int_{t_0}^t \left(f(s) - \frac{1}{2}\sigma^2(s)\right) + \int_{t_0}^t \sigma(s)dw(s)\right] \quad \text{and} \quad t \in J,$$

where $f = \lambda + \gamma$. This completes the solution process of the illustration.

Example 6.5.1. Let us consider Example 6.2.3. We choose $V(t, x) = \frac{1}{2}x^2$. In this case,

$$LV(t, x(t)) = \frac{x^2}{1+t} + \frac{1}{2}\sigma^2(t, x). \tag{6.5.25}$$

(a) In order to apply Comparison Theorem 6.5.1, we need to impose the following conditions:

$$V_x(t, x)\sigma(t, x) = x\sigma(t, x), \tag{6.5.26}$$

$$\sigma^2(t, x) \le \gamma(t)x^2 \quad \text{and} \quad 2x\sigma(t, x) = b(t)x^2, \tag{6.5.27}$$

where γ and b are continuous functions defined on R into R. From (6.2.41), (6.5.25)–(6.5.27), and by imitating the argument used in Illustration 6.5.1, we have

$$\begin{cases} dV(t, x) = LV(t, x)dt + G(t, V(t, x))dw(t), \\ dV(t, x) \le g(t, V(t, x)) + G(t, V(t, x))dw(t), \end{cases} \tag{6.5.28}$$

where

$$g(t, V(t, x)) = \left(\frac{2}{1+t} + \gamma(t)\right)V(t, x), \quad G(t, V(t, x)) = b(t)V(t, x), \tag{6.5.29}$$

and a stochastic comparison differential equation:

$$du(t) = \left(\frac{2}{1+t} + \gamma(t)\right)u\,dt + \sigma(t)u\,dw(t), \quad \text{for } t \ge t_0, \quad u(t_0) = u_0. \tag{6.5.30}$$

By applying Theorem 6.5.1, we have the following estimate for the solution process of (6.2.41):

$$\|x(t)\|^2 \le \|x_0\|^2 \exp\left[\int_{t_0}^t \left(f(s) - \frac{1}{2}\sigma^2(s)\right) + \int_{t_0}^t \sigma(s)dw(s)\right], \quad \text{for } t \ge t_0, \tag{6.5.31}$$

where $f(t) = \frac{2}{1+t} + \gamma(t)$.

(b) On the other hand, if we apply Theorem 6.5.2, we can drop the relation (6.5.26) and relax the condition (6.5.27) to get

$$\sigma^2(t, x) \leq \gamma(t)x^2. \tag{6.5.32}$$

In this case, we have

(i) $\quad LV(t, x) \leq \left(\dfrac{2}{1+t} + \gamma(t)\right)V(t, x); \tag{6.5.33}$

(ii) a deterministic comparison differential equation,

$$du(t) = \left(\frac{2}{1+t} + \gamma(t)\right)u\,dt, \quad u(t_0) = u_0. \tag{6.5.34}$$

From the application of Theorem 6.5.2, we have the following mean square estimate for the solution process of (6.2.41):

$$E\|x(t)\|^2 \leq E[\|x_0\|^2]\exp\left[\int_{t_0}^t f(s))\right], \tag{6.5.35}$$

where f is as defined in (6.5.31).

Observation 6.5.2

(i) Example 6.5.1 exhibits the strength and weakness of the condition $V_x(t, x)\sigma(t, x) = G(t, V(t, x))$. Theorem 6.5.2 is applicable to a wider class of differential equations of the type (6.1.1). On the other hand, Theorem 6.5.1 is restricted to a special class of partial differential equations determined by $V_x(t, x)\sigma(t, x) = G(t, V(t, x))$, for some G. There is a tradeoff between these conditions.

(ii) Of course, there is very close relationship between the energy function method for finding the closed-form solution to the scalar version of (6.1.1) and the comparison method for finding estimates for the evolution process of (6.1.1).

(iii) Both methods assume that the IVP (6.1.1) has solutions. Moreover, the solution process of the transformed (reduced) stochastic differential equation can be considered to be a very special estimate for the solution to the original stochastic differential equations. In both cases, we need to find $LV(t, x)$ and $V_x(t, x)\sigma(t, x)$. However, for Theorem 6.5.2, $V_x(t, x)\sigma(t, x) = G(t, V(t, x))$ is not required.

Exercises

1. Let $V(t, x) = x^2$ and let $x(t)$ be any solution to the differential equation. Using Theorem 6.5.1 (if possible), find an estimate for the solution to the following differential equations with initial data (t_0, x_0):

   (a) $dx = \left[\cos tx - \dfrac{x}{1+\sin^2 x}\right]dt + \sigma(t)x\,dw(t);$

(b) $dx = \left[2x - \frac{6x}{1+\exp[-x]}\right]dt + 3x\,dw(t);$

(c) $dx = [2x - x(1 + \exp[1 + \sin x])]dt + 2x\,dw(t).$

2. Let $B = \text{diag}(d_1, \ldots, d_i, \ldots, d_n)$ be a diagonal matrix. Let f be a smooth function. Let $V(t, x) = x^T x$. Applying Theorem 6.5.1, find an estimate for the solution process of the following equation:

$$dx = f(t, x)dt + Bx\,dw(t).$$

3. Let us consider the following stochastic differential equations:

$$dx = f(t, x)dt + \sigma(t, x)dw(t), \quad x(t_0) = x_0,$$

where f and σ are smooth functions defined on $R \times R$ into R to assure the existence of a solution process. Moreover, $\sigma \neq 0$, $\frac{\partial}{\partial t}\sigma$, and $\frac{\partial}{\partial x}\sigma$ are continuous. Using $V(t, x) = \int^x \frac{ds}{\sigma(t,s)}$:

(a) Find $LV(t, x)$ and $V_x(t, x)\sigma(t, x)$.
(b) Under what conditions, on can prove or disprove the following:

   (i) $LV(t, x) \leq g(t, V(t, x))$ for some smooth function g, and
   (ii) $V_x(t, x)\sigma(t, x) = dw(t)$.

   (Hint: Chapter 3, energy function method.)
(c) If (b) is valid, then which of the comparison results is applicable?
(d) If the answer to the question in (c) is Theorem 6.5.1, then prove or disprove that under the assumption of (b), every nonlinear stochastic scalar differential equation has a stochastic comparison differential equation of the type

$$du = g(t, u) + b(t)dw(t), \quad u(t_0) = u_0,$$

or one may even be able find a closed-form solution to the given nonlinear stochastic differential equation by solving the stochastic comparison differential equations.

(e) Find an estimate for the given stochastic differential equation.
(f) Using (a)–(e), what kind(s) of significant conclusion(s) can you draw?

4. To what extent can one generalize Exercise 3 to a system of nonlinear stochastic differential equations of the type (6.1.1)?

6.6 Variational Comparison Method

In this section, we generalize the comparison theorems presented in Section 6.5. This generalization is based on the ideas of two nonlinear methods, namely the method of nonlinear variation of constants parameters and the classical Lyapunov's second method. We referred to this as the "hybrid dynamic method." By employing the concept of the Lyapunov function and differential inequalities of Section 6.5, we present a very general variational comparison theorem. This result connects the solution processes of the dynamic system (6.1.1) with the maximal solution to the corresponding comparison dynamic equation (6.4.1) and a solution to an auxiliary dynamic system, either (6.3.1) or (6.3.2). For simplicity, we frequently use (6.3.2). As a byproduct of this, several auxiliary comparison results are formulated. These results generalize and extend the existing results.

Theorem 6.6.1. *Assume that all the hypotheses of Theorem 6.3.1 are satisfied. Further assume that:*

$(\mathbf{H_{6.6.1}})$(i) *for $LV(s, z(t, s, x))$ in (6.3.4), $g(t, u)$ and $G(t, u)$ in (6.4.1) satisfy the following inequalities*

$$\begin{cases} LV(s, z(t, s, x)) \leq g(s, V(s, z(t, s, x))), \\ \dfrac{\partial}{\partial x}V(s, z(t, s, x))\Phi(t, s, x)\sigma(s, x) = G(s, V(s, z(t, s, x))), \end{cases} \tag{6.6.1}$$

where $z(t, s, x)$ is the solution process to the auxiliary system (6.3.2) through (s, x);

$(\mathbf{H_{6.6.1}})$(ii) *let $r(t) = r(t, t_0, u_0)$ be the maximal solution to the comparison differential equation (6.4.1) existing for $t \geq t_0$;*

$(\mathbf{H_{6.6.1}})$(iii) *let $x(s) = x(s, t_0, x_0)$ be a solution to (6.1.1) existing for $t \geq t_0$.*

Then,

$$V(t, x(t, t_0, x_0)) \leq r(t, t_0, u_0), \quad \text{for } t \geq t_0, \tag{6.6.2}$$

provided that

$$V(t, z(t, t_0, x_0)) \leq u_0. \tag{6.6.3}$$

Proof. For $t_0 \leq s \leq t$, we imitate the proof of Theorem 6.3.1 and arrive at (6.3.5), i.e.

$$dV(s, z(t, s, x(s))) = LV(s, z(t, s, x(s)))ds$$
$$+ \frac{\partial}{\partial x}V(s, z(t, s, x(s)))\Phi(t, s, x(s))\sigma(s, x(s))dw(s). \tag{6.6.4}$$

From (6.6.1) and (6.6.4), we have

$$\begin{cases} dV(s, z(t, s, x(s))) = LV(s, z(t, s, x(s)))ds + G(s, V(s, z(t, s, x(s))))dw(s), \\ dV(s, z(t, s, x(s))) \leq g(s, V(s, z(t, s, x(s))))ds + G(s, V(s, z(t, s, x(s))))dw(s), \end{cases}$$
$$\tag{6.6.5}$$

provided that $\Delta s > 0$ and $\Delta s = ds$. Now, we set

$$m(s) = V(s, z(t, s, x(s))), \quad m(t_0) = V(t, z(t, t_0, , x_0)). \tag{6.6.6}$$

From (6.6.6), (6.6.5) reduces to

$$\begin{cases} dm(s) = LV(s, z(t, s, x(s)))ds + G(s, m(s))dw(s), \quad m(t_0) = V(t, z(t, t_0, x_0)), \\ dm(s) \leq g(s, m(s))ds + G(s, m(s))dw(s). \end{cases}$$
$$\tag{6.6.7}$$

Under the assumption of the theorem as well as relations (6.6.3) and (6.6.7), Theorem 6.4.1 is applicable. Thus, repeating the argument used in Theorem 6.5.1, the proof of the theorem can be completed. The details are left to the reader. $\square$

Theorem 6.6.2. *Assume that all hypotheses of Theorem* 6.6.1 *remain true except that the relation* $\frac{\partial}{\partial x} V(s, z(t, s, x(s)))\Phi(t, s, x(s))\sigma(s, x(s)) = G(s, V(s, z(t, s, x(s))))$ *in* (6.6.1) *is replaced by the following:*

($H_{6.6.1}$)(d) *g in* (6.4.1) *is concave in u for fixed t;*
($H_{6.6.1}$)(e) *The expected value of $E[V(t, x(t))]$ exists;*
($H_{6.6.1}$)(f) *$E[V(t, z(t, t_0, x_0))] \le u_0$.*

Then,

$$E[V(t, x(t))] \le r(t), \quad for\ t \ge t_0, \tag{6.6.8}$$

where $r(t) = r(t, t_0, u_0)$ is the maximal solution of the system of nonlinear deterministic comparison differential equations (6.5.13).

Proof. The proof of this is parallel to the proof of Theorem 6.5.2 and its deterministic version (Chapter 6 of Volume 1). The details are left as an exercise. $\square$

Observation 6.6.1. We remark that the presented comparison theorems generalize the existing comparison theorems in a systematic and unified way. For example:

(i) If an auxiliary rate function $a = 0$, then (6.6.3) reduces to $V(t, x_0) \le u_0$.
(ii) If an auxiliary rate function $a = A(t)z$, then (6.6.3) reduces to $V(t, \Phi(t, t_0)x_0) \le u_0$, and so on.

Illustration 6.6.1. We consider the mathematical Illustration 6.2.1 and choose $V(t, x) = \frac{1}{2}x^2$. Let us assume that the auxiliary stochastic differential equation (6.3.1) is scalar and linear with drift and diffusion coefficients $a(t)z$ and $\kappa(t)z$. Examine the feasibility of applying at most two theorems: Theorems 6.6.1 and 6.6.2.

Solution process. First, we compute $LV(s, z(t, s, x))$. Using the necessary information in Illustration 6.2.1, we compute

$$dV(s, (z, (t, s, x))) = \frac{1}{2}dz^2(t, s, x)$$

$$= z(t, s, x)dz(t, s, x) + \frac{1}{2}(dz(t, s, x))^2$$

$$= z(t, s, x)\left[\partial_s z(t, s, x) + \partial_{x_0} z(t, s, x) + \partial_{x_0 s}^2 z(t, s, x) + \frac{1}{2}\frac{\partial^2}{\partial z_0 \partial z_0} z(t, s, x)\right]$$

$$+ \frac{1}{2}(dz(t, s, x))^2 \quad \text{(from Observations 2.5.9 and 4.6.7)}$$

$$= z(t, s, x)\Phi(t, s)[[-a(s) + \kappa^2(s)]x\, ds - \kappa(s)x\, dw(s)) + dx$$

$$- \kappa(s)[\sigma(s)x + q(s, x)](dw(t))^2] + \frac{1}{2}(dz(t, s, x))^2$$

$$= z(t, s, x)\Phi(t, s)[[[-a(s) + \kappa^2(s) + f(s) - \kappa(s)\sigma(s)]x$$

$$+ p(s, x) - \kappa(s)q(s, x)]ds + [(-\kappa(s) + \sigma(s))x(s) + q(s, x)]dw(s)]$$

$$+ \Phi^2(t, s)[(-\kappa(s) + \sigma(s))x + q(s, x)]^2 ds$$

$$= z(t, s, x)\Phi(t, s)[[(f(s) - a(s) + \kappa^2(s) - \kappa(s)\sigma(s))x$$

$$+ p(s, x) - \kappa(s)q(s, x)]ds + [(\sigma(s) - \kappa(s))x(s) + q(s, x)]dw(s)]$$

$$+ \frac{1}{2}\Phi^2(t, s)[(\sigma(s) - \kappa(s))x + q(s, x)]^2 ds$$

$$= LV(s, z(t, s, x))ds + z(t, s, x)\Phi(t, s)[(\sigma(s) - \kappa(s))x(s) + q(s, x)]dw(s).$$

$$(6.6.9)$$

Here,

$$LV(s, z(t, s, x))$$

$$= z(t, s, x)\Phi(t, s)[(f(s) - a(s) + \kappa^2(s) - \kappa(s)\sigma(s))x + p(s, x) - \kappa(s)q(s, x)]ds$$

$$+ \frac{1}{2}\Phi^2(t, s)[(\sigma(s) - \kappa(s))x + q(s, x)]^2 ds$$

$$= (f(s) - a(s) + \kappa^2(s) - \kappa(s)\sigma(s))z(t, s, x)\Phi(t, s)x\, ds$$

$$+ z(t, s, x)\Phi(t, s)(p(s, x) - \kappa(s)q(s, x))ds$$

$$+ \frac{1}{2}\Phi^2(t, s)[(\sigma(s) - \kappa(s))x + q(s, x)]^2 ds$$

$$\leq e(s)V(s, z(t, s, x)), \tag{6.6.10}$$

where

$$\begin{cases} f_1(s) = 2(f(s) - a(s) + \kappa^2(s) - \kappa(s)\sigma(s)), \\ 2(p(s, x) - \kappa(s)q(s, x)) \leq f_2(s)x^2, \\ [(\sigma(s) - \kappa(s))x + q(s, x)]^2 \leq \gamma(s)x^2, \\ e(s) = f_1(s) + f_2(s) + \gamma(s); \end{cases} \tag{6.6.11}$$

$$V_x(s, y(t, s, x(s))\Phi(t, s, x(s))\sigma(s, x(s))$$

$$= z(t, s, x)\Phi(t, s)[(\sigma(s) - \kappa(s))x(s) + q(x, s)]$$

$$= G(s, V(s, z(t, s, x(s)))) \tag{6.6.12}$$

if and only if $q(t, x) = \eta(t)x$ (linear function in x). $G(s, u) = \nu(t)u$, $g(s, u) = e(s)u$, where $\nu(t) = \sigma(t) - \kappa(t) + \eta(t)$. Thus, Theorem 6.6.1 is applicable to the perturbed stochastic differential equation (6.2.25) if and only if q is linear in x. In this case, the estimate (6.6.2) on the solution process (6.2.25) is

$$V(x(t)) \leq V(z(t)) \exp\left[\int_{t_0}^t (e(s) - \nu^2(s))ds + \int_{t_0}^t \nu(s)dw(s)\right], \quad \text{for } t \geq t_0.$$

$$(6.6.13)$$

From the definition of V, we have

$$|x(t, t_0, x_0)|^2 \leq |z(t, t_0, z_0)|^2 \exp\left[\int_{t_0}^t (e(s) - \nu^2(s))ds + \int_{t_0}^t \nu(s)dw(s)\right]. \quad (6.6.14)$$

(ii) For any $q(t, x)$ that satisfies (6.6.11), we apply Theorem 6.6.2, and obtain the following mean square estimate:

$$E[V(x(t))] \leq E[V(z(t))] \exp\left[\int_{t_0}^t e(s)ds\right]. \quad (6.6.15)$$

From the definition of V, we have

$$E[|x(t, t_0, x_0)|^2] \leq E[|z(t, t_0, z_0)|^2] \exp\left[\int_{t_0}^t e(s)ds\right]. \quad (6.6.16)$$

This completes the solution procedure of the illustration.

Observation 6.6.2

(i) The estimates (6.6.13) and (6.6.15) justify the statements about the dependence of the solution processes of dynamic processes in the chemical, biological, engineering, medical, physical, and social sciences described by:

(a) The state dynamic $x(t)$.

(b) The Auxiliary dynamic process $z(t)$ (it characterizes the externally applied goal-oriented adjustment process).

(c) The comparison dynamic process (it characterizes the information determination analytic approximation method).

(ii) For $f(t) = a(t)$ and $\sigma(t) = \kappa(t)$, (6.6.13), (6.6.15), and (6.6.11) reduce to

$$V(x(t)) \leq V(z(t)) \exp\left[\int_{t_0}^t (e(s) - \nu^2(s))ds + \int_{t_0}^t \nu(s)dw(s)\right], \quad \text{for } t \geq t_0,$$

$$(6.6.17)$$

$$E[V(x(t))] \leq E[V(z(t))] \exp\left[\int_{t_0}^t e(s)ds\right], \quad (6.6.18)$$

$$\begin{cases} 2(p(s, x) - \sigma(s)q(s, x)) \leq f_2(s)x^2, \\ 2[q(s, x)]^2 \leq \gamma(s)x^2, \\ e(s) = f_2(s) + \gamma(s), \end{cases} \quad (6.6.19)$$

respectively. Furthermore, from Illustration 6.6.1, several special cases can be made.

Example 6.6.1. We consider Example 6.2.1. We assume that $a(t,z) = f(t,z) = -\frac{1}{2}z^3$, and pick $V(t,x) = \frac{1}{2}x^2$. Examine the feasibility of applying at most two theorems: Theorems 6.6.1 and 6.6.2.

Solution process

(a) Recalling the expressions for $z(t,t_0,x_0)$, $\frac{\partial}{\partial z_0}z(t,t_0,z_0)$ and $\frac{\partial^2}{\partial z_0 \partial z_0}z(t,t_0,z_0)$, we
   compute

$$LV(s,z(t,s,x(s))) = \frac{1}{2}\frac{\sigma^2(s,x(s))(1-3x^2(t-s))}{[1+x^2(s)(t-s)]^3}$$

$$\leq \frac{1}{2}\frac{\sigma^2(s,x(s))}{[1+x^2(s)(t-s)]^3} \leq \gamma(s)\frac{1}{2}\frac{x^2(s)}{[1+x^2(s)(t-s)]}.$$

$$(6.6.20)$$

This is true due to the fact that $1 \leq [1+x^2(s)(t-s)]$. We examine the expression

$$V_x(s,y(t,s,x(s)))\Phi(t,s,x(s))\sigma(s,x(s)) = z(t,s,x(s)\Phi(t,s,x(s)))\sigma(s,x(s))$$

$$= \frac{x(s)\sigma(s,x(s))}{[1+x^2(s)(t-s)]^2}$$

$$= G(s,V(s,z(t,s,x(s))))$$

if and only if $\sigma(t,x) = \sigma(t)x$ (linear function in x). $G(s,u) = \sigma(t)u$, $g(s,u) = \gamma(s)u$. Thus, Theorem 6.6.1 is applicable to the stochastic perturbed stochastic differential equation (6.2.32) if and only if σ is linear in x. In this case, the estimate (6.6.2) for the solution process (6.2.32) is

$$V(x(t)) \leq V(z(t))\exp\left[\int_{t_0}^t (\gamma(s) - \sigma^2(s))ds + \int_{t_0}^t \sigma(s)dw(s)\right], \quad \text{for } t \geq t_0,$$

(b) For any $\sigma(t,x)$ that satisfies (6.6.20), we apply Theorem 6.6.2, and obtain the following mean square estimate:

$$E[V(x(t))] \leq [V(z(t))]\exp\left[\int_{t_0}^t \gamma(s)ds\right]. \qquad (6.6.21)$$

This completes the solution procedure of the example.

Exercises

1. It is assumed that the solution process of the following stochastic differential equations exists on $J = [t_0, t_0 + \alpha)$. For the following pairs of perturbed and unperturbed nonlinear stochastic/deterministic differential equations, establish the feasibility of applying either of the at least one of the comparison theorems (Theorems 6.6.1 and 6.6.2). [Hint: Use $V(t,x) = \frac{1}{2}x^2$.]

(a) $\begin{cases} dz = a(t)z\,dt + \kappa(t)z\,dw(t), \quad z(t_0) = z_0 \text{ (auxiliary equation)}, \\ dx = \left[-\frac{\cos t}{2(\sin t+2)}x^3 + p(t,x)\right]dt + \sigma(t)x\,dw(t), \quad x(t_0) = x_0, \\ dy = -\frac{\cos t}{2(\sin t+2)}y^3\,dt, \quad y(t_0) = x_0, \end{cases}$

where $p \in C[J \times R, R]$.

(b) $\begin{cases} dz = a(t)z\,dt + \kappa(t)z\,dw(t), \quad z(t_0) = z_0 \text{ (auxiliary equation)}, \\ dx = \left[f(t)x - \frac{1}{4}x^5 + p(t,x)\right]dt + [\sigma(t)x + q(t,x)]dw(t), \quad x(t_0) = x_0, \\ dy = -\frac{1}{4}y^5\,dt, \quad y(t_0) = x_0, \end{cases}$

where $p, q \in C[J \times R, R]$.

(c) $\begin{cases} dz = -\frac{1}{4}z^5\,dt, \quad z(t_0) = x_0 = z_0 \text{ (auxiliary equation)}, \\ dx = \left[f(t)x - \frac{1}{4}x^5 + p(t,x)\right]dt + [\sigma(t)x + q(t,x)]dw(t), \quad x(t_0) = x_0, \\ dy = -\frac{1}{4}y^5\,dt, \quad y(t_0) = x_0, \end{cases}$

where $p, q \in C[J \times R, R]$.

(d) $\begin{cases} dz = \frac{z}{1+t}\,dt, \quad z(t_0) = z_0 \text{ (auxiliary equation)}, \\ dx = \left[\frac{1}{1+t}x + p(t,x)\right]dt + [\sigma(t)x + q(t,x)]dw(t), \quad x(t_0) = x_0, \\ dy = \frac{y}{1+t}\,dt, \quad y(t_0) = x_0, \end{cases}$

where $p, q \in C[J \times R, R]$.

(e) $\begin{cases} dz = -\frac{a(t)}{2n}z^{2n+1}\,dt, \quad z(t_0) = z_0 \text{ (auxiliary equation)}, \\ dx = \left[-\frac{a(t)}{2n}x^{2n+1} + p(t,x)\right]dt + q(t,x)dw(t), \quad x(t_0) = x_0, \\ dy = -\frac{1}{2n}y^{2n+1}\,dt, \quad y(t_0) = x_0, \end{cases}$

where $p, q \in C[J \times R, R]$.

2. Is Theorem 6.6.1 applicable to the stochastic nonlinear differential equations in Exercise 1? If the answer is "Yes," then determine an estimate for a solution process.

3. Is Theorem 6.6.2 applicable to the stochastic nonlinear differential equations in Exercise 1? If answer is "Yes," then determine the estimates for the solution process.

6.7 Linear Hybrid Systems

A stochastic hybrid dynamic system is a pair of interconnected dynamic subsystems operating under both continuous and discrete times. For example, a continuous dynamic process is interrupted by the discrete time events. These types of systems provide mathematical models for interconnected dynamic phenomena evolving under different measure chains with state dependent discrete events.

In this section, we motivate stochastic hybrid dynamic phenomena by presenting a dynamic process in the medical and biological dynamic processes. In addition, we present a few elementary examples to illustrate the role and scope of the stochastic modeling of dynamic processes that are operating on different timescales.

This section briefly presents the following well-known and well-established problem: the "dosage problem" [81, 90]. The common conceptual understanding of this problem can be applied to any problem in the agricultural, biological, chemical, engineering, medical, physical, and social sciences. For example, the term "dose" is limited not only to the measured quantity of a therapeutic chemical/physical agent but also to a portion of an experience or knowledge applied to problem-solving processes in any field. The main goal of this problem is to determine the correct dose of the substance/knowledge/information/experience and the interdosage time so that the predetermined objective is achieved.

Illustration 6.7.1 (drug prescription problem). This problem is a continuation of the deterministic version of the drug prescription problem (Illustration 6.8.1 of Volume 1). We continue to answer two of the most important questions in the context of the randomly varying environmental perturbations. For easy reference, the two questions are:

(i) What dosage of the drug is to be prescribed?
(ii) How frequently should the drug be administered?

It is known that most of the drug concentration needs to be within the safe and effective bounds or zones. Moreover, the frequency of the drug administration must produce the maximum benefit.

Following the development of the deterministic version of the answers to these questions (Illustration 6.8.1 of Volume 1: notations, definitions, and results PDP1–PDP7), we arrive at

$$dc = -\epsilon c\, dt + \sigma x\, dw(t), \quad c(0) = c_0. \tag{6.7.1}$$

This is based on the diffusion process caused by inhomogeneity and the condition of the cell membrane. Due to the changes in ionic conductance, the coefficient of the permeability of the membrane is subject to the influence of random fluctuations. These random perturbations provide insights into the opening and closing of the kinetics of the ionic channels. The main ionic species are sodium (Na^+), potassium (K^+), and chloride (Cl^-) ions. The resting membrane potential of the membrane depends on the concentrations of Na^+, K^+, Cl^-, the pH value, and the permeability. For $t \geq 0$, $w(t)$ and c_0 are independent.

By imitating the development (Illustration 6.8.1 of Volume 1) and using the methods of solving stochastic differential equations (Section 2.3 of Chapter 2), we have

$$c(t) = c_0 \exp\left[-\left(\epsilon + \frac{1}{2}\sigma^2\right)t + \sigma w(t)\right], \quad \text{for } t \geq 0. \tag{6.7.2}$$

The right-hand side expression gives the concentration of the drug in the plasma at any time $t \geq 0$.

In this case, the discrete dynamic of the drug administration process is described by

$$c(k) = \exp\left[-\left(\epsilon + \frac{1}{2}\sigma^2\right)T + \sigma w(T)\right]c(k-1) + c_0, \quad c(0) = c_0, \quad \text{for } k \geq 1.$$

$$(6.7.3)$$

The first term on the right-hand side represents the residual of the drug concentration between the kth and the $(k-1)$th drug dose administration process over a time interval of length T for any $k \geq 1$. Thus, the amount of drug assimilated during the interval of length T is $c(k-1)\left(1 - \exp\left[-\left(\epsilon + \frac{1}{2}\sigma^2\right)T + \sigma w(T)\right]\right)$.

The solution process of the discrete time drug dose iterative process (6.7.3) determines the drug accumulation in the plasma under the repeated dosages. It is

$$c(k) = \exp[-\alpha(T,1,k)]c_0 + \exp[-\alpha(T,1,k-1)]c_0 + \cdots$$

$$+ \exp[-\alpha(T,1,2)]c_0 + \exp[-\alpha(T,1,1)]c_0 + c_0$$

$$= \exp[-\alpha(T,1,1)][1 + \exp[-\alpha(T,2,2)] + \exp[-\alpha(T,2,3)] + \cdots$$

$$+ \exp[-\alpha(T,2,k)]]c_0 + c_0$$

$$= \exp[-\alpha(T,1,1)]R(k)c_0 + c_0, \quad (6.7.4)$$

where $\alpha(T,\ell,i) = \sum_{j=\ell}^{i}\left[\left(\epsilon + \frac{1}{2}\sigma^2\right)\Delta T_j - \sigma\Delta_k w(jT)\right]$, $\Delta_k w(jT) = w((k+1-j)T) - w((k-j)T))$, $\Delta T_j = T_j - T_{j-1}$, $\Delta T_j = T_j - T_{j=1}$, for $i = 1, 2, \ldots, k, \ldots$, and

$$R(k) = 1 + \exp[-\alpha(T,2,2)] + \exp[-\alpha(T,2,3)] + \cdots + \exp[-\alpha(T,2,k)].$$

From this, we note that $\exp[-\alpha(T,1,1)]R(k)c_0$ is the aggregate residual of the concentration of the drug due to the first k successively repeated doses over the time interval $[0, kT]$. We further note that for $j = 1, 2, \ldots, k$, $\Delta_k w(jT)$'s are independent and identical normally distributed $N(0, T)$ random variables (Definition 1.2.12 and Observation 1.2.11) and $E(\exp[\Delta_k w(jT)] = \exp[T]$. Thus,

$$E[R(k) \mid F_k] = 1 + E[\exp[-\alpha(T,2,2)]] + E[\exp[-\alpha(T,2,3)]] + \cdots$$

$$+ E[\exp[-\alpha(T,2,k)]]$$

$$= 1 + \exp[-\epsilon T] + \exp[-2\epsilon T] + \cdots + \exp[-\epsilon(k-1)T]$$

$$= \frac{1 - \exp[-\epsilon kT]}{1 - \exp[-\epsilon T]}.$$

Therefore, (6.7.4) reduces to

$$E[c(k) \mid F_k] = \exp[-\alpha(T,1,1)]\frac{(1 - \exp[-\epsilon kT])c_0}{1 - \exp[-\epsilon T]} + c_0. \quad (6.7.5)$$

Determination of the dosage schedule. From the expression of $c(k)$ (6.7.5), we have

$$\lim_{k \to \infty} E[c(k) \mid F_k] = \lim_{k \to \infty} [\exp[-\alpha(T, 1, 1)] R(k) c_0 + c_0]$$

$$= \lim_{k \to \infty} \exp[-\alpha(T, 1, 1)] \frac{(1 - \exp[-\epsilon k T]) c_0}{1 - \exp[-\epsilon T]} + \lim_{k \to \infty} [c_0]$$

$$= \frac{\exp[-\alpha(T, 1, 1)]}{1 - \exp[-\epsilon T]} c_0 + c_0. \tag{6.7.6}$$

We note that the above expression is valid for any arbitrary interdosage time $T > 0$. From a choice of c_0, we choose the interdosage time T so that

$$D = \frac{\exp[-\alpha(T, 1, 1)]}{1 - \exp[-\epsilon T]} c_0 + c_0 = \frac{\exp[-\alpha(T, 1, 1)]}{1 - \exp[-\epsilon T]} c_0 + \frac{1 - \exp[-\epsilon T]}{1 - \exp[-\epsilon T]} c_0.$$

Hence,

$$D(1 - \exp[-\epsilon T]) = (\exp[-\alpha(T, 1, 1)] + 1 - \exp[-\epsilon T]) c_0$$

$$= (\exp[-\alpha(T, 1, 1)] + 1 - \exp[-\epsilon T])(D - d)$$

$$= D(1 - \exp[-\epsilon T] + \exp[-\alpha(T, 1, 1)])$$

$$- d(1 - \exp[-\epsilon T] + \exp[-\alpha(T, 1, 1)]),$$

which implies that

$$\frac{D}{d} = \exp[\alpha(T, 1, 1)](1 - \exp[-\epsilon T] + \exp[-\alpha(T, 1, 1)]).$$

We solve the above equation with respect to T:

$$\ln \left[\frac{D}{d} \right] = \alpha(T, 1, 1) + \ln(1 - \exp[-\epsilon T] + \exp[-\alpha(T, 1, 1)])$$

$$= \alpha(T, 1, 1) + \ln(1 - \exp[-\epsilon T] + \exp[-\alpha(T, 1, 1)])$$

$$= \left(\epsilon + \frac{1}{2}\sigma^2 \right) T - \sigma \Delta_k w(T) + \ln(1 - \exp[-\epsilon T] + \exp[-\alpha(T, 1, 1)]).$$

Thus, the interdosage time (schedule) is determined by

$$T = \frac{1}{\epsilon + \frac{1}{2}\sigma^2} \ln \left[\frac{D}{d} \right] - \frac{\sigma \Delta_k w(T) + \ln(1 - \exp[-\epsilon T] + \exp[-(1 + \frac{1}{2}\sigma^2)T + \sigma w(T)])}{\epsilon + \frac{1}{2}\sigma^2}$$

$$\tag{6.7.7}$$

Dosage prescription and schedule. Following the argument used in the deterministic modeling of the drug prescription and schedule [80], we summarize that the drug dosage prescription and the interdosage time (the time interval between the consecutive dosages) are

$$c_0 = D - d,$$

$$E[T] = \frac{1}{\epsilon + \frac{1}{2}\sigma^2} \ln\left(\frac{D}{d}\right) - \frac{E[\ln(1 - \exp[-\epsilon T] + \exp[-(1 + \frac{1}{2}\sigma^2)T + \sigma\Delta_k w(T)])}{\epsilon + \frac{1}{2}\sigma^2}.$$

Moreover, $E[T] \geq \frac{1}{\epsilon + \frac{1}{2}\sigma^2} \ln(\frac{D}{d})$.

Conclusions

(i) From (6.7.6) and $c_0 = D - d$, we note that

$$D = \frac{\exp[-\alpha(T, 1, 1)]}{1 - \exp[-\epsilon T]} c_0 + c_0 = \frac{\exp[-\alpha(T, 1, 1)]}{1 - \exp[-\epsilon T]} c_0 + D - d,$$

and hence

$$d = \frac{\exp[-\alpha(T, 1, 1)]}{1 - \exp[-\epsilon T]} c_0 = \frac{c_0}{\exp[\alpha(T, 1, 1)](1 - \exp[-\epsilon T])}$$

$$= \frac{c_0}{\exp[\frac{1}{2}\sigma^2 T - \sigma\Delta_k w(T)]\exp[\epsilon T] - 1)}.$$

(ii) From (i), $\frac{d}{c_0}$ is close to 0 if the interdosage time is very large. Therefore, the intermediate aggregate residual concentration at the kth dose is $R(k)$, and hence

$$0 < E[R(k)|F_T] < \frac{c_0}{\exp[\frac{1}{2}\sigma^2 T - \sigma\Delta_k w(T)]\exp[\epsilon T] - 1)} = \frac{E[\exp[-\alpha(T, 1, 1)]]}{1 - \exp[-\epsilon T]} c_0.$$

$$(6.7.8)$$

Moreover,

$$0 < E[R(k)] < \frac{\exp[-\epsilon T]}{1 - \exp[-\epsilon T]} c_0 = \frac{c_0}{\exp[\epsilon T] - 1}. \qquad (6.7.9)$$

(iii) If T is chosen in such a way that $\exp[\epsilon T]$ is close to 1, then $\exp[\epsilon T] - 1$ is positive. In this case, $\frac{d}{c_0}$ is significantly greater than 1. Therefore, the kth dosage concentration $c(k)$ increases as k increases. Thus, from SPDP3, the drug concentration in the blood will be very close to c_0 due to the dose. As a result of this and from (i), the drug concentration in the blood will be oscillating between D and $\frac{\exp[-\alpha(T, 1, 1)]}{1 - \exp[-\epsilon T]} c_0 = \frac{c_0}{\exp[\frac{1}{2}\sigma^2 T - \sigma\Delta_k w(T)](\exp[\epsilon T] - 1)}$.

Observation 6.7.1

(i) From Illustration 6.7.1, the mathematical formulation of the drug prescription problem is:

$$\begin{cases} dx = -\epsilon c\, dt + \sigma x\, dw(t), & c(t_k) = c_k, \quad t \in [kT, (k+1)T), \text{ for } k \geq 0, \\ c(k) = \exp\left[-\left(\epsilon + \frac{1}{2}\sigma^2\right)T + \sigma\Delta_k w(T)\right] c(k-1) + c_0, & c(0) = c_0. \end{cases}$$

$$(6.7.10)$$

(ii) From (i), it is clear that the continuous time dynamic process outlined in SPDP3 is perturbed by the discrete time event (drug administration) which is described by the stochastic drug administration process. Thus, the dynamic process of the drug prescription problem is modeled by the interconnected discrete and continuous time stochastic processes. Such dynamic processes are called *stochastic hybrid dynamic processes*.

(iii) We further remark that the description of stochastic hybrid model in (i) is linear in both discrete and continuous time dynamic processes. Therefore, it is called a stochastic linear hybrid system.

General problem formulation. In our subsequent discussion, we present a few examples to show the nature of the stochastic hybrid dynamic system. Let us consider the following nonlinear stochastic hybrid dynamic system:

$$du = g(t, u)dt + \Lambda(t, u)dw(t), \quad u(t_k) = u_k \quad t \neq t_k, \quad k \in I(0, \infty),$$

$$u_k = \psi(u_{k-1}(t_k^-, t_{k-1}, u_{k-1}), k), \quad u(t_0) = u_0, \quad k \in I(1, \infty), \qquad (6.7.11)$$

where we assume:

$(H_{6.7.1})(i)$ $g : R_+ \times R^N \to R^N$ are continuous on $[t_k, t_{k+1}) \times R^N$ and for each (t, u), $\lim g(t, u) = g(t_{k+1}^-, \bar{u})$ as $(t, u) \to (t_{k+1}^-, \bar{u})$;

$(H_{6.7.1})(ii)$ Λ is defined on $R_+ \times R_+^N$ into R^{Nm} and statisfies $(H_{6.4})$ of Theorem 6.4.1;

$(H_{6.7.1})(iii)$ $\psi : R_+ \times R_+^N \to R_+^N$ is a Borel-measurable function;

$(H_{6.7.1})(iv)$ $w(t)$ is an m-dimensional normalized Wiener process and it is independent of u_0.

From Observation 6.7.1, we further recall that $r(t, t_0, u_0)$ is defined by

$$r(t) = r(t, t_0, u_0) = \begin{cases} r_0(t, t_0, r_0), & 0 \leq t < t_1, \\ r_1(t, t_1, r_1), & t_1 \leq t < t_2, \\ \cdots \quad \cdots & \cdots \quad \cdots \\ r_{n-1}(t, t_{n-1}, r_{n-1}), & t_{n-1} \leq t < t_n, \\ \cdots \quad \cdots & \cdots \quad \cdots . \end{cases}$$

Moreover, the solution process is described by the following iterative process:

$$\begin{cases} r(t, t_{k-1}, u_{k-1}) = r_{k-1}(t, t_{k-1}, u_{k-1}), & t_{k-1} \leq t < t_k, \\ u_k = \psi(r_{k-1}(t_k^-, t_{k-1}, u_{k-1}), k), & u(t_0) = u_0, \quad k \in I(1, \infty). \end{cases} \qquad (6.7.12)$$

Observation 6.7.2. From (6.7.12), we note that for each $k \in I(1, \infty)$, $(r_{k-1}(t), u_k)$ is a solution process of (6.7.11) on $t_{k-1} \leq t < t_k$, where $r_{k-1}(t) = r(t, t_{k-1}, u_{k-1})$, and $u_k = u(k) = u(k, 0, u_0)$. It is determined by the one-step cyclic (Gauss–Seidel) iterative procedure [100].

We present now a few examples of simple versions of the scalar impulsive dynamic system (6.7.1), and its explicit solution processes.

Example 6.7.1. Let $g(t, u) = \lambda(t)u$ and $\Lambda(t, u) = \Lambda(t)u$. In this case, the switching rule ψ in (6.7.11) is $\psi(u) = \delta(k - 1)u$. The one-step cyclic solution process is expressed as

$$
\begin{cases}
r(t, t_{k-1}, u_{k-1}) = u_{k-1} \exp\left[\int_{t_{k-1}}^t \left(\lambda(s) - \frac{1}{2}\Lambda^2(s)\right) ds + \int_{t_{k-1}}^t \Lambda(s)dw(s)\right], \\
u_k = \delta(k - 1)u_{k-1} \exp\left[\Delta(t_k, t_{k-1}, w(t_k), w(t_{k-1})\right],
\end{cases}
$$

(6.7.13)

for $t_k \le t < t_{k+1}$, $k \in I(0, \infty)$; $\psi(k-1)$, $u(t_{k-1}) = \delta(k-1)u_{k-1}$, and $u(t_0) = u_0$. Here $\Delta(t, t_{k-1}, w(t_k), w(t_{k-1})) = \int_{t_{k-1}}^{t_k} (\lambda(s) - \frac{1}{2}\Lambda^2(s))ds + \int_{t_{k-1}}^{t_k} \Lambda(s)dw(s)$. The overall solution process is further explicitly represented by

$$
r(t, t_0, u_0) = \prod_{t_0 < t_k < t} \psi(u_k) \exp\left[\int_{t_{k-1}}^t \left(\lambda(s) - \frac{1}{2}\Lambda^2(s)\right) ds\right.
$$

$$
\left. + \int_{t_{k-1}}^t \Lambda(s)dw(s)\right] u_0 \quad \text{for all } t \ge t_0,
$$

$$
= \left(\prod_{k=1}^{n+1} d(k-1)\right) \exp\left[\sum_{k=1}^n \Delta(t_k, t_{k-1}, w(t_k), w(t_{k-1}))\right.
$$

$$
\left. + \Delta(t, t_k, w(t), w(t_k))\right] u_0,
$$

(6.7.14)

for all $t \ge t_0$, where $\nu(t) = [(\lambda(t) - \frac{1}{2}\Lambda^2(t)]$.

6.8 Stochastic Hereditary Systems

The growth rates of biological, chemical, compartmental, engineering, physical, and social systems depend on both the present and past history of the state of the system. The time delays are due to the time it takes to respond to the random external/internal environmental changes in dynamic processes. Mathematical models of stochastic hereditary dynamic systems are described by the following types of systems of stochastic functional or delay differential equations:

$$
dx = f(t, x_t)dt + \sigma(t, x_t)dw(t), \quad x_{t_0} = \varphi_0,
$$

(6.8.1)

where m column vectors of σ, $f \in C[R_+ \times C^n \times R, R^n]$, $t_0 \in [0, \infty)$; for $\tau > 0$, $C^n = C[[-\tau, 0], R^n]$ denotes the space of continuous functions with the domain $[-\tau, 0]$ and the range in R^n; x_t is an element of C^n defined by $x_t(\theta) = x(t + \theta)$, $-\tau \le \theta \le 0$; for $-\tau \le \theta \le 0$, $\varphi_0(\theta)$ is an n-dimensional random vector defined on a complete probability space $(\Omega, \mathcal{F}, P)$ representing an initial past history of the state of the system, and it is sample-continuous; $w(t) = (w_1(t), w_2(t), \dots, w_m(t))^T$ is an m-dimensional normalized Wiener process with independent increments; $w(t)$ are $\mathcal{F}_t$-measurable for all $t \ge t_0$ and $\mathcal{F}_t$ is an increasing family of sub-σ-algebras of $\mathcal{F}$;

$\varphi_0(\theta)$ and $w(t)$ are mutually independent for each $t \geq t_0$; it is assumed that σ and f are smooth enough to insure the existence of a solution process $x(t) = x(t_0, \varphi_0)(t)$ of (6.8.1).

Employing the concept of the Lyapunov function and deterministic functional differential inequalities [86, 88, 91–93], we present a very general variational comparison theorem. The result connects solutions processes of the dynamic system (6.8.1) with the maximal solution to the corresponding comparison dynamic equation and a solution to an auxiliary dynamic system. As a byproduct of this, some auxiliary comparison results are formulated. These results generalize and extend the existing results.

For $V \in C[[-\tau, \infty) \times R^n \times R, R_+^N]$, and $\frac{\partial}{\partial t}V$, $\frac{\partial}{\partial x}V$, and $\frac{\partial^2}{\partial x \partial x}V$ exist and are continuous on $[-\tau, \infty) \times R^n$; $\frac{\partial^2}{\partial x \partial x}V$ is locally Lipschitzian in z for each (t,s), $t_0 \leq s \leq t$. From this and the hypotheses of Theorem 6.3.1 in the context of (6.8.1), and using the logic from the proof of Theorem 6.3.1, we have

$$dV(s, \varphi, z(t, s, \varphi(0))) = LV(s, \varphi, z(t, s, \varphi(0)))$$
$$+ \frac{\partial}{\partial x}V(s, \varphi, z(t, s, \varphi(0)))\sigma(s, \varphi)dw(s), \quad (6.8.2)$$

where

$$LV(s, \varphi, z(t, s, \varphi(0)))$$
$$= \frac{\partial}{\partial s}V(s, \varphi, z(t, s, \varphi(0))) + \frac{1}{2} \operatorname{tr}\left(\frac{\partial^2}{\partial x \partial x}V(s, \varphi, z(t, s, \varphi(0)))\Theta(t, s, \varphi)\right)$$
$$+ \frac{\partial}{\partial x}V(s, \varphi, z(t, s, \varphi(0))) \left[\frac{1}{2} \operatorname{tr}\left(\frac{\partial^2}{\partial x \partial x}z(t, s, \varphi(0))\Lambda(s, \varphi)\right)\right.$$
$$+ \left. \Phi(t, s, \varphi(0))f(s, \varphi(0))\right], \quad (6.8.3)$$

$\Lambda(s, \varphi) = \sigma(s, \varphi)\sigma^T(s, \varphi)$ and $\Theta(t, s, \varphi) = \Phi(t, s, \varphi(0))\Lambda(s, \varphi)\Phi^T(t, s, \varphi(0))$.

Theorem 6.8.1. *Assume that all the hypotheses of Theorem 6.3.1 are satisfied with the modification in its time domain of V from R_+ to $[-\tau, \infty)$. Further assume that:*

(H$_{6.8.1}$)(i) *$LV(s, z(t, s, x))$ in (6.8.3) satisfies the inequality*

$$LV(s, \varphi, z(t, s, \varphi(0))) \leq g(s, V(s, z(t, s, \varphi(0))), V_s), \quad (6.8.4)$$

where $V_s = V(s + \theta, z(t, s + \theta, \varphi(s + \theta)))$ for $\theta \in [-\tau, 0]$;

(H$_{6.8.1}$)(ii) *$g \in C[R_+ \times R_+^N \times C_+^N, R_+^N]$, $g(t, u, \eta)$ is concave in (u, η) for each t, quasi-monotone nondecreasing in u for each $(t, \eta) \in R_+ \times C_+^N$, and increasing in η for each $(t, u) \in R_+ \times R_+^N$;*

(H$_{6.8.1}$)(iii) $r(t_0, \eta_0)(t) = r(t)$ *is the maximal solution to the comparison system of functional differential equations*

$$du = g(t, u, u_t), \quad u_{t_0} = \eta_0, \tag{6.8.5}$$

existing for $t \geq t_0$;

(H$_{6.8.1}$)(iv) *for any solution process, $x(t_0, \varphi_0)(t) = x(t)$ of (6.8.1), $E[V(t, x(t))]$ exists for $t \geq t_0$ and*

$$E[V(t_0 + \theta, z(t, s + \theta, \varphi(\theta)))] \leq \eta_0(\theta), \tag{6.8.6}$$

for all $\theta \in [-\tau, 0]$.

Then,

$$E[V(t, x(t) \mid x(t_0)) = \varphi(0)] \leq r(t_0, \eta_0)(t), \quad t \geq t_0. \tag{6.8.7}$$

Proof. The proof of this result is left as an exercise. $\qquad\square$

To illustrate the scope and the significance of Theorem 6.8.1, we first present a corollary. This will be helpful in studying the behavior of the system of comparison functional differential equations.

Corollary 6.8.1. *Assume that the hypotheses of Theorem 6.8.1 are satisfied. In addition, assume that g satisfies*

$$\sum_{k=1}^{N} d_k g_k(s, V(s, z(t, s, \varphi(0))), V_s)$$

$$\leq \alpha(s)\nu(s, z(t, s, \varphi(0))) + \beta(s)\int_{-\tau}^{0} \nu(s + \theta, z(t, s + \theta, \varphi(\theta)))d_\theta\beta(s, \theta) \tag{6.8.8}$$

where $d_k > 0$; $\alpha, \beta \in C[J, R]$, $\beta \geq 0$, $\beta(s, \theta)$ is a function of bounded variation [169], $d_\theta\beta(s, \theta)$ is a non-negative Stieltjes measure on $[-\tau, 0]$, for $t_0 \leq s \leq t$, and

$$\nu(s, z(t, s, \varphi(0))) = \sum_{k=1}^{N} d_k V_k(s, z(t, s, \varphi(0))). \tag{6.8.9}$$

Then,

$$L\nu(s, z(t, s, \varphi(0))) \leq \ell(s, \nu_s), \quad where$$

$$\ell(s, (\nu_s) = \alpha(s)\nu(s) + \beta(s)\int_{-T}^{0} \nu(s + \theta))d\theta\beta(s, \theta). \tag{6.8.10}$$

Moreover,

$$E[\nu(t, x(t)) \mid \varphi_0(t_0) = x_0] \leq r(t_0, \eta_0)(t) \tag{6.8.11}$$

whenever

$$\nu_{t_0} \leq \eta_0,$$

where $r(t_0, \eta_0)(t)$ is the maximal solution to the following linear system of functional differential equations:

$$du = \ell(s, \nu_s)ds, \quad u_{t_0} = \eta_0. \tag{6.8.12}$$

Proof. The proof of the inequality (6.8.11) follows from the uniqueness of the solution to (6.8.12) and the application of Theorem 6.8.1. This completes the proof of the corollary. $\qquad\square$

In the following, we provide a simple example that illustrates the usefulness of the presented results.

Example 6.8.1. We consider a very simple scalar version of (6.8.1):

$$dx = \left(-\frac{1}{2}x^2 + f(t)\right)x\,dt + b(t)x(t-\tau)dw(t), \quad x_{t_0} = \varphi_0, \tag{6.8.13}$$

where $a, b \in C[R, R]$, $\tau > 0$ is a stochastic time delay defined on the complete probability space $(\Omega, \mathcal{F}, P)$, and it is independent of the random process $w(t)$ and the initial state process φ_0. Furthermore, $\tau \in [\nu, \mu]$ with probability 1, where $\nu \geq 0$ and $\mu > 0$. This assumption on τ characterizes the "reaction time delay" in the hereditary system with past memory μ [88, 101].

(a) Let F be a distribution function of a random variable τ defined on a complete probability space $(\Omega, \mathcal{F}, P)$. Let us note that

$$\int_{-\infty}^{\infty} dF(\theta) = \int_{\nu}^{\mu} dF(\theta) = 1$$

and it is obvious that (6.8.13) can be rewritten as

$$dx = \left(-\frac{1}{2}x^2 + f(t)\right)x\,dt + b(t)\int_{\nu}^{\mu} x(t-\theta)dF(\theta)dw(t), \quad x_{t_0} = \varphi_0, \tag{6.8.14}$$

which implies that

$$dx = \left(-\frac{1}{2}x^2 + f(t)\right)x\,dt - b(t)\int_{-\mu}^{-\nu} x(t+\theta)dF(-\theta)dw(t), \quad x_{t_0} = \varphi_0. \tag{6.8.15}$$

(b) Let F in (a) be defined by

$$F(\theta) = \begin{cases} 0, & \text{if } \theta < \nu, \\ \frac{\theta-\nu}{\mu-\nu}, & \text{if } \nu \leq \theta \leq \mu, \\ 1, & \text{if } \mu < \theta. \end{cases}$$

In this case, (6.8.15) reduces to

$$dx = \left(-\frac{1}{2}x^2 + f(t)\right)x\,dt + \frac{b(t)}{\mu-\nu}\int_{-\mu}^{-\nu} x(t+\theta)d\theta\,dw(t), \quad x_{t_0} = \varphi_0. \tag{6.8.16}$$

For this example, we consider the following scalar version of the auxiliary differential equation (6.3.2):

$$dz = -\frac{1}{2}z^3 dt, \quad z(t_0) = z_0. \tag{6.8.17}$$

We note that $f(t, z) = -\frac{1}{2}z^3$ is twice differentiable with respect to z. For details, see Example 6.2.1. We choose a Lyapunov-like function: $V(t, x) = \frac{1}{2}|x|^2$. In this case, the expressions in (6.8.3) reduce to

$$f(s, \varphi) = -\frac{1}{2}\varphi^3(0) + f(s)\varphi(0), \quad \sigma^2(s, \varphi) = \left[b(s) \int_{-\mu}^{-\nu} \varphi(\theta) dF(-\theta) \right]^2,$$

$$\Theta(t, s, \varphi) = \left[b(s) \int_{-\mu}^{-\nu} \varphi(\theta) dF(-\theta) \right]^2 [1 + x^2(s)(t-s)]^{-3}, \quad \text{for } t_0 \le s \le t,$$

$$LV(s, \varphi, z(t, s, \varphi(0)))$$

$$= \frac{1}{2} \text{tr} \left(\frac{\partial^2}{\partial x \partial x} V(s, z(t, s, \varphi(0))) \Theta(t, s, \varphi) \right)$$

$$+ \frac{\partial}{\partial z} V(s, z(t, s, \varphi(0))) \left[\frac{1}{2} \text{tr} \left(\frac{\partial^2}{\partial z_0 \partial z_0} z(t, s, \varphi(0)) \Lambda(s, \varphi) \right) \right.$$

$$\left. + \Phi(t, s, \varphi(0)) f(s)\varphi(0) \right]$$

$$= \frac{1}{2} \left[b(s) \int_{-\mu}^{-\nu} \varphi(\theta) dF(-\theta) \right]^2 [1 + \varphi^2(0)(t-s)]^{-3}$$

$$+ \varphi(0)[1 + \varphi^2(0)(t-s)]^{-1/2} \left(f(s)\varphi(0)[1 + \varphi^2(0)(t-s)]^{-3/2} \right.$$

$$\left. - \frac{3}{2}(t-s)\varphi(0)[1 + \varphi^2(0)(t-s)]^{-5/2} \left[b(s) \int_{-\mu}^{-\nu} \varphi(\theta) dF(-\theta) \right]^2 \right)$$

$$= [1 + \varphi^2(0)(t-s)]^{-2} f(s)\varphi^2(0) + \frac{1}{2} \left(1 - 3\varphi^2(0)(t-s) \right)$$

$$\times [1 + \varphi^2(0)(t-s)]^{-3} \left[b(s) \int_{-\mu}^{-\nu} \varphi(\theta) dF(-\theta) \right]^2 \right)$$

$$\le 2f(s)\frac{1}{2}z^2(t, s, \varphi(0)) + \frac{1}{2}[1 + \varphi^2(0)(t-s)]^{-2} \left[b(s) \int_{-\mu}^{-\nu} \varphi(\theta) dF(-\theta) \right]^2.$$

From this, the Cauchy–Bunyakovski–Schwarz inequality [169], and the nature of the solution to the auxiliary differential equation (6.2.31), we have

$$LV(s, \varphi, z(t, s, \varphi(0))) \le 2f(s)V(s, z(t, s, \varphi(0))) + b^2(s)\|V_s\|_0, \tag{6.8.18}$$

where $\max_{\theta \in [-\mu, -\nu]} V(s + \theta, z(t, s + \theta, \varphi(\theta))) = \|V_s\|_0$. The corresponding comparison functional differential equation is linear like in (6.8.12):

$$du = 2f(s)u + \frac{1}{2}b^2(s)\|u_t\|_0, \quad u_{t_0} = \eta_0. \tag{6.8.19}$$

Thus, by the application of Theorem 6.8.1, we have

$$E[V(x(t))] \le E|V(z(t))|^2 \exp\left[\int_{t_0}^t (2f(s) + \mu b^2(s))ds\right], \quad \text{for } t \ge t_0. \tag{6.8.20}$$

Observation 6.8.1. Using the minimal class of functions [80, 86, 91–93], the maximal solutions of (6.8.12) and (6.8.19) can be further estimated by the ordinary comparison differential equation with rate coefficients $(\alpha(s) + \beta(s))$ and $(2f(s) + \mu b^2(s))$, respectively. Because of this, the estimates (6.8.11) and (6.8.20) are reduced to

$$E\left[\sum_{k=1}^N d_k V_k(t, x(t_0, \varphi_0)(t))) \mid x_{t_0} = \varphi_0\right]$$

$$\le \sum_{k=1}^N d_k V_k(t, z(t)) \exp\left[\int_{t_0}^t (\alpha(s) + \tau \beta(s))ds\right], \quad \text{for } t \ge t_0, \tag{6.8.21}$$

$$E[|x(t)|^2] \le E|z(t)|^2 \exp\left[\int_{t_0}^t (2f(s) + \mu b^2(s))ds\right], \quad \text{for } t \ge t_0. \tag{6.8.22}$$

Moreover, the estimates (6.8.21) and (6.8.22) depend on the time delays (τ and μ). Further, it provides information about the effects of hereditary on the dynamic processes, and a partial solution to the problem of "hereditary versus nonhereditary."

Exercises

1. Prove Theorem 6.8.1.
2. Prove Corollary 6.8.1.
3. Using the concept of the minimal class of functions [80, 86, 91–93], justify the estimate (6.8.21). Moreover, show that

$$r(t_0, \eta_0)(t) \le \sum_{k=1}^N d_k V_k(t, z(t)) \exp\left[\int_{t_0}^t (\alpha(s) + \tau \beta(s))ds\right], \quad \text{for } t \ge t_0.$$

4. Using the concept of the minimal class of functions [80], justify the estimate (6.8.21). Moreover, show that

$$r(t_0, \eta_0)(t) \le \frac{1}{2}|z(t)|^2 \exp\left[\int_{t_0}^t (2f(s) + \mu b^2(s))ds\right], \quad \text{for } t \ge t_0,$$

6.9 Qualitative Properties of Solution Processes

In this section, a mathematical model for the inflation-unemployment process under stochastic environmental perturbations is presented. The development of the mathematical model stems from the basic deterministic model described in Section 6.10, [80] (IUP1–IUP3). It is assumed that φ_{11}, the speed of adjustment of the expected rate of inflation, is deterministic, while φ_{22}, the speed of adjustment of the rate of unemployment, is perturbed by Gaussian white noise. Based on this, the differential equation relative to the expected rate of inflation π remains unchanged, and the differential equation relative to the unemployment rate U is subject to stochastic perturbations. Now, it is modified, as

$$dU = (-\varphi_{22} + \xi(t))(m - p - g)U\,dt, \tag{6.9.1}$$

where $\xi(t)$ is Gaussian white noise with mean zero and covariance $(\sigma(t, m - p - g))^2 \delta(t - s)$; δ is the Dirac delta function, and σ is a continuous function defined on $J \times R = [t_0, \infty) \times R$, and it satisfies the growth condition:

$$\sigma^2(t, -y)(\beta + |y|) \leq \bar{\sigma}^2. \tag{6.9.2}$$

This assumption shows that the level of employment is assumed to be randomly affected by monetary policy. In view of this change, Equation (6.9.1) can be written as an Itô–Doob-type stochastic differential equation:

$$dU = -\varphi_{22}(m - p - g)U\,dt + \sigma(t, m - p - g)(m - p - g)U\,dw(t), \tag{6.9.3}$$

where $w(t)$ is a normalized Wiener process (Definition 1.2.12).

Applying Theorem 1.3.2 to $\ln U$ and using (6.9.3), we obtain the expression

$$d\ln U = [-\varphi_{22}(m - p - g) - \frac{1}{2}\sigma^2(t, m - p - g)(m - p - g)^2]dt$$
$$+ \sigma(t, m - p - g)(m - p - g)dz(t). \tag{6.9.4}$$

Repeating the argument used in the development of deterministic modeling of the inflation-unemployment process (Chapter 6 [80]), applying Theorem 1.3.2 to p, and using (6.9.4), we have the following Itô–Doob-type stochastic differential equation:

$$dp = -\beta d\ln U + h\,d\pi$$
$$= -\beta \left[\left[-\varphi_{22}(m - p - g) - \frac{1}{2}\sigma^2(t, m - p - g)(m - p - g)^2 \right] dt \right.$$
$$\left. + \sigma(t, m - p - g)(m - p - g)dz(t) \right] + h\varphi_{11}(p - \pi)dt$$
$$= \left[(-\beta\varphi_{22} + h\varphi_{11})p - h\varphi_{11}\pi + \beta\varphi_{22}(m - g) \right.$$

$$+ \frac{1}{2}\beta\sigma^2(t, m - p - g)(m - p - g)^2 \Bigg] dt$$

$$- \beta\sigma(t, m - p - g)(m - p - g)dz(t). \tag{6.9.5}$$

Under the above-stated conditions and development, the deterministic mathematical model (Section 6.10 [80]) takes the following form:

$$\begin{cases} d\pi = \varphi_{11}(p - \pi)dt, \\ dp = \Big[(-\beta\varphi_{22} + h\varphi_{11})p - h\varphi_{11}\pi + \beta\varphi_{22}(m - g) \\ \quad + \frac{1}{2}\beta\sigma^2(t, m - p - g)(m - p - g)^2 \Big] dt \\ \quad - \beta\sigma(t, m - p - g)(m - p - g)dz(t). \end{cases} \tag{6.9.6}$$

The equilibrium state of (6.9.6) is determined by solving the system of algebraic equations

$$\begin{cases} 0 = \varphi_{11}(p - \pi)dt, \\ 0 = \Big[(-\beta\varphi_{22} + h\varphi_{11})p - h\varphi_{11}\pi + \beta\varphi_{22}(m - g) \\ \quad + \frac{1}{2}\beta\sigma^2(t, m - p - g)(m - p - g)^2 \Big] dt \\ \quad - \beta\sigma(t, m - p - g)(m - p - g)dz(t). \end{cases} \tag{6.9.7}$$

Moreover, its solution is denoted by (p^*, π^*). Again, $p^* = \pi^* = m - g$. Setting $x = [x_1, x_2]^T$, where $x_1 = \pi - \pi^*$, $x_2 = p - p^*$, system (6.9.6) is reduced to

$$\begin{cases} d\pi = [-\varphi_{11}x_1 + \varphi_{11}x_2]dt, \\ dp = \Big[-h\varphi_{11}x_1 + (-\beta\varphi_{22} + h\varphi_{11})x_2 + \frac{1}{2}\beta\sigma^2(t, -x_2)x_2^2\Big]dt \\ \quad + \beta\sigma(t, -x_2)x_2 dz(t). \end{cases} \tag{6.9.8}$$

Hence,

$$dx = [Ax + p(t, x)]dt + q(t, x)dz(t), \quad x(t_0) = x_0, \tag{6.9.9}$$

where A is defined (Section 6.10 of Volume 1); and $p(t, x)$ and $q(t, x)$ are defined as

$$p(t, x) = \Big[0, \frac{1}{2}\beta\sigma^2(t, -x_2)x_2^2\Big]^T \quad \text{and} \quad q(t, x) = [0, \beta\sigma(t, -x_2)x_2]^T, \tag{6.9.10}$$

respectively.

To investigate the stability properties of the stochastic mathematical model of the inflation-unemployment process (6.9.9), we choose the Lyapunov-like function V, defined by

$$V(x) = d_1 x_1^2 + d_2 x_2^2, \tag{6.9.11}$$

where d_1 and d_2 are arbitrary positive real numbers.

Applying Theorem 1.3.2 to V in the context of (6.9.9), $\mathcal{L}V$ is determined as

$$\mathcal{L}V(x) = V_x(x)[Ax + p(t,x)] + \frac{1}{2}(V_{xx}(x)q(t,x)q^T(t,x)), \qquad (6.9.12)$$

which can be estimated:

$$\mathcal{L}V(x) \le d_1[-2a_{11}ax_1^2 + 2|a_{12}||x_1x_2|]$$
$$+ d_2[-2a_{22}x_2^2 + 2|a_{21}||x_1x_2| + \beta x_2^2(\beta + |x_2|)\sigma^2(t,-x_2)].$$

Using some elementary algebraic rearrangements, simplifications, and the estimate (6.9.2) on σ, the above inequality can be reduced to

$$\mathcal{L}V(x) \le d_1[-2a_{11}ax_1^2 + |a_{12}||x_1|^2 + |a_{12}||x_2|^2]$$
$$+ d_2[-2a_{22}x_2^2 + |a_{21}||x_1|^2 + |a_{21}||x_2|^2 + \beta\bar{\sigma}^2 x_2^2]$$
$$\le \mu(G)V(x), \qquad (6.9.13)$$

where G is a 2×2 matrix defined by

$$G = \begin{bmatrix} -2a_{11} + |a_{12}| & |a_{12}| \\ |a_{21}| & -2a_{22} + |a_{21}| + \beta\bar{\sigma}^2 \end{bmatrix}, \qquad (6.9.14)$$

and $\mu(G)$ is the logarithmic norm of G defined in Definition 6.6.1 (Section 6.6 [80]).

Now, we are ready to state and prove the mean square stability result.

Theorem 6.9.1. *Assume that the logarithmic norm of the matrix G satisfies the following condition:*

$$\mu(G) \le -\nu, \quad \text{for some positive number } \nu. \qquad (6.9.15)$$

Then, the equilibrium state (p^*, π^*) of the stochastic system (6.9.6) is exponentially mean-square-stable. Moreover, the estimate on the rate of decay of the solution process of (6.9.6) is ν.

Proof. By employing the differential inequalities and Compassion Theorem 6.6.2, in particular Observation 6.6.1(i), we have

$$E[V(x(t))] \le [V(x_0)]\exp[-\nu(t-t_0)], \quad t \ge t_0.$$

Using algebraic manipulations, the above inequality is rewritten as

$$d_m E[\|x(t)\|^2] \le d_M E[\|x_0\|^2]\exp[-\nu(t-t_0)], \quad t \ge t_0.$$

Hence,

$$E[\|x(t)\|^2] \le M E[\|x_0\|^2]\exp[-\nu(t-t_0)], \quad t \ge t_0, \qquad (6.9.16)$$

where $L = \frac{d_M}{d_m}$.

The proof of the theorem can be completed by following the standard argument used in the literature [8, 56, 98]. $\qquad\qquad\square$

6.10 Critical Point Theory: Stochastic Versus Deterministic

In this section, a few elementary examples showing the random perturbation effects in the critical point theory [165] of a two-dimensional autonomous dynamic vector field are presented. These examples exhibit the role and scope of mathematics in real world problems. Moreover, they show the significance of the two-dimensional dynamic vector field under randomly varying environmental perturbations. This would also motivate one to undertake a further study of critical point theory under the influence of random perturbations. In addition, the examples raise several issues, in particular "stochastic versus deterministic" [112] in the modeling of dynamic processes. This important problem is one of the authors' current research projects.

Points Versus Focal Point 6.10.1. Let's consider a degenerate two-dimensional deterministic dynamic vector field,

$$dy = Ay \, dt, \tag{6.10.1}$$

where

$$A = \begin{bmatrix} 0 & 0 \\ 0 & 0 \end{bmatrix}.$$

In this case, the fundamental matrix solution (Chapter 4 of Volume 1) of (6.10.1) is

$$\Phi_d(t) = \begin{bmatrix} 1 & 0 \\ 0 & 1 \end{bmatrix}. \tag{6.10.2}$$

Therefore, in this case, all points, are critical points, and all trajectories reduce to the points.

Now, we consider the Itô–Doob-type stochastic perturbed system relative to (6.10.1) as

$$dy = Ay \, dt + By \, dw(t), \tag{6.10.3}$$

where dy stands for the Itô–Doob-type stochastic differential of y, the matrix A is as defined in (6.10.1), and the matrix B is defined by

$$B = \begin{bmatrix} 0 & 1 \\ -1 & 0 \end{bmatrix}.$$

Moreover, it is trivial that $AB = BA$.

Therefore, by following the solution procedure in Section 4.3, the fundamental matrix solution process of (6.10.3) is

$$\Phi(t) \equiv \Phi_s(t) = \exp\left[\frac{1}{2}t\right] \begin{bmatrix} \cos w(t) & \sin w(t) \\ -\sin w(t) & \cos w(t) \end{bmatrix}. \tag{6.10.4}$$

From this, it is easy to conclude that the critical point is the origin, and it is the *focal point* (focus).

From this, we can conclude that the Itô–Doob-type stochastic perturbation has caused to generate of the focal point.

Node Versus Focus 6.10.2. Next, we consider a two-dimensional deterministic dynamic vector field described by the deterministic system of differential equations

$$dx = Ax\,dt, \tag{6.10.5}$$

where

$$A = \begin{bmatrix} 1 & 0 \\ 0 & 1 \end{bmatrix}.$$

In this case, the fundamental matrix solution (Chapter 4 of Volume 1) of (6.10.5) is

$$\Phi_d(t) = \exp[t] \begin{bmatrix} 1 & 0 \\ 0 & 1 \end{bmatrix}. \tag{6.10.6}$$

Therefore, in this case, the critical point is the origin, and the origin is the *node* (unstable).

Now, we consider the Itô–Doob-type stochastic perturbed system relative to (6.10.5):

$$dy = Ay\,dt + By\,dw(t), \tag{6.10.7}$$

where dy stands for the Itô–Doob-type stochastic differential of y, the matrix A is as defined in (6.10.5), and the matrix B is as defined in (6.10.3). It is obvious that $AB = BA$. In this case, by following the solution procedure of Section 4.4, the fundamental matrix solution process of (6.10.7) is

$$\Phi(t) = \exp\left[\frac{3}{2}t\right] \begin{bmatrix} \cos w(t) & \sin w(t) \\ -\sin w(t) & \cos w(t) \end{bmatrix}. \tag{6.10.8}$$

From this, it is easy to conclude that the critical point is the origin and is also the *focal point* (focus). The focus is unstable.

From this, we conclude that the Itô–Doob-type stochastic perturbation has forced to transfer of the critical point "node" to the critical point "focus."

Node Versus Center 6.10.3. In the following, we consider a two-dimensional deterministic dynamic vector field described by

$$dx = Ay\,dt, \tag{6.10.9}$$

where

$$A = \begin{bmatrix} -\frac{1}{2} & 0 \\ 0 & -\frac{1}{2} \end{bmatrix}.$$

In this case, the fundamental matrix solution (Chapter 4 of Volume 1) of (6.10.9) is

$$\Phi_d(t) = \exp\left[-\frac{1}{2}t\right]\begin{bmatrix} 1 & 0 \\ 0 & 1 \end{bmatrix}. \tag{6.10.10}$$

Therefore, the critical point is the origin and the origin is the *node* (stable).

Now, we consider the Itô–Doob-type stochastic perturbed system relative to (6.10.9):

$$dy = Ay\,dt + By\,dw(t), \tag{6.10.11}$$

where dy stands for the Itô–Doob-type stochastic differential of y, the matrix A is as defined in (6.10.9) and the matrix B is as defined in (6.10.3). Again, it is clear that $AB = BA$.

In this case, by following the solution procedure in Section 4.4, the fundamental matrix solution process of (6.10.11) is

$$\Phi(t) = \begin{bmatrix} \cos w(t) & \sin w(t) \\ -\sin w(t) & \cos w(t) \end{bmatrix}. \tag{6.10.12}$$

From this, it is easy to conclude that the critical point is the origin, and it is the *center* (focus).

From this, we conclude that the Itô–Doob-type stochastic perturbation has caused to generate the critical point "center" from the stable "node."

Center Versus Focus 6.10.4. Below, we consider a two-dimensional deterministic dynamic vector field described by

$$dx = Ax\,dy, \tag{6.10.13}$$

where

$$A = \begin{bmatrix} 0 & 1 \\ -1 & 0 \end{bmatrix}.$$

In this case, the fundamental matrix solution to (6.10.13) (Chapter 4 of Volume 1) is

$$\Phi_d(t) = \begin{bmatrix} \cos t & \sin t \\ -\sin t & \cos t \end{bmatrix}. \tag{6.10.14}$$

Therefore, the critical point is the origin, and the origin is the *center*.

Now, we consider the Itô–Doob-type stochastic perturbed system relative to (6.10.13):

$$dy = Ay\,dt + By\,dw(t), \tag{6.10.15}$$

where dy stands for the Itô–Doob-type stochastic differential of y, the matrix A is as defined in (6.10.13), and the matrix B is defined by

$$B = \begin{bmatrix} 1 & 0 \\ 0 & 1 \end{bmatrix}.$$

We observe that $AB = BA$.

In this case, by following the solution procedure in Section 4.4, the fundamental matrix solution process of (6.10.15) is

$$\Phi(t) = \exp\left[-\frac{1}{2}t + w(t)\right] \begin{bmatrix} \cos t & \sin t \\ -\sin t & \cos t \end{bmatrix}. \tag{6.10.16}$$

From this, we conclude that the critical point is the origin, and it is the focal point (focus).

From this, we conclude that the Itô–Doob-type stochastic perturbation has caused the transfer of the critical point "node" to the critical point "focus."

6.11 Statistical Properties and Effects of Random Perturbations

In this section, we outline a few statistical properties of the solution processes of stochastic differential equations of the Itô–Doob type. In particular, we find the mean and variance of the solution processes. Using the mean and variance, we shed light on the effects of the random perturbations in the modeling of dynamic processes. In order to meet the goal of this section, we need the following, elementary result.

Lemma 6.11.1. *Let Y be a normally distributed random variable $\mathcal{N}(\mu, \sigma^2)$ with mean μ and variance σ^2. Then, for every $p > 0$,*

$$E[(\exp[Y])^p] = E[\exp[pY]] = \exp\left[\mu p + \frac{p^2\sigma^2}{2}\right]. \tag{6.11.1}$$

Proof. First, we note that

$$py - \frac{(y-\mu)^2}{2\sigma^2} = \frac{2p\sigma^2 y - (y-\mu)^2}{2\sigma^2} = \frac{-(y-\mu-p\sigma^2)^2 + 2p\mu\sigma^2 + p^2\sigma^4}{2\sigma^2}$$

$$= -\frac{(y-\mu p - \sigma^2)^2}{2\sigma^2} + \mu p + \frac{p^2\sigma^2}{2}. \tag{6.11.2}$$

Using (6.11.2), we determine the following:

$$E[(\exp[Y])^p] = E[\exp[pY]] \quad \text{(by the laws of exponents)}$$

$$= \frac{1}{\sqrt{2\pi\sigma^2}} \int_{-\infty}^{\infty} \exp\left[py - \frac{(y-\mu)^2}{2\sigma^2}\right] dy \quad \text{(by Definition 1.1.16)}$$

$$= \exp\left[\mu p + \frac{p^2\sigma^2}{2}\right] \frac{1}{\sqrt{2\pi\sigma^2}} \int_{-\infty}^{\infty} \exp\left[\frac{-(y-\mu-p\sigma^2)^2}{2\sigma^2}\right] dy.$$

The conclusion of the lemma immediately follows. $\qquad\square$

Knowing the nature of the Wiener process and Itô–Doob calculus [7, 8, 32, 42, 43, 52, 98], we find the statistical properties of the solution processes of some Illustrations, Examples, and the mathematical models (predator–prey and diabetes mellitus). It is assumed that the initial state data is independent of the Wiener process.

Mean of the Solution Process 6.11.2. We examine both single- and multi-state dynamic systems described by the stochastic differential equations.

A. Single-state dynamic process. The means of the solution processes of the dynamic processes described in Illustrations 2.3.3 and 2.4.2 are:

1. Illustration 2.3.3. We recall Illustration 2.3.3 (stochastic version of Lambert's law) as it relates to the mathematical model of the dynamic process of photon absorption under a randomly varying environment. It is described by

$$dn(x) = -kn(x)dx + \sigma n(x)dw(x), \quad n_0 = n(0) = \frac{N_0}{A}. \tag{6.11.3}$$

Its closed-form solution process is

$$I(x) = I_0 \exp\left[-\left(k + \frac{1}{2}\sigma^2\right)x + \sigma w(x)\right], \tag{6.11.4}$$

where $I(x)$ is the intensity of the light beam that is directly proportional to the photon flux $\frac{n(x)}{A}$, and I_0 is independent of the Wiener process w. This is the stochastic version of Lambert's law. Applying Lemma 6.11.1, the mean ($p = 1$) of the solution $I(x)$ is

$$E[I(x)] = E[I_0]E\left[\exp\left[-\left(k + \frac{1}{2}\sigma^2\right)x + \sigma w(x)\right]\right]$$

$$= E[I_0]\exp\left[-\left(k + \frac{1}{2}\sigma^2\right)x\right]E\left[\exp[\sigma w(x)]\right]$$

$$= E[I_0]\exp\left[-\left(k + \frac{1}{2}\sigma^2\right)x\right]\exp\left[\frac{1}{2}\sigma^2 x\right]$$

$$= E[I_0]\exp[-kx]. \tag{6.11.5}$$

On the other hand, using the argument presented in Observations 2.1.1, 2.1.3, and 2.1.5, the corresponding mean differential equation is

$$dm(t) = -km(t)dx, \quad E[n_0] = E\left[\frac{N_0}{A}\right], \tag{6.11.6}$$

where $E[I(x)] = m(t)$. Its solution process is

$$m(t) = m_0 \exp[-kx]. \tag{6.11.7}$$

This is indeed Lambert's law [44, 152].

Conclusion 6.11.1

(i) We note that the mean of the solution process (6.11.4) and the solution to the mean differential equation (6.11.6) are equal.

(ii) The deterministic version of Lambert's law implies that as the depth or thickness of the medium increases, the intensity of the light decreases. In the case of the stochastic version of Lambert's law, the randomness accelerates the decrease in the light intensity by the rate $\frac{1}{2}\sigma^2$. This means that as the magnitude of the fluctuation increases, the intensity of the light further decreases.

2. Illustration 2.4.2. From Illustration 2.4.2 (the birth–death and immigration processes), we have

$$dx = (fx + p)dt + (\sigma x + q)dw(t), \quad x(t_0) = x. \tag{6.11.8}$$

Its closed-form solution process is

$$x(t) = \exp\left[\left(f - \frac{1}{2}\sigma^2\right)(t - t_0) + \sigma(w(t) - w(t_0))\right] x_0$$
$$+ \frac{q}{\sigma}\left(\exp\left[\left(f - \frac{1}{2}\sigma^2\right)(t - t_0) + \sigma(w(t) - w(t_0))\right] - 1\right)$$
$$+ \frac{\sigma p - fq}{\sigma}\int_{t_0}^t \exp\left[\left(f - \frac{1}{2}\sigma^2\right)(t - s) + \sigma(w(t) - w(s))\right] ds. \tag{6.11.9}$$

Using Observation 1.2.1(i), Definition 1.1.16, and Theorem 1.1.1, the calculation of the mean $(p = 1)$ of the solution process $x(t)$ is

$$E[x(t)]$$
$$= E[x_0]E\left[\exp\left[\left(f - \frac{1}{2}\sigma^2\right)(t - t_0) + \sigma(w(t) - w(t_0))\right]\right]$$
$$+ \frac{q}{\sigma}\left(E\left[\exp\left[\left(f - \frac{1}{2}\sigma^2\right)(t - t_0) + \sigma(w(t) - w(t_0))\right] - 1\right]\right)$$
$$+ \frac{\sigma p - fq}{\sigma}E\left[\int_{t_0}^t \exp\left[\left(f - \frac{1}{2}\sigma^2\right)(t - s) + \sigma(w(t) - w(s))\right] ds\right]$$
$$= E[x_0]\exp[f(t - t_0)] + \frac{q}{\sigma}[\exp[f(t - t_0)] - 1]$$
$$+ \frac{\sigma p - fq}{\sigma}\int_{t_0}^t \exp[(f(t - s)]ds \quad \text{(by Lemma 6.11.1 and integral interchange)}$$
$$= E[x_0]\exp[f(t - t_0)] + \frac{q}{\sigma}[\exp[f(t - t_0)] - 1]$$
$$+ \frac{\sigma p - fq}{\sigma}\exp[ft]\left[-\frac{1}{f}\exp[-fs]\right]\Bigg|_{t_0}^t \quad \text{(by definite integration)}$$

$$= E[x_0] \exp[f(t - t_0)] + \frac{q}{\sigma} [\exp[f(t - t_0)] - 1]$$

$$+ \frac{\sigma p - fq}{\sigma} \exp[ft] \left[-\frac{1}{f} \exp[-ft] + \frac{1}{f} \exp[-ft_0] \right]$$

$$= E[x_0] \exp[f(t - t_0)] + \frac{q}{\sigma} [\exp[f(t - t_0)] - 1]$$

$$+ \left[\frac{p}{f} - \frac{q}{\sigma} \right] [-1 + \exp[f(t - t_0)]] \quad \text{(by simplifying)}$$

$$= E[x_0] \exp[f(t - t_0)] + \frac{p}{f} [-1 + \exp[f(t - t_0)]] \quad \text{(by simplifying).} \tag{6.11.10}$$

Again, on the other hand, by imitating the argument presented in Observations 2.1.1, 2.1.3, and 2.1.5, the mean differential equation associated with (6.11.8) and its solution are

$$dm = (fm + p)dt, \quad E[x_0] = m_0, \tag{6.11.11}$$

$$m(t) = \exp[f(t - t_0)]m_0 + \frac{p}{f}(\exp[f(t - t_0)] - 1), \tag{6.11.12}$$

respectively.

Conclusion 6.11.2

(i) We note that the mean of the solution process (6.11.9) and the solution to the mean differential equation (6.11.11) are equal.

(ii) Depending on the sign of the expected intrinsic growth rate (birth rate–death rate) f and the expected size of the species in a birth–death and immigration process, the size of the population either approaches the deterministic steady state $\left(-\frac{p}{f} \right)$ size (if $f < 0$), or it approaches ∞ (explosion) ($f > 0$) irrespective of the magnitude of the expected/deterministic/permitted/legal immigration rate of population p.

(iii) In the presence of the randomly varying environment and depending on the magnitude of fluctuations $\frac{1}{2}\sigma^2$ and the sign of $\left(f - \frac{1}{2}\sigma^2 \right)$, the size of population either converges or diverges from its random steady state. Moreover,

   (a) If $f < \frac{1}{2}\sigma^2$, then the size of the population approaches its random steady-state size.

   (b) If $f > \frac{1}{2}\sigma^2$, then the size of the population grows infinitely large as time increases.

(iv) Under condition (a), randomness can strengthen or generate the convergence, and under condition (b), the population grows unboundedly, which is not practically possible.

(v) Certain special cases can further simplify discussions (a) and (b) and can provide useful information. For example, $\frac{p}{f} = \frac{q}{\sigma}$. In this situation, the randomness preserves the steady state, and the effects of randomness are more obvious.

B. Multistate dynamic processes. Now, we present the mean of multivariate dynamic processes described in Example 4.3.3 (predator–prey model), Example 4.4.5, and Illustration 4.5.4 (diabetes mellitus model and Example 4.4.5 [80]).

1. Example 4.3.3 (predator–prey model). We consider the predator–prey mathematical variational model,

$$dN = CN\, dt + \sigma CN\, dw(t), \quad N(t_0) = N_0, \tag{6.11.13}$$

and its solution representation (Procedure 4.4.4, Example 4.3.3, and Example 4.4.5 [80]),

$$\Phi(t, w(t), 0)x_0 = \exp\left[\frac{\beta^2\sigma^2}{2}t\right]$$

$$\times \begin{bmatrix} \cos\theta(\Delta t, \Delta w(t)) & -\eta\sin\theta(\Delta t, \Delta w(t)) \\ \frac{1}{\eta}\sin\theta(\Delta t, \Delta w(t)) & \cos\theta(\Delta t, \Delta w(t)) \end{bmatrix} N_0, \tag{6.11.14}$$

where $\theta(\Delta t, \Delta w(t)) = \beta(t - t_0) + \sigma(w(t) - w(t_0))$, $\Delta t = t - t_0$, and $(w(t) - w((t_0)) = \Delta w(t)$; $\eta = \frac{\alpha_1}{\beta_1}\sqrt{\frac{\beta_0}{\alpha_0}}$.

Based on the above presentation, we recognize that the problem of finding the mean of the solution reduces to the problem of computing the improper integrals. This problem is left to the reader as an exercise. We caution that the closed-form integrals are not always feasible.

Again, applying Observations 4.1.1 and 4.1.2, the mean differential equation (6.11.13) and its closed-form solution are

$$dm = Cm\, dt, \quad E[N_0] = m_0, \tag{6.11.15}$$

$$m(t) = \begin{bmatrix} \cos\beta(t - t_0) & -\eta\sin\beta(t - t_0) \\ \frac{1}{\eta}\sin\beta(t - t_0) & \cos\beta(t - t_0) \end{bmatrix} m_0, \tag{6.11.16}$$

respectively.

The comparison of $E[N(t)]$ and $m(t)$ is left as an exercise (if the integral is computable in the closed form).

Conclusion 6.11.3. We observe that the sample paths of the population process and the solution to the mean differential equations are periodic; the two have the same critical point. In the case of (6.11.13), the critical point is an unstable focus; on the other hand, the critical point of (6.11.15) is the center. This implies that the randomness has destroyed the center and generated an unstable focal point.

2. Example 4.4.5. We recall that the closed-form solution process of Example 4.4.5 is

$$\Phi(t, w(t), 0)x_0 = e^{3t+\alpha(t,w(t))} \begin{bmatrix} \phi_{11}(t, w(t), \psi) & \phi_{12}(t, w(t), \psi) \\ \phi_{21}(t, w(t), \psi) & \phi_{22}(t, w(t), \psi) \end{bmatrix} x_0, \qquad (6.11.17)$$

where $\psi = \arctan\left(\frac{1}{2}\right)$,

$$\phi \equiv \phi(t, w(t)) = -2\sigma^2 t + \sigma w(t), \quad 3t + \alpha(t, w(t)) = 3\left(1 - \frac{\sigma^2}{2}\right)t + 2\sigma w(t).$$

by using the closed-form solution, it is not so obvious to compute $E[x(t)]$. The issue is an improper integration problem. We will not compute the integral, but we recommend that the reader make an effort to compute the integral as an exercise.

On the other hand, applying Observations 4.1.1 and 4.1.2, the mean differential equation is

$$dm = Am\, dt, \quad E[x_0] = m_0, \qquad (6.11.18)$$

where the matrix A is as given in Example 4.4.5 [80]. The solution of this mean system of differentials is

$$m(t) = e^{3t} \begin{bmatrix} \sqrt{5}\sin(t + \psi) & -\sqrt{5}\sin t \\ \sqrt{5}\sin t & -\sqrt{5}\sin(t + \psi) \end{bmatrix} m_0. \qquad (6.11.19)$$

Conclusion 6.11.4. In the absence of the closed-form knowledge about the mean of the solution process (6.11.17), we compare the solution process (6.11.17) with the solution to the mean differential equation (6.11.18). It is clear that both types of systems of differential equations have the origin as the critical point.

(i) In the case of a system of mean differential equations, it is an unstable focal point.
(ii) With respect to the system of stochastic differential equations in Example 4.4.5 and depending on the sign of $\left(1 - \frac{\sigma^2}{2}\right)$, it has the following three cases:

(a) stable focal point $(2 < \sigma^2)$ or
(b) unstable focal point $(2 > \sigma^2)$ or
(c) center $(2 = \sigma^2)$.

Obviously, the stable focal and center points are generated by the effects of random perturbations.

3. Illustration 4.5.4 (diabetes mellitus model). For the sake of simplicity, we examine the homogeneous system of differential equations corresponding to the diabetes detection model (Illustration 4.5.4). Again, we recall the closed-form solution expression of the homogeneous system associated with Illustration 4.5.4,

$$dx = (Ax)dt + (Bx)dw(t), \quad x(t_0) = x_0, \tag{6.11.20}$$

with its solution,

$$\Phi(t, w(t), 0)x_0 = \exp[\alpha_0(t - t_0)] \begin{bmatrix} \phi_{11}(t, w(t), \psi) & \phi_{12}(t, w(t), \psi) \\ \phi_{21}(t, w(t), \psi) & \phi_{22}(t, w(t), \psi) \end{bmatrix} x_0. \tag{6.11.21}$$

Once more, the computation of the mean $E[x(t)]$ is left as an exercise to the reader.

By repeating the above argument (Observations 4.1.1 and 4.1.2), the mean differential equation and its closed-form solution are

$$dm = Am\,dt, \quad E[x_0] = m_0, \tag{6.11.22}$$

$$m(t) = \exp\left[-\frac{k_{11} + k_{22}}{2}(t - t_0)\right] \begin{bmatrix} \phi_{11}^d(t - t_0) & \phi_{12}^d(t - t_0) \\ \phi_{21}^d(t - t_0) & \phi_{22}^d(t - t_0) \end{bmatrix} m_0 = \Phi_d(t, 0)m_0, \tag{6.11.23}$$

respectively, where $\psi = \arctan\left[\frac{a_{11} - a}{\beta}\right]$,

$$\phi_{11}(t - t_0) = \sqrt{\frac{k_{12}k_{21}}{k_{12}k_{21} - (k_{11} - k_{22})^2}} \cos \varphi_d(t, \psi),$$

$$\phi_{12}(t - t_0) = \frac{k_{12}}{\beta} \sin \varphi_d(t),$$

$$\phi_{21}(t - t_0) = \frac{k_{21}}{\beta} \sin \varphi_d(t),$$

$$\phi_{22}(t - t_0) = \sqrt{\frac{k_{12}k_{21}}{k_{12}k_{21} - (k_{11} - k_{22})^2}} \cos \varphi_d(t, \psi),$$

$\alpha = -\frac{k_{11} + k_{22}}{2}$, and $\Phi_d(t, 0)$ is a normalized fundamental matrix solution to (6.11.22).

Conclusion 6.11.5. Conclusions parallel to Conclusion 6.11.3 can be reformulated. In this case, the solution processes of (6.11.20) and (6.11.22) have the same behavior. In fact, both of them have stable focal points. In this case, randomness accelerates the rate of convergence.

Observation 6.11.1. Employing Lemma 6.11.1 and the elementary integration techniques, the higher moments $(r > 0)$ of the solution processes of illustrations and examples are computable in closed form. For instance, the rth moments of the solution process of Illustrations 2.3.3 is easy to compute.

$$E[|I(x)|^r] = E[|I_0|^r] \exp\left[r\left(-k + \frac{1}{2}\sigma^2(r - 1)\right)x\right].$$

However, in general, the rth moments of the solution processes of the illustrations and examples similar to Illustration 2.4.2 are not easily computable: for example, in the case of Illustration 2.4.2, we have

$$E[|x(t)|^r] = E\left[\left|\exp\left[\left(f - \frac{1}{2}\sigma^2\right)(t - t_0) + \frac{1}{2}\sigma^2(w(t) - w(t_0))\right]x_0\right.\right.$$
$$+ \frac{q}{\sigma}\left(\exp\left[\left(f - \frac{1}{2}\sigma^2\right)(t - t_0) + \sigma(w(t) - w(t_0))\right] - 1\right)$$
$$\left.\left.+ \frac{\sigma p - f q}{\sigma}\int_{t_0}^t \exp\left[\left(f - \frac{1}{2}\sigma^2\right)(t - s) + \sigma(w(t) - w(t_0))\right]ds\right|^r\right].$$

Depending on $r > 0$, using the concavity/convexity and Jensen's inequality [98, 169], one can estimate the rth moments.

Covariance and Variance of the Solution Process 6.11.3. The covariance of the solution process describes the measure of the effects of the single random perturbation on the single/multiple interacting components of the dynamic process. We recall that the variance of the random variable Y measures (spread or dispersion) the expected value of the square of the deviation of Y from its expected value. The concept of covariance of two random variables is a measure of the dependence between them. Of course, if the random variables are independent, the covariance is identically zero. However, if the covariance is zero, the random variables are called uncorrelated. Let us illustrate the computation of the variance of the solution process by computing the variance of the solution process of Illustration 2.3.3.

From Definition 1.1.19, we observe that $\text{Var}(X) = E[X^2] - (E[X])^2$, and compute

$$\text{Var}(I(x)) = (E[(I(x) - x(t))^2]) = E[|I(x)|^2] - |m(t)|^2$$
$$= E[|I_0|^2]\exp[(-2k + \sigma^2)x] - m_0^2\exp[-2kx] \qquad \text{(by Lemma 6.11.1)}$$
$$= \exp[-2kx][E[|I_0|^2]\exp[\sigma^2 x] - m_0^2] \qquad \text{(by simplifying)}$$
$$= \exp[-2kx][\exp[\sigma^2 x](E[|I_0|^2] - m_0^2)$$
$$+ m_0^2(\exp[\sigma^2 x] - 1] \qquad \text{(by the additive inverse)}$$
$$= \text{Var}(I_0)\exp[(-2k + \sigma^2)x]$$
$$+ m_0^2(\exp[\sigma^2 x] - 1)\exp[-2kx] \qquad \text{(by Definition 1.1.19)}.$$

Further analysis and results can be found in Ref. 96.

Passive and Active Random Effects 6.11.4. The passive and the active randomness depend on the basic characteristic of the dynamic processes in science and engineering [96].

(i) The dynamic processes described in Illustrations 2.3.3 and 4.5.4 are considered to be *passive* with regard to random perturbations. This is due to fact that the behavior of the state of the dynamic process is unchanged under random perturbations. In this case, in fact, it accelerates the inherent character of the process. Of course, this type of random effects can produce undesirable outcomes. For a satisfactory maintenance of the performance of the process, the process needs to be monitored, periodically.

(ii) The dynamic process presented in Illustration 2.4.2 is considered to be *conditionally passive and active* with regard to random perturbations. This is based on the sign of the coefficient of the exponent $\left(f - \frac{1}{2}\sigma^2\right)$ and the magnitude of spread or dispersion of the random environmental fluctuations (σ^2). In this case, the behavior of the state of the dynamic process is influenced by the magnitude of the random internal or external perturbations. In fact, it addresses:

(a) the acceleration;
(b) the deceleration;
(c) stopping of the existing state of the action process.

This generates the domain of decomposition of the parametric set associated with stochastic disturbances. The domain of decomposition of the random environmental parametric set provides a tool for the decision-making process.

(iii) The dynamic outlined in Example 4.3.3 is referred to as *partially active* with regard to the random environmental perturbations. This is characterized by partially maintaining the behavior with a simultaneously increasing activation process and the magnitude of fluctuations. Example 4.3.3 exhibits the change from the center of the critical point to the unstable focal point. Moreover, the magnitude of the amplitude increases as the magnitude of the disturbance increases.

(iv) The dynamic process described in Example 4.4.5 is considered to be *active* in nature. The presence of random disturbance in the behavior of the dynamic process is completely altered. With regard to this example, in the absence of randomness, the critical point $(0,0)$ is unstable. However, in the presence of random fluctuations, the unstable system can be stabilized by the choice of the magnitude of random variations. For further details, see Conclusion 6.11.4.

(v) In summary, the random perturbations can alter the qualitative properties of the evolution of the dynamic process. For example:

(a) Stabilizing and destabilizing (Illustration 2.3.3 and Example 4.4.5).
(b) Convergence versus divergence (Illustration 2.4.2).
(c) Effects on the amplitude of oscillatory behavior (predator–prey and insulin–glucose control modes).

This leads to a fundamental problem, "stochastic versus deterministic" (or "uncertainty versus certainty"), in dynamic processes under the randomly varying environment.

Exercises

Find the mean and variance of the following state representations of the dynamic processes described in Illustrations and Examples:

1. Illustration 2.3.1: Chemical kinetics under a random environment.
2. Illustration 2.3.2: Arterial pulse–diastolic phase under random fluctuations.
3. Illustration 2.4.1: Newton's law of cooling under random thermal agitation.
4. Illustration 2.4.3: Arterial pulse–systolic phase under random fluctuations.
5. Illustration 2.4.4: Cell membrane.
6. Illustration 2.4.5: Diffusion process under a random environment.
7. Illustration 2.4.6: Central nervous system.

6.12 Notes and Comments

The content of this chapter provides a short and limited synopsis of many attributes of the dynamic processes that can be incorporated in mathematical modeling. This is exhibited by the limited usage of mathematical analysis. Again, our citations are rather limited but well-focused. This is due to the fact that a large amount of literature has been developed in engineering, mathematics and other sciences over the past 60+ years. Section 6.1 briefly outlines the fundamentals of the model validation of nonlinear, nonstationary dynamic processes. These results are based on class notes of the second author and existing graduate-level books [80, 98]. Section 6.2 contains the most current version of the method of variation of parameters for nonlinear and nonstationary systems [27, 75, 98, 112–114]. The generalized method of variation of parameters is adapted from Refs. 83, 90–93, 112 in Section 6.3. For its further extensions, see Refs. 94 and 95. The formulation of basic differential inequalities, the Lyapunov function, and comparison results of Sections 6.4 and 6.5 are based on Ref. 94. The ideas and results regarding the variational comparison method of Section 6.6 stem from Refs. 83, 90–93. For further extensions, see Refs. 4–6, 15–20, 46, 47, 51, 99, 102–124. An emerging area of research, i.e. modeling of hybrid dynamic systems, is introduced in Section 6.7. It is adpated from Ref. 81, and for its further extensions and generalizations see Refs. 18, 72, 73, 94, 95, 123, 124, 176, 177. A brief presentation of the scope of hereditary effects exhibited in Section 6.8 is adapted from Refs. 88, 91–93. For further details, see Refs. 22, 85, 86. In Section 6.10, an illustration of a qualitative property, namely stability of the equilibrium state of the inflation-unemployment process, is adapted from Refs. 3, 35, 36, 46, 50, 97, 147, 149, 196. The effects of the random perturbations in the context of critical points are exhibited by the examples in Section 6.10. In Section 6.11, the statistical properties of the solution process and the solution process of the mean of differential equations are outlined via examples. The issue of "stochastic versus deterministic" is also illustrated, explicitly in Refs. 4–6, 15, 16, 51, 72, 73, 75, 77, 84, 86, 88, 93, 96, 98, 99, 102–105, 107, 112, 119.

Bibliography

[1]   Ackerman, Eugene (1962). *Biophysical Sciences* (Prentice-Hall, Englewood Cliffs).

[2]   Allen, Edward (2007). *Modeling with Ito Stochastic Differential Equations* (Springer).

[3]   Alogoskoufis, George S. and Smith, Ron (1991). The Phillips curve, the persistence of infation, and the Lucas critique: evidence from exchange-rate regimes, *The American Review of Economics* **81**, 5, pp. 1254–1275.

[4]   Anabtawi, M. J. and Ladde, G. S. (2005). Dynamics of fluid flows under Markovian structural perturbations, *Mathematical and Computer Modelling* **42**, pp. 967–976.

[5]   Anabtawi, M. J. and Ladde, G. S. (2000). Convergence and stability analysis of system of partial differential differential equations under Markovian structural perturbations I: Vector Lyapunov-like functions, *Stochastic Analysis and Applications* **18**, pp. 493–524.

[6]   Anabtawi, M. J. and Ladde, G. S. (2000). Convergence and stability analysis of system of partial differential differential equations under Markovian structural perturbations II: Vector Lyapunov-like functionals, *Stochastic Analysis and Applications* **18**, pp. 671–696.

[7]   Applebaum, David (2004). *Levy Processes and Stochastic Calculus* (Cambridge University Press).

[8]   Arnold, Ludwig (1974). *Stochastic Differential Equations* (John Wiley & Sons, New York).

[9]   Arrow, K. J. and Hahn, F. H. (1971). *General Equilibrium Analysis* (Holden-Day, San Francisco).

[10] Baily, N. T. J. (1957). *The Mathematical Theory of Epidemics* (Hafner, New York).

[11] Batschelet, E. (1971). *Introduction to Mathematics for Life Scientists* (Springer-Verlag, New York).

[12] Bartlett, M. S. (1960). *Stochastic Population Models in Ecology and Epidemiology* (Methuen, London).

[13] Bolie, Victor W. (1961). Coefficients of normal blood glucose regulation, *Journal of Applied Physiology* **16**, pp. 783–788.

[14] Canavos, George C. (1984). *Applied Probability and Statistical Methods* (Little, Brown and Company, Boston).

[15] Chandra, J., Ladde, G. S. and Lakshmikantham, V. (1983). Stochastic analysis of compressible gas lubrication slider bearing problem, *SIAM Journal on Applied Mathematics* **43**, pp. 1174–1186.

[16] Chandra, J. and Ladde, G. S. (1983/84). On roughness effects in a compressible lubrication problem, *Recent Developments in Applied Mathematics* (Rensselaer, New York) pp. 8–12.

[17] Chandra, J., Ladde, G. S. (2012). Dynamic stochastic models of social networks, ABSTRACTS: 1079-34-120, Aus Tampa Spring Meeting.

[18] Chandra, J. and Ladde, G. S. (2004). Stability analysis of stochastic hybrid systems, *International Journal of Hybrid Systems* **4**, pp. 179–198.

[19] Chandra, J. and Ladde, G. S. (2010). Collective behavior of multiagent network dynamic systems under internal and external random perturbations, *Nonlinear Analysis: Real World Applications* **11**, pp. 1330–1344.

[20] Chandra, J., Ladde, G. S. and Sirisaengtaksin, O. (1988). On multitime method for large-scale filtering, *International Journal of Systems Science* **19**, pp. 1579–1604.

[21] Chandrasekhar, S. (1943). Stochastic problems in physics and astrology, *Review of Modern Physics*, pp. 1–89.

[22] Chang, Mou-Hsiung (2006). *Stochastic Control of Hereditary Systems and Applications* (Springer-Verlag, New York).

[23] Cope, F. W. (1976). Derivation of Weber–Fechner law and Lowenstein equation, *Bulletin of Mathematical Biology* **38**, pp. 111–118.

[24] Cope, F. W. (1970). The solid-state physics of electron and ion transport in biology, *Advances in Biological and Medical Physics* **13**, pp. 1–42.

[25] Crow, J. F. and Kimura, M. (1971). *Introduction to Population Genetics* (Harper and Row, New York).

[26] Defares, J. G. and Sneddon, L. N. (1961). *An Introduction to the Mathematics of Medicine and Biology* (Year Book Medical Publishers, Chicago).

[27] Deo, S. G. and Lakshmikantham, V. (1998). *Method of Variation of Parameters for Dynamic Systems* (Gordon and Breach Science, Amsterdam).

[28] Doob, J. L. (1953). *Stochastic Processes* (Wiley, New York).

[29] Dold, A. and Eckmann, B. (eds.) (1972). *Stability of Stochastic Dynamical Systems; Proceedings of the International Symposium* **294**, University of Warwick, July 10–14, 1972 (Springer-Verlag, New York).

[30] Dublin, Neil (1976). *A Stochastic Model for Immunological Feedback in Carcinogenesis* **9** (Springer-Verlag, New York).

[31] Dynkin, E. G. (1965). *Markov Processes* (Springer-Verlag, Berlin).

[32] Elliott, Robert (1982). *Stochastic Calculus and Applications* (Springer-Verlag, New York).

[33] Etheridge, Alison (2002). *A Course in Financial Calculus* (Cambridge University Press).

[34] Foster, Lisa D., Saha, Nirzhar, Ross, Mark, Ladde, G. S. and Wang, P. (2008). Using frequency analysis to determine wetland hydroperiod, *Neural, Parallel, and Scientific Computations* **16**, pp. 17–34.

[35] Friedman, Milton (1968). The role of monetary policy, *The American Economic Review* **58**, 1, pp. 1–17.

[36] Friedman, Milton (1977). Noble Lecture: inflation and unemployment, *Journal of Political Economy* **85**, 3, pp. 1–17.

[37] Gandolfo, Giancarlo (1972). *Mathematical Methods and Models in Economic Dynamics* (North-Holland, Amsterdam).

[38] Gardiner, C. W. (1985). *Handbook of Stochastic Methods for Physics, Chemistry and the Natural Sciences*, 2nd ed. (Springer-Verlag, New York).

[39] Gatewood, Lael C., Ackerman, Eugene, Rosevear, John W. and Molnar, George D. (1970). Modeling blood glucose Dynamic, *Behavioral Science* **15**, 1, pp. 72–86.

[40] Gavalas, G. R. (1968). *Nonlinear Differential Equations of Chemical Reacting Systems* (Springer-Verlag, New York).

[41] Gerasimov, Ya., Dreving, V., Eremin, E., Kiselev, A., Lebedev, V., Panchenkov, G. and Shlygin, A. (1974). *Physical Chemistry*, Vol. 1 and 2, (MIR, Moscow).

[42]   Gikhman, I. I. and Skorokhod, A. V. (1969). *Introduction to the Theory of Random Processes* (W. B. Saunders, Philadelphia).

[43]   Gikhman, I. I. and Skorokhod, A. V. (1972). *Stochastic Differential Equations* (Springer-Verlag, New York).

[44]   Gnedenko, B. (1973). *The Theory of Probability*, 2nd ed. (MIR, Moscow).

[45]   Goel, N. S. and Richter-Dyn, N. (1974). *Stochastic Models in Biology* (Academic, New York).

[46]   Golec, J. and Ladde, G. S. (1989). Euler-type approximation for systems of stochastic differential equations, *Journal of Applied Mathematics and Simulation* **2**, pp. 239–249.

[47]   Golec, J. and Ladde, G. S. (1990). Averaging principle and systems of singularly perturbed stochastic differential equations, *Journal of Mathematical Physics* **31**, pp. 1116–1123.

[48]   Goodman, L. A. (1967). On the age–sex composition of the population that would result from given fertility and mortality conditions, *Demography* **4**, pp. 423–441.

[49]   Gowen, John W. (1964). Effects of x-Rays of different wavelengths on viruses, Chap. 39, *Statistics amd Mathematics in Biology* (Editors: Oscar Kempthorne *et al.*) (Hafner, New York), pp. 495–510.

[50]   Greenwood, J. and Huffman, G. W. (1987). A dynamic model of inflation and unemployment, *Journal of Monetary Policy* **19**, pp. 203–228.

[51]   Griffin, Byron L. and Ladde, G. S. (2004). Qualitative properties of stochastic iterative processes under random structural perturbations, *Mathematics and Computers in Simulations* **67**, pp. 181–200.

[52]   Grigoriu, Mircea (2002). *Stochastic Calculus* (Birkhauser, Boston).

[53]   Gutfreund, H. (1972). *Enzymes: Physical Principles* (John Wiley and Sons, New York).

[54]   Guyton, Arthur C. (1971). *Basic Human Physiology: Normal Function and Mechanisms of Disease* (W. B. Saunders, Philadelphia).

[55]   Halliday, David, Resnick, Robert and Walker, Jearl (2001). *Fundamentals of Physics*, 6th ed. (John Wiley and Sons, New York).

[56]   Hasminskii, R. Z. (1980). *Stochastic Stability of Differential Equations* (Sijthoff & Noordhoff, Rockville, Maryland).

[57]   Holden, Arun V. (1976). *Models of the Stochastic Activity of Neurones*, Vol. 12 (Springer-Verlag, New York).

[58]   Homans, G. S. (1950). *The Human Group* (Harcourt Brace, New York).

[59]   Huxley J. S. (1932). *Problems of Relative Growth* (Mathuen, London).

[60]   Ito, K. (1951). On stochastic differential equations, *Mem. Math. Soc.*, No. 4.

[61]   Ito, K. (1951). On a formula concerning stochastic differential, *Nagoya Mathemathical Journal* **55**.

[62]   Jacquez, John A. (1972). *Compartmental Analysis in Biology and Medicine* (Elsevier, New York).

[63]   Johnson, Frank H., Eyring, Henry and Stover, Besty Jones (1974). *The Theory of Rate Processes in Biology and Medicine* (John Wiley and Sons, New York).

[64]   Kapur, J. N. (1985). *Mathematical Models in Biology and Medicine* (Affiliated East–West Press, New Delhi, India).

[65]   Kats, Ya and Martynyuk, A. A. (2002). *Stability and Stabilization of Nonlinear Systems with Random Structure* (Taylor and Francis, New York).

[66]   Kendall, Davis G. (1949). Stochastic processes and population growth, *Journal of Royal Statistical Society, Series B* **11**, pp. 230–282.

[67]   Kimura, Motto (1964). Diffusion models in population genetics, *Journal of Applied Probability* **1**, pp. 177–232.

[68]   Kimura, M, Kallianpur and Hida. T. (eds.) (1987). *Stochastic Methods in Biology — Proceedings*, Nagoya, Japan, 1985, Vol. 70 (Springer-Verlag, New York).

[69]   Kimura, Motto and Ohta, Tomoko (1971). *Theoretical Aspects of Population Genetics* (Princeton University Press).

[70]   Kirby, Roger D., Ladde, Anil G. and Ladde, G. S. (2010). Stochastic Laplace transform with applications, *Communications in Applied Analysis* **14**, pp. 373–392.

[71]   Kloeden, P. E. and Platen, E. (1992). *Numerical Solution of Stochastic Differential Equations* (Springer-Verlag, New York).

[72]   Korzeniowski, Andrzej and Ladde, G. S. (2009). Modeling of hybrid network dynamics under random perturbations, *Nonlinear Analysis: Hybrid Systems* **3**, pp. 143–149.

[73]   Korzeniowski, Andrzej and Ladde, G. S. (2010). Random network with interacting nodes, *Neural, Parallel and Scientific Computations* **18**, pp. 333–342.

[74]   Kramers, H. A. (1940). Brownian motion in a field of force and the diffusion model of chemical reactions, *Physica VII*, 4.

[75]   Kulkarni, R. and Ladde, G. S. (1979). Stochastic stability of short-run market equilibrium: a comment, *The Quarterly Journal of Economics*, pp. 731–735.

[76]   Ladde, A. G. (1992). Selecting the best baseball cards for investment, *Mathematical and Computer Modelling* **16**, pp. 135–142.

[77]   Ladde, A. G. and Ladde, G. S. (2007). Dynamic processes under random environmental perturbations, *Bulletin of the Marathwada Mathematical Society* **8**, pp. 96–123.

[78]   Ladde, Anil G. and Ladde, G. S. (2010). Stochastic modeling analysis and applications, *International Encyclopedia of Statistical Sciences* (Editor: Miodrag Lovric) (Springer), pp. 1526–1531.

[79]   Ladde, Anil G. and Ladde, G. S. (2010). Determinant functions and applications to stochastic differential equations, *Communications in Applied Analysis* **14**, pp. 409–434.

[80]   Ladde, Anil G. and Ladde, G. S. (2012). *An Introduction to Differential Equations I: Deterministic Modeling, Methods, and Analysis* (World Scientific Publishing Company, Singapore).

[81]   Ladde, Jay G. and Ladde, G. S. *Mathematical Modeling of Drug Prescription and Scheduling Process* (reprint).

[82]   Ladde, G. S. (1974). *An Introduction to Biomathematics I; Lecture Notes* (The State University of New York at Potadam).

[83]   Ladde, G. S. (1975). Variational comparison theorem and perturbation of nonlinear systems of differential equations, *Proceedings of the American Mathematical Society* **52**, pp. 181–187.

[84]   Ladde, G. S. (1977). Competitive processes II: stability of random systems, *Journal of Theoretical Biology* **68**, pp. 331–354.

[85]   Ladde, G. S. (1978). *Stability of General Systems in Biological, Physical, and Social Sciences*, Applied General Systems Research (Editor: George J. Klir) (Plenum, New York), pp. 575–587.

[86]   Ladde, G. S. (1981). Stochastic stability analysis of model ecosystems with time-delay, *Differential Equations and Applications in Ecology, Epidemics and Population Problems* (Editors: Stavros N. Busenberg and Kenneth L. Cooke) (Academic, New York), pp. 215–228.

[87]　Ladde, G. S. (1991). Modeling of dynamic systems by Itô-type systems of stochastic differential equations, *Integral Methods in Science and Engineering — 90* (Editors: A. Haji-Sheikh, Constantin Corduneanu, John L. Fry, Tseng Huang and Fred R. Payne) (Hemisphere, Washington, D.C.), pp. 63–74.

[88]　Ladde, G. S. (2003). Stabilizing and oscillizing hereditary and random structural perturbations effects on multispecies processes, *Proceedings of Conference on Nonlinear Systems: Modeling, Simulation and Applications* (Editor: S. B. Agase) (The Publication of Science College, Nanded, India), pp. 1–20.

[89]　Ladde, G. S. (2004). Stability of stochastic distributed parameter large-scale control systems under random structural perturbations, *Dynamic of Continuous, Discrete and Impulsive Systems Series A: Mathematical Analysis* **11**, pp. 233–254.

[90]　Ladde, G. S. (2005). Variational comparison method and stochastic time series analysis, *Mathematical and Computational Models — NCMC* 2005 (Editors: R. Arumuganathan and R. Nadarajan) (Allied, New Delhi), pp. 16–40.

[91]　Ladde, G. S. (2006). Large-scale stochastic hereditary systems under Markovian structural perturbations I: Variational comparison theorems, *Journal of Applied Mathematics and Stochastic Analysis* **2006**, Article ID 19871, 11 pp. JAMSA/ 19871.

[92]　Ladde, G. S. (2006). Large-scale stochastic hereditary systems under Markovian structural perturbations II: Qualitative analysis of isolated systems, *Journal of Applied Mathematics and Stochastic Analysis* **2006**, Article ID 67268, 14 pp. JAMSA/67268.

[93]　Ladde, G. S. (2006). Large-scale stochastic hereditary systems under Markovian structural perturbations III: Qualitative analysis, *Journal of Applied Mathematics and Stochastic Analysis* **2006**, Article ID 24643, 10 pp. JAMSA/24643.

[94]　Ladde, G. S. (2005). Hybrid dynamical inequalities and applications, *Dynamical Systems and Applications* **14**, pp. 481–514.

[95]　Ladde, G. S. (2008). Stochastic systems: A class of hybrid systems, *Abstract of Papers Presented to the American Mathematical Society* **29**, 1, p. 173.

[96]　Ladde, G. S. *Competitive Cooperative Process in Biological, Engineering, Medical, Physical and Social Sciences* (in progress).

[97]　Ladde, G. S. *Stochastic Stability of Inflation–Unemployment Process; Lecture Notes*.

[98]　Ladde, G. S. and Lakshmikantham, V. (1980). *Random Differential Inequalities* (Academic, New York).

[99]　Ladde, G. S. and Lakshmikantham, V. (1980). Competitive–cooperative processes and stability of diffusion systems, *Applied Stochastic Processes* (Editor: G. Adomian) (Academic, New York), pp. 83–108. MR 83j 60084.

[100]　Ladde, G. S., Lakshmikantham, V. and Vatsala, A. S. (1985). *Monotone Iterative Techniques for Nonlinear Differential Equations* (Pitman Advanced Publishing Program, Boston).

[101]　Ladde, G. S., Lakshmikantham, V. and B. G. Zhang (1987). *Oscillation Theory of Differential Equations with Deviating Arguments* (Marcel Dekker, New York).

[102]　Ladde, G. S. and Lawrence, Bonita (1995). Stability and convergence of large-scale stochastic approximation procedures, *International Journal of Systems Science* **26**, pp. 595–618.

[103]　Ladde, G. S. and Lawrence, Bonita (2001). Stability and convergence of stochastic approximation procedures under Markovian structural perturbations, *Dynamic Systems and Applications* **10**, pp. 145–175.

[104] Ladde, G. S. and Lawrence, Bonita (2003). On joint probability density functions of discrete time processes, *Mathematics and Computer in Simulation* **63**, pp. 629–650.

[105] Ladde, G. S. and Lawrence, Bonita (2004). Stability and convergence of large-scale stochastic approximation under Markovian structural perturbations, *Differential Equations and Dynamical Systems* (Editor: D. Bhauguna), (Narosa, New Delhi), pp. 25–48.

[106] G. S. Ladde and N. G. Medhin (1995). Derivation of optimality conditions for a stochastic control problem, *Stochastic Analysis and Applications* **13**, pp. 165–176.

[107] Ladde, G. S., Medhin, N. G. and Sambandham, M. (2003). Error estimates for random boundary value problems with applications to a hanging cable problem, *Mathematical and Computer Modelling* **38**, pp. 1037–1050.

[108] Ladde, G. S. and Okonkwo, Zephyrinus C. (1997). Itô-type stochastic differential systems with abstract Volterra operators, *Dynamic Systems and Applications* **6**, pp. 461–468.

[109] Ladde, G. S., Peterson, S. and Sambandham, M. (2003). Comparison of actual and analytic computational error estimates to Ito-type stochastic differential equations, *International Journal of Computational and Numerical Analysis and Applications* **4**, pp. 189–199.

[110] Ladde, G. S. and Robinson, J. V. (1981). A stochastic version of Turing's cell morphogenetic model, *Biomathematics and Cell Kinetics* (Editor: M. Rotenberg) (Elsevier/North-Holland: Biomedical Press, Amsterdam, The Netherlands), pp. 349–356.

[111] Ladde, G. S. and Robinson, J. V. (1984). Stability of limiting distributions of one and two species stochastic population models, *Mathematical Modelling* **5**, pp. 331–338.

[112] Ladde, G. S. and Sambandham, M. (2004). *Stochastic Versus Deterministic Systems of Differential Equations* (Marcel Dekker, New York).

[113] Ladde, G. S. and Sathananthan, S. (1992). Stability of Lotka–Volterra model, *Mathematical and Computer Modelling* **16**, pp. 99–107.

[114] Ladde, G. S., Sathananthan, S. and Pirapakaran, R. (1995). Numerical treatment of random population models, *Proceedings of Neural, Parallel and Scientific Computations* **1**, pp. 257–260.

[115] Ladde, G. S. and Siljak, D. D. (1975). Stability of multispecies communities in randomly varying environments, *Journal of Mathematical Biology* **2**, pp. 165–178.

[116] Ladde, G. S. and Siljak, D. D. (1975). Connective stability of large-scale stochastic systems, *International Journal of Systems Science* **6**, pp. 713–721.

[117] Ladde, G. S. and Siljak, D. D. (1975). Stochastic stability and instability of model ecosystems, *Proceedings of the 6th International Federation of Automatic Control World Congress* (The Publication of IFAC, Boston, Massachusetts), pp. 55.4:1–7.

[118] Ladde, G. S. and Siljak, D. D. (1983). Multi-parameter singular perturbations of linear systems with multiple time scales, *Automatica* **19**, pp. 385–394.

[119] Ladde, G. S. and Siljak, D. D. (1983). Multiplex control systems: stochastic stability and dynamic reliability, *International Journal of Control* **28**, pp. 515–524.

[120] Ladde, G. S. and Sirisaengtaksin, O. (1988). Near-optimum regulators for stochastic singularly perturbed systems, *Stochastic Analysis and Applications* **6**, pp. 11–79.

[121] Ladde, G. S. and Sirisaengtaksin, O. (1989). Large-scale stochastic singularly perturbed systems, *Mathematics and Computers in Simulation* **31**, pp. 31–40.

[122] Ladde, G. S. and Smith, M. S. (1989). Processing of prefiltered GPS data, *IEEE Transactions on Aerospace and Electronic Systems* **25**, pp. 711–728.

[123] Ladde, G. S. and Wu, Ling (2009). Stochastic modeling and statistical analysis on the stock price processes, *Nonlinear Analysis: Theory and Methods* **71**, pp. e1203–e1208.

[124] Ladde, G. S. and Wu, Ling (2010). Development of nonlinear stochastic models by using stock price data and basic statistics, *Neural, Parallel and Scientific Computations* **18**, pp. 269–282.

[125] Langevin, P. (1908). *Comptes Rendus* **146**, p. 530.

[126] Latham, J. L. (1969). *Elementary Reaction Kinetics*, 2nd ed. (Butterworth, London).

[127] Leithold, Louis (1976). *The Calculus with Analytic Geometry*, 3rd ed. (Harper and Row, New York).

[128] Levins, Richard (1970). Extinction, some mathematical questions in biology, lecture on mathematics in life sciences, *The American Mathemathical Society* **2**, pp. 75–108.

[129] Loewenstein, W. R. (1961). Excitation and inactivation in a receptor membrane, *Annals. of the New York Academy of Sciences* **94**, pp. 510–534.

[130] Lotka, A. J. (1956). *Elements of Mathematical Biology* (Dover, New York).

[131] Ludwig, Donald (1974). *Stochastic Population Theories*, Vol. 3 (Springer-Verlag, New York).

[132] Maruyama, Takeo (1977). *Stochastic Problems in Population Genetics*, Vol. 17 (Springer-Verlag, New York).

[133] Mattson, H. F. (1993). *Discrete Mathematics with Applications* (John Wiley and Sons, New York).

[134] May, Robert M. (1973). *Stability and Complexity in Model Ecosystems* (Princeton University Press).

[135] McDonald, D. A. (1960). *Blood Flow in Arteries* (Arnold, London).

[136] Mullins, Jr., E. R. and Rosen, David (1971). *Probability and Calculus* (Bogden and Quigley, New York).

[137] Murdick, R. C. (1971). *Mathematical Models in Marketing* (Intext Educational, Toronto).

[138] Needham, J. (1934). Chemical heterogony and the group-plan of animal growth, *Biol. Rev.* **9**, pp. 79–109.

[139] Nelson, Edward (1967). *Dynamical Theories of Brownin Motion* (Princeton University Press).

[140] Nicosia, F. M. (1966). *Consumer Decision Processes: Marketing and Advertising* (Prentice-Hall, Englewood, California, New Jersey).

[141] Oksendal, B. (1985). *Stochastic Differential Equations: An Introduction with Applications* (Springer-Verlag, New York).

[142] Ornstein, L. S. (1919). On the Brownian motion, physics, *Proceedings of the Section of Sciences*, Vol. XXI (1st Part, Nos. 1–5) (Johannes Muller, Amsterdam, The Netherland), pp. 96–108.

[143] Pannetier, G. and Souchay, P. (1967). *Chemical Kinetics* (Elsevier, New York).

[144] Pedjeu, Jean-Claude and Ladde, G. S. (2012). Stochastic fractional differential equations: modeling, methods and analysis, chaos, solitons and Fractals: Nonlinear Sciences and Nonequilibrium and Complex Phenomina, Volume 45, pp. 279–293.

[145] Pedjeu, Jean-Claude and Ladde, G. S. (2012). A class of higher order stochastic differential equation, Dynamic Systems and Applications (in press).

[146] Phelps, Edmund S. (1967). Phillips curves, expectations of inflation and optimal unemployment over time, *Economica* **34**, pp. 254–281.

[147] Phelps, Edmund S. (1970). Money wage dynamics and labor market equilibrium, *Microeconomic Foundation of Employment and Inflation Theory* (Editor: E. S. Phelps, (Norton, New York).

[148] Pielou, E. C. (1969). *An Introduction to Mathematical Biology* (John Wiley & Sons, New York).

[149] Phillips, A. W. (1958). The relation between unemployment and the rate of change of money wage rates in the United Kingdom, 1861–1957, *Econamica* **25**, pp. 283–299.

[150] Poland, John (1987). A modern fairy tale, *The American Mathemathical Monthly* **94**, 3, pp. 291–295.

[151] Pollard, J. H. (1973). *Mathematical Models for the Growth of Human Population* (Cambridge University Press).

[152] Riggs, Douglas Shepard (1963). *The Mathematical Approach to Physiological Problems* (The M.I.T. Press, Cambridge, Massachusetts).

[153] Randall, James, E. (1958). *Elements of Biophysics*, 2nd ed. (Year Book Medical Publishers, Chicago).

[154] Rashevsky, N. (1960). *Mathematical Biophysics* (Dover, New York).

[155] Rashevsky, N. (1961). *Mathematical Principles in Biology and Their Applications* (Charles C. Thomas, Springfield, Illinois).

[156] Rescigno, Aldo and Beck, James S. (1972). *Compartments, Foundations of Mathematical Biology*, Vol. II: *Cellular Systems* (Editor: R. Rosen) (Academic) pp. 255–322.

[157] Rescigno, A. and Segre, G. (1966). *Drug and Tracer Kinetics* (Blaisdell, Waltham, Massachusetts).

[158] Ricciardi, Luigi M. (1977). *Diffusion Processes and Related Topics in Biology*, Vol. 14 (Springer-Verlag, New York).

[159] Rosen, Robert (1967). *Optimality Principles in Biology* (Plenum, New York).

[160] Rosen, Robert (1970). *Dynamical System Theory in Biology* (John Wiley & Sons, New York).

[161] Rosen, Robert (1972). *Foundations of Mathematical Biology*, Vol. III (Editor: Robert Rosen (Academic).

[162] Ross, Sheldon M. (1996). *Stochastic Processes*, 2nd ed. (John Wiley & Sons, New York).

[163] Ross, Sheldon M. (1972). *Introduction to Probability Models* (Academic, New York).

[164] Roughgarden, J. (1979). *Theory of Population Genetics and Evolutionary Ecology: An Introduction* (MacMillian, New York).

[165] Roxin, Emilio, O. (1972). *Ordinary Differential Equations* (Wadsworth, Belmont, California).

[166] Rubinow, S. I. (1975). *Introduction to Mathematical Biology* (John Wiley & Sons, New York).

[167] Rudnic, Joseph and Gaspari, George (2004). *Elements of the Random Walk* (Cambridge University Press).

[168] Rudin, W. (1976). *Principles of Mathematical Analysis*, 3rd ed. (MaGraw-Hill, New York).

[169] Royden, H. L. (1988). *Real Analysis*, 3rd ed. (Prentice-Hall).

[170] Ryan, Francis J. (1964). Analysis of populations of mutating bacteria, Chap. 15, *Statistics and Mathematics in Biology* (Editors: Oscar Kempthorne *et al.* (Hafner, New York), pp. 217–225.

[171]   Saaty, Thomas L. (1968). *Mathematical Models of Arms Control and Disarmament: Applications of Mathematical Structures in Politics* (John Wiley & Sons, New York).

[172]   Sagawa, K., Lie, R. K., Schaefer, J. and Frank, O. (1850). Die Grundform des Arterielle Pluses, *Zeitschrift fur Biologie* **37** (1899), pp. 483–526: A translation by K. Sagawa *et al.*, *Journal of Molecular Cell Cardiology* **22** (1990), pp. 253–277.

[173]   Saha, Nirzhar, Ross, Mark, and Ladde, G. S. (2008). Dynamic modeling of root water uptake using soil moisture data, *Neural, Parallel, and Scientific Computations* **16**, pp. 105–124.

[174]   Sampath, G. and Srinivasan, S. K. (1977). *Stochastic Models for Spike Trains of Single Neurons*, Vol. 16 (Springer-Verlag, New York).

[175]   Segel, Inwin H. (1975). *Enzyme Kinetics: Behavior and Analysis of Rapid Equilibrium and Steady-State Enzyme Systems* (Wiley Interscience, New York).

[176]   Siu, D. P. and Ladde, G. S. (2011). Stochastic hybrid system with nonhomogeneous jumps, *Nonlinear Analysis: Hybrid Systems* **5**, pp. 591–602.

[177]   Siu, D. P. and Ladde, G. S. A multivariate stochastic hybrid model with switching coefficients and jumps: solution and distribution, *Journal of Probability and Statistics*, Volume 2011 ID 720614, 20 pages, 2011. Doi: 115/2011/720614.

[178]   Siljak, Dragoslav D. (1978). *Large-Scale Dynamic Systems: Stability and Structure* (North-Holland, New York).

[179]   Smith, J. M. (1974). *Models in Ecology* (Cambridge University Press).

[180]   Simon, H. A. (1957). *Models of Man* (Wiley, New York).

[181]   Simpson, George Gayland, Roe, Anne and Lewontin, Richard C. (1960). *Quantitative Zoology* (Harcourt, Brace and Company, New York).

[182]   Soong, T. T. (1973). *Random Differential Equations in Science and Engineering* (Academic, New York).

[183]   Stevens, Brian (1965). *Chemical Kinetics* (Franklin, Englewood, New Jersey).

[184]   Solow, Robert M. (1987). *Growth Theory: An Exposition* (Oxford University Press, New York).

[185]   Solow, Robert M. (1956). A contribution to the theory of economic growth, *The Quarterly Journal of Economics* **70**, pp. 65–94.

[186]   Tapiero, Charles S. (1990). *Applied Stochastic Models and Control in Management* (Elsevier, Amsterdam, The Netherlands).

[187]   Teissier, Georges (1960). Relative growth: the physiology of crustacea (Editor: Talbot H. Waterman), Vol. 1, *Metabolism and Growth* (Academic, New York).

[188]   Teorell, T. (1937). Kinetics of distribution of substances administered to the Body — I, *Archives Inter. de Pharmaco. et de Therapie* **57**, pp. 205–225.

[189]   Teorell, T. (1937). Kinetics of distribution of substancees administered to the Body — II, *Archives Inter. de Pharmaco. et de Therapie* **57**, pp. 226–240.

[190]   Uhlenbeck, G. E. and Ornstein, L. S. (1930). On the theory of the Brownian motion, *Physical Review* **36**, pp. 823–841.

[191]   Van Egmond, A. A. J., Groen, J. J. and Jongkees, L. B. W. (1949). The mechanics of the circular canal, *Journal of Physiology* **110**, pp. 1–17.

[192]   Wang, Ming Chen and Uhlenbeck, G. E. (1945). On the theory of Brownian motion, *Reviews of Modern Physics* **17**, pp. 323–342.

[193]   Whittaker, R. H. and Levin, S. A. (1975). *Niche Theory and Applications* (Halsted, Chichester).

[194]   Waltman, Paul (1974). *Deterministic Threshold Models in the Theory of Epidemics* (Springer-Verlag, New York).

[195] Wong, Eugene (1971). *Stochastic Processes in Information and Dynamical Systems* (McGraw-Hill, New York).

[196] Yeung, D. (1995). Employment adjustment noise and the extected rate of inflation in a simple inflation-unemployment model, *Stochastic Annalysis and Applications* **13**, 1, pp. 125–135.

[197] The Constitution of the United States of America, U.S. Government Printing Office, Washington, D.C. (1976).

[198] Funk & Wagnalls New Encyclopedia (Editor-in-Chief: Robert S. Phillips) Vol. 7 (Funk & Wagnalls, 1975).

Index